ANNUAL REVIEW OF GENETICS

ANNUAL REVIEW OF GENETICS

VOLUME 23, 1989

ALLAN CAMPBELL, *Editor*
Stanford University, Stanford

BRUCE S. BAKER, *Associate Editor*
Stanford University, Stanford

IRA HERSKOWITZ, *Associate Editor*
The University of California, San Francisco

ANNUAL REVIEWS INC.   4139 EL CAMINO WAY   PO BOX 10139   PALO ALTO, CALIFORNIA 94303-0897

ANNUAL REVIEWS INC.
Palo Alto, California, USA

International Standard Serial Number: 0066–4197
International Standard Book Number: 0–8243–1223-6
Library of Congress Catalog Card Number: 63-8847

∞ The paper used in this publication meets the minimum requirements of American National Standard for Information Sciences—Permanence of Paper for Printed
Library Materials, ANSI Z39.48-1984.

Typesetting by Kachina Typesetting Inc., Tempe, Arizona; John Olson, President
Typesetting Coordinator, Janis Hoffman

PRINTED AND BOUND IN THE UNITED STATES OF AMERICA

Herschel Roman

DEDICATION

Herschel Lewis Roman (1914–1989) was the founding editor of Annual Review of Genetics, a position in which he served for 20 years. After his retirement as editor in 1984, he continued to be accessible for valued advice and counsel to the present editor. He also authored a prefatory chapter for the 1986 volume that includes some highlights from his own research career[1].

As a graduate student with Lewis J. Stadler at the University of Missouri, Roman carried out his first genetic research, on the cytogentics of the B chromosomes in maize. Upon assuming an academic post at the University of Washington, he found the climate in Seattle less suitable for maize research than that in Columbia and soon switched his attention to the then embryonic field of yeast genetics. He continued to work on yeast for the rest of his career. His early work established a rigorous foundation for yeast genetics. He established the Genetics Department at Seattle as a center of yeast research and played a key role in promoting yeast as a prime experimental system for molecular genetics of eukaryotes. He deserves much of the credit for the vitality of contemporary yeast genetics. A major theme of his research was the relationship between gene conversion and genetic recombination and his interest in and contributions to this area continued up to the time of his death. His contributions were recognized by many honors, including membership in the National Academy of Sciences, the Gold Medal of the Emil Christian Hansen Foundation, and the Thomas Hunt Morgan Medal of the Genetics Society of America.

Roman established the Department of Genetics at the University of Washington and served as chairman for many years. He did much to develop that Department to reflect his own high scientific standards and broad interests across the entire field of genetics. Education and training in genetics had high priority for him. The whole department was imbued with love of and commitment to genetics, and graduate students left his department with an indelible mark.

During his later years, his physical activities were restricted as the result of a severe stroke suffered in 1976. This did not prevent him from continuing to serve for some years as editor of Annual Review of Genetics nor from directing an active research program.

Herschel will be remembered with great affection by many former students of the Department and by his scientific colleagues and friends, including the present editors. We dedicate this volume of Annual Review of Genetics to his memory.

1. Roman, H. 1986. The early days of yeast genetics: a personal narrative. *Annu. Rev. Genet.* 20:1–12

Annual Review of Genetics
Volume 23 (1989)

CONTENTS

SOME RELATED ARTICLES IN OTHER *ANNUAL REVIEWS*

From the *Annual Review of Biochemistry,* Volume 58, 1989

DNA Topoisomerase Poisons as Antitumor Drugs, *L. F. Liu*
DNA Conformation and Protein Binding, *A. A. Travers*
Animal Virus DNA Replication, *M. D. Challberg and T. J. Kelly*
Molecular Mechanisms of Transcriptional Regulation in Yeast, *K. Struhl*
Eukaryotic Transcriptional Regulatory Proteins, *P. F. Johnson and S. L. McKnight*
Mutational Effects on Protein Stability, *T. Alber*
Gene Conversion and the Generation of Antibody Diversity, *L. J. Wysocki and M. L. Gefter*

From the *Annual Review of Biophysics and Biophysical Chemistry,* Volume 18, 1989

Toward a Unified Model of Chromatin Folding, *J. Widom*
Physical Studies of Protein-DNA Complexes by Footprinting, *Thomas D. Tullius*

From the *Annual Review of Cell Biology,* Volume 5, 1989

The Chloroplast Chromosomes in Land Plants, *M. Sugiura*
Initiation of Eukaryotic DNA Replication In Vitro, *B. Stillman*
Communication Between Mitochondria and the Nucleus in the Regulation
 of Cytochrome Genes in the Yeast *Saccharomyces cerevisiae, S. L. Forsburg
 and L. Guarente*
Origin and Evolution of Mitochondrial DNA, *M. W. Gray*
RNA Editing, *L. Simpson*

From the *Annual Review of Entomology,* Volume 34, 1989

Expression of Foreign Genes in Insects Using Baculovirus Vectors, *Susumu Maeda*

From the *Annual Review of Immunology,* Volume 7, 1989

V-Region Connectivity in T-Cell Repertoires, *P. Pereiras, A. Bandeira, A.
 Coutinho, M.-A. Marcos, M. Toribio, and C. Martinez-A.*
Immunogenetics of Human Cell Surface Differentiation, *W. J. Rettig and L. J. Old*

From the *Annual Review of Medicine,* Volume 40, 1989

Genetic Changes in the Pathogenesis of Lung Cancer, *M. J. Birrer and J. D. Minna*

(*continued*)

From the *Annual Review of Microbiology,* Volume 43, 1989

ANNUAL REVIEWS INC. is a nonprofit scientific publisher established to promote the advancement of the sciences. Beginning in 1932 with the *Annual Review of Biochemistry*, the Company has pursued as its principal function the publication of high quality, reasonably priced *Annual Review* volumes. The volumes are organized by Editors and Editorial Committees who invite qualified authors to contribute critical articles reviewing significant developments within each major discipline. The Editor-in-Chief invites those interested in serving as future Editorial Committee members to communicate directly with him. Annual Reviews Inc. is administered by a Board of Directors, whose members serve without compensation.

For the convenience of readers, a detachable order form/envelope is bound into the back of this volume.

Sewall Wright

Annu. Rev. Genet. 1989. 23:1–18

SEWALL WRIGHT'S CONTRIBUTIONS TO PHYSIOLOGICAL GENETICS AND TO INBREEDING THEORY AND PRACTICE

Elizabeth S. Russell

The Jackson Laboratory, 600 Main Street, Bar Harbor, Maine 04609

CONTENTS

INTRODUCTION

Throughout Sewall Wright's long and varied scientific careers, stretching from graduate days at the Bussey Institute of Harvard, through ten years collecting and analyzing data on inbred guinea pigs at the USDA in Beltsville,

1

0066-4197/89/1215-0001$02.00

29 years of teaching and research at the University of Chicago, and finally on to 33 years of fruitful thinking and writing at the University of Wisconsin, he consistently made fundamental conceptual and practical contributions.

Sewall Wright's work on evolution is widely known from his long controversy with Oxford's R. A. Fisher, and from his four volume summarization, *Evolution and Population Genetics* (44–49), brought together during his years at the University of Wisconsin. We already have an excellent biography (7) that emphasizes Wright's work on evolution. In the course of writing this biography, Provine spent many informal and highly profitable hours consulting directly with Wright, which brings to his presentation a welcome ring of truth.

I was delighted when I was asked to prepare, for the *Annual Review of Genetics,* a "short biography of the first half of Sewall Wright's life." I wanted to accept the challenge, but hesitated. What could I add to Provine's excellent account? How could a scientific biography be limited to the first half of anyone's life? This objection is especially pertinent with Wright since, once he got into an interesting area, he never let it go! Pigmentation genetics and analysis of results of inbreeding, both of which he began to study very early, were both subjects of papers written more than 50 years later! I decided to concentrate on Wright's times at the Bussey and the USDA, and the first part of his time at the University of Chicago, and to reminisce about Wright as a teacher, concentrating on the two fields of mammalian genetics that I know best: physiological genetics and characterization and use of inbred strains.

Before I plunge in, I want to express my deep gratitude to Wright's nephew, Christopher Wright, who loaned me a series of fascinating letters, recently rediscovered, written by Wright between 1912 and 1933, mostly to his mother (13).

This is also an appropriate place to make some generalizations about Wright's personality and character.

1. Although Wright did not enter readily into everyday chitchat, he loved discussions and maintained numerous deep friendships throughout his life, most of which began from scientific connections. Some (C. C. Little, Harrison Hunt) dated from the Bussey. For a couple of years while at the USDA, he shared an apartment with Paul Popenoe, then editor of the *Journal of Heredity,* and developed firm connections with experts in livestock breeding. While at Chicago he greatly enjoyed discussions with biology faculty companions during long walks on the Indiana dunes. One important feature of his work on evolution was his friendship, including field trips, with Dobzhansky. His best and very much appreciated friend late in life was Jim Crow.

2. Wright was a very conscientious, hard-working man, who kept excellent records. His special love was trying to convert those records into forms

that could be analyzed mathematically. He also threw himself whole heartedly into analyzing research problems of all the many scientists who came to him for help.

3. Wright was an instinctual mathematician, having learned the basics from his father while an undergraduate at Lombard College. When tackling a new problem, he always went back to first principles. Perhaps because of this approach, he stuck firmly to his conclusions. Provine quotes him as saying, "I acquired some facility in translating questions into mathematical symbolism and solving as best I could." Provine continues, "He was not primarily interested in pure mathematics, but in devising ways to analyze problems quantitatively, and became adept at teaching himself new mathematics" (7).

   While I was his graduate student at Chicago, he often became completely engrossed, in the midst of a course lecture, in deriving a new formula. Usually none of his students could follow his derivation, nor even see most of what he had written on the blackboard.

4. Wright was a great believer in Occam's razor, or the principle of parsimony: Always cut assumptions to the bare minimum, adopting the simplest hypothesis concordant with the facts. And, please, "Don't go the long way around Robin Hood's barn."

5. At least up to 1940, Wright prided himself on reading and digesting all publications in genetics, and presented neat summaries to his students, usually referring to a stack of notes on 3x5 cards.

THE BUSSEY INSTITUTE

Wright's training as a mammalian geneticist started soon after the rediscovery of Mendel's laws and the reorganization (1909) of the Bussey Institute of Harvard University to be a center for graduate studies in genetics. Its director, W. E. Castle, was an early leader of mammalian genetics, and his associate, E. M. East, was an excellent botanical geneticist. Together they provided a very stimulating environment for a group of excellent graduate students, including C. C. Little, E. C. MacDowell, and J. A. Detlefson, working on mammals, and E. W. Sinnott and R. A. Emerson, working with plants. All of these became real scientific leaders. There were excellent seminars and discussions made especially lively by the fact that Castle and East, both ardent believers in Mendelism, did not completely agree on other genetic questions. Students were encouraged to contribute to the discussions.

When Wright arrived in 1912, Castle and students had already demonstrated Mendelian inheritance of several distinct coat-color differences in rabbits, mice and guinea pigs. In 1911 new and different stocks from South

America had been added to the guinea pig colony. By 1912, Castle himself was deeply involved in selection experiments on spotting patterns in hooded rats. Clarence Cook Little, who had begun to work with Castle even before the Bussey became a genetics institute, was by 1912 enthusiastically working on multiple factor inheritance of cancer susceptibility in mice. Detlefson, who was to leave soon, was in charge of the guinea pig colony. In 1912, the guinea pig colony contained new, not yet understood color variations, brought in from Peru. Castle immediately assigned to Wright responsibility for maintenance of the guinea pig colony, with freedom to analyze all genetic differences. Guinea pigs, with small litters and a long gestation period, are not the easiest mammals for genetic analysis, but Wright managed to make critical tests to assign all the new variability to the proper allelic series. Each of these new recessive diluting mutants could have been like "blue dilution" in the rabbit, or pink-eyed dilution in the mouse, or members of the albino series, or could belong to a new, previously unknown series. At this time, only one multiple allelic series had been described. One important contribution made by Wright at the Bussey was his demonstration that the guinea pig albino series was a second multiple allelic series that contained at least four distinct alleles.

Wright was also much interested in problems of selection and inbreeding. The importance of Johannsen's pure self-fertilized lines of Phaseolus beans, and of the work of G. H. Shull and E. M. East on inbred and F_1 hybrid corn, were becoming recognized, but there was considerable doubt among animal breeders about the possibility and utility of brother-sister inbreeding for development of animal stocks. As I've heard informally from his co-students C. C. Little and H. D. Fish, Wright's mathematical mind led him to contribute very effectively to local discussions of inbreeding. Wright was interested both in continuing to work on guinea pig coat-color genetics and in studying mammalian inbreeding and inbred lines.

Early in 1915, Dr. George Rommel, Chief of the USDA Bureau of Animal Industry, came to Castle seeking a geneticist who could tackle analysis of a very large body of data already accumulated at the USDA during nine years of inbreeding guinea pigs. Before 1906, non-inbred guinea pigs were maintained at USDA to supply research needs of government workers. Rommel, with considerable astuteness, started 35 brother-sister mated inbred lines in 1906, to test feasibility of inbreeding animals. While some of these lines died out, others survived, though their characteristics had never been studied. Castle felt that Wright's diligence and mathematical approach were ideal for analysis of this data, and highly recommended Wright for the position. Wright accepted Dr. Rommel's job offer and in the summer of 1915 packed up representatives of the special guinea pig coat-color stocks and took off for the USDA in Washington and Beltsville.

PERSONAL LIFE AT THE USDA

Before considering Wright's scientific work at the USDA, I want to talk a bit about developments in his personal life during those ten years. He arrived from academia as a bachelor, first living in a rented room, then sharing an apartment with Paul Popenoe. After a time he joined the Cosmos Club and found it a good place for lively discussions. He maintained scientific friendship contacts by attending and presenting papers at AAAS and American Naturalist meetings. This activity included getting to know members of the new Drosophila genetic group, especially Sturtevant and Morgan.

In the summer of 1920, he spent all of his 30-day annual vacation at the Station for Experimental Evolution at Cold Spring Harbor. C. C. Little was there, and Wright's good friends Phineas and Anna Rachel Whiting, and a young woman, Louise Williams, who had obtained a master's degree in Zoology with Harold Fish, one of Wright's friends at the Bussey. Miss Williams was now an instructor at Smith College, but had come to Cold Spring Harbor for the summer to take care of Fish's rabbit colony. The first evening, when Wright and Miss Williams met at dinner, Louise told him she wished he had not been "just another geneticist," but rather the new animal caretaker who was supposed to clean her rabbit cages. He volunteered to help her next day, a good start to a deepening friendship that included conversations, evening strolls, and flower gathering. She was somewhat lame from a congenital hip dislocation, but this did not keep her from walking, somewhat slowly, for miles. "She had an attractive gaiety and a good sense of humor, and was sympathetic with anyone in trouble . . . I was so much in love with her at the end of my vacation . . . that I proposed on the last evening" (7). They were married at her home in Ohio, in September, 1921. After the wedding they had a canoe-trip honeymoon above the falls on the Potomac. During his last four years at the USDA, he was a happily married man, with two sons (Dick and Bob) born by 1925. Their third child was a daughter (Betty) born in Chicago in 1929. Louise Wright was a truly lovely woman, always welcoming to students. Sewall Wright, as usual, by following his instincts, had made an excellent choice.

WORK ACCOMPLISHMENTS AT THE USDA

Wright found the atmosphere at the USDA very different from the Bussey. For quite a while he was very busy with both scientific and not-so-scientific assignments. In addition to animal caretakers, he worked with livestock breeders, and developed quite an interest in breeds of horses, cattle, pigs, and sheep and went as a part of his job to the yearly International Livestock

Show. He also received and answered many inquiries from farmers. He liked and appreciated his boss, Dr. Rommel, chief of the Bureau of Animal Industry, who helped Wright to get new metal animal cages. One early job that amused Wright (and me) was working out the market value of a "lot of very fancy cattle," which were quarantined for a year because they got hoof and mouth disease. The government had figured these cattle to be worth $70,000, the cattlemen, $186,000. Wright worked out a scientific scheme of evaluation that took their ancestry into account, "a modification of Galton's law. I don't know if the USDA will use my figures. Their problem probably really calls for a diplomat rather than a student of heredity" (13).

However odd these "real world" assignments may seem, Wright's ten years at the USDA were extremely productive. His contributions to methodology for quantitative genetic improvement of livestock breeds made him greatly revered by both theoretical and practical agriculturalists. He was called on frequently to solve quantitative analytic problems for the whole Agriculture Division; perhaps his most famous assistance there was his study, involving use of path coefficients, of "Corn and Hog Correlations" (37). By 1921 members of the Agriculture Division had persuaded him to offer courses in genetics and statistics for government workers. In the early years, he wrote two series of papers for the *Journal of Heredity,* these providing a simple, concise background for a wide audience. The first such series, covering ten species, was on color inheritance in mammals (16–26). Later he published five papers on Systems of Mating (27–31), which became the "bible" on genetic improvement of livestock. This series involved early use of the concept of path coefficients, tracing all pathways of influence between two individuals, and evaluating the contribution of each path to the total variance.

I am limiting my detailed scientific discussion of his work at the USDA to the developing area of physiological genetics, and to his studies analyzing and characterizing guinea pig inbred strains.

Studies of Coat-Color Genetics

Wright found it hard to prepare his Bussey thesis for publication, approximately one and a half years after its original composition. He found many points on which his opinion had changed, and it was difficult to make alterations so that the new text (15) harmonized with his new views (13). He kept collecting more data on the same and new genetic color combinations. Soon after arriving at USDA, he developed a series of grades for each of the color types, by matching each guinea pig's color(s) to a series of standard skins that marked barely perceptible increases in color intensity. He assigned genotypes through a combination of color-grade and progeny test. Many of the guinea pigs were tricolored, with patches of black-sepia, red-yellow, and white. This pattern was particularly useful since the observer could grade both sepia and yellow spots on the same individual, which obviously had only one genotype. Sepia grades went from 3 to 14 (intense black), red/yellow grades from 1 to 13 (intense red), and pink-eyed sepia from one (barely distinguishable from white) to 12 (light sepia, almost identical to dark-eyed sepia grade 5). He really had no direct measurement of the amount of melanin deposited in the hair. The only pigment granules he actually saw were in a few histological sections prepared in collaboration with Harrison Hunt. Using the color-grading methodology, he wrote three interim papers (20, 35, 36), a collaborative paper (6), and a big summary paper (39). These papers dealt almost entirely with the effects of five allelic series affecting quality and intensity of pigment over the whole body. He did not study the agouti series, which puts a red-yellow band on a sepia hair shaft. In the guinea pig, the extension locus affects quality of pigment, with two alleles controlling production of eumelanin (dominant E-) vs phaeomelanin (e/e) on the whole body, and a third allele (e^p/e^p) producing both colors in different areas (spots or brindling).

The multiple allelic albino series (C, c^k, c^d, c^r, c^a) affects intensity of both black-sepia and red-yellow. The C allele is always top dominant, and the c^a the lowest recessive. The effects of c^k are somewhat different on sepia and yellow, and the red-eyed c^r gene substitution does not reduce sepia pigment intensity in $E/\text{-}c^r/c^r$, but completely eliminates all color in $e/e\ c^r/c^r$. This very different ordering of albino allele affects in combination with sepia vs yellow is very unusual.

Two allelic series, B/b and P/p, alter eumelanic E-pigment, but have no effect on phaeomelanin.

One series, F/f, affects only red-yellow pigment, f/f reducing red to yellow. The only place where f/f affects eumelanin is in combination with pink-eyed dilution. $E/\text{-}\ C/\text{-}\ f/f\ p/p$ is very pale cream.

In the discussion section of his summary paper (39), he speaks of tyrosine as a possible chromogen, and tyrosinase as the probable product of the albino locus. He develops a "paper scheme" that suggests how the gene series *might* interact, but limits his conclusions as follows: "First, the immediate product (I) of action of genes of the C series has different thresholds of effectiveness depending on whether an accessory substance (II) (product of E-), necessary for any sort of sepia or brown, is present or not and whether or not II, if present, is modified by the P-series; second, that above these thresholds there is competition between the precursor of yellow (I alone) and the precursor of sepia and brown (I–II). Factors B and F must be assigned effects following, or otherwise without effect upon, the above threshold and competition effects" (40).

Early in these studies, Wright said, "coat patterns in guinea pigs, and doubtless in other animals, must be determined by a complex of causes of very diverse kinds (19).

Wright concluded that products of different genes must interact with each other. Certain genes must act before or after others, and two different gene products may compete for the same chromogen. When this was written, only Wright and Goldschmidt (4) were really thinking about what genes actually do.

Effects of Inbreeding on the Original Brother-Sister Inbred Lines

In 1906, a total of 35 lines were set up. Twenty-three of these survived through enough successive generations of full-sib matings to be called "lines," or "inbred families." Seventeen of these were still doing well in 1915. By that time, it had been clearly established that guinea pigs *could* be inbred! However, no one had carefully analyzed the effects of this inbreeding, nor studied the characteristics established. Rommel's original idea and Wright's special qualities formed a very happy combination. Wright found excellent records on over 30,000 guinea pigs from 23 families maintained over ten years in a single colony, with good information on date of birth, maternal age at birth, litter size, mortality at birth and between birth and weaning, birth weights, and number of litters. These were all available to a geneticist who loved to look for correlations and patterns. Quotes from a series of letters in spring, 1916, show his attitude: "My work is getting very interesting; every correlation determined suggests the desirability of finding another. I find all kinds of relationships between rate of growth and size of litters both complex and hard to interpret. It is exceedingly difficult to distinguish cause from effect, and from effects of a common cause" (13).

The system even contained a control outbred stock, with no matings closer than between second cousins. In this colony, there were strong environmental influences, both annual and seasonal, that affected both inbred and outbred stocks.

During 13 years of inbreeding, from 1906 to 1920, "there was an average decline in vigor in all characteristics," "most marked in frequency and size of litters" and "ability to raise large litters has fallen off much more than ability to raise small litters" (33).

Experimental inoculation with tuberculosis had shown that, on the average, the inbreds were inferior to their outbred counterparts in disease resistance (32, 50).

There was both a decrease in vigor during inbreeding and a differentiation among families as to the most affected elements of vigor. "Each family came to be characterized by a particular combination of traits, usually involving strength in some respects with weakness in others." "Crosses between different inbred families resulted in marked improvement over both parental stocks in every respect, due allowance having been made for the effects of size of litter on the other characters" (34).

Differentiation among Inbred Families

Even more interesting (to me) than the decline in vigor was the differentiation Wright observed among families. Each family produced a variety of colors in early generations, but "as time went on, different colors automatically became fixed in different lines" (34). Intensity of color and amount of white in spotting patterns also differed markedly between stocks. "The 23 families . . . have automatically become so differentiated from each other that a new litter could easily be recognized at a glance as belonging to its particular family" (34).

Many of the distinguishing traits of a particular family were not seen in all individuals within that family, since incidence was affected by environmental as well as genetic influences. Wright's method of path coefficients, tracing all possible connections between individuals, provided a means to evaluate the contributions of heredity and environment to expression of each observed character. Variability of expression persisted even after genetic homozygosity had been achieved, but there was no longer any correlation between expression in parent and in offspring.

Wright's analysis of tricolor coat patterns (26) provides an early, neat example using his "threshold" concept and his new method of path coefficients. He measured the amount of white in the color patterns of tricolored guinea pigs from three inbred families and the outbred control stock B, and attributed observed variance in amount of white to contributions from heredity (H^2), environment common to littermates (E^2) and "accidents of development" (D^2). Wright assumed that each area of a guinea pig's coat required at least a minimum "threshold" level of influences favoring color before pigment could be formed. The "sum of favorable and unfavorable influences could be arranged in a graded series," roughly fitted to a normal bell-shaped curve, whose mean was genetically fixed, surrounded by plus and minus environmental variations. Pigment was deposited only where the curve covered areas above threshold level. In the three inbred families, individual guinea pigs varied in proportion of white, but the overall mean for family 39 was 20% (left skew), for family 35 was near 50%, and for family 13, 85% (right skew). The distribution curve for the outbred stock B looked very much like that for family 35, with approximately 50% white. For the two families with skewed distributions, Wright determined medians, and upper and lower quartile values, then transformed the scale so that the distribution approximated a normal curve, whose variance and standard deviation he determined. Using these values, he calculated H^2, E^2, and D^2 for each stock. In the three inbred families, H^2 was practically zero, since there was essentially no genetic variation within strain. In family 35, with over 50% white in tricolors, less than 3% could be attributed to heredity, 5% to environment common to littermates, and 92% to accidents of development. In the very similar looking outbred stock B, the causes of variation were very different: 42% of the variance in stock B could be attributed to the large amount of hereditary difference within the stock, and 58% to accidents of development. My summary is: identical appearance does not necessarily mean genetic identity, and genetic identity does not guarantee identical appearance.

A later study of family 35 (38) showed correlation between amount of white and increasing maternal age. The coats of offspring of 3-month old mothers averaged 60% white, while offspring of 21- to 46-month old mothers averaged 73% white. Since 3-month females had usually achieved only one half of their adult weight, Wright felt the correlation between proportion of white with advancing age of the mother probably implied some kind of growth competition between mother and offspring.

Genetic Monsters

One type of trait that Wright found that differed among inbred families was the tendency to produce low but characteristic numbers of some specific developmental anomaly. The defects were not just culminations of the deleterious effects of inbreeding, since the affected strains were usually among the healthiest. The first "monstrosity" observed, otocephaly, was very rare, with 82 cases among almost 40,000 guinea pigs. There was a genetic basis for susceptibility to the trait, since 50 of these cases appeared among the 3253 individuals in family 13. Otocephaly varied from slight reduction of space between the ears (grade 1) to complete absence of the head (grade 12). "In grades 6 to 12 the primary feature . . . was progressive arrest of the brain" (49). Except for the head defect, the rest of the body was near normal in size, plump, and well developed. "The genetic basis determines an individual, not a maternal character . . . By elimination it is concluded that the main factor is probably chance delay, or other irregularity, in implantation, acting at a particular moment in ontogeny, to produce the resulting temporary arrest in development" (50). Otocephaly seems to have been a threshold character in family 13.

Guinea pigs normally have only three toes on each hind foot. The relatively mild condition, atavistic polydactyly, or return to the ancestral rodent type with four toes per hind foot, was found in a fair proportion of individuals in several different lines. Within one line, incidence of four toes was determined by environmental influences, but between lines, differences in susceptibility appeared to be controlled by 3 or 4 pairs of genes (41).

Quite independently, a different form of polydactyly appeared, dependent upon presence of a single semi-dominant gene *(Px)* (40, 41). Heterozygous

Px/px individuals often developed thumbs as well as little toes. Homozygous *Px/Px* individuals developed paddle-shaped feet, each with "7–12 undifferentiated digits, and other severe developmental anomalies."

Advances in Understanding of Inbred Lines

Wright's careful studies and development of mathematical methods for analyzing characteristics of inbred strains led to increased interest in genetic differences, reduced prejudice against inbred lines, and encouraged their use by both agriculturists and biomedical researchers. His studies of vital statistics confirmed that vigor declined during inbreeding, but showed differences between strains in the particular nature of deleterious effects. Usually F_1 hybrids between strains were much more vigorous than were pure lines.

The importance of using particular inbred strains depends upon genetically influenced characters differing between strains. Wright found many cases of such differentiation, and developed excellent methods for studying characteristics affected by internal and external environmental factors as well as heredity. His own classification of types of characters fixed in inbred strains (49) has been very helpful to me in thinking about the power of inbreeding, and I hope it may also be useful to other researchers:

1. Complete fixation: all individuals within a strain appear identical, as in quality of coat color or nature of histocompatibility genes.
2. Fixation of average grade: "There are characters affected by nongenetic factors in addition to genetic ones" where "the former are of the nature of irregularities of development unrelated to tangible external conditions." One can think of myriad examples; he cites proportion of white in tricolors.
3. Fixation of ranking: Although weight is much influenced by tangible environmental factors, each strain appears to have a fixed degree of genetic susceptibility, so that ranking, but not values, remains constant under varying conditions.
4. Fixation in reaction to particular conditions. Determination of vigor, or disease resistance, or many other characters depends on susceptibility to different insults in different strains. In these cases, what is fixed is a specific susceptibility, not the final character.

Physiological Genetic Concepts

From the very beginning, Wright saw that genes acted within cells, and that they must interact with other genes and with environmental influences. His first studies of action of coat-color genes demonstrated this kind of interdependence. Later he deduced that pathways of interdependence were important in expression of variable characters in inbred strains. In these cases it was usually difficult to sort out effects of particular genes, but analysis of inbred strain characteristics broadened the range of processes on which genetic factors could be seen to have effects. The genetic monsters, otocephaly and single-gene-induced polydactyly, both seemed to involve gene-induced changes in the rate of development at critical times and in particular places in early embryonic development.

In early 1925, Dr. Frank R. Lillie, head of the University of Chicago Zoology Department, offered Wright a position as associate professor, with responsibility for research and graduate courses in genetics. By this time, the Wrights were planning to build a home in the Washington area and research at the USDA was progressing very well. They weren't at all sure that they wanted to move. Wright wrote to Lillie of his concerns, including need for guinea-pig housing and for ample time for writing and research. Then, in February, 1925, Wright learned that President Coolidge, attempting to cut expenditures, had dismissed E. D. Ball, director at research at the USDA. Wright approved highly of Ball, who had benefited his research program extensively. As soon as Wright learned of Ball's dismissal, he wrote to Lillie, showing much more interest in the Chicago offer, and before long it was agreed that Wright would move to the University of Chicago in January, 1926.

Wright looked forward to interacting with embryologists at Chicago, and to increasing his attack on theoretical, rather than practical, problems.

THE UNIVERSITY OF CHICAGO ZOOLOGY DEPARTMENT

Analysis of development was a central interest in the Chicago Zoology Department. Its first chairman, C. O. Whitman, and its second chairman, Frank R. Lillie, were both embryologists. Charles M. Child proposed and promoted the concept that dominant apical to subordinate basal physiological gradients controlled development and differentiation in planaria. Carl R. Moore analyzed mammalian endocrinological development and differentiation. For seven years after Wright arrived, B. J. Willier, a leading amphibian embryologist, was at Chicago, and they became close friends. After Willier left, Victor Hamburger and Paul Weiss, both experimental embryologists, joined the Chicago department. All of these individuals were lively, driving people, devoted to their own research topics at a time when developmental biology was advancing rapidly, with much interest in Spemann's induction phenomena. These developmentalists thought of genes as influencing superficial characters and, at least at first, did not seem too eager to think about gene action in early development.

Other research fields were represented on the faculty. W. C. Allee was a fine ecologist with special interest in animal aggregations. He managed to be very active in spite of being confined to a wheelchair. Alfred Emerson was an entomologist, ecologist, zoogeographer, and expert on termites. H. H. Newman, who organized the undergraduate biology courses, was interested in one phase of genetics, study of characteristics of identical versus fraternal twins, reared together or apart. Dr. H. H. Strandskov, who had been Newman's student, particularly in human genetics, became a member of the faculty after 1931.

All of the zoology courses, and some of the research, took place in the stone Gothic-style Hull building on the main campus. This was two short blocks from the Whitman Laboratory, where both staff and students in genetics and ecology, plus some of those in endocrinology, had laboratory space. They all worked well together. Between 1925 and 1936, there was relatively little contact between zoology and the nearby Billings Laboratories of The University of Chicago Medical School.

Wright's office, on the ground floor of Whitman, was right next to the free-standing guinea-pig house. It was always interesting to hear the chirping voices of the guinea pigs when Wright rattled his keys, preparatory to entering their quarters.

Wright's Development at Chicago

During his first ten years at Chicago, Wright's career advanced rapidly. He was more than ever asked to present important papers, including one for the major session on Physiological Genetics at the International Genetics Congress in 1932, in Ithaca, New York. He was president of that Congress. He became a very strong proponent of Physiological Genetics, and came more and more to introduce his ideas with graphs and diagrams, showing first "the relation of factors," environmental and genetic, "to any developmental process of a vertebrate, occurring at a particular time and place." Usually this was followed by a "diagram illustrating the chain of processes relating to the immediate physiological action of a gene to characters at different levels."

Wright's accomplishments and abilities came to be very widely appreciated during his first decade at Chicago. He was elected to the American Philosophical Society in 1932, to the National Academy of Sciences in 1934, and served as vice-president (1933) and president (1934) of the Genetics Society of America. He served on many editorial boards and committees, and continued to receive even more requests for scientific advice, and through all of this, he continued to record data on his guinea pigs and to write papers!

In 1935 Harvard University made a very attractive offer to Wright, and the University of Chicago responded by offering him even more. The zoology faculty unanimously wrote him a letter, urging him to stay at Chicago. I remember his talking with his graduate students about the problem, trying to make up his mind. The school atmosphere and lifestyle of Chicago pleased both Wrights, and they decided to stay. Everyone was delighted.

WRIGHT AS A TEACHER

Wright took very seriously his responsibility for teaching all of Chicago's graduate genetics courses. I feel my courses with Wright were invaluable. Besides Fundamental Genetics, where I was teaching assistant rather than student, I had *two* biometry courses, each presented formally, although there were only two or three students. I couldn't possibly take satisfactory notes on all that he presented, but I absorbed a great deal about probability, normal curves, threshold characters, transformations of scale, variance, means, standard deviations, heritability, and even path coefficients, though I still can't really use that method effectively. The two most exciting courses, which drew a larger attendance, were Physiological Genetics and Evolution. In Physiological Genetics he discussed the development of ideas and concepts, and reviewed practically all recent genetic publications, putting them into appropriate perspective. He discussed what was known at that time (1934) about the chemistry of chromosomes, and the basic problem of the nature of the gene and its duplication. At that stage, he leaned toward the protein component as the site of gene specificity.

The evolution course was also exciting. His long controversy with R. A. Fisher about the processes of evolution, and the nature and significance of dominance, had just begun. I agree with Paul Scott (personal communication) that we all learned from that controversy: "Treat your opponent with respect, but state your own viewpoint firmly."

The majority of Wright's graduate students tackled research problems related to Wright's studies of guinea pigs, and used animals from his research colony, but he left us pretty much on our own in developing methodology. His first doctoral student, John Paul Scott, analyzed the embryonic development of the polydactylous monsters. He did a very credible job, first producing a "Normentafel" of guinea-pig embryology, then comparing "pollex" monsters with normal littermates. Beginning at 18½ days, almost all organs of the "pollex" embryos showed a much higher than normal rate of mitoses, with abnormal morphogenesis, and by 26 days almost all of these embryos were dead. Scott called these 8 days the critical period. I hope he won't mind my saying this, but aside from concern with genetics, the most apparent carryover from his thesis to later research was use of the term "critical period."

Herman Chase studied the distribution of sepia, yellow, and white areas on tricolor guinea pigs with highly variable total amounts of white. Certain regions were much more apt than others to be colored. Genetic factors influenced the amount of white on the body as a whole, and non-genetic factors acted locally. The intermediate allele e^p was not completely dominant over e, nor was S completely dominant over s (1, 2).

William L. Russell studied reactions of frozen sections of guinea pig hair follicles and of pigment cells in the basal epidermis to exposure to buffered solutions of 3,4-dioxyphenylalanine, or dopa. For each genotype tested, the graded responses in hair follicles agreed very well with the intensity of pigment in yellow hairs of corresponding e/e genotypes, but not at all well with intensity of sepia (9). Especially noticeable was the complete absence of dopa reaction in both dark sepia $E/-$ $c^r c^r$ and white e/e c^r/c^r. The reactions in epidermal pigment cells corresponded with colors in the eyes.

Benson Ginsburg studied dopa reactions using extracts from guinea pig skin, and obtained results corresponding with yellow hair pigment-color grades (3).

Attempts to extract and measure concentrations in solutions of pigment from hair of different genotypes were made by myself (8) and Gertrude Heidenthal (5). Both of us were successful with solutions of yellow pigment in dilute alkali. Sepia and brown melanins were very insoluble. Both of us dissolved hair keratin with 6N hydrocloric acid, and obtained a precipitate of eumelanin. I tried weighing this, and oxidizing it with potassium permanganate. Heidenthal was more successful in making a melanin solution by long boiling in strong alkali. Conversion curves were developed for evaluating the relative amounts of pigment corresponding to the color grades and to hair genotypes.

Although Wright's students carried out experiments largely on their own, the genotypes and color grades of our guinea pigs had been assigned by Wright. The writing up of our results involved very long, fair, but piercing sessions with Wright.

In spite of his busyness, Wright enjoyed having students go with him to lectures and seminars in other departments. I particularly enjoyed going with him to hear Maud Slye tell about cancer in her mouse colony. She kept complete pedigrees, but never inbred her animals and found many different kinds of cancer irregularly distributed in her colony. She believed the tendency to all kinds of cancer was inherited as one simple recessive that might be transmitted hidden for many generations. Her lecture was a good demonstration of the need for background genetic control, and for testing your assumptions.

All of Wright's students learned a great deal about genetics, and continued to do genetic research after leaving Chicago. Scott and Ginsburg worked on genetics in social behavior in dogs. The rest of us worked with mice. Chase continued with pigment and with skin and hair. William Russell became an expert on induced mutation rates. My physiological genetic research came to center on anemias and other hematological characters.

THE WANING OF THE GUINEA PIG

Although Wright continued his research with guinea pigs until he left Chicago in 1955, his central interest came to be evolution in natural populations, particularly Drosophila. More examples, and more varied types of mutations were found in mice, and linkage maps were developed. The characterization and supply of a great variety of inbred mouse strains and mutants was also important in providing tools for research. Obviously, it was less expensive to work with mice than with guinea pigs.

Some, but not all, of his five favorite guinea pig inbred families were maintained at NIH after he moved to Wisconsin in 1955. This loss of inbred families seems very regrettable to me, especially as the early guinea pig studies of inbreeding and of gene-actions contributed so much to the development of mammalian genetics. We can not even fall back on the excuse that there was no published record of guinea pig genetics. As I worked on this review, I thought how useful the dark sepia $E/- c^r c^r$ vs. the white $e/e c^{cr}$ guinea pigs might be for today's molecular study of enzymes and enzyme reactions in pigmentation. I very much doubt if any one anywhere can find a live c^r/c^r guinea pig!

I would like to end by expressing the great debt of The Jackson Laboratory to Sewall Wright for his careful study of inbreeding and for his example and his training of many Jackson Laboratory staff members. Earl L. Green, the second director of the Jackson Laboratory, was a postdoctoral student of Wright, and four staff members (J. Paul Scott, William L. Russell, Elizabeth S. Russell, and Willys K. Silvers) were Wright's doctoral students. In 1954, Wright gave the important closing paper, *Summary of Patterns of Mammalian Gene-Action* (43) at The Jackson Laboratory's 25th Anniversary Symposium, and he was our special honored guest at the Laboratory's 50th Anniversary Symposium in 1979.

Wright was a wonderful man, and a truly excellent pioneer geneticist.

Literature Cited

1. Chase, H. B. 1939a. Studies on the tricolor pattern of the guinea-pig I. The relations between different areas of the coat in respect to the presence of color. *Genetics* 24:610–21
2. Chase, H. B. 1939b. Studies on the tricolor pattern of the guinea-pig II. The distribution of black and yellow as affected by white-spotting and by imperfect dominance in the tortoise shell series of alleles. *Genetics* 24:622–43
3. Ginsburg, B. 1944. The effects of the major genes controlling coat colors in the guinea-pig on the dopa oxidase activity of skin extracts. *Genetics* 29:176–98
4. Goldschmidt, R. 1938. *Physiological Genetics*. New York: McGraw-Hill. 375 pp.
5. Heidenthal, G. 1939. A colorimetric study of genic effect on guinea pig coat color. *Genetics* 25:197–214
6. Hunt, H. R., Wright S. 1918. Pigmentation in guinea-pig hair. *J. Hered.* 9:178–81
7. Provine, W. B. 1986. *Sewall Wright and Evolutionary Biology*. Chicago: Univ. Chicago Press
8. Russell, E. S. 1939. A quantitative study of genic effects on guinea pig coat color. *Genetics* 24:332–55.

9. Russell, W. L. 1939. Investigation of the physiological genetics of hair and skin color in the guinea pig by means of the copa reaction. *Genetics* 24:645–67.

10. Scott, J. P. 1937. The embryology of the guinea pig I. A table of normal development. *Am. J. Anat.* 60:397–432

11. Scott, J. P. 1937. The embryology of the guinea pig III. The development of the polydactylous monster. A case of growth acceleration at a particular period by a semi-dominant gene. *J. Exp. Zool.* 77:123–57

12. Scott, J. P. 1938. The embryology of the guinea-pig II. The development of the polydactylous monster. A new teras produced by the genes PxPx. *J. Morphol.* 62:299–321

13. Wright, C. 1989. Loan of rediscovered letters from S. Wright to family, 1912–1933

14. Wright, S. 1915. The albino series of allelomorphis in guinea-pig. *Am. Nat.* 49:140–48

15. Wright, S. 1916. An intensive study of the inheritance of color and other coat characters in guinea-pigs, with especial reference to graded variation. *Carn. Inst. Wash. Publ.* 241:59–121

16. Wright, S. 1917. Color inheritance in mammals. *J. Hered.* 8:224–35

17. Wright, S. 1917. Color inheritance in mammals II. The mouse. *J. Hered.* 8:373–78

18. Wright, S. 1917. Color inheritance in mammals III. The rat. *J. Hered.* 8:426–30

19. Wright, S. 1917. Color inheritance in mammals IV. The rabbit. *J. Hered.* 8:473–75

20. Wright, S. 1917. Color inheritance in mammals V. The guinea-pig. *J. Hered.* 8:476–80

21. Wright, S. 1917. Color inheritance in mammals VI. Cattle. *J. Hered.* 8:561–64

22. Wright, S. 1917. Color inheritance in mammals VII. The horse. *J. Hered.* 8:561–64

23. Wright, S. 1918. Color inheritance in mammals IX. The dog. *J. Hered.* 9:89–80

24. Wright, S. 1918. Color inheritance in mammals X. The cat. *J. Hered.* 9:178–81

25. Wright, S. 1918. Color inheritance in mammals XI. Man. *J. Hered.* 9:227–40

26. Wright, S. 1920. The relative importance of heredity and environment in determining the piebald pattern of guinea-pigs. *Proc. Natl. Acad. Sci. USA* 6:320–32

27. Wright, S. 1921. Systems of mating I. The biometric relation between parent and offspring. *Genetics* 6:111–23

28. Wright, S. 1921. Systems of mating II. The effects of inbreeding on the genetic composition of a population. *Genetics* 6:124–43

29. Wright, S. 1921. Systems of mating III. Assortative mating based on somatic resemblance. *Genetics* 6:144–61

30. Wright, S. 1921. Systems of mating IV. The effects of selection. *Genetics* 6:162–66

31. Wright, S. 1921. Systems of mating V. General considerations. *Genetics* 6:168–78

32. Wright, S. 1922. The effects of inbreeding and crossbreeding of guinea-pigs I. Decline in vigor. *USDA Bull. #1090,* pp. 1–36

33. Wright, S. 1922. The effects of inbreeding and crossbreeding on guinea-pigs II. Differentiation among inbred families. *USDA Bull. #1090,* pp. 37–63

34. Wright, S. 1922. The effects of inbreeding and crossbreeding on guinea-pigs III. Crosses between highly inbred families. *USDA Bull. #1121*

35. Wright, S. 1923. Two new color factors in the guinea-pig. *Am. Nat.* 57:42–51

36. Wright, S. 1925. The factors of the albino series of guinea-pigs and their effects on black and yellow pigmentation. *Genetics* 10:223–60

37. Wright, S. 1925. Corn and hog correlations. *USDA Bull. #1300*

38. Wright, S. 1926. Effects of age of parents on characteristics of the guinea-pig. *Am. Nat.* 60:552–69

39. Wright, S. 1927. The effect in combination of the major color-factors of the guinea-pig. *Genetics* 12:530–69

40. Wright, S. 1934. The results of crosses between inbred strans of guinea-pigs differing in number of digits. *Genetics* 19:537–51

41. Wright, S. 1934. Polydactylous guinea pigs: Two types respectively heterozygous and homozygous in the same mutant gene. *J. Hered.* 25:359–62

42. Wright, S. 1934. Physiological and evolutionary theories of dominance. *Am. Nat.* 68:25–53

43. Wright, S. 1954. Summary of patterns of mammalian gene action. *J. Nat. Cancer Inst.* 15:837–51

44. Wright, S. 1968. *Evolution and the Genetics of Populations, Vol. 1. Genetic and Biometric Foundations.* Chicago: Univ. Chicago Press

45. Wright, S. 1969. *Evolution and the Genetics of Populations, Vol. 2. Genetic and Biometric Foundations.* Chicago: Univ. Chicago Press

46. Wright, S. 1977. *Evolution and the Genetics of Populations, Vol. 3. Genetic and Biometric Foundations.* Chicago: Univ. Chicago Press
47. Wright, S. 1977. *Evolution and the Genetics of Populations, Vol. 4. Genetic and Biometric Foundations.* Chicago: Univ. Chicago Press
48. Wright, S., Chase, H. B. 1936. On the genetics of the spotted pattern of the guinea-pig. *Genetics* 21:758–78
49. Wright, S., Eaton, O. N. 1929. The persistence of differentiation among inbred families of guinea-pigs. *USDA Techn. Bull. #103*
50. Wright, S., Lewis, P. 1921. Factors in the resistance of guinea-pigs to tuberculosis with especial regard to inbreeding and heredity. *Am. Nat.* 55:20–50
51. Wright, S., Loeb, L. 1927. Transplantation and individuality differentials in inbred families of guinea-pigs. *Am. J. Pathol.* 3:251–85

Annu. Rev. Genet. 1989. 23:19–36

GENETICS OF ALCOHOLISM

Eric J. Devor and C. Robert Cloninger

Department of Psychiatry, Washington University School of Medicine, St. Louis, Missouri 63110

CONTENTS

INTRODUCTION

The genetics of alcoholism is at a pivotal point in its history. Decades of study of alcoholics, their families, and controls have revealed an increasingly sophisticated clinical picture in which distinct subtypes are recognized on the basis of discrete symptoms, associated psychiatric and behavioral entities, and baseline personality differences (11). Moreover, familial/genetic studies indicate that these distinct subtypes are subject to different forms and intensities of genetic control (35). Today, attention is turning toward the new clinical classifications of alcoholics to collect samples that are as homogeneous as possible. The use to which the samples will be put is in genetic linkage studies. Thus, the years of effort that have gone into establishing the hypothesis that genetic influences in alcoholism do indeed exist have set the stage for new efforts to detect and map alcoholism-susceptibility genes.

0066-4197/89/1215-0019$02.00

In this review, we develop and document the emergence of the current view of the genetics of alcoholism. We also examine evidence implicating a wide-range of neurophysiological and biochemical traits as potential candidates for the major loci imparting increased risk for alcoholism. Finally, we touch briefly upon the methods and strategies being employed in the search for those loci.

AN HISTORICAL PERSPECTIVE

The use of alcoholic beverages by man is an ancient phenomenon. As long ago as the Paleolithic era, some tens of thousands of years ago, fermented fruit juices, fermented grain, and fermented honey were available for consumption (82). Probably even then man first began to notice that individuals who tended to drink too much had close relatives who exhibited the same behavior. Thus, the idea that alcoholism "runs in families" is an old one. Both Aristotle and Plutarch believed that alcoholics "begat" other alcoholics (36). The ancient Egyptians not only recorded evidence of drunkenness, but also produced the oldest known temperance tract (82). In the Middle Ages, Chaucer repeatedly used the word "dronkelew" to indicate that dependence upon alcohol was a mental illness, organic, and heritable (37). However, it was not until 1849 that the term "alcoholism" was first used by a Swedish public health official (37).

The notion that alcoholism is a disease and not simply a "weakness" has its roots in the writings of nineteenth century physicians such as Benjamin Rush and Thomas Trotter (37). Hospitals dedicated to the treatment of alcoholics were opened in the United States beginning in the 1840s and alcoholism, or "inebriety," became a subject of serious medical study in the latter half of the nineteenth century (36). Systematic studies of the relatives of alcoholics began with the pioneering work of Long (58) who reported that of all of the "inherited disease" observed in the offspring of alcoholics, 21% was due directly to alcoholism in the parents. Other similar studies suggested that alcoholism was a major cause of mentally defective offspring. Shuttleworth & Beach (94) noted that 19% of the "idiots" in a paupers' home, and 13% in a charity house had at least one alcoholic parent. Goodwin (36) notes that the largest study of its kind, by McNicholl (62), found that 53% of 6,624 children were retarded to some degree compared with only 10% of 13,000 children of abstainers.

In his classic early study of familiality (22), Crothers reported over 70% of the more than 4,000 alcoholics whom he had treated over a 35-year period to have a positive family history of alcoholism. Over the past 80 years, this result has been replicated time and time again (21). In fact, only one negative report, a controversial and flawed study by Elderton & Pearson (27), has detracted from this basic observation. Dahlberg & Stenberg (23) noted that one-fourth of the alcoholics they studied had a positive family history. Among first-degree male relatives of alcoholics the risk was 20–25% and among first-degree female relatives it was fully 5%. These risks were four to five times greater than those in the general population at that time.

As Goodwin (36) has observed, "familial" does not necessarily mean "hereditary." Jellinek & Joliffe (51) were the first to suggest that alcoholism was a heterogeneous disease, having both familial and nonfamilial forms. The authors claimed at the time that there was no evidence for a genetic component in either form. However, over the past thirty years increasingly sophisticated and careful studies have produced evidence that Jellinek & Jolliffe were, at once, right and wrong. Indeed, alcoholism is heterogeneous and there are both familial and nonfamilial cases, but there is also a strong genetic influence on many cases of the familial form of alcoholism.

FAMILIAL/GENETIC STUDIES OF ALCOHOLISM

Studies of the genetics of complex traits traditionally proceed on three fronts: twins, adoptions, and families. Evidence of genetic predisposition, where it exists, results primarily from the first two types of study whereas the "architecture" of genetic predisposition derives from family studies.

Twin Studies

The appeal of twin studies is that they afford a way to assess the relative strengths of environmental and genetic components of phenotypic variance. Since monozygotic (MZ) twins are genetically identical and dizygotic (DZ) twins share only one half of their genes on average, a greater concordance among MZ than DZ twins is expected for a trait under strong genetic control. From the exact differences between MZ and DZ concordances an estimate on heritability can be obtained.

Several twin studies of alcoholism, or of traits relevant to alcoholism, have been carried out since 1960. The most widely cited of these is the Finnish investigation by Partanen of 902 male twin-pairs (70). Three major components of drinking behavior were identified: frequency of drinking, amount consumed, and loss of control. Only loss of control failed to provide evidence of genetic determinants. Moderate heritabilities of 0.39 and 0.36 were estimated for frequency and amount of drinking respectively. A study by Kaprio et al of 11,500 Finnish twin-pairs of both sexes (54) and a study by Jonsson & Nilsson of 7,500 Swedish twin-pairs of both sexes (52) also yielded heritability estimates in the 0.3–0.4 range for amount and frequency of alcohol consumption.

The earliest of the twin studies, by Kaij (53), was based upon 174 Swedish male pairs. The primary focus was on alcohol abuse. A similarly focused study of 850 American twin-pairs was conducted by Loehlin (57). Heritability estimates from both studies were in the range suggesting that one-third of the variance in alcohol-related traits is under direct genetic control. Among all of the extant twin studies, however, only Kaprio et al (54) reported sex-specific heritabilities. The difference between an estimated male heritability of 0.37 and an estimated female heritability of 0.25 for alcohol intake was significant.

Two twin studies of clinically defined alcoholism have been reported. Hrubek and Omenn (45) investigated 15,924 male twin-pairs ascertained through Veterans Administration records. They found a significantly higher concordance for disease among MZ than DZ twin-pairs. Further, MZ twins were significantly more concordant for secondary effects such as alcoholic psychosis and liver cirrhosis. Murray et al reached a very different conclusion (65); from a preliminary analysis of 56 twin-pairs, they observed virtually no difference in concordance. While this study is preliminary, it does represent the only result at odds with the consensus of the small pool of twin studies. The majority of twin-study results lead to a conclusion that a genetic predisposition to various aspects of alcohol abuse and alcoholism does exist.

Adoption Studies

Disentangling genetic and nongenetic influences through the study of adoptees relies on the assumption that a genetic trait present in biological parents will still be expressed in adoptees, regardless of the genotypic status and environmental circumstances of the adoptive parents. Thus, adoption studies are sensitive indicators of genetic effects. Roe carried out the first adoption study of alcoholism in the United States in the early 1940s (81). The sample was small, 49 adopted children of whom 27 had an alcoholic biologic parent, but none had an alcoholic adoptive parent. Alcohol abuse was not significantly greater among the children of the alcoholic biological parents than among controls, but there was a slight excess.

Several larger and more thorough adoption studies have been conducted since. Goodwin et al (38–40) reported on a systematic investigation of 133 male and 96 female adoptees in Denmark. 55 male and 49 female adoptees had one or both biological parents with alcoholism. Moreover, all adoptees were separated from their biological families within six weeks of birth and were adopted into unrelated, nonalcoholic families. After matching a control sample of adoptees for several social, economic, and demographic variables, analyses of the samples revealed that some 5% of the control adoptees developed alcoholism later in life, compared to 18% of the adoptees with a positive biological family history. Significant, too, was the observation that the majority of the excess was restricted to male adoptees since only two of the 49 family-history-positive adopted females and two of 47 control females became alcoholic. Also, while alcoholic male adoptees exhibited specific symptoms such as loss of control, hallucinations, and morning drinking, none of the four alcoholic female adoptees did so.

A small-scale American adoption study by Cadoret & Gath (9) added little to the systematic work of Goodwin and colleagues. However, the large Swedish adoption study of Bohman, Cloninger, and colleagues (5, 6, 14–16, 95, 96) has added a great deal of knowledge regarding genetic influences in alcoholism. The Swedish adoption sample included all 862 male and 913 female offspring born to single mothers in Stockholm between 1930 and 1949 and adopted away at an early age by nonrelatives. The Swedish health-care and other registration systems were able to provide data on alcohol abuse, criminality, occupation, and medical and social history on all 1775 adoptees, their adoptive families, and their biological parents. This remarkable amount of data has led to a series of results that go far beyond mere confirmation of a genetic component in alcoholism. Indeed, three classes of alcoholism have been identified, with distinct clinical characteristics and a different balance of genetic and environmental determinants. They have been named milieu-limited or Type I alcoholism, male-limited or Type II alcoholism, and anti-social behavior disorder with alcohol abuse (11, 18, 24). Relevant clinical characteristics of the two major subtypes of alcoholism are summarized in Table 1.

Milieu-limited, Type I, alcoholism occurs both in men and women and is distinguished by adult onset (after age 25), a less severe and often treatable course, as well as few, if any, alcohol-related social problems such as fighting or arrests. In contrast, the male-limited, Type II, subtype occurs predominantly in males, has an early age of onset (well before age 25), a severe

TABLE 1   Characteristics of Type I versus Type II alcoholism

Distinguishing Characteristic	Type I	Type II
Alcohol-related problems:		
Age of onset (years)	After 25	Before 25
Spontaneous alcohol seeking		
(inability to abstain)	Infrequent	Frequent
Alcohol-related legal problems		
(fighting, arrests)	Infrequent	Frequent
Psychological dependence		
(loss of control)	Frequent	Infrequent
Guilt and fear		
(about dependence)	Frequent	Infrequent
Personality traits:		
Novelty seeking	Low	High
Harm avoidance	High	Low
Reward dependence	High	Low

and intractable course, and many alcohol-related problems. An important aspect of the families of Type II alcoholics is that female siblings do not abuse alcohol but rather show an abnormally high prevalence of low-frequency somatization disorder (95). Low-frequency somatization is characterized by recurrent complaints of headache, backache, vague abdominal pain, and an early age of onset. A second, more severe form of somatization, termed high frequency (15), was found in families in which the male relatives abuse alcohol and have histories of violent criminal behavior.

Cross-fostering analyses within the Swedish adoption data indicated that, among males, risk of Type I alcoholism increased as a function of family history *and* adoptive environment (14, 18). Conversely, Type II alcoholism appeared to be highly heritable from father to son regardless of environmental background (18). There was a nine-fold increase in risk to sons of Type II fathers compared to other fathers in both Type II environment and all other adoptive environments. Similar studies of female adoptees showed that the Type I adopted daughters were much more likely to abuse alcohol than were Type II adopted daughters (18). This result is consistent with the view that Type II alcoholism is predominantly a male phenomenon. Too, the high-frequency somatization disorder appears to be quite heritable in much the same way as Briquet's syndrome in the families of sociopaths (18).

Family Studies

Schuckit et al (86, 90) conducted an investigation of 164 half-sibs of 69 alcoholics. Among the probands 60 were male and 9 female and the half-sibs who became alcoholic were most often male. Concordance was highest if the shared parent was the alcoholic regardless of whether or not they shared an alcoholic home environment. Therefore, the best predictor of alcoholism among half-sibs is the existence of a shared alcoholic parent.

Two early family studies of alcoholics hinted at the heterogeneity the adoption studies would later find. Frances et al (32) reviewed family data on 2215 male alcoholics treated in United States Navy centers. They found that those alcoholics who had an affected first-degree relative did worse in treatment than did those with no family history. Moreover, among those having the positive family history there was a positive correlation between prognosis and number of affected relatives. The following year, Pennick et al (73) added two more aspects to the familial patterns of alcoholism; familial alcoholics tended to have a younger age-of-onset and the course of disease was more severe.

Beginning in 1977, a detailed study of the families of alcoholics has been underway in St. Louis as part of the Washington University Alcoholism Research Center. Pedigrees have been obtained on 286 hospitalized alcoholics. Analyses of these pedigrees revealed that the families of male alcoholics were heterogeneous. Some families appeared to be like the families of the female probands (Type I-like), whereas others had the appearance of Type II, male-only families (12, 35). Heritability estimates for alcoholism in Type I families were 0.21, but in the Type II families it was a very high 0.88 (12, 35). These estimates are consistent with those from the Stockholm Adoption Study (14, 18).

Segregation analysis of the St. Louis families (35) confirmed both the significant heterogeneity among male alcoholics and a significantly different genetic "architecture." The highly heritable Type II alcoholism was found to be under the control of a single major gene effect while the less heritable Type I form was more consistent with a polygenic influence (35).

Personality and Alcoholism

The Stockholm Adoption Study revealed the existence of distinct subtypes of alcoholic-Type I and Type II. The family/pedigree studies in the St. Louis Alcoholism Research Center have confirmed and extended this result (18, 35). In addition to the medical, behavioral, and genetic features that distinguish the two subtypes currently recognized, there is a consistent and reproducible personality dimension that itself has a strong genetic component. Studies of the personalities of male-limited, Type II alcoholics show them to be impulsive and sensation-seeking individuals. Cloninger (10, 11) advanced a general theory involving three discrete heritable dimensions of personality termed "novelty seeking," "harm avoidance," and "reward dependence." Individual differences across the three dimensions are largely independent of each other (10, 11).

Studies of male alcoholics showed that Type I alcoholics tend to have passive-dependent or "anxious personalities (12, 13). They are high in reward dependence: emotionally dependent, sentimental, and sensitive to social cues. They are also high in the harm-avoidance dimension: cautious, apprehensive, inhibited, and shy. Conversely, they are low in novelty seeking: rigid, reflective, orderly, and loyal. By contrast the Type II alcoholic, as noted, is impulsive and antisocial to a degree. They are low in reward dependence: detached, emotional, emotionally cool, tough-minded, and self-willed. They are also low in harm avoidance: confident, care-free, energetic and uninhibited. Finally, they are characteristically high in novelty seeking: exploratory, excitable, disorderly, and distractable. These dimensions are summarized in Table 1. While individual alcoholics display the full range of personality traits seen in the general population, these general subtype classes do differ quantitatively from the general population and from each other as groups. The importance of these differences beyond their value in classification must await specification of the role genetics plays in each dimension.

THE SEARCH FOR A BIOLOGICAL MARKER

In the face of the evidence of familial/genetic studies of alcoholism favoring the view that a genetic predisposition exists for at least some forms of the disease, a reproducible, reliable, and diagnostic marker has become the Holy Grail of alcoholism research. Attempts to produce such a marker have led investigators toward a vast array of plausible and implausible candidates. Clearly, such a marker will be a biological trait or gene product that either directly participates in the pathophysiology of the disease or is closely linked to a trait that does (88). While the presence of a marker for genetic predisposition to alcoholism has often been assumed regardless of prior alcohol consumption, there is currently no reason to believe that alcoholism might not be a very subtle phenomenon triggered by alcohol consumption or even triggering that consumption (102).

Electrophysiological Markers of Alcoholism

Ethanol ingestion has an effect upon the central nervous system. Two measures of CNS function are spontaneous brain activity via the electroencephalogram (EEG) and elicited activity via the evoked-potential (EP) or event-related potential (ERP). In order to determine if any CNS electrophysiological markers of predisposition to alcoholism exist, it is critical to study heritable traits in which alcohol-related or alcohol-induced effects are known (34).

Particular patterns of resting EEG in both men and women have been associated with compulsive drinking (34). The EEG patterns are known to be heritable and involve poorly synchronized slow wave, or alpha, activity (76). Male alcoholics and their sons have shown evidence of significantly decreased slow wave activity on background cortical EEG. However, following alcohol ingestion there was a greater increase of slow wave activity and decrease in fast wave activity among the alcoholics than among controls (74).

Similarly, intriguing results have been obtained in EP studies of alcoholics. Begleiter and colleagues have documented a phenomenon involving a positive polarity brain wave seen 300 milliseconds after an external stimulus is applied. Among alcoholics this brain wave, termed P300, is significantly smaller in amplitude compared to controls (75). Moreover, this diminished P300 response was present in the nondrinking preadolescent sons of alcoholics (3). After controlling for any alcohol-induced effects, Begleiter and colleagues concluded that P300 amplitude is a genetically influenced trait that may serve as a biological marker for alcoholism (3).

A similarly significant reduction in the N430 component of ERP was reported for high-risk males having alcoholic fathers (85). Here, however, the latency of the N430 component was influenced by the amount of alcohol consumed.

Biochemical Markers of Alcoholism

The search for biochemical markers proceeds on two fronts. One focuses on specific biochemical processes, products, and catalysts while the other relies upon the possible serendipitous finding. In the latter category, a number of studies have chosen one of the favorite targets for association—the HLA system on chromosome 6p. However, the majority of these studies have been concerned not with alcoholism itself, but with secondary end-organ damage such as alcoholic cirrhosis or pancreatitis (24). Association studies of alcoholic cirrhosis, for example, have variously indicated significance for the antigens HLA-B8 (2), HLA-Bw40 (4), HLA-B13 (63), HLA-A2, and HLA-DR3 (8), and HLA-DR2 (104). Rada et al invoked a protective role (77) for the haplotype HLA-B8/B12. For alcoholic pancreatitis both HLA-B40 (29) and HLA-Bw39 (112) have been implicated.

One association study (20) has used clinically defined alcoholism itself as the trait of interest. A sample of 30 alcoholics was compared to two samples of controls for 11 HLA-A antigens, 16 HLA-B antigens, and 9 HLA-DR antigens. Among these, significant associations were seen for increased HLA-B40 and HLA-D3 frequencies and for decreased HLA-DR4 frequencies. The authors concluded that their findings, particularly for the HLA-DR antigens, are consistent with the view that alcoholism has a significant genetic component. Gilligan & Cloninger (34) note that all such associations must be assessed carefully. The number of antigens screened and the sample sizes can have a marked influence on the chances of finding a positive association; independent verification is essential and for HLA associations such confirmation has not been forthcoming.

One of the most widely discussed and cited markers in alcohol-association studies is platelet monoamine oxidase (MAO). The genetics of MAO activity is as yet unclear. Rice et al (79, 80) presented evidence that activity levels are influenced by a single major gene locus. Cloninger et al (17), on the other hand, found evidence for five components in the distribution of platelet MAO activity. This latter finding is consistent with a multigenic system of control or, at least, a multiple allele system. Numerous studies have reported a consistently lowered MAO activity in the platelets of alcoholics compared to controls (1, 28, 59, 91, 103, 111). Moreover, von Knorring (109) found that platelet MAO activities were consistently lower in the more genetically determined Type II alcoholics than in the Type 1 alcoholics, who were more like controls. While MAO activity might seem to be a good potential marker, it may well not be specific enough. Significant MAO activity differences have been reported among chronic schizophrenics (113).

Another platelet enzyme system that has proved to be a potential marker in alcoholism is adenylate cyclase (AC). Tabakoff et al (103) showed that adenylate cyclase activity, whether stimulated by guanine nucleotides, cesium fluoride, or prostaglandin E_1, was significantly lowered in a sample of 95 alcoholics compared to 33 control subjects. In particular, the cesium fluoride-stimulated AC activity correctly classified three-fourths of alcoholics and controls. In addition, these differences appeared not to be due to alcohol ingestion but rather were consistent with inherent, possibly inherited, characteristics (43, 103).

Other enzymes that have shown significant differences between alcoholics and controls are arylsulfatase A (46) and phospholipase D (64). In the latter case, a unique metabolite of ethanol called phosphatidylethanol was present in higher amounts in lymphocytes of males with a personal and family history of alcoholism than in the lymphocytes of controls. Synthesis of phosphatidylethanol is mediated by phospholipase D and the activity of this enzyme has been suggested as the differential factor. Like HLA associations, however, assessment of the significance of such findings must await independent verification.

Schuckit & Gold (89) have cited several instances in which the presence of ethanol has revealed significant differences between alcoholics and non-alcoholics that may be due to inherent genetic factors. The hormones prolactin (92) and cortisol (93) have both shown highly suggestive differences among individuals at high risk for alcoholism versus controls. Moreover, when data for both hormones are combined and reassorted on a family-history positive (FHP)/family-history negative (FHN) criterion, clear differences were observed indicative that hormonal changes in general may serve as a marker. To underscore this point, Topel (108) and Swartz et al (101) each presented evidence of altered hormone processing and responsiveness among alcoholics. Tope (108) reported both baseline deficits and altered precursor processing for β-endorphin in chronic alcoholics. The cleavage products β-endorphin and ACTH of the precursor molecule proopiomelanocortin (POMC) were present in significantly different proportions in the CSF of alcoholics compared to controls. The level of β-endorphin was lowered while ACTH levels were elevated in the alcoholics. Resting epinephrine levels as well as stress-induced epinephrine release were also lowered in nonalcoholics with a positive family history of alcoholism.

Clearly, a wide-range of potential markers exists for alcoholism and while some of these are alcohol-induced or influenced, none has so far been in the direct metabolic pathways of ethanol in the body. The first two enzyme systems in the oxidative pathway of ethanol are the alcohol dehydrogenases (ADH) and the aldehyde dehydrogenases (ALDH). To date, no unequivocal evidence exists to suggest that the rate of ethanol metabolism, as determined by the multiple isozyme forms of ADH and/or ALDH, has any influence on the risk of becoming alcoholic (84, 87, 106).

In recent years a secondary oxidative pathway for ethanol metabolism has revealed a suggestive property with respect to alcoholism. Lieber and colleagues defined the microsomal ethanol oxidizing system (MEOS) that is known to account for nearly all non-ADH oxidative metabolism of ethanol (56). Properties of MEOS indicate that an ethanol-induced cytochrome P450-dependent monoxygenase is driving the system. This cytochrome P450 is now known as P450IIE1 (98). It was found that MEOS plays a major role in the development of ethanol tolerance following chronic alcohol consumption (56). While it has not yet been systematically investigated, MEOS could in some way be involved in the build-up of tolerance characteristic of Type I alcoholics (11).

MAPPING ALCOHOLISM TO THE HUMAN GENOME

The rise of molecular biology and its incorporation into virtually every aspect of biological and medical research has resulted in dramatic advances in understanding and bodes well for the future. Botstein et al (7) ambitiously proposed a high resolution genetic map of the entire human genome. Today, much of this map is being realized (26, 44). As the techniques involved become more and more commonplace, hundreds of informative DNA probes are becoming available. Further, genetic analysis theory is being translated into practical, usable applications. These advances are making it feasible to map genes responsible for virtually any trait or disorder. Thus, it is to molecular biology, gene mapping, and linkage analysis that research on the genetics of alcoholism is turning. The task is enormous, the pitfalls many, and the cost high, yet the potential payoff is staggering—a fundamental understanding of the genetics of an addiction and a realistic program of intervention based upon that understanding.

The attempt to map risk genes for alcoholism, or for any trait or disease, must proceed under a paradigm mindful of four aspects of a genetic linkage study to avoid if the enterprise is to have a chance of success. These aspects are:

1. Unreliable, unreproducible, or inappropriate diagnostic criteria,
2. Improper ascertainment and/or uninformative families,
3. Uninformative genetic probes or errors in analysis of probe results,
4. Misspecification of the transmission model used in the linkage analysis.

One misstep along the way could doom a linkage study no matter how good the other components might be.

The diagnosis of alcoholism now relies on a set of criteria outlined by Goodwin & Guze (37). The diagnosis is given if an individual manifests three or more behaviors such as excessive ingestion of alcohol, persistent but unsuccessful desire to control ingestion, large amount of time and effort spent in obtaining alcohol, disruption of social, occupational, or recreational activities due to alcohol, classical withdrawal symptoms necessitating ingestion. Other more specific criteria have also been used in many studies (31). These criteria, coupled with the advent of subtyping, have proven to be quite reliable and reproducible.

Once appropriate diagnostic criteria for alcoholism are in place, the next step is to collect a sample of alcoholics and their families using those criteria that will give a reasonable chance for success in mapping. For Mendelian traits this process is quite straightforward (19, 67), but for a complex trait such as alcoholism it is not. To begin, the mode of transmission is unknown, though a major gene effect is likely for Type II alcoholic families (35). When the mode of transmission is not known, the method of sib-pair analysis and its modern refinements (42, 71, 100, 107) can be used to find a linked marker. Gershon & Goldin (33) showed that if the risk gene is a recessive then twenty-five pairs of affected sibs are required to detect linkage. When the gene is a dominant that number increases to forty-five pairs. Even so, the ease of ascertainment and computation makes sib-pair and sib-trio methods attractive (19).

If the mode of transmission can be assumed to be a single major locus, then the classical method of ascertaining large families with multiple affected members is optimal. Gershon & Goldin (33) estimated that eight to ten moderate-sized pedigrees (averaging fifteen members) would be sufficient regardless of the underlying genetic architecture. However, even though the pedigree method has tremendous advantages, such as high statistical power and the ability to provide genetic distance estimates, it suffers greatly if there is etiologic heterogeneity (13, 33). Weeks & Lange (110) have advanced a compromise method in which affected family members are used regardless of degree of relationship. The utility of this approach is still being evaluated. In the Washington University Alcoholism Research Center we have chosen an ascertainment scheme that hedges our bets. Families are initially ascertained through affected sib-pairs and then all nuclear family members are included. A third step follows in which the pedigree is extended only through affected relatives. This makes the families useful by any of the methods cited above.

Utility of the families is, however, only as good as the genetic markers used to type them. Three strategies are commonly employed (24). The first and most attractive is the candidate-gene approach. Numerous candidate genes have been offered and some were previously discussed as biochemical markers. Others have been found through the linkage studies already done. These include the enzyme esterase D (105) and the MNS blood group (41). A list of the available information on potential candidate loci for alcoholism is presented in Table 2. Failure to find linkage to candidate loci makes the second strategy optimal. This second strategy is the so-called "shotgun" method (24).

As noted, a large number of highly informative markers have been produced and are being mapped to the human genome (26, 66, 72). These probes may be employed in groups to systematically scan the entire genome for linkage. The last, and least attractive, method is "random-mapping" using the hypervariable, or "minisatellite," DNA probes (24). These probes initially held great promise for rapidly screening the equivalent of 50–60 conventional probes (47–49). However, two critical drawbacks are now known; linkage data are difficult to generate and are not additive across families, and, second, these probes are subject to very high spontaneous mutation rates (50). Thus, the first two strategies hold the best promise for future linkage and mapping studies of alcoholism.

Finally, sophisticated linkage-analysis applications are becoming generally

TABLE 2 A Listing of potential candidate genes in alcoholism

Candidate Gene	Cloned[a]	Chromosome Localization[a]	RFLP	PIC[b]	References
Alcohol dehydrogenase 1	yes	4q21-25			
Alcohol dehydrogenase 2	yes	4q21-25	Rsa I	.27	97
			Pvu II	.37	69
Alcohol dehydrogenase 3	yes	4q21-25	Msp I	.38	97
Alcohol dehydrogenase 4	no	4q21-25	Xba I	.37	97
Alcohol dehydrogenase 5	yes	4q21-25			97
Aldehyde dehydrogenase 1	yes	9q21			78
Aldehyde dehydrogenase 2	yes	12q24	EcoRI		97
Aldehyde dehydrogenase 3	yes	17			83
Aldehyde dehydrogenase 4	no	—			
Formaldehyde dehydrogenase	no	4q21-25			97
Cytochrome P450IIE1	yes	10	Taq I	.16	61
Fatty acid ethyl ester synthase	no	—			24
Esterase D	yes	13q14	Apa I	.27	99
Transketolase	no	—			24
2,3-butanediol dehydrogenase	no	—			24
Monoamine oxidase A	yes	Xp11-21	EcoRV	.37	68
Monoamine oxidase B	yes	Xp11-21			
Propiomelanocortin	yes	2p23	Rsa I	.35	
			Sst I	.27	30
Adenylate cyclase	no	—			
Phospholipase D	no	—			64
Arylsulfatase A	yes	22q13-qter			46
MNS blood group	no	4q28-31			60
HLA-A,B,C,DR	yes	6p21	Many	.21-.97	72

[a] Compiled from HGM9 (72) and cited references.
[b] Polymorphic Information Content (see reference 7).

available. Success of analysis depends upon the correct specification of parameters such as ascertainment and allelic penetrances (19, 33, 55, 67). While the sib-pair methods are to some extent immune to these problems, there is the attendant loss of power to consider when designing the ascertainment scheme. Each of the four areas discussed is being addressed by investigators in all branches of genetics and the solutions found will have a profound influence on the future studies of alcoholism.

CONCLUSIONS

In the past few years the study of the genetics of alcoholism has reached a level of sophistication and understanding that make it possible for the first time to realistically discuss the search for the genes responsible. Over the next few years several laboratories will be engaged in genetic linkage studies that take advantage of the parallel advances in DNA technology and genetic analysis. At least some of the pathophysiological processes involved in the disease of alcoholism will surely become known in detail.

Literature Cited

1. Alexopoulos, G. S., Lieberman, K. W., Frances, R., Stokes, P. E. 1981. Platelet MAO during the alcohol withdrawal syndrome. *Am. J. Psychiatry* 138:1254–55
2. Bailey, R. J., Krasner, N., Eddleston, A. L., Williams, R., Tee, D. E. H., et al. 1976. Histocompatibility antigens, autoantibodies, and immunoglobulins in alcoholic liver disease. *Br. Med. J.* 2:727–29
3. Begleiter, H., Porjesz, B., Bihari, B., Kissin, B. 1984. Event-related brain potentials in boys at risk for alcoholism. *Science* 227:1493–96
4. Bell, H., Nordhagen, R. 1978. HLA antigens in alcoholics with special reference to alcoholic cirrhosis. *Scand. J. Gastroenterol.* 15:453–56
5. Bohman, M., Sigvardsson, S., Cloninger, C. R. 1981. Maternal inheritance of alcohol abuse. Cross-fostering analysis of adopted women. *Arch. Gen. Psychiatry* 38:965–69
6. Bohman, M., Cloninger, C. R., Sigvardsson, S., von Knorring, A.-L. 1987. The genetics of alcoholisms and related disorders. *J. Psychiatr. Res.* 21:447–52
7. Botstein, D., White, R. L., Skolnick, M., Davis, R. W. 1980. Construction of a genetic linkage map in man using restriction fragment length polymorphisms. *Am. J. Hum. Genet.* 32:314–31
8. Bron, B., Kubski, D., Widmann, J. J., von Fliedner, V., Jeannet, M. 1982. Increased frequence of DR3 antigen in alcoholic hepatitis and cirrhosis. *Hepatogastroenterology* 29:183–86
9. Cadoret, R. J., Gath, A. 1978. Inheritance of alcoholism in adoptees. *Br. J. Psychiatry* 132:252–58
10. Cloninger, C. R. 1986. A unified biosocial theory of personality and its role in the development of anxiety stress. *Psychiatric Dev.* 3:167–222
11. Cloninger, C. R. 1987. Neurogenetic adaptive mechanisms in alcoholism. *Science* 236:410–16
12. Cloninger, C. R. 1987. Recent advances in family studies of alcoholism. In *Genetics and Alcoholism,* ed. H. W. Goedde, D. P. Agarwal, pp. 47–60. New York: Liss
13. Cloninger, C. R. 1989. Clinical heterogeneity in families of alcoholics. In *Genetic Aspects of Alcoholism,* ed. K. Küanmaa, B. Tabakoff, T. Saito, pp. 55–65. Helsinki: Finn. Found. Alcohol Studies
14. Cloninger, C. R., Bohman, M., Sigvardsson, S. 1981. Inheritance of alcohol abuse. Cross-fostering analysis of adopted men. *Arch. Gen. Psychiatry* 38:861–68
15. Cloninger, C. R., Sigvardsson, S., von Knorring, A.-L., Bohman, M. 1984. An adoption study of somatoform disorders:

II. Identification of two discrete somatoform disorders. *Arch. Gen. Psychiatry* 41:863–71

16. Cloninger, C. R., Bohman, M., Sigvardsson, S., von Knorring, A.-L. 1985. Psychopathology in adopted-out children of alcoholics. The Stockholm Adoption Study. In *Recent Developments in Alcoholism*, ed. M. Galanter, pp. 37–51. New York: Plenum

17. Cloninger, C. R., von Knorring, L., Oreland, L. 1985. Pentameric distribution of platelet monoamine oxidase activity. *Psychiatry Res.* 15:133–43

18. Cloninger, C. R., Sigvardsson, S., Gilligan, S. B., von Knorring, A.-L., Reich, T., Bohman, M. 1989. Genetic heterogeneity and the classification of alcoholism. In *Alcohol Research From Bench to Bedside.* ed. E. Gordis, B. Tabakoff, M. Linnoila. pp. 3–16. New York: Haworth

19. Conneally, P. M. 1989. Criteria for optimal linkage studies in alcoholism. See Ref. 13, pp. 259–64

20. Corsico, R., Pessino, O. L., Morales, V., Jmelninsky, A. 1988. Association of HLA antigens with alcoholic disease. *J. Stud. Alcohol* 49:546–50

21. Cotton, N. S. 1979. The familial incidence of alcoholism. *J. Stud. Alcohol* 40:89–116

22. Crothers, T. D. 1909. Heredity in the causation of inebriety. *Br. Med. J.* 2:659–61

23. Dahlberg, G., Stenberg, S. 1934. *Alkoholismen som Samhalls Problem.* Stockholm: Oskar Eklunds

24. Devor, E. J., Reich, T., Cloninger, C. R. 1988. Genetics of alcoholism and related end-organ damage. *Semin. Liver Dis.* 8:1–11

25. Deleted in proof

26. Donis-Keller, H., Green, P., Helms, C., Cartinhour, S., Weiffenbach, B., et al. 1987. A genetic linkage map of the human genome. *Cell* 51:319–37

27. Elderton, E., Pearson, K. 1910. A first study of the influence of parent's alcoholism on the physique and ability of the offspring. *Eugenics Laboratory Memoir X.* London: Cambridge Univ. Press

28. Faraj, B. A., Lenton, J. D., Kutner, M., Camp, V. M., Stammers, T. W., et al. 1987. Prevalence of low monoamine oxidase function in alcoholism. *Alcohol Clin. Exp. Res.* 11:464–67

29. Fauchet, R., Genetet, B., Gosselin, M., Gastard, J. 1979. HLA antigens in chronic alcoholic pancreatitis. *Tissue Antigens* 13:163–66

30. Feder, J., Gurling, H. M. D., Darby, J., Cavalli-Sforza, L. L. 1985. DNA restriction fragment analysis of the proopiomelanocortin gene in schizophrenia and bipolar disorders. *Am. J. Hum. Genet.* 37:186–94

31. Feighner, J. P., Robins, E., Guze, S. B., Woodruff, A. R., Winokur, G., Munoz, R. 1972. Diagnostic criteria for use in psychiatric research. *Arch. Gen. Psychiatry* 26:57–63

32. Frances, R. J., Bucky, S., Alexopoulos, G. S. 1984. Outcome study of familial and nonfamilial alcoholism. *Am. J. Psychiatr.* 141:1469–71

33. Gershon, E. S., Goldin, L. R. 1987. The outlook for linkage research in psychiatric disorders. *J. Psychiatr. Res.* 21:541–50

34. Gilligan, S. B., Cloninger, C. R. 1987. The genetics of alcoholism. In *Human Metabolism of Alcohol*, ed. K. Crow, R. D. Bott. Boca Raton, Florida: CRC Press. In press

35. Gilligan, S. B., Reich, T., Cloninger, C. R. 1987. Etiologic heterogeneity in alcoholism. *Genet. Epidemiol.* 4:395–414

36. Goodwin, D. W. 1987. Genetic influences in alcoholism. *Adv. Intern. Med.* 32:283–98

37. Goodwin, D. W., Guze, S. B. 1989. *Psychiatric Diagnosis*, 4th ed. Oxford: Oxford Univ. Press

38. Goodwin, D. W., Schulsinger, F., Hermansen, L., Guze, S. B., Winokur, G. 1973. Alcohol problems in adoptees raised apart from alcoholic biological parents. *Arch. Gen. Psychiatry* 28:238–43

39. Goodwin, D. W., Schulsinger, F., Knop, J., Mednick, S. A., Guze, S. B. 1977. Alcoholism and depression in adopted-out daughters of alcoholics. *Arch. Gen. Psychiatry* 34:751–75

40. Goodwin, D. W., Schulsinger, F., Knop, J., Mednick, S. A., Guze, S. B. 1977. Psychopathology in adopted and non-adopted daughters of alcoholics. *Arch. Gen. Psychiatry* 34:1005–9

41. Hill, S. Y., Aston, C., Rabin, B. 1988. Suggestive evidence of genetic linkage between alcoholism and the MNS blood group. *Alcohol Clin. Exp. Res.* 12:811–14

42. Hodge, S. E. 1984. The information contained in multiple sibling pairs. *Genet. Epidemiol.* 1:109–22

43. Hoffman, P. L., Lee, J. M., Saito, T., Willard, B., DeLeon-Jones, F., et al. 1989. Platelet enzyme activities in alcoholics. See Ref. 13, pp. 95–106

44. Howard Hughes Medical Institute, Salt

Lake City, Utah, 1987. *Linkage Maps of Human Chromosomes* (unpubl.), 86 pp.

45. Hrubek, Z., Omenn, G. S. 1981. Evidence of genetic predisposition to alcoholic cirrhosis and psychosis: Twin concordances for alcoholism and its biologic end-points by zygosity among male veterans. *Alcohol Clin. Exp. Res.* 5:207–15

46. Hulgalkar, A. R., Nora, R., Manowitz, P. 1984. Arylsulfatase A variants in patients with alcoholism. *Alcohol Clin. Exp. Res.* 8:337–41

47. Jeffreys, A. J., Wilson, V., Thein, S. L. 1985. Hypervariable "minisatellite" regions in human DNA. *Nature* 314:67–73

48. Jeffreys, A. J., Wilson, V., Thein, S. L. 1985. Individual-specific "fingerprints" of human DNA. *Nature* 316:76–9

49. Jeffreys, A. J., Wilson, V., Thein, S. L., Weatherall, D. J., Ponder, B. A. J. 1986. DNA "fingerprints" and segregation analysis of multiple markers in human pedigrees. *Am. J. Hum. Genet.* 39:11–24

50. Jeffreys, A. J., Royle, N. J., Wilson, V., Wong, Z. 1988. Spontaneous mutation rates to new length alleles at tandem-repetitive hypervariable loci in human DNA. *Nature* 332:278–81

51. Jellinek, E. M., Jolliffe, N. 1940. Effects of alcohol on the individual: Review of the literature of 1939. *Quart. J. Stud. Alcohol.* 1:110–81

52. Jonsson, E., Nilsson, T. 1968. Alkohlkonsumption has monozygota och dizygota tvillingar. *Nord. Hyg. Tidskr.* 49:21–25

53. Kaij, L. 1960. *Studies of the Etiology and Sequels of Abuse of Alcohol.* Lund, Sweden: Univ. Lund Press

54. Kaprio, J., Sarna, S., Koskenvuo, M., Rantasalo, I. 1978. *The Finnish Twin Registry: Baseline Characteristics.* Section II. History of symptoms and illness, use of drugs, physical characteristics, smoking, alcohol and physical activity. Helsinki: Kansanterveystieteen Julkaisuja

55. Kidd, K. K. 1987. Research design considerations for linkage studies of affective disorders using recombinant DNA markers. *J. Psychiat. Res.* 21:551–57

56. Lieber, C. S. 1987. Effects of chronic alcohol consumption on the metabolism of ethanol. See Ref. 12, pp. 161–72

57. Loehlin, J. C. 1972. An analysis of alcohol-related questionnaire items from the National Merit Twin Study. *Ann. NY Acad. Sci.* 197:117–20

58. Long, J. F. 1879. Use and abuse of alcohol. *Trans. Med. Soc. N.C.* 26:87–100

59. Major, L. F., Goyer, P. F., Murphy, D. L. 1981. Changes in platelet monoamine oxidase activity during abstinence. *J. Stud. Alcohol* 42:1052–57

60. Mattei, M. G., London, J., Rahuel, C., d'Auriol, L., Colin, Y., et al. 1987. Chromosome localization by in situ hybridization of the gene for human erythrocyte glycophrin to region 4q28–31. *Cytogenet. Cell Genet.* 46:658

61. McBride, O. W., Umeno, M., Gelboin, H. V., Gonzalez, F. J. 1987. A Taq I polymorphism in the human P450IIE1 gene on chromosome 10 (CYP2E). *Nucleic Acids Res.* 15:10071

62. MacNicholl, T. A. 1905. A study of the effects of alcohol on school children. *Q. J. Inebriation* 27:113–17

63. Melendez, M., Vargas-Tank, L., Fuentes, C., Armas-Mercio, R., Castillo, D., et al. 1979. Distribution of HLA histocompatibility antigens, ABO blood groups and Rh antigens in alcoholic liver disease. *Gut* 20:288–90

64. Mueller, G. C., Fleming, M. F., LeMahieu, M. A., Lybrand, G. S., Barry, K. J. 1988. Synthesis of phosphatidylethanol—a potential marker for adult males at risk for alcoholism. *Proc. Natl. Acad. Sci. USA* 85:9778–82

65. Murray, R. M., Clifford, C., Burling, H. M. D., Tophan, A., Clow, A., Bernadt, M. 1983. Current genetic and biological approaches to alcoholism. *Psychiatr. Dev.* 2:179–92

66. Nakamura, Y., Leppert, M., O'Connell, P., Wolff, R., Holm, T., et al. 1987. Variable number of tandem repeat (VNTR) markers for human gene mapping. *Science* 235:1616–22

67. Ott, J. 1985. *Analysis of Human Genetic Linkage.* Baltimore: The Johns Hopkins Univ. Press

68. Ozelius, L., Hsu, Y.-P., Bruns, G., Powell, J. F., Chen, S., et al. 1988. Human monoamine oxidase gene (MAOA): chromosome position (Xp21-p11) and DNA polymorphism. *Genomics* 3:53–58

69. Pandolfo, M., Smith, M. 1988. A Pvu II RFLP in the human ADH3 gene. *Nucleic Acids Res.* 16:11857

70. Partanen, J., Bruun, K., Markanen, T. 1966. *Inheritance of drinking behavior.* Helsinki: Finn. Found. Alcohol Studies

71. Payami, H., Thomson, G., Louis, E. J. 1984. The affected sib method. III. Selection and recombination. *Am. J. Hum. Genet.* 36:352–62

72. Pearson, P. L., Kidd, K. K., Willard, M. F., Track, R., Consiglio, D., et al. 1987. Report of the Committee on Human Gene Mapping by Recombinant DNA Techniques. *Cytogenet. Cell Genet.* 46:390–566

73. Penick, E. C., Read, M. R., Crowley, P. A., Powell, B. J. 1978. Differentiation of alcoholics by family history. *J. Stud. Alcohol.* 39:1944–48

74. Pollock, B. E., Volavka, J., Goodwin, D. W. 1983. The EEG after alcohol administration in men at risk for alcoholism. *Arch. Gen. Psychiatry* 40:857–61

75. Porjesz, B., Begleiter, H. 1983. Brain dysfunction and alcohol. In *The Pathogenesis of Alcoholism,* ed. B. Kissin, H. Begleiter, pp. 415–83. New York: Plenum

76. Propping, P., Kruger, J., Mark, N. 1981. Genetic predisposition in alcoholics and their relatives. *Hum. Genet.* 59:51–59

77. Rada, R. T., Knodell, R. G., Troup, G. M., Kellner, R., Hermanson, S. M., Richards, M. 1981. HLA frequences in cirrhotic and noncirrhotic male alcoholics: a controlled study. *Alcohol Clin. Exp. Res.* 5:188–91

78. Raghunathan, L., Hsu, L. C., Klisak, I., Sparkes, R. S., Yoshida, A., Mohandas, T. 1988. Regional localization of the human genes for aldehyde dehydrogenase-1 and aldehyde-dehydrogenas-2. *Genomics* 2:267–69

79. Rice, J., McGuffin, P., Shaskan, E. G. 1982. A commingling analysis of platelet monoamine oxidase activity. *Psychiatry Res.* 7:325–35

80. Rice, J., McGuffin, P., Goldin, L. R., Shaskan, E. G., Gershon, E. S. 1984. Platelet monoamine oxidase (MAO) activity: Evidence for a single major locus. *Am. J. Hum. Genet.* 36:36–43

81. Roe, A. 1944. The adult adjustment of children of alcoholic parents raised in foster homes. *Q. J. Stud. Alcohol* 5:378–93

82. Rouché, B. 1960. *Alcohol.* New York: Grove

83. Santisteban, I., Povey, S., West, L. F., Parrington, J. M., Hopkinson, D. A. 1985. Chromosome assignment, biochemical and immunological studies on a human aldehyde dehydrogenase, ALDH3. *Am. Hum. Genet.* 49:87–100

84. Saunders, J. B., Williams, R. 1983. The genetics of alcoholism: is there an inherited susceptibility to alcohol-related problems? *Alcohol* 18:189–217

85. Schmidt, A. L., Neville, H. J. 1985. Language processing in men at risk for alcoholism: an event-related potential study. *Alcohol* 2:529–33

86. Schuckit, M. A. 1972. Family history and half-sibling research in alcoholism. *Ann. NY Acad. Sci.* 197:121–25

87. Schuckit, M. A. 1984. Subjective responses to alcohol in sons of alcoholics and controls. *Arch. Gen. Psychiatry* 41:879–84

88. Schuckit, M. A. 1985. Trait (and state) markers of a predisposition to psychopathology. In *Physiological Foundations of Clinical Psychiatry,* ed. L. L. Judd, P. Groves, Philadelphia: Lippincott

89. Schuckit, M. A., Gold, E. O. 1988. A simultaneous evaluation of multiple markers of ethanol/placebo challenges in sons of alcoholics and controls. *Arch. Gen. Psychiatry* 45:211–16

90. Schuckit, M. A., Goodwin, D. W., Winokur, G. A. 1972. A study of alcoholism in half-siblings. *Am. J. Psychiatry* 128:1132–36

91. Schuckit, M. A., Shaskan, E., Duby, J. 1982. Platelet MAO activities in the relatives of alcoholics and controls. *Arch. Gen. Psychiatry* 39:137–40

92. Schuckit, M. A., Parker, D. C., Rossman, L. R. 1983. Prolactin responses to ethanol in men at elevated risk for alcoholism and controls. *Biol. Psychiatry* 18:1153–59

93. Schuckit, M. A., Gold, E., Risch, C. 1987. Plasma cortisal levels following ethanol in sons of alcoholics and controls. *Arch. Gen. Psychiatry* 44:942–45

94. Shuttleworth, G. E., Beach, F. 1900. Idiocy and imbecility. In *A System of Medicine by Many Writers,* ed. T. C. Allbutt, pp. 223–247. New York: MacMillan

95. Sigvardsson, S., von Knorring, A.-L., Bohman, M., Cloninger, C. R. 1984. An adoption study of somatoform disorders: I. The relationship of somatization to psychiatric disability. *Arch. Gen. Psychiatry* 41:853–62

96. Sigvardsson, S., Cloninger, C. R., Bohman, M. 1985. Prevention and treatment of alcohol abuse: Use and limitations of the high risk paradigm. *Soc. Biol.* 32:185–94

97. Smith, M. 1986. Genetics of human alcohol and aldehyde dehydrogenases. *Adv. Hum. Genet.* 14:249–90

98. Song, B.-J. Gelboin, H. V., Park, S.-S., Yang, C. S., Gonzalez, F. J. 1986. Complementary DNA and protein sequences of ethanol-inducible rat and hu-

man cytochrome P-450s. *J. Biol. Chem.* 261:16689–97

99. Squire, J., Dryja, T. P., Dunn, J., Goddard, A., Hoffman, T., et al. 1986. Cloning of the esterase D gene: A polymorphic gene probe closely linked to the retinoblastoma locus on chromosome 13. *Proc. Natl. Acad. Sci. USA* 83:6573–77

100. Suarez, B. K., Rice, J. P., Reich, T. 1978. The generalized sib-pair IBD distribution: Its use in the detection of linkage. *Am. Hum. Genet.* 42:87–94

101. Swartz, C. M., Drews, V., Cadoret, R. 1987. Decreased epinephrine in familial alcoholism. *Arch. Gen. Psychiatry.* 44:938–41

102. Tabakoff, B., Hoffman, P. L. 1989. Genetics and biological markers of risk for alcoholism. See Ref. 13, pp. 127–42

103. Tabakoff, B., Hoffman, P. L., Lee, J. M., Saito, T., Willard, B., DeLeon-Jones, F. 1988. Differences in platelet enzyme activity between alcoholics and nonalcoholics. *New Engl. J. Med.* 318:134–39

104. Tait, B. D., Mackay, I. R. 1982. HLA and alcoholic cirrhosis. *Tissue Antigens* 19:6–10

105. Tanna, V. L., Wilson, A. F., Winokur, G., Elston, R. C. 1988. Possible linkage between alcoholism and esterase-D. *J. Stud. Alcohol* 49:472–76

106. Thacker, S. B., Veech, R. L., Vernon, A. A., Rutstein, D. D. 1984. Genetic and biochemical factors relevant to alcoholism. *Alcohol. Clin. Exp. Res.* 8:375–83

107. Thomson, G. 1986. Determining the mode of inheritance of RFLP-associated diseases using the sib-pair method. *Am. J. Hum. Genet.* 39:207–21

108. Topel, H. 1988. Beta-endorphin genetics in the etiology of alcoholism. *Alcohol* 5:159–65

109. von Knorring, A.-L., Bohman, M., von Knorring, L., Oreland, L. 1985. Platelet MAO activity as a biology marker in subgroups of alcoholism. *Acta Psychiatr. Scand.* 72:51–58

110. Weeks, D. E., Lange, K. 1988. The affected-pedigree-member method of linkage analysis. *Am. J. Hum. Genet.* 42:315–26

111. Wiberg, A., Gottfries, C.-G., Oreland, L. 1977. Low platelet monoamine oxidase activity in human alcoholics. *Med. Biol.* 55:181–86

112. Wilson, J. S., Gossat, D., Tait, A., Rouse, S., Juan, X. J., Pirola, R. C. 1984. Evidence for an inherited predisposition to alcoholic pancreatitis: A controlled HLA typing study. *Dig. Dis. Sci.* 29:272–30

113. Wyatt, R. J., Potkin, S. G., Murphy, D. L. 1979. Platelet monoamine oxidase activity in schizophrenia: a review of the data. *Am. J. Psychiatry* 136:377–85

Annu. Rev. Genet. 1989. 23:37–69

MECHANISMS THAT CONTRIBUTE TO THE STABLE SEGREGATION OF PLASMIDS

Kurt Nordström

Department of Microbiology, Uppsala University, The Biomedical Center, Box 581, S-751 23 Uppsala, Sweden

Stuart J. Austin

Laboratory of Chromosome Biology, BRI-Basic Research Program, NCI-Frederick Cancer Research Facility, Frederick, Maryland 21701

CONTENTS

0066-4197/89/1215-0037$02.00

PLASMIDS CARRY GENES FOR THEIR STABLE INHERITANCE

All the essential genetic information in a bacterial cell is located on one circular, double-stranded DNA molecule, the chromosome. In addition, most bacteria carry other DNA molecules, called plasmids. These replicate in harmony with their host cells. They may contribute to the phenotype of the host but are not essential for cell viability.

During the life cycle of a cell, two processes take place; growth (increase in mass) and cell division (increase in number). Similarly, two events can be recognized in the life cycle of a replicon (chromosome or plasmid); replication (increase in number) and partition (distribution to the daughter cells at cell division). Both events are of utmost importance for the maintenance of plasmids in bacterial populations. Therefore, plasmids control replication and partition. In addition, they often adopt other strategies that increase the fidelity of inheritance.

Natural bacterial plasmids are inherited with a very high degree of stability. The loss rate of low-copy-number plasmids can be as low as 10^{-7} per cell division (57). This implies the existence of mechanisms that control the distribution of the plasmids at cell division. This realization spurred the search for plasmid mutants that had lost their stability of inheritance. Such mutants were first described in the early 1970s (31). Yoshikawa (91), in an elegant analysis, used bacteriophage P1 transduction to create deletions in the IncFII plasmid R100 that was integrated into the *Escherichia coli* chromosome. This defined a region, RepB, that "was not necessary for replication but needed for stable replication" (maintenance).

In the early 1980s different laboratories deliberately used restriction endonucleases to create deletions and to define basic replicons (the smallest part of a plasmid able to replicate). Researchers observed that these basic replicons were unstably inherited and that this instability was caused by the lack of various functions present on the wild-type plasmid. In this way, regions called *stb* (plasmid R100/NR1, (52)), and *par* (plasmid pSC101, (51); plasmid R1,

(58)) were described. Subsequently, regions that stabilize plasmids have been found in many systems. However, a closer analysis of different plasmids revealed that some of the stability functions did not involve real partition functions, i.e. functions that actively distribute the plasmid molecules to the daughter cells, but affected other aspects of plasmid life cycles, such as the resolution of oligomers formed by recombination or the killing of plasmid-free cells by plasmid-coded functions. Moreover, it was found that a single plasmid can make use of several such functions. Thus the extreme fidelity of maintenance found in naturally occurring plasmids appears to be due to the concerted efforts of multiple systems. This article reviews these processes and attempts to explain how they work together to achieve stable maintenance. To do this, we first discuss plasmid maintenance from a theoretical point of view before moving into a short discussion of real systems. This paper does not attempt an encyclopaedic coverage of the literature but rather discusses key concepts and areas of research.

THEORETICAL CONSIDERATIONS

Natural plasmids often encode several different systems that help to maintain them stably. These are encoded in separable sections of the plasmid genome, hence they can be progressively deleted. Actually, only one segment, the basic replicon, is absolutely essential for normal plasmid replication. A basic replicon is the smallest part of a plasmid able to replicate with the normal copy number of its parent plasmid. All basic replicons currently described consist of an origin of replication and a negative-feedback control system that responds to cell growth in such a way as to limit the propagation of the plasmid to a characteristic copy number.

The "Ideal" Basic Replicon

A hypothetical "ideal" basic replicon has the following characteristics:

1. A replication control system that responds rapidly to any variations in the copy numbers in newborn cells such that there are exactly $2n$ in each cell at the time of cell division.
2. Random distribution of the copies of the plasmid to the daughter cells at cell division. Hence, each individual plasmid molecule has an equal chance to end up in either of the two daughter cells. Consequently, plasmid-free cells will appear with a probability that is dependent upon the copy number of the plasmid. This probability is denoted L_{th} (Loss frequency, theoretical) and is given by the following equation:

$$L_{th} = (1/2)^{2n}$$

1.

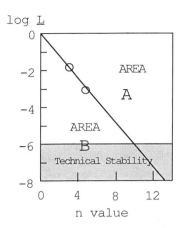

Figure 1 The frequency of formation of plasmid-free cells (*L*) as a function of the copy number (*n*) of the plasmid. The line shows $L_{th} = (½)^n$, i.e. the *L* value for an ideal basic replicon. Areas A and B represent worse-than-random and better-than-random segregation, respectively. The circles show L_{ex} of the basic replicon of plasmid R1 (left) and a copy mutant of plasmid R1 (right) (58). Plasmids with *L* values of less than 10^{-6} are considered to be stable as measurement of loss levels becomes technically unfeasible at lower values and mutational instability is likely to limit the efficiency of any system at about this value.

Equation 1 gives a straight line in the semilogarithmic plot $L_{th} = f(n)$ shown in Figure 1.

Ideal basic replicons with $n \geq 10$ would be stably inherited in practice since *L* values below 10^{-6} cannot readily be measured and this range is approaching that in which mutations affecting plasmid maintenance occur. Thus, at such a high copy number, an ideal basic replicon might well be able to function as a viable plasmid in nature. However, in practice, basic replicons deviate from ideal behavior. The experimental loss frequency L_{ex} is larger than L_{th}; i.e. it falls in area A of Figure 1. These deviations from ideal behavior occur because neither of the assumptions made for ideal behavior is completely valid.

Effect of Variations in Copy Number

The first assumption is not strictly correct because the replication controls of real basic replicons do not respond in such a way as to give every cell precisely 2*n* copies of the plasmid at the time of cell division. In practice there is a spread in the copy number. This tends to increase the value of L_{ex} into area A of Figure 1. However, unless the spread is very broad, this will only cause a small increase in loss frequency (57, 58).

The Unit of Segregation may be More than One Plasmid Genome

A far more significant deviation from ideal behavior is that, in practice, not all copies of a basic replicon are free to diffuse randomly to daughter cells. The actual unit of segregation may be larger than a single plasmid copy, moving the L_{ex} value of the basic replicon further into area A of Figure 1. This can cause a considerable increase in loss frequency. Two different mechanisms may contribute to this effect:

1. Recombination gives rise to dimers and other oligomeric forms and hence reduces the number of partitioned units. This effect can be eliminated by use of Rec⁻ host cells. Unfortunately, Rec⁻ strains are not ideal experimental systems as they often have aberrant growth characteristics.

2. Noncovalent clustering or compartmentalization of plasmids reduces the number of partitioned units. It is possible that some basic replicons are unable to diffuse freely because a complex is formed at the synthesis site or copies subsequently clump into larger units.

Plasmid Systems that Affect the Stability of Segregation

Naturally occurring plasmids make use of a basic replicon but also contain one or more additional elements to achieve a high degree of stability. These elements act in one of two ways. The first group of systems act by minimizing the deviations from the ideal state of the basic replicon so that the experimental loss frequency approaches the theoretical ideal value L_{th} more closely. We refer to these elements as "helper elements" as they help the random distribution of the basic replicon. Although they can substantially improve the maintenance of the replicon, they are not able to reduce the loss frequency below the value of L_{th}. Hence the L_{ex} remains in area A of Figure 1. The second group of elements work by actively improving the distribution of the plasmid copies in the population. These systems, that we refer to as "active elements", can allow the replicon to achieve L_{ex} values smaller than L_{th} (area B, Figure 1).

For high-copy-number plasmids, values of L_{ex} that lie close to L_{th} may be quite sufficient for stable inheritance. Thus high-copy-number plasmids may, and probably do, make use of just a basic replicon and one or more helper elements that aid its random distribution. However, low-copy-number plasmids, although they can—and often do—benefit from these aids, must have an active element. This follows as a general principle because these plasmids have high values for L_{th} and cannot achieve stability unless L_{ex} is lower and lies well within area B of Figure 1.

HELPER ELEMENTS THAT AID RANDOM DISTRIBUTION OF THE BASIC REPLICON

Site-Specific Recombination Systems

As all copies of a given plasmid in a cell are homologous with each other, they can recombine to form plasmid oligomers. *E. coli* cells transformed with monomeric forms of small, multicopy plasmids grow to form populations in which as much as 50% of the plasmid DNA can be in oligomeric forms (39). As plasmid-replication controls do not compensate for oligomer formation, the number of segregating units of the basic replicon is reduced. In the case of high-copy-number plasmids, the loss frequency is increased because random distribution of the units is less effective. Multimer formation in low-copy-number plasmids can also have a marked effect as some dividing cells will contain only a single multimeric plasmid, making any partition impossible. Thus all plasmids can benefit from some mechanism to overcome this problem. In those cases studied so far, the mechanism used is a site-specific recombination system. By greatly increasing the rate of interplasmid recombination, any multimeric forms are rapidly resolved into monomers that can then be partitioned properly (Figure 2). These site-specific recombination systems fall into two general classes: in the first (e.g. the *lox-cre* recombination system of the P1 plasmid (6)), the recombination site and the recombinase enzyme are both plasmid-encoded, and in the other (e.g. the *cer-xer* system of ColE1 (79)), the recombinase is host-encoded.

All natural plasmids probably make use of some type of highly active

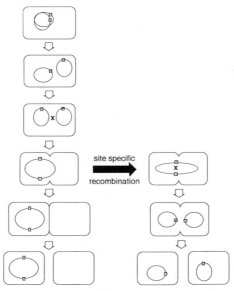

site specific
recombination

Figure 2 Effect of a site-specific recombination system on the inheritance of a plasmid. Generalized recombination tends to make dimers and other multimers from plasmid copies. This limits the availability of separate molecules for partition. This can be a particularly severe problem with low-copy-number plasmids and can cause frequent loss events. Natural plasmids use one or more highly active site-specific recombination systems to promote a rapid interchange of dimer to monomer forms. Such systems ensure that monomers are always available for partition, thus solving the problem. Left pathway: generalized recombination between daughter plasmids gives rise to dimers. Plasmid loss occurs. Right pathway: highly active site-specific recombination resolves the dimers and allows proper partition.

site-specific recombination system. Cases supporting this point are steadily accumulating. Many plasmids carry transposable elements, of which some, such as Tn*1000* (γ-δ), Tn*21*, and Tn*3*, encode highly active site-specific recombination systems of their own that resolve co-integrate intermediates during transposition (25). These systems probably are equally proficient at resolving oligomers. Thus, the common association of transposons with plasmids may benefit both the transposon by providing a mobile replicon for its maintenance, and the plasmid by aiding its stability.

Anticlumping systems

Some basic replicons are much less stably maintained than would be expected from a random distribution of copies, even under conditions where multimer formation is blocked. These replicons may form noncovalent aggregates that prevent their free diffusion. The non-random distribution of pSC101 derivatives has been attributed to aggregate formation and the action of the pSC101 *par* locus to dissociation of plasmid aggregates (83), although recent data present an alternative explanation (see below). Since the existence of such "anti-clumping" activities cannot be excluded at present, this possibility should nevertheless be considered when analyzing new plasmid stability functions.

Functions that Improve Copy Number Distribution

As pointed out above, basic replicons in practice have a spread in the copy number of individual cells prior to division, due to the nature of their copy-number control systems. The cause and extent of this spread for InFII plasmids is understood (57, 87). The spread is considerable, but has a very modest effect on the value of L_{ex} (57, 58). However, other basic replicons may have sufficiently broad distributions of copy number to cause large loss frequencies. If so, stabilizing functions may exist that act by improving the characteristics of the replication control system and narrowing the copy-number distribution. The *par* region of pSC101 may encode such a function (see below). There is strong evidence that the *Staphylococcus* plasmid pT181 has a function that works in this way (34).

ACTIVE SYSTEMS: BETTER-THAN-RANDOM SEGREGATION

Killer Systems

In 1975, Koyama & Yura (43) pointed out that accuracy of plasmid maintenance could, in principle, be achieved by an "addiction" mechanism; if cells that lose an established plasmid die, the population would never contain viable cured cells. L_{ex} would be drastically reduced and fall within area B of

Figure 1. Such systems have in fact been described in the F plasmid (68) and R1 (24), The F *ccd* system was originally thought to operate by delaying cell division of cells with too few copies of the plasmid, whereas it clearly acts by killing cured cells (38). These systems appear to work on a "poison and antidote principle." The plasmids produce a killer substance and an agent that blocks its activity. When the plasmid is lost the blocking agent decays, but the killer substance persists in the cytoplasm and kills the cell. Both the killer and blocking agents of F are probably proteins (68), whereas for R1 the killer is a messenger RNA for a toxic protein and the blocking agent is a small antisense RNA that blocks expression of the mRNA (70).

Killer systems are incompatibility determinants, i.e. they contribute to the activity that prevents two similar plasmids from being stably maintained in the same cell. If a killer system is cloned into a standard vector such as pBR322, the cloned DNA exerts a destabilizing effect on any plasmid that makes use of the same killer system to aid its stability (68). This follows because the cloned DNA, which is itself stably maintained, produces the blocking agent. Thus, when the resident plasmid is lost through some error in replication or partition, the killer substance fails to kill the cell and plasmid-free cells accumulate in the population.

True Partition Systems

A true partition system selectively moves DNA molecules to ensure that each daughter cell receives at least one copy. Assuming that the replicon and its accessory systems achieve a copy number of at least two free plasmids in every dividing cell, an accurate partition system would give perfect plasmid maintenance without any killing of cells. Strong evidence supports the presence of active partition elements in plasmids F, P1, and the *incFII* plasmids NR1 and R1 (2, 23, 67, 80). All naturally occurring, low-copy-number plasmids are likely to have such a system, in our view, given the obvious selective advantage involved.

The mechanism of active partition remains an interesting problem, and models for this process and supporting evidence are discussed below. However, that some systems do indeed promote active partition now seems reasonably well established. The evidence for this is most straightforward in the case of very low-copy-number plasmids such as P1 and F. The DNA regions involved are well defined and in each case consist of a *cis*-acting recognition site and open reading frames for two essential proteins (1, 55). Plasmids have been constructed that consist only of these elements and a well-defined, low-copy-number, minimal replicon (2). The plasmids have loss frequencies far lower than the L_{th} values calculated for the replicon, with L_{ex} values well into area B of Figure 1. Thus the elements actively promote a nonrandom distribution of copies. Moreover, the plasmids have little or no effect on the growth rate of cells harboring them. They are not, therefore, acting by a cell-killing mechanism that would have a very considerable effect at such low copy numbers. Active partitioning remains as the only viable explanation. Apparently, these systems must promote the active segregation of plasmid copies in a way analogous to the segregation of chromosomes by mitosis in higher cells.

As a general rule, true partition systems are also incompatibility determinants (59). This has important implications for the way in which copies are chosen for partition and is discussed at length below.

HOW CAN SYSTEMS THAT STABILIZE THE INHERITANCE OF PLASMIDS BE IDENTIFIED?

The preceding theoretical discussion indicates the existence of several systems that contribute to the stability of inheritance of any given plasmid. How can these individual elements be defined and described?

Firstly, the systems that contribute to the stability of inheritance of plasmids seem in general to be independent of each other. Each one is encoded by a cassette of DNA; cassettes from different plasmids may be combined to give stable replicons. These cassettes are the targets of the mutations and deletions that are the starting points for analysis. The phenotype of a mutant that has lost one of the several systems contributing to the stability of inheritance is an increased loss rate. Mapping such loci indicates the presence of a stabilizing cassette in the vicinity, but the phenotype does not define the type of function that has been lost. Nor is the severity of the phenotype likely to be an indicator of mechanism, as the presence of other cassettes can mask the full capabilities of the lost function. Therefore the function must generally be studied in isolation, which unfortunately requires a careful genetic analysis to determine the element's limits and to ensure that only one cassette is present in the defined region. It is then possible to study the capabilities of the element by joining it to a defined minimal replicon whose properties are well understood. With such constructs, the phenotype caused by the element can point to its likely function.

The degree of instability, the L value as well as the copy number (n) must be measured. A comparison of the L values of the basic replicon with and without the stabilizing function in a $L = f(n)$ diagram (see Figure 1) indicates the type of system present. Functions that aid random distribution can only give a stability approximately fitting the line $L = (\frac{1}{2})^{2n}$ in Figure 1. Only the active systems are able to move the stability into the B region of Figure 1.

There are significant technical difficulties in determining the L value. The experimental loss frequency L_{ex} cannot be determined directly. Rather, the kinetics are determined of the growth of the plasmid-free, the plasmid-carrying, and the total populations. This will make the result dependent upon the relative growth rate of the plasmid-free (P^-) and the plasmid-carrying (P^+) cells. Normally the plasmid-free cells grow more rapidly than the plasmid-carrying cells. This must be corrected for. Differences in growth rate will express themselves as nonlinear curves on plots of $\log [P^+/(P^+ + P^-)] = f(t)$ (59). Measurements may give very misleading L values if determinations are made at two points only, especially if the loss has been substantial. Even worse is to dilute the culture and let it grow into the stationary phase before analysis, since virtually nothing is known about replication under such circumstances. Hence, only by extrapolating back to initial rates can L_{ex} be determined with precision.

As explained below, most active systems exert incompatibility effects towards their parental plasmids, whereas other systems never do. This effect can be determined with isolated elements provided the basic replicon used is itself compatible with the parent plasmid. The combination of stability and incompatibility phenotypes is usually sufficient to determine whether or not an active or helper element is present.

The two types of active systems, partition and killer systems, are easy to distinguish from each other. The latter has deleterious effects on the growth rate and causes dead cells to accumulate. However, only when very low-copy-number replicons are used will the effects be easily measurable. Killer systems have been readily identified by cloning the cassette onto a temperature-sensitive (*rep*-ts) replicon and determining what happens after a shift to nonpermissive temperature (24, 38). Once the copy number has dropped to one per cell, a killer system will ultimately result in death of half the progeny in each generation. This will completely stop the viable cells from increasing (Figure 3). In this type of test, a true partition system should not affect growth and plasmid-free cells should rapidly accumulate.

To distinguish between systems that aid random distribution may not be as simple. Site-specific recombination systems are readily detected as their stabilizing effect is suppressed in *recA* hosts (6). Also, it is fairly easy to perform direct assays for recombinational mobilization of the plasmid from one strain to another (6, 79). Any remaining systems (which may include replication control enhancers and anti-clumping systems) are difficult to classify as the existence and properties of such systems are still uncertain.

EXAMPLES OF PLASMIDS

The plasmid-maintenance functions of the several plasmids of *E. coli* that we discuss do not by any means represent a complete list of those for which information is available, nor is *E. coli* the only organism in which such studies have been carried out. Important studies have also been carried out

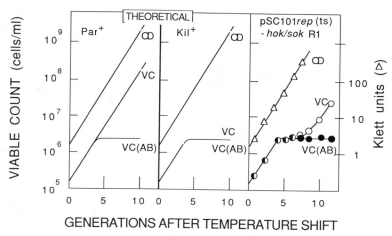

Figure 3   Effect of plasmid partition and killer functions on growth of the host bacteria under conditions where the plasmid does not replicate. The two left panels show a theoretical experiment in which a *rep(ts)* replicon carrying either a *par* or a *kil* function are grown exponentially at permissive temperature and at zero time are shifted to non-permissive temperature. The right panel shows the result of an experiment with plasmid pSC101*rep(ts)* carrying the *hok/sok* region of plasmid R1 (K. Nordström, unpublished). Although some cured cells do survive and eventually grow out of the population, the *hok/sok* system is clearly of the killer (Kil[+]) type rather than the Par[+] type. OD=optical density. VC-viable counts without selection. VC(AB)=viable counts on antibiotic-containing medium selecting for cells that retain the plasmid.

with plasmids in *Staphylococcus* species for example (63). However, the *E. coli* systems have been the subject of extensive investigation and illustrate well the range of systems likely to be encountered.

The F Plasmid

The sex factor F is maintained as a plasmid with one or two copies per host chromosome, yet loss of the plasmid from growing cells is too infrequent to be readily detectable (17, 69).

F is a prime example of a plasmid that achieves a high degree of stability by combining the effects of multiple functional cassettes. The plasmid encodes three replication systems and at least five other elements that are likely to contribute to maintenance stability (Figure 4). Many studies have concentrated on miniplasmids consisting of the 6.5 kb-*Eco*RI fragment of the 100 kb-F genome linked to an antibiotic resistance marker. These plasmids are maintained at low copy number and are stably inherited (49, 81). This small region consists of a densely packed cluster of plasmid-maintenance elements. A core region contains the *ori-2* (also referred to as *oriS*) origin with its associated regulatory elements and the *repE* gene that produces an essential initiator protein (40). This is followed directly by the active partition element (variously named the *sop* region or F *par* (2, 68)). An additional origin, referred to as *ori-1* or *oriV*, which may in fact be the origin most frequently used by the wild-type plasmid (40), lies to the left of this core. This origin also requires the *repE* product in addition to at least two other protein products of the region. The "extended" replicon that includes both origins and the functions required to operate them is known as the *repF1A* region (8; Figure 4). Interdigitated with the replication functions of the *repF1A* region are two other plasmid-maintenance functions. Closely associated with the *ori-1* origin is a site-specific recombination site (*rsfF*) (66). The product of the *D* gene that is situated some 0.5 kb to the right of the site acts as a specific recombinase and is also required for the activity of *ori-1* (45). This recombinase activity is

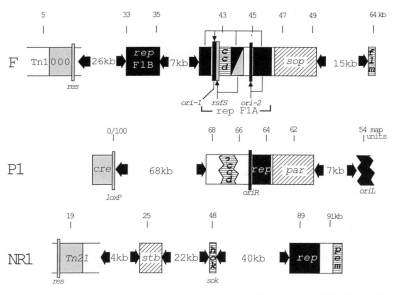

Figure 4 Genetic map of the functions that contribute to the stability of inheritance of the low-copy-number plasmids F, P1 and NR1 (IncFII). Black boxes are replicons or parts thereof. Stippled boxes are site-specific recombination systems known or likely to be involved in multimer resolution. Diagonally shaded boxes are true partition systems. Horizontally shaded boxes are killer systems. Some key *cis*-acting sites are shown as taller boxes. Boxes of uncertain extent and position are shown with jagged boundaries. The existence of the *ccd* box of P1 marked with a question mark is not fully established. Note that most functions can be regarded as discrete functional cassettes. The *rep* FIA region of plasmid F is exceptional as it contains two replicons and two other functional systems in an interdigitated array, with some of the functions being shared. Lines and arrows link the elements of these "split cassettes". The co-ordinates shown are approximate and are conventional for maps of the entire plasmids. The units of the F and R1 maps are kilobase-pairs whereas the P1 map units are approximately 0.9 kb long.

highly effective at resolving dimeric plasmids that carry the site (45, 66). The system appears to aid plasmid stability (S. J. Austin, unpublished). Also in the vicinity are the *ccd* genes, which were originally thought to act by controlling cell division (67), but actually have a cell-killing function (38). The region contains open reading frames for two small proteins of which one, CcdB, is capable of killing the cell while the other, CcdA, appears to act as an antagonist (see above).

Not all relevant functions are clustered in this one *Eco*RI fragment. Although the replication functions that appear to drive the wild-type plasmid are clustered here, an additional region capable of autonomous replication (*Eco*RI fragment 7) has also been described (46). Yet another replication region lies in the vicinity of F coordinate 2.5 (*Eco*R1 fragment 12). This consists of an IncFII-type replicon similar to that of plasmids R1 and NR1. However, this is not active in F as it is interrupted by the Tn*1000* (γ-δ) transposon (30). At least one additional plasmid-stability element lies outside the F5 region. This is the *flm* locus, located at map position F63.65–64.11 in the region that is first transferred during F conjugation (48). The locus is structurally and functionally homologous to the *hok-sok* region of plasmid R1, and acts by cell killing. Furthermore, the site-specific resolvase system of the transposon Tn*1000* (γ-δ) that is present in the wild-type F plasmid (26) most likely acts as an additional dimer-resolution system to aid plasmid stability.

In all then, the F plasmid encodes at least three replication systems and five systems that promote plasmid stability. It illustrates well the concept of multiple cassettes that aid plasmid maintenance. However, not all of the cassettes are functionally independent: the *ori-1* and *ori-2* replication systems share a key component in both requiring the E protein for function; both *ori-2* and the *rsfF* recombination site require the D protein. Thus the concept of functional cassettes, although useful, has some limitations.

The P1 Plasmid

P1 is a temperate bacteriophage of *E. coli*. It is unusual among coliphages in that its prophage is a low-copy-number plasmid (69). The plasmid-maintenance genes are separable from phage-propagation genes, and can be used to construct low-copy-number miniplasmids with properties that are very similar to those of other low-copy-number plasmids (75). The P1 plasmid is lost at a rate of less than one in 10^6 cell-division events (6). The plasmid is present at slightly less than one copy per host chromosome equivalent; probably the lowest of any plasmid so far studied (69). Miniplasmids consisting of some 10 kb of the 100 kb-P1 genome contain sufficient information for stable plasmid maintenance at low copy number (4). A single segment of some 4.5 kb contains the relevant genes. These consist of adjacent replication and partition cassettes that are organized in a fashion similar to the region containing the *ori*-2 and partition cassettes in plasmid F (Figure 4), although very little if any DNA homology is evident between the two plasmids.

The information present in the *rep-par* region has been studied extensively. However, in other respects wild-type P1 is rather understudied. Nevertheless, like F, P1 clearly has several additional systems that can aid plasmid maintenance (Figure 4). At least two additional replication systems are present that are capable of driving plasmid replication. One lies to the right of the core-maintenance region and may be involved in the lytic replication of P1 phage (77). The other has not been precisely located, but is clearly distinct from the other two (19).

Distal to the core maintenance region is the site-specific recombination site *loxP* that defines the ends of the P1 physical map and the *cre* gene that encodes a specific recombinase. This system promotes plasmid stability by resolving multimers (6). Indirect evidence suggests that P1 also has a killer system. Plasmids containing a region that lies just to the left of the replicon (Figure 4) are capable of inducing SOS functions (11), an effect shared with similar plasmids that contain F *ccd* (68).

The Plasmids R1/NR1

Plasmids R1 and NR1 are the best studied members of the IncFII group. The term IncFII was used because its conjugal transfer system is similar to that of F. In other respects, such as replication control, the IncFII plasmids are very different from the IncFI plasmids typified by F. A complete physical and genetic map of NR1 was recently published (88). Plasmids R1 and NR1 are present at low copy number with slightly more than one copy per chromosome equivalent. The copy number increases as growth rate is decreased (15). The basic replicon consists of some 2 kb of DNA and is governed by two control systems of which one, the interaction of CopB protein with the RepA promoter, appears to be a back-up system to speed up replication should the copy number become too low (57, 61, 87).

R1 and NR1 are very stably inherited with a loss rate of less than 10^{-7} per cell division (57). This is due to the combined action of several independent stability cassettes encoded from several widely separated loci (Figure 4).

The NR1 *stb* (R1 *parA*) cassette appears to encode a true partition function (23, 80). It resembles that of F and P1, having two essential open reading frames and a *cis*-acting locus (80; Figure 5). Unlike F and P, however, the *cis*-acting site is upstream of the open reading frames (54). Little is known at present about the biochemistry of *stb/parA* function.

The *parB* or *hok/sok* cassette of R1 encodes a killer function that kills a high proportion of any cells that fail to inherit a plasmid copy (9, 24). The mechanism is relatively well understood and involves the activity of a small

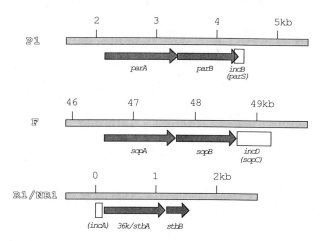

Figure 5 Genetic organization of the *par* region of plasmids F, P1, and R1/NR1(IncFII). Arrows indicate open reading frames for essential proteins. The open boxes show the regions containing the *cis*-acting partition sites for the systems. These are associated with incompatibility determinants in P1, F and R1. The NR1 region does not appear to be an incompatibility determinant. The *stbA* gene of NR1 is referred to as the 36k-open reading frame in R1. The *stbB* gene of NR1 probably has a counterpart in R1 but this remains to be established. The co-ordinates given are those of the P1 plasmid maintenance map of Abeles et al (1); the conventional F physical map and the NR1 *stb* map of Tabuchi et al (80).

antisense RNA to block the expression of a messenger RNA for a toxic product (70).

Recently, an additional cell-killing system, *pemK/pemI*, was discovered in plasmid NR1 (referred to as R100 by the authors; 82). Its cassette is closely linked to the basic replicon (Figure 4). It appears to be an analog of the F *ccd*-killing system. An equivalent function, *parD*, is present in plasmid R1, but in this case its activity seems to be repressed in the wild-type plasmid, although it can be activated by mutation (10). Although the *pemK/pemI* system clearly contributes to plasmid stability, it is uncertain whether *parD* contributes to the stability of the wild-type R1 plasmid despite the identical sequences of the two regions (10, 82). There is no information available to indicate whether IncFII plasmids make use of site-specific recombination systems for multimer resolution, although this seems likely on theoretical grounds. Multimers of R1 are not normally detected. As these plasmids carry transposons and transposon fragments, multimer resolution is likely carried out by these regions. Both NR1 and R1 carry the transposon Tn*21* that has an active resolvase system closely related to that of Tn*1000* and Tn*3* (14). The Tn*21* resolvase system probably plays a role in the stability of these plasmids and we have included it on the map of plasmid-maintenance cassettes in Figure 4.

The pSC101 Plasmid

Plasmid pSC101 has a moderately high copy number of approximately 15 per average cell (12). As such, it may not need stability functions. However, a region called *par,* closely linked to the basic replicon, is implicated in stability (51). When completely deleted, maintenance is highly unstable even though the average copy number is maintained (83). This may reflect clumping of copies or, as now seems more likely, a very considerable spread of the copy numbers in individual cells. Partial deletion of the locus gives Par^+ mutants with the Cmp (competition) phenotype. These are preferentially lost when competing with the wild-type in incompatibility tests (83), also probably because of a defect in replication fidelity. The *par* region was recently shown to be a specific binding site for DNA gyrase (86). Thus the replication abnormalities of *par* mutants may be due to a defect in DNA topology. A region with similar genetic properties has been identified in the *Staphylococcus* plasmid pT181 (21). Deletions in this region seem to affect the timing of replication by reducing affinity of the replicon for the initiator protein RepC.

The ColE1 Plasmid

ColE1 is representative of a group of high-copy-number plasmids and is probably the best understood of all replicons. In addition, it contains a recombination site (*cer*) that resolves multimeric forms (79). Cloning vehicles such as pBR322 that are closely related to ColE1 lack the *cer* site and are prone to accumulate as multimers. This leads to a measurable instability despite their high copy numbers. The recombinase for the *cer* site (*xer*) is host-encoded (72). No other maintenance-related functions appear to be present, although, as explained below, the production of the colicin encoded by the plasmid can provide a means of selecting against plasmid-free cells by killing them from without.

THE PARTITION PROCESS

The Structure of Partition Loci

Elements that meet the criteria that we have defined for active partition functions have been characterized in the F (67), P1 (2) and the closely related IncFII plasmids R1 (23) and NR1 (80). The basic structures of these units are illustrated in Figure 5. The general organization of the F and P1 regions is very similar; both contain two open reading frames and a downstream *cis*-acting site (1, 55). The protein products of both open reading frames are essential for partition activity (5, 18). The proteins can act in *trans* to the partitioned plasmid as long as the cognate *cis*-acting site is present (50, 67). The *cis*-acting sites of both systems act as incompatibility determinants, although in both cases functional truncated sites can be produced that have altered incompatibility properties (13, 47). The structures of the sites in F and P1 are remarkably different; the former consisting of eleven 43-base pair imperfect direct repeats, and the latter of a 13-base pair perfect palindrome with a 3-base pair spacer and flanking bases (50, 55).

The smaller of the two Par proteins in the F system, SopB, appears to be a DNA-binding protein that recognizes its cognate partition site (33). In the P1 case, the Par proteins have been purified, and the ParB protein was found to bind specifically to the P1 partition site in vitro (13). Recognition of the partition site probably involves this specific ParB binding.

The partition region of the IncFII plasmid NR1 (*stb*) is organized rather differently (Figure 5). It also has two open reading frames but the *cis*-acting site is immediately *upstream* of them. The homologous system of R1 (*parA*) may have a greatly shortened second open reading frame (23), although the two sequences would be reconciled if a single base-sequence error is assumed (80). The region containing the *cis*-acting sites of these plasmids has an extraordinary base composition, with one strand being almost exclusively composed of C and A bases. The region contains a small palindrome but no extensive direct repeats (23, 80). These sequences also contain the promoter for expression of the open reading frames and the expression is autoregulated (23). One or both of the protein products likely binds here both to regulate expression and to promote partition, although no direct evidence as to the function of the IncFII proteins is available at present.

Randomization during Active Partition—Partition-mediated Incompatibility

Novick & Hoppensteadt (65) pointed out that true partition loci might act as incompatibility determinants. At the time of their writing, it was clear that copies of several plasmid types were withdrawn from a random pool for replication (7, 28, 71), and partition might therefore also involve withdrawal of copies from a free pool. If so, an incoming plasmid with the same partition specificity would compete in the process, tending to displace the resident plasmid from the pool (Figure 6). Furthermore, if much of the partition machinery is assumed to be *trans*-acting, and the key *cis*-acting element a specific partition-recognition site, it follows that a cloned partition site would act as an incompatibility determinant. In general, this concept has proved to be true. Quantitative incompatibility experiments support the concept (60) and the *cis*-acting partition sites of F and P1 are both associated with in-compatibility determinants (47, 50, 55). Moreover, studies with hybrid plasmids have proven that the target mechanism that is sensitive to incompatibility is species-specific and resides in the partition mechanism itself (2). However, the mapping of an incompatibility determinant to a putative partition locus does not always identify a *cis*-acting partition site. Cloned regions that overexpress certain plasmid-encoded partition proteins can also exert incompatibility effects (3, 20, 44, 54). Presumably the concentrations of these products are critical for the systems to function. Moreover, careful study of the incompatibility properties of the P1 and F partition sites reveals unexpected complexities that are mentioned below and the *cis*-acting site of one partition system, that of plasmid NR1, appears not to exert incompatibility at all (54). These observations are important when assessing the viability of models for partition.

Models for Partition

The existence of partition-mediated incompatibility provides strong evidence that the copies are free in the cytoplasm at some stage in the cell-cycle. This follows as a plasmid introduced from the outside can compete for maintenance with the resident copies and displace them from the growing population. However, during active partition, copies are constrained to specific positions in the cell. Moreover, the system must sense where the plasmids are. If one plasmid (or group of plasmids) goes to one nascent cell, another must be directed to the other daughter cell. Plasmids must somehow be recognized as different members of a pair or of a pair of groups whose members have different fates. A seemingly simple way to achieve this would be an "exclusive site" process. Each cell would possess a unique and highly effective pair of binding sites for the plasmid DNA. Once a plasmid binds to one of the two sites, another plasmid has no choice but to bind to the other. This type of model cannot be excluded but it seems unattractive as it requires

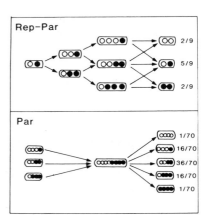

Figure 6 Plasmid incompatibility as a consequence of randomization during replication and partition. The open and full circles represent two differently marked plasmids; in the Rep-Par case, the two plasmids species have the same replication and partition systems; in the Par case, two different replicons have the same Par system.

the existence of a separate, exclusive pair of host sites for each specific type of plasmid.

An alternative model would be if copies could be recognized by binding to themselves rather than to exclusive sites. Self-paired plasmids could then be subject to a process very similar to mitosis in higher cells; the pair would attach to one of a number of equivalent host sites on a particular host structure that pulls or pushes the copies towards opposing poles of the dividing cell (Figure 7) (3). The host sites need not be exclusive to the plasmid type but could be shared by a number of different replicons. In this model, incompatibility could be explained by similar plasmids sharing the same type of partition site and thus forming heterologous pairs, leading to the random assortment of incoming and resident plasmids.

Observations on the properties of the P1-partition site are most readily explained by the prepairing model: Deletion of a small region gives rise to a site that still functions but has altered incompatibility properties. It exerts incompatibility against plasmids that use the same mutated site, but not against the wild-type (13). Such sites may have changed in a way such that they can pair with themselves but not with the wild-type. This alteration appears to be paralleled by a change in the structure of a protein complex at the site, due at least in part to the loss of a binding site for the host IHF protein (13, 20). In the wild-type case, binding of IHF appears to alter the configuration of the structure so that a different incompatibility specificity results. These observations are difficult to reconcile with the exclusive host-site

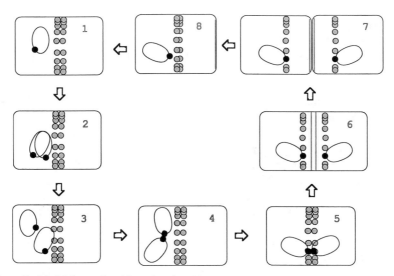

Figure 7   Model for equipartition of a plasmid.

model as a change in incompatibility in this model implies that a completely different host site is recognized by the wild-type and the deletion mutant; a most unlikely situation in our view.

An important feature of the exclusive site model is that it can only affect the distribution of a single pair of plasmids. As most plasmids have more than two copies per cell, the excess copies would be randomly distributed, a process we refer to as one-pair partition (57). By contrast, the self-pairing model has the potential to promote equipartition; any number of paired copies could be properly distributed, giving an equal number (or near-equal with an odd number) of plasmids in each daughter cell (3).

Equipartition or One-Pair-Partition?

It would be most useful to be able to distinguish between one-pair and equipartition. In practice this is not easy, mainly because the copy-control systems of replicons tend to compensate for any inequalities in copy distribution in such a way that they are not perpetuated for more than a generation or so. There are several experimental systems that theoretically could distinguish between the two alternative modes. However, to date none has given a definitive answer. Answers have been sought from the following sources:

1. The curing kinetics obtained by shifting a population of cells carrying a temperature-sensitive (*rep(ts)*) derivative of a plasmid to a temperature that does not allow plasmid replication. Hashimotoh-Gotoh & Ishii (32) reported that a *rep(ts)* derivative of plasmid pSC101 gave rise to plasmid-free cells earlier than predicted by equipartition models. However, if variations are considered in copy number as well as in cell age at division, curing curves of the type they reported could be obtained even if equipartition were the mode of partition used (56).

2. Plasmid loss rates with defined replicons. The known spread in copy number of plasmids such as R1 should give some instability even if provided with a perfect partition system. The instability should be more pronounced with one-pair than with equipartition. However, the predicted difference between the two modes of partition is fairly small and small deviations in the behavior of the replication control system might interfere with the result (57). Furthermore, the copy number has to be known very accurately. Hence, this approach is probably not workable.

3. Incompatibility between plasmids with compatible basic replicons but the same *par* function. The two modes of partition should give different results. However, after careful analysis, we believe that at present the differences are too small and the uncertainties too great to interpret such experiments with confidence. The theoretical basis for this conclusion is given in the appendix.

4. Kinetics of copy-number change in *oriC* plasmids. Plasmids derived

from the *E. coli* chromosomal origin *oriC* do not seem to have any copy-number control; rather each copy replicates once every cell cycle (42). The copy-number distribution rapidly broadens with time due to unequal segregation. This broadening is much slower if the *oriC* plasmid also carries the F partition system. (K. Rasmussen & M. R. Jensen, personal communication). The results obtained when the F partition system is present do not fit with one-pair-partition but are not totally consistent with equipartition either.

Host Structures in Partition

Stable partition requires selective movement of DNA molecules to daughter cells. Which structure is involved in this process? The replicon model by Jacob et al (37) grew out of early analysis of this problem. In their model, replicons (chromosomes and plasmids) were assumed to be continuously bound to sites along the equator of the cell. For each replicon type a specific site was postulated. DNA-replication was assumed to involve replication of the site also. The synthesis of invaginating membrane material for the cell septum occurred between the two elements of these dimeric sites, carrying the attached replicons into opposite halves of the dividing cell. Division of the site was assumed to be essential both for partition and initiation of the next round of replication. The replication control aspect of this model is clearly wrong as Par⁻ plasmids replicate without problem. Furthermore, the specificity may cause difficulties; there are many more than 20 different plasmid-replicon types in *E. coli* (64). Finally, random spread in replication as well as random assortment during partition (see above) do not easily fit into this beautiful, classic theory. However, the idea of multiple equatorial partition sites that divide and are moved apart by envelope growth is still a viable and attractive explanation for the motive force that drives partition.

In principle, exclusive membrane binding sites (as in one-pair partition models) could be located at any specific point in the nascent cell. Attachment of replicons to polar sites is possible for example. However, the prepairing type of model envisions attachment of the paired partition sites to a single location, a condition better satisfied by equatorial binding sites.

Gustafsson et al (29) showed that Par⁺ R1 and pSC101 but not Par⁻ deletion derivatives of these plasmids are present in the outer membrane fraction isolated from a neutral sucrose gradient. However, this observation has not yet been followed up by further, more detailed studies. It may be noted that plasmid pSC101, whose *par* function may not be involved in active partition (see above), also mediates membrane binding.

The studies of Gruss & Novick (27) provide an intriguing clue to the location of plasmid DNA in the cell. They reported that, in spite of a fairly high copy number (20 or more), plasmids are frequently lost during formation and regeneration of protoplasts of *Staphylococcus aureus*. This would indicate that the plasmid molecules exist as sequestered pools in the protoplasts and that these pools might be associated in some way with outer layers of the envelope of the cell. Furthermore, compatible plasmids were cured independently, whereas incompatible ones were cured jointly. This finding might suggest that each replicon forms its own pool or is located in its own compartment.

The involvement of the host chromosome has also been proposed in plasmid partition. This idea reduces the problem of plasmid partition to that of distributing the chromosome itself to the daughter cells (see below). The bacterial chromosome can be isolated in a highly compact form (1600–2400 S), the so-called folded chromosome (78, 89) where the DNA is held together by noncovalently linked RNA and protein. Low-copy-number plasmids like F or R1 cosediment with the folded chromosome (41), whereas most of the copies of high-copy-number plasmids do not (53). The association seems to be with the chromosome rather than with membrane (90). Kline & Miller (41) also suggested that plasmid association with the folded chromosome coupled plasmid replication with replication of that part of the chromosome to which the plasmid associates. That plasmids replicate throughout the whole cell cycle (7, 28, 71) invalidates this idea, however.

The Host Chromosome

Does the chromosome need a Par function? One possibility is that the sheer size of the chromosome excludes random partition. The idea might perhaps be rephrased as placing the membrane invagination/septum between the daughter nucleoids rather than placing the latter on each side of the septum. Chromosome replication, segregation, and septum localization are highly coordinated both in time and space. Mutations in at least four genes, *parA, gyrB, dnaG* (16, 62), and *parD* (36), prevent completed chromosomes from separating. In the *gyrB* case this seems to result from catenation of the DNA (74). A *parD* mutant replicates its DNA normally but is unable to separate the daughter molecules (36). However, the number of cell divisions is unaffected. If the chromosomes have not separated, the septum is formed between the cell pole and the non-separated DNA, supporting the idea that chromosome partition is normally a passive process. If septum formation were an active process taking place in the middle of the cell, non-separated chromosomes would prohibit septum formation rather than cause the septum to be misplaced.

EUKARYOTIC AND PROKARYOTIC REPLICONS

What are the differences between prokaryotic and eukaryotic chromosomes? Prokaryotic and eukaryotic chromosomes seem to share some properties that are different from those of bacterial plasmids. (*a*) There is no correction during replication for differences in copy number—all chromosomes replicate only once during the cell cycle, as also seems true for minichromosomes of *E. coli* (42) and eukaryotic plasmids. Hence, there is no randomization during replication. In general, the sister chromatids of eukaryotic chromosomes seem to remain associated prior to mitosis, so that pairing is predetermined and limited to coproducts of replication. There is no randomization during mitosis either. Thus there appears to be no counterpart of partition-mediated incompatibility in eukaryotes.

An interesting parallel to prokaryotic replicons has emerged from studies of the 2-micron circle replicon in yeast. This plasmid actively maintains a high, stable copy number in the yeast nucleus by a novel mechanism involving site-specific recombination. The replicon is governed by normal cellular controls that allow initiation once per generation. However, additional copies can be generated by a site-specific inversion event in the replicative intermediate. This event is subject to a feedback control that represses expression of the recombinase in response to increasing copy number (73). In addition, the plasmid encodes a distribution system that allows copies to distribute to the daughter nucleus (85). When this system is inactivated, the copies often all remain in the mother cell during budding. Two plasmid-encoded proteins and a *cis*-acting site are needed. Attachment of the copies to random sites on the nuclear envelope may be involved (85).

STRATEGIES FOR SUCCESSFUL PLASMID MAINTENANCE

We have seen that large plasmids contain a large number of cassettes for different types of plasmid-maintenance systems. Probably, none of these systems is highly accurate when acting alone, although in combination they achieve virtually complete stability. For plasmids such as F, the functional relationships between these individual activities can be deduced. The multiple replication systems are each subject to negative feedback that is sensitive to copy number. This results in a constant, low copy number in each cell at the time of division. The existence of multiple pathways for replication is likely to increase the reliability of initiation. In some cells, plasmid dimers are produced by generalized recombination, blocking proper segregation. These are rapidly resolved by one of the two site-specific recombination systems. Again, the reliability of this step is probably increased by the existence of more than one pathway for resolution. The active partition system then segregates the plasmid monomers into daughter cells. Occasionally, a division event will give rise to one cell without a plasmid due to errors in replication, resolution, or partition. Most of these cells will be killed by the combined action of the two killing systems. The result is a population in which plasmid-free cells are virtually undetectable.

This type of strategy is highly successful. Precise numbers are not available for F, but the L_{ex} value for plasmid R1 is $<10^{-7}$ per cell division (57). Nordström & Aagaard-Hansen (57) calculated the expected degree of stability of plasmid R1 based on the known properties of the basic replication-control system. They found that some dividing cells must exist with only a single copy. The calculations showed that the L value should be 10^{-4} to 10^{-3} even if a perfect partition system were assumed. The discrepancy could be accounted for by (a) an auxilliary replication-control circuit (the copB circuit) that is activated at low copy number; and (b) the hok/sok system that kills plasmid-free cells with high efficiency. The basic replicon alone is lost with a frequency of about 1.5% per cell generation. The par system alone reduces the loss rate to about 10^{-4}, as does the hok/sok system alone (24). The contribution of multimer resolution was not considered in this analysis. All the constructs used presumably have some such system. In combination, the R1 elements combine to achieve what amounts to total stability.

Neither in F nor R1 do the killer systems act as the primary mechanism for plasmid stability. Rather they are used to improve the efficiency of the active partition systems by killing only those few cells that lose the plasmid due to errors in the replication or partition controls. This can be understood on theoretical grounds; these systems kill a proportion of the population and therefore adversely affect growth rate of the host. Basic replicons that have relatively low copy numbers have high L_{th} values. Plasmids using these replicons in conjunction with a killer system would be a significant liability to the host unless some true partition system were also present to reduce the loss frequency.

The killing of plasmid-free cells is a highly successful and widely used strategy in the plasmid world. Many plasmids carry genes for toxic substances, bacteriocins. When released into the medium, these proteins kill bacteria by several different mechanisms. To protect the (bacteriocinogenic) producer strains, the plasmid codes for an immunity substance that interferes with the toxic molecule. Hence, the bacteriocinogenic strain has a selective advantage over related strains that do not carry the plasmid. These cells can either be those that have lost the plasmid or that have never had it. In such a scenario, the killer substances discussed in previous sections (e.g. Hok/Sok of IncFII plasmids) can be regarded as bacteriocins that kill from within rather than from without. However, these systems are inactive against cells that never had the plasmid or cells that escape killing. Therefore, true bacteriocins may be more effective survival agents in natural, mixed populations.

Plasmids may be regarded as selfish DNA. Hence, it is crucial for a plasmid (a) not to affect its host seriously, e.g. by amounting to a large fraction of the cell's DNA, (*b*) to be maintained stably, (*c*) to compete favorably with cells that do not carry the plasmid, and (*d*) to be able to spread to plasmid-free cells. As we have seen, plasmids can be classified into two general classes. Small plasmids have few maintenance cassettes and rely on higher copy numbers to achieve stability. Large plasmids have multiple cassettes that, in concert, can achieve great stability at low copy number. The relationship between size, genetic complexity, and copy number is a logical consequence of the nature of the bacterial cell. Bacteria seem to possess a trimmed, minimal genome. Any increase in the genome reduces the growth rate. As large plasmids are present in low copy number and small plasmids in higher copy numbers, the plasmid content typically amounts to 2–5% of the total DNA irrespective of the size of the plasmid. Hence, the burden of carrying a plasmid is minimized. Evolution produces two divergent strategies in the plasmid world to achieve this low burden. The first trend is toward a minimal genome with few cassettes and therefore the necessity of a high copy number. The other trend selects for the acquisition of multiple cassettes that allows and necessitates the development of low-copy-number replication. The latter strategy allows other large blocks of genetic information for transfer functions etc to be carried, whereas the former does not. The ability to colonize other cell populations by transfer is an important one as it counteracts the small but inevitable loss of the plasmid from the population in long-term growth. It is interesting that some small, high-copy-number plasmids have solved their transfer dilemma by acting as parasites on their low-copy-number counterparts; they have acquired origins of conjugal transfer that take up very little space on the genome because they use the transfer products of large plasmids present in the same cell. Many small plasmids can neither transfer themselves nor be mobilized by known conjugative plasmids. However, it is likely that transformation and generalized transduction by phages such as P1 play a role in these cases. Self-contained transfer systems, for instance the conjugal systems of plasmids F and R1, and the phage transfer system of P1, are costly in terms of the bulk of the required genes. By employing a whole battery of maintenance cassettes these plasmids have achieved stability at such low copy numbers that their very large size (about 100 kb) can be tolerated. The organization of individual plasmid-stability elements as self-contained cassettes allows them to be reassorted in different combinations. The same is true of cassettes that carry information for drug resistance, specialist metabolic functions, and pathogenicity. Comparative studies of different plasmids provide ample evidence that the individual cassettes rather than the plasmids themselves are the basic units on which evolution builds. The activity of the transposable elements so often associated with plasmids and the susceptibility of plasmids to undergo frequent recombination events ensure a dynamic interchange of cassettes that constantly creates new plasmid types to challenge the environment. It is this extraordinary flexibility that ensures their success.

CONCLUSION

Plasmids are inherited with a very high degree of stability. Plasmid-free cells are formed at a frequency of less than one per 10^6 cell divisions even for low-copy-number plasmids (2–10 plasmid molecules per average cell). Plasmids use several fundamentally different systems to achieve this stability. These systems are coded for by casettes of DNA that build up the plasmids. The replication frequency is controlled such that deviations from the average copy number are rapidly corrected. Rescue systems speed up replication at too low copy numbers or compensate for failure of the normal system. Basic replicons are randomly distributed between the daughter cells at cell division. The spontaneous loss frequency is increased because plasmid dimers, trimers, etc. are formed by recombination. To counteract this, plasmids carry genetic information for site-specific recombination systems that resolve these oligomers. The stability of inheritance is improved beyond that obtained by random segregation of monomers by systems that actively partition the plasmids at cell division. Finally, plasmids carry genes coding for proteins that kill from within those cells that do not receive any plasmid copy at cell division. None of the systems is perfect, but by using several of them, even low-copy-number plasmids are virtually never lost. High-copy number plasmids do not require as many different systems to obtain a very high degree of stability of inheritance.

ACKNOWLEDGMENTS

We express our sincere gratitude to Christina Pellettieri for excellent secretarial work during the preparation of the manuscript. S. A. by invitation of the Science Faculty was visiting professor at Uppsala University. Research grants to K. N. from the Swedish Natural Science Research Council and the Swedish Cancer Society are gratefully acknowledged.

APPENDIX

PLASMID INCOMPATIBILITY AND DISCRIMINATION BETWEEN EQUIPARTITION AND ONE-PAIR-PARTITION

Randomization during Partition—Partition Incompatibility

Plasmid replication is random. It spreads over the whole cell cycle and some plasmid molecules replicate more than once during one cell cycle whereas others do not replicate at all during that time period (7, 28, 71). Apart from a short eclipse period, newly replicated plasmid molecules have the same probability as nonreplicated ones to replicate (28). If the plasmids in a cell happen to have different markers that are not related to replication, e.g. antibiotic resistances that are located within regions not essential for replication, the randomness in replication causes distortion in the distribution between the two versions of the plasmid. Since replication cannot correct these distortions, pure lines will inevitably form (Figure 1) (60, 65). This phenomenon is called *incompatibility* and is a logical consequence of the replication system.

The rate of formation of pure lines from cells carrying a heteroplasmid population is a function of the copy number of the plasmid (59, 60). Quantitative incompatibility experiments have revealed that the rate of formation of pure clones from a heteroplasmid population is consistent with randomization not only in replication but also in partition (60). This predicts that compatible replicons carrying the same *par* function should be incompatible. This prediction is generally true (59, 60). Hence, incompatibility is an important phenotype expressed by most *par* systems.

However, quantitatively, par incompatibility is only evident at very low copy numbers. This is because the replication control system will compensate for distorted copy-number distributions. This is evident from the following comparison: Assume that two distinguishable variants (A and B) of the same replicon carry the same *par* region. All cells are assumed to contain exactly n plasmid copies at birth and $2n$ at cell division. Random replication will then yield equal numbers of cells with the following ratios between variants A and B: $(2n-1)/1$, $(2n-2)/2$ . . . $2/(2n-2)$, and $1/(2n-1)$ (see Figure 6 in the main paper and reference (59)). If these copies are equipartitioned at cell division

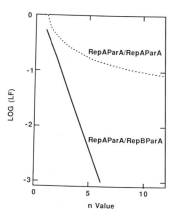

Figure 1 Theoretical incompatibility experiment involving two differently marked plasmids that have the same Rep and Par functions (dotted curve) or that have compatible replicons with the same Par function (solid curve). The copy number is the same for both plasmids in the latter case. The replication control system is assumed to settle the copy number in all cells at exactly $2n$ at cell division. The plasmid molecules are supposed to be equipartitioned at cell division. For diagram, see Figure 6 (main paper).

(i.e. each daugher cell receives exactly n copies), the rate of appearance of pure lines (LF) will be (for details see (59)):

$$LF = 2n/(n + 1)(2n - 1))$$

At high n values LF approaches $1/n$.

If two different replicons (C and D) (both with a copy number of n) have the same *par* region, equipartition would result in:

$$LF = (2n)!(2n)!/(4n)!$$

Figure 1 shows $LF = f(n)$ for the two cases. It is apparent that the two situations lead to very different results, and that the partition incompatibility is weak except at rather low values of n.

Equipartition or One-Pair-Partition?

What are the kinetics of controlled partition? There are two major alternatives, *equipartition* and *one-pair-partition* (57). In the former case, each daughter cell receives half the number of plasmid copies or, at odd copy number, one daughter receives one plasmid copy more than the other daughter cell. During one-pair-partition, only one pair is equipartitioned, whereas the rest of the plasmid copies are randomly distributed.

One way to discriminate between equipartition and one-pair-partition would be to use incompatibility. If two different compatible replicons (C and D) have the same *par* function, have the copy numbers m and n, respectively, and are partitioned by equipartition, the loss rate will be

$$LF_D = (2m)!(m+n)!/((m-n)!(2m+2n)!)$$

If $m \gg n$, LF_D will approach $(\frac{1}{2})^{2n}$, i.e. the L value of a Par^- mutant.

For one-pair-partition the LF value would be

$$LF_C = \frac{2m^2 + 4mn - m}{(m + n)(2m + 2n - 1)}(\frac{1}{2})^{2m}$$

$$LF_D = \frac{2n^2 + 4mn - n}{(m + n)(2m + 2n - 1)}(\frac{1}{2})^{2n}$$

For $m \gg n$, LF_C approaches $(\frac{1}{2})^{2n}$, i.e. the L value of a Par^- plasmid.

A comparison between the behavior of the two modes of partition in an incompatibility test is shown in Figure 2; in this case n and m are assumed to

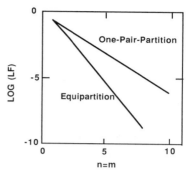

Figure 2 Theoretical incompatibility test involving two differently marked plasmids that have the same Rep and Par system. The copy number is the same for both plasmids. The copy number is assumed to be the same for both plasmids and is exactly $2n$ in all cells at cell division. Two modes of partition are considered, equipartition and one-pair partition.

be equal. The two modes of partition give clearly different results. However, in that part of the diagram where there is a big difference, the *LF* value is very low. In the region where the *LF* value is easily determined, the difference is smaller, which makes a correct determination of copy number critical. Figure 3 shows the *LF* values for $n = 3$ (the copy number of the IncFII plasmid R1) as a function of m. In this figure, the experimental value from such a test (60) has been included. The data are in best agreement with equipartition. However, this type of test would be useful in discarding at least one of the modes of partition discussed only if true copy number can be determined with a high degree of accuracy. Furthermore, the exact kinetics of the copy-number-control system has to be included in the calculations. Plasmid R1, for example, replicates according to the $+n$ mode of replication (57), presumably with a Poissonian-spread around n. In such a system, the curves of Figure 3 will be moved upwards.

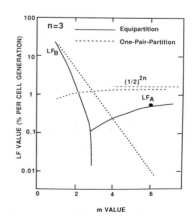

Figure 3 Incompatibility test involving two differently marked compatible replicons that have the same Par system. The copy number for the two plasmids is supposed to be exactly $2m$ and $2n$ at cell division, respectively. The copy number n is assumed to be 3 for plasmid A and m for plasmid B. The solid line and the dotted line show the calculated results for equipartition and one-pair-partition, respectively. The filled circle shows the result of an experiment with plasmid R1 ($n=3$) and plasmid RSF1010 carrying the *par* region of plasmid R1 (60).

Literature Cited

1. Abeles, A. L., Friedman, S. A., Austin, S. J. 1985. Partition of unit-copy miniplasmids to daughter cells. III. The DNA sequence and functional organization of the P1 partition region. *J. Mol. Biol.* 85:261–72

2. Austin, S., Abeles, A. 1983. Partition of unit-copy miniplasmids to daughter cells. I. P1- and F-miniplasmids contain discrete, interchangeable sequences sufficient to promote equipartition. *J. Mol. Biol.* 169:353–72

3. Austin, S., Abeles, A. 1983. Partition of unit-copy miniplasmids to daughter cells. II. The partition region of miniplasmid-P1 encodes an essential protein and a centromere-like site at which it acts. *J. Mol. Biol.* 169:373–87

4. Austin, S., Hart, F., Abeles, A., Sternberg, N. 1982. Genetic and physical map of a P1-miniplasmid. *J. Bacteriol.* 152:63–71

5. Austin, S., Wierzbicki, A. 1983. Two mini-F encoded proteins are essential for equipartition. *Plasmid* 10:73–81

6. Austin, S., Ziese, M., Sternberg, N. 1981. A novel role for site-specific recombination in maintenance of bacterial replicons. *Cell* 25:729–36

7. Bazaral, M., Helinski, D. R. 1970. Replication of a bacterial plasmid and an episome in *Escherichia coli. Biochemistry* 9:399–406

8. Bergquist, P. L., Saadi, S., Maas, W. K. 1986. Distribution of basic replicons having homology with RepFIA, Rep-FIB, and RepFIC among IncF group plasmids. *Plasmid* 15:19–34

9. Boe, L., Gerdes, K., Molin, S. 1987. Effects of genes exerting growth inhibition and plasmid stability on plasmid maintenance. *J. Bacteriol.* 169:4646–50

10. Bravo, A., de Torrontegui, G., Diaz, R. 1987. Identification of components of a new stability system of plasmid R1, ParD, that is close to the origin of replication of this plasmid. *Mol. Gen. Genet.* 210:101–10

11. Capage, M. A., Scott, J. R. 1983. SOS induction by P1-Km miniplasmids. *J. Bacteriol.* 155:473–80

12. Cohen, S. N., Chang, A. C. Y. 1973. Recircularization and autonomous replication of a sheared R-factor DNA segment in *E. coli* transformants. *Proc. Natl. Acad. Sci. USA* 70:1293–97

13. Davis, M. A., Austin, S. J. 1988. Recognition of the P1 plasmid centromere analog involves binding of the parB protein and is modified by a specific host fact. *EMBO J.* 7:1881–88

14. de la Cruz, F., Grinsted, J. 1982. Genetic and molecular characterization of Tn21, a multiple resistance transposon from R100.1. *J. Bacteriol.* 151:222–28

15. Engberg, B., Nordström, K. 1975. Replication of R-Factor R1 in *Escherichia coli* K-12 at different growth rates. *J. Bacteriol.* 123:179–86

16. Fairweather, N. F., Orr, E., Holland, B. 1980. Inhibition of deoxyribonucleic acid gyrase: effects on nucleic acid synthesis and cell division in *Escherichia coli* K-12. *J. Bacteriol.* 142:153–61

17. Frame, R., Bishop, J. O. 1971. The number of sex factors per chromosome in *E. coli. Biochem. J.* 121:93–103

18. Friedman, S. A., Austin, S. J. 1988. The P1 plasmid-partition system synthesizes two essential proteins from an autoregulated operon. *Plasmid* 19:103–12

19. Froehlich, B. J., Watkins, C., Scott, J. R. 1986. IS1-dependent generation of high copy number replicons from P1ApCm as a mechanism of gene amplification. *J. Bacteriol.* 166:609–17

20. Funnell, B. E. 1988. Mini-P1 plasmid partitioning; excess ParB protein destabilizes plasmids containing the centromere parS. *J. Bacteriol.* 170:954–60

21. Gennaro, M. L., Novick, R. P. 1986. cmp, a *cis*-acting plasmid locus that increases interaction between replication origin and initiator protein. *J. Bacteriol.* 168:160–66

22. Gerdes, K., Larsen, J. E. L., Molin, S. 1985. Stable inheritance of plasmid R1 requires two different loci. *J. Bacteriol.* 161:292–98

23. Gerdes, K., Molin, S. 1986. Partitioning of plasmid R1. Structural and functional analysis of the parA locus. *J. Mol. Biol.* 190:269–79

24. Gerdes, K., Rasmussen, P. B., Molin, S. 1985. Unique type of plasmid maintenance function: Post segregational killing of plasmid free cells. *Proc. Natl. Acad. Sci. USA* 83:3116–20

25. Grindley, N. D., Lauth, M. R., Wells, R. G., Wityk, R. J., Salvo, J. J., Reed, R. R. 1982. Transposon-mediated site-specific recombination: identification of three binding sites for resolvase at the *res* sites of gamma delta and Tn3. *Cell* 30:19–27

26. Grindley, N. D., Reed, R. R. 1985. Transpositional recombination in prokaryotes. *Annu. Rev. Biochem.* 54:863–96

27. Gruss, A., Novick, R. 1986. Plasmid instability in regenerating protoplasts of

Staphylococcus aureus is caused by aberrant cell division. *J. Bacteriol.* 165: 878–83

28. Gustafsson, P., Nordström, K. 1975. Random replication of the stringent plasmid R1 in *Escherichia coli* K-12. *J. Bacteriol.* 123:443–48

29. Gustafsson, P., Wolf-Watz, H., Lind, L., Johansson, K-E., Nordström, K. 1983. Binding between the *par* region of plasmids R1 and pSC101 and the outer membrane fraction of the host bacteria. *EMBO J.* 2:27–32

30. Guyer, M. S. 1978. The γ-δ sequence of F is an insertion sequence. *J. Mol. Biol.* 126:347–65

31. Hashimoto, H., Mitsuhashi, S. 1970. Segregation of R factors. In *Progress in Antimicrobial and Anticancer Chemotherapy. Proc. 6th Int. Congr. Chemotherapy*, ed. H. Umezawa, pp. 545–51. Tokyo: Univ. Tokyo Press

32. Hashimotoh-Gotoh, T., Ishii, K. 1982. Temperature-sensitive plasmids are passively distributed during cell division at non-permissive temperature: A new model for replicon duplication and partitioning. *Mol. Gen. Genet.* 187:523–25

33. Hayakawa, Y., Murotsu, T., Matsubara, K. 1985. Mini-F protein that binds to a unique region for partition of mini-F plasmid DNA. *J. Bacteriol.* 163:349–54

34. Highlander, S. K., Novick, R. P. 1987. Plasmid repopulation kinetics in *Staphylococcus aureus*. *Plasmid* 17:210–21

35. Deleted in proof

36. Hussain, K., Begg, K. J., Salmond, G. P. C., Donachie, W. D. 1987. ParD: a new gene coding for a protein required for chromosome partitioning and septum localization in *Escherichia coli*. *Mol. Microbiol.* 1:73–81

37. Jacob, F., Brenner, S., Cuzin, F. 1963. On the regulation of DNA replication in bacteria. *Cold Spring Harbor Symp. Quant. Biol.* 28:329–48

38. Jaffe, A., Ogura, T., Hiraga, S. 1985. Effects of the ccd function of the F plasmid on bacterial growth. *J. Bacteriol.* 163:841–49

39. James, A. A., Morrison, P. T., Kolodner, R. 1982. Genetic recombination of plasmid DNA. Analysis of the effect of recombination-deficient mutants on plasmid recombination. *J. Mol. Biol.* 160:411–30

40. Kline, B. C. 1985. A review of mini-F plasmid maintenance. *Plasmid* 14:1–16

41. Kline, B. C., Miller, J. R. 1975. Detection of nonintegrated plasmid deoxyribonucleic acid in the folded chromosome of *Escherichia coli*: Phys-

iochemical approach to studying the unit of segregation. *J. Bacteriol.* 121: 165–72

42. Koppes, L., von Meyenburg, K. 1987. Nonrandom minichromosome replication in *Escherichia coli* K-12. *J. Bacteriol.* 169:430–33

43. Koyama, A. H., Yura, T. 1975. Plasmid mutations affecting self-maintenance and host growth in *Escherichia coli*. *J. Bacteriol.* 122:80–88

44. Kusukawa, N., Mori, H., Kondo, A., Hiraga, S. 1987. Partitioning of the F plasmid: Overproduction of an essential protein for partition inhibits plasmid maintenance. *Mol. Gen. Genet.* 208: 365–72

45. Lane, D., de Feyter, R., Kennedy, M., Phua, S-H., Semon, D. 1986. D-protein of mini F plasmid acts as a repressor of transcription and as a site-specific resolvase. *Nucleic Acids Res.* 14:9713–28

46. Lane, D., Gardner, R. C. 1979. Second *Eco*R1 fragment of F capable of self-replication. *J. Bacteriol.* 139:141–51

47. Lane, D., Rothenbuehler, R., Merrillat, A-M., Aiken, C. 1987. Analysis of the F plasmid centromere. *Mol. Gen. Genet.* 207:406–12

48. Loh, S. M., Cram, D. S., Skurray, R. A. 1988. Nucleotide sequence and transcriptional analysis of a third function (Flm) involved in F-plasmid maintenance. *Gene* 66:259–68

49. Lovett, M. S., Helinski, D. R. 1976. Method for the isolation of the replication region of a bacterial replicon: Construction of a mini-F'Km plasmid. *J. Bacteriol.* 127:982–89

50. Martin, K. A., Friedman, S. A., Austin, S. J. 1987. Partition site of the P1 plasmid. *Proc. Natl. Acad. Sci. USA* 84: 8544–47

51. Meacock, P. A., Cohen, S. N. 1980. Partitioning of bacterial plasmids during cell division: A *cis*-acting locus that accomplishes stable plasmid inheritance. *Cell* 20:529–42

52. Miki, T., Easton, A. M., Rownd, R. H. 1980. Cloning of replication incompatibility and stability functions of R plasmid NR1. *J. Bacteriol.* 141:87–99

53. Miller, J., Manis, J., Kline, B., Bishop, A. 1978. Nonintegrated plasmid-folded chromosome complexes. *Plasmid* 1: 273–83

54. Min, Y-N., Tabuchi, A., Fan, Y-L., Womble, D. D., Rownd, R. H. 1988. Complementation of mutants of stability locus of IncFII plasmid NR1: Essential functions of the trans-acting *stbA* and *stbB* gene. *J. Mol. Biol.* 204:345–56

55. Mori, H., Kondo, A., Ohshima, A.,

Ogura, T., Hiraga, S. 1986. Structure and function of the F-plasmid genes essential for partitioning. *J. Mol. Biol.* 192:1–15

56. Nordström, K. 1984. Equipartition and other modes of partition: on the interpretation of curing kinetics using *rep(ts)* plasmids. *Mol. Gen. Genet.* 198: 185–86

57. Nordström, K., Aagaard-Hansen, H. 1984. Maintenance of bacterial plasmids: Comparison of theoretical calculations and experiments with plasmid R1 in *Escherichia coli. Mol. Gen. Genet.* 197:1–7

58. Nordström, K., Molin, S., Aagaard-Hansen, H. 1980. Partioning of plasmid R1 in *Escherichia coli.* 1. Kinetics of loss of plasmid derivatives deleted of the *par* region. *Plasmid* 4:215–27

59. Nordström, K., Molin, S., Aagaard-Hansen, H. 1980. Partitioning of plasmid R1 in *Escherichia coli.* II. Incompatibility properties of the partitioning system. *Plasmid* 4:332–49

60. Nordström, K., Molin, S., Aagaard-Hansen, H. 1981. Plasmid R1 incompatibility. In *Molecular Biology, Pathogenicity, and Ecology of Bacterial Plasmids,* ed. S. B. Levy, R. Clowes, E. L. Koenig, pp. 291–301. New York: Plenum

61. Nordström, K., Molin, S., Light, J. 1984. Control of replication of bacterial plasmids: genetics, molecular biology, and physiology of the plasmid R1 system. *Plasmid* 12:71–90

62. Norris, V., Alliotte, T., Jaffé, A., D'Ari, R. 1986. DNA replication termination in *Escherichia coli parB* (a *dnaG* allele), *parA,* and *gyrB* mutants affected in DNA distribution. *J. Bacteriol.* 168:494–504

63. Novick, R. P. 1987. Plasmid incompatibility. *Microbiol. Rev.* 51:381–95

64. Novick, R. P., Clowes, R. C., Cohen, S. N., Curtiss, R. III, Datta, N., Falkow, S. 1976. Uniform nomenclature for bacterial plasmids: a proposal. *Bacteriol. Rev.* 40:168–89

65. Novick, R. P., Hoppensteadt, F. C. 1978. On plasmid incompatibility. *Plasmid* 1:421–34

66. O'Connor, M. B., Malamy, M. H. 1984. Role of the F-factor *oriV1* region in *recA*-independent illegitimate recombination. Stable replicon fusions of the F-derivative pOX38 and pBR322-related plasmids. *J. Mol. Biol.* 175:263–84

67. Ogura, T., Hiraga, S. 1983. Partition mechanisms of F plasmid: Two plasmid gene-encoded products and a cis-acting region are involved in partition. *Cell* 32:351–60

68. Ogura, T., Hiraga, S. 1983. Mini-F plasmid genes that couple host-cell division to plasmid proliferation. *Proc. Natl. Acad. Sci. USA* 80:4784–88

69. Prentki, P., Chandler, M., Caro, L. 1977. Replication of the prophage P1 during the cell cycle of *Escherichia coli. Mol. Gen. Genet.* 152:71–76

70. Rasmussen, P. B., Gerdes, K., Molin, S. 1987. Genetic analysis of the *parB+* locus of plasmid R1. *Mol. Gen. Genet.* 209:122–28

71. Rownd, R. 1969. Replication of a bacterial episome under relaxed control. *J. Mol. Biol.* 44:387–402

72. Sherratt, D. J., Summers, D. K., Boocock, M., Stirling, C., Stewart, G. 1986. Novel recombination mechanisms in the maintenance and propagation of plasmid genes. *Banbury Report 24: Antibiotic Resistance Genes: Ecology, Transfer, and Expression,* pp. 263–73. New York: Cold Spring Harbor Lab.

73. Som, T., Armstrong, K. A., Volkert, F. C., Broach, J. R. 1988. Autoregulation of 2-micron circle gene expression provides a model for maintenance of stable plasmid copy levels. *Cell* 15:27–37

74. Steck, T. R., Drlica, K. 1984. Bacterial chromosome segregation: evidence for DNA gyrase involvement in decatenation. *Cell* 36:1081–88

75. Sternberg, N., Austin, S. 1983. Isolation and characterization of P1 minireplicons, λ-P1:5R and λ-P1:5L. *J. Bacteriol.* 153:800–12

76. Sternberg, N., Hamilton, D. 1981. Bacteriophage P1 site-specific recombination I. Recombination between *loxP* sites. *J. Mol. Biol.* 150:467–86

77. Sternberg, N., Hoess, R. 1983. The molecular genetics of bacteriophage-P1. *Annu. Rev. Genet.* 17:123–54

78. Stonington, O., Pettijohn, D. 1971. The folded genome of *Escherichia coli* isolated in a protein-DNA-RNA complex. *Proc. Natl. Acad. Sci. USA* 68:6–9

79. Summers, D. K., Sherratt, D. J. 1984. Multimerization of high copy number plasmids causes instability: ColE1 encodes a determinant essential for plasmid monomerization and stability. *Cell* 36:1097–103

80. Tabuchi, A., Min, Y-N., Kim, C. K., Fan, Y-L., Womble, D. D., Rownd, R. H. 1988. Genetic organization and nucleotide sequence of the stability locus of IncFII plasmid NR1. *J. Mol. Biol.* 202:511–25

81. Timmis, K., Cabello, F., Cohen, S. 1975. Cloning, isolation, and characterization of replication regions of complex plasmid genomes. 1975. *Proc. Natl. Acad. Sci. USA* 72:2242–46

82. Tsuchimoto, S., Ohtsubo, H., Ohtsubo, E. 1988. Two genes, *pemK* and *pemI*, responsible for stable maintenance of resistance plasmid R100. *J. Bacteriol.* 170:1461–66

83. Tucker, W. T., Miller, C. A., Cohen, S. N. 1984. Structural and functional analysis of the par region of the pSC101 plasmid. *Cell* 38:191–201

84. Uhlin, B. E., Nordström, K. 1985. Preferential inhibition of plasmid replication in vivo by altered DNA gyrase activity in *Escherichia coli*. *J. Bacteriol.* 162:855–57

85. Volkert, F. C., Broach, J. R. 1986. Site-specific recombination promotes plasmid amplification in yeast. *Cell* 15:541–50

86. Wahle, E., Kornberg, A. 1988. The partition locus of plasmid pSC101 is a specific binding site for DNA gyrase. *EMBO J.* 7:1889–95

87. Womble, D. D., Rownd, R. H. 1986. Regulation of IncFII plasmid DNA replication. A quantitative model for control of plasmid NR1 replication in the bacterial cell division cycle. *J. Mol. Biol.* 192:529–48

88. Womble, D. D., Rownd, R. H. 1988. Genetic and physical map of plasmid NR1: Comparison with other IncFII antibiotic resistance plasmids. *Microbiol. Rev.* 52:433–51

89. Worcel, A., Burgi, E. 1972. On the structure of the folded chromosome of *Escherichia coli*. *J. Mol. Biol.* 71:127–47

90. Worcel, A., Burgi, E. 1974. Properties of a membrane-attached form of the folded chromosome of *Escherichia coli*. *J. Mol. Biol.* 82:91–105

91. Yoshikawa, M. 1974. Identification and mapping of the replication genes of an R factor, R100-1, integrated into the chromosome of *Escherichia coli* K-12. *J. Bacteriol.* 118:1123–31

Annu. Rev. Genet. 1989. 23:71–85

MAIZE TRANSPOSABLE ELEMENTS

A. Gierl and H. Saedler

Abteilung Molekulare Pflanzengenetik, Max-Planck-Institut für Züchtungsforschung, 5000 Köln 30, Federal Republic of Germany

P. A. Peterson

Department of Agronomy, Iowa State University, Ames, Iowa 50011

CONTENTS

INTRODUCTION

The concept of transposable elements was developed by Barbara McClintock forty years ago. These genetic units move within genomes and frequently can be phenotypically recognized when they integrate into a gene. Characteristically, these insertion mutations are somatically unstable, i.e. excision of the element from the locus restores gene activity in this and all progeny cells. Such variegated phenotypes have appeared in many life forms and were attributed to transposable elements. Soon after the molecular isolation of transposons from *E. coli* (33, 73) it became apparent that transposable

71

0066-4197/89/1215-0071$02:00

elements are natural components of most genomes, because they have been found in representatives of virtually all organisms.

In general, transposable elements can be grouped into two classes; the first comprises elements with terminal inverted repeats (TIRs) that are attached to a region encoding functions necessary for transposition: the second class is represented by the retrotransposons. The latter category can be divided into retrovirus-like elements flanked by long terminal repeats, and nonviral retrotransposons terminating in a stretch of poly(A) residues of variable length. In maize, the *Bs1* (32) element represents the first subdivision while the *Cin4* elements (70) belongs to the nonviral retrotransposons. The coding regions in *Bs1* and *Cin4* share homology with viral reverse transcriptases.

Since the genetics of maize transposable elements (13, 16, 55) and various aspects of their molecular biology (10, 14, 15, 21, 47, 71) have recently been reviewed, this article focuses on the mechanism and the regulation of transposition of elements with TIRs. The biological relevance of transposable elements is considered by describing their prevalence in different maize populations.

Of the ten transposable element systems that have been genetically identified in maize (listed in 55), two have contributed most of the molecular data. These are the *Enhancer (En)* or *Suppressor-mutator (Spm)* system and the *Activator (Ac)* system. With a certain preference for *En/Spm,* these two systems are used as the principle examples to outline the current status in the molecular analysis of transposition.

MECHANISM OF TRANSPOSITION

Structural Features of En/Spm and Ac

The *En-* and *Spm*-transposable element systems were identified independently by Peterson (52) and McClintock (40). Both genetically (53) and molecularly (36, 50, 51) *En* and *Spm* are virtually identical. The *En/Spm* element is 8.3 kb in length, has 13-bp perfect TIRs, and causes a 3-bp duplication of the target sequences upon insertion (22, 51, 67). It encodes at least two products (50), termed tnpA and tnpB (20). The 69-kD tnpA protein is encoded by a 2.5-kb transcript and the structure of the corresponding gene has been determined (50) (Figure 1). TnpB is encoded by a 6-kb transcript and its structure is still hypothetical, although judging from the hybridization pattern, it is probably encoded, at least in part, by the open reading frames (ORF1, ORF2) that are present in the first intron of tnpA (Figure 1). TnpA mRNA is about 100 times more abundant than tnpB mRNA (50). Transcription of tnpA (50), and probably of tnpB, is initiated at position 209 at the left end of the element (20). Hence, these two transcripts overlap and represent alternative splice products of a precursor transcript. In addition to the structurally intact, transposition-competent *En/Spm* element, numerous defective elements have been isolated (22, 36, 65, 68, 80). These elements represent internal deletion derivatives of *En/Spm* and transpose only in the presence of the active intact element. Because of this responsiveness, defective and active elements constitute a two-component system, consisting of the receptor (defective) element and the autonomous (intact) element (see below).

The *Ac* element was originally identified by McClintock as a locus required to activate chromosome breakage at a site where a particular *Dissociation (Ds)* element is located (37). *Ac* was subsequently shown to be a transposable element (38, 39). The *Ac* element is 4.6 kb in length (45, 57, 58), has 11-bp imperfect TIRs (one mismatch at the terminus) and causes an 8-bp duplication of target sequences upon insertion. The single product of *Ac* is encoded by a 3.5-kb mRNA and consists of 807 amino acids (34). As with the *En/Spm* family, defective elements *(Ds)* have been isolated that transpose only in the presence of *Ac*. In contrast to the defective *En/Spm* elements, the *Ds* elements represent a rather heterogeneous group. Common to all *Ds* elements is the 11-bp TIR, which, unlike that of *Ac*, is in perfect form. Some *Ds* elements are simple internal deletion derivatives of *Ac*. Others, like *Ds2* (43), share only rudimentary homology with *Ac* or have little more than the 11-bp TIR in common with *Ac,* as in the case of *Ds1* (79). Not all *Ds* elements induce the *Ac*-dependent chromosome breakage by which the *Ac-Ds* family was identified. Chromosome-breaking *Ds* elements seem to be of the "double *Ds*" type: one copy of a 2-kb *Ds* element (internal deletion derivative of *Ac*) is inserted in inverted orientation into another copy of the same *Ds* element (11, 83; for review, see 10, 14).

Models for Transposition

In principle, a transposable element consists of a region encoding the transposase and attached to it are sequences that serve as substrates for transposition. Excision and reinsertion of an element is accomplished by interaction of the transposase and possibly other cellular factors with the substrate sites. Transposition of *En/Spm* and *Ac* seems to be a nonreplicative process, though it might be linked to chromosomal replication (26, 27). In any case, the element excises from a donor site and inserts at a target site (25), often to linked sites on the same chromosome (26, 48).

Figure 1 Structural organization of *En/Spm*. Hatched boxes represent the highly structured terminal repetitive regions as shown in Figure 2. Open boxes indicate the exons of tnpA, translation of which commences within the second exon. ORF1 and ORF2 are indicated by the arrows in the first intron of tnpA.

DNA-sequence analysis of several revertant genes that occur after transposable elements are excised revealed that although excision is rather imprecise (3, 57, 69, 75, 79, 83) it still follows certain rules (4, 64). As a consequence of excision, altered sequences, "footprints," are left behind at the site of excision. According to the model of Saedler & Nevers (64), transposition takes place by a "cut-and-paste" mechanism. The transposase recognizes the ends of the element and introduces staggered nicks at the ends of the target-site duplication, in the same fashion as postulated for reinsertion. Footprints result from the action of DNA-repair enzymes on the protruding single-stranded fringes at the excision site and at the element in the "transposition complex." According to Coen et al's model (4) the cutting activity for excision and reinsertion is different. A staggered nick, of a size corresponding to the target-site duplication, is formed only at the recipient site, whereas the element at the donor site is released by cutting more precisely at the ends of the element. Excision generates two hairpin structures at the site of excision. Footprints are generated by resolution of these hairpins and ligation of the products. Both models suggest an intermediate free form of the element that is not chemically linked to chromosomal DNA. Such an extrachromosomal copy has been found for the *Mu1* transposable element (78) of *Z. mays.*

The observation that transposition occurs most frequently to linked positions has been explained in a recent model (62). This model implies a physical association between donor and recipient site. Cutting occurs first only at one end of the element, followed by ligation of the free end into the recipient site. Subsequently the other end of the element is released from the donor site and joined to the recipient site. Cutting could occur by either of the two mechanisms explained above. This model denies that a free copy of the element is a "normal" intermediate of transposition.

Substrate Specificity of Transposition

In general, the interaction of element-encoded products with their substrates is very specific. This has been tested by making use of the fact that most maize-transposable element systems exist in one of two forms: The autonomous element that encodes all functions for transposition and the nonautonomous (receptor or defective element) that usually represent internal deletion derivatives of the autonomous element. Since the defective elements have retained the substrate sites for transposition, they can transpose in the presence of the active autonomous element. Therefore, family relationships can be determined by testing whether a particular autonomous element can trigger the excision of a variety of defective elements (belonging also to other families). Using this technique, ten separate transposable element systems have been defined in maize (for a detailed list see 55). Furthermore, the identity of transposable element systems that have been isolated independently, such as *En* and *Spm* (53) or *Ac* and *Mp,* was thereby established (2).

Some unexpected system relationships have also been observed: *Ac* triggers the excision of normal *Ds*, *Ds1*, and *Ds2* elements. However, an element termed *Ac2* interacts predominantly only with *Ds2* (59). *Ac* does not interact with the receptor elements of the *Ubiquitous (Uq)* system, although *Uq* triggers *Ds1* excision (E. E. O. Caldwell, personal communication). Clarification of this situation awaits molecular characterization of *Ac2* and *Uq*.

Substrates for Transposition

The termini of *En/Spm* are made up of two domains: a 13-bp perfect TIR and subterminal regions containing several direct and inverse repetitions of a 12-bp sequence motif (20, 36, 50). These terminal regions comprise 180 bp at the 5' end and 300 bp at the 3' end of En/Spm (Figure 2). Deletion of the outermost nucleotide of the 13-bp TIR drastically reduces the excision frequency of this element (65). Partial deletion of the 12-bp motifs in the subterminal region also correlates with decreased excision rates (56, 68, 80). The extent of the deletions seems to be proportional to the amount of reduction (Figure 2). The entire deletion of the 12-bp motifs at one end of the element completely abolishes excision ability (A. Gierl, unpublished data). In summary, the above observations strongly suggest that the 13-bp TIRs and the subterminal region containing the 12-bp motifs both make up the substrates for excision in the *En/Spm* system.

Subterminal regions also seem to be involved in the excision of *Ac*. The analysis of deletion derivatives of *Ac* in transgenic tobacco (7) revealed that internal sequences (between position 44 and 181) are required for excision. The other determinant for *Ac* excision is the 11-bp TIR. The deletion of the four terminal nucleotides at one end of *Ac* abolishes excision (30), yet the two terminal nucleotides of *Ac* are not complementary, in contrast to *Ds*, as mentioned above. This implies that the terminal base is not necessary for

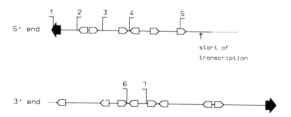

Figure 2 Terminal structures of *En/Spm*. The 200 bp of the 5' terminus and the 300 bp of 3' terminus of *En/Spm* are given at the top and bottom of the figure, respectively. The filled arrows indicate the 13-bp TIR. The open arrows represent the 12-bp sequence motifs to which binding of tnpA protein was observed. The start of transcription is indicated. The deletion breakpoints of several *En/Spm* deletion derivatives are given in sequential order corresponding to the following alleles: 1) *bz-m13* state CS6 (65); 2) *a2-m1* state class II (A. Gierl, unpublished); 3) *a1-m1* state 1112 (80); 4) *bz-m13* state CS9, (65); 5) *a1-m1* state 5719A (68); 6) *a1-m2* state 7997 (36); 7) *a1-m2* state 7995 (36).

recognition by the transposase. In summary, similar to *En/Spm*, the 11-bp TIRs and subterminal sequences are required for *Ac* transposition.

The transposition of *Ds1* in the presence of *Ac* conflicts with the above interpretation. As already mentioned, *Ds1* has the 11-bp TIRs but is devoid of internal *Ac* sequences. *Ac*-mediated *Ds1* excision was recently shown unambiguously to occur in transgenic tomato (35), where there were no other maize elements like *Uq* to obscure the picture. It is not understood how the structural *Ac* requirements for excision are substituted by *Ds1* sequences. Since *Ds1* is also excised in the presence of *Uq* (see above), analysis of the *Uq* system may clarify this problem.

Analysis of En/Spm *and* Ac-*Encoded Functions*

The biochemical characterizations of *En/Spm*-encoded functions has only recently started. cDNA of tnpA, the more abundant product, was expressed in *E. coli* and shown to encode a DNA-binding protein (20). It recognizes the 12-bp sequence motifs in the subterminal regions of *En/Spm* (Figure 2), which are defined as *cis*-determinants for transposition (see above). The property of tnpA to bind to the termini of *En/Spm* has led to the conclusion that tnpA represents the *suppressor* component (19, 20), which has been genetically defined for the *En/Spm* system (41). The suppressor function can be monitored with certain alleles in which the insertion of a small defective *En/Spm* element reduces, but does not eliminate, gene expression. However, when an element that expresses tnpA is present in the same genome, this residual gene expression is suppressed. *Suppressor* function seems to be mediated by bound tnpA that blocks transcription readthrough (20, 22). When *suppressor* action was tested transiently in tobacco protoplasts, binding of tnpA to various combinations of the 12-bp motif had different consequences for suppression (S. R. Grant, personal communication).

Binding of tnpA can also have an opposite effect on gene expression. The expression of the *a1-m2* allele is activated in the presence of *En/Spm* (41a). In this case, the insertion of the element disrupts the promotor region of the *A1* gene (72), thereby inhibiting gene expression. However, in the presence of *En/Spm*-encoded functions *A1* expression is partially restored. Binding of element-encoded functions to the element in the *A1* promoter has been suggested as activating this promoter (36, 66, 72).

The fairly symmetric array of the tnpA-binding sites (Figure 2) at both termini also suggests a functional role for tnpA in transposition. The lack of tnpA-binding sites at one end of the element abolished excision (see above). However, tnpA alone is not sufficient to execute excision, as can be concluded from the analysis of mutant *En/Spm* elements like *Spm-w-8011* and *En-2*. These elements still express tnpA, but are defective in promoting excision. In these elements sequences from ORF1 and ORF2 are deleted (19, 36).

These latter findings indicate that a second function is required for transposition, which is derived at least in part from ORF1 and ORF2. This function is probably encoded by the 6-kb mRNA and has been termed tnpB (20). TnpB has not been analysed biochemically, due to the low abundance of its transcript. Its functional role may, however, be predicted based on a comparison of *En/Spm* related elements from other plant species.

The *Tam1* element of *Antirrhinum majus* (3, 76), the *Tgm* element of *Glycine max* (60, 61), and *En/Spm* can be grouped into one family, based on the property that these elements produce a 3-bp target-site duplication upon insertion and share nearly identical 13-bp TIRs (3, 51, 61, 67). Because of the terminal similarities, these elements have been termed "CACTA"-elements (3).

The structural organization of the "CACTA"-elements is strikingly similar (Figure 3). *Tam1* encodes two transcripts, a major 2.5-kb mRNA and a less abundant 5-kb mRNA. The predominant smaller transcripts of *En/Spm* (tnpA) and *Tam1* are not related in sequence. In contrast, the larger transcripts (tnpB), containing sequences from the open reading frames present in the center of *En/Spm* and *Tam1*, share significant homology at the amino-acid level (76). Hitherto no gene structure has been established for *Tgm* and the structure of the fully intact element has not yet been reported. Nevertheless, in some *Tgm* elements an open reading frame was found that also shares homology with *En/Spm* (61) and *Tam1* (Figure 3). The conservation of these polypeptide sequences indicates a common function of the encoded products. Since the 13-bp TIRs are also conserved within the "CACTA"-elements, it has been speculated (20, 76) that the putative product (tnpB) may interact with the 13-bp TIRs and may in fact represent the function that cleaves at the element's termini.

In conclusion, two functions are associated with transposition of *En/Spm:* tnpA, which may stabilize the transposition complex by binding to the motifs in the subterminal region, and tnpB, which may accomplish endonucleolytic cleavage next to the element's ends after interacting with the 13-bp TIRs.

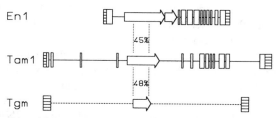

Figure 3   Comparison of the structural organization of *En1 (Z. mays), Tam1 (A. majus)* and *Tgm (G. max)*. Hatched boxes represent the terminal repetitive regions. Open boxes indicate the exons of the most abundant product (tnpA). The homology of a 1-kb region of the open reading frames is indicated (amino acid level).

The conserved features found in the "CACTA"-family raise the question of how these elements have evolved. Common to all members seems to be an ancestral endonuclease (tnpB) and its substrate in the form of the TIRs. However, the elements from individual species may have gained a second function (tnpA), each unrelated to the other. The putative DNA-binding motifs in the subterminal regions of En/Spm, Tam1, and Tgm are also nonhomologous. Analysis of other members of the "CACTA"-family might elucidate more about the evolution of these elements, some of which, such as Tam2 of A. majus (81) MPI1 of Z. mays (84) and Pis1 of Pisum sativum (74), have been identified.

In contrast to En/Spm, only one product is encoded by Ac (34). Expression of the corresponding cDNA in transgenic tobacco (7) demonstrated that this product suffices to transpose Ds. The Ac product was also expressed in insect cells by a baculo virus vector (39) and shown to bind to Ac DNA (R. Kunze & P. Starlinger, personal communication). DNA-sequence analysis of the TAM3 element of A. majus (76) has identified this element as a relative of Ac. Both elements have 7 bp of their 11-bp TIR in common and cause a 8-bp target-site duplication. The putative protein encoded by Tam3 shares small regions of fairly high homology (50–65%) with the Ac protein (76). One of these regions also shows some homology to a segment of the hobo element from Drosophila (34, 77). Tpc1, another Ac/Tam3-like element, has been isolated as an insertion into an allele of the chalcone synthase gene of Petroselinum crispum (31), suggesting the evolution here of another transposable-element family from a common precursor.

REGULATION OF En/Spm TRANSPOSITION

Transposition is tightly linked to mutation and should therefore be regulated to a level that is not deleterious for the cell. This is illustrated for example by the phenomena of hybrid dysgenesis in Drosophila (12) or by the high mutability associated with the Mutator (Mu) transposable element system of Z. mays. In active Mu lines the mutation frequency is increased to levels of 30–50 times above the spontaneous mutation rate (63). The Mu system in these lines avoids the deleterious consequences of this high rate, by frequently changing to an inactive state. (2a, 3a, 82).

There are several mechanisms by which transposition can be regulated and adjusted to acceptable levels. First, the En/Spm promoter is rather weak. The abundance of tnpA mRNA in poly (A) RNA preparations is about 10^{-6} (H. Cuypers, personal communication). Secondly, the mRNAs of tnpA and tnpB seem to be derived from one precursor transcript. TnpB mRNA is the minor product of this maturation process. TnpB might therefore be rate-limiting for transposition in maize. An inverse situation has been reported for En in transgenic potato (18). In this case the tnpB transcript is the more abundant product, and tnpA might be rate-limiting. This may also indicate that alternative splicing might be regulated in maize.

En/Spm *is Subject to Negative Regulation*

An active *En/Spm* can lapse into inactivity (41, 17). Inactivity has been correlated with increased levels of C-methylation of the element (5). In particular, methylation sites have been investigated in a GC-rich region of the first exon of tnpA and in the region upstream of the site of transcription initiation (1). In active *En/Spm* elements these regions are undermethylated relative to the rest of the element (1). A high degree of methylation is also correlated with low rates of transcription (1). The consequences of methylation for *En/Spm* activity have been recently reviewed in detail (15).

A completely different type of negative regulation was detected when the inhibitory effect on *En/Spm* transposition of the *En-I102* element was analysed (8). This internal deletion derivative of *En* expresses an aberrant polypeptide (tnpR) that shares homology with both tnpA and tnpB. TnpR represses transposition, probably by competitive inhibition of tnpA and/or tnpB function (8). Since there are about 50–100 *En/Spm* homologs (mainly deletion derivatives) distributed throughout the maize genome (67), products encoded by these elements could feasibly modulate *En/Spm* transposition.

En-Spm *is also Subject to Positive Regulation*

A positive regulator for *En/Spm* gene expression has been postulated (46), based on McClintock's observation that inactive elements can be transiently or heritably reactivated by an active element (42). The finding that *Spm-w-8011,* which probably only encodes tnpA (see above), is also capable of *trans*-activating inactive elements (1) suggests that tnpA is the activator. The mechanism of activation has been considered in several models (15, 20, 36), and since tnpA binds to the region upstream of the *En/Spm* specific transcription initiation site (20), bound tnpA protein very likely provides a shelter against methylation of the promoter region and thus prevents inactivation of the element. This implies that the presence of tnpA is essential to maintain its own expression. Overexpression of element-encoded functions could be avoided by the fact that the innermost tnpA-biding site at the left end of *En/Spm* overlaps with the "TATA"-box of the element's promoter (20). Binding to this site might result in repression of the promoter. In this sense tnpA would be an autoregulator.

Furthermore, it was shown in vitro that binding of tnpA is reduced when the binding motif is methylated (20). If for unknown reasons tnpA concentration drops to levels too low to protect these binding sites, the promoter region would therefore become methylated and the element would lapse into inactivity. In vitro tests have shown that *hemi*-methylated motifs have a higher binding affinity to tnpA than *holo*-methylated forms (20). The reactivation of inactive, methylated elements in the presence of an active element would probably occur during replication when the binding motifs are undermethylated.

BIOLOGICAL RELEVANCE OF TRANSPOSABLE ELEMENTS

While the molecular details of transposable elements are of interest to the molecular biologist, how significant is their contribution to biological processes in nature? Are transposable elements confined to the geneticist's nursery?

Several studies determined the distribution of active elements in native maize lines, apart from a genetic nursery. By the use of an *En/Spm* receptor allele, crosses were made with tribal Indian corn. *En/Spm* was present in both the Quapaw and Kiowa Tribal Indian lines from Oklahoma (54). Corn from the Cuña Indian tribe from a remote region in northwestern Columbia near the Panama Canal contained both *En/Spm* and *Dt* (23), in addition to the *Fcu* system (24). The functional, active *Bg* element has been found in maize varieties from Italy, Chile and Morocco (44). These results provide evidence for the universal distribution of these elements in maize.

The next question is whether transposable elements also occur in maize-breeding populations that are selected for quantitative traits, such as yield. In populations such as Iowa Stiff Stalk Synthetic (BSSS) or Lancaster Shure Crop, only the *Mrh* and *Uq* elements were uncovered at a significant frequency (56). To pursue this question the BSSS population was tested in greater detail. BSSS-derived inbreds are the source for 42% of the hybrids used in the U.S. corn belt (9). BSSS was established by intercrossing of 16 elite lines. The progeny were tested and the best were intercrossed in successive cycles of selection (28). This repetitive procedure led to consistent improvement of the population (Figure 4). Although *Uq* was only detected in one of the 16 progenitor lines (6), it persisted in the BSSS population and the number of *Uq* increased concomitantly with the cycles of selection (K. R. Lamkey, personal communication). However, when inbred lines were established from the various selection cycles, such as B14, B37, B73, and B84 (Figure 4), *Uq* was not longer detected (6). These results seem to indicate that under moderate selection pressure *Uq* is present in the BSSS population and that there is no selection against *Uq* activity. In contrast, *Uq* is lost in the process of homogenization of the genome by inbreeding, which is a selection against variability. Whether this also indicates that *Uq* contributes to the genetic flexibility responsible for population improvement awaits additional experimentation.

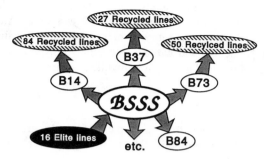

Figure 4 Development of inbred lines from the Iowa Stiff Stalk Synthetic (BSSS) population. 16 elite lines were intercrossed to make up the original BSSS (BSSSCo) population (28). BSSS in the figure represents a continuum from the original BSSSCo population through cycles of selection spanning 30–40 years of population improvement. At different cycles inbred lines have been established illustrated by the B lines. B14 and B37 came out of the original BSSSCo population, B73 from cycle 5 and B84 from cycle 7, respectively. Single traits were incorporated into these inbreds by crossing a B line to a local line, for example, in Minnesota for earliness. With such crosses and using the B line as one parent in one or two backcrosses, recycled lines such as CM105, AG32, A634, A635, etc were developed. As shown, 84 recycled lines originate from B14, illustrating the superior value of this BSSS-derived line. Though the figure only illustrates U.S. corn belt-developed recycled lines, the presence of these lines in Spain, Italy, Yugoslavia, Pakistan, and Zambia demonstrates the worldwide importance of the BSSS population.

A second example for transposable-element activity in breeding populations is represented by the long-term Illinois oil- and protein-selection program. From one original elite line four separate populations were formed by selecting for low or high oil and for low or high protein levels, respectively. In these highly selected breeding populations, originating from a narrow genetic base, no activity of Uq was detectable. However, after 50 cycles selection was reversed, e.g. from high to low protein, and Uq activity appeared in the reverse protein populations (6).

Again, as with inbreds, Uq activity was lost when selection was only for one specific trait. Reversal of this selection may create a condition in which more variation is allowed to appear and might even be necessary for successful selection into the opposite direction. How this procedure activated Uq (49) is not understood. Transposable elements are known to be activated by "stress" or "genomic shock," such as viral infection, chromosome breakage, irradiation, and crossings (reviewed in 47, 55). Whether there is a common mechanism in all these activation phenomena remains to be elucidated. Molecular cloning of Uq will provide the tools to follow the activation of Uq and to gain insight into the putative effects of Uq action.

In summary, given the limiting data available thus far, activity of transposable elements might be a condition to provide the flexibility for populations to adapt or to respond successfully to environmental changes. Molecular probes for many plant transposable elements are now available. Analysis of the prevalance of transposable elements and their role in plant breeding could thus be verified independently and extended to other transposable element systems. In this regard it might also be worthwhile to examine more natural plant populations.

ACKNOWLEDGMENTS

We thank several colleagues for communicating results prior to publication. We are grateful to F. Salamini, B. Scheffler, Zs. Schwarz-Sommer, and P. Starlinger for helpful comments on the manuscript.

Literature Cited

1. Banks, J. A., Masson, P., Fedoroff, N. 1988. Molecular mechanisms in the developmental regulation of the maize *Suppressor-mutator* transposable element. *Genes Dev.* 2:1364–80
2. Barclay, P. C., Brink, R. A. 1954. The relationship between modulator and activator in maize. *Proc. Natl. Acad. Sci. USA* 40:1118
2a. Bennetzen, J. L. 1987. Covalent DNA modification and the regulation of Mutator element transposition in maize. *Mol. Gen. Genet.* 208:45–51
3. Bonas, U., Sommer, H., Saedler, H. 1984. The 17 kb Tam1 element of *Antirrhinum majus* induces a 3 bp duplication upon integration into the chalcon synthase gene. *EMBO J.* 3:1015–19
3a. Chandler, V., Walbot, V. 1986. DNA modification of a maize transposable element correlates with loss of activity. *Proc. Natl. Acad. Sci. USA* 83:1767–71
4. Coen, E. H., Carpenter, R., Martin, C. 1986. Transposable elements generate novel spatial patterns of gene expression in *Antirrhinum majus*. *Cell* 47:285–96
5. Cone, K. C., Burr, F. A., Burr, B. 1986. Molecular analysis of the maize regulatory locus *C1*. *Proc. Natl. Acad. Sci. USA* 83:9631–35
6. Cormack J. B., Cox, D. F., Peterson, P. A. 1988. Presence of the transposable element *Uq* in maize breeding material. *Crop Sci.* 28:941–44
7. Coupland, G., Baker, B., Schell, J., Starlinger, P. 1988. Characterization of the maize transposable element *Ac* by internal deletions. *EMBO J.* 7:3653–59
8. Cuypers, H., Dash, S., Peterson, P. A., Saedler, H., Gierl, A. 1988. The defective En-1102 element encodes a product reducing the mutability of the En/

Spm transposable element system of *Zea mays*. *EMBO J.* 7:2953–60
9. Darrah, L. L., Zuber, M. S. 1986. 1985 United States farm maize germplasm base and commercial breeding strategies. *Crop. Sci.* 26:1109–13
10. Döring, H.-P., Starlinger, P. 1986. Molecular genetics of transposable elements in plants. *Annu. Rev. Genet.* 20:175–200
11. Döring, H.-P., Tillman, E., Starlinger, P. 1984. DNA sequence of the maize transposable element *Dissociation*. *Nature* 307:127–30
12. Engels, W. R. 1983. The P family of transposable elements in *Drosophila*. *Annu. Rev. Genet.* 17:315–44
13. Fedoroff, N. 1983. Controlling elements in maize. In *Mobile Genetic Elements*, ed. J. Shapiro, pp. 1–63. New York: Academic
14. Fedoroff, N. 1989. Maize transposable elements. In *Mobile DNA*, ed. M. Howe, D. Berg. Washington, DC:ASM Press. In press
15. Fedoroff, N. 1989. Maize transposable elements and development. *Cell* 56:181–91
16. Fincham, J. R. S., Sastry, G. R. K. 1974. Controlling elements in maize. *Annu. Rev. Genet.* 8:15–50
17. Fowler, R. G., Peterson, P. A. 1978. An altered state of a specific En regulatory element induced in a maize tiller. *Genetics* 90:761–82
18. Frey, M., Tavantzis, S. M., Saedler, H. 1989. The maize *En-1/Spm* element transposes in potato. *Mol. Gen. Genet.* In press
19. Gierl, A., Cuypers, H., Lütticke, S., Pereira, A., Schwarz-Sommer, Zs., et al. 1988. Structure and function of the

En/Spm transposable element system of *Zea mays:* identification of the Suppressor component of En. In *Proc. Internatl. Symp. on Plant Transposable Elements.* pp. 115–19. ed. O. Nelson. New York: Plenum

20. Gierl, A., Lütticke, S., Saedler, H. 1988. *TnpA* product encoded by the transposable element En-1 of *Zea mays* is a DNA binding protein. *EMBO J.* 7:4045–53

21. Gierl, A., Saedler, H. 1989. Transposition in plants. In *Nucleic Acids and Molecular Biology,* ed. F. Eckstein, D. M. J. Lilley. Heidelberg: Springer. In press

22. Gierl, A., Schwarz-Sommer, Zs., Saedler, H., 1985. Molecular interactions between the components of the En-I transposable element system of *Zea mays. EMBO J.* 4:579–83

23. Gonella, J., Peterson, P. A. 1975. The presence of En among some maize lines from Mexico, Columbia, Bolivia and Venezuela. *Maize Gen. Coop. Newsl.* 49:73

24. Gonella, J. A., Peterson, P. A. 1977. Controlling elements in a tribal maize from Columbia: *Fcu,* a two-unit system. *Genetics* 85:629–45

25. Greenblatt, I. M. 1968. The mechanism of *Modulator* transposition in maize. *Genetics* 58:585–97

26. Greenblatt, I. M. 1984. A chromosomal replication pattern deduced from pericarp phenotypes resulting from movements of the transposable element, *Modulator,* in maize. *Genetics* 108:471–85

27. Greenblatt, I. M., Brink, R. A. 1962. Twin mutations in medium variegated pericarp maize. *Genetics* 47:489–501

28. Hallauer, A. R., Russel, W. A., Smith, O. S. 1983. Quantitative analysis of Iowa Stiff Stalk Synthetic. *Stadler Symp.* 15:83–104

29. Hauser, C. H., Fusswinkel, H., Li, J., Oellig, C., Kunze, R., et al. 1989. Overproduction of the protein encoded by the maize transposable element *Ac* in insect cells by a baculo virus vector. *Mol. Gen. Genet.* In press

30. Hehl, R., Baker, B. 1989. Induced transposition of *Ds* by a stable *Ac* in crosses of transgenic tobacco plants. *Mol. Gen. Genet.* In press

31. Herrmann, A., Schulz, W., Hahlbrock, K. 1988. Two alleles of the single copy chalcone synthase gene in parsley differ *ba* a transposon-like element. *Mol. Gen. Genet.* 212:93–98

32. Johns, M. A., Mottinger, J., Freeling, M. 1985. A low copy number, copialike transposon in maize. *EMBO J.* 4:1093–102

33. Jordan, E., Saedler, H., Starlinger, P. 1968. 0° and strong polar mutations in the *gal* operon ar insertions. *Mol. Gen. Genet.* 102:353–63

34. Kunze, R., Stochaj, U., Lauf, J., Starlinger, P. 1987. Transcription of transposable element *activator (Ac)* of *Zea mays L. EMBO J.* 6:1555–63

35. Lassner, M. W., Palys, J. M., Yoder, J. I. 1989. Genetic transactivation of dissociation elements in transgenic tomato plants. *Mol. Gen. Genet.* In press

36. Masson, P., Surosky, R., Kingsbury, J., Fedoroff, N. V. 1987. Genetic and molecular analysis of the *Spm*-dependent *a-m2* alleles of the maize a locus. *Genetics* 177:117–37

37. McClintock, B. 1948. Mutable loci in maize. *Carnegie Inst. Washington Yearb.* 47:155–69

38. McClintock, B. 1949. Mutable loci in maize. *Carnegie Inst. Washington Yearb.* 48:142–54

39. McClintock, B. 1951. Mutable loci in maize. *Carnegie Inst. Washington Yearb.* 50:174–81

40. McClintock B. 1954. Mutations in maize and chromosomal aberrations in neurospora. *Carnegie Inst. Washington Yearb.* 53:254–260

41. McClintock, B. 1961. Further studies on the *Suppressor-mutator* system of control of gene action in maize. *Carnegie Inst. Washington Yearb.* 60:469–76

41a. McClintock, B. 1962. Topographical relations between elements of control systems in maize. *Carnegie Inst. Washington Yearb.* 61:448–61

42. McClintock, B. 1971. The contribution of one component of a control system to versatility of gene expression. *Carnegie Inst. Washington Yearb.* 70:5–17

43. Merckelbach, A., Döring, H.-P., Starlinger, P. 1986. The aberrant *Ds* element in the *Adh1-2F11::Ds* allele. *Maydica* 31:109–22

44. Montanelli, C., Di Fonzo, N., Marotta, R., Motto, M., Soave, C., Salamini, F. 1984. Occurrence and behavior of the components of the *o2-m(r)-Bg* system of maize controlling elements. *Mol. Gen. Genet.* 197:209–18

45. Müller-Neumann, M., Yoder, J. I., Starlinger, P., 1984. The DNA sequence of the transposable element of *Ac* of *Zea mays L. Mol. Gen. Genet.* 198:19–24

46. Nevers, P., Saedler, H. 1977. Transposable genetic elements as agents of gene

instability and chromosome rearrangements. *Nature* 268:109–15

47. Nevers, P., Shepherd, N. A., Saedler, H. 1986. Plant transposable elements. *Adv. Bot. Res.* 12:102–203

48. Nowick, E. M., Peterson, P. A. 1981. Transposition of the enhancer controlling element system in maize. *Mol. Gen. Genet.* 183:440–48

49. Pan, Y.-B., Peterson, P. A. 1988. Spontaneous activation of quiescent *Uq* transposable elements during endosperm development in *Zea mays. Genetics* 119: 457–464

50. Pereira, A., Cuypers, H., Gierl, A., Schwarz-Sommer, Zs., Saedler, H. 1986. Molecular analysis of the En-Spm transposable element system of *Zea mays. EMBO J.* 835–41

51. Pereira, A., Schwarz-Sommer, Zs., Gierl, A., Peterson, P. A., Saedler, H. 1985. Genetic and molecular analysis of the Enhancer (En) transposable element system of *Zea mays. EMBO J.* 4:17–23

52. Peterson, P. A. 1953. A mutable pale green locus in maize. *Genetics* 38:682–83

53. Peterson, P. A. 1965. A relationship between the *Spm* and *En* control systems in maize. *Am. Nat.* 44:391–98

54. Peterson, P. A. 1986. Mobile elements in maize: a force in evolutionary and plant breeding processes. *Stadler Genet. Symp.* 17:47–78

55. Peterson, P. A. 1987. Mobile elements in plants. €RC *Crit. Rev. Plant Sci.* 6:105–208

56. Peterson, P. A., Salamini, F. 1986. Distribution of active transposable elements among important corn breeding populations. *Maydica* 31:163–72

57. Pohlman, R., Fedoroff, N., Messing, J. 1984. The nucleotide sequence of the maize controlling element *Activator. Cell* 37:635–42

58. Pohlman, R., Fedoroff, N., Messing, J. 1984. Correction: nucleotide sequence of *Ac. Cell* 39:417

59. Rhoades, M. M., Dempsey, E. 1983. Further studies on two-unit mutable systems found in our high-loss studies and on the specificity of interaction of responding and controlling events. *Maize Genet. Coop. Newslett.* 57:14–17

60. Rhodes, P. R., Vodkin, L. O. 1985. Highly structured sequence homology between an insertion element and the gene in which it resides. *Proc. Natl. Acad. Sci. USA* 82:493–97

61. Rhodes, P. R., Vodkin, L. O. 1988. Organization of the *Tgm* family of transposable elements in soybean. *Genetics* 120:597–604

62. Robbins, T. P., Carpenter, R., Coen, E. S. 1989. A chromosome rearrangement suggests that donor and recipient sites are associated during Tam3 transposition in *Antirrhinum majus. EMBO J.* 8:5–13

63. Robertson, D. S. 1978. Characterization of a mutator system in maize. *Mutat. Res.* 51:21–28

64. Saedler, H., Nevers, P. 1985. Transposition in plants: a molecular model. *EMBO J.* 4:585–90

65. Schiefelbein, J. W., Raboy, V., Kim, H. Y., Nelson, O. E. 1988. Molecular characterization of Suppressor-mutator (Spm)-induced mutations at the bronze-1 locus in maize: the bz-m13 alleles. In *Proc. Internatl. Symp. Plant Transposable Elements.* ed. O. Nelson, pp. 261–78. New York: Plenum

66. Schwarz-Sommer, Zs. 1987. The significance of plant transposable elements in biological processes. In *Results and Problems in Cell Differentiation.* ed. W. Hennig, pp. 213–21. Berlin/Heidelberg: Springer

67. Schwarz-Sommer, Zs., Gierl, A., Klösgen, R. B., Wienand, U., Peterson, P. A., Saedler, H. 1984. The Spm (En) transposable element controls the excision of a 2 kb DNA insert at the Wx-m8 allele of *Zea mays. EMBO J.* 3:1021–28

68. Schwarz-Sommer, Zs., Gierl, A., Berntgen, R., Saedler, H. 1985. Sequence comparison of "states" of a1-m1 suggests a model of Spm (En) action. *EMBO J.* 4:2439–43

69. Schwarz-Sommer, Zs., Gierl, A., Cuypers, H., Peterson, P. A., Saedler, H. 1985. Plant transposable elements generate the DNA sequence diversity needed in evolution. *EMBO J.* 4:591–97

70. Schwarz-Sommer, Zs., Leclercq, L., Göbel, E., Saedler, H. 1987. Cin4, an insert altering the structure of the A1 gene in Zea mays, exhibits properties of nonviral retrotransposons. *EMBO J.* 6:3873–80

71. Schwarz-Sommer, Zs., Saedler, H. 1988. Transposons and retrotransposons in Plants: analysis and biological relevance: In *Transposition,* ed. A. J. Kingsman, S. M. Kingsman, K. F. Chater pp. 343–54. Cambridge: Cambridge Univ. Press

72. Schwarz-Sommer, Zs., Shepherd, N., Tacke, E., Gierl, A., Rhode, W., et al. 1987. Influence of transposable elements on the structure and function of the A1 gene of Zea mays. *EMBO J.* 6:287–94

73. Shapiro, J. A. 1969. Mutations caused by the insertion of genetic material into

the galactose operon of *Escherichia coli.* *J. Mol. Biol.* 40:93–105

74. Shirsat, A. H. 1988. A transposon-like structure in the 5' flanking sequence of a legumin gene from *Pisum sativum. Mol. Gen. Genet.* 212:129–33

75. Sommer, H., Carpenter, R., Harrison, B. J., Saedler, H. 1985. The tranposable element Tam3 of *Antirrhinum majus* generates a novel type of sequence alteration upon excision. *Mol. Gen. Genet.* 199:255–31

76. Sommer, H., Hehl, R., Krebbers, E., Piotrowiak, R., Lönnig, W.-E., Saedler, H. 1988. Transposable elements of *Antirrhinum majus.* See Ref. 65, pp. 227–36

77. Streck, R. D., MacGaffey, J. E., Beckendorf, S. K. 1986. The structure of hobo transposable elements and their insertion sites. *EMBO J.* 5:3615–23

78. Sundaresa, V., Freeling, M. 1987. An extrachromosomal form of the *Mu* transposons of maize. *Proc. Natl. Acad. Sci. USA* 84:4924–28

79. Sutton, W. D., Gerlach, W. L., Schwartz, D., Peacock, W. J. 1984. Molecular analysis of Ds controlling element mutations at the Adh1 locus of maize. *Science* 223:1265–68

80. Tacke, E., Schwarz-Sommer, Zs., Peterson, P. A., Saedler, H. 1986. Molecular analysis of "states" of the *A1* locus of *Zea mays. Maydica,* 31:83–91

81. Upadhyaya, K. C., Sommer, H., Krebbers, E., Saedler, H. 1985. The paramutagenic line niv-44 has a 5 kb insert, Tam2, in the chalcone synthase gene of *Antirrhinum majus. Mol. Gen. Genet.* 199:201–7

82. Walbot, V. 1986. Inheritance of Mutator activity in *Zea mays* as assayed by somatic instability of the *bz2-mu1* allele. *Genetics* 114:1293–312

83. Weck, E., Courage, U., Döring, H.-P., Federoff, N., Starlinger, P. 1984. Analysis of *sh-m6233,* a mutation induced by the transposable element Ds in the sucrose synthase gene of *Zea mays. EMBO J.* 3:1713–16

84. Weydemann, U., Wienand, U., Niesbach-Klösgen, U., Peterson, P. A., Saedler, H. 1988. Cloning of the transposable element Mpi1 from c2-m3. *Maize Genet. Coop. Newslett.* 64:48

Annu. Rev. Genet. 1989. 23:87–120

RECOMBINATIONAL CONTROLS OF rDNA REDUNDANCY IN *DROSOPHILA*

R. Scott Hawley and Craig H. Marcus

Department of Molecular Genetics, Albert Einstein College of Medicine, Bronx, New York 10461

CONTENTS

INTRODUCTION

The presence of tandemly repeated sequences is a characteristic of the eukaryotic genome. Moreover, although specific situations exist in which cer-

0066-4197/89/1215-0087$02.00

tain repeated sequences can be quantitatively altered, tandemly repetitive sequences normally exhibit remarkable quantitative and qualitative stability (67, 119). In contrast, tandemly repeated sequences are virtually absent in prokaryotic genomes. This striking difference in genome organization is vividly illustrated by the observation that when a tandemly repeated gene cluster from a eukaryote is placed in a prokaryote by recombinant DNA techniques, or when a segment of the bacterial chromosome is tandemly duplicated by genetic means, the redundancy is quite unstable and rapidly eliminated (1, 16). These observations suggest that eukaryotic organisms possess one or more genetic control mechanism(s) that function to maintain normal redundancy and to alter that redundancy under certain conditions.

The tandemly repeated 18S and 28S ribosomal RNA genes (rDNA) in *Drosophila melanogaster* have long served as a useful experimental system in which questions regarding the control of gene redundancy can be addressed. The well-documented stability of these genes in genetically normal individuals, the existence of known copy-number variants and of both somatic and germline processes that can alter redundancy, and, finally, the ability to detect changes in copy number at the phenotypic level provide an ideal system to study genetic events that can either increase or decrease rDNA redundancy.

In this review we consider the genetic basis of a number of processes known to result in changes in rDNA copy number. Specifically, we consider several recombinational phenomena that result in stable and heritable changes in rDNA redundancy: unequal recombination between sister chromatids (rDNA magnification), rDNA-mediated recombination between the X and Y chromosomes, and intrachromosomal rDNA recombination induced by the *Rex* mutation.

Other processes produce nonheritable increases or changes in somatic rDNA redundancy. These include rDNA compensation, a somatic process that results in an overall increase in rRNA-gene copy number in rDNA deficient flies (115, 116) and "independent polytenization," a process that results in amplification of rDNA in the polytene chromosomes of the larval salivary gland chromosomes (7–9, 26, 28–30, 32, 109). However, because our focus is on the role that recombination plays in the maintenance of gene copy number they are not considered here.

Throughout this review we are guided by several basic questions. First, to what extent do recombinational processes act to control rDNA copy number in wild-type Drosophila? Second, if any of these processes is induced by an rDNA deficiency and therefore can be considered as a compensatory mechanism, then by what mechanism(s) is an rDNA deficiency sensed and then corrected? Third, we explore whether the systems controlling rDNA copy number in flies represent mechanisms unique to rDNA, or rather, are they general mechanisms which function to control other aspects of gene structure or stability? We first present a brief summary of the genetics and molecular biology of rRNA genes is Drosophila.

THE BASIC STRUCTURE AND GENETICS OF THE rDNA

In wild-type *Drosophila melanogaster* males there are two clusters of tandemly repeated 18S and 28S rRNA genes (rDNA). Each of these tandem arrays comprises a single nucleolus organizer (see 40, 49, 61 for recent reviews). One nucleolus organizer is located in the proximal heterochromatin of the *X* chromosome and the other on the short arm of the *Y* chromosome (2). The *X* and *Y* chromosomal rRNA genes are organized as a large tandem array of rRNA transcription units separated by nontranscribed spacers (57, 85). Although variations have been observed in the number of rRNA genes possessed by a single nucleolus organizer, wild-type *X* and *Y* chromosomes usually contain approximately 150–225 rRNA genes (25, 69, 95, 116).

There are three major classes of rRNA genes that differ by the presence or absence of insertions in the 28S gene (40, 41, 85, 124). The first class is the uninterrupted and therefore functional 11-kb gene. A second class is composed of repeats containing a 0.5–6.5-kb insert known as Type I (124, 126). The third class of repeats possesses a different insertion, Type II (22). The Type I and Type II insertion sites are located 51bp apart within the 28S coding region but their presence is not coupled (102). Insert and non-insert bearing rDNA genes are interspersed, possibly not at random, within the rDNA arrays (39, 47, 121, 123). Although these insertion elements are predominantly located within the rDNA, copies of these insertion sequences are also found at other sites within the genome (21, 84).

X and *Y* chromosomal rDNA differ from each other in several important respects. First, 65–70% of the *X* chromosomal rRNA genes contain Type I insertion sequences (45, 125), whereas the *Y* contains few, if any, such insert-bearing genes (121, 125). Second, Yagura et al (131) showed that the 18S RNA transcribed from *X* rDNA differs from that of the *Y* by at least one base substitution. Third, Coen et al (19) found that the 5' end of the nontranscribed spacer is different in *X* and *Y* rDNA. Finally, the two arrays also differ in terms of size and frequency of long spacer length variants (50), and *X*-linked rDNA spacers are in general much more similar to each other than they are to *Y*-linked spacers (127). These data indicate that rDNA recombination between the *X* and *Y* chromosomes occurs at low frequency. However, such exchanges do occur and are the subject of a later section.

Genetically these two ribosomal gene clusters correspond to the *X* and *Y* chromosomal *bobbed (bb)* loci (99). Mutations at these loci are known as *bobbed* mutations and correspond to partial deficiencies of the rRNA genes

(96). These mutants range in severity from very weak alleles to lethal mutations (bb^l) according to the extent of deletion. In the absence of a bb or bb^+ homolog, X-chromosomal bb alleles comprised of 90–110 genes generally exhibit a strong pleiotropic phenotype that includes extremely short and thin bristles, etching of the abdominal cuticle, and lengthened developmental time. All of these aspects of the bobbed phenotype reflect a decrease in the cell's ability to produce ribosomes, and thus a general deficiency in the capacity for protein synthesis. Arrays bearing substantially fewer than 90 genes are usually lethal in the absence of a bb or bb^+ homolog (96, 116). Complete deficiencies (bb^o) and near-complete deficiencies (bb^-) exist for both the X and Y chromosomal bb loci.

Since the X and Y *bobbed* loci contribute to rRNA production in an additive fashion, the combined presence of approximately 150 rRNA genes is sufficient to produce a wild-type individual. This is true either when 150 or more genes are present on one homolog and the other chromosome carries few or no genes, or when both homologs possess an intermediate number of rRNA genes.

In this review, we refer extensively to two chromosomes that carry complete of near-complete rDNA deficiencies. The $In(1)sc^{4L}sc^{8R}$ chromosome, referred to as sc^4sc^8, carries a deficiency for most of the X chromosomal heterochromatin, including the rDNA. This chromosome carries no detectable rRNA genes and therefore is designated as bb^o. Many of the studies discussed also rely on an rDNA-deficient Y chromosome known as Ybb^-. Although this chromosome was originally thought to carry a nearly wild-type number of nonfunctional rRNA genes (90), Tartof (116) has shown that the Ybb^- chromosome is deficient for at least 80% of the rDNA usually found on a wild-type Y chromosome. According to Endow (27), the Ybb^- chromosome contains approximately 10–12 rRNA genes of which at least some are interrupted by Type II insertions and are therefore nonfunctional. The Ybb^- chromosome exerts a weak antimorphic effect on the bobbed phenotype (3) such that bb/Ybb^- males are phenotypically more severely bobbed than are $bb/0$ males. The basis of this effect is not well understood. These two deficiencies provide the necessary tools for detecting those changes in rDNA redundancy that result in alterations in the bobbed phenotype.

rDNA MAGNIFICATION

In otherwise genetically normal individuals, rDNA redundancy is generally quite stable (117, 118). For example, by combining the data from published controls it may be concluded that the spontaneous reversion rate of the bb^2 allele, which carries approximately 110 rDNA repeats, to bb^+ is less than one in twenty thousand. Despite this impressive stability, Ritossa (91) demonstrated that alterations in X chromosomal rDNA redundancy occur at high frequencies in the germlines of males carrying a *bobbed-deficient* Y chromosome (Ybb or Ybb^-). These changes in rDNA redundancy may be observed as reversions of bb to bb^+ and are the result of heritable increases in rDNA redundancy (91, 117, 118). The process that allows or facilitates these increases in rDNA redundancy is known as rDNA magnification.

Tartof recently presented an historical perspective on the analysis of magnification (120). Here we discuss the genetic characteristics of rDNA magnification, as well as the evidence that the primary mechanism of rDNA magnification is unequal rDNA exchange between sister chromatids. We also consider two hypotheses regarding the mechanisms by which magnification is induced or allowed. According to the first, magnification is mediated by an rDNA-specific recombination system induced by a deficiency of rDNA (31, 92). The second hypothesis holds that magnification is independent of rDNA copy number, but rather is executed by general cellular recombination or repair systems and is induced or allowed by failed or abnormal pairing between the X and Y chromosomes (47). We present evidence that both models have merit, and consider the circumstances under which each mechanism operates.

Genetic Assays for rDNA Magnification

MAGNIFICATION IN MALES   There are two commonly used methods for assaying rDNA magnification in the germlines of Drosophila males. A single generation assay for magnification was presented by Tartof (117, 118), and is referred to as the sc^4sc^8 assay (see Figure 1). In this assay, bb/Ybb^- males are crossed to sc^4sc^8/dl-49 females (dl-49 denotes $In(1)dl$-49, a bb^+ balancer chromosome used only to maintain the sc^4sc^8 chromosome in stock). Because the sc^4sc^8 chromosome carries a complete deficiency for bb (bb^o), bb to bb^+ reversion may be observed by examining the phenotype of the X/sc^4sc^8 daughters. In the sc^4sc^8 assay, the frequency of X chromosomal magnification is characteristic of the specific *bobbed* allele being tested (117, 118), and is influenced by age, autosomal background, and phenotypic severity with respect to bobbed (64).

For purposes of illustration we consider the use of this assay to measure the magnification of a commonly used bb allele known as bb^2 (116). Like all or most alleles of bb, the bb^2 mutation is stable in the germlines of otherwise genetically normal males. Therefore, in crosses of bb^2/Ybb^+ males to sc^4sc^8/dl-49 females, all the X/sc^4sc^8 daughters are phenotypically bobbed. However, when bb^2/Ybb^- males are crossed to sc^4sc^8/dl-49 females, approximately 18% of the X/sc^4sc^8 daughters are phenotypically bobbed-plus, while the remainder still exhibit a severe bobbed phenotype (117, 118). The phenotypically bobbed-plus daughters carry bb^+ revertants of the bb^2 mutation that are associated with increases in rDNA redundancy, and are stable even after testcrosses involving several generations of out-crossing (117, 118).

Some bb^2/Ybb^- males produce large numbers of revertant progeny, while others produce few or none (117, 118). Indeed the frequency of magnification in a single fly can be as high as 80–90% (117, 118). The extensive clustering of magnification events among the progeny of single males has led Tartof (117, 118) to suggest that the initial magnifying event can occur at low frequency in both meiotic and premeiotic cells. He (117, 118) proposed that premeiotic magnification events in the germline can be amplified by cell proliferation and clonal selection and thus produce large clusters of revertant-bearing progeny. By analyzing the rRNA repeat-class composition of bb^+ revertants produced by a single male, Terracol (122) demonstrated that clusters of magnified progeny are in fact the consequence of single premeiotic magnification events.

Ritossa (91) in the course his original studies, introduced a different assay for magnification in males, referred to as the Ybb^- assay. In this assay bb/Ybb^- males are backcrossed to *attached-X/Ybb^-* females and the X/Ybb^- sons are scored with respect to the bobbed phenotype. This set of backcrosses is then repeated for several generations and the frequency of bobbed-plus males is scored at each generation (see Figure 1). Ritossa (91) and others (10, 48) observed that a large fraction of the X/Ybb^- sons were less severely bobbed than their fathers and were often phenotypically bobbed-plus. This amelioration of the bobbed phenotype continued over several generations of backcrossing to attached-X/Ybb^- females.

Ritossa (91) and others (10, 48) have also shown that only a minority of the X chromosomes carried by phenotypically bobbed-plus X/Ybb-males after one or two generations of magnification carry a stable bb^+ revertant; the majority retest as strongly bobbed following association with a bb^+ homolog for several generations. After many generations under magnifying conditions however, most if not all bobbed-plus progeny carry stable and heritable bb^+ alleles (10, 48, 91, 94).

Although both the sc^4sc^8 and the Ybb^- assays depend on placing X chromosomes derived from bb/Ybb^- males opposite an rDNA-deficient homolog (bb^0 or bb^-), they produce different results in both the frequency and the heritability of the magnification process. More specifically, in the sc^4sc^8 assay the majority of the X/sc^4sc^8 progeny are still strongly bobbed, while the remainder appear as bobbed-plus progeny bearing a stable and heritable X chromosomal bb^+ revertant. In contrast, although the majority of X/Ybb^- progeny produced in the Ybb^- assay are less severely bobbed that their bb/Ybb^- fathers, upon retesting the majority of these males are shown to carry unmagnified bb alleles. As discussed in a later section, the paradoxical differences between these two assays were resolved by demonstrating that many

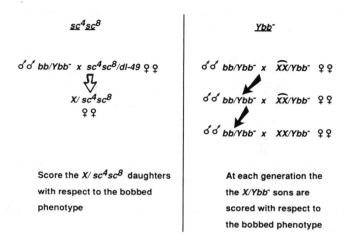

Figure 1 Two commonly used assays for rDNA magnification in Drosophila males (see text for explanation).

of the phenotypically bobbed-plus offspring observed in the Ybb^- assay arise as a consequence of autosomal suppressors of the bobbed phenotype, and not as the result of bb to bb^+ reversion (73).

For purposes of exposition we have focused solely on X chromosomal magnification. However, the rDNA of Ybb chromosomes can also be magnified in assays similar to those just described (10, 72). The frequency of magnification varies according to the Ybb allele; some Y chromosomes show very high levels of magnification (10, 94), while others magnify poorly, if at all (128). The bases of these differences are not understood.

MAGNIFICATION IN FEMALES   Komma & Endow (53, 54) used an assay similar to the sc^4sc^8 assay to demonstrate and characterize magnification in Drosophila females. In one example, magnification is assayed in the progeny of phenotypically bobbed females of the genotype $bb/attached$-$XYbb^-$ (the symbol $attached$-$XYbb^-$ denotes an rDNA-deficient $attached$-XY chromosome derivative of the Ybb^- chromosome). As with the two male assays, magnification was monitored by placing the X chromosomes derived from these females opposite an rDNA-deficient homolog (i.e. by crossing the indicated females to $attached$-$XYbb^-/Ybb^+$ males and scoring the resulting $X/attached$-$XYbb^-$ daughters with respect to bobbed). Phenotypically bobbed-plus offspring carrying stable bb^+ revertants were recovered at low frequencies (1–2%). Magnification was also demonstrated in females bearing a free Ybb^- chromosome, but did not occur in the absence of a Y chromosome. The significance of the requirement for a Y chromosome for magnification in females is discussed in a later section.

This low frequency of female magnification at least partially reflects the absence of the clusters of revertant progeny so commonly observed when assaying male magnification. As Komma & Endow pointed out (53), both the lack of clusters and the low frequency of magnification are presumably a consequence of differences between germline development in males and females. In males, where all the products of definitive stem-cell divisions become functional gametes, premeiotic events can produce large clusters of magnified progeny; however, in females, where the same number of definitive stem-cell divisions produce only a single oocyte, the products of premeiotic magnification events are frequently lost. Despite these numerical differences, Komma & Endow's data (53) clearly demonstrate that the magnification process can occur in the germlines of both sexes and result in stable and heritable reversion at the *bobbed* locus.

Pseudo-reversion Mimics rDNA Magnification

In order to explain the discrepant results regarding frequency and heritability produced by the sc^4sc^8 and Ybb^- assays for male magnification, Marcus et al (73) used both assays to examine the magnification of a single well-characterized bb allele, bb^2 (116). To eliminate any subjective bias in scoring the bobbed phenotype, they used as their metric the length of the posterior scutellar bristles. For both assays, their data confirm the observations made by previous investigators. In the sc^4sc^8 assay they observed a low frequency of magnified progeny (10%), most or all of which carried a stable bb^+ revertant; in the Ybb^- assay, they saw a very high frequency of phenotypically bobbed-plus progeny (47%), only a fraction of which could be attributed to stable bb^+ revertants. The majority of the phenotypically bobbed-plus progeny obtained in the first generation using the Ybb^- assay did not carry a bb^+ revertant, and retested as severely bobbed after only one generation over a bb^+ homolog.

Marcus et al (73) explained these apparently contradictory data by demonstrating that a substantial fraction of the phenotypically bobbed-plus male progeny observed in the Ybb^- assay arose by a process or mechanism not revealed by the sc^4sc^8 assay: namely, the accumulation of recessive autosomal suppressors of the bobbed phenotype present in the *attached-X/Ybb$^-$* stock. The existence of such modifiers was initially suggested by Lindsley & Grell (59), but rejected by Atwood (3) and Harford (43). However, Marcus et al (73) demonstrated that at least two unrelated stocks of X/Ybb^- males and *attached X/Ybb$^-$* females carry autosomal modifiers that strongly suppress the bobbed phenotype, allowing for the recovery of bobbed-plus pseudo-revertants. These modifying loci can fully ameliorate the bobbed phenotype of bb/Ybb^- males whose fathers were nonmagnifying bb^2/Ybb^+ males, and even allow substantial phenotypic reversion when bb to bb^+ genetic reversion is blocked by magnification-defective mutations (73).

Therefore, a substantial fraction of the apparent genetic instability observed in the Ybb^- assay is due to the impossibility of phenotypically distinguishing between males bearing magnified bb^+ alleles and bb/Ybb^- pseudo-revertants. Autosomal modifying loci also explain the excess of phenotypically bobbed-plus progeny observed using the Ybb^- assay compared to the sc^4sc^8 assay. This difference is in large part due to the homozygosis of magnifiers in the Ybb^- assay that result in a high frequency of pseudo-revertant progeny. Since such modifiers are common in stocks in which the females carry a Ybb^- chromosome (C. H. Marcus, unpublished observation), the differences in magnification frequencies observed by previous investigators using different assays may often reflect the action of these modifiers.

Finally, the combined actions of phenotypic suppression and magnification at the *bobbed* locus explain the observation that most, if not all, of these bobbed-plus progeny examined after several generations of magnification carry heritable bb^+ alleles. For example, at each generation the sources of bobbed-plus progeny are (*a*) sons of fathers already carrying a bb^+ revertant, (*b*) bb^+ bearing sons from magnifying bb/Ybb^- males, and (*c*) bb/Ybb^- males made phenotypically bobbed-plus by the modifying loci. Thus, with each generation the fraction of true revertant progeny should increase until, by later generations, the entire population of males should be genotypically bb^+/Ybb^-. This is exactly what is observed.

Magnification is a Stable and Heritable Process

The heritability and stability of X chromsomal bb^+ reversion has been clearly demonstrated in studies in which pseudo-reversion does not occur. For example, Tartof (117, 118) addressed the question of instability by examining the progeny of bobbed-plus X/bb^o daughters produced by bb^2/Ybb^- males. Each of these daughters was then crossed to bb^o/Ybb^+ males and the X/bb^o daughters were examined with respect to the bobbed phenotype. All of the revertants tested in this manner for stability produced only bobbed-plus progeny in the next generation. Similar results were obtained by Hawley & Tartof (45–47) and by Hawley et al (44). These workers have also observed that similar if not identical frequencies of magnification are obtained regardless of whether the magnified X chromosomes are recovered directly (i.e. by placing them opposite a bb^o bearing homolog) or maintained for one or more generations opposite a bb^+ homolog (i.e. a normal Y chromosome) and then assayed by crossing to appropriate tester females. Tartof (117) also demonstrated that when such males carry a magnified bb^+X chromosome and a normal Y chromosome, they transmit only bb^+-bearing X chromosomes to their progeny. These results argue strongly that bb^+ revertants are genetically stable even in the absence of selection.

Although a few examples of apparent instability have been reported, these instances are usually confined to cases where the magnified offspring exhibited incomplete phenotypic reversion; instability is only rarely observed in the descendants of strongly bobbed-plus individuals (62, 73). Therefore, it is possible that much of the purported instability is due to the reduced penetrance of weak alleles of bb that arise as a consequence of magnification. Locker (62) also observed rare reversions of strong bb^+ loci kept opposite a bb^+ balancer chromosome for a few generations to bb^{rl} alleles. These exceptions remain unexplained, but their rarity coupled with the general stability of reversion events, argues convincingly that most, if not all, magnification events are both stable and heritable.

The first insight into the mechanism by which such revertants arise came from the discovery that the same bb/Ybb^- males that produce magnified X chromosomes also produce gametes with a substantially reduced number of rRNA genes. This phenomenon was termed rDNA reduction.

rDNA Reduction

Reduction was originally defined as the production of *bobbed-reduced lethal* (bb^{rl}) mutations (i.e. X chromosomes carrying a reduced number of rRNA genes) by bb/Ybb^- males. The observation that both reduction and magnification occurred in the germlines of magnifying males was initially made by Locker & Prud'homme (64), confirmed by Tartof (118), and extended to Y chromosomal magnification by Malva and coworkers (72). The basic experiment in these studies involves crossing magnifying males to appropriate tester females, such that the father's X or Y chromosome is recovered without selection opposite a bb^+ homolog. For example, by crossing bb^2/Ybb^- males to *attached-X/Ybb$^+$* females, the X chromosome donated by the father can be recovered opposite a Ybb^+ chromosome donated by the mother. The status of the bb allele carried by each of these sons is then assayed by individually crossing each male to females carrying a bb^o homolog. In those cases where the tester males carried a bb^{rl} mutation, X/bb^o daughters do not survive. Males carrying a bb^+ revertant are also easily detected because they produce only bobbed-plus X/bb^o daughters. Using this methodology, Tartof (118) demonstrated that reduction to bb^{rl} occurs at a frequency of approximately 3% in the germlines of bb^2/Ybb^- males.

X chromosomal reduction occurs in the germlines of both bobbed and bobbed-plus males (i.e. bb^1/Ybb^-, bb^+/Ybb^-, or $bb/y^+bb^+Ybb^-$) and requires the presence of either the Ybb^- chromosome or its derivatives (47, 83, 117, 118). Similarly, Y chromosomal reduction can also occur in the germlines of bobbed and bobbed-plus males (i.e. bb/Ybb or bb^+/Ybb) (72).

Unlike magnification events, reduction shows no evidence of clustering and occurs at a frequency at least tenfold less than that observed for magnification (44, 47, 64, 72, 117, 118). Despite these differences, Tartof (118) used two lines of evidence to argue that magnification and reduction are likely to be reciprocal products of the same process. First, the occurrence of the bb^{rl} chromosomes is correlated with the occurence of bb^+ revertants in the same germline. Second, the rDNA content of a bb^{rl} allele and of a bb^+ revertant, both derived from bb^2, have been measured and found to be 20 and 218 rRNA genes, respectively, while the parental bb^2 possesses approximately 110 rRNA genes. To quote Tartof (118), "Thus, the 90 rRNA genes lost by reduction is roughly equal to the number of rRNA genes gained by magnification." Tartof (118) further proposed that the observed excess of bb^+ revertants is due to both germline selection against bb^{rl}/Ybb^- cells, which will be grossly deficient in rDNA content, and to the possibility that bb^{rl} chromosomes can undergo subsequent magnification events in the same germline. After discussing two possible mechanisms for the magnification process, we will return to the relationship between magnification and reduction and present further evidence that they are in fact the reciprocal parts of the same process.

Possible Mechanisms of Magnification

In the two decades since its discovery, two general models have been proposed to explain the magnification process. Ritossa (10, 92, 97) initially proposed a model of magnification in which the bobbed phenotype causes extra rDNA synthesis and in which the newly synthesized rRNA genes can reintegrate into the chromosome. According to Ritossa (94), "The extra rDNA is free in the majority of the somatic cells, but can be integrated into the chromosome in the germ line. Integration is visualized as due to crossing over between the chromosome and circularized extra copies of rDNA." He proposed that these integration events resulted in a stable increase in rDNA copy number and thus in phenotypic reversion. This hypothesis was initially used to explain (*a*) the large increases in the rDNA redundancy observed in males of the genotype bb/Ybb^- (10, 23, 63, 94); (*b*) the gradual reduction in severity of the bobbed phenotype over several generations; and (*c*) the apparent instability of bb to bb^+ reversion during the first few generations of magnification. This hypothesis is discussed in detail below.

The rDNA redundancy in bb/Ybb^- males frequently exceeds that observed in genetically wild-type bb^+/Ybb^+ males. Indeed, at least one group has presented electron-microscopic evidence for the physical existence of the postulated rDNA episomes in the testes of magnifying flies (42). These increases in rDNA redundancy, however, are not restricted to males undergoing magnification and more likely reflect the effect of rDNA compensation that allows for disproportionate increases in rDNA redundancy in the somatic tissues. Compensation occurs when an X chromosomal rDNA locus is present in a single copy (i.e. in $X/0$ or X/Ybb^- males or in X/sc^4sc^8 females) and is restricted to the X chromosomal rDNA (115, 116). Tartof (115) has demonstrated that compensation occurs in numerous genotypes, including X/Ybb^-, $X/0$, and X/sc^4sc^8, but cannot ameliorate the bobbed phenotype. In fact, phenotypically bobbed bb^2/sc^4sc^8 females possess more rRNA genes than do bb^+/bb^+ wild-type females (115). The increase in rDNA redundancy observed under compensating conditions varies between genotypes and is also influenced by autosomal background (24, 116). Most importantly, compensation is nonheritable. Therefore the observed increases in rDNA redundancy in males undergoing magnification are easily explained as a compensatory response, and do not require the effect of episomal replication and reintegration to be invoked.

Ritossa's model also cleverly explained the apparent instability of bb reversion, as well as the gradual reduction in the severity of the bobbed phenotype over several generations (10, 48, 71, 91). According to Ritossa (94), "Some rDNA is integrated into the chromosome and some 'goes along' with it in some other manner, but is ultimately integrated." Thus, the model invokes attachment without integration to explain instability, and eventual integration to explain the gradual amelioration to the bobbed phenotype. However, as explained in the preceding section, these phenomena are largely a consequence of pseudo-reversion arising from the accumulation of autosomal modifiers and not of changes at the bb locus. Indeed, as already discussed, considerable evidence exists that magnification events are stable and can occur within a single generation.

Finally, the excision and reintegration model is also inconsistent with the failure of ring chromosomes to magnify (discussed in next section) and the evidence that the X and Y rDNA loci are maintained independently (19, 50, 121, 131). Thus, the Ritossa model, although crucial to the development of the field, is inconsistent with most of the available data.

Alternatively, Tartof (117, 118) proposed that rDNA magnification-reduction results from unequal mitotic and meiotic exchanges involving the rDNA of sister chromatids. Such recombination events would lead to the production of two sister chromatids differing in rRNA gene number: one magnified chromatid with a greater number and the other a reduced chromatid with fewer rDNA-tandem repeats that are borne by either parental chromatid. This hypothesis is supported by the following observations: (a) both rDNA magnification and reduction occur in the germlines of individual bb/Ybb^- males; (b) neither magnification nor reduction is the consequence of recombination events involving the X and Y chromosomes; (c) rDNA reduction and magnification are both reversible and apparently reciprocal events; and (d) magnification does not occur or is severely inhibited in males bearing a ring-X chromosome, a condition in which recombination between sister chromatids would result in the creation of inviable dicentric ring chromosomes. As Tartof pointed out (117), "The virtue of this hypothesis rests on the fact that it is consistent with all observations regarding the *bb* locus, does not require special ad hoc assumptions, is readily testable and relies on fundamental orthodox genetic principles."

This magnification model has been referred to as the unequal sister-chromatid exchange (USCE) hypothesis. Unfortunately, the term sister-chromatid exchange has come to imply both postreplicational recombination between sister chromatids and exchange events occurring at replication forks themselves (52). Because the former process corresponds more closely to that depicted by Tartof's model, we refer instead to "recombination between sister chromatids."

Unequal Recombination between the Bobbed Loci of Sister Chromatids as the Mechanism of rDNA Magnification

The unequal sister-chromatid recombination hypothesis is supported by two separate lines of evidence. First, *bb*-bearing ring-X chromosomes are incapable of magnification. Second, when the confounding effects of selection and clonal amplification are eliminated, magnification and reduction occur at equal frequencies, and both events are stable and heritable.

The behavior of ring-X chromosomes under magnifying conditions provides a critical test of the unequal sister-chromatid recombination hypothesis. Sister-chromatid recombination within a ring chromosome produces dicentric chromatids that either will be broken or lost at the succeeding mitotic anaphase, or at anaphase II of meiosis (76, 77). Therefore, if the unequal sister-chromatid recombination hypothesis is correct, then ring-*bb* chromosomes should be incapable of both magnification and reduction. Tartof (118) examined magnification in the germlines of ring-*bb*/*Ybb*⁻ males over four generations of magnification. Despite scoring more than 1500 offspring not a single magnified ring-*bb*⁺ chromosome was obtained. The failure of ring-*bb*/*Ybb*⁻ males to produce *bb*⁺ magnificants has been confirmed by Endow et al (34), and extended to magnification in females by Komma & Endow (53).

Furthermore, Endow et al (34) have presented direct genetic evidence that magnifying conditions cause ringe chromosome loss (presumably as a consequence of the formation and subsequent loss of dicentric rings). They have shown that in males undergoing magnification, ring-X chromosomes are transmitted less efficiently that are their *Ybb*⁻ homologs, resulting in an excess of male progeny. In contrast, the ring-X and Y chromosomes are transmitted with equal efficiency in males not undergoing magnification. The decrease in the fraction of ring-X chromosomes transmitted by magnifying males must be due to chromosome loss during meiosis because premeiotic loss events are lethal, and thus would not bias the sex ratio. That the ring-loss events are the consequence of recombination events involving the rDNA is suggested by the finding that no sex ratio distortion is observed in the progeny of ring-X $bb°/Ybb$ males, even though the Ybb chromosome is an efficient inducer of magnification.

Finally, Endow et al (34) have presented cytological evidence that $R(1)bb/Ybb^-$ males produce a higher frequency of aberrant ring structures at anaphase II of meiosis. These abberrant ring structures were predominantly dicentric and interlocked ring chromosomes, but broken, lagging, or interconnected ring chromosomes were also observed. In these experiments pupal testes of $R(1)bb/Ybb^-$ and $R(1)bb/Ybb^+$ males were examined for the presence of second meiotic division figures. The ring-bearing figures, which constituted half the total, were then scored for the presence of abnormal ring structures. $R(1)bb/Ybb^-$ males produced abnormal anaphase II figures at a frequency of 11.5% (71% of which were dicentrics), compared to a value of 2.4% observed in $R(1)bb/Ybb^+$ controls. The dicentrics observed in nonmagnifying males likely arise as a consequence of spontaneous recombination events between sister-chromatids occurring outside of the rDNA. Although spontaneous recombination between sister chromatids has not been observed in normal mitotic cells (38), dicentric rings have been reported in both meiotic and mitotic cells carrying ring chromosomes (37). In summary, these data show that the number of dicentric ring chromosomes is increased under magnifying conditions, and are consistent with an increase in recombination between the rDNA of sister chromatids in magnifying males.

Although magnification is blocked in $R(1)bb$ chromosomes, Endow et al (34) have reported the recovery of at least one reduced ring chromosome among the progeny of $R(1)bb/Ybb^-$ males crossed to bb/bb females. Nineteen daughters bearing the putative reduced ring chromosomes were recovered; of these four were fertile. Further test-crossing showed that one of these females did in fact carry a $R(1)bb^{r1}$ derivative of $R(1)bb$. Endow et al (34) ascribe these exceptions to intrachromatid exchanges that result in nonreciprocal excision events. However, given the high frequency of ring chromosome breakage observed at anaphase II in of $R(1)bb/Ybb^-$ males, and the high frequency of subsequent zygotic lethality, it is possible that the origin of these exceptions is more complex. This point needs clarification and further study.

In summary, the cytological observations of Endow et al (34) strongly support the hypothesis that those magnification events that occur meiotically are the consequence of recombination between sister chromatids. Moreover, the failure of $R(1)bb/Ybb^-$ to produce any bb^+ progeny in males (34, 118), and in females (53) would seem to suggest that sister-chromatid recombination is the underlying chromosomal process for magnification whenever it occurs (44, 47). Our only reservation on this point is that it remains to be shown that a given *bb* mutation carried by a ring chromosome will be capable of magnifying when carried by a rod chromosome. At least some *bb* alleles, especially those associated with heterochromatic rearrangements, are notoriously refractory to premeiotic magnification, even on rod chromosomes (62, 64, 89; R. S. Hawley, unpublished data). Moreover, Gatti (37) has argued that the ability of an individual ring chromosome to undergo sister-chromatid recombination "is determined by factors, in addition to size, intrinsic to its structure." Therefore, the failure of a ring-borne *bb* allele to magnify might possibly reflect either an unusual property of that *bb* allele or a peculiarity in the structure of the ring chromosome, such as a heterochromatic rearrangement. Until a given ring chromosome can be placed through several cycles of ring opening and reclosure and tested for magnification at each step, the simple result that *R(1)bb* chromosomes do not magnify does not strictly prove that premeiotic rDNA magnification occurs as a consequence of recombination between sister chromatids.

The second part of the Tartof hypothesis postulates that the observed excess of sperm bearing bb^+ chromosomes relative to sperm bearing bb^{rl} chromatids is a consequence of both selection in the germline for cells with normal or near normal rDNA redundancy and the opportunity for bb^{rl} chromatids to undergo further cycles of magnification. This hypothesis predicts that magnified chromatids and reduced chromatids will be recovered at equal frequencies under conditions where magnification to bb^+ would not provide a germline cell with a selective advantage. It also predicts that the magnification events should increase in frequency and become more highly clustered when a bobbed phenotype is imposed.

This prediction has been tested by using a series of derivatives of the Ybb^- chromosome to which varying amounts of X-chromosomal rDNA have been appended (47). These chromosomes are designated as: bb^+Ybb^- (a Ybb^- chromosome to which the X chromosomal bb^+ locus has been appended), $bbYbb^-$ (a Ybb^- chromosome to which the X chromosomal *bb* locus has been appended), and $bb^{rl}Ybb^-$ (a reduced-lethal derivative of bb^+Ybb^-). When bb^2/bb^+Ybb^-, $bb^2/bbYbb^-$, and $bb^2/bb^{rl}Ybb^-$ males were compared, the frequencies of magnification of the bb^2 chromosome were 0.6%, 1.8%, and 10.7%, respectively (47). Therefore the frequency of magnification increases with the severity of the bobbed phenotype. Moreover, large clusters of magnified progeny were produced only in the presence of the bobbed phenotype (i.e. by $bb^2/bb^{rl}Ybb^-$ males). However, magnification events that occurred as single events, rather than in clusters, were produced with similar frequencies by bb^2/bb^+Ybb^-, $bb^2/bbYbb^-$, and $bb^2/bb^{rl}Ybb^-$ males.

The frequency of reduction (approximately 1.0%) was also similar for all three classes of males (47; R. S. Hawley unpublished data). Indeed, among the progeny of bb^2/bb^+Ybb^- males, bb^+ and *reduced-lethal*-bearing chromosomes were recovered at equal frequency. Therefore, as Tartof predicted, removing the selective advantage possessed by bb^+ magnificants in an otherwise bb germline creates a situation where the frequency of magnification equals the frequency of reduction. Once a bobbed phenotype is imposed, however, both the excess of magnification events relative to reduction events and the clustering of magnification events within single germlines can be restored.

Tartof's prediction that bb^+ revertants will enjoy a growth advantage in an otherwise phenotypically bobbed germline has been directly tested by pole-cell transplantation experiments (D. Wright & R. S. Hawley, unpublished data). In these experiments bb/Ybb^+ pole cells were transplanted into bb/Ybb^- recipients and vice versa in order to create mosaic germlines. When bb/Ybb^+ pole cells were injected into bb/Ybb^- recipients, large clusters of donor-derived sperm were obtained. However, when bb/Ybb^- pole cells were transplanted into bb/Ybb^+ recipients, donor-derived sperm were not observed. Although only a small number of mosaic animals were analyzed, these data do support the existence of germline selection or competition with respect to the bobbed phenotype.

Therefore, in addition to providing a simple and aesthetically pleasing model for both magnification and reduction, the unequal sister-chromatid recombination hypothesis has been able to withstand a number of direct tests. This is especially true for meiotic magnification events where direct cytological observations are available. The data are less compelling for premeiotic events, and alternative hypotheses can still be considered in this case.

The Bobbed Phenotype and the Induction of rDNA Magnification

Endow & Atwood (31) have argued, "Unlike other systems of gene amplification, magnification is induced by a deficiency in the genes that amplify; the subsequent gene increase corrects the original deficiency." Indeed, the standard assays for rDNA magnification in males require testing the progeny of phenotypically bobbed males for the production of bb^+-bearing sperm, and considerable evidence suggests a strong correlation between the severity of the phenotype and the frequency of magnification. Endow & Atwood (31) have further claimed that "magnification does not occur in the absence of an rDNA deficiency." Nonetheless, at least four lines of evidence suggest that although the bobbed phenotype is required for the induction of premeiotic magnification events, it is not essential for the induction of meiotic magnification or reduction in males.

First, although reduction does not occur in bb/Ybb^+ males, it is observed in the germlines of phenotypically bobbed-plus bb^+/Ybb^- males (83, 117).

Indeed, the frequency of reduction is unaffected by the presence or absence of the bobbed phenotype (47, 72, 83, 117). Therefore, if magnification and reduction are reciprocal events, then magnification must also be occurring in the germlines of these males.

Second, Hawley & Tartof (47) have used a series of bb and bb^+ derivatives of the standard Ybb^- chromosome to analyze the role of the bobbed phenotype in the control of rDNA magnification. They observed that both magnification and reduction events occur in the progeny of bb/bb^+Ybb^- or $bb/bbYbb^-$ males, and that those magnification events that occur as single events rather than in clusters (presumably meiotic magnification events) occur with similar frequencies in bb/bb^+Ybb^-, $bb/bbYbb^-$, and bb/Ybb^- males. Indeed, as noted above, among the progeny of bb/bb^+Ybb^- males, bb^+ and reduced-lethal-bearing chromosomes are recovered at equal frequency. This observation demonstrates that meiotic magnification events are not a consequence of the bobbed phenotype.

Third, other structurally abnormal Ybb^+ chromosomes (such as $B^sv^+Ybb^+y^+$), which are unrelated to Ybb^-, can also induce X chromosomal magnification, albeit at much reduced frequencies (47). Magnification can also be observed among the progeny of bb/Ybb^- males carrying a bb^+ bearing duplication of the X chromosome (31).

Finally, that meiotic magnification is independent of the bobbed phenotype is confirmed by the analysis of Ybb^--induced loss of ring-X chromosomes (34). As described in the preceding section, the decrease in the fraction of ring-X chromosomes transmitted by magnifying males must be due to chromosome loss during meiosis, since premeiotic loss events are lethal and would not bias the sex ratio. The presence of the Ybb^- chromosome, which creates magnifying conditions, causes significant ring loss regardless of whether the ring-X is bb or bb^+. Therefore, like bb to bb^+ reversion events, ring-X loss is independent of the bobbed phenotype and requires only the presence of the Ybb^- chromosome.

These experiments demonstrate that the induction of meiotic magnification and reduction events in males is not related to rDNA dosage, but appears rather to be induced by the presence of a structurally abberant Y chromosome. Endow & Atwood (31) have claimed that the requirement for an rDNA deficiency is absolute even at meiosis, but that "the Ybb^- chromosome is constitutive for magnification." However, hypotheses that imbue the Ybb^- chromosome with special properties cannot explain the observation that other structurally abnormal Y chromosomes, such as $B^sv^+Ybb^+y^+$ which carries a bb^+ allele, also induce low frequencies of meiotic magnification and reduction (47).

In contrast to meiotic magnification events, whose induction is independent of the bobbed phenotype, premeiotic magnification events are observed only in phenotypically bobbed males. Premeiotic events are observed as large clusters of bb^+ progeny that appear in vast excess to their bb^{rl} sibs. Because of the size of these clusters and their distribution among bb/Ybb^- individuals, they necessarily reflect an early magnification event that probably arises during the mitotic or syncytial stage of germline development. However, the expected clusters of bb^+ are not produced in bb/bb^+Ybb^- germlines, nor did Hawley & Tartof (47) observe large clusters of bb^{rl} alleles that would be expected from phenotypically bobbed-plus cells that are genotypically bb^{rl}/bb^+Ybb^-.

The fact that these clusters of bb^+ X chromosomes are not produced in phenotypically bobbed-plus germlines is curious, given the small number of progenitor cells comprising the Drosophila germline (60). The male germline develops from three to seven pole cells. These cells then undergo three asynchronous mitoses to generate approximately 37 cells that then migrate into the embryo to form the presumptive gonad. Approximately ten to fourteen of these cells then go on to form the adult stem-cell population located in the apical region of the testis. Stem cells then divide mitotically to produce a new stem cell and a primary spermatogonium. The primary spermatogonium goes on to divide in a synchronous and syncytial manner to produce a sixteen-cell cyst. Finally, each nucleus within the cyst undergoes two meiotic divisions to ultimately yield a bundle of 64 spermatids.

Given this scheme of spermatogenesis, if magnification did occur premeiotically in bb/bb^+Ybb^- germlines, it should produce at least some clusters of bb^+ or bb^{rl} alleles even in the absence of selection. The failure to recover such clusters, given the large number of males examined, leads to the conclusion that early magnification events occur only in the presence of the bobbed phenotype. This conclusion is strengthened by the observation that the ability of various Y chromosomes to induce premeiotic magnification is strictly correlated with the severity of the bobbed phenotype.

Although the bobbed phenotype is necessary for the induction of premeiotic magnification events, we do not know whether it is sufficient. A structural aberration on the Y chromosome, such as that possessed by $B^s v^+ Y bb^+ y^+$ or Ybb^-, or a specific Y chromosomal site may also be required. Indeed, Komma & Endow (53, 54) have used segmental aneuploidy to demonstrate that a locus on the tip of the long arm of the Y chromosome is necessary for normal levels of magnification in females. Preliminary evidence (cited in 54) suggests that this region of the Y chromosome may also be required for magnification in males.

Data from our own laboratory further demonstrate that a deficiency in the capacity for normal levels of protein synthesis, as manifest by the bobbed phenotype, is not sufficient to induce magnification. This is to say that

bb/Ybb^+ males bearing deficiencies for the 5S rRNA genes, which are phenotypically indistinguishable from bb/Ybb^- males, do not exhibit detectable levels of magnification (L. Benjamin & R. S. Hawley, unpublished data). Similarly, bb/Ybb^+ males carrying a *Minute* mutation known to encode a ribosomal protein (55) also fail to undergo magnification. Therefore, either premeiotic magnification requires the presence of a structurally aberrant Y chromosome in addition to a general deficiency in protein synthetic capacity, or the cell's ability to induce magnification in the presence of the bobbed phenotype reflects its ability to directly sense a deficiency in rDNA redundancy.

Komma & Endow (53, 54) have also argued that the bobbed phenotype is required for magnification in females. However, their conclusion is based only on the observation that magnification does not occur in $bb/sc^4sc^8/Ybb^+$ females. Unfortunately, given that Ybb^+ chromosomes do not induce magnification in males, this experiment does not prove that an rDNA deficiency is necessary to produce magnification. Until these studies are repeated using an efficient inducer of magnification, such as a bb^+Ybb^- chromosome, the role of the bobbed phenotype in female magnification will remain unclear.

In summary, these data suggest that the requirement for the bobbed phenotype is limited to magnification events that occur early in spermatogenesis and thus produce large clusters of revertants. In contrast, magnification events occurring late in spermatogenesis, probably at meiosis, are independent of the bobbed phenotype. Both bobbed-dependent and bobbed-independent magnification events appear to depend on the presence of an rDNA-deficient or otherwise structurally abnormal Y chromosome. As shown in the following section, male meiotic and premeiotic magnification events also differ in their response to mutations that impair the magnification process.

Mutants that Affect rDNA Magnification

Genes required for the execution of the premeiotic versus meiotic magnification events can be delineated by the analysis of repair and/or recombination deficient mutations that also inhibit rDNA magnification. For example, the *mei-41* and *mus-101* mutations appear to specifically inhibit those premeiotic magnification events in males that would normally produce large clusters of revertants (44, 46, 73). Single magnification events, however, occur at normal frequencies in males bearing these mutations, and, at least for *mei-41*, there is no effect on the frequency of magnification observed in males bearing $bbYbb^-$ chromosome (44). (It is not possible to assay reduction in *mei-41*-bearing males because nonmagnifying *mei-41* males also produce X-chromosomal bb mutations at high frequencies (46)). The ability of *mei-41* and *mus-101* mutations to specifically inhibit premeiotic magnification further supports the argument that premeiotic and meiotic magnification are under separate control.

In contrast, both clustered and single magnification events are inhibited in males bearing *mus-108* (44). Magnification is also abolished in *mus-108 bb/bbYbb⁻* and *mus-108 bb/bb⁺Ybb⁻* males (44, R. S. Hawley, unpublished observations). Therefore, the *mus-108* mutation blocks both premeiotic and meiotic magnification events, supporting the hypothesis that early and late magnification events do share some functions. Since the sole available allele of *mus-108* also blocks rDNA reduction (R. S. Hawley, unpublished observations), this mutation may well define a general component of the magnification-reduction process, such as the exchange event. That aberrant exchange events may still be occurring in the germlines of these males is suggested by the observation of Hawley et al (44) that significant changes in the repeat-class composition of the *X*-chromosomal rDNA can occur within the germlines of *mus-108 bb/Ybb⁻* males.

mus-108, mei-41, and *mus-101* have been defined as postreplication-repair defective on the basis of their inability to synthesize normal-length DNA on a UV-damaged template (11, 13–15). Although the mechanism of postreplication repair in eukaryotes is unclear, in prokaryotes this process is dependent on recombination between sister chromatids (103, 104). If a similar mechanism operates in eukaryotic cells, it might explain the effects of such repair-defective mutations on the magnification process. We must note, however, that the *mus-108, mei-41,* and *mus-101* mutations are pleiotropic and also affect such processes as recombination and mitotic chromosome stability (4, 5, 15). Hence, although the correlation of the magnification defect with the repair phenotype is striking, the possibility that the inhibition of magnification is a consequence of a block in some other cellular process cannot be discounted.

Contradictory data have been obtained on the possible involvement of the *mei-9* locus in rDNA magnification. The *mei-9* locus is defined by alleles whose phenotypes include decreased recombination, decreased somatic sister-chromatid exchange, a defect in excision repair, and mutagen-sensitivity (4–6, 11, 12, 36, 37, 74, 81). Although Polito et al (88) have reported that magnification is impaired or prevented in *mei-9ᵃbb/Ybb⁻* males, normal levels of magnification are observed when another allele *(mei-9ᵇ)* is tested (44, 46, 73). Endow & Atwood (31) have argued that these observations reflect either a difference in penetrance between *mei-9ᵃ* and *mei-9ᵇ*, or that the two alleles affect different components of a multifunctional protein. Until more alleles are tested, the basis for the role of the *mei-9* alleles in magnification remains obscure.

A Model for the Control of rDNA Magnification

As previously discussed, rDNA magnification appears to be under two levels of control. The evidence for this assertion is based on both the differential requirement of meiotic and premeiotic events with respect to the bobbed phenotype and on the finding of mutations such as *mei-41* and *mus-101* that specifically inhibit premeiotic magnification.

At one level, the germline can adjust its abundancy of rDNA in response to the bobbed phenotype early in development at the mitotic or syncytial stages. We propose that the control of these early magnification events results from the differences in rRNA transcription observed in phenotypically bobbed flies. Although Type I and Type II containing rRNA genes appear to be transcriptionally inactive under normal conditions (65, 66, 68), they appear to be transcribed in phenotypically bobbed individuals. Moreover, Labella et al (56) have shown that the transcription of Type II bearing rRNA genes is amplified 16-fold in males undergoing magnification when compared to nonmagnifying control females. Moreover, Eickbush & coworkers (16a, 130) have shown that the Type II insert of the silk moth *Bombyx mori* encodes a double-stranded endonuclease which when expressed in *E. coli* can specifically cleave the 28S rRNA gene at the site of the Type II insertion as well as at the 5'28S/Type II junction (Y. Xiong & T. H. Eickbush, personal communication). The Type II insertion sequences of *Bombyx mori* and *melanogaster*, have identical structures and appear to encode the same proteins (J. Jakubzak, Y. Xiong & T. H. Eickbush, personal communication). Therefore, one could imagine that the Drosophila Type II insert also encodes an rDNA-specific endonuclease.

deCicco & Glover (23) and Stark & Wahl (110) have speculated that nicks or gaps generated by insertion elements may be involved in magnification. More specifically, we propose that Type II insertion-bearing genes are preferentially transcribed under magnifying conditions allowing the production of an rDNA-specific endonuclease. The double-strand gaps produced by this nuclease then serve as the substrates for repair/replication/recombination enzymes that produce the observed increases in rDNA redundancy.

This model is consistent with both the observed dependence of magnification on the bobbed phenotype and with the observation that premeiotic magnification events are inhibited by mutations such as *mei-41*, *mus-101* and *mus-108* that are deficient in DNA repair (44, 46). The model also explains the observation that under magnifying conditions *mei-41* alleles induce both deletions and interchanges involving the X rDNA at a high frequency (46). Following Hawley & Tartof (46), we speculate that these aberrations may well arise as a consequence of the abnormal repair of nicks or gaps generated by the gene product of Type II insertion elements that would normally serve as substrates or intermediates in the magnification process. Although several aspects of this model are highly speculative, it provides a useful paradigm for the induction of premeiotic magnification events.

Both Type II and Type I elements show similarity in structure and coding capacity to a class of retrotransposons, known as non-LTR transposons (16b). Thus, the rDNA-specific endonuclease produced by these elements can be thought of as part of the integrase-function of these retrotransposons (T. H. Eichbush, personal communication). According to our model, the enhanced transcription of Type II-bearing rRNA genes in the presence of the bobbed phenotype, allows the organism to conscript a transposon-integration function into the service of a gene-specific recombination process.

A second level of control of magnification occurs late in spermatogenesis, probably at about the time of meiosis. The occurrence of bb^+ and bb^{rl} products in phenotypically bobbed-plus germlines demonstrates that meiotic magnification events are neither a consequence of the bobbed phenotype nor of the deficiency for rDNA carried by Ybb^-. Therefore, the induction of at least meiotic magnification and reduction events is unrelated to rDNA dosage but rather appears to be the consequence of Y chromosomal structural anomalies in the region of the rDNA.

Recently, McKee & Lindsley (78) have presented evidence that the X and Y chromosomal rRNA genes define the meiotic pairing sites that ensure proper sex-chromosome segregation at meiosis in males. We propose that one function or property of rDNA-mediated sex chromosome pairing in Drosophila males is to maintain proper sister-chromatid alignment within the nucleolus organizer. Indeed, McClintock (75) has shown that unequal sister-chromatid alignment is a common consequence of failed meiotic pairing in corn. According to our model, in the presence of Y chromosomes carrying *bobbed* deficiencies, or that are otherwise structurally abnormal, improper or failed pairing allows chromatid slippage, and hence unequal alignment of sister chromatids. We further propose that it is the presence of such improper chromatid alignments that induces the observed sister-chromatid exchange (perhaps as a means of ensuring sister-chromatid cohesion). Thus, unlike premeiotic magnification, meiotic magnification cannot be thought of as a compensatory response to an rDNA deficiency, but rather as an accident of improper chromosome behavior.

The Significance of rDNA Magnification

The most surprising feature of rDNA magnification in Drosophila is the degree to which this process is controlled in at least two separate ways. The first level of control, exerted in premeiotic cells, allows magnification in response to an rDNA deficiency. In this sense, magnification may indeed be viewed as a locus-specific compensatory mechanism. Attempts to find similar mechanisms acting upon other tandemly repeated genes, such as those encoding histones or 5S rRNA, have so far been unsuccessful. Moreover, the "sensing-system" that facilitates premeiotic magnification is specific to a deficiency for the genes involved and is not triggered by other mutations that phenocopy the bobbed phenotype. On the other hand, meiotic magnification and reduction events occur at normal frequencies even in the absence of the bobbed phenotype. These events seem to require only the presence of a structurally abnormal Y chromosome for their induction. Therefore, meiotic magnification does not represent a compensatory process, but rather apparently reflects a by-product of failed or abnormal pairing.

Both early and late magnification events can be inhibited by repair or recombination-deficient mutations. This suggests that many of the events of magnification depend largely on general cellular repair/recombination systems rather than on a specific rDNA recombination system. Therefore, it seems likely that the underlying basis of magnification is a DNA repair process involving recombination between sister chromatids. To the extent that such a process is general, then magnification-like events may provide a general method for altering the dose of tandemly reiterated genes. If this is the case, then premeiotic rDNA magnification represents the prototype of this process, in that a deficiency in copy number causes locus-specific recombinational repair. Those exchanges that increase copy number can then be amplified by germline selection.

Although intrachromosomal exchanges at the Bar locus represent the only other confirmed example of unequal sister-chromatid exchanges occuring at a tandemly repeated locus in Drosophila (86), the occurrence of unequal recombination between the rDNA of sister chromatids has been elegantly documented in prokaryotes and in yeast (1, 87, 114).

INTERCHROMOSOMAL rDNA RECOMBINATION

In the preceeding section we have considered the role that recombination between sister chromatids plays in the control of rDNA copy number. In this section we consider the effects of exchange between the two X chromosomes and between the X and Y chromosomes.

rDNA Mediated Recombination between the X and Y Chromosomes

Meiotic recombination is generally absent in Drosophila males, however recombination between the X and Y chromosomes is occasionally observed at a low frequency ($\sim 10^{-3}$–10^{-4}) (80, 111). Although Cooper (20) noted that the X could pair with either arm of the Y chromosome, virtually all reported cases of X-Y recombination have involved the base of Y^s (46, 58, 70, 79, 80,

98, 108, 127). Two lines of evidence demonstrate that these exchanges result from recombination between the rRNA genes on the X and Y chromosomes. First, the finding that recombination between bb^+-bearing X and Y chromosomes can yield bb or even bb^l combinants suggests that most if not all X-Y recombination occurs within the rDNA and that X-Y pairing is frequently unequal, at least in those cases where recombination occurs (70). Second, molecular evidence that X-Y interchange results from recombination between the X and YrDNA chromosomal clusters has been provided by Coen & Dover (18), Komma & Endow (53), and Williams et al (127).

Despite assertions to the contrary (82), a number of studies have demonstrated that X-Y interchange occurs at similarly low frequencies regardless of the orientation of the X chromosomal rDNA with respect to the centromere (46, 70). This suggests that rDNA clusters of either (or both) the X and Y chromosome contain genes in both orientations. This conclusion has recently been confirmed by analyzing rDNA recombination events involving two nucleolus organizers borne by the same X chromosome (101). These exchanges were induced by a maternal effect mutation known as *Rex* that induces intrachromosomal rDNA-exchange events (100). *Rex*-induced exchanges were observed resulting from exchanges between the rDNA clusters in both possible orientations. This observation furthers argues that the rDNA of one or both clusters must contain both tandem and reverse repeats of rRNA genes.

The timing of spontaneous X-Y interchange remains an unanswered question. In a recent study, Williams et al (127) observed no clustering among 17 crossovers recovered among 82,595 progeny, leading them to conclude that these recombination events are probably meiotic. However, in an experiment reported by Komma & Endow (33), X-Y interchange exhibits substantial evidence of clustering, suggesting that such interchanges can occur early in germline development (i.e. 20 out of 22 interchanges were recovered in a single bottle). Ritossa and coworkers (93, 98) have also argued that most if not all X-Y interchanges occuring in males undergoing magnification were meiotic in origin. However, a reanalysis of Ritossa's data (93), which considers the total number of recombinants produced by each male, reveals a significant departure from the expectations of a Poisson distribution ($x^2 = 15.25$, P<0.01). Moreover, the cytogenetic arguments of Ritossa et al (98) hinge on the incorrect assumption that the rDNA on the X and Y chromosomes exists in opposite polarities. Their observations seem more consistent with a premeiotic origin of X-Y interchange. Extensive clustering has also been observed in other studies of X-Y interchange (58, 70, 80). Indeed, Lindsley considered X-Y interchange in males to be primarily a premeiotic process (58). In summary, there is evidence for both premeiotic and meiotic X-Y

interchange, suggesting that exchange might occur at any point in germline development.

The relationship of *X-Y* interchange to the bobbed phenotype also remains unclear. At a low but measurable frequency ($\sim 10^{-3}$), *bb* or nearly *bb*$^{+}$ derivatives of *bb*^{l}X chromosome can be recovered from among the progeny of *bb*$^{l}/Ybb$ (magnifying) males, a process known as "one step" magnification (3, 92). Although these events could be explained by successive cycles of sister-chromatid recombination within the germline, Endow & Komma (33) have demonstrated that most, if not all, of such reversion events are the consequence of *X-Y* interchange. *X-Y* interchanges were found among the progeny of magnifying (*bb*$^{l}/Ybb$) males (33). Ritossa has also reported a hundredfold increase in the frequency of *X-Y* interchange in phenotypically bobbed males relative to that observed in a bobbed-plus control cross (93). However, as noted above, in both sets of experiments the observed frequencies of *X-Y* interchange are elevated by the occurrence of clustering and possibly by premeiotic selection, and thus the actual frequency of *X-Y* interchange events in magnifying males may be substantially lower than reported by these authors. These concerns make it difficult to determine to what extent the process of *X-Y* interchange is enchanced by the bobbed phenotype. It is at least possible that these recombinational events are in some way related to the magnification process. The nature of such a connection is obscured in that the magnified *bb*$^{+}$ chromosomes produced by *bb/Ybb*$^{-}$ males rarely, if ever, are associated with *X-Y* interchange.

Little is known regarding the genetic basis or control of *X-Y* interchange. Hawley & Tartof (46) have noted a large increase in the frequency of *X-Y* interchange in *mei-41 bb*$^{+}$*/Ybb*$^{-}$ males when compared to *mei-41 bb*$^{+}$*/Ybb*$^{+}$, *bb*$^{+}$*/Ybb*$^{-}$, or *bb*$^{+}$*/Ybb*$^{+}$ males. The frequency of these events is unaffected by the presence or absence of a bobbed phenotype, and the *X-Ybb*$^{-}$ interchanges observed in these experiments were recovered as single events rather than in clusters. There are at least two significant differences between the *X-Ybb*$^{-}$ interchanges induced by *mei-41* and those described above. First, although most or all of these interchanges involved the *X* chromosomal rDNA, the *Y* chromosomal breakpoints were apparently distributed at random on both arms of the *Ybb*$^{-}$ chromosome. Second, these same males also produce *X*-chromosomal rDNA deletions at very high frequencies. These results suggest that the observed *X-Ybb*$^{-}$ interchanges are not the consequence of simple rDNA recombination, but rather that both the deletions and the interchanges reflect the aberrant repair of damage to the *X* chromosomal rDNA. Therefore, it seems likely that these *X-Ybb*$^{-}$ interchanges are unrelated to rDNA recombination events that underlie spontaneous *X-Y* interchange.

rDNA-Mediated Recombination between the X Chromosomes in Females

In genetically normal Drosophila females, heterochromatic intervals, including the X chromosomal rDNA, exhibit extremely low levels of recombination (105, 106). A recent study showed the frequency of heterochromatic exchange to be approximately 10^{-4}, and most of this exchange to involve the rDNA (127). Higher levels of recombination, however, have been observed when assayed in females in which the rDNA has been moved distally and in which substantial amounts of flanking heterochromatin have been removed (105, 106). The frequency of these exchanges ranged from 0.41%–0.02%, depending on genetic background (107). The frequency of rDNA exchange in females is unaffected by the presence or absence of a Y chromosome or by the bobbed phenotype (107). For this reason it seems unlikely that rDNA exchange in females represents a mechanism of rDNA copy control, rather it seems more prudent to view this process as an aberrant result of normal repair processes.

Spontaneous rDNA Recombination and the Separate Maintenance of the Two Nucleolus Organizers

The role of X-Y interchange in the co-evolution of the X and Y rDNA arrays also remains unresolved. Frankham et al (35) and Gillings et al (39) have observed that *bobbed* mutations generated by selection for reduced bristle length result from spontaneous X-Y interchange. This led them to hypothesize that such exchanges are a major force in maintaining the coevolution of the X and Y chromosomal arrays. In support of this hypothesis, Gillings et al (39) have demonstrated that recombinant $X \cdot Y^L$ chromosomes can be found in wild populations; and that in stocks homozygous for $X \cdot Y^L$, such chromosomes can subsequently lose all or part of the Y^L arm by an undetermined mechanism. However, based on their estimates of the frequencies of spontaneous X-Y and X-X rDNA recombination, Williams et al (127) have concluded that "the patterns of variation within and between the two sex chromosomes cannot be explained solely as a product of reciprocal recombination." Thus, despite considerable work on this system, the role of interchromosomal rDNA recombination in either the maintenance of rDNA redundancy or the homogenization of repeat classes within and between arrays remains unresolved.

THE Rex MUTATION AND ZYGOTIC rDNA RECOMBINATION

The *Rex* mutation exerts a dominant maternal effect that induces or allows intrachromatid recombination between two separated blocks of rDNA on a single paternally derived X chromosome (100, 113). *Rex* was detected because it caused or allowed mitotic intrachromosomal exchange in a paternally derived attached-XY chromosome bearing two sets of rRNA genes (100, 113). *Rex* induces these events at frequencies of 1–8%, generally before the S-phase of the first mitotic division (113). Although most of the analysis of *Rex* has concentrated on assaying exchange in paternally derived chromosomes, at least one tester chromosome has been shown to undergo *Rex*-induced exchange even when transmitted by the mother (112). The use of a number of X chromosomes bearing heterochromatic duplications as target chromosomes has shown that the site of *Rex*-induced exchange is the rDNA (101, 113). Indeed, the ability of *Rex* to induce exchanges when the two rDNA clusters are in parallel or inverted orientations has been used to demonstrate that at least some rRNA genes exist in both orientations within a given nucleolus organizer (101).

Rex maps within the X heterochromatin, probably within the rDNA, and is most likely a neomorph (100). The *Rex* mutation was found on an X chromosome bearing a very severe allele of *bb* that is refractory to magnification (R. S. Hawley, unpublished data). However, X chromosomes that carry a less severe *bb* allele but still exert a strong *Rex* phenotype have been constructed by intrachromosomal exchange (R. Rasooly & L. G. Robbins, personal communication).

Curiously enough, *Rex*-induced exchanges between bb^+ rDNA clusters on an attached-XY chromosome produce a much larger number of *bb*-bearing recombinants than would be predicted on the basis of unequal exchange (100). This observation has now been extended to an X chromosome bearing an rDNA duplication, and suggests that *Rex* induces both exchanges and deletions.

Such observations suggest that the *Rex* mutation may allow or induce the formation of nicks and or gaps in the rDNA prior to or during the S-phase of the first mitotic division. We propose that both the deletions and the exchanges result from the repair of this damage. Moreover, the observation that bb^+ revertants of paternally derived bb^2 chromosomes have been obtained in crosses of bb^2/Ybb^+ males to *Rex*-bearing females (R. S. Hawley, unpublished data) suggests that *Rex* may also facilitate recombination between sister chromatids.

We have proposed that premeiotic magnification is induced by an rDNA-specific recombinase encoded by the Type II insertion element. Perhaps the Rex phenotype also results from the inclusion of such a recombinase activity in the egg. One could then imagine that the neomorphic *Rex* mutation that maps to the rDNA is, in fact, an altered insert-bearing rRNA gene whose activity is greatly enhanced during oogenesis. Whether or not this model is correct, the analysis of the *Rex* mutation argues that the recombinational activity of the rDNA in the early embryo can be controlled by factors present in embryonic cytoplasm.

SUMMARY AND NEW DIRECTIONS

We have described recombinational processes of the rDNA that involve exchange at seemingly all possible levels (i.e. intrachromatid exchanges, sister-chromatid exchanges, exchanges between homologs, and possibly even exchanges involving amplified rDNA molecules). Nonetheless, these processes are either very rare, as in the case of interchromosomal recombination, or only occur in the presence of certain genotypes, as in the case of magnification. Indeed, recombination within the rDNA is usually prevented. Although the basis of this suppression is unclear, in yeast the suppression of mitotic rDNA recombination results from the combined action of DNA topoisomerases I and II (17). We presume that similar mechanisms act to inhibit rDNA recombination in Drosophila, and that recombination occurs only in the context of recombinational repair processes.

We have also presented the argument that the induction of such repair processes may be induced by nicks or gaps generated by the activity of Type I or Type II insertion sequences. The preferential transcription of insert-bearing rRNA genes in rDNA-deficient animals may provide a sensing system by which premeiotic magnification can be induced and normal rDNA redundancy restored.

The study of rDNA recombination has long been complicated by the problems inherent in studying several hundred rRNA genes at the same time. Thus it has been impossible to define or isolate recombinational junction sequences or to inquire about specific sites within the rDNA that facilitate exchange. A new era in these studies began when Karpen et al (51) achieved P-element-mediated transformation of a single rRNA gene, and demonstrated that single rRNA genes can be expressed at a high rate even when inserted into euchromatic sites. This technique opens the door to studies in which constructs carrying one or more copies of all or part of an rRNA gene can be inserted into the Drosophila genome and their recombinational activity assayed in a variety of genotypes. Such experiments seem likely to provide the Drosophila rDNA geneticist with the ability to construct the sort of marked rDNA repeating units that will definitively demonstrate the relative importance of various classes of exchange events.

It seems to us then, that the future of this subject lies less in the further elaboration of the often confusing, and frequently contentious, formal genetics of the *bobbed* locus and more in the molecular analysis of the exchange events themselves, and of the molecular biology of rDNA insertions.

ACKNOWLEDGMENTS

This work was supported by a grant to R. S. H. from the Eukaryotic Genetics Program of the National Science Foundation (DCB-8815749). R. S. H. is also the recipient of a Faculty Research Award from the American Cancer Society (FRA-324) and of an Irma T. Hirschl Monique-Weill Caulier Career Scientist Award. We also thank our colleagues Dr. David Wright, Dr. Leonard Robbins, Dr. Ellen Swanson, Ms. Rebeckah Rasooly, and Ms. Laura Benjamin for permission to cite unpublished data. We are also grateful to Ms. Donna Lombardi for heroic secretarial assistance. Finally we thank Dr. Jeanne Mosca-Hawley and Ms. Tara Hawley for their patience, kindness, and endurance during the course of this effort.

Literature Cited

1. Anderson, R. P., Roth, J. R. 1977. Tandem genetic duplications in phage and bacteria. *Annu. Rev. Microbiol.* 31:473–505

2. Appels, R., Hilliker, A. J. 1982. The cytogenetic boundaries of the rDNA region within heterochromatin of the *X* chromosome of *Drosophila melanogaster* and their relationship to the male meiotic pairing site. *Genet. Res.* 39:149–156

3. Atwood, K. C. 1969. Some aspects of the bobbed problem in *Drosophila*. *Genetics* 61:319–28 (Suppl.)

4. Baker, B. S., Boyd, J. B., Carpenter, A. T. C., Green, M. M., Nguyen, T. D., et al. 1976. Genetic controls of meiotic recombination of somatic DNA metabolism in *Drosophila melanogaster*. *Proc. Natl. Acad. Sci. USA* 73:4140–43

5. Baker, B. S., Carpenter, A. T. C. 1972. Genetic analysis of sex chromosomal meiotic mutants in *Drosophila melanogaster*. *Genetics* 71:255–86

6. Baker, B. S., Carpenter, A. T. C., Ripoll, P. 1978. The utilization during mitotic cell division of loci controlling meiotic recombination and disjunction in *Drosophila melanogaster*. *Genetics* 90:531–78

7. Beckingham, K., Thompson, N. 1982. Under-replication of rDNA cistrons in polyploid nurse cell nuclei of *Calliphora erythrocephala*. *Chromosoma* 87:177–96

8. Belikoff, E. J., Beckingham, K. 1985. Both nucleolar organizers are replicated in dipteran polyploid tissues: a study at the level of individual nuclei. *Genetics* 111:325–36

9. Belikoff, E. J., Beckingham, K. 1985. A stochastic mechanism controls the relative replication of equally competent ribosomal RNA gene sets in individual dipteran polyploid nuclei. *Proc. Natl. Acad. Sci. USA* 82:5045–5049

10. Boncinelli, E., Graziani, F., Polito, L., Malva, C., Ritossa, F. 1972 rDNA magnification at the bobbed locus of the *X* chromosome in *Drosophila melanogaster*. *Cell Differ.* 1:133–42

11. Boyd, J. B., Golino, M. D., Nguyen, T. D., Green, M. M. 1976. Isolation and characterization of *X*-linked mutants of *Drosophila melanogaster* which are sensitive to mutagens. *Genetics* 84:485–506

12. Boyd, J. B., Golino, M. D., Setlow, R. B. 1976. The *mei-9ᵃ* mutant of *Drosophila melanogaster* increases mutagen sensitivity and decreases excision repair. *Genetics* 84:527–44

13. Boyd, J. B., Setlow, R. B. 1976. Characterization of postreplication repair in mutagen-sensitive strains of *Drosophila melanogaster*. *Genetics* 84:507–26

14. Boyd, J. B., Shaw, K. E. S. 1982. Postreplication repair defects in mutants of *Drosophila melanogaster*. *Mol. Gen. Genet.* 186:289–94

15. Brown, T. C., Boyd, J. B. 1982. Abnormal recovery of DNA replication in ultraviolet-irradiated cell cultures of *Drosophila melanogaster* which are defective in DNA repair. *Mol. Gen. Genet.* 183:363–68

16. Brutlag, D., Fry, F., Nelson, T., Hung, P. 1977. Synthesis of hybrid bacterial plasmids containing highly repeated satellite DNA. *Cell* 10:509–19

16a. Burke, W. D., Calalang, C. C., Eickbush, T. H. 1987. The site specific ribosomal DNA insertion element type II of *Bombyx mori* and of *Drosophila melanogaster* (R2B$_m$) *Proc. Natl. Acad. Sci. USA* 32:649–54

17. Christman, M. F., Dietrich, F. S., Fink, G. R. 1988. Mitotic recombination of the rDNA of *S.* cerevisiae is suppressed by the combined action of DNA topoisomerases I and II. *Cell* 55:413–25

18. Coen, E. S., Dover, G. A. 1983. Unequal exchanges and the co-evolution of *X* and *Y* chromosomal rDNA arrays in *Drosophila melanogaster*. *Cell* 33:849–55

19. Coen, E. S., Thoday, J. M., Dover, G. 1982. Rate of turnover of structural variants in the rDNA gene family of *Drosophila melanogaster*. *Nature* 295:564–68

20. Cooper, K. W. 1964. Meiotic conjunctive elements not involving chiasmata. *Proc. Natl. Acad. Sci. USA* 52:1248–55

21. Dawid, I. B., Long, E. O., Dinocera, P. P., Pardue, M. L. 1981. Ribosomal insertion-like elements in *Drosophila melanogaster* are interspersed with mobile sequences. *Cell* 25:399–408

22. Dawid, I. B., Wellauer, P. K., Long, E. O. 1978. Ribosomal DNA in *Drosophila melanogaster*. I. Isolation and characterization of cloned fragments. *J. Mol. Biol.* 126:749–68

23. DeCicco, D. V., Glover, D. M. 1983. Amplification of rDNA and Type I sequences in *Drosophila* males deficient in rDNA. *Cell* 32:1217–25

24. Dutton, F. L., Krider, H. M. 1984. Factors influencing disproportionate replication of the ribosomal RNA cistrons *Drosophila melanogaster*. *Genetics* 107:395–404

25. Dutton, F. L., Krider, H. M. 1984. Ribosomal RNA cistrons of *X*-chromosomes clonally derived from *Drosophila melanogaster* laboratory populations, redundancy, organization and stability. *Genetics* 107:405–21

26. Endow, S. A. 1980. On ribosomal gene compensation in *Drosophila*. *Cell* 22:145–55

27. Endow, S. A. 1982. Molecular characterization of ribosomal genes on *Ybb*-chromosome of *Drosophila melanogaster*. *Genetics* 102:91–99

28. Endow, S. A. 1982. Independent replication of ribosomal genes in *Drosophila*. In *Gene Amplification*, ed. R. T. Schimke, pp. 115–19. Cold Spring Harbor, NY: Cold Spring Harbor Lab.

29. Endow, S. A. 1982. Polytenization of the ribosomal genes on the *X* and *Y* chromosomes of *Drosophila melanogaster*. *Genetics* 100:375–85

30. Endow, S. A. 1983. Nucleolar dominance of polytene cells of *Drosophila*. *Proc. Natl. Acad. Sci. USA* 80:4427–31

31. Endow, S. A., Atwood, K. C. 1988. Magnification gene amplification by an inducible system of sister chromatid exchange. *Trends Genet.* 4:348–51

32. Endow, S. A., Glover, D. M. 1979. Differential replication of ribosomal gene repeats in polytene nuclei of *Drosophila*. *Cell* 17:597–605

33. Endow, S. A., Komma, D. J. 1986. One-step and stepwise magnification of a *bobbed-lethal* chromosome in *Drosophila melanogaster*. *Genetics* 114:511–23

34. Endow, S. A., Komma, D. J., Atwood, K. C. 1984. Ring chromosomes and rDNA magnification in *Drosophila*. *Genetics* 108:969–83

35. Frankham, R., Briscoe, D. A., Nurthen, R. K. 1980. Unequal crossing over at the rDNA tandon as a source of quantitative genetic variation. *Genetics* 95:727–42

36. Gatti, M. 1979. Genetic control of chromosome breakage and rejoining in *Drosophila melanogaster*. *Proc. Natl. Acad. Sci. USA* 76:1377–81

37. Gatti, M. 1982. Sister chromatid exchanges in *Drosophila*. In *Sister Chromatid Exchange*, ed. S. Wolff, pp. 267–97. New York: Wiley

38. Gatti, M., Santini, G., Pimpinelli, S., Olivieri, G. 1979. Lack of spontaneous sister chromatid exchanges in somatic cells of *Drosophila melanogaster*. *Genetics* 91:255–74

39. Gillings, M. R., Frankham, R., Spiers, J., Whalley, M. 1987. *X-Y* exchange and coevolution of the *X* and *Y* rDNA arrays in *Drosophila melanogaster*. *Genetics* 116:241–51

40. Glover, D. M. 1981. The rDNA of *Drosophila melanogaster*. *Cell* 26:297–98

41. Glover, D. M., Hogness, D. S. 1977. A novel arrangement of the 18S and 28S sequences in a repeating unit of *Drosophila melanogaster*. *Cell* 10:167–76

42. Graziani, F., Caizzi, R., Gargano, S. 1977. Circular ribosomal DNA during ribosomal magnification in *Drosophila melanogaster*. *J. Mol. Biol.* 112:49–63

43. Harford, A. G. 1974. Ribosomal gene magnification in *Drosophila:* a chromosomal change. *Genetics* 78:887–96

44. Hawley, R. S., Marcus, C. H., Cameron, M. L., Schwartz, R. L., Zitron, A.

E. 1985. Repair defective mutants inhibit rDNA magnification in *Drosophila* and discriminate between meiotic and premeiotic magnification events. *Proc. Natl. Acad. Sci. USA* 82:8095–99

45. Hawley, R. S., Tartof, K. D. 1983. The ribosomal DNA of *Drosophila melanogaster* is organized differently from that of *Drosophila hydei*. *J. Mol. Biol.* 163:499–503

46. Hawley, R. S., Tartof, K. D. 1983. The effect of *mei-41* on rDNA redundancy in *Drosophila melanogaster*. *Genetics* 104: 63–80

47. Hawley, R. S., Tartof, K. D. 1985. A two-stage model for the control of rDNA magnification. *Genetics* 109:691–700

48. Henderson, A., Ritossa, F. 1970. On the inheritance of rDNA of magnified *bobbed* loci in *Drosophila melanogaster*. *Genetics* 66:463–73

49. Hilliker, A. J., Appels, R., Schalet, A. 1980. The genetic analysis of *Drosophila melanogaster* heterochromatin. *Cell* 21:607–19

50. Indik, Z. K., Tartof, K. D. 1980. Long spacers of ribosomal genes in *Drosophila melanogaster*. *Nature* 284:420–70

51. Karpen, G. H., Schaefer, J. E., Laird, C. D. 1988. A Drosophila rRNA gene located in euchromatin is active in transcription and nucleolus formation. *Genes Dev.* 2:1745–63

52. Kato, H. 1977. Mechanisms for sister chromatid exchanges and their relation to the production of chromosome aberrations. *Chromosoma* 59:179–91

53. Komma, D. J., Endow, S. A. 1986. Magnification of the ribosomal genes in female *Drosophila melanogaster*. *Genetics* 114:859–74

54. Komma, D. J., Endow, S. A. 1987. Incomplete *Y* chromosome promote rDNA magnification in male and female *Drosophila*. *Proc. Natl. Acad. Sci. USA* 84:2382–86

55. Kongsuwan, K., Yu, Q., Vincent, A., Frisardi, M. C., Rosbash, M., et al. 1985. A *Drosophila minute* gene encodes a ribosomal protein. *Nature* 317:555–58

56. Labella, T., Vicari, L., Manzi, A., Graziani, F. 1983. Expression of rDNA insertions during rDNA magnification in *Drosophila melanogaster*. *Mol. Gen. Genet.* 190:487–93

57. Laird, C. D., Wilkinson, L. E., Foe, V. E., Chooi, W. Y. 1976. Analysis of chromatin-associated fiber arrays. *Chromosoma* 50:169–92

58. Lindlsey, D. L. 1955. Spermatogonial exchange between the *X* and *Y* chromosomes of *Drosophila melanogaster*. *Genetics* 40:24–44

59. Lindsley, D. L., Grell, E. H. 1968. Genetic variations of *Drosophila melanogaster*. *Carnegie Inst. Wash. Publ.* 627

60. Lindsley, D. L., Tokuyasu, K. T. 1980. Spermatogenesis. In *The Genetics and Biology of Drosophila*, ed. M. Ashburner, T. R. F. Wright, IId:225–94. London: Academic

61. Lindsley, D. L., Zimm, G. 1985. The genome of *Drosophila melanogaster*. Part 1: Genes A–K. *Dros. Inf. Serv.* 62:108

62. Locker, D. 1976. Instability at the bobbed locus following magnification in *Drosophila melanogaster*. *Mol. Gen. Genet.* 143:261–68

63. Locker, D., Marrakechi, M. 1977. Evidence for an excess of rDNA in testis of *Drosophila melanogaster* male underjoins rDNA magnification. *Mol. Gen. Genet.* 154:249–54

64. Locker, D., Prud'homme, N. 1973. Study of several factors producing a variation of the reversion frequencies at the bobbed locus in *Drosophila melanogaster*. *Mol. Gen. Genet.* 124: 11–19

65. Long, E. O., Collins, M., Kiefer, B. I., Dawid, I. B. 1981. Expression of the ribosomal DNA insertions in *bobbed* mutants of *Drosophila melanogaster*. *Mol. Gen. Genet.* 182:377–84

66. Long, E. O., Dawid, I. B. 1979. Expression of ribosomal insertion in *Drosophila melanogaster*. *Cell* 18:1185–96

67. Long, E. O., Dawid, I. B. 1980. Repeated genes in eukaryotes. *Annu. Rev. Biochem.* 49:727–64

68. Long, E. O., Rebbert, M. L., Dawid, I. B. 1980. Structure and expression of ribosomal RNA genes of *Drosophila melanogaster* interrupted by Type II insertions. *Cold Spring Harbor Symp. Quant. Biol.* 45:667–72

69. Lyckegaard, E. M. S., Clark, A. G. 1989. Ribosomal DNA and stellate gene copy number variation on the *Y* chromosome of *Drosophila melanogaster*. *Proc. Natl. Acad. Sci. USA* 89:1944–48

70. Maddern, R. H. 1981. Exchange between the ribosomal RNA genes of *X* and *Y* chromosomes in *Drosophila melanogaster* males. *Genet. Res.* 38:1–7

71. Malva, C., Graziani, F., Polito, L., Boncinelli, E., Ritossa, F. 1972. Check of gene number during the process of rDNA magnification. *Nature New Biol.* 239:135–36

72. Malva, C., LaVolpe, A., Gargiulo, G. 1980. Comparison between rDNA mag-

nification and *bb* lethal mutation frequencies in *Drosophila melanogaster. Mol. Gen. Genet.* 180:511–15

73. Marcus, C. H., Zitron, A. E., Wright, D. A., Hawley, R. S. 1986. Autosomal modifiers of *bobbed* phenotype are a major component of the rDNA magnification paradox in *Drosophila melanogaster. Genetics* 113:305–19

74. Mason, J. M., Green, M. M., Shaw, K. E. S., Boyd, J. B. 1981. Genetic analysis of *X*-linked mutagen-sensitive mutants of *Drosophila melanogaster. Mutat. Res.* 81:329–43

75. McClintock, B. 1933. The association of non-homologous parts of chromosomes in the mid-prophase of meiosis in *Zea mays. Z. Zellforsch Mikrosk. Anat.* 19:191–237

76. McClintock, B. 1938. The production of homozygous deficient tissues with mutant characteristics by means of the aberrant mitotic behavior of ring-shaped chromosomes. *Genetics* 23:315–76

77. McClintock, B. 1941. The association of mutants with homozygous deficiencies in *Zea mays. Genetics* 26:542–71

78. McKee, B., Lindsley, D. L. 1987. Inseparability of *X*-Heterochromatic functions responsible for *X:Y* pairing, meiotic drive, and male fertility in *Drosophila melanogaster. Genetics* 116:399–407

79. Muller, H. J. 1948. The construction of several types of *Y* chromosomes. *Dros. Inf. Serv.* 22:73–74

80. Neuhaus, M. H. 1937. Additional data on crossing over between the *X* and *Y* chromosomes of *Drosophila melanogaster. Genetics* 32:333–39

81. Nguyen, T. D., Boyd, J. B. 1977. The *meiotic-9 (mei-9)* mutants of *Drosophila melanogaster* are deficient in repair replications of DNA. *Mol. Gen. Genet.* 158:141–47

82. Palumbo, G., Caizzi, R., Ritossa, F. 1973. Relative orientation with respect to the centromere of ribosomal RNA genes of the *X* and *Y* chromosomes of *Drosophila melanogaster. Proc. Natl. Acad. Sci. USA* 70:1883–85

83. Palumbo, G., Endow, S. A., Hawley, R. S. 1984. Reduction of wild-type *X* chromosomes with the *Ybb*-chromosome of *Drosophila melanogaster. Genet. Res.* 43:93–98

84. Peacock, W. J., Appels, R., Endow, S. A., Glover, D. 1981. Chromosomal distribution of the major insert in *Drosophila melanogaster* 28S rRNA genes. *Genet. Res.* 37:209–14

85. Pellegrini, M., Manning, J., Davidson, N. 1977. Sequence arrangement on the rDNA of *Drosophila melanogaster. Cell* 10:213–24

86. Peterson, H. M., Laughnan, J. R. 1963. Intrachromosomal exchange at the *Bar* locus in *Drosophila. Proc. Natl. Acad. Sci. USA* 50: 126–33

87. Petes, T. D. 1980. Unequal meiotic recombination within tandem arrays of yeast ribosomal DNA genes. *Cell* 19: 765–74

88. Polito, L. C., Cavaliere, D., Zazo, A., Furia, M. 1982. A study of rDNA magnification phenomenon in a repair-recombination deficient mutant in *Drosophila melanogaster. Genetics* 102:39–48

89. Procunier, D., Tartof, K. D. 1978. A genetic locus having *trans* and contiguous *cis* functions that control the disproportionate replication of ribosomal RNA genes in *Drosophila melanogaster. Genetics* 88:67–79

90. Ritossa, F. M. 1968. Non-operative DNA complementary to ribosomal RNA. *Proc. Natl. Acad. Sci. USA* 59: 1124–31

91. Ritossa, F. M. 1968. Unstable redundancy of genes for ribosomal RNA. *Proc. Natl. Acad. Sci. USA* 60:509–16

92. Ritossa, F. M. 1972. A procedure for magnification of lethal deletions of genes for ribosomal RNA. *Nature New Biol.* 240:109–111

93. Ritossa, F. 1973. Crossing-over between *X* and *Y* chromosomes during ribosomal DNA magnification in *Drosophila melanogaster. Proc. Natl. Acad. Sci. USA* 70:1950–54

94. Ritossa, F. M. 1976. The bobbed locus. In *The Genetics and Biology of Drosophila*, ed. E. Novitski, M. Ashburner, 1B:801–49. New York: Academic

95. Ritossa, F. M., Atwood, K. C., Lindsley, D. L., Spiegelman, S. 1966. On the chromosomal distribution of DNA complementary to ribosomal and soluble RNA. *Natl. Cancer Inst. Monogr.* 23: 449–72

96. Ritossa, F. M., Atwood, K. C., Spiegelman, S. 1966. A molecular explanation of the bobbed mutants of *Drosophila* as partial deficiencies of "ribosomal" DNA. *Genetics* 54:819–34

97. Ritossa, F., Malva, C., Boncinelli, E., Graziani, F., Polito, L. 1971. The first steps of rDNA magnification of DNA complementary to ribosomal RNA in *Drosophila melanogaster. Proc. Natl. Acad. Sci. USA* 68:1580–84

98. Ritossa, F., Scalenghe, F., DiTuri, N., Contini, A. M. 1974. On the cell stage

of *X-Y* recombination in *Drosophila. Cold Spring Harbor Symp. Quant. Biol.* 38:483–90

99. Ritossa, F. M., Spiegelman, S. 1965. Localization of DNA complementary to ribosomal RNA in the nucleolus organizer region of *Drosophila melanogaster. Proc. Natl. Acad. Sci. USA* 53:737–45

100. Robbins, L. G. 1981. Genetically induced mitotic exchange in the heterochromatin of *Drosophila melanogaster. Genetics* 99:443–59

101. Robbins, L. G., Swanson, E. E. 1988. *Rex*-induced recombination implies bipolar organization of the ribosomal RNA genes of *Drosophila melanogaster. Genetics* 120:1053–59

102. Roiha, H., Miller, J. R., Woods, L. C., Glover, D. M. 1981. Arrangements and rearrangements of sequences flanking the two types of rDNA insertions of *Drosophila melanogaster. Nature* 290:749–53

103. Rupp, W. D., Howard-Flanders, P. 1968. Discontinuities in the DNA synthesis in an excision-defective strain of *Escherichia coli* following ultraviolet irradiation. *J. Mol. Biol.* 31:291–304

104. Rupp, W. D., Wilde, C. E., Reno, D. C., Howard-Flanders, P. 1971. Exchanges between DNA strands in ultraviolet-irradiated *Escherichia coli. J. Mol. Biol.* 61:25–41

105. Schalet, A. 1969. Exchanges at the bobbed locus of *Drosophila melanogaster. Genetics* 63:133–53

106. Schalet, A. 1972. Crossing over in the major heterochromatic region of the *X* chromosome in normal and inverted sequences. *Dros. Inf. Serv.* 48:111–13

107. Schalet, A., Lefevre, G. 1976. The proximal region of X-chromosome. See Ref. 94, pp. 847–902

108. Siderov, B. N. 1940. The causes of mosaicism in aberrations connected with breaks in the inert chromosome regions in *Drosophila melanogaster. Bull. Biol. Med. Exp. URSS* 9:10–12

109. Spradling, A., Orr-Weaver, T. 1987. Regulation of DNA replication during Drosophila development. *Annu. Rev. Genet.* 21:373–403

110. Stark, G. R., Wahl, G. M. 1984. Gene amplification. *Annu. Rev. Biochem.* 53:447–91

111. Stern, C., Doan, D. 1936. A cytogenetic demonstration of crossing over between the *X* and *Y* chromosomes in the male of *Drosophila melanogaster. Proc. Natl. Acad. Sci. USA* 32:649–54

112. Swanson, E. E. 1984. *Parameters of the* Rex *phenotype in Drosophila melanogaster.* PhD thesis. Mich. State Univ.

113. Swanson, E. E. 1987. The responding site of the *Rex* locus of *Drosophila melanogaster. Genetics* 115:271–76

114. Szostak, J. W., Wu, R. 1980. Unequal crossing over in the ribosomal DNA of *Saccharomyces cerevisiae. Nature* 284:426–30

115. Tartof, K. D. 1971. Increasing the multiplicity of ribosomal RNA in *Drosophila melanogaster. Science* 171:294–97

116. Tartof, K. D. 1973. Regulation of ribosomal RNA genes in *Drosophila melanogaster. Genetics* 73:57–71

117. Tartof, K. D. 1974. Unequal sister chromatid exchange and disproportionate replication as mechanisms regulating ribosomal RNA gene redundancy. *Cold Spring Harbor Symp. Quant. Biol.* 38:491–500

118. Tartof, K. D. 1974. Unequal mitotic sister chromatid exchange as the mechanism of ribosomal RNA gene magnification. *Proc. Natl. Acad. Sci. USA* 71:1272–76

119. Tartof, K. D. 1975. Redundant genes. *Annu. Rev. Genet.* 9:355–85

120. Tartof, K. D. 1988. Unequal crossing over then and now. *Genetics* 120:1–6

121. Tartof, K. D., Dawid, I. B. 1976. Similarities and differences in the structure of *X* and *Y* chromosomal rRNA genes in *Drosophila. Nature* 263:27–70

122. Terracol, R. 1987. Differential magnification of rDNA gene types in *bobbed* mutants of *Drosophila melanogaster. Mol. Gen. Genet.* 208:168–76

123. Terracol, R., Prud'homme, N. 1986. Differential eliminations of rDNA genes in *bobbed* mutations of *Drosophila melanogaster. Mol. Cell Biol.* 6:1023–1031

124. Wellauer, P. K., Dawid, I. B. 1977. The structural organization of ribosomal DNA in *Drosophila melanogaster. Cell* 10:193–212

125. Wellauer, P. K., Dawid, I. B., Tartof, K. D. 1978. *X* and *Y* chromosomal ribosomal DNA of *Drosophila:* comparison of spacers and insertions. *Cell* 14:269–78

126. White, R. L., Hogness, D. S. 1977. R loop mapping of 18S and 28S sequences in the long and short repeating units of *Drosophila melanogaster* rDNA. *Cell* 10:177–92

127. Williams, S. M., Kennison, J. A., Robbins, L. B., Strobeck, C. 1989. Reciprocal recombination and the evolution of

the ribosomal gene family of *Drosophila melanogaster*. *Genetics*. In press

128. Williamson, J. H., Procunier, J. D., Church, R. B. 1973. Does the rDNA in the Y chromosome of *Drosophila melanogaster* magnify? *Nature New Biol.* 243:190–91

129. Deleted in proof

130. Xiong, Y., Eickbush, T. H. 1988. Functional expression of a sequence-specific endonuclease encoded by the retrotransposon R2B$_m$. *Cell* 55:235–46

131. Yagura, T., Yagura, M., Muramatso, M. 1979. *Drosophila melanogaster* has different ribosomal RNA sequences on X and Y chromosomes. *J. Mol. Biol.* 133:533–47

Annu. Rev. Genet. 1989. 23:121–39

THE MOLECULAR GENETICS OF SELF-INCOMPATIBILITY IN *BRASSICA*

J. B. Nasrallah and M. E. Nasrallah

Section of Plant Biology, Division of Biological Sciences, Cornell University, Ithaca, New York 14853

CONTENTS

INTRODUCTION

The evolution of multicellular organisms, plants included, was no doubt accompanied by the development of cell-cell recognition and communication. These processes are crucial to the ability of the organism to distinguish between self and non-self, and to undergo fertilization. Sexual recognition processes have been described in a large number of taxa. Among the most

0066-4197/89/1215-0121$02.00

extensively studied are those operating in the mating systems of many ascomycetes, slime molds and algae, and in gametic recognition in mammals. In all of these cases, recognition between the appropriate cell types allows only legitimate cellular unions and thus ensures the integrity and continuity of the species. It is reasonable to assume that similar recognition processes operate between sperm cells and ovules during fertilization in plants. Plants, however, have evolved an additional mechanism that prevents self-fertilization and promotes outbreeding. This phenomenon is known as self-incompatibility (SI) and interferes with the normal development of a pollen grain and its ultimate ability to deliver the male gametic nucleus to the female gametic nucleus of the ovule. In a successful pollination, the pollen grain lands on the stigma and then must literally invade and grow through the diploid tissue of the stigma, then the style, and finally enter the ovary (see Figure 1). In plants that exhibit SI, the interactions between pollen and pistil lead to the inhibition of self-pollen development and hence the failure to produce seed.

The ultimate fate of a pollen grain is dependent on a series of events involving cell-cell recognition, followed by signal transduction and cellular response. We expect these complex interactions to be mediated by cell surface macromolecules present on pistil and pollen. This review does not attempt to cover the entire field of SI research since various aspects have been extensively reviewed (7, 16, 25, 29). Primary emphasis is placed on the initial recognition process and its genetic control. After a brief description of the distribution and diversity of SI systems, we detail the SI system that operates in the crucifer family, the Brassicaceae, (which includes such plants as broccoli, cabbage, kale, mustards, oil seed crops and radish). Specifically, the cellular and genetic basis of pollen recognition in *Brassica* is reviewed in light of recent data derived from the molecular genetic analysis of the interaction between pollen and stigma.

The Diversity of Self-Incompatibility Systems

SI systems occur widely among flowering plants, but only a few have been subjected to genetic analysis. Different plant families vary extensively with

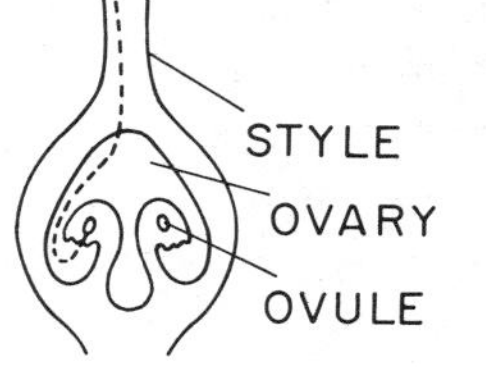

Figure 1   Schematic diagram of a plant pistil. The dotted line indicates the pollen tube.

respect to the genetic control and cytological manifestation of the phenomenon. As a result, SI systems have been classified into three basic groups (15). We include a summary of the major features of the various SI groups in Table 1.

Briefly, SI systems in which incompatibility phenotype is associated with a number of morphological differences in the flower (heteromorphic systems) can be distinguished from SI systems with no associated morphological differences (homomorphic systems). A further distinction can be made among the homomorphic group, in gametophylic systems, the SI phenotype of pollen is determined by the genotype of the gametophyte (that is, by the pollen grain itself); in sporophytic systems the SI phenotype of pollen is determined by the genotype of the sporophyte (that is, by the plant that produced the pollen) (Figure 2). For many species, genetic control is mediated by a single locus termed the S locus, which has multiple alleles. Incompatibility results if the same S allele is active in pollen and pistil. Genetic control by more than one locus has also been described in some species, such as the two-locus gametophytic system of grasses (31). Species also differ in the site at which pollen behavior is inhibited. In gametophytic systems, inhibition occurs

Table 1 Inheritance and general features of self-incompatibility in four families

Classification	Inheritance	Alleles	Allelic interactions	Inhibition
Heteromorphic: Primulaceae (33)	one locus	2 ss = pin Ss = thrum	dominant/sporophytic control of pollen reaction	pollen germination and/or arrest of pollen tubes in stigma or style
Homomorphic: Solanaceae (11) Gametophytic	one locus (S)	many (S_1-S_n)	codominant in style/ pollen reaction determined by haploid gametophyte	pollen tubes inhibited in stylar transmitting tract following 7–8 hr normal growth
Graminae (31) Gametophytic	two loci (S and Z)	many (S_1-S_n) (Z_1-Z_n)	as in Solanaceae	pollen tubes inhibited at or close to stigma surface within 2–10 min of germination
Cruciferae (3) Sporophytic	one locus (S)	many (S_1-S_n) 50 known	dominant or codominant (pollen reaction determined by diploid sporophyte)	pollen or pollen tubes inhibited at the stigma surface within 30 min of contact

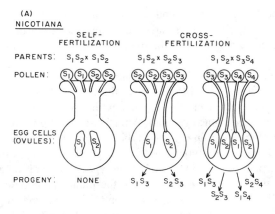

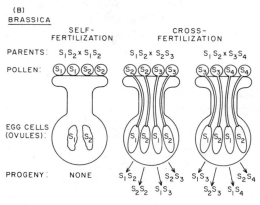

Figure 2 Genetic control of self-incompatibility in *Nicotiana* and *Brassica*. In both species, a pollination is incompatible when the S alleles (S_1, S_2. . . . S_n) expressed in pollen and pistil are matched. Three examples are shown to illustrate the behavior of pollen from different genotypes on an S_1S_2 pistil. (A). In the gametophytically controlled *Nicotiana* system, the phenotype of a pollen grain is determined by its haploid genotype, and codominant expression of S alleles in the pistil is the rule. When an S_1S_2 plant is self-pollinated or crossed to plants of the same genotype, the growth of all pollen (S_1 and S_2) is inhibited in the style. In a cross with pollen from an S_2S_3 plant, only S_2 pollen is inhibited. In a cross with pollen from an S_3S_4 plant, no alleles are matched and all pollen grains are functional.

(B). In the sporophytically controlled *Brassica* system, the phenotype of a pollen grain is determined by the diploid genotype of its parent. Allelic interactions of codominance and dominance-recessiveness occur in stigma and pollen. In the examples shown, S_1 and S_2 exhibit codominance. As diagrammed for *Nicotiana,* in a self-pollination all pollen grains are inhibited, and in a cross with pollen from an S_3S_4 plant all pollen grains are functional. However, in a cross with pollen from an S_2S_3 plant where S_3 is dominant to S_2, all pollen grains are S_3 in phenotype, the cross if fully compatible, and four classes of progeny are produced including S_2S_2.

when the pollen tube has grown into the style for a number of species including *Nicotiana* and other Solanaceae. Inhibition of pollen at or close to the stigma surface has been reported in self-incompatible rye and other grasses (31). In sporophytic systems, inhibition of pollen at the stigma surface appears to be the rule.

It has been suggested that the basic difference between gametophytic and sporophytic systems may be the timing or site of gene action during pollen development. As a result of its developmental history, pollen exhibits both gametophytically and sporophytically determined characteristics. Sporophytic control, obeying the rules of diploid genetics, is thought to result if the relevant genes are expressed pre-meiotically in the pollen mother cells or in the cells of the sporophytic tissues surrounding the developing pollen. Gametophytic determination on the other hand is a consequence of the action of the haploid genome of the pollen grain itself.

THE GENETIC AND CELLULAR BASIS OF SELF-INCOMPATIBILITY IN CRUCIFERS

Sporophytic Control by a Single Polymorphic S Locus

SI was first observed in radish *(Raphanus sativus)* in a report published in 1920 (57). Not until the early fifties, however, did the study of SI in crucifers begin in earnest with Bateman's analysis of the inheritance patterns of SI in *Iberis* and his interpretation of the results in *Raphanus* obtained by A. G. Brown at the John Innes Institute (3). Bateman explained the complex inheritance patterns by invoking the action of one locus with multiple alleles that exhibited co-dominant, dominant, or recessive allelic interactions. He proposed further that the pollen reaction was controlled sporophytically. The hypothesis of sporophytic control was verified in the kale cultivar of *B. oleracea* (60), radish (51) and several other cultivars of *B. oleracea* and *B. campestris*.

The number of alleles at the S locus was estimated to be 22 in *Iberis* populations (3) and 25–34 in *R. raphanistrum* (52). Forty-one different alleles were enumerated in cultivated crops of *B. oleracea* (45, 46), and the total number of alleles isolated to date in this species appears to have reached a plateau at 50–60 alleles. Complex allelic interactions are a feature of the sporophytic SI systems (Figure 1): in some cases, one allele exhibits dominance over another. In other cases, there is a mutual weakening of the alleles. The range of allelic interactions observed appears to vary with the species under study (2, 52, 61). Overall, 60% of allelic pairs are estimated to exhibit co-dominance in crucifers, in contrast to species with gametophytic self-incompatibility such as *Nicotiana* in which co-dominance is the absolute rule.

The occurrence of co-dominant interactions in gametophytic and sporophytic systems has favored a unifying model of plant SI. In this model, recognition of self is proposed to be an "oppositional" system with inhibition arising from the interaction of identical alleles (29).

Site of Self-Pollen Inhibition

Sears concluded from microscopic examination of *R. sativus* that in-compatibility (sporophytic, in this case) was due to the failure of pollen tubes to invade the stigma (53). The crucifer stigma has at its surface a layer of elongated cells called papillar cells or papillae. Transmission electron micros-copy (8, 23, 26) and fluorescence microscopy (16) show that a pollen grain interacts with the apical or subapical region of a single papillar cell. In a successful pollination, pollen first adheres to the papillar cell surface, hy-drates, germinates, and the emerging pollen tube penetrates the cuticle and grows within the papillar cell wall (8, 12). In an incompatible pollination, the inhibition response is very rapid and occurs soon after pollen land on the stigma. Incompatible pollen generally fail to germinate. Germination may occur in some lines, but the resulting pollen tubes grow abnormally and fail to invade the stigmatic papillar cell wall. In the latter case, pollen tubes grew through the outermost cuticle layer and were inhibited in the cellulose-pectin layer of the papillar cell wall (23). On the papillar cell side, a distinctive rejection response is viualized as deposition of the β-1,3-glucan callose at the site of contact with incompatible pollen (16). The papillar cells of the stigma surface thus constitute the barrier to incompatible pollen development in crucifers. If this barrier is penetrated by pollen tubes, no further obstacles lie in their journey to the ovary (47, 53). Self-incompatibility in members of this family can therefore be viewed as a highly specific cell-cell recognition between a pollen grain and the papillar cell that it contacts.

The Developmental Regulation of the Self-Incompatibility Response

To understand how recognition processes between pollen and stigma operate, it is important to consider how the two structures develop and when the competence for self-recognition is acquired. *Brassica* flowering buds show a developmental progression with the mature open flowers toward the base and the young immature buds toward the tip of the inflorescence. This progression is sufficiently predictable to allow the classification of buds into groups of equivalent age for subsequent biochemical and molecular analyses (36, 38). Reliable criteria for this classification are the size of the bud and its position along the inflorescence relative to the opened flowers.

The diploid somatic tissues of the pistil complete their morphological differentiation early in bud development. Stigmatic papillae appear as well-defined cells in buds at 5–7 days before flower opening or anthesis. Pollination analysis has shown that stigmas from immature buds up to 2 days prior to anthesis and earlier allow normal development of self-pollen and therefore cannot discriminate between self and non-self pollen. The self-incompatibility response is acquired in stigmas only at 1–2 days prior to anthesis (50, 55). Thus, while the components necessary to support normal pollen development are expressed early in pistil development and thereafter remain active and stable, it can be inferred that the relevant recognition molecules are expressed or become active late in stigma development. Because self-pollination of bud stigmas results in the production of a full seed complement, this is a well-established and convenient method to produce and maintain S-allele homozygotes.

Pollen maturation and development from the sporogenous tissue of the anther is basically the same in crucifers as in other dicots (25). The pollen develops in anther locules surrounded by a sporophytic layer of nurse cells, the tapetum. Pollen mother cells undergo meiosis to form haploid pollen grains that reach maturity in buds at one day prior to anthesis, and mature pollen grains are shed as the anthers dehisce in the open flower. Pollen maturation is accompanied by changes in the sporophytic tissues of the anther. In particular, the cells of the tapetum become disorganized and dissolve. Electron microscopic studies indicate that material from the tapetum is transferred onto the outer layer of the pollen wall, the sculptured exine (17). Fractions derived from the pollen exine elicited the typical callose response in stigmatic papillae (9, 16). These observations have led Heslop-Harrison to hypothesize that sporophytic control of SI is a result of S-gene expression in the tapetum (16). However, other hypotheses have been proposed to explain sporophytic control. Pandey, for example, has suggested that a pollen grain produced by an S-allele heterozygote carries both allelic specificities because the S gene is expressed in the diploid pollen mother cells before meiosis (48).

THE MOLECULAR ANALYSIS OF SELF-INCOMPATIBILITY IN *BRASSICA*

Classical and Molecular Tools for the Study of Self-Incompatibility

SI phenotype is scored by two methods: the number of seeds is determined after maturation, or pollen-tube growth is directly monitored by microscopy. One of the more rapid methods is UV fluorescence microscopy, whereby pollinated stigmas are fixed and stained with decolorized aniline blue (24). Microscopic examination of a few pollinated stigmas is usually sufficient to determine if the pollen reaction is compatible or incompatible.

Genetic analysis of self-incompatibility in crucifers has been greatly aided by the identification in stigma extracts of proteins that were first recognized as S-allele specific antigens (42, 54) and later as S-locus specific glycoproteins (SLSG) (18, 35, 43, 50). SLSGs exhibit extensive polymorphisms that are most easily detected on isoelectric focusing (IEF) gels. Following sodium dodecyl sulfate-polyacrylamide gel electrophoresis (SDS-PAGE) of stigma extracts, SLSGs migrate as a complex of several closely spaced molecular-weight species differing by approximately 2,000 daltons (35). The average molecular size of the SLSG band complexes varies with the S-allele analyzed and ranges from 57 to 65 kd. Genetic data from our laboratory and from that of Hinata and colleagues have demonstrated a perfect correlation of the segregation of these SLSGs and corresponding S-alleles. Because of this strict correspondence and because expression of SLSG by two alleles is codomi-nant, screening segregates by SLSG polymorphism is very useful in assigning S-genotypes.

The identification of S-gene products is further enhanced by the use of specific antibody probes directed against purified SLSG (22). When these antibodies are used in combination with an enzyme-linked detection system to stain blots of stigma proteins resolved by SDS-PAGE, the sensitivity is such that SLSG from one stigma or a fraction of a stigma is easily detected. This method makes it much easier not only to see the SLSG polymorphism and consequently verify the S-genotype, but also to analyze the variation in SLSG levels and investigate SLSG function.

SLSG patterns can only be identified at the flowering stage, and the time required for screening therefore depends on plant generation time. Recently, cDNA sequences encoding SLSG from *Brassica* (38) and the genes encoding them (SLG genes) were isolated (39). The use of these cDNA probes has revealed extensive restriction fragment length polymorphisms (RFLPs) when genomic DNA from several different S-allele homozygotes was analyzed. As with SLSG polymorphism, this DNA polymorphism cosegregates with the corresponding S-alleles, and in all cases analyzed to date, the S locus geno-type can be correctly inferred from the pattern of the genomic restriction fragments (38, 41). These DNA markers provide convenient and rapid analy-sis of segregating populations and circumvents the time-consuming task of performing diallel pollinations. S-related DNA patterns can be deduced from genomic DNA samples prepared rapidly from a small amount of leaf tissue. Screening can therefore be accomplished very early at the seedling stage in order to establish putative incompatibility groups.

Temporal and Spatial Patterns of Expression

CELL-TYPE SPECIFICITY   SLSGs are detected in stigmas but not in styles, ovaries, or seedlings (36). The accumlation of SLSG correlates with the shift from self-compatibility to self-incompatibility in the developing stigma (18,

36, 40, 44, 50). Metabolic labeling of excised stigmas with ^{35}S-methionine indicated that SLSGs are synthesized early in the self-compatible bud stage (36). The synthesis rate increases sharply in stigmas from self-incompatible buds and reaches a maximum at approximately 1 day prior to anthesis in correlation with the onset of the self-incompatibility response. At its highest level, synthesis of SLSG is 5% of the synthesis of total soluble proteins in the stigma of some genotypes. RNA blot analysis using a cDNA encoding SLSG demonstrated that SLSG mRNA exhibits the same tissue-specific and temporal pattern of expression as the corresponding protein (38). In situ hybridization of pistil sections with ^{3}H-labeled cDNA probes or with ^{35}S-labeled single-stranded RNA probes has further shown that S-transcripts are exclusively localized in the papillar cells of the stigma surface and cannot be detected in any other cell layer of the pistil (39).

The analysis of S-gene expression in anthers is much less advanced. The fact that an SI response results from the activity of the same S-allele in pollen and stigma has suggested that pollen inhibition is based on the interaction of identical S-molecules that form inhibitory complexes (28). However, analysis of pollen proteins has not detected molecules with the electrophoretic and immunological properties of stigma SLSG. Evidence for the expression of S sequences in anther tissue has been obtained by the analysis of RNA isolated from anthers at various stages of development (J. B. Nasrallah, in preparation). The intensity of hybridization to SLSG-cDNA obtained in anthers is several hundredfold lower than that obtained with stigma RNA. The expression of S sequences in anthers occurs during a relatively narrow developmental window, in the post-meiotic anthers of buds at 3 and 4 days prior to anthesis. S-transcripts cannot be detected in premeiotic anthers, in mature anthers containing mature pollen, or in pollen. This temporal pattern of expression is in agreement with Heslop-Harrison's hypothesis of S-gene expression in the tapetum (16), but formal proof of this hypothesis must await additional studies.

IMMUNOLOCALIZATION OF SLSG IN THE STIGMA   The subcellular distribution of SLSG in the stigmatic papillae was analyzed by indirect immunogold labeling with monoclonal antibodies specific for a protein epitope of SLSG (22). Labeling with gold particles was first detected in stigmas from self-compatible buds at 2 to 3 days before anthesis, especially over the region of the cell wall adjacent to and along the plasma membrane. In the mature self-incompatible buds and open flowers, abundant SLSG were found in the cell wall of stigmatic papillae. In the cytoplasm, immunogold particles were detected over the rough endoplasmic reticulum (ER) and Golgi apparatus, indicating that the export of SLSG to the cell wall may follow a conventional pathway of secretion, proceeding with synthesis in the rough ER, followed by processing in the Golgi apparatus and secretion into the wall by fusion of vesicles with the plasma membrane. The papillar cell wall therefore becomes essentially a reservoir of SLSG in the self-incompatible stigma. In cases where self-pollen fail to germinate, SLSG may be released from the wall to react with pollen. When germination is allowed, SLSGs are precisely at the location in the wall where the inhibition of pollen-tube development occurs.

The Structure of SLSG and the Basis of Allelic Specificity

The extreme polymorphism exhibited by SLSG can be reasonably assumed to reflect the variability required for specific pollen recognition. Thus clues to the basis of allelic specificity have been sought in the structure of the SLSG molecule. The contribution of both the carbohydrate and protein moieties to allelic recognition should be considered, since data from a number of recognition systems suggest that cellular interactions may be mediated by the oligosaccharide or protein components of surface molecules. For example, cell adhesion mediated by fibronectin is dependent on a tetrapeptide (49), while oligosaccharides have been implicated in the adhesion between sperm and egg in the mouse (14).

THE OLIGOSACCHARIDE CHAINS OF SLSG   The most detailed analysis of the carbohydrate moiety of SLSG has been performed in *B. campestris* (58). SLSG isolated from S_8 homozygotes was estimated to carry 5 (58) to 7 (59) oligosaccharide side chains, all apparently clustered in the N-terminal region of the molecule.

The oligosaccharides were determined to have a neutral sugar composition of xylose: fucose: mannose in the proportions of 4.1: 4.9: 12.1 mol/mol protein (58). The N-glycosidic linkages involve the amide group of asparagine and the anomeric carbon of N-acetylglucosamine. The N-linked oligosaccharides are of the complex glycan type and have a $Man_3(GlcNAc)_2$ core structure. Two structural forms of glycans have been deduced: form B has the basic $Man_3(GlcNAc)_2$ structure with an attached xylose and fucose residue; Form A, $Man_3(GlcNAc)_3$, has the same structure as form B but contains an additional N-acetylglucosamine. Form A is considered to be the mature form and is the main carbohydrate component of SLSG. Both glycan forms are of common occurrence in plant glycoproteins (13, 58). The A structure has been found in a seed protease inhibitor and in several lectins, and structures A and B are components of laccase. Furthermore, since the complex glycans of SLSGs derived from the S_9- and S_{12}-alleles also had the same structures (58), it was concluded that the glycans of SLSG are not responsible for the specificity of the pollen-stigma interaction in *Brassica*. However, comparative nucleotide sequence analyses in *B. oleracea* have shown that each S allele specifies a distinct pattern of potential N-glycosylation sites (see next section), and the involvement of these glycans in determining S-allele specificity cannot therefore be ruled out.

THE POLYPEPTIDE MOIETY OF SLSG   Amino acid sequence information has been obtained for SLSG from *B. oleracea* by the nucleotide sequence analysis of cDNA clones encoding these molecules (37), and for SLSGs from *B. campestris* by direct amino acid analysis of peptide fragments (59). The nucleotide sequence of SLSG cDNA derived from the S_6 homozygote of *B. oleracea* predicts a polypeptide of 436 amino acids. Amino acid hydropathy profiles reveal a 31-amino acid hydrophobic signal peptide at the N-terminus of the polypeptide, in keeping with the localization of SLSG in the papillar cell wall. Comparative analysis of cDNA clones encoding SLSG from four closely related alleles of *B. oleracea* has revealed 90% sequence similarity at the DNA level (38; C. F. Aquadro, M. E. Nasrallah & J. B. Nasrallah, in preparation). The mature SLSG polypeptides (405 amino acids) predicted from these sequences show substantial differences. A distinct distribution of potential N-glycosylation sites is predicted for each of the SLSGs examined to date (Figure 3). In addition, these polypeptides differ along their length by many substitutions and by deletions and insertions of groups of amino acids. Based on the distribution of these allele-encoded differences, the SLSG polypeptide can be divided into three regions (Figure 3): the amino-terminal region comprising amino-acid residues 1–181 is relatively conserved with approximately 80% conserved positions; a second region extending roughly from residue 182 to residue 268 is the most variable with only approximately 52% conserved positions; finally, the carboxy-terminal region with eleven invariant cysteines contains approximately 78% conserved positions.

The three basic structural domains of SLSG may each correspond to a functional domain. Although the limits of these domains are bound to change as more data become available, the carboxy-terminal domain with its eleven conserved cysteine residues may be involved in intrachain disulfide-bond formation or in bond formation with other stigma molecules or with pollen incompatibility components. On the other hand, the hypervariable region of the molecule could conceivably impart allelic specificity, either in itself or in conjunction with the distinctive array of potential N-glycosylation sites exhibited by each allele.

The S-Multigene Family

In contrast to the genetic prediction of a single gene involved in determining incompatibility specificity, blot analysis of *Brassica* genomic DNA probed with SLSG-cDNA reveal multiple bands of hybridization that indicate the presence of several sequences related to the SLSG structural gene (38, 39). The occurrence of approximately 12 genomic regions with homology to the

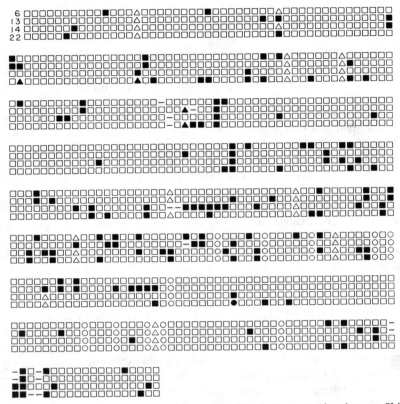

Figure 3 Diagramatic representation of the amino-acid sequence comparison between SLSGs encoded by the S_6, S_{13}, S_{14} and S_{22} alleles of *B. oleracea*. The sequence starts with residue 1 of the mature protein. Potential N-glycosylation sites are indicated by triangles, cysteine residues by circles, and any other amino acids by squares. Unfilled symbols indicate residues that are conserved in the four alleles, and filled symbols indicate variable amino acid residues. Dashes are gaps introduced to maximize sequence alignment.

SLSG-cDNA was demonstrated by the isolation of the corresponding genomic clones from *Brassica* (10, 39). These regions homologous to S show varying degrees of DNA sequence similarity to the SLSG-cDNA ranging from 60% to 100%. The analysis of the S-multigene family is in its early stages. To date, only two expressed gene members have been analyzed in detail (27, 39; and see below). For the remaining genes, the question of linkage relationships, expression, and function still have to be addressed. As for other multigene families, this analysis will require the development of gene-specific probes capable of differentiating between individual gene members.

MOLECULAR ORGANIZATION OF THE S LOCUS   The S locus, although behaving as a single genetic locus, appears to have a complex organization. Genomic sequences homologous to S and flanking regions isolated to date contain over 200 kb of DNA. However, none of the genomic clones contain more than one region homologous to S, nor do the restriction maps show any overlap in the flanking regions (10). The actual size of the S locus depends on how many of the S-related copies map to this region. The linkage of at least some of these copies to the S locus has been established genetically by the cosegregation of self-incompatibility phenotype with the RFLPs characteristic of the segregating alleles (34, 38, 41). At least one of the S-related genes is not linked to the S locus (27).

The S-related genes exhibit varying degrees of divergence from the SLSG structural gene. DNA blot analysis of different S-allele homozygotes indicated that, in general, the restriction fragments that hybridize most strongly to the SLSG-cDNA probe cosegregate with the corresponding S-alleles in genetic crosses and are therefore derived from the S-locus region. In higher eukaryotes, linked members of multigene families have been suggested to coevolve as a result of frequent gene conversion (21). Sequence homogeneity may therefore have been maintained in the self-incompatibility gene copies clustered at the S-locus region, whereas other copies have diverged because they reside elsewhere in the genome. Furthermore, clusters of related sequences are thought to promote unequal crossing-over and the production of polymorphisms, as in the case of the mouse H-2 region (56). Similar mechanisms may likewise be important in the generation of the extensive genetic polymorphisms and the evolution of the self-incompatibility system.

TWO EXPRESSED MEMBERS OF THE S-GENE FAMILY   Two members of the S-multigene family have been shown to be expressed. One of these genes is the SLSG structural gene, the expression and properties of which have been reviewed above, and will be referred to as the SLG gene. The other gene has been designated SLR1 for S-locus related. Molecular analysis has revealed similarities and also profound differences in the properties of the two genes (27). They share a common structure and have coding regions that are uninterrupted by introns. They also show the same general pattern of expression in the stigma: their transcripts show maximal levels at one day prior to anthesis, and are exclusively localized in the stigma papillar cells. Significant differences between the two genes are revealed by sequence and genetic analyses. In contrast to the extensive sequence variability associated with SLSG, the analysis of SLR1 sequences isolated from three different S-allele homozygotes has demonstrated that this gene is highly conserved and encodes identical proteins in strains that differ in their S-alleles. In addition, genetic analysis of segregating populations of *B. oleracea* and *B. campestris* has demonstrated the independent segregation of SLR1 DNA polymorphisms from SLG polymorphisms and SI phenotype (27, 34). The SLR1 gene cannot therefore be a determinant of allelic specificity.

Classical genetic studies did not reveal the existence of the SLR1 gene. The demonstration that this additional member of the S gene family is also expressed suggests the existence of a family of related proteins in the *Brassica* flower. The primary translation products of SLG and SLR1 exhibit 70% sequence similarity and share common structural features that may be the hallmark of the self-incompatibility gene family products. These features include the presence of a signal peptide (suggesting that these molecules may have similar subcellular localization), glycosylation (SLSG is glycosylated and the putative SLR1 product is predicted to be glycosylated), and, notably, the same precise arrangement of eleven cysteine residues at the carboxy terminus of the molecule.

The function of the predicted SLR1 protein product and its involvement in pollen recognition is not known. The obvious differences in selective pressure operating on SLG and SLR1 to generate diversity in the first case and to maintain identity in the second must be related to the very different biological functions of the two gene products. While molecular data, together with extensive genetic evidence, have assigned the SLG gene and its SLSG product a central role in the discrimination between "self" and "non-self" pollen, the predicted SLR1 glycoprotein has now been identified but nothing is known of its possible role. The extreme conservation of SLR1 sequences even across *Brassica* species, and their expression to high levels even in self-compatible strains in which the SLG gene is either not functional or not expressed, suggests for the SLR1 protein product a fundamental role in pollination events in *Brassica*.

EVOLUTIONARY CONSIDERATIONS

The Origin Of S-Alleles

Of the large number of S-alleles in *Brassica*, only a handful have been analyzed at the molecular level. Although the evolutionary relationship of the various S-alleles is not known, these can be classified into different groups based on a number of criteria. By classical genetic analysis, they have been arranged into a complex dominance series based on their genetic interactions with other alleles in heterozygotes (61). More recently, monoclonal antibodies produced against SLSG purified from a *B. oleracea* S_6-homozygote have allowed a classification of S-alleles based on antigenic relatedness (M. E. Nasrallah, unpublished information; 22). One monoclonal antibody that recognizes SLSG via a protein epitope reacts with SLSG from a wide variety of S-genotypes of *Brassica*, including *B. oleracea*, *B. campestris*, *B. napus*, and of *Raphanus,* including *R. sativus* and *R. raphanistrum*. Within *B. oleracea,* an interesting epitope polymorphism was uncovered by this antibody: out of twenty S-allele homozygotes tested, sixteen (including the S_6-, S_{13}- and S_{14}-alleles discussed above) reacted positively, and four others, the S_2-, S_4-, S_5- and S_{15}-homozygotes, did not. In this latter group, S_2, S_5 and S_{15} were previously described as being low on the dominance scale and generally recessive in pollen (61).

DNA and amino acid sequence data provide additional criteria for allelic classification. Two distinct subsets have been identified in *B. oleracea*. One subset includes the S_6-, S_{13}-, S_{14}- and S_{22}-alleles that exhibit over 90% overall DNA sequence similarity to one another (37; and C. F. Aquadro, M. E. Nasrallah, & J. B. Nasrallah, in preparation). Another subset includes the pollen-recessive S_2-, S_5- and S_{15}-alleles that as a group also share over 90% DNA similarity with one another. However, these pollen-recessives show less than 70% overall DNA sequence similarity relative to the first subset of alleles (C. H. Chen & J. B. Nasrallah, in preparation). This degree of sequence divergence is remarkably high. Future research should determine if the SLSGs and the corresponding specificities encoded by these two subsets of alleles arose by sequence divergence of the same gene member or if they are specified by different linked members of the S-multigene family.

Examination of levels and patterns of DNA sequence variability among four alleles belonging to one subset of alleles from *B. oleracea* reveals a remarkably ancient and diverse evolutionary history of polymorphism (C. F. Aquadro, M. E. Nasrallah, & J. B. Nasrallah, in preparation). Average nucleotide divergence at synonymous sites (that do not change amino acids) between alleles is 21.4%, ranging from 15.2 to 27.6% for the 433–436 codons compared. However, this variation is distributed in a highly nonrandom fashion across the gene and with respect to nonsynonymous nucleotide sites (that change amino acids). Approximately 18% of synonymous sites differ between alleles. Assuming a substitution rate of 5×10^{-9}/site/year/lineage, a value calculated for one plant nuclear gene (63) and close to the average rate of 4.65×10^{-9} calculated for vertebrate nuclear genes (30), it appears that the alleles diverged more than 21 million years ago. Nonsynonymous sites show remarkable variation across the gene, with one 80bp segment in the central region of the protein differing by as much as 28% on a per nonsynonymous site basis. Two alleles in fact show 13 nonsynonymous differences and no synonymous differences in this region, directly implicating natural selection for protein diversity as a cause of the tremendous allelic variation at this locus. These data combined with the great age of the alleles found in present populations are consistent with the expectations of population genetic models of SI where new antigenic alleles have a strong selective advantage and are maintained polymorphic in the population for long periods of time (32, 64). Analysis of the patterns of substitutions along the chromo-somes also provides strong evidence that intragenic recombination has contributed to the generation of allelic diversity.

The Origin of Self-Compatibility

The analysis of self-compatible *Brassica* strains has shown that self-compatibility results from the action of genes unlinked to the S locus, usually involving loss of the incompatibility response in the stigma but not in the pollen (19, 20, 40, 41, 62). In studies of the self-compatible yellow sarson, a natural cultivar of *B. campestris*, little or no activity of the S locus was reported and self-compatibility was attributed to the action of an unlinked recessive gene designated "m" (20). In other self-compatible strains, self-compatibility was correlated with a reduction of SLSG in the stigma, and caused by the action of unlinked suppressor *(sup)* genes showing recessive (41), incompletely dominant (40), or dominant inheritance (62). Self-compatible strains are therefore generated from self-incompatible strains by a mechanism that involves a reduction of SLSG levels in the stigma as a result either of decreased protein or transcript stability, or of down-regulation of the SLSG structural gene itself. In addition, other mutations in genes responsible for the post-translational modification of SLSG, its activation, or correct subcellular targeting may also lead to self-compatibility.

The ancestral nature of self-compatibility in plant populations and the evolution of active S-alleles by successive mutations of an ancestral self-fertile allele has been discussed (5, 6). However, it follows from the occurrence of single-gene mutations to self-compatibility that self-compatible species can also result from secondary loss of SI. In the crucifer family, a number of species including *Arabidopsis thaliana* have no reported incidence of SI, but nevertheless contain multiple S-related sequences (39). We postulate that while some of these sequences may represent non-functional remnants of an S-gene system, others, exemplified by the SLR1 gene, may fulfil a basic pollination-related function. In view of its ubiquitous nature, we further suggest that this SLR1 gene represents the ancestral gene from which SI genes have evolved in crucifers. Duplication of the SLR1 gene might have produced the other members of the S-gene family, and at least one of these duplicated copies, freed from the selective forces operating on SLR1, diverged and became the first functional SI gene.

CONCLUDING REMARKS

Molecular genetic approaches have contributed significantly to our understanding of the phenomenon of self-incompatibility in *Brassica*. The available molecular data have, however, shed little light on the relationships between sporophytic and gametophytic systems. In *Nicotiana alata*, a species with gametophytic SI, polymorphic glycoproteins associated with S-alleles and their corresponding cDNAs have also been identified (1), but no similarities have been uncovered in DNA sequence, in predicted overall protein structure, or in gene copy number between the *Brassica* system and the *Nicotiana* (4). Intriguing results have emerged from recent plan transformation experiments in which the *Brassica* SLG gene was introduced into *N. tabacum* (H. Moore and J. B. Nasrallah, in preparation). In transgenic tobacco, the *Brassica* SLG gene is expressed in the transmitting cells of the style. This pattern of expression is identical to that described for the S-associated gene and its product in *N. alata* (1). It is interesting that the *Brassica* and *Nicotiana* genes appear to have similar patterns of expression in cells that lie in the path of the developing pollen tubes.

By using a combination of genetics, biochemistry, cell biology, and molecular biology, the mechanism of pollen recognition should be elucidated in the future. Already, plant transformation experiments are showing promise in the analysis of various aspects of the SI response. Plants stably transformed with SLSG-encoding sequences placed under the control of plant constitutive promoters are likely to facilitate the isolation of large quantities of the protein, and consequently the biochemical analysis of SLSG and the development of in vitro bioassays for SLSG activity. Most informative will be *Brassica* transformation experiments in which SLG genes cloned from a number of S-homozygous genotypes are introduced into recipient lines carrying different S-alleles. The results of these experiments should determine if the SLG gene is sufficient to impart an incompatibility response typical of the donor genotype on pollen as well as on the stigma. Only then will the hypothesis of single-gene control of allelic specificity be verified, and new insights into the mechanism of pollen recognition be gained.

Literature Cited

1. Anderson, M. A., Cornish, E. C., Mau, S.-L., Williams, E. G., Hogart, R., et al. 1986. Cloning of cDNA for a stylar glycoprotein associated with expression of self-incompatibility in *Nicotiana alata*. *Nature* 321:38–44
2. Bateman, A. J. 1954. Self-incompatibility systems in angiosperms. II. *Iberis amara*. *Heredity* 8:305–32
3. Bateman, A. J. 1955. Self-incompatibility systems in angiosperms. III. Cruciferae. *Heredity* 9:52–68
4. Bernatzky, R., Anderson, M. A., Clarke, A. E. 1988. Molecular genetics of self-incompatibility in flowering plants. *Dev. Genet.* 9:1–12
5. Charlesworth, D. 1988. Evolution of homomorphic sporophytic self-incompatibility. *Heredity* 60:445–53
6. Charlesworth, D., Charlesworth, B. 1979. The evolution and breakdown of S-allele systems. *Heredity* 43:41–55
7. deNettancourt, D. 1977. *Incompatibility in Angiosperms*. Berlin/Heidelberg/New York: Springer 230 pp.
8. Dickinson, H. G., Lewis, D. 1973. Cytochemical and ultrastructural differences between intraspecific compatible and incompatible pollinations in *Raphanus*. *Proc. R. Soc. London Ser. B* 183:21–28
9. Dickinson, H. G., Lewis, D. 1973. The formation of tryphine coating the pollen grains of *Raphanus*, and its properties relating to the self-incompatibility system. *Proc. R. Soc. London Ser. B* 184:148–65
10. Dwyer, K. G., Chao, A., Cheng, B.,

Chen, C.-H., Nasrallah, J. B. 1989. The *Brassica* self-incompatibility multigene family. *Genome*. In press

11. East, E. M. 1940. The distribution of self-sterility in flowering plants. *Proc. Am. Philos. Soc.* 82:449–518

12. Elleman, C. J., Willson, C. E., Sarker, R. H., Dickinson, H. G. 1988. Interaction between the pollen tube and the stigmatic cell wall following pollination in *Brassica oleracea. New Phytol.* 109:111–17

13. Faye, L. 1988. Structure, biosynthesis, and function of plant glycoprotein glycans. *Curr. Top. Plant Biochem. Physiol,* 7:62–82

14. Flornan, H. M., Wassarman, P. M. 1985. O-linked oligosaccharides of mouse egg ZP3 account for its sperm specificity. *Cell* 41:313–24

15. Gibbs, P. E. 1986. Do homomorphic and heteromorphic self-incompatibility systems have the same sporophytic mechanism? *Plant Syst. Evol.* 154:285–323

16. Heslop-Harrison, J. 1975. Incompatibility and the pollen stigma interaction. *Annu. Rev. Plant Physiol.* 26:403–25

17. Heslop-Harrison, J., Knox, R. B., Heslop-Harrison, Y. 1974. Pollen-wall proteins. Exine-held fractions associated with the incompatibility response in Cruciferae. *Theor. Appl. Genet.* 44:133–37

18. Hinata, K., Nishio, T. 1978. Stigma proteins in self-incompatible *Brassica campestris* L. and self-incompatible relatives, with special reference to S-allele specificity. *Jpn. J. Genet.* 53:27–33

19. Hinata, K., Okasaki, K. 1986. Role of the stigma in the expression of self-incompatibility in crucifers in view of genetic analysis. In *Biotechnology and Ecology of Pollen*, ed. D. L. Mulcahy, G. B. Mulcahy, E. Ottaviano, pp. 185–190. Berlin/New York/London: Springer-Verlag

20. Hinata, K., Okasaki, K., Nishio, T. 1983. Gene analysis of self-compatibility in *Brassica campestris* var yellow sarson (a case of recessive epistatic modifier). In *Proc. 6th Intern. Rapeseed Conf.*, Paris, I:354–59

21. Hood, L., Campbell, J. H., Elgin, S. C. R. 1975. The organization, expression and evolution of antibody genes and other multigene families. *Annu. Rev. Genet.* 9:305–53

22. Kandasamy, M. K., Paolillo, D. J., Nasrallah, J. B., Faraday, C. D., Nasrallah, M. E. 1989. The S locus specific glycoproteins of *Brassica* accumulate in the cell wall of developing stigma papillae. *Dev. Biol.* 134:462–72

23. Kanno, T., Hinata, K. 1969. An electron microscopic study of the barrier against pollen-tube growth in self-incompatible Cruciferae. *Plant Cell Physiol.* 10:213–16

24. Kho, Y. O., Baer, J. 1968. Observing pollen tubes by means of fluorescence. *Euphytica* 17:298–302

25. Knox, R. B. 1984. Pollen-pistil interactions. In *Encyclopaedia of Plant Physiology*, ed. H. F. Linskens, J. Heslop-Harrison, 17:508–92. Berlin/Heidelburg/New York: Springer

26. Kroh, M. 1964. An electron microscopic study of the behavior of Cruciferae pollen after pollination. In *Pollen Physiology and Fertilization*, ed. H. F. Linskens, pp. 221–24. Amsterdam: North Holland

27. Lalonde, B., Nasrallah, M. E., Dwyer, K. D., Chen, C. H., Barlow, B., Nasrallah, J. B. 1989. A highly conserved *Brassica* gene with homology to the S-locus specific glycoprotein structural gene. *Plant Cell* 1:249–58

28. Lewis, D. 1964. A protein dimer hypothesis on incompatibility. In *Genet. Today. Proc. 11th Int. Cong. Genet., The Hague, 1963,* ed. S. J. Geerts, 3:656–63

29. Lewis, D. 1979. Genetic versatility of incompatibility in plants. *N. Z. J. Bot.* 17:637–44

30. Li, W.-S., Luo, C.-C., Wu, C.-I. 1985. Evolution of DNA sequences. In *Molecular Evolutionary Genetics*, ed. R. J. MacIntyre. pp. 1–94. New York/London: Plenum

31. Lundqvist, A. 1954. Studies on self-sterility in rye, *Secale cereale* L. *Hereditas* 40:278–94

32. Maruyama, T., Nei, M. 1981. Genetic variability maintained by mutation and overdominant selection in finite populations. *Genetics* 98:441–59

33. Mather, K. 1950. The genetical architecture of heterostyly in *Primula sinensis. Evolution* 4:340–52

34. Nasrallah, J. B. 1989. Molecular genetics of self-incompatibility in *Brassica.* In *Plant Reproduction: From Floral Induction to Pollination,* ed. E. M. Lord, G. Bernier. *Am. Soc. Plant Physiol. Sym.* Ser. 1:156–64

35. Nasrallah, J. B., Nasrallah, M. E. 1984. Electrophoretic heterogeneity exhibited by the S-allele specific glycoproteins of *Brassica. Experientia* 40:279–81

36. Nasrallah, J. B., Doney, R. C., Nasrallah, M. E. 1985. Biosynthesis of glycoproteins involved in the pollen-stigma

interaction of incompatibility in developing flowers of *Brassica oleracea* L. *Planta* 165:100–7

37. Nasrallah, J. B., Kao, T. H., Chen, C. H., Goldberg, M. L., Nasrallah, M. E. 1987. Amino-acid sequence of glycoproteins encoded by three alleles of the S locus of *Brassica oleracea*. *Nature* 326:617–19

38. Nasrallah, J. B., Kao, T. H., Goldberg, M. L., Nasrallah, M. E. 1985. A cDNA clone encoding an S-locus specific glycoprotein from *Brassica oleracea*. *Nature* 318:263–67

39. Nasrallah, J. B., Yu, S. M., Nasrallah, M. E. 1988. Self-incompatibility genes of *Brassica oleracea*: expression, isolation and structure. *Proc. Natl. Acad. Sci. USA* 85:5551–55

40. Nasrallah, M. E. 1974. Genetic control of quantitative variation in self-incompatibility proteins detected by immunodiffusion. *Genetics* 76:45–50

41. Nasrallah, M. E. 1989. The genetics of self-incompatibility in *Brassica* and the effects of suppressor genes. See Ref. 34, pp. 146–55

42. Nasrallah, M. E., Wallace, D. H. 1967. Immunogenetics of self-incompatibility in *Brassica oleracea* L. *Heredity* 22:519–27

43. Nasrallah, M. E., Wallace, D. H., Savo, R. M. 1972. Genotype, protein, phenotype relationships in self-incompatibility of *Brassica*. *Genet. Res.* 20:151–60

44. Nishio, T., Hinata, K. 1982. Comparative studies on S-glycoproteins purified from different S-genotypes in self-incompatible *Brassica* species. I. Purification and chemical properties. *Genetics* 100:641–47

45. Ockendon, D. J. 1974. Distribution of self-incompatibility alleles and breeding structure of open-pollinated cultivars of Brussel sprouts. *Heredity* 33:159–71

46. Ockendon, D. J. 1982. An S-allele survey of cabbage (*Brassica oleracea* var *capitata*). *Euphytica* 31:325–31

47. Ockendon, D. J., Gates, P. J. 1975. Growth of cross- and self-pollen tubes in the styles of *Brassica oleracea*. *New Phytol.* 75:155–60

48. Pandey, K. K. 1970. Time and site of the S-gene action, breeding systems and relationships in incompatibility. *Euphytica* 19:364–72

49. Piershbacher, M. D., Ruoslahti, E. 1984. Cell attachment activity of fibronectin can be duplicated by small synthetic fragments of the molecule. *Nature* 309:30–33

50. Roberts, I. N., Stead, A. D., Ockendon, D. J., Dickinson, H. G. 1979. A glycoprotein associated with the acquisition of the self-incompatibility system by maturing stigmas of *Brassica oleracea*. *Planta* 146:179–83

51. Sampson, D. R. 1957. The genetics of self-incompatibility in the radish. *J. Hered.* 48:26–29

52. Sampson, D. R. 1964. A one-locus self-incompatibility system in *Raphanus raphanistrum*. *Can. J. Genet. Cytol.* 6:435–45

53. Sears, E. R. 1937. Cytological phenomena connected with self-sterility in the flowering plants. *Genetics* 22:130–81

54. Sedgley, M. A. 1974. Assessment of serological techniques for S-allele identification in *Brassica oleracea*. *Euphytica* 23:543–52

55. Shivanna, K. R., Heslop-Harrison, Y., Heslop-Harrison, J. 1978. The pollen-stigma interaction: bud pollination in the Cruciferae. *Acta Bot. Neerl.* 27:107–19

56. Steinmetz, M., Winot, A., Minard, K., Hood, L. 1982. Clusters of genes encoding mouse transplantation antigens. *Cell* 28:489–98

57. Stout, A. B. 1920. Further experimental studies on self-incompatibility in hermaphroditic plants. *J. Genet.* 9:85–129

58. Takayama, S., Isogai, A., Tsukamoto, C., Ueda, Y., Hinata, K., et al. 1986. Structure of carbohydrate chains of S-glycoproteins in *Brassica campestris* associated with self-incompatibility. *Agric. Biol. Chem.* 50:1673–76

59. Takayama, S., Isogai, A., Tsukamoto, C., Ueda, Y., Hinata, K., et al. 1987 Sequences of S-glycoproteins, products of the *Brassica campestris* self-incompatibility locus. *Nature* 326:102–4

60. Thompson, K. F. 1957. Self-incompatibility in narrow-stem kale, *Brassica oleracea* var *acephala*. I. Demonstration of a sporophytic system. *J. Genet.* 55:45–60

61. Thompson, K. F., Taylor, J. P. 1966. Non-linear dominance relationships between S alleles. *Heredity* 21:345–62

62. Thompson, K. F., Taylor, J. P. 1971. Self-compatibility in kale. *Heredity* 27:459–71

63. Wolfe, K. H., Li, W.-H., Sharp, P. M. 1987. Rates of nucleotide substitution vary greatly among plant mitochondrial, chloroplast and nuclear DNAs. *Proc. Natl. Acad. Sci. USA* 84:9054–58

64. Wright, S. 1939. The distribution of self-sterility alleles in populations. *Genetics* 24:538–52

Annu. Rev. Genet. 1989. 23:141–61

ADENOVIRUS E1A PROTEIN PARADIGM VIRAL TRANSACTIVATOR

Jane Flint[1] and Thomas Shenk[2]

Department of Biology[1],[2] and Howard Hughes Medical Institute [2], Princeton University, Princeton, New Jersey 08544

CONTENTS

VIRAL TRANSACTIVATING PROTEINS

It is ten years since the recognition of the critical role played by an E1A protein in the regulation of adenovirus transcription. Viruses carrying mutations that impaired or prevented expression of E1A gene products were shown to produce much reduced concentrations of viral early mRNAs (7, 42), the result of inefficient transcription (70). Such transactivation has proved to be widespread among viral systems. Viral proteins that enhance the rate of

0066-4197/89/1215-0141$02.00

transcription or interact with DNA sequences or proteins within promoter domains include the adenovirus E1A protein, the herpes simplex virus α-TIF (also known as ICP25, VP16 or Vmw65) and ICP4 (also known as Vmw175) proteins, the pseudorabies virus immediate early protein, the Epstein-Barr virus BZLF1 protein, the papilloma virus full-length E2 protein, and SV40 and polyoma virus large-T antigens. Many other viral proteins increase the level of target gene expression, but have not been definitively proven to function at the level of transcription. These include the adenovirus E1B-21kd polypeptide, herpes simplex virus ICP0 and ICP27 proteins, cytomegalovirus immediate early proteins, Epstein-Barr virus BMLF1 and BRLF1 proteins, varicella zoster virus 140kd protein, papillomavirus E7 protein, SV40 small-t antigen, hepatitis B virus X protein, human retrovirus tat proteins, bovine leukemia virus p34 protein, and a mouse mammary tumor virus 3' orf polypeptide.

The proteins that have been clearly shown to function at the level of transcription can be divided into two groups based on whether they exhibit sequence-specific DNA binding. The herpes simplex virus ICP4 protein (16, 63), Epstein-Barr virus BZLF1 protein (17), the papilloma virus full-length E2 protein (87), and SV40 and polyoma virus large-T antigens (21) bind specific DNA sequence motifs, whereas the herpes simplex virus α-TIF protein (49, 62), the pseudorabies virus immediate early protein (1, 2) and the adenovirus E1A protein do not. Several observations suggest such a classification does not define two mechanistic classes of activator proteins. First, although SV40 T antigen can bind to DNA in a sequence-specific fashion, its transactivation function does not require direct interaaction of the protein with DNA (23). Second, although the herpes simplex virus α-TIF protein cannot bind to DNA, it does interact at a defined sequence motif through an interaction with cellular DNA-binding proteins (63). Finally, although the adenovirus E1A and pseudorabies virus immediate early proteins do not exhibit sequence-specific DNA-binding activity, they do bind to DNA apparently independent of sequence (10, 12). Thus, DNA binding is not likely to provide a reliable basis for functional groupings of transactivating proteins. As yet, we know too little about the mechanisms underlying transactivation to guess with any confidence whether the known activators mediate the process via the same or different mechanisms.

The E1A protein is by far the most extensively studied of the viral transactivators, in part because it is not only a transcriptional regulatory protein but also on oncogene product. However, the ability to promote oncogenic transformation and transcriptional induction appear to be separable and distinct activities of the protein (reviewed in 67).

The transactivation activity of the E1A protein was discovered because of its critical role in productive infection. The most striking property to emerge from subsequent studies has been the wide range of promoters on which the E1A protein can act. All adenoviral promoters that have been examined respond to the E1A protein. Activation occurs for promoters that are introduced into E1A protein-expressing cells as part of the adenoviral genome, in plasmids, or following integration into the cellular genome (reviewed in 6). Thus, E1A protein transactivation is not restricted to promoters carried by an adenoviral genome. Transactivable promoters share no common sequence elements (reviewed in 6, 19), implying that transactivation is not mediated via interaction of the protein with a specific, 'E1A-response' promoter element. Both cellular promoters and those of other viruses are transactivated (reviewed in 6). The generality with which the E1A protein can act is further emphasized by its ability to stimulate transcription by RNA polymerase III (25, 38, 50). Such observations have led to the seemingly inevitable conclusion that the E1A proteins act by a mechanism that must be both indirect and general.

In this review we focus on the mechanism underlying the promiscuous transactivation activity of the E1A protein.

STRUCTURAL AND GENERAL PROPERTIES OF E1A-PROTEINS

During the early phase of productive adenovirus infection, two major, differentially spliced E1A mRNAs are made (Figure 1). The 12S and 13S mRNAs encode 243 and 289 amino acid (R) proteins, respectively. These proteins share N- and C-terminal sequences and differ only in a 46R internal, cys-rich segment unique to the larger E1A protein (Figure 1). The amino acid sequence of this domain is conserved among distantly related human adenoviruses and has been termed conserved region 3(CR3) (67). Although the 243R E1A protein has been reported to possess a low level of transactivation activity in certain circumstances (18, 53, 90), the 289R E1A protein is largely responsible for this function (reviewed in 6). The transactivation activity of the 243R E1A protein will not, therefore, be considered further.

Several mutational studies have confirmed the importance of the unique region of the 289R E1A protein to its activation activity that is implied by failure of the smaller protein to transactivate efficiently. Mutations within segments shared by the two E1A proteins, including two conserved regions in the first exon (CR1 and CR2, Figure 1), induce little or no change in transactivation activity. Conversely, mutations in CR3, the unique segment of the 289R protein, impair transactivation (reviewed in 67). Remarkably, a synthetic peptide that comprises the unique 46R segment of the larger E1A protein and three adjacent conserved amino acids from the N-terminus of Exon 2 (Figure 1) stimulates transcription from several adenovirus promoters

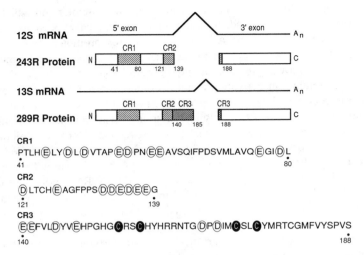

Figure 1 Diagram of E1A mRNAs and polypeptides and amino acid sequence of conserved protein domains. 12S and 13S mRNA exons are represented by lines, introns by caret symbols, and poly(A) sequences by An. The 243 and 289 amino acid (243R) polypeptide encoded by the 12S and 13S mRNAs, respectively, are designated by rectangles. Polypeptide domains that are conserved between different adenovirus serotypes are indicated as conserved regions 1, 2, and 3 (CR1, CR2, CR3) above the rectangles, and the amino acid numbers of the first and last residue comprising each conserved region are indicated below the rectangles. N and C mark the N-terminal and C-terminal domains of the polypeptides, respectively. The amino acid sequences of the conserved domains are displayed in the bottom portion of the figure. Acidic residues within these sequences are marked by open circles and cysteines comprising the metal-binding region in CR3 are marked by shaded circles. CR3 is the transactivating domain.

(28, 59). This result implies that these 49 amino acids of the E1A protein are sufficient to induce transactivation and that they can function autonomously. However, it seems certain that other regions of the protein normally contribute. The peptide is less effective, on a molar basis, than full-length 289R E1A protein as a transactivator of either RNA polymerase II or RNA polymerase III transcription (S. Data, P. K. Chatterjee, S. J. Flint & M. L. Harter, in preparation), and an acidic sequence that spans the junction between CR2 and CR3 (Figure 1) plays a role in transcriptional activation (58).

The E1A proteins are acidic, particularly the N-terminal 140R segment that contains 34 negatively charged residues out of a total of 185 arranged in several clusters (Figure 1). In this respect, the E1A protein resembles many other eukaryotic proteins that activate transcription (reviewed in 72, 80). The E1A proteins are phosphorylated at several sites, but no one phosphorylation appears to be essential to transactivation or oncogenic transformation (14, 76, 78). Phosphorylation clearly contributes to the extensive heterogeneity of E1A protein species synthesized in infected cells (36). The significance of such heterogeneity has not been established.

The unique segment (CR3) of the 289R E1A protein contains one copy of a class II Zn^{2+}-finger motif (Figure 1), a primary sequence characteristic of DNA-binding domains of some eukaryotic proteins that was first recognized in TFIIIA (4, 63). The 289R E1A protein binds one mole Zn^{2+}/mole protein and point mutations that alter the sequence of its Zn^{2+}-finger motif impair or prevent Zn^{2+} binding and transactivation (13), indicating that the Zn^{2+}-finger domain of the 289R E1A protein plays a critical role in this process. In other proteins, Zn^{2+}-finger motifs of the form present in the E1A protein mediate sequence-specific DNA binding (reviewed in 4). The 289R E1A protein also binds to double-stranded DNA (10), but no sequence specificity has been detected. Although the DNA-binding domain of the 289R E1A protein includes the Zn^{2+}-finger (10), it has not yet been demonstrated to mediate binding of the E1A protein to DNA.

CELLULAR TRANSCRIPTION FACTORS: TARGETS FOR E1A-MEDIATED ACTIVATION

The central issue to emerge from surveys of the range of E1A protein activity concerns the mechanism by which it can act to stimulate transcription from a diverse array of promoters: responsive promoters share no common sequence elements and indeed may be transcribed by either RNA polymerase II or III. Thus, studies of the mechanism of transactivation have focussed on identifying the cellular transcription components that mediate transcription from individual adenovirus promoters and elucidating alterations in the activities of these components that are induced by the E1A protein.

TATA-binding Proteins

The E1B promoter is one of the simplest in the viral genome (Figure 2). Analysis of mutations within the E1B promoter in vivo established that two sequences, a TATA-box and a consensus Sp1-binding sequence, are the principal elements of the E1B promoter (70, 93). Far upstream sequences that modulate the efficiency of E1B transcription in the absence of a functional Sp1-binding site and bind cellular factors have also been described (70). Of the promoter-proximal elements, only the TATA sequence appears to be essential for transactivation. Mutations in the TATA element eliminated the activation potential of the E1B promoter, whereas mutations at other sites, including the Sp1-consensus sequence, did not (93). E1A protein-mediated stimulation of transcription from the E1B promoter is indifferent to spacing between the two major promoter elements (92). This implies that the E1A protein affects the TATA-binding factor directly, rather than facilitating interactions between factors bound at the promoter. The TATA motif is also important to E1A protein transactivation of some cellular genes (29, 69, 82) and, under some conditions, the adenovirus major late promoter (55). The

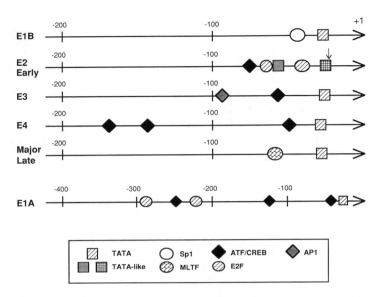

Figure 2 Diagram of adenovirus promoter elements and factor binding sites. Adenovirus promoters are represented by the solid horizontal lines on which the arrowheads designate the major sites of transcription initiation. Promoter elements and binding sites for cellular factors are indicated by the symbols shown in the figure. The vertical arrow above the E2 early promoter marks the minor initiation site. Binding sites for additional factors, whose roles in transcription or E1A protein-transactivation are less well-established, are described in the test. Note that the E1A transcriptional control region is shown on a different scale.

pseudorabies virus immediate early protein (a viral transactivating protein with properties similar to those of E1A) also stimulates transcription in vitro via a TATA-binding factor (2).

TATA elements are recognized by general, DNA-binding proteins such as TFIID (75, 79). These motifs are present in most RNA polymerase II promoters characterized to date and might reasonably account for the E1A protein's ability to stimulate transcription so generally. This, however, is not the case. Transactivation of some viral early promoters does not depend on their TATA elements. Moreover, Simon and colleagues (82) have demonstrated that certain TATA elements, including that of the SV40 early promoter, are refractory to the E1A protein, whereas others, such as that of a human hsp70 gene, are responsive.

MLTF

Like the E1B promoter, the adenovirus major late promoter, which is active during the early phase of infection and responsive to E1A protein transactivation in vivo (reviewed in 19), is constructed of two principal elements, a

TATA box and an upstream element (UE) (Figure 2). These elements were initially identified by detailed mutational analysis (6, 19) and subsequently shown to bind the general factor TFIID (75) and a gene-specific factor, the major late transcription factor (MLTF) (8, 64, 79), respectively. The major late promoter is particularly active when assayed in vitro and has been widely used as a model promoter. Transcription from the major late promoter can be substantially stimulated by E1A proteins present in infected cell extracts (3, 20, 54), by bacterially synthesized E1A protein (3, 87), or by a 49R synthetic peptide containing CR3 (28, 59).

Interestingly, the major late promoter elements critical for transactivation depend on experimental conditions. The factor that is responsible for activating transcription directed by the major late promoter in whole cell extracts prepared from cells harvested late after infection is found in the same fraction as TFIID (TATA-binding factor) after one chromatographic step (55). As stimulation of major late transcription under these conditions does not require promoter sequences upstream of the TATA element, TFIID has been suggested to play an important role in transactivation of major late transcription (55). By contrast, maximal stimulation of major late transcription by the 289R E1A protein present in nuclear extracts prepared from cells harvested during the early phase of infection, or by exogenous, bacterially synthesized E1A protein, depends on the UE sequence, and thus MLTF (3). Deletion of the UE sequence has also been reported to impair transactivation of the major late promoter in transfected cells (56). An MLTF-binding site also contributes to the E1A protein inducibility of the adeno-associated virus P5 promoter and can confer sensitivity to E1A protein-transactivation on a heterologous promoter comprising the initiation site and TATA motif of the SV40 early promoter (9).

E2F

The E2 transcription unit is transcribed in infected cells from two separate promoters. One, designated E2 early, is active during both the early and late phases, whereas the second, E2 late, is activated only during the late phase of the viral replication cycle (reviewed in 6, 19). The E2 late promoter has been reported to be refractory to 289R E1A protein transactivation (51). By contrast, the E2 early promoter is efficiently transactivated and has probably been the most intensively studied as a model for transactivation. The E2 early 'promoter' in fact comprises two overlapping transcriptional control regions that direct initiation at major ($+1$) and minor (-26) sites (Figure 2). It is clear from several mutational studies (reviewed in 6, 19) that sequences positioned approximately 30bp upstream of the two initiation sites (Figure 2) determine the sites of initiation of E2 early transcription. It is not clear why the element centered on -56 (TTAAATTT) is weaker than that at -28 to -23

(TTAAGA), despite its better match to the canonical TATA sequence. This property might be a function of the cellular factors that bind at the two sites, which have not been identified, or simply the result of too close a positioning (Figure 2) of the -56 element to sites at which other cellular factors essential for E2 transcription bind (7).

Genetic and biochemical studies have identified sites for two such factors, termed E2 factor (E2F) and activating transcription factor (ATF; originally E2A-EF) (40, 45, 46, 50, 68, 83, 84, 98). Duplicated sites for E2F are present within the sequences -71 to -53 and -49 to -33, as well as in the E1A transcriptional control region (Figure 2). Zajchowski et al (98) reported that a mutation in the -45 region reduces transcription within infected cells from both E2 early start sites to about 10% of normal levels and highly purified E2F stimulates in vitro transcription from a promoter containing two E2F binding sites (95). However, other mutations in the E2F-binding region induce no significant change in E2 transcription, as judged by a transfection assay (68). The more upstream sequence (-82 to -66) bound by ATF is clearly essential for efficient E2 early transcription from either initiation site (68, 98). Segments of E2 early promoter DNA that contain the binding sites for these two cellular factors can activate E2 early transcription and display properties typical of enhancer elements (40, 41).

Despite the contribution of cellular factors that bind E2 early upstream promoter elements to transcription from both the $+1$ and -26 initiation sites, the two E2 early transcriptional control regions exhibit different requirements for activation by E1A protein. Transactivation of transcription from the -26 cap site is severely impaired by mutation of the TATA-like sequence at -56 (98). Neither this mutation nor mutation of the sequence at -28 to -23 inhibit transactivation of transcription from the major E2 early initiation site at $+1$. Indeed, it is well-established that transactivation of transcription from the major E2 early initiation site does not depend upon any single promoter element. Rather, all mutations that depress E1A protein-induced transcription also impair basal transcription, and vice versa (40, 44, 68, 69). It can, therefore, be concluded that the E1A protein can transactivate such E2 early transcription in the absence of any one of the E2 early promoter elements described above.

Despite this conclusion, much attention has been given to a role for E2F in transactivation. This was the first cellular protein identified as binding to the E1 early promoter, and its discovery was accompanied by the demonstration of a remarkable increase in the DNA-binding activity of this factor in adeno-virus-infected cells (45, 46). Indeed, in these initial experiments very little E2F DNA-binding activity could be detected in uninfected HeLa cell extracts, although such activity has subsequently been detected (35, 84). This discrepancy can probably be explained by both the nature of the nonspecific DNA used in DNA-binding assays (84) and the nature of the DNA-binding probe. Using a DNA fragment containing a single E2F binding site, Hardy & Shenk (35) readily detected E2F activity in uninfected as well as infected cells. By contrast, a 'double-site' probe containing two E2 early E2F sites, analogous to E2 early promoter fragments used by Nevins and colleagues, detected induction of novel E2F-containing complexes. Several lines of evidence, including the nonlinear dependence of formation of the infected cell-specific complex on E2F concentration, the greatly increased stability of the double-site complex formed with infected cell E2F and the sensitivity of formation of this complex to both the spacing and orientation of the two E2F binding sites, established that E2F from infected cells, but not from uninfected cells, binds to the double site probe cooperatively (35). Thus, while E2F is present in readily detectable quantities in uninfected HeLa cells, adenovirus infection induces a profound change in its interaction with the E2 early promoter. Neither the mechanism of cooperative binding nor that responsible for the altered behavior of infected cell E2F is yet understood. However, altered E2F-binding activity appears to be induced by a mechanism operating post-translationally (74).

The report that the altered E2F activity characteristic of infected cells required production of the 289R E1A protein (45–47, 74), the reported induction of this alteration between three and five hours after infection (74), and the presence of E2F sites in a second E1A protein-inducible promoter— that of the E1A gene itself (47), provided strong, circumstantial evidence that E2F played an important role in E1A-protein-mediated transactivation of E2 early transcription. Nevertheless, more recent evidence suggests that the adenovirus-induced modification of E2F activity is unlikely to contribute to E1A-mediated transactivation. The infected cell-specific form of E2F is not present in cells infected by a mutant virus from which virtually the entire E4 transcription unit is deleted (33). Conversely, infected cell-specific E2F was formed in cells infected by mutant viruses unable to produce E1A proteins (33). The E4 function required for induction of cooperatively binding E2F has been mapped to ORF 6/7 (M. Marton, S. Bain & T. Shenk, submitted), which encodes a polypeptide of 17kd. Cells infected by viruses from which this ORF has been completely or partially deleted produce no altered E2F, but activation of E2 early transcription appears to take place normally. These findings provide strong evidence that the E4 protein-dependent modification of E2F is not necessary for E1A protein transactivation of E2 early transcription.

Whether E2F, despite its modification by action of the E4-17kd-polypeptide, can contribute to E1A protein transactivation of E2 early transcription has not been clearly established. Note in this context that no single E2 promoter element appears essential to transactivation, a property suggesting that E2F sites alone should be able to confer sensitivity to E1A protein transactivation. Such a construct has not been tested. Thus, the question of how E2 early transcription is stimulated by the E1A protein remains. More attention will likely now be paid to factors binding to the element at -28 to -23 or the far upstream element, -82 to -66 (Figure 2). The latter is of particular interest: it is bound by ATF, a cellular factor that is either identical, or closely related, to the cyclic AMP response element binding protein (CREB). This factor also binds to the E3, E4, and E1A promoters and is activated by cAMP.

ATF (CREB)

Analysis of the E3 and E4 control regions has further focussed attention on ATF. These promoters each comprise a TATA element preceded by binding sites for several transcription factors (Figure 2). The E3 binding site for ATF lies adjacent to the TATA element, at -68 to -44 (39). Deletion of this region reduces tenfold the efficiency of E3 transcription in vitro (52). A sequence closely related to the ATF consensus is present between positions -102 and -81, a region also important for E3 promoter activity (24, 89). However, careful competition experiments have established that the factor AP1 has a higher affinity than ATF for this E3 promoter element (39). The sequence between -233 and -111 relative to the E3 cap site also contains an E3 promoter element and a binding site for NF1 (39). However, deletion of this entire region decreases the efficiency of E3 transcription by only two- to fourfold (89).

As with the E2 promoter, analysis of E3 transcription from deleted promoters in the presence of E1A proteins has failed to identify promoter elements that are specific for transactivation: the TATA element, ATF, and AP1 binding sites are all important for transcription activated by E1A protein and for basal transcription (24, 39, 52, 89).

The E4 promoter lies at the extreme right-hand end of the genome and from positions -227 to -329 comprises the right-end copy of the 103 base-pair inverted terminal repetition. However, deletion of the terminal repetition induces only two- to threefold decreases in the efficiency of E4 transcription assayed by in vitro transcription or transfection and such mutated promoters remain sensitive to transactivation (11, 26, 31, 32, 49). The E4 promoter contains three binding sites for ATF between the initiation site and the terminal repetition (Figure 2; 49, 50). Several lines of evidence indicate that these sites, and thus ATF, play important roles in both basal and E1A protein-induced E4 transcription. First, the 5' boundary of a region whose deletion decreases basal or transactivated transcription has been located between -179 and -158 (26, 49, 57). Second, E4 promoter fragments containing the two more upstream ATF sites restored activity to E4 promoters deleted to -100 or beyond, and conferred E1A protein-inducibility on heterologous promoters (27, 49). Third, DNA fragments containing the pair of ATF binding sites at −172 to −135 or the single promoter-proximal ATF site efficiently competed for basal E4, but not major late transcription in vitro. Finally, two copies of the promoter-proximal ATF binding sites placed at −138 in a deleted E4 promoter restored full promoter activity in vitro and in vivo (49). While such results illustrate the role of ATF in basal and transactivated E4 transcription, the contribution of individual sites has not been analyzed in detail, e.g. by systematic mutation of individual or multiple ATF binding sites in an otherwise wild-type background. A similar approach will be required to clarify the roles of other E4 promoter sequences that regulate the levels of basal and E1A protein-induced E4 transcription (11, 27).

The identification of sequences important for basal and E1A protein-activated transcription of the E2, E3, and the E4 promoters led to the recognition that these, and the E1A transcriptional control region itself, each contained one or more copies of the consensus sequence ^A_TCGTCA. Cellular factors binding to such sites present in the E2, E3, and E4 promoters were individually identified (39, 49, 83) and subsequently shown to possess identical DNA-binding specificities and transcriptional activities (50). The consensus sequences for ATF binding sites in adenoviral promoters are identical to those of a cellular factor, CREB, previously recognized as binding to promoter elements mediating cAMP induction of such genes as somatostatin, c-*fos,* and fibronectin (64, 66, 81). Binding-competition experiments, as well as comparison of the patterns of methylation protection on viral ATF and cellular CREB sites, established that CREB and ATF represent the same, or very closely related, factors (34, 39, 57, 60). Indeed, proteins binding to the cellular and viral elements copurify through two chromatographic steps (34) and appear as major polypeptides of 45kDa apparent molecular weight upon UV-light-induced crosslinking of complexes formed on viral ATF or cellular CREB binding sites (60). Moreover, the viral ATF sites mediate cAMP-induced stimulation of transcription from viral promoters.

Increases in intracellular cAMP concentration activate cAMP-dependent protein kinases (77). The catalytic subunit of one, protein kinase A, is required for cAMP-inducible transcription (30) and may mediate the increased phosphorylation of CREB that is observed when intracellular cAMP levels are raised (69, 94). The identity of CREB binding sites of cAMP-inducible cellular genes and adenoviral ATF binding sites suggested that viral promoters might also respond to cAMP. Indeed, transcription from the E4 and E1A promoters in cAMP-responsive mouse cells infected by adenovirus is stimulated when increases in the intracellular concentration of cAMP are induced (15), as is transcription from E2, E3, E4, or E1A promoters in transfected, cAMP-responsive rat or human cells (57, 78). The lack of response observed in cell lines deficient in cAMP-dependent protein kinase A

(15, 78) indicates that such transcriptional induction is probably mediated by the same pathway as induction of cAMP-responsive cellular genes. In transfected rat pheochromocytoma PC12 cells lacking the cAMP-dependent kinase, expression of the 289R E1A protein stimulated E2 or E3 expression to levels similar to those observed in wild-type cells that had been treated with forskolin to induce cAMP levels (78). Moreover, expression from these promoters in cells exposed to both the 289R E1A protein and the cAMP inducer were not significantly increased above the level observed with either inducer alone, leading to the suggestion that the E1A protein is capable of directly or indirectly modifying ATF (CREB) to circumvent lack of cAMP-dependent protein kinase A activity (78). In adenovirus-infected mouse cells, by contrast, a substantial degree of synergy between cAMP- and E1A protein-induction of transcription from the E4 promoter has been observed (15). The E1A protein and cAMP inducers stimulated transcription from the E4 promoter ten- and fivefold, respectively. But exposure to the two inducers simultaneously resulted in a 200-fold increase in transcription (15). Transcription of the E1A gene, whose transcriptional control region also contains ATF (CREB) binding sites (Figure 2), responds in the same manner (15).

EivF and E4F

The E4 promoter is also bound by additional cellular factors. One of these, termed EivF, was purified by virtue of its ability to activate the E4 promoter in an in vitro transcription system and by its binding to the sequence GTGACGT, closely related to the ATF (CREB)-binding site consensus (11). Whether EivF is related to ATP (CREB) or plays a role in transactivation of E4 transcription has not been established. Another factor, termed E4F, has also been described (74): it binds to the −53 to −46 region of the E4 promoter and is significantly increased in activity in infected cells. Although the E4F binding site, GACGTAAC, overlaps the promoter proximal ATF (CREB) binding site, ATGACGT, no binding sites for E4F were detected in other viral promoters (74), indicating that E4F and CREB/ATF are distinct factors. Neither the role of E4F in E4 transcription nor the significance of its increased binding activity in adenovirus-infected cells has been investigated.

TFIIIC

Basal level transcription of the adenovirus VA RNA genes requires TFIIIB, TFIIIC, and RNA polymerase III. The rate-limiting activity needed for their transcription is TFIIIC (22, 38, 97), which binds to the B box of the intragenic control region. The E1A protein changes the activity of TFIIIC (97) to stimulate transcription of VA RNA genes both in vivo and in vitro (5, 25, 37, 38). Two forms of TFIIIC-VA RNA promoter DNA complex can be detected by band-shift assay, and one of these increases in amount in response to adenovirus infection (31). The two complexes can be interconverted by acid phosphatase treatment (37), suggesting that adenovirus infection influences the phosphorylation state of TFIIIC. Presumably, the E1A protein is responsible for this alteration in TFIIIC, although this conclusion must remain tentative until a role for other viral gene products is ruled out.

MECHANISMS OF E1A-MEDIATED ACTIVATION

The promiscuity with which the 289R E1A protein transactivates transcription and the failure to identify promoter elements specifically mediating transactivation by the E1A protein inevitably led to the view that the E1A protein modulates transcription indirectly. Transactivation has, therefore, been generally considered to result from E1A protein-induced increases in the concentration or activity of cellular transcription factors (reviewed in 6, 43). The ability of E1A proteins to interact physically with cellular proteins (36, 96), including specific transcription factors (K. Magire & R. Weinmann, personal communication), is also consistent with an indirect role for the E1A protein. Nevertheless, it now seems likely that the E1A protein participates more directly in transactivation than was recognized previously.

Increased Concentration of Cellular Factors

Models of transactivation invoking increased concentrations of transcription factors (reviewed in 6) merely replace the question of how the E1A protein stimulates transcription from viral promoters with that of how the transactivator, acting transcriptionally or post-transcriptionally, increases the amount of cellular factors. Thus, they provide no mechanistic explanation of transactivation; rather, several lines of evidence argue against this class of model. The concentrations of several cellular factors that can participate in transactivation, such as ATF (CREB) and MLTF, are unchanged in adenovirus-infected cells (3, 39, 49, 84), the increased concentration of TFIIIC originally described in adenovirus-infected cells (97) has more recently been ascribed to activation of pre-existing TFIIIC (37), and exogenous 289R E1A protein or peptide stimulate in vitro transcription by RNA polymerase II or III in extracts prepared from uninfected cells (28, 59, 87).

Modification of Cellular Factors

The notion that the E1A protein modulates the activity of cellular transcription factors via post-translational modification such as phosphorylation has considerable appeal. This type of regulatory mechanism has clear precedent in cellular systems (e.g. heat shock: 57, 86; CREB: 65, 94), the E1A protein is itself a phosphoprotein, and ATF (CREB), whose activity is thought to be regulated by phosphorylation (65, 94), can contribute to transactivation. Such a mechanism could also explain the ability of the E1A protein to work with or upon both general factors that mediate RNA polymerase II or RNA polymerase III transcription (TFIID and TFIIIC, respectively) as well as gene-specific factors (ATF(CREB), AP1). Nevertheless, it now appears that the best documented modification of the activity of an RNA polymerase II transcription factor in adenovirus-infected cells, that of E2F DNA binding activity, is the result of action of an E4 gene product (33). The phosphorylation of TFIIIC that accompanies its activation in adenovirus-infected cells (37), as well as alterations of factors that bind the E4 promoter (11, 73), could represent modifications induced directly or indirectly by the E1A protein. The argument for direct involvement of the E1A protein in modification of TFIIIC rests on the observation that the change is not detected in cells infected by *dl*312, which cannot express the 289R E1A protein. Under the conditions of infection employed, it is possible that no other viral early proteins were made at significant levels, because of the lack of transactivator (7, 42). Thus, it is not at present clear whether the phosphorylation of TFIIIC requires a viral product whose expression is regulated by the 289R E1A protein or if it is a direct result of E1A protein activity.

It has also been reported that cellular factors that bind to elements in the E2, E3, and E4 promoters involved in transactivation, such as ATF/(CREB) and AP1, exhibit no changes in DNA-binding activity following adenovirus infection (39, 49, 83). Such observations do not preclude the possibility that the transactivator induces activities of these proteins other than specific DNA-binding, or modifications that cannot be detected in the band-shift assay. Additional experiments are required to examine the transcriptional activity and interactions at the promoter of cellular factors isolated from infected cells, define unambiguously the transactivator's role in inducing transcription factor modifications, and establish the potential of the 289R E1A protein to modulate the activity and physical state of cellular transcription factors.

Function at the Promoter

Recent evidence suggests that the E1A protein may stimulate transcription through its presence at promoters as part of transcriptional initiation complexes.

Lillie & Green (58) have recently demonstrated that the 289R E1A protein possesses an acidic transcriptional activating region and a segment directing the transactivator to the promoter, properties typical of several euykaryotic factors that activate transcription (72, 80). A hybrid protein comprising residues 121–233 of the 289R E1A protein fused to the DNA-binding domain of the yeast transcriptional activator GAL4 transactivated transcription from E4, E1B, and MMTV promoters more efficiently when these promoters contained GAL4 binding sites. As the segment of the GAL4 protein present in the fusion protein lacks sequences required for activation of transcription, this result implies that binding the E1A transactivator to promoters, or binding it more efficiently, increases its transactivation activity. The activities displayed by a series of truncated E1A proteins identified amino acids 138–149 as essential for transactivation of promoters that contained or lacked GAL4 binding sites. This region of the E1A protein is highly acidic (Figure 1) and could, indeed, be replaced by an acidic activating sequence of the HSV-1 α-TIF protein with no loss of function. A fusion protein from which amino acids 180–223 were deleted activated expression only from promoters containing GAL4 binding sites, implying that the primary role of the 180–223R segment of the transactivation domain is to bring the transactivator to the promoter (58). The segment 180–223 contains a highly acidic and unusual repeated amino acid sequence $(EP)_6$ (amino acids 189–200), as well as one of the most basic regions of the protein (amino acids 202–216). Clearly, more subtle mutations will be required to define the functions of these potentially interesting primary sequences at and beyond the C-terminus of the transactivation domain.

The view that the E1A protein functions at the promoter is also supported by the in vitro transactivation of major late transcription by the 289R E1A protein present in infected cell extracts or bacterially synthesized E1A protein (3). Maximal transactivation by either form of the protein requires interaction of MLTF with the major late promoter. Although neither an increase in the affinity of MLTF for UE nor modification of MLTF could be detected in the presence of the 289R E1A protein, the specific binding rate of MLTF was accelerated. Moreover, both the enhanced transcription from the major late promoter and accelerated binding of MLTF to UE DNA characteristic of extracts prepared 8–9 hr after adenovirus infection, but not basal activities, were inhibited by monoclonal antibodies specific for the E1A protein (3). As this result implies that active E1A protein must be present during transcription or binding reactions, the simplest explanation of the altered binding properties of MLTF is that the E1A protein directly facilitates its specific binding to UE DNA, via interaction with the factor, the DNA, or both. Indeed, efficient transactivation by exogenous E1A protein depended on its simultaneous interaction with template DNA and transcriptional components present in nuclear extracts. The failure of the E1A protein to enhance major late transcription efficiently when added to reactions after initiation complexes had been permitted to form (3) and the greater quantities of such complexes formed in the presence of the protein (20, 54) suggest that the primary role of the transactivator in such in vitro systems is to facilitate the formation of productive initiation complexes. Taken together, the results of these in vitro experiments with endogenous or exogenous E1A protein provide additional evidence that the 289R E1A protein is itself a participant in the sequence of reactions that culminates in transactivation of major late transcription.

These recent experiments make a strong case for function of the E1A protein at the promoter and suggest the basis for a model of transactivation that can account for the most puzzling aspect of the activity of the E1A protein, its promiscuity of action. Such a model considers the 289R E1A protein as a participant in the reactions by which transcription from responsive promoters is induced. Although Lillie & Green (58) suggested an indirect interaction of the transactivator with the promoter, it seems reasonable to suppose that the double-stranded DNA binding activity of the protein (10) also contributes to its interaction with promoters. Thus, the interaction probably involves both protein-protein and protein-DNA interactions. Once the protein is bound to a promoter, its acidic activating regions (for example, see 58) might facilitate the binding of one or more specific or general factors. Alternatively, promoter-bound E1A protein might alter the interactions among the factors mediating basal transcription or modify such factors to facilitate the rate-limiting step in initiation of transcription.

These possible functions of the E1A protein need not be mutually exclusive. The most important quantitative effects of the transactivator might be on DNA-binding activity of a specific factor at one promoter but on factor-factor interactions at a second promoter, or under different conditions of transcription. This view is suggested by the repeated observation that no one promoter element, and thus transcription factor, is absolutely essential for transactivation of most viral promoters. The most straightforward interpretation of this property is that the greatest effect of the E1A protein will be manifest at that step in initiation of transcription that is rate-limiting under each particular set of in vivo or in vitro conditions.

The very broad range of the adenovirus transactivator could then be explained by (a) its ability to bind to DNA nonspecifically, and thus to any accessible promoter with which it comes into contact; (b) its ability to interact with a variety of cellular proteins, and (c) the presence, particularly in the N-terminal exon, of several clusters of negatively charged residues that could provide acidic, activating surfaces suited to different cellular factors.

The model proposing that the E1A protein acts directly at the promoter to mediate transactivation predicts both that the DNA-binding of the transactivator is critical and that the protein is associated, at least transiently, with assembling transcription complexes or complexes containing individual transcription factors and DNA. While such predictions have not yet been tested, indirect support for this class of model is provided by analogy with the pseudorabies virus immediate early protein. This protein, like the 289R E1A protein and other viral transactivators, stimulates transcription quite generally. Detailed analysis of the role of this protein in transactivating in vitro transcription from the major late promoter has established that its primary function is to faciliate the assembly of transcription complexes by increasing the efficiency of formation of a complex between TFIID and the promoter (2). Interestingly, this complex forms slowly in vitro in the absence of immediate early protein and is readily competed by binding of either nonspecific proteins

(2) or histones (91) to the template. The immediate early protein binds directly to promoters whose transcription it stimulates and oligomers that compete for immediate early protein binding preferentially inhibit transcriptional stimulation (12). Thus, the promoter-binding and transactivation functions of the protein appear to be related.

SUMMARY AND PERSPECTIVES

Transactivator proteins have been described for many different viruses. The paradigm for such transactivators is the adenovirus 289R E1A protein, which activates transcription of many different promoters. It accomplishes this by altering the activity from a variety of transcription factors that play a role in RNA polymerase II-directed (TATA binding proteins, MLTF, ATF-CREB) or RNA polymerase III-directed (TFIIIC) transcription.

Several different models for the mechanism of E1A protein activation can be proposed and they are not mutually exclusive. The most attractive and best-supported of these is that the protein functions at the promoter as part of the transcription initiation complex. Definitive proof of this model awaits the isolation of a protein complex formed on an E1A-inducible promoter element that contains the E1A protein, and the demonstration that the presence of the transactivator protein in the complex enhances its ability to function in the initiation of transcription. If this model proves to be correct, the next order of business will be the exploration of the molecular events mediated by the E1A protein within the initiation complex that lead to an enhanced rate of transcriptional initiation.

ACKNOWLEDGMENTS

We thank our colleagues who kindly provided us with reprints and preprints of their work. We gratefully acknowledge the competent secretarial assistance of Elena Chiarchiaro and Sharon Doherty and the preparation of figures by Yumi Kasai.

Literature Cited

1. Abmayr, S. M., Feldman, L. D., Roeder, R. G. 1985. *In vitro* stimulation of specific RNA polymerase II-mediated transcription by the pseudorabies virus immediate early protein. *Cell* 43:821–29
2. Abmayr, S. M., Workman, J. L., Roeder, R. G. 1988. The pseudorabies immediate early protein stimulates *in vitro* transcription by facilitating TFIID:protein interactions. *Genes Dev.* 2:542–53
3. Albin, R. L., Harter, M. L., Flint, S. J. 1989. The adenovirus 289R E1A protein plays a direct role in transactivation of transcription from the Ad2 ML promoter. Submitted.
4. Berg, J. M. 1986. Potential metal-binding domains in nucleic acid binding proteins. *Science* 232:429–552
5. Berger, S. L., Folk, W. R. 1985. Differential activation of RNA polymerase III-transcribed genes by the polyomavirus enhancer and the adenovirus E1A gene products. *Nucleic Acids Res.* 13:1413–28
6. Berk, A. J. 1986. Adenovirus promoters

and E1A transactivation. *Annu. Rev. Genet.* 20:45–79

7. Berk, A. J., Lee, F., Harrison, T., Williams, J., Sharp, P. A. 1979. A pre-carly adenovirus 5 gene product regulates synthesis of early viral messenger RNAs. *Cell* 17:935–44

8. Carthew, R. W., Chodosh, L. A., Sharp, P. A. 1985. An RNA polymerase II transcription factor binds to an upstream element in the adenovirus major late promoter. *Cell* 43:439–48

9. Chang, L.-S., Shi, Y., Shenk, T. 1989. Adeno-associated virus P5 promoter contains an adenovirus E1A-inducible element and a binding site for the major late transcription factor. *J. Virol.* 63: In press

10. Chatterjee, P. K., Bruner, M., Flint, S. J., Harter, M. L. 1988. DNA-binding properties of an adenovirus 289R E1A protein. *EMBO J.* 7:835–41

11. Cortes, P., Buckbinder, L., Leza, M. A., Rak, N., Hearing, P., Merino, A., Reinberg, D. 1988. EivF, a factor required for transcription of the adenovirus E1V promoter, binds to an element involved in E1a-dependent activation and cAMP induction. *Genes Dev.* 2:975–90

12. Cromlish, W. A., Abmayr, S. M., Workman, J. L., Horikoshi, M., Roeder, R. G. 1989. Transcriptionally active immediate early protein of pseudorabies virus binds to specific sites on class II gene promoters. *J. Virol.* 63:1869–76

13. Culp, J. S., Webster, L. C., Friedman, D. J., Smith, C. L., Huang, W. J., et al. 1988. The 289-amino acid E1A protein of adenovirus binds zinc in a region that is important for trans-activation. *Proc. Natl. Acad. Sci. USA* 85:6450–54

14. Dumont, D. J., Tremblay, M. L., Branton, P. E. 1989. Phosphorylation at serine 89 induces a shift in gel mobility but has little effect on the function of adenovirus type 5 E1A proteins. *J. Virol.* 63:987–91

15. Engel, D. A., Hardy, S., Shenk, T. 1988. cAMP acts in synergy with E1A protein to activate transcription of the adenovirus early genes E4 and E1A. *Genes Dev.* 2:1517–28

16. Faber, S. W., Wilcox, K. W. 1986. Association of the herpes simplex virus regulatory protein ICP4 with specific nucleotide sequences in DNA. *Nucleic Acids Res.* 11:1475–89

17. Farrell, P., Rowe, D., Rooney, C. Kouzarides, J. T. 1989. Epstein-Bar virus BZLF1 trans-activator specifically binds to consensus Ap1 site and is related to *c-fos. EMBO J.* 8:127–32

18. Ferguson, B., Krippl, B., Andrisani, O., Jones, N., Westphal, H. 1985. E1A 13S and 12S mRNA products made in *Escherichia coli* both function as nucleus-localised transcription activators but do not directly bind DNA. *Mol. Cell. Biol.* 5:2653–61

19. Flint, S. J. 1986. Regulation of adenovirus mRNA formation. *Adv. Virus Res.* 31:169–228

20. Flint, S. J., Leong, K. 1986. Enhanced transcription activity of HeLa cell extracts infected with adenovirus type 2. *Cancer Cells* 4:137–46

21. Fried, M., Prives, C. 1986. The biology of simian virus 40 and polyomavirus. *Cancer Cells* 4:1–16

22. Fuhrman, S. A., Engelke, D. A., Geiduschek, E. P. 1984. HeLa cell RNA polymerase III transcription factors. *J. Biol. Chem.* 259:1934–43

23. Gallo, G. J., Galinger, G., Alwine, J. C. 1988. Simian virus 40 T antigen alters the binding characteristics of specific simian DNA-binding factors. *Mol. Cell. Biol.* 8:1648–56

24. Garcia, J., Wu, F., Gaynor, R. 1987. Upstream regulatory regions required to stabilize binding to the TATA sequence in an adenovirus early promoter. *Nucleic Acids Res.* 15:8367–85

25. Gaynor, R. B., Feldman, L. T., Berk, A. J. 1985. Viral immediate early proteins activate transcription of class III genes. *Science* 230:447–50

26. Gilardi, P., Perricaudet, M. 1984. The E4 transcriptional unit of Ad2: far upstream sequences are required for its transactivation by E1A. *Nucleic Acids Res.* 12:7877–88

27. Gilardi, P., Perricaudet, M. 1986. The E4 promoter of adenovirus type 2 contains an E1A dependent cis-acting element. *Nucleic Acids Res.* 14:9035–49

28. Green, M., Loewenstein, P. M., Pusztai, R., Symington, J. S. 1988. An adenovirus E1A protein domain activates transcription *in vivo* and *in vitro* in the absence of protein synthesis. *Cell* 53:921–26

29. Green, M. R., Treismann, R., Maniatis, T. 1983. Transcriptional activation of cloned human β-globin genes by viral immediate-early gene products. *Cell* 35:137–48

30. Grove, J. R., Price, D. J., Goodman, H. M., Avruch, J. 1986. Recombinant fragment of protein kinase inhibitor blocks cyclic AMP-dependent gene transcription. *Science* 238:530–33

31. Hanaka, S., Nishigaki, T., Sharp, P. A., Handa, H. 1987. Regulation of *in vitro* and *in vivo* transcription of early

region IV of adenovirus type 5 multiple *cis*-acting elements. *Mol. Cell. Biol.* 7:2578–87

32. Handa, H., Sharp, P. A. 1984. Requirement for distal upstream sequences for maximal transcription *in vitro* of early region IV or adenovirus. *Mol. Cel. Biol.* 4:791–98

33. Hardy, S., Engel, D., Shenk, T. 1989. An adenovirus early region 4 product is required for induction of the infection-specific form of cellular E2F activity. *Genes Devel.* In press

34. Hardy, S., Shenk, T. 1988. Adenovirus E1A gene products and adenosine 3',5'-cyclic monophosphate activate transcription through a common factor. *Proc. Natl. Acad. Sci. USA* 85:4171–75

35. Hardy, S., Shenk, T. 1989. E2F from adenovirus-infected cells binds cooperatively to DNA containing two properly oriented and spaced recognition sites. *Mol. Cell. Biol.* Submitted

36. Harlow, E., Whyte, B., Franza, B. R. Jr., Schley, C. 1986. Association of adenovirus early region 1A proteins with cellular polypeptides. *Mol. Cell. Biol.* 6:1579–89

37. Hoeffler, W. K., Kovelman, R., Roeder, R. G. 1988. Activation of transcription factor IIIC by the adenovirus E1A protein. *Cell* 53:907–20

38. Hoeffler, W. K., Roeder, R. G. 1985. Enhancement of RNA polymerase III transcription by the E1A gene product of adenovirus. *Cell* 41:955–63

39. Hurst, H. C., Jones, N. C. 1987. The identification of factors that interact with the E1A-inducible adenovirus E3 promoter. *Genes Dev.* 1:1132–46

40. Imperiale, M. J., Hart, R. P., Nevins, J. R. 1985. An enhancer-like element in the adenovirus E2 promoter contains sequences essential for uninduced and E1A-induced transcription. *Proc. Natl. Acad. Sci. USA* 82:381–85

41. Jalinot, P., Kedinger, C. 1986. Negative regulatory sequences in the E1A inducible enhancer of the adenovirus-2 early promoter. *Nucleic Acids Res.* 14:2651–69

42. Jones, N., Shenk, T. 1979. An adenovirus type 5 early gene function regulates expression of other early viral genes. *Proc. Natl. Acad. Sci. USA* 76:3665–69

43. Jones, N. C., Rigby, P. W. J., Ziff, E. B. 1988. Trans-acting protein factors and the regulation of eukaryotic transcription: lessons from studies on DNA tumor viruses. *Genes Dev.* 2:267–81

44. Kingston, R. E., Kaufman, R. J.,

Sharp, P. A. 1984. Regulation of transcription of the adenovirus EII promoter by E1A gene products: absence of sequence specificity. *Mol. Cell. Biol.* 4:1970–77

45. Kovesdi, I., Reichel, R., Nevins, J. R. 1986. E1A transcription-induction: enhanced binding of a factor to upstream promoter sequences. *Science* 231:719–22

46. Koveski, I., Reichel, R., Nevins, J. R. 1986. Identification of a cellular factor involved in E1A transactivation. *Cell* 45:219–28

47. Kovesdi, I., Reichel, R., Nevins, J. R. 1987. Role of an adenovirus E2 promoter-binding factor in E1A-mediated coordinate gene control. *Proc. Natl. Acad. Sci. USA* 84:2180–84

48. Kristie, T. M., Roizman, B. 1984. Separation of sequences defining basal expression from those conferring α gene recognition within the regulatory domains of herpes simplex virus 1 α genes. *Proc. Natl. Acad. Sci. USA* 81:4065–69

49. Lee, K. A. W., Green, M. R. 1987. A cellular transcription factor E4F1 interacts with an Ela-inducible enhancer and mediates constitutive enhancer function *in vitro*. *EMBO J.* 6:1345–53

50. Lee, K. A. W., Hai, T. Y., SivaRaman, L., Thimmappaya, B., Hurst, H. C., et al. 1987. A cellular transcription factor ATF activates transcription of multiple E1A-inducible adenovirus early promoters. *Proc. Natl. Acad. Sci. USA* 84: 8355–59

51. Leff, T., Chambon, P. 1986. Sequence-specific activation of transcription by adenovirus E1A products is observed in HeLa but not in 293 cells. *Mol. Cell. Biol.* 6:201–8

52. Leff, T., Corden, J., Elkaim, R., Sassone-Corsi, P. 1985. Transcriptional analysis of the adenovirus-5 EIII promoter: absence of sequence specificity for stimulation by E1A gene products. *Nucleic Acids Res.* 13:1209–21

53. Leff, T., Elkaim, R., Goding, C. R., Jalinot, P., Sassone-Corsi, P., et al. 1984. Individual products of the adenovirus 12S and 13S mRNAs stimulate viral E2A and E3 expression at the transcriptional level. *Proc. Natl. Acad. Sci. USA* 81:4381–85

54. Leong, K., Berk, A. J. 1986. Adenovirus early region 1A protein increases the number of template molecules transcribed in cell-free extracts. *Proc. Natl. Acad. Sci. USA* 83:5844

55. Leong, K., Brunet, L., Berk, A. J. 1988. Factors responsible for the higher transcriptional activity of extracts of

adenovirus infected cells fractionate with the TATA box transcription factor. *Mol. Cell. Biol.* 8:1765–74

56. Lewis, E. D., Manley, J. L. 1985. Control of adenovirus late promoter expression in two human cell lines. *Mol. Cell. Biol.* 5:2433–42

57. Leza, M. A., Hearing, P. 1988. Cellular transcription factor binds to adenovirus early region promoters and to a cyclic AMP response element. *J. Virol.* 62:3003–13

58. Lillie, J. W., Green, M. R. 1989. Transcription activation by the adenovirus E1a protein. *Nature* 338:39–44

59. Lillie, J. W., Loewenstein, P. M. ,Green, M. 1987. Functional domains of adenovirus type 5 E1a proteins. *Cell* 50:1091–100

60. Lin, Y. S., Green, M. R. 1988. Interaction of a common transcription factor ATF, with regulatory elements in both E1a- and cyclic AMP-inducible promoters. *Proc. Natl. Acad. Sci. USA* 85:3396–400

61. McKnight, J. L. C., Kristie, T. M., Roizman, B. 1987. Binding of the virion protein mediating α gene induction in herpes simplex virus 1-infected cells to its cis site requires cellular proteins. *Proc. Natl. Acad. Sci. USA* 84:7061–65

62. Michael, N., Spector, D., Mavromara-Nazos, P. 1988. The DNA-binding properties of the major regulatory protein α4 of herpes simplex viruses. *Science* 239:1531–34

63. Miller, J., McLachlan, A. D., Klug, A. 1985. Repetitive zinc-binding domains in the protein transcription factor IIIA from Xenopus oocytes. *EMBO J.* 4:1609–14

64. Miyamoto, N. G., Moncolin, V., Egly, J. M., Cambon, P. 1985. Specific interaction between a transcription factor and the upstream element of the adenovirus-2 major late promoter. *EMBO J.* 4:3563–70

65. Montminy, M. R., Bilezikjian, L. M. 1987. Binding of a nuclear protein to the cyclic AMP response element of the somatostatin gene. *Nature* 328:175–78

66. Montminy, M. R., Sevarino, K. A., Wagner, J. A., Mondel, G., Goodman, R. H. 1986. Identification of a cyclic AMP responsive element within the rat somatostatin gene. *Proc. Natl. Acad. Sci. USA* 83:6682–86

67. Moran, E., Mathews, M. B. 1987. Multiple functional domains in the adenovirus E1A gene. *Cell* 48:177–78

68. Murthy, S. C. S., Bhat, G. P., Thimmappaya, B. 1985. Adenovirus EIIA early promoter: transcription control elements and induction by the viral pre-early EIA gene, which appears to be sequence independent. *Proc. Natl. Acad. Sci. USA* 82:2230–34

69. Nabel, G. J., Rice, S. A., Knipe, D. M., Baltimore, D. 1988. Alternative mechanisms for activation of human immunodeficiency virus enhancer in T cells. *Science* 239:1299–301

70. Nevins, J. R. 1981. Mechanism of activation of early viral transcription by the adenovirus E1A gene product. *Cell* 26:213–20

71. Parks, C. L., Banerjee, S., Spector, D. J. 1988. Organization of the transcriptional control region of the E1b gene of adenovirus type 5. *J. Virol.* 62:54–67

72. Ptashne, M. 1988. How eukaryotic transcriptional activators work. *Nature* 335:683–89

73. Raychauduri, P., Rooney, R., Nevins, J. R. 1987. Identification of an E1A inducible cellular factor that interacts with regulatory sequences within the adenovirus E4 promoter. *EMBO J.* 6:4073–81

74. Reichel, R., Kovesdi, T., Nevins, J. R. 1987. Developmental control of a promoter-specific factor that is also regulated by the E1A gene product. *Cell* 48:501–6

75. Reinberg, D., Horikoshi, M., Roeder, R. G. 1987. Factors involved in specific transcription by mammalian RNA polymerase II: functional analysis of initiation factors IIA and IID and identification of a new factor operating downstream of the initiation site. *J. Biol. Chem.* 262:3322–30

76. Richter, J. D., Slavicek, J. M., Schneider, J. F., Jones, N. C. 1988. Heterogeneity of adenovirus type 5 E1A proteins: multiple serine-phosphorylations induce slow-migrating electrophoretic variants but do not affect E1A-induced transcriptional activation or transformation. *J. Virol.* 62:1948–55

77. Ross, E. M., Gilman, A. G. 1980. Biochemical properties of hormone-sensitive adenylate cyclase. *Annu. Rev. Biochem.* 49:533–64

78. Sassone-Corsi, P. 1988. Cycli AMP induction of early adenovirus promoters involves sequences required for E1A trans-activation. *Proc. Natl. Acad. Sci. USA* 85:7192–96

79. Sawadago, M., Roeder, R. G. 1985. Interaction of a gene-specific transcription factor with the adenovirus major late promoter upstream of the TATA box region. *Cell* 43:165–75

80. Sigler, P. B. 1988. Transcription activa-

tion. Acid blobs and negative noodles. *Nature* 333:210–12

81. Silver, B. J., Bokar, J. A., Virgin, J. B., Vallen, E. A., Milsted, A., Nelsen, J. 1987. Cyclic-AMP regulation of the human glycoprotein hormone α-subunit gene is mediated by an 18 base-pair element. *Proc. Natl. Acad. Sci. USA* 84:2198–202

82. Simon, M. C., Fisch, T. M., Benecke, B. J., Nevins, J. R., Heintz, N. 1988. Identification of multiple, functionally distinct TATA elements, one of which is the target in the hsp70 promoter for E1A regulation. *Cell* 52:723–29

83. SivaRaman, L., Subramarian, S., Thimmappaya, B. 1986. Identification of a factor in HeLa cells specific for an upstream transcriptional control sequence of an E1A-inducible adenovirus promoter and its relative abundance in infected and uninfected cells. *Proc. Natl. Acad. Sci. USA* 83:5914–18

84. SivaRaman, L., Thimmappaya, B. 1987. Two promoter-specific host factors interact with adjacent sequences in an E1A-inducible adenovirus promoter. *Proc. Natl. Acad. Sci. USA* 84:6112–16

85. Sorger, P. K., Lewis, M. J., Pelham, H. R. B. 1987. Heat-shock factor is regulated differently in yeast and HeLa cells. *Nature* 329:81–84

86. Spalholz, B. A., Byrne, J. C., Howley, P. M. 1988. Evidence for cooperativity between E2 binding sites in E2 transregulation of bovine papillomavirus type 1. *J. Virol.* 62:3143–50

87. Spangler, R., Bruner, M., Dalie, B., Harter, M. L. 1987. Activation of adenovirus promoters by the adenovirus E1A protein in cell-free extracts. *Science* 237:1044–46

88. Tsukamoto, A., Ponticelli, A., Berk, A. J., Gaynor, R. B. 1986. Genetic mapping of a major site of phosphorylation in adenovirus type 2 E1A proteins. *J. Virol.* 59:14–22

89. Weeks, D. L., Jones, N. C. 1985. Adenovirus E3-early promoter: sequences required for activation by E1A. *Nucleic Acids Res.* 13:5389–402

90. Winberg, G., Shenk, T. 1984. Dissection of overlapping functions within the adenovirus type 5 E1A gene. *EMBO J.* 3:1907–12

91. Workman, J. L., Abmayr, S. M., Cromlish, W. A., Roeder, R. G. 1988. Transcriptional regulation by the immediate early protein of pseudorabies virus during *in vitro* nucleosome assembly. *Cell* 55:211–19

92. Wu, L., Berk, A. 1988. Constraints on spacing between transcription factor binding sites in a simple adenovirus promoter. *Genes Dev.* 2:403–11

93. Wu, L., Rosser, D. S. E., Schmidt, M. G., Berk, A. J. 1987. A TATA box implicated in E1a transcriptional activation of a simple adenovirus 2 promoter. *Nature* 326:512–15

94. Yamamoto, K. K., Gonzalez, G. A., Biggs, W. H., Montminy, M. R. 1988. Phosphorylation-induced binding and transcriptional efficiency of nuclear factor CREB. *Nature* 334:494–98

95. Yee, A. S., Raychauduri, P., Jakoi, L., Nevins, J. R. 1989. The adenovirus-inducible factor E2F stimulates transcription after specific DNA binding. *Mol. Cell. Biol.* 9:575–85

96. Yee, S.-P., Branton, P. 1985. Detection of cellular proteins associated with human adenovirus type 5 early region 1A polypeptides. *Virology* 147:141–53

97. Yosinaga, S., Dean, N., Han, M., Berk, A. J. 1986. Adenovirus stimulation of transcription by RNA polymerase III: evidence for an E1A-dependent increase in transcription factor IIIC concentration. *EMBO J.* 5:343–54

98. Zajchowski, D. A., Boeuf, H., Kedinger, C. 1985. The adenovirus-2 early EIIa transcription unit possesses two overlapping promoters with different sequence requirements for E1a-dependent stimulation. *EMBO J.* 4:1293–300

Annu. Rev. Genet. 1989. 23:163–98

GENETICS OF PROTEOLYSIS IN ESCHERICHIA COLI*

Susan Gottesman

Laboratory of Molecular Biology, National Cancer Institute, Bethesda, Maryland 20892

CONTENTS

INTRODUCTION

Protein turnover has been considered an important part of the cell's metabolism, yet its role in regulating the availability of cellular functions in bacterial systems has received relatively little attention until recently. Over the last ten years it has become increasingly apparent that turnover of particular proteins, under specific conditions, can play as central a role as the transcriptional and translational regulatory mechanisms. This review summarizes our current knowledge of this particular role of proteases in *Escherichia coli,* with attention to the energy-dependent proteases responsible for the initial cleavages in most of these interesting degradative processes.

Proteases have two general functions: (*a*) responsibility for protein processing to form functional proteins from precursors, and (*b*) destruction of functional and non-functional proteins. This review focuses on the protein degradative role, with emphasis on the proteases that initiate degradation. The roles of both proteases and peptidases in completing the degradative process begun by the initiating proteases are discussed only in summary form, as an update of other reviews on the subject (52, 53, 98, 102).

Proteins are degraded to allow the recycling of the components of improperly made proteins, to provide a source of amino acids under starvation conditions, and to regulate the amount of functional protein. The degradation of both abnormal and normal short-lived proteins is often energy-dependent. The last ten years have seen a rapid expansion in our understanding of the nature of energy-dependent proteolysis in both eukaryotic and prokaryotic cells. The in vivo dependence on energy can be attributed to proteases that are activated by or totally dependent on ATP for proteolysis. However, the known ATP-dependent proteases do not account for all energy-dependent degradation that goes on in cells.

In addition to energy-dependent proteases, cells contain many distinct energy-independent proteases (48, 98). Finally, many surveys with general protease subtrates may not be sufficient to detect those proteases with very specific substrate preferences, or proteases expressed only under specific growth conditions. Thus, our current catalog of proteases may still represent only a small proportion of those present and playing important roles in the cell.

Roles for Proteases in Regulatory Circuits

Regulation of protein synthesis is widespread, and in bacteria such as *E. coli* much of that regulation takes place at the level of transcription. Since most *E. coli* proteins have half-lives much longer than the doubling time of cells (101), the shut-off of synthesis leaves the cell with a population of active protein molecules until their concentration is reduced by growth of the cells and dilution. When proteins are part of a developmental program, or are part of an emergency-response system that may require activities inconsistent with normal growth, it may become essential, as part of the developmental progression or recovery from stress, to rid the cell rapidly of some proteins. I discuss a number of such cases below. In these cases, rapid turnover of the proteins can help to regulate precisely the availability of the protein, or degradation can be part of the developmental or recovery pathway.

I define a *constitutively unstable* protein as one whose half-life is less than the doubling time of the cell under all conditions tested. Therefore, such a protein will be depleted more rapidly by degradation than by dilution. A *conditionally unstable* protein is one whose half-life under many conditions may be quite long, but whose rapid degradation is triggered by particular physiological conditions.

Substrates for Proteolysis in E. coli

The study of protein degradation can begin with either the substrates or the proteases; given either one, the other can usually be identified.

UNSTABLE LAMBDA PROTEINS   One hint of the possible range of proteins that might be unstable in *E. coli* and other organisms is provided by a survey of the role of protein turnover for the well-studied bacteriophage lambda. Lambda regulators cl, cII, and N are all subject to proteolysis; cII and N are quite unstable under most tested conditions (54), while cl is a conditionally unstable protein, degraded rapidly only after DNA damage (86, 147; and see below). In addition, the lambda DNA-replication initiation protein, O, is rapidly turned over (54). Thus far, different proteases have been implicated in the turnover of each of these proteins; cl instability is dependent on RecA, cII turnover seems to involve Hfl, N is subject to degradation by Lon, and the protease responsible for O degradation has not yet been identified (54, 63, 86). This small sample indicates the importance of protein turnover and the multiplicity of degradative pathways in *E. coli*.

CIS-ACTING PROTEINS   A variety of *cis*-acting proteins have been described in which it is difficult (although frequently not impossible) to provide the function from a chromosome other than the one on which it works. Recent experiments with the *cis*-acting Tn903 transposase provide an excellent example of the role of proteases in regulating protein availability (N. Grindley, personal communication). The transposase protein turns over with a half-life of less than 5 minutes in wild-type cells. *lon* mutations, which eliminate a major ATP-dependent protease (see below), increase the half-life at least twentyfold. The transposase also becomes able to work in *trans* in *lon* mutant hosts (C. Derbyshire, M. Kramer, & N. Grindley, manuscript in preparation).

Other *cis*-acting proteins, including lambda O protein and Mu transposase, are either chemically or functionally unstable (54, 110). Many work stochiometrically at DNA sites. One can speculate that a stochiometric requirement for the presence of such proteins at a nearby site, combined with a high affinity for that site and instability when not complexed to the site, may explain the preference for *cis*-action. If instability of the protein is in fact an important component of the *cis*-action, selection of mutations that allow a *cis*-acting protein to function more readily in *trans* may lead to the identification of the relevant protease.

REC A-DEPENDENT PROTEOLYSIS

Probably the first recognized role for a proteolytic event in the regulation of gene expression in *E. coli* was the discovery that the induction of lambda prophage after DNA damage involved the cleavage of the lambda cI repressor in a RecA-dependent fashion (120). This cleavage is dependent on the host RecA function; *recA* mutants are defective in both DNA recombination and the induction of prophage lambda and the cellular DNA damage repair response, called the SOS response. The cellular LexA repressor is also cleaved after DNA damage, allowing the induction of a number of DNA-damage repair systems, including the *uvrABC* genes, *umuDC, sulA,* and other genes of unidentified function. This circuit has been well reviewed by several authors (see 86, 147). I review here the outlines of the circuit and discuss the role of the proteolytic activity, with emphasis on recent findings.

Induction of the SOS system provides an example of conditionally unstable proteins. LexA protein and lambda repressor are normally stable proteins that act to repress cellular repair functions and lambdalytic operons, respectively. After DNA damage, these stable proteins suddenly become unstable and are destroyed quickly enough to lead to rapid induction of the repressed genes. As the DNA damage is repaired, the half-life of the LexA protein gradually returns to that of the normal stable protein (85).

Cleavage of LexA and cI repressors can be demonstrated in vitro, in the presence of RecA, ATP (or a nonhydrolyzable analogue), and polynucleotide (29, 30). Cleavage occurs between an Ala and Gly in the hinge region between the two domains of the proteins. The role of the RecA protein in this process is complicated by Little's observation (84) that the activity necessary for the cleavage resides in the substrates themselves. At high pHs, both lambda repressor and LexA are capable of autodigestion at the same site as that seen in the RecA-dependent reaction. Recent analysis of mutants defective in both the autodigestion and RecA-dependent degradation suggests that amino acids in the carboxy-terminal domain, including a serine at amino acid 119 of LexA (conserved in the lambdoid repressors), are directly involved in the reaction (131).

RecA appears to have a second role in the SOS response in addition to cleavage of the LexA repressor. Recent work in several laboratories indicates that RecA is needed for cleavage of UmuD; UmuD is necessary for UV-induced mutagenesis (16, 106, 126, 147). Cleavage is necessary for activation of UmuD (106). The possibility that UmuD might be a substrate for cleavage was first suggested by the similarity in the protein sequence between the caboxy-terminal domains of UmuD, LexA, and the lambdoid repressors (112). Since residues in this domain of LexA have been implicated in mediating both autodigestion and RecA-stimulated cleavage (131), the sequence similarities implied that UmuD might also be able to participate in the same types of reactions (112). As in the case of LexA and lambda repressor, the UmuD protein can autodigest, and probably carries the active site for proteolysis in its C-terminal half (16, 106). Thus, UmuD requires two RecA-dependent protein cleavages to allow it to function: one to allow its synthesis from a LexA repressed promoter, and a second to convert it to a functional form.

The conditional digestions of LexA and UmuD in a RecA-dependent reaction provide an extreme example of the manner in which specific targets are recognized by proteases. In this case, the targets themselves have evolved to provide much of the proteolytic machinery. The activation of this potential protease occurs when RecA interacts with a signal generated by damaged DNA, presumably oligonucleotides or stretches of single-stranded DNA. The "protease" can only work on those substrates capable of participating actively in their own cleavage.

THE LON SYSTEM

In 1973 Bukhari & Zipser identified mutations that allowed intragenic complementation between fragments of β-galactosidase (15). The original fragments, formed when *amber* termination mutations were isolated in the *lacZ* gene, were highly unstable in vivo. Because the suppressing mutations increased the stability of the fragments, they called the new mutations *degT,* for *deg*radation of early *t*ermination fragments (15). Apte et al identified mutations, called *degR,* that seemed to specifically suppress the instability of polypeptide chains lacking the amino-terminus of β-galactosidase (4). Both *degR* and *degT* mutant hosts were shown to carry mutations in a locus that had previously been called either *lon* or *capR* (57, 127). Previously isolated *lon* and *capR* mutations were able to suppress the instability of β-galactosidase fragments sufficiently to allow intramolecular complementation. *lon(capR)* mutations make cells UV-sensitive and allow increased synthesis of capsular polysaccharide (62, 90). The demonstration that the *lon* locus codes for an ATP-dependent protease (20, 25), called Protease La by Goldberg, and the identification of constitutively unstable proteins subject to Lon proteolysis, now provides a unified picture of the role of Lon in the cell.

UV sensitivity: the Lon phenotype

In 1964, Howard-Flanders and coworkers first observed that one class of UV-sensitive mutations had the property of forming *long*, non-septated filaments and dying after UV treatment (62). They called these mutations *lon*. Second site suppressors of the UV sensitivity of *lon* mutations can be isolated at two sites in *E. coli*, called *sulA* and *sulB* (also called *sfiA* and *sfiB*) (42, 43, 55, 65). The UV-sensitivity of *lon* mutations was shown to depend on the induction of the SOS system; George et al (43) found that *lon* cells are inviable after the SOS system is triggered by temperature-activatable mutations in the *recA* gene (*tif* mutations). This lethality (and associated filamentation) can be blocked either by *lexA* mutations (alleles unable to be cleaved and therefore uninducible), or mutations in genes that they called *sfiA* and *sfiB*. George et al hypothesized that *sfiA* or *sfiB* might code for a protease-sensitive cell-division inhibitor (43). This model has proven to be correct.

sulA (sfiA) is a typical gene of the SOS global control circuit. It has a highly efficient promoter, overlapped by a LexA binding site (99). Repression of the transcription of the gene by LexA has been demonstrated both in vivo and in vitro (64, 99). In vivo, induction occurs after DNA-damaging treatments or in appropriate mutant hosts that induce the SOS response (64). Induction of SulA, in the absence of any other part of the SOS response, is sufficient to inhibit cell division (66, 124). While cells continue to elongate and DNA continues to be made and distributed along the elongated filament, no septa are formed to divide daughter cells.

The product of the *sulA* gene is a protein of 169-amino acids that is highly unstable in wild-type cells under all conditions tested (100). *lon* mutations increase the half-life of SulA from less than 2 minutes to more than 20 minutes (100). In addition, the in vivo degradation of SulA has recently been demonstrated to be energy dependent, as expected for Lon-dependent proteolysis (D. Canceill & O. Huisman, manuscript in preparation; 144). Preliminary results suggest that purified SulA can be degraded by Lon protease in a purified system (D. Canceill, O. Huisman & M. Maurizi, manuscript in preparation).

The observations suggest a straightforward model for the UV sensitivity of *lon* mutants and reveal the role of the Lon protease in the normal aftermath of an SOS response. SulA synthesis is induced after any sort of DNA damage. While SulA is being rapidly synthesized, it accumulates in amounts sufficient to inhibit cell division, despite its rapid degradation by Lon. Therefore, there is a transient inhibition of cell division in lon^+ cells as part of the SOS response. After DNA damage has been repaired, SulA synthesis is shut off, as the LexA repressor reaccumulates in cells. At this point, SulA concentrations rapidly fall in wild-type cells, which are able to degrade SulA, and the cells resume synthesis of septa and continue cell division. In *lon* mutant cells, however, SulA remains active and continues to inhibit cell division. Consequently, the cells elongate and the characteristic long, non-septated filaments are formed. It is unclear at what stage this elongation becomes lethal to the cells, but blocking filamentation (with *sul* mutations) is sufficient to make *lon* cells UV-resistant. There is no other detectable defect in DNA repair or mutagenesis in *lon* mutants.

The probable target of SulA action is the product of the *ftsZ* gene, an essential cell-division gene. Evidence for this is provided primarily by Lutkenhaus' finding that *sulB* mutations are alleles of *ftsZ* (87). Inactivation of FtsZ with temperature-sensitive mutations results in the formation of long, non-septated filaments with the DNA distributed throughout the filament (reviewed in ref. 33). Overproduction of FtsZ can overcome inhibition by SulA (88). More direct evidence of a SulA/SulB interaction has been provided by the observation that the turnover of SulA can be slowed in the presence of wild-type FtsZ protein (72). Finally, overproduction of the *ftsZ* product in *lon*⁺ cells can lead to minicell formation, suggesting a key role for FtsZ in determining the frequency of septum formation (149). Therefore, we imagine that SulA interacts stochiometrically with FtsZ to block its action. In the presence of Lon, the SulA/FtsZ interaction is sufficiently reversible to allow degradation of SulA and subsequent return of cell division. Whatever its role in cell division, cells can temporarily tolerate lack of functional FtsZ: long filaments made in the absence of functional FtsZ can resolve when FtsZ becomes available (89). At some point, this process becomes irreversible and cells die.

This model for Lon degradation of SulA is an excellent example of an emergency response function that is useful to the cell during DNA repair but lethal for normal long-term growth. The rapid turnover of SulA provides a simple mechanism for regulating the action of SulA via regulation of its synthesis. The Lon protease, whose synthesis is not regulated as part of the SOS system (see below), is always available to degrade SulA soon after it is made. Only under conditions where SulA synthesis remains at a high level will SulA be active. Presumably, this transient inhibition of cell division allows DNA repair to proceed before chromosomes are segregated to daughter cells. Unfortunately, data to support this presumption have been difficult to obtain; *lon sulA* cells show no increased UV sensitivity compared to wild-type cells.

Although Lon is not essential for bacterial growth under standard growth conditions, *lon* mutants are poised to exhibit a variety of growth defects: transferring *lon*⁻ cells from simple to complex medium (42), increasing Tn10 transposition (119), or exposing the cells to even very low doses of UV light causes lethal filamentation. It is extremely difficult to introduce dam^- mutations, which increase the basal level of SOS gene expression slightly (31, 113), into lon^- hosts; the double mutants grow poorly and accumulate suppressing mutations (S. Gottesman, unpublished observations). Lysogenization by phage P1 is also decreased more than 10^3-fold in lon^- hosts (55, 65, 139). All these phenotypes are abolished by the introduction of either $sulA$ or $sulB$ mutations (42, 55, 65). A low level expression of the $sulA$ gene, sufficient to cause lethal filamentation, is therefore the likely explanation for failure to grow in all of these situations. Both expression of the Tn10 transposase (119) and P1 infection (32) do in fact lead to SOS induction.

Lambda lysogeny is also reduced in lon mutants (146, 148), and Markovitz and coworkers have found an increase in lysogenization in sul derivatives (42). Although other explanations for the decrease in lambda lysogeny can be offered (see below), some transient SOS induction may also play a role.

Given the difficulties facing the lon^- cell, it is interesting that $E.\ coli$ B seems to be a natural lon mutant (35, 152); the B/r radiation-resistant derivatives commonly used carry a sul mutation (70, 152).

Capsule Overproduction: The CapR Phenotype

A. Markovitz (90) first isolated $capR(lon)$ mutants of $E.\ coli$ as cells with aberrant regulation of capsular polysaccharide synthesis. lon mutants overproduce by 10- to 50-fold the colanic acid capsular polysaccharide. Markovitz demonstrated that some enzymes necessary for synthesis of colanic acid are overproduced in lon mutant cells (91) and identified two other loci leading to high levels of capsular polysaccharide synthesis, $capS$ and $capT$. Neither mutation was well-characterized or precisely mapped. It is unclear if these loci are the same as the regulatory genes described below. (For a summary of early work on capsule regulation see ref. 91.)

If Lon acts as a protease to regulate capsular polysaccharide synthesis, overproduction of capsular polysaccharide in lon mutants can most easily be explained by postulating that a positive regulator of capsule synthesis is a substrate for the Lon protease. This has in fact been demonstrated: RcsA, a positive regulator of transcription of the genes necessary for capsule synthesis, is unstable in lon^+ cells (half-life of less than 5 min) and is significantly more stable in lon^- cells (half-life greater than 20 min) (142). Therefore, the extra accumulation of the RcsA protein in lon^- mutants may explain the overproduction of capsule in lon cells.

Markovitz's initial observations on the increased synthesis of some enzymes in lon mutants led us to look directly for transcriptional fusions to genes involved in capsule synthesis. Using the Mu-lac transposition vector (18), we isolated and mapped a series of insertions that reduced or abolished

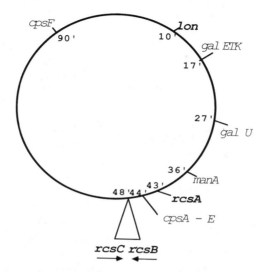

Figure 1 A genetic map of markers implicated in synthesis of capsular polysaccharide synthesis. The genes in bold type are regulatory genes.

capsule synthesis in *lon* mutants; the genes inactivated by insertion were called *cps,* and at least six different complementation groups were defined (145). Five of these map in a cluster at 43', close to a region in which *non* mutations (non-mucoid) had previously been mapped (115), and next to the *rfb* locus, implicated in synthesis of lipopolysaccharide in *Salmonella* and *E. coli* (6, 123). The sixth locus, *cpsF,* maps near 90' and is leaky for capsule synthesis. All the fusions that showed any function (Lac$^+$) were regulated in a similar fashion; they were expressed at high levels in *lon*$^-$ cells and at low levels in *lon*$^+$ cells. The genes involved in capsule synthesis are shown in Figure 1. Only the *cps* genes are regulated by *lon* and the *rcs* functions.

Using *cps-lac* fusions to distinguish between regulatory mutations and structural gene mutations for capsule synthesis, regulatory mutations were identified in three loci, in addition to *lon* (56). Loss-of-function mutations in either *rcsA*(41') or *rcsB*(44') abolish synthesis from the *cps* genes in *lon*$^-$ hosts (56; Table 1 and Figure 1); some mutations in *rcsC* (close to *rcsB* at 44') increase synthesis in *lon*$^+$ hosts (Table 1). At high temperatures, the synthesis of colanic acid and the expression of the *cps-lac* fusions is reduced dramatically in *lon* mutants, but not in *rcsC* mutants (Table 1). All of these mutations are recessive to single copies of the wild-type gene, introduced on transducing phage. Dominant mutations at the *rcsA* locus have also been identified; these mutations, termed *rcsA**, increase synthesis of *cps-lac* in *lon*$^+$ hosts (56).

Although the roles of *rcsA, rcsB,* and *rcsC* have not been fully determined,

Table 1   Regulation of capsular polysaccharide synthesis

Genotype	Expression of cps-lac fusion[a] (units β-galactosidase)	
	30°C	42°C
lon^+	8	1
$lon\Delta$	482	6
$lon\Delta$ $rcsA$::ΔTn10	1	1
$lon\Delta$ $rcsB$::ΔTn10	1	1
lon^+ $rcsC$137	415	238

[a] Cells were grown in minimal glucose medium.

the genetic and sequence analysis of these genes suggests the following model
for regulation of capsular polysaccharide synthesis. RcsA is an unstable
protein with a half-life of 5 min in lon^+ hosts (142). It is apparently normally
limiting for capsule synthesis, since introducing even a second copy on an F'
plasmid or on a transducing phage causes a noticeable increase in the expression of capsule and of the *cps* genes (128, 142). Therefore, the removal of the
protease (Lon) responsible for its degradation significantly increases the
accumulation of RcsA and the subsequent synthesis of colanic acid. The RcsA
protein is a very basic protein of 23,500 mol wt, with no strong resemblance
to know DNA-binding proteins (A. Torres-Cabassa, D. Gutnick, V. Stout,
M. Maurizi & S. Gottesman, manuscript in preparation). RcsA analogs with
extensive homology with the *E. coli* gene have been identified in *Klebsiella*
(2) and in *Erwinia stewartii* (143; D. Coplin, personal communication). In
Erwinia, the RcsA analogue has been implicated in the synthesis of capsular
polysaccharides that share some structural properties with colanic acid. The
Erwinia capsule is necessary for pathogenesis in corn; *rcsA* mutants of
Erwinia are avirulent (143). From the conservation of its function, RcsA
could be a general component of many capsule regulatory circuits.

The similarities in DNA sequence of the positive regulator *rcsB* (V. Stout
& S. Gottesman, in preparation) with effectors from two-component regulatory systems suggest that it is also such an effector. These two-component
regulatory systems have been identified throughout the bacterial world (121).
In the best studied case, nitrogen regulation, the effector is phosphorylated by
a sensor in response to nitrogen deprivation (118). After such modification, it
is able to act as a positive regulator of transcription for the target genes. The
genetics of RcsB action are consistent with its playing a similar role in the
positive regulation of transcription from the *cps* promoters (14). If so, the best
candidate for a sensor to transmit environmental information to RcsB is the
rcsC product. This 104,000 mol wt protein crosses the inner membrane, and

has a cytoplasmic domain that shares homology with other sensors (V. Stout & S. Gottesman, manuscript in preparation). Mutations defective in temperature regulation of capsule synthesis map in this gene (Table 1), suggesting that temperature is one component of the environmental information read by the RcsB/C system.

RcsB, in contrast to RcsA, is a stable protein (14). However, if it works like its analogue in the nitrogen-regulation system, NtrC, its activity may be regulated by unstable phosphorylation in response to environmental cues. Thus, the instability of the RcsA protein and of RcsB activity may provide two points for the precise control of capsule synthesis.

Such tight and complex regulation inevitably raises questions about the function of the capsule, the conditions under which it must be synthesized and under which it is dispensable. Unfortunately, complete answers are not yet available. Capsule clearly can provide a physical barrier and camouflage for bacterial cells (28, 37), but the colanic acid capsule has not been shown to provide an advantage in protecting cells from phagocytosis. Given the function of RcsA in a plant pathogen, and the temperature-sensitive expression of colanic acid, increased colanic acid synthesis may more reasonably be imagined as a reaction to stresses in the outside environment. One demonstrated function for capsule is a protection of E. coli cells from dehydration; cells that are overproducing capsule survive overnight drying by a factor of 10^3-fold better than those that cannot produce capsule (S. Gottesman, unpublished observations). One reason for cells to rapidly decrease capsule synthesis when it is no longer needed may simply be the enormous energy loss in synthesizing vast amounts of this polysaccharide. Certainly cells that are synthesizing large amounts of capsule or constitutively expressing the cps genes seem to accumulate secondary suppressing mutations at a rapid rate (S. Gottesman, unpublished observations). It is interesting in this context to note that E. coli B, which carries a lon mutation, has a secondary mutation that suppresses capsule overproduction (34). Many standard E. coli K12 strains have also lost the capacity to synthesize capsule (S. Gottesman, unpublished observations).

Lambda Growth in lon Mutants

Bacteriophage lambda lysogenizes poorly in lon mutants, and lambda N protein is degraded by the Lon protease. In addition, lon⁻ lysogens carrying a prophage with the temperature-inducible cI857 repressor commit more rapidly to lytic growth after short incubations at high temperatures (54). Although stabilization of the N protein, which promotes lytic growth, may be sufficient in principle to explain these results, the observation that N⁻nin mutants still show this rapid commitment to lytic growth in lon⁻ hosts suggests that stabilization of N cannot be the only explanation (54). The observation that lambda cII protein, a positive regulator of lysogeny, is more unstable in lon mutants (54) may also explain these results. Loss of the Lon protease must act indirectly to stimulate cII degradation, possibly by an effect on the *hfl* protease that has been implicated in cII turnover (see below).

The sensitivity of N protein to degradation by Lon has provided the best available model for studying the in vitro activity of Lon protease. In vitro, purified N is rapidly degraded by Lon in the presence of ATP (94). The protein is cut at six sites; no clear sequence requirements can be deduced. Mutations in the sites should help determine the important aspects of these sequences. A comparison between N and the RcsA and SulA targets for Lon degradation shows some similar sequences in these proteins, although no one sequence is shared by all targets.

Abnormal Proteins as Targets for Lon: The Deg Phenotype

As mentioned above, one of the first observations linking *lon* to proteolysis was the selection of *lon* mutants as sufficient to stabilize nonsense and restart fragments of β-galactosidase (4, 15). Such fragments are degraded much more rapidly than the parental intact molecule. Puromycin treatment leads to premature termination of translation; the resulting polypeptide-chain fragments are also subject to rapid degradation (47). The turnover of these fragments is slowed but not stopped in *lon* mutants (76, 94).

The abnormally high rate of degradation of protein fragments can be attributed to their inability to fold properly and, therefore, to the exposure of internal, protease-sensitive sites on their surfaces. If true, other mutations that may lead to unfolding or improper folding should also cause sensitivity to proteolysis. In fact, many temperature-sensitive mutations, other types of missense mutations, and proteins containing amino acid analogs can all be unstable (47, 57, 156). Removal of the *lon* protease by mutation is sometimes sufficient to suppress the temperature-sensitive phenotype of some mutations. Thus, phages with temperature-sensitive mutations in the lambda-replication initiation protein, O, were shown to grow at normally non-permissive temperatures in *lon* mutants (47). A temperature-sensitive mutation in the sigma subunit of RNA polymerase is also suppressed in *lon* mutants, and the suppression correlates with an increase in the half-life of the protein (59).

Substitution of canavanine for arginine during protein synthesis increases protein turnover dramatically; *lon* mutations slow the degradation by about 50%, although the remaining turnover is still energy-dependent (44, 76, 95). If individual proteins are examined after a pulse of canavanine, many of the bands visible by one-dimensional gel electrophoresis are unstable in *lon*$^+$ cells and are stabilized in *lon* mutants (93). Therefore, at least by this crude measure, most proteins made unstable by this substitution are substrates for Lon, while most wild-type proteins in *E. coli* are very stable.

Finally, both foreign proteins and some *E. coli* proteins cloned and over-produced in *E. coli* are frequently unstable (53, 98). Overproduction of one subunit of Integration Host Factor (IHF) leads to instability of the over-produced subunit (103). A similar phenomenon has been observed in both bacterial and animal cells, where mutations in one component of a multi-protein complex can result in instability for the remaining components (80, 105). Presumably this instability of "proteins without partners" is another manifestion of the instability of abnormal proteins. Overproduction may also promote differences in folding due to different synthesis conditions. In some, but hardly all cases, the instability is lessened in *lon* mutant hosts. The instability of abnormal proteins is discussed further below.

Thus, in vitro, Lon recognizes rather specific sites in N, as well as in the small protein substrates insulin and glucagon (94). Other proteins, some of which are unstable in vivo, are not degraded by Lon. Denaturation some-times, but not always, increases a protein's susceptibility to proteolysis (94, 151). In contrast to its narrow specificity in vitro, the ability of Lon to degrade abnormal proteins in vivo is relatively broad. Its substrates include many protein fragments and missense proteins. This paradox in Lon recognition of its substrates remains a major problem in understanding the in vivo role of energy-dependent proteases such as Lon. One possible insight into the prob-lem, discussed below, is provided by recent experiments on mutants defective in the heat-shock proteins.

Regulation of Lon Synthesis and Activity

HEAT-SHOCK REGULATION OF LON SYNTHESIS   Although Lon is easily detectable in normally growing cells, its synthesis increases transiently when cells are subjected to a heat shock (44, 114). The promoter region of *lon* contains a consensus sequence for the heat-shock sigma factor (24, 41). In addition to Lon, other proteases seem to be heat-shock induced, as judged by the proteolysis defect of cells with mutations in *htpR,* the heat-shock sigma factor (see below). Abnormal protein overproduction can induce the heat-shock response (46). Therefore, one role of Lon may well be to help the cell deal with denatured proteins resulting from stress conditions such as heat shock. The connections between the heat-shock response, abnormal proteins, and protein turnover are discussed more fully below.

There is no evidence that the heat-shock response is required for Lon to degrade its normal substrates, such as SulA, RcsA, and N, at optimal rates: degradation occurs at low as well as high temperatures, and *htpR* mutants do not have UV-sensitivity or capsule overproduction phenotypes (8; S. Gottes-man, unpublished observations). Overproduction of Lon substrates such as SulA does not induce either *lon* synthesis or a heat-shock response (D. Canceill, personal communication). The overproduction of SulA can in fact titrate the availability of Lon sufficiently to increase the synthesis from RcsA-dependent *cps-lac* fusions (E. Dervyn, D. Canceill & O. Huisman, manuscript in preparation). If induction of Lon synthesis occurred in this situation, such titration would seem unlikely. This difference in the ability of overproduced abnormal proteins and overproduced natural substrates to induce the heat-shock response may suggest a fundamantal difference in how these normal substrates and abnormal proteins are recognized by the cell (see below).

There is thus far no evidence for in vivo regulation of Lon activity by changes in ATP levels. ATP is rapidly hydrolyzed by Lon to ADP in vitro, and ADP acts to inhibit protease activity (96, 97). Goldberg and coworkers have presented a model for Lon action in which Lon protease activity, normally inhibited in vivo by bound ADP, is activated by substrate binding and release of ADP (96, 97, 151). DNA has also been demonstrated to stimulate Lon activity in vitro (19, 26, 154), but the significance of these observations for in vivo activities is unclear.

REGULATION OF LON ACTIVITY BY PHAGE T4   An unexpected control of Lon activity occurs after infection with phage T4. It had long been observed that protein fragments and other abnormal proteins are stable after T4 infection (129). Recently, Simon and coworkers have cloned a T4 function, called *pin*, capable of making cells phenotypically Lon⁻ when present in single copy. Pin interferes specifically with Lon protease by an unknown mechanism (130). Presumably, T4 has proteins that would otherwise be Lon substrates but are protected by the *pin* inhibition. Analysis of the behavior of T4 *pin* mutants may point to these substrates.

MUTATION IN LON AS A CONTROL OF ACTIVITY: SPECULATION   Nearly everyone who has worked with *E. coli* has been struck by the high frequency with which mucoid derivatives of normally non-mucoid strains arise on plates left out on the bench, during selections for mutants resistant to various phage or other harsh treatments, and under a variety of other conditions. This same instability is evident in the rapid accumulation of mutations that increase expression of *cps-lac* fusions. These *cps-lac* strains have acquired mutations predominantly in the *lon* gene, as judged by mapping and their UV-sensitive phenotype (S. Gottesman, unpublished observations). Cells carrying *lon* on multi-copy plasmids are also subject to a high frequency of *lon* inactivation (45, 155; M. Maurizi, unpublished observations); most are IS insertions at one site early in the gene.

These data suggest that one possible mechanism for reducing the availability of Lon is by inactivation of its gene by IS insertion. If true, it will be interesting to determine if there is an increase in the *lon*-mutation frequency under certain conditions. High levels of capsule synthesis are a significant energy drain during rapid growth, but capsule may serve to protect the cell from drying and from harm during periods of starvation. Mutation of a fraction of a given population to *lon⁻* could conceivably lead to sufficient synthesis of mucoidy to protect the rest of the population; if so, any disadvantage for the individual cell in being *lon⁻* will not affect the rest of the population when conditions for growth improve. The demonstration of a dependence on SulA-dependent inhibition of cell division could also require testing under starvation conditions.

While a role for IS insertion in the control of *lon* expression is purely speculative at this point, genetic instability has been identified as a mode of gene regulation of various capsular polysaccharides, including alginate synthesis by *Pseudomonas* (39), exopolysaccharide synthesis in the marine bacteria *Pseudomonas atlantica* (11), synthesis of the Vi capsular antigen by *Citrobacter* (107) and capsule synthesis in *H. influenzae* (78).

MAP LOCATION AND GENE DOSAGE: ANOTHER REGULATORY MECHANISM? The *lon* gene is located at 10' on the genetic map of *E. coli*; its substrates, SulA and RcsA, are coded for by genes located clockwise of *lon* at 22' and 43', respectively. Increasing the gene dosage of *rcsA⁺* even moderately, with a F' carrying *rcsA*, is sufficient to make the cell mucoid (128). Is it significant that replication of the host chromosome will always increase Lon gene dosage before increasing the amounts of SulA or RcsA, which lie closer to the termination region (6)? If so, moving the *rcsA* gene to a site close to the origin, or moving *lon* farther away, may increase capsule synthesis in this very delicately balanced system. If, in the absence of additional inducing signals or mutations, capsule synthesis is tied to cell division, such a gene-dosage arrangement may be one way to balance synthesis and degradation.

Distribution of Lon and its Substrates

Although only the Lon ATP-dependent protease from *E. coli* has been studied in detail, the conservation of Lon substrates throughout gram-negative bacteria suggests that very similar systems operate in many organisms.

The *sulA* gene, for instance, has been cloned and sequenced from four bacterial hosts (40). The SulA homologs are well-conserved throughout three-fourths of the protein length. These SulA proteins are not active in *E. coli*, suggesting that the diverged regions may specify interaction with FtsZ. A broader survey by Southern hybridization indicates that *sulA* analogues are present in other *Enterobacteriaceae* (27). This conservation may imply both the existence of similar SOS systems in these organisms and conservation of the target FtsZ protein. Proteins with the ability to cross-react with *E. coli*

FtsZ have been detected in essentially all bacteria tested, including gram-positives (27).

Homologs of RcsA, the unstable positive regulator for capsule synthesis, have been found in other gram-negative bacteria. The conservation of the amino acid sequences of these proteins and the ability of RcsA to act in heterologous hosts suggests that the target of RcsA is also conserved throughout these different organisms, presumably to regulate different capsules.

Lon protease itself has been purified from *Salmonella typhimurium*, as well as from *E. coli* (36). Although Rupprecht & Markovitz (122) detected homologous DNA in various gram-negative bacteria and the archaebacterium *Halobacterium* by Southern hybridization with *lon*-containing plasmids, the significance of these results with respect to Lon is clouded by the discovery that the gene immediately 3' to *lon*, and presumably present on their plasmids, codes for a subunit of the highly conserved HU protein (73). Many bacteria produce proteins of similar size to Lon that cross-react with Lon antibody (Y. Katayama, in preparation). A monoclonal antibody to a human heat-shock protein, hsp90, cross-reacts with *E. coli* Lon, raising the possibility that this type of ATP-dependent protease is indeed very widespread (79). ATP-dependent proteolysis in mammalian cells clearly plays important roles, and work thus far on these less well-defined proteases suggests many parallels between Lon and these mammalian proteolytic systems (12, 49, 116).

Mechanism of Lon Action

Studies on the mechanism of Lon action have begun to shed significant light on the unique biochemistry of energy-dependent proteolysis, but this review only highlights some of the major conclusions, with attention to their relevance to the in vivo role of Lon.

Lon is composed of four 87,000 dalton subunits. It behaves as a serine protease in its sensitivity to inhibitors such as diisopropyl fluorophosphate (150). The protein sequence of Lon, predicted from the DNA sequence, does not contain sequences similar to the serine active sites of other serine proteases (24). This may not be surprising in view of Lon's unique requirement for ATP. A consensus nucleotide binding site has been identified in the middle of the protein (24).

In vitro degradation by Lon of short fluorogenic peptides, low molecular weight proteins (glucagon), and some high molecular weight proteins (casein) can proceed with nonhydrolyzable ATP analogs, whereas degradation of other high molecular weight proteins (denatured BSA) requires hydrolysis of ATP (50). Degradation of N, a physiological substrate for Lon, proceeds at about half of the maximal rate in vitro when the nonhydrolyzable analog AMPPNP is substituted for ATP (94).

Studies on the kinetics of ATP usage by Menon & Goldberg (97) suggest the existence of two different types of ATP sites per tetramer of Lon protease. Binding to the two high affinity sites activates degradation of peptides and some protein substrates, while rapid cleavage of large proteins requires ATP binding to both of the two high affinity sites and two additional low affinity sites. The binding of ATP to the first two sites is belived to change the affinity of the protein for ATP, creating the low affinity sites. There is no evidence for phosphorylation of either protein substrates or Lon protease (50); the observation of protease activity with nonhydrolyzable ATP analogs is consistent with the idea that phosphate transfer is not an intrinsic requirement for cleavage of peptide bonds by this protease.

Goldberg has proposed that ATP serves as an allosteric effector of Lon activity, based on a variety of studies. Since ATP is quickly hydrolyzed to ADP, and ADP is not released from the protein until a substrate protein (but not a peptide) is bound, his model envisions a cycle of activation and inactivation in which the Lon protease will only be activated by substrate binding. Thus, protein-substrate binding also acts as an allosteric effector, causing release of ADP (96, 151). If so, then Lon's specificity of degradation may depend on the specificity of binding to this postulated allosteric activation site. This model still leaves unexplained the ability of Lon to catalytically degrade proteins such as N, using non-hydrolyzable ATP analogs.

HEAT-SHOCK RESPONSE AND PROTEOLYSIS

Essentially all cells that have been examined respond to increases in temperature by a rapid but transient increase in the synthesis of a number of proteins. This response, called the heat-shock response, is mediated in *E. coli* by a heat-shock sigma factor, sigma 32, that recognizes the promoters of heat-shock genes (60, 104). Many heat-shock genes are conserved between prokaryotes and eukaryotes. Although their functions are still under intensive investigation, some heat-shock proteins have been implicated in the in vivo unfolding and folding of proteins. In fact, abnormal proteins seem to provide one pathway for induction of the heat-shock response. Given that a major consequence of high temperature is the denaturation of proteins, the heat-shock response may help the cell to deal with this stress by providing ways to handle denatured protein, either through degradation or renaturation.

Heat-Shock Induction by Abnormal Proteins

In both prokaryotic and eukaryotic cells, the accumulation of abnormal proteins, which frequently are subject to proteolysis, leads to induction of the heat-shock response (3, 46). Goff & Goldberg demonstrated that incorporation of canavanine into protein or overproduction of a foreign unstable protein induces the heat-shock response and increases expression of *lon* (46). These effects are dependent on the *htpR* gene product. Parsell & Sauer provide a further insight into this phenomenon (109a). Using a series of mutant derivatives of the lambda repressor N-terminal domain, they demonstrated a correlation between in vitro unfolding and the in vivo ability to induce a heat-shock response. A particularly interesting finding is that unfolded but stable derivatives were more effective as inducing signals than some mutants that were subject to proteolysis. This suggests that the cell has a mechanism for recognizing unfolded proteins per se, and that unfolding and sensitivity to degradation are not necessarily coincident.

DEGRADATION OF ABNORMAL PROTEINS IN HEAT-SHOCK MUTANTS
Thus, abnormal proteins, which are frequently but not always subject to proteolysis, induce heat shock. The ability to induce the heat-shock response also seems necessary for proteolysis of many abnormal proteins. *htpR* mutants, which are defective in the heat-shock sigma factor and therefore cannot increase heat-shock protein synthesis at high temperatures, are defective in proteolysis of canavanine-containing proteins and puromycyl peptides, as well as the degradation of the X90 fragment of β-galactosidase (8, 44). The proteolytic defect in *htpR* mutants is more severe than that found in a *lon* mutation. Therefore, at least one proteolytic system other than *lon* must be missing in *htpR* mutants.

DnaK, a major *E. coli* heat-shock protein, is highly homologous to the eukaryotic hsp70 class of heat-shock proteins. In *E. coli, dnaK* has been implicated in recovery from heat shock (140). *dnaK* mutants have higher basal levels of heat-shock proteins and synthesis of heat-shock proteins persists in these mutants. Proteolysis of a number of mutant proteins, including LacZX90, is elevated at low temperatures in a *dnaK* mutant; this increase is found even in *lon* mutants (76, 136). These observations complement those found with the *htpR* mutants by suggesting that increasing heat-shock proteins other than Lon can increase proteolysis.

SIGMA 32 AS A SUBSTRATE FOR PROTEOLYSIS   One apparent substrate for these heat-shock dependent proteases is sigma 32, the *htpR* product and regulator of the response. Sigma 32 has a half-life of one minute during steady-state growth at both low and high temperatures (135, 141). After a heat shock, the half-life increases to eight minutes for a few minutes, and accumulation of sigma 32 increases more than tenfold. Within ten minutes, the amount of sigma 32 has begun to drop (135). Since the rate of synthesis of heat-shock proteins is dependent on the level of sigma 32 (61, 135), this transient increase explains the transient increase in rates of synthesis of these proteins. As heat-shock proteins accumulate, increased proteolytic activity increases degradation of sigma 32, causing a decrease in synthesis of the heat-shock proteins. Consistent with this model is the observation that mutations in *dnaK* both increase the half-life of sigma 32 and extend the heat-shock response (140, 141).

A remaining question is whether the kinetics of induction of heat shock by unfolded protein can be considered adequate to account for the rapid onset of the heat-shock response. If unfolding in response to heat is in fact the primary inducer and if the mechanism of this induction is competition for the proteolytic system that degrades sigma 32, will random unfolding of cellular proteins compete rapidly and efficiently enough to account for the instantaneous increase in sigma 32 levels? An alterative to the partial unfolding of many proteins would be one or more "thermometer" proteins, with high affinity for the recognition element in the degradative pathway for unfolded proteins. Such proteins would presumably unfold gradually over a wide temperature range. The thermometer itself need not be degraded, since Parsell & Sauer's work demonstrates that degradation is not a necessity for heat-shock induction (109a). The thermometer should not itself be induced by heat shock; if it were, one would not expect a transient induction response. If a single such thermometer protein exists, mutants that eliminate its synthesis or that cannot be recognized by the system may be unable to induce heat shock; mutations that lead to unfolding at low temperatures may be constitutive for the heat-shock response. In cells in which abnormal proteins are overproduced, these proteins may substitute for the thermometer proteins in competing for some element of the system. The recognition elements, if overproduced, should also block induction of the heat-shock response.

A model in which competition for sigma 32 turnover explains all aspects of heat-shock induction may well be too simplistic. In addition to changes in sigma 32 half-life during induction, changes in transcription and translation rates have been observed (61). Therefore, multiple distress signals may lead to various responses, all leading to the transient increase in sigma 32 levels. It would be interesting to determine if overproduction of an unfolded protein leads only to stabilization of sigma 32 or also to changes in expression.

PROTEIN DEGRADATION BY HEAT-SHOCK PROTEASES   If *htpR* is necessary for proteolysis of abnormal proteins, some of the heat-shock proteins themselves might be expected to be proteases. As mentioned above, Lon is in fact a heat-shock protein, but Lon-dependent degradation is not sufficient to account for all the *htpR*-dependent protein degradation (8, 44). In fact, mutations in many heat-shock genes lead to defects in proteolysis. Cells mutant in *dnaJ* and *groEL* are defective for degradation of puromycyl peptides; *dnaJ* mutants are also defective for LacZX90 degradation (136). *dnaK* mutants, which have increased degradation of some missense proteins at low temperatures, are defective for degradation of these same proteins at high temperature, and are defective for degradation of canavine-containing proteins and puromycyl peptides at all temperatures (76, 136). *grpE* mutants show a pattern similar to that seen with *dnaK* (136). Therefore, one must postulate that each of these proteins is an essential part of a complex proteolytic system or is in the pathway to induction of such a system. Such a multi-component protease has not been directly demonstrated, but is reminiscent of reports of complex proteolytic systems in eukaryotes (12, 92, 116). Some of these proteins or their eukaryotic analogs have the property of promoting the unfolding of proteins, and are referred to as unfoldases (111). The eukaryotic analogue of DnaK is involved in the posttranslational unfolding and insertion of proteins into membranes. An unfoldase could be a component of a system for making proteins sensitive to a protease. What remains unclear is how such proteins, which are abundant in cells, protect cells against heat shock. Most are essential cell proteins (see ref. 104). Most appealing is the idea that by helping to unfold proteins, heat-shock proteins protect normal proteins from abnormal refolding and allow them to refold on a proper pathway. If so, the problem of recognizing the abnormal (to be degraded) from the normal (to be refolded) remains.

The paradoxical observations on the phenotypes of *dnaK* and *grpE* mutants for protein turnover may reflect the involvement of these proteins in heat-shock induction and recovery. The increase in degradation of LacZX90 or certain LacI proteins at low temperatures (76, 136) may be ascribed to the increased synthesis of the putative "heat-shock proteases" including, but not limited to, Lon in *dnaK* mutants (140). When the degradation of the canavanine-containing proteins and puromycyl peptides is studied, the synthesis of these abnormal proteins is in itself sufficient to induce the heat-shock response (46). Therefore, in this case one is comparing the effect of *dnaK* and *grpE* mutants in a heat-shock induced background. The results suggest that, in addition to an involvement in recovery from heat shock, *dnaK* and *grpE* must also participate directly or indirectly in the degradation of abnormal proteins.

LON DEGRADATION OF ABNORMAL PROTEINS DURING HEAT SHOCK
Either *lon* mutations or certain mutations in the heat-shock proteins (for instance, *dnaJ*) increase the half-life of abnormal substrates like lacZX90 from less than 5 minutes to more than 30 minutes (136). Such results indicate that either degradation of the protein by Lon involves DnaJ as a necessary component, or that Lon and DnaJ are influencing some other protease's ability to act. By a similar argument, the stabilization of many abnormal proteins in *htpR* mutants suggests that heat-shock proteins are important components of the proteases that degrade these abnormal proteins. Yet *htpR* mutants and *dnaJ* mutants do not have the general *lon* phenotypes, and

therefore presumably are still able to degrade SulA and RcsA (8; S. Gottesman, unpublished observations). One possible explanation for this inconsistency is a model in which Lon has two modes of in vivo protein degradation. One of them, responsible for the rapid and specific degradation of the Lon-specific substrates, requires only Lon, ATP and the substrate. The other, responsible for the degradation of abnormal proteins, involves the participation of other cellular proteins, most notably DnaJ. DnaJ might be acting as an auxiliary unfoldase in this case, to make an abnormal protein susceptible to Lon degradation. Alternatively, it may interact directly with Lon, as it seems to with DnaK and GrpE to promote lambda replication (71, 81). By this model, Lon could degrade abnormal proteins that would not otherwise act as substrates when the heat-shock response is induced. Since canavanine incorporation induces the heat-shock response, the ability of Lon to participate in degradation of most canavanine-containing proteins would reflect this second, more promiscuous activity. A prediction of such a model is that the defects of *dnaJ* and *lon* mutants for degradation of abnormal proteins will not be additive, while *dnaJ* mutations will have no effect on Lon's ability to degrade RcsA or SulA.

CLP, A TWO-COMPONENT ATP-DEPENDENT PROTEASE

Cells that are devoid of Lon because of deletions or insertions early in the gene still show ATP-dependent degradation of casein in vitro and energy-dependent degradation of a number of unstable proteins, including abnormal proteins, in vivo (95). Purification of an energy-dependent casein degradation activity led to the identification of the two-component Clp protease of *E. coli* (74, 75), (called Ti by Hwang et al (67, 68)). The Clp protease contains equimolar amounts of the two polypeptide components ClpA (81,000 dalton mol wt) and ClpP (23,000 daltons) (67, 68, 74, 75; M. Maurizi & S.-H. Kim, in preparation). The ClpA subunit binds ATP and shows substrate-stimulated ATPase activity (68, 74). The ClpP subunit has peptidase activity (153) and contains the protease active site (68; M. Maurizi, unpublished observations). Neither subunit alone is active in protein degradation, and the degradation of casein by the combined subunits is absolutely dependent on ATP (67, 75).

Using antibody to the purified subunits, we have been able to identify bacteriophage clones containing the *clpA* and *clpP* subunits, map the genes, and begin the process of analyzing their in vivo function. *clpA* maps at 19' on the *E. coli* genetic map. Inactivation of the gene by either insertion or deletion is not deleterious to the cell under normal growth conditions, either in the presence or absence of Lon. Cells that are mutant for Lon have reduced degradation of abnormal canavanine-containing proteins, as discussed above;

introduction of *clpA* mutations into these cells reduces further the energy-dependent degradation of these polypeptides (74).

Sequencing of the *clpA* structural gene has revealed that ClpA has two ATP-consensus binding sites similar to the single observed ATP-consensus site in Lon (M. Maurizi, W. Clark & S. Gottesman, manuscript in preparation). In addition, the sequence allowed the detection through searches of sequence banks of a well-conserved homolog of *clpA* in *E. coli*. The protein coded for by this gene, sequenced and referred to as ORF-BG by C. Squires (personal communication) differs from ClpA for a stretch of 165 amino acids at the amino-terminus, and contains a stretch of 127 amino acids in the middle of the protein, between the two ATP-binding sites, which is not found in ClpA. For the two conserved stretches (of 235 and 150 amino acids), the proteins share more than 85% similarity or identity. In addition, other proteins with homology to this ClpA-like protein have been identified in bacteria, parasites, and plant systems, suggesting that Clp-like proteases may be widespread (C. Squires, personal communication). It is tempting to speculate that ORF-BG participates in another ATP-dependent protease system in *E. coli*.

The structural gene for the ClpP subunit maps far from *clpA*, at 10' on the *E. coli* map. The gene is close to that for *lon*, but there is no evidence of its synthesis being closely tied to that of Lon (M. Maurizi, W. Clark & S. Gottesman, manuscript in preparation). The *clpP* sequence does not reveal any extensive homologies to either Lon or other proteases for which sequence information is available. The ClpP subunit can be labeled with serine protease inhibitors such as diisopropyl-fluorophosphate (68; M. R. Maurizi, in preparation), and a low level of peptidase activity can be detected with the isolated ClpP protein (153).

The observation of peptidase activity for the ClpP subunit is reminiscent of both Lon, which has a peptidase activity in the absence of ATP cleavage, and the RecA-independent autocleavage of LexA and other substrates. These three conditionally energy-dependent proteases are all serine proteases; all have reactions detectable in the absence of metabolic energy under special conditions. Formally, RecA and LexA can be considered a two-component protease like Clp, in which the ATPase regulatory subunit (RecA or ClpA) is separate from the protease subunit (LexA or ClpP).

Substrates for Clp Proteolysis

Although the in vitro identification and purification of Clp has used degradation of casein as an assay, the in vivo activities of Clp still remain elusive. ClpA protein itself is unstable in vivo and in vitro and a ClpA-lacZ protein fusion, containing the first 42 amino acids of ClpA, is quite unstable in wild-type strains. Interestingly, the turnover of this fusion protein is significantly slowed in *clpA* mutants (S. Gottesman, W. Clark & M. Maurizi, manuscript in preparation). Thus, ClpA may itself be a target for Clp-dependent proteolysis. Protein and operon fusions were surveyed for expression changes when *clpA⁺* is introduced into a *clpA* mutant host. Fusions of *lacZ* to two positions in the *rbs* (ribose) operon were identified (S. Gottesman, W. Clark, & M. Maurizi, manuscript in preparation). Thus some specific fusions are substrates for Clp; in both the *clp-lac* and the *rbs-lac* cases, *lon* has little or no effect on the stability of these fusions.

Regulation of Clp Synthesis

In the absence of either phenotypes or known natural substrates for the Clp protease (other than ClpA itself), an investigation of the regulation of Clp synthesis may give some insight into the in vivo role of Clp. Clp is not a heat-shock protein (74) and thus differs fundamentally from Lon in its synthesis pattern. Using *clp-lac* transcriptional and translational fusions, we have found that *clpA* expression is regulated similarly to the *pepT* peptidase of *Salmonella* (134; S. Gottesman, W. Clark & M. Maurizi, manuscript in preparation). Synthesis increases abruptly as the cell density reaches 0.4–0.5 O.D. units; synthesis can be increased at lower density by growing cells anaerobically. Thus the substrates for Clp degradation presumably become subject to degradation when cells are growing anaerobically. Such substrates could include proteins synthesized aerobically that interfere with anaerobic growth, or specific, anaerobically incuded proteins with short half-lives. Finally, given the observation that the tripeptidase *pepT* has similar regulation, Clp and PepT may act together to generate available amino acids for growth under anaerobic conditions when better carbon and nitrogen sources are not available.

Lon's activities are consistent with a role in bacterial growth in environments outside the mammalian gut. Capsule synthesis is stimulated during growth at low temperature and protects cells from drying. RcsA analogs are found in plant pathogens and soil microorganisms. SulA synthesis and degradation in response to DNA damage may well be a stress encountered most in the outside environment, and the filamentation response seems to be dependent on aerobic growth (1, 42). Clp, on the other hand, can most easily be pictured as having a role in the degradation of proteins to amino acids in the anaerobic conditions of the mammalian gut.

ALP-DEPENDENT PROTEOLYSIS

Although the Clp protease was initially identified biochemically in cells devoid of Lon, new proteases can also be identified by their ability to substitute for Lon in vivo. Trempy & Gottesman (144) have described the identification of a gene, named *alp*, that, when present on high copy number plasmids, can suppress the capsule overproduction and UV-sensitivity phenotype of *lon* mutants. The Alp function increases the degradation of SulA in *lon* mutants; the increased degradation is energy-dependent. Alp function is independent of either *clpA* or *orf*BG, and is present in *htpR* mutant hosts (J. E. Trempy & S. Gottesman, manuscript in preparation). *alp* chromosomal mutants are viable, and no specific natural substrate for Alp has yet been identified.

Sequencing of the alp region reveals an open reading frame for a 70 amino acid protein. Given the size of the other known energy-dependent proteases (Table 2), the region cloned on the plasmid probably represents either a regulator that increases synthesis of a protease or the rate-limiting component of a larger protease. Identification of both the protease and its natural substrates may help clarify the in vivo role of this system.

ENERGY-DEPENDENT PROTEASES: SUBSTRATE RECOGNITION

How cells discriminate between cellular garbage—proteins to be discarded as rapidly as possible—and the majority of functional proteins remains an

Table 2 Comparison of ATP-dependent proteases

Protease	Map location (min.)	Subunit size (daltons)	Subunits/ protein	ATP Sites/ subunit[d]	Protease type	Regulation of Synthesis
Lon	10 (470kb)[b]	87,000	4	1	Serine	Heat shock
(La)[a]						
(CapR)						
Clp	ClpA 19 (938kb)	84,000	4	2		Anaerobiosis
(Ti)	ClpP 10 (465kb)	21,000	10–12	0	Serine	
RecA	58 (2835)	37,800	[filament][e]	1		
&						
LexA[c]	93 (4310)	22,700			Serine	DNA damage
UmuD[c]	26 (1248)	15,000				

[a] Alternate names.
[b] Values in parentheses are coordinates from the physical map of *E. coli* (ref. 77).
[c] Alternative substrates; active sites for RecA (see text).
[d] Based on sequence comparisons.
[e] Active when bound on DNA.

intriguing problem. For eukaryotic cells, this is frequently a question of localization; targeting proteins to the lysosomes is tantamount to targeting them for destruction. Other eukaryotic degradative systems involve tagging with ubiquitin, a precursor for cytoplasmic energy-dependent degradation. According to recent work, the ubiquitin tagging rules may depend on the N-terminal amino acid sequence of the target protein, and the availability of a ubiquitin-conjugating residue (lysine) at an appropriate distance from the end (5, 51, 117). Given that the ubiquitin-sensitive N-terminal residues are rarely found naturally at the N-terminus of proteins, the initial cleavages that may lead to subsequent recognition by the ubiquitinating system become the basic targetting system.

There is no evidence thus far that prokaryotic cells contain ubiquitin, or a parallel tagging mechanism for proteins targetted for destruction. For a protease like RecA, the recognition problem is bypassed by the presence of the protease active site in the substrate itself. Clearly this is not a solution for more general proteases that degrade abnormal proteins.

The ability of unfolded proteins to induce heat shock, and the instability of some missense proteins and many protein fragments implicate unfolding as at least part of the recognition signal for proteolysis. In fact, experiments by Parsell & Sauer (109) suggest that, for mutant derivatives of a given protein, there is a direct relationship between the degree of unfolding observed in vitro and the susceptibility to degradation observed in vivo. For at least one of the unstable, unfolded substrates tested, *htpR* but not *lon* mutants prevented turnover (109).

For a given protein subject to proteolysis in vivo, substantial evidence implies that other signals may be important in addition to unfolding. In particular, either recognition or degradation itself may be influenced by the carboxy-terminus of a protein. Bowie & Sauer selected revertants of an unstable mutant phage regulatory protein, Arc, and isolated derivatives that increase stability by adding a carboxy-terminal extension to the protein (13). The same carboxy-terminal extension can stabilize unstable derivatives of another protein, the N-terminal domain of lambda repressor (13). Since proteins with these same carboxy-terminal extensions are effective in inducing the heat-shock response, the recognition that leads to induction must be independent, or comprise only one component, of the process of degradation. The Tn903 transposase, which is rapidly turned over in vivo in a Lon-dependent fashion, is also stabilized by fusion of β-galactosidase at the carboxy terminus (C. Derbyshire, M. Kramer, & N. Grindley, manuscript in preparation).

The ability of carboxy-termini to determine stability, even for an unfolded protein, seems one sensible way to allow the degradation of one class of abnormal proteins, those whose synthesis is interrupted before completion.

This reversal of Varshavsky's N-end rule for substrate recognition may only be pertinent to one class of proteases in *E. coli*, since a ClpA-LacZ fusion protein, carrying only the amino-terminal 42 amino acids of ClpA, is degraded in a Clp-dependent fashion, as is the ClpA protein itself.

ENERGY-INDEPENDENT PROTEOLYSIS

Although most of this review has concentrated on the energy-dependent cytoplasmic proteases of *E. coli*, many protease activities detected in extracts are in fact energy-independent (48, 98). Many of these are membrane-associated or periplasmic (48, 98, 108). Genetic identification of these activities is proceeding, and three proteases for which there is recent progress are discussed here. Those for which there is both genetic and biochemical evidence of protease activity are listed in Table 3.

Hfl and the Degradation of cII Protein

The lysis-lysogeny decision for bacteriophage lambda is regulated by the availability of cII protein, a positive regulator for both lambda repressor synthesis and Int synthesis. cII is a highly unstable protein, with a half-life of less than 2 minutes (54). Its instability seems to depend, at least in part, on the host functions coded for by the *hflA* and *hflB* loci. Loss-of-function mutations at either locus increase lambda lysogenization (*hfl:* *h*igh *f*requency *l*ysogenization) and increase the half-life of cII. Double mutants have an additive effect on cII degradation and lysogenization, suggesting the existence of two pathways for cII degradation (10, 63). Mutations in the phage cIII protein also have an effect on both cII degradation and lysogenization: cIII seems to act to protect cII from degradation.

The degradation of cII has recently been demonstrated in vitro, using purified proteins coded for by the *hflA* locus. This reaction is sensitive to serine protease inhibitors but does not require energy (22). Since this reaction is still significantly slower than that seen in vivo, some component of the reaction may be missing. A regulatory component could possibly convert this reaction to an efficient, energy-dependent one. The in vivo energy-dependence of cII degradation has not been examined.

In vivo, cII degradation can be inhibited by serine protease inhibitors in *hflA*$^+$ hosts; in *hflA*$^-$ hosts, the residual degradation of cII is sensitive to NEM. It is not yet clear whether the *hflB* products are responsible for the NEM-sensitive degradation of cII (63).

The dramatic effects of the *hfl* mutants and cIII on lysogenization (twenty-fold increase in lysogenization in *hflA* mutants and 100-fold decrease in lysogenization in *cIII* mutants) are reflected in only modest (twofold) effects on cII turnover (63). Given the very rapid turnover of this protein and the

Table 3 Some energy-independent proteases of *Escherichia coli*[a]

Protease	Genetic locus	Map position (min.)	Subunit size (daltons)	Localization	Substrates	Comments	References
HflA	hflK hflC	94.5	46,000 37,000	Cytoplasmic	cII	Turbid λ plaques	9, 22
Protease III (Pi)[b]	ptr	61	110,000	Periplasmic	Insulin, small proteins	Deletions are viable	23, 38, 138
Protease IV	sspA	n.d.[c]	67,000	Cytoplasmic membrane	Signal peptides		69
OmpT (Protease VII)[b]	ompT	12.5	35,500	Outer membrane	Paired basic residues	Active in extracts	58, 125, 137

[a] Only those proteases with demonstrated in vitro activity and identified structural gene(s) are listed.
[b] Alternate name.
[c] n.d.: not determined.

precise balance in which the lysis-lysogeny switch seems to reside, this difference may be sufficient. However, some experiments on cII turnover suggest that another important factor in the level of cII in the cell is the rapidity with which degradation starts (10). In hfl^- cells, which show high level lysogeny, there is a lag before degradation of cII begins; no such lag is seen under conditions leading to lytic growth (hfl^+). This lag may represent an essential processing step (for example, at the N-terminus; 10) before rapid proteolysis can commence; if so, the implication of the genetics is that cIII, $hflA$ and $hflB$ may all influence this rate-limiting step.

The rapid turnover of cII is reminiscent of the instability of sigma 32. It is interesting that cIII, which protects cII from degradation, also induces the heat-shock response and increases the half-life of sigma 32, even though hfl mutants do not affect heat-shock induction (7). This induction plays a peculiar role with respect to lambda physiology: although the heat-shock $groE$, $dnaJ$, K, and $grpE$ genes are all necessary for lambda replication and morphogenesis, these proteins are primarily necessary for phage committed to the lytic response. Presumably, cIII induction of these proteins is not essential for lytic growth, since cIII mutants grow well lytically. The protection by cIII of both sigma 32 and cII from degradation suggests that cIII may act as a general inhibitor of proteolysis.

A role for $hflA$ for E. coli itself has not yet been found: null mutants exhibit no growth defects (9). $hflA$ mutants have increased synthesis of a set of E. coli proteins detected by two-dimensional gel electrophoresis; some of these proteins are normally somewhat unstable (21). It is not yet clear whether these proteins represent unstable protein substrates of the Hfl system or intermediates in the degradation of unstable proteins.

Deg P. mutants of E. coli (133) fail to cleave a membrane-bound Tsr-PhoA fusion protein. $degP$ mutants also fail to degrade other membrane-associated fusion proteins but have no effect on cytoplasmic fusions. Insertion mutations in the same gene, called $htrA$ by Georgopoulos and coworkers, were isolated for their inability to grow at temperatures above 42°C (83). The $degP(htrA)$ gene codes for a 48,000 dalton membrane protein (82). While the synthesis of $degP$ increases sharply at high temperature, the increase is independent of the sigma 32 heat-shock regulator. Instead, the promoter of this gene may be recognized by sigma E, a high temperature regulator found to act at the $htpR$ promoter as well (82).

DegP may thus serve as a high temperature clean-up system for the membrane and periplasm. The temperature sensitivity of the mutations, which is increased when abnormal protein substrates are overproduced, may reflect the lethality of junk protein clogging the membrane (132). Alternatively, either a specific essentially cleaved substrate may exist, or DegP may help to generate amino acids for cell use.

Omp T

Several workers have observed the rapid processing of proteins in extracts of *E. coli*. Ada, UvrB, T7 RNA polymerase, φX174 head protein, and the enterobactin receptor, among others, all undergo relatively limited proteolysis (see, for instance, 17). Recent work has identified the protease responsible for this degradation as the outer membrane-associated Protease VII or OmpT product (58, 125, 137). The purified protein is able to cut between paired basic residues. Mutations in *ompT* grow well; the gene can be deleted from *E. coli*. Therefore, although the function of this protein is still unclear, mutations in it may be particularly useful to workers trying to purify unprocessed proteins from *E. coli* extracts.

CONCLUSIONS

Protein turnover and the specific degradation of both naturally unstable proteins and abnormal proteins are closely intertwined with the rest of the cell's regulatory circuitry. The ubiquitous heat-shock response seems closely tied to the production of abnormal proteins in the cell and to their destruction by degradation. The RecA protease serves to initiate the SOS response to DNA damage, both by destroying the LexA repressor and by activating the UmuD product. Lon also participates in recovery from the SOS response by degrading the cell division inhibitor SulA, and regulates capsular polysaccharide synthesis via degradation of RcsA. These proteolytic events are all initiated with energy-dependent cleavages. The existence of other energy-dependent proteases with rather different roles is suggested by the discovery of the anaerobically induced ClpA protease and its mysterious sibling, coded for by *orfBG*. While proteases can participate in the induction of some of these stress responses, as it does with RecA cleavage of LexA, an even more general role for protein degradation will likely be found in the recovery from many stress responses, when proteins made under one condition are troublesome under the recovery conditions. The role of Lon for recovery from SOS induction and the increase in sigma 32 degradation after induction of the heat-shock proteases may be two examples of this phenomenon. The mechanism of recognition of these energy-dependent proteases for their substrates still remains relatively obscure.

ACKNOWLEDGMENTS

I am grateful to all my scientific colleagues who generously provided unpublished information. I thank members of my laboratory and M. Maurizi for their comments on the manuscript and their many other contributions to both the science and my thinking about it. I thank Michael Gottesman for his comments on the manuscript and his tolerance. I would like to dedicate this review to my late colleage, collaborator and friend Olivier Huisman, who died in 1988. His enthusiasm for science and his wealth of ideas about new ways to approach the issue of proteolysis will be missed.

Literature Cited

1. Adler, H. I., Carrasco, A., Crow, W., Gill, J. S. 1981. Cytoplasmic membrane fraction that promotes septation in an *Escherichia coli lon* mutant. *J. Bacteriol.* 147:326–32
2. Allen, P., Hart, C. A., Saunders, J. R. 1987. Isolation from *Klebsiella* and characterization of two *rcs* genes that activate colanic acid capsular biosynthesis in *Escherichia coli. J. Gen. Microbiol.* 133:331–40
3. Ananthan, J., Goldberg, A. L., Voellmy, R. 1986. Abnormal proteins serve as eukaryotic stress signals and trigger the activation of heat shock genes. *Science* 232:522–24
4. Apte, B. N., Rhodes, H., Zipser, D. 1975. Mutation blocking the specific degradation of reinitiation polypeptides in *E. coli. Nature* 257:329–31
5. Bachmair, A., Varshavsky, A. 1989. The degradation signal in a short-lived protein. *Cell* 56:1019–32
6. Bachmann, B. J. 1987. Linkage map of *Escherichia coli* K-12, Edition 7. In *Escherichia coli and Salmonella typhimurium: Cellular and Molecular Biology,* ed. F. C. Neidhardt, J. L. Ingraham, K.B. Low, B. Magasanik, M. Schaechter, H. E. Umbarger. pp. 807–76 Am. Soc. Microbiol.: Washington, DC
7. Bahl, H., Echols, H., Straus, D. B., Court, D., Crowl, R., Georgopoulos, C. P. 1987. Induction of the heat shock response of *E. coli* through stabilization of sigma 32 by the phage lambda cIII protein. *Genes Dev.* 1:57–64
8. Baker, T. A., Grossman, A. D., Gross, C. A. 1984. A gene regulating the heat shock response in *Escherichia coli* also affects proteolysis. *Proc. Natl. Acad. Sci. USA* 81:6779–83
9. Banuett, F., Herskowitz, I. 1987. Identification of polypeptides encoded by an *Escherichia coli* locus (*hflA*) that governs the lysis-lysogeny decision of bacteriophage lambda. *J. Bacteriol.* 169:4076–85
10. Banuett, F., Hoyt, M. A., McFarlane, L., Echols, H., Herskowitz, I. 1986. *hflB*, a new *Escherichia coli* locus regulating lysogeny and the level of bacteriophage lambda cII protein. *J. Mol. Biol.* 187:213–24
11. Bartlett, D. H., Wright, M. E., Silverman, M. 1988. Variable expression of extracellular polysaccharide in the marine bacterium *Pseudomonas atlantica* is controlled by genome rearrangement. *Proc. Natl. Acad. Sci. USA* 85:3923–27
12. Bond, J. S., Butler, P. E. 1987. Intracellular proteases. *Annu. Rev. Biochem.* 56:333–64
13. Bowie, J. U., Sauer, R. T. 1989. Identification of C-terminal extensions that protect proteins from intracellular proteolysis. *J. Biol. Chem.* 264:7596–602
14. Brill, J. A., Quinlan-Walshe, C., Gottesman, S. 1988. Fine-structure mapping and identification of two regulators of capsule synthesis in *Escherichia coli* K-12. *J. Bacteriol.* 170:2599–611
15. Bukhari, A. I., Zipser, D. 1973. Mutants of *Escherichia coli* with a defect in the degradation of nonsense fragments. *Nature* 243:238–41
16. Burckhardt, S. E., Woodgate, R., Scheuermann, R. H., Echols, H. 1988. UmuD mutagenesis protein of *Escherichia coli:* Overproduction, purification, and cleavage by RecA. *Proc. Natl. Acad. Sci. USA* 85:1811–15
17. Caron, P. R., Grossman, L. 1988. Potential role of proteolysis in the control of UvrABC incision. *Nucleic Acids Res.* 16:10903–12
18. Casadaban, M. J., Cohen, S. N. 1980. Lactose genes fused to exogenous promoters in one step using a *Mu-lac* bacteriophage in vivo probe for transcriptional control sequences. *Proc. Natl. Acad. Sci. USA* 76:4530–33
19. Charette, M. F., Henderson, G. W., Doane, L. L., Markovitz, A. 1984. DNA stimulated ATPase activity of the Lon (CapR) Protein. *J. Bacteriol.* 158:195–201
20. Charette, M., Henderson, G. W., Markovitz, A. 1981. ATP hydrolysis-dependent activity of the *lon(capR)* protein of *E. coli* K12. *Proc. Natl. Acad. Sci. USA* 78:4728–32
21. Cheng, H. H., Echols, H. 1987. A class of *Escherichia coli* proteins controlled by the *hflA* locus. *J. Mol. Biol.* 196:737–40

22. Cheng, H. H., Muhlrad, P. J., Hoyt, M. A., Echols, H. 1988. Cleavage of the cII protein of phage lambda by purified HflA protease: control of the switch between lysis and lysogeny. *Proc. Natl. Acad. Sci. USA* 85:7882–86

23. Cheng, Y.-S. E., Zipser, D., Cheng, C.-Y., Roiseth, S. J. 1979. Isolation and characterization of mutations in the structural gene for protease III *(ptr.) J. Bacteriol.* 140:125–30

24. Chin, D. T., Goff, S. A., Webster, T., Smith, T., Goldberg, A. L. 1988. Sequence of the *lon* gene in *Escherichia coli:* A heat-shock gene which encodes the ATP-dependent protease La. *J. Biol. Chem.* 263:11718–28

25. Chung, C. H., Goldberg, A. L. 1981. The product of the *lon(capR)* gene in *Escherichia coli* is the ATP-dependent protease, protease La. *Proc. Natl. Acad. Sci. USA* 78:4931–35

26. Chung, C. H., Goldberg, A. L. 1982. DNA stimulates ATP-dependent proteolysis and protein-dependent ATPase activity of protease La from *Escherichia coli. Proc. Natl. Acad. Sci. USA* 79: 795–99

27. Corton, J. C., Ward, J. E. J., Lutkenhaus, J. 1987. Analysis of cell division gene *ftsZ (sulB)* from gram-negative and gram-positive bacteria. *J. Bacteriol.* 169:1–7

28. Costerton, J. W., Irvin, R. T., Cheng, K.-J. 1981. The role of bacterial surface structures on pathogenesis. *Crit. Rev. Microbiol.* 8:303–38

29. Craig, N. L., Roberts, J. W. 1980. *E. coli* recA protein-directed cleavage of phage-lambda repressor requires polynucleotide. *Nature* 83:26–29

30. Craig, N. L., Roberts, J. W. 1981. Function of nucleoside triphosphate and polynucleotide in *Escherichia coli* recA protein directed cleavage of phage-lambda repressor. *J. Biol. Chem.* 256: 8039–44

31. Craig, R. J., Arraj, J. A., Marinus, M. G. 1984. Induction of damage inducible (SOS) repair in *dam* mutants of *Escherichia coli* exposed to 2-aminopurine. *Mol. Gen. Genet.* 194:539–40

32. D'Ari, R., Huisman, O. 1982. DNA replication and indirect induction of the SOS response in *Escherichia coli. Biochimie* 64:623–27

33. Donachie, W. D., Robinson, A. C. 1987. Cell division: parameter values and the process. See Ref. 6, pp. 1578–93

34. Donch, J., Greenberg, J. 1970. Protection against lethal effects of ultraviolet light on *Escherichia coli* by capsular polysaccharide. *Mutat. Res.* 10:153–55

35. Donch, J., Greenberg, J. 1968. Genetic analysis of *lon* mutants of strain K-12 of *Escherichia coli. Mol. Gen. Genet.* 103: 105–15

36. Downs, D., Waxman, L., Goldberg, A. L., Roth, J. 1986. Isolation and characterization of *lon* mutants in *Salmonella typhimurium. J. Bacteriol.* 165: 193–97

37. Dudman, W. F. 1977. The role of surface polysaccharides in natural environments. In *Surface Carbohydrates of the Prokaryotic Cell,* ed. I. Sutherland, pp. 357–414. London: Academic

38. Dykstra, C. C., Kushner, S. R. 1985. Physical characterization of the cloned protease III gene from *Escherichia coli* K-12. *J. Bacteriol.* 163:1055–59

39. Flynn, J. A. L., Ohman, D. E. 1988. Cloning of genes from mucoid *Pseudomonas aeruginosa* which control spontaneous conversion to the alginate production phenotype. *J. Bacteriol.* 170:1452–60

40. Freudl, R., Braun, G., Honore, N., Cole, S. T. 1987. Evolution of the enterobacterial *sulA* gene: a component of the SOS system encoding an inhibitor of cell division. *Gene* 52:31–40

41. Gayda, R. C., Stephens, P. E., Hewick, R., Schoemaker, J. M., Dreyer, W., J., Markovitz, A. 1985. Regulatory region of the heat shock-inducible *capR (lon)* gene: DNA and protein sequences. *J. Bacteriol.* 162:271–75

42. Gayda, R. C., Yamamoto, L. T., Markovitz, A. 1976. Second-site mutations in *capR(lon)* of *Escherichia coli* K-12 that prevent radiation sensitivity and allow bacteriophage lambda to lysogenize. *J. Bacteriol.* 127:1208–16

43. George, J., Castellazzi, M., Buttin, G. 1975. Prophage induction and cell division in *E. coli*. III. *sfiA* and *sfiB* restore division in *tif* and *lon* strains and permit the mutator properties of *tif*. *Mol. Gen. Genet.* 140:309–32

44. Goff, S. A., Casson, L. P., Goldberg, A. L. 1984. Heat shock regulatory gene *htpR* influences rates of protein degradation and expression of the *lon* gene in *Escherichia coli. Proc. Natl. Acad. Sci. USA* 81:6647–51

45. Goff, S. A., Goldberg, A. L. 1987. An increased content of protease La, the *lon* gene product, increases protein degradation and blocks growth in *Escherichia coli. J. Biol. Chem.* 262:4508–15

46. Goff, S. A., Goldberg, A. L. 1985. Production of abnormal proteins in *E. coli*

stimulates transcription of *lon* and other heat shock genes. *Cell* 41:587–95

47. Goldberg, A. L. 1972. Degradation of abnormal proteins in *Escherichia coli*. *Proc. Natl. Acad. Sci. USA* 69:422–26

48. Goldberg, A. L., Sreedhara Swamy, K. H., Chung, C. H., Larimore, F. S. 1983. Proteases of *Escherichia coli*. *Methods Enzymol.* 80:680–702

49. Goldberg, A. L., Voellmy, R., Chung, C. H., Menon, A. S., Desautels, M., et al. 1985. The ATP dependent pathway for protein breakdown in bacteria and mitochondria. In *Intracellular Protein Catabolism*, pp. 33–45. New York: Liss

50. Goldberg, A. L., Waxman, L. 1985. The role of ATP hydrolysis in the breakdown of proteins and peptides by Protease La from *Escherichia coli*. *J. Biol. Chem.* 260:12029–34

51. Gonda, D. K., Bachmair, A., Wunning, I., Lane, W. S., Varshavsky, A. 1988. Universality and structure of the N-end rule of protein turnover. In *The Ubiquitin System: Current communications in Molecular Biology*, ed. M. J. Schlesinger, pp. 97–105. Cold Spring Harbor, NY: Cold Spring Harbor Lab.

52. Gottesman, S. 1987. Regulation by proteolysis. See Ref. 6, pp. 1308–12

53. Gottesman, S. 1989. Minimizing proteolysis in *E. coli*: Genetic solutions. *Methods Enzymol.*, In press

54. Gottesman, S., Gottesman, M., Shaw, J., Pearson, M. L. 1981. Protein degradation in *E. coli*: The *lon* mutation and bacteriophage lambda N and cII protein stability. *Cell* 24:225–33

55. Gottesman, S., Halpern, E., Trisler, P. 1981. Role of *sulA* and *sulB* in filamentation by *lon* mutants of *Escherichia coli* K-12. *J. Bacteriol.* 148:265–73

56. Gottesman, S., Trisler, P., Torres-Cabassa, A. S. 1985. Regulation of capsular polysaccharide synthesis in *Escherichia coli* K12: Characterization of three regulatory genes. *J. Bacteriol.* 162:1111–19

57. Gottesman, S., Zipser, D. 1978. The Deg phenotype of *Escherichia coli lon* mutants. *J. Bacteriol.* 133:844–51

58. Grodberg, J., Dunn, J. J. 1988. *ompT* encodes the *Escherichia coli* outer membrane protease that cleaves T7 RNA polymerase during purification. *J. Bacteriol.* 170:1245–53

59. Grossman, A. D., Burgess, R., Walter, W., Gross, C. 1983. Mutations in the *lon* gene of *E. coli* K12 phenotypically suppress a mutation in the sigma subunit of RNA polymerase. *Cell* 32:151–59

60. Grossman, A. D., Erickson, J. W., Gross, C. A. 1984. The *htpR* gene product of *E. coli* is a sigma factor for heat shock promoters. *Cell* 39:383–90

61. Grossman, A. D., Straus, D. B., Walter, W. A., Gross, C. A. 1987. Sigma 32 synthesis can regulate the synthesis of heat shock proteins in *Escherichia coli*. *Genes Dev.* 1:179–84

62. Howard-Flanders, P., Simson, E., Theriot, L. 1964. A locus that controls filament formation and sensitivity to radiation in *Escherichia coli* K12. *Genetics* 49:237–46

63. Hoyt, M. A., Knight, D. M., Das, A., Miller, H. I., Echols, H. 1982. Control of phage lambda development by stability and synthesis of cII protein: Role of the viral cIII and host *hflA*, *himA* and *himD* genes. *Cell* 31:565–73

64. Huisman, O., D'Ari, R. 1981. An inducible DNA-replication-cell division coupling mechanism in *E. coli*. *Nature* 290:797–99

65. Huisman, O., D'Ari, R., George, J. 1980. Further characterization of *sfiA* and *sfiB* mutations in *Escherichia coli*. *J. Bacteriol.* 144:185–91

66. Huisman, O., D'Ari, R., Gottesman, S. 1984. Cell division control in *Escherichia coli*: specific induction of the SOS SfiA protein is sufficient to block septation. *Proc. Natl. Acad. Sci. USA* 81:4490–94

67. Hwang, B. J., Park, W. J., Chung, C. H., Goldberg, A. L. 1987. *Escherichia coli* contains a soluble ATP-dependent protease (Ti) distinct from protease La. *Proc. Natl. Acad. Sci. USA* 84:5550–54

68. Hwang, B. J., Woo, K. M., Goldberg, A. L., Chung, C. H. 1988. Protease Ti, a new ATP-dependent protease in *Escherichia coli*, contains protein-activated ATPase and proteolytic functions in distinct subunits. *J. Biol. Chem.* 263:8727–34

69. Ichihara, S., Suzuki, T., Suzuki, M., Mizushima, S. 1986. Molecular cloning and sequencing of the *sppA* gene and characterization of the encoded protease IV, a signal peptide peptidase, of *Escherichia coli*. *J. Biol. Chem.* 261:9405–11

70. Johnson, B. F., Greenberg, J. 1975. Mapping of *sul*, the suppressor of lon in *Escherichia coli*. *J. Bacteriol.* 122:570–74

71. Johnson, C., Chandrasekhar, G. N., Georgopoulos, C. 1989. *Escherichia coli* DnaK and GrpE heat shock proteins interact both in vivo and in vitro. *J. Bacteriol.* 171:1590–96

72. Jones, C. A., Holland, I. B. 1985. Role of the SfiB (FtsZ) protein in division inhibition during the SOS response in *E. coli:* FtsZ stabilizes the inhibitor SfiA in maxicells. *Proc. Natl. Acad. Sci. USA* 82:6045–49

73. Kano, Y., Wada, M., Nagase, T., Imamoto, F. 1986. Genetic characterization of the gene *hupB* encoding the HU-1 protein of *Escherichia coli. Gene* 45:37–44

74. Katayama, Y., Gottesman, S., Pumphrey, J., Rudikoff, S., Clark, W. P., Maurizi, M. R. 1988. The two-component ATP-dependent Clp protease of *Escherichia coli:* purification, cloning, and mutational analysis of the ATP-binding component. *J. Biol. Chem.* 263:15226–36

75. Katayama-Fujimura, Y., Gottesman, S., Maurizi, M. R. 1987. A multiple-component, ATP-dependent protease from *Escherichia coli. J. Biol. Chem.* 262:4477–85

76. Keller, J. A., Simon, L. D. 1988. Divergent effects of a *dnaK* mutation on abnormal protein degradation in *Escherichia coli. Mol. Microbiol.* 2:31–41

77. Kohara, Y., Akiyama, K., Isono, K. 1987. The physical map of the whole *E. coli* chromosome: Application of a new strategy for rapid analysis and sorting of a large genomic library. *Cell* 50:495–508

78. Kroll, J. S., Maxon, E. R. 1988. Capsulation and gene copy number at the *cap* locus of *Haemophilus influenzae* Type b. *J. Bacteriol.* 170:859–64

79. Latchman, D. S., Chan, W. L., Leaver, C. E. L., Patel, R., Oliver, P., La Thangue, N. B. 1987. The human Mr90,000 heat shock protein and the *Escherichia coli* Lon protein share an antigenic determinant. *Comp. Biochem. Physiol. B* 87:961–67

80. Lazarides, E., Moon, R. T. 1984. Assembly and topogenesis of the spectrin-based membrane skeleton in erythroid development. *Cell* 37:354–56

81. Liberek, K., Georgopoulos, C., Zylica, M. 1988. Role of the *Escherichia coli* DnaK and DnaJ heat shock proteins in the initiation of bacteriophage lambda DNA replication. *Proc. Natl. Acad. Sci. USA* 85:6632–36

82. Lipinska, B., Sharma, S., Georgopoulos, C. 1988. Sequence analysis and regulation of the *htrA* gene of *Escherichia coli:* a sigma 32-independent mechanism of heat-inducible transcription. *Nucleic Acids Res.* 16:10053–67

83. Lipinska, B., Fayet, O., Baird, L., Georgopoulos, C. 1989. Identification, characterization, and mapping of the *Escherichia coli htrA* gene, whose product is essential for bacterial growth only at elevated temperatures. *J. Bacteriol.* 171:1574–84

84. Little, J. W. 1984. Autodigestion of LexA and phage lambda repressors. *Proc. Natl. Acad. Sci. USA* 81:1375–79

85. Little, J. W. 1983. The SOS regulatory system: Control of its state by the level of RecA protease. *J. Mol. Biol.* 167:791–808

86. Little, J. W., Mount, D. W. 1982. The SOS regulatory system of *Escherichia coli. Cell* 29:11–22

87. Lutkenhaus, J. F. 1983. Coupling of DNA replication and cell division: *sulB* is an allele of *ftsZ. J. Bacteriol.* 154:1339–46

88. Lutkenhaus, J., Sanjanwala, B., Lowe, M. 1986. Overproduction of FtsZ suppresses sensitivity of *lon* mutants to division inhibition. *J. Bacteriol.* 166:756–62

89. Maguin, E., Lutkenhaus, J., D'Ari, R. 1986. Reversibility of SOS-associated division inhibition in *Escherichia coli. J. Bacteriol.* 166:733–38

90. Markovitz, A. 1964. Regulatory mechanisms for synthesis of capsular polysaccharide in mucoid mutants of *Escherichia coli* K12. *Proc. Natl. Acad. Sci. USA* 51:239–46

91. Markovitz, A. 1977. Genetics and regulation of bacterial capsular polysaccharide biosynthesis and radiation sensitivity. In *Surface Carbohydrates of the Prokaryotic Cell,* ed. I. Sutherland, pp. 415–62. London: Academic

92. Matthews, W., Tanaka, K., Driscoll, J., Ichihara, A., Goldberg, A. L. 1989. Involvement of the proteasome in various degradative processes in mamalian cells. *Proc. Natl. Acad. Sci. USA* 86:2597–601

93. Maurizi, M. R., Katayama, Y., Gottesman, S. 1988. Selective ATP-dependent degradation of proteins in *Escherichia coli.* In *The Ubiquitin System. Current communications in Molecular Biology,* ed. M. J. Schlesinger, pp. 147–54. Cold Spring Harbor, NY: Cold Spring Harbor Lab.

94. Maurizi, M. R. 1987. Degradation in vitro of bacteriophage lambda N protein by Lon protease from *Escherichia coli. J. Biol. Chem.* 262:2696–703

95. Maurizi, M. R., Trisler, P., Gottesman, S. 1985. Insertional mutagenesis of the *lon* gene in *Escherichia coli: lon* is dispensable. *J. Bacteriol.* 164:1124–35

96. Menon, A. S., Goldberg, A. L. 1987. Protein substrates activate the ATP-dependent protease La by promoting nucleotide binding and release of bound ADP. *J. Biol. Chem.* 262:14929–34

97. Menon, A. S., Goldberg, A. L. 1987. Binding of nucleotides to the ATP-dependent protease La from *Escherichia coli*. *J. Biol. Chem.* 262:14921–28

98. Miller, C. G. 1987. Protein degradation and proteolytic modification. See Ref. 6, pp. 680–91

99. Mizusawa, S., Court, D., Gottesman, S. 1983. Transcription of the *sulA* gene and repression by LexA. *J. Mol. Biol.* 171: 337–43

100. Mizusawa, S., Gottesman, S. 1983. Protein degradation in *Escherichia coli:* The *lon* gene controls the stability of the SulA protein. *Proc. Natl. Acad. Sci. USA* 80:358–62

101. Mosteller, R. D., Goldstein, R. V., Nishimoto, K. R. 1980. Metabolism of individual proteins in exponentially growing *Escherichia coli*. *J. Biol. Chem.* 255:2524–32

102. Mount, D. W. 1980. The genetics of protein degradation in bacteria. *Annu. Rev. Genet.* 14:279–319

103. Nash, H. A., Robertson, C. A., Flamm, E., Weisberg, R. A., Miller, H. I. 1987. Overproduction of *Escherichia coli* integration host factor, a protein with nonidentical subunits. *J. Bacteriol.* 169:4124–27

104. Neidhardt, F. C., VanBogelen, R. A. 1987. Heat shock response. See Ref. 6, pp. 1334–45

105. Nishi, K., Schnier, J. 1988. The phenotypic suppression of a mutation in the gene *rplX* for ribosomal protein L24 by mutations affecting the *lon* gene product for protease LA in *Escherichia coli* K12. *Mol. Gen. Genet.* 212:177–81

106. Nohmi, T., Battista, J. R., Dodson, L. A., Walker, G. C. 1988. RecA-mediated cleavage activates UmuD for mutagenesis: Mechanistic relationship between transcriptional derepression and posttranslational activation. *Proc. Natl. Acad. Sci. USA* 85:1816–20

107. Ou, J. T., Baron, L. S., Rubin, F. A., Kopecko, D. J. 1988. Specific insertion and deletion of insertion sequence 1-like DNA element causes the reversible expression of the virulence capsular antigen Vi of *Citrobacter freundii* in *Escherichia coli*. *Proc. Natl. Acad. Sci. USA* 85:54402–05

108. Palmer, S. M., St. John, A. C. 1987. Characterization of a membrane-associated serine protease in *Escherichia coli*. *J. Bacteriol.* 169:1474–79

109. Parsell, D. A., Sauer, R. T. 1989. The structural stability of a protein is an important determinant of its proteolytic susceptibility in *E. coli*. *J. Biol. Chem.* 264:7590–95

109a. Parsell, D. A., Sauer, R. T. 1989. Induction of a heat shock-like response in *E. coli* by an unfolded protein depends on the level, not the degradation, of the protein. *Genes Dev.* In press

110. Pato, M. L., Reich, C. 1982. Instability of transposase activity: evidence from bacteriophage Mu DNA replication. *Cell* 29:219–25

111. Pelham, H. R. B. 1986. Speculations on the functions of the major heat shock and glucose-regulated proteins. *Cell* 46:959–61

112. Perry, K. L., Elledge, S. J., Mitchell, B. B., Marsh, L., Walker, G. C. 1985. *umuDC* and *mucAB* operons whose products are required for UV light- and chemical-induced mutagenesis: UmuD, MucA, and LexA proteins share homology. *Proc. Natl. Acad. Sci. USA* 82: 4331–35

113. Peterson, K. R., Wertman, K. F., Mount, D. W., Marinus, M. G. 1985. Viability of *Escherichia coli* K-12 DNA adenine methylase (*dam*) mutants requires increased expression of specific genes in the SOS regulon. *Mol. Gen. Genet.* 201:14–19

114. Phillips, T. A., VanBogelen, R. A., Neidhardt, F. C. 1984. *lon* gene product of *Escherichia coli* is a heat shock protein. *J. Bacteriol.* 159:283–87

115. Radke, K. L., Siegel, E. C. 1971. Mutation preventing capsular polysaccharide synthesis in *Escherichia coli* K-12 and its effect on bacteriophage resistance. *J. Bacteriol.* 106:432–37

116. Rechsteiner, M. 1987. Ubiquitin-mediated pathways for intracellular proteolysis. *Annu. Rev. Cell Biol.* 3:1–30

117. Reiss, Y., Kaim, D., Hershko, A. 1988. Specificity of binding of NH2-terminal residue of proteins to ubiquitin-protein ligase. *J. Biol. Chem.* 263:2693–98

118. Reitzer, L. J., Magasanik, B. 1987. Ammonia assimilation and the biosynthesis of glutamine, glutamate, aspartate, asparagine, L-alanine, and D-alanine. See Ref. 6, pp. 302–20

119. Roberts, D., Kleckner, N. 1988. Tn10 transposition promotes RecA-dependent induction of a lambda prophage. *Proc. Natl. Acad. Sci. USA* 85:6037–6041

120. Roberts, J. W., Roberts, C. W. 1975. Proteolytic cleavage of bacteriophage lambda repressor in induction. *Proc. Natl. Acad. Sci. USA* 72:147–51

121. Ronson, C. W., Nixon, B. T., Ausubel,

F. M. 1987. Conserved domains in bacterial regulatory proteins that respond to environmental stimuli. *Cell* 49:578–81

122. Rupprecht, K. R., Markovitz, A. 1983. Conservation of *capR(lon)* DNA of *Escherichia coli* K-12 between distantly related species. *J. Bacteriol.* 155:910–14

123. Sanderson, K. E., Hurley, J. A. 1987. Linkage map of *Salmonella typhimurium*. See Ref 6, pp. 877–918

124. Schoemaker, J. M., Gayda, R. C., Markovitz, A. 1984. Regulation of cell division in *Escherichia coli:* SOS induction and cellular location of the SulA protein, a key to *lon*-associated filamentation and death. *J. Bacteriol.* 158:551–61

125. Sedgwick, B. 1989. In Vitro proteolytic cleavage of the *Escherichia coli* Ada protein by the *ompT* gene product. *J. Bacteriol.* 171:2249–51

126. Shinagawa, H., Iwasaki, H., Kato, T., Nakata, A. 1988. RecA protein-dependent cleavage of UmuD protein and SOS mutagenesis. *Proc. Natl. Acad. Sci. USA* 85:1806–10

127. Shineberg, B., Zipser, D. 1973. The *lon* gene and degradation of β-galactosidase nonsense fragments. *J. Bacteriol.* 116:1469–71

128. Silverman, M., Simon, M. 1973. Genetic analysis of flagellar mutants in *Escherichia coli. J. Bacteriol.* 113:105–13

129. Simon, L. D., Tomczak, K., St. John, A. C. 1978. Bacteriophages inhibit degradation of abnormal proteins in *E. coli. Nature* 275:424–28

130. Skorupski, K., Tomaschewski, J., Ruger, W., Simon, L. D. 1988. A bacteriophage T4 gene which functions to inhibit *Escherichia coli* Lon protease. *J. Bacteriol.* 170:3016–24

131. Slilaty, S. N., Little, J. W. 1987. Lysine-156 and serine-119 are required for LexA repressor cleavage: A possible mechanism. *Proc. Natl. Acad. Sci. USA* 84:3987–91

132. Strauch, K. L., Johnson, K., Beckwith, J. 1989. Characterization of *degP,* a gene required for proteolysis in the cell envelope and essential for growth of *Escherichia coli* at hight temperature. *J. Bacteriol.* 171:2689–96

133. Strauch, K. L., Beckwith, J. 1988. An *Escherichia coli* mutation preventing degradation of abnormal periplasmic proteins. *Proc. Natl. Acad. Sci. USA* 85:1576–80

134. Strauch, K. L., Lenk, J. B., Gamble, B. L., Miller, C. G. 1985. Oxygen regulation in *Salmonella typhimurium. J. Bacteriol.* 161:673–80

135. Straus, D. B., Walter, W. A., Gross, C. A. 1987. The heat shock response of *E. coli* is regulated by changes in the concentration of sigma 32. *Nature* 329:348–91

136. Straus, D. B., Walter, W. A., Gross, C. A. 1988. *Escherichia coli* heat shock gene mutants are defective in proteolysis. *Genes Dev.* 2:1851–58

137. Sugimura, K., Nishihara, T. 1988. Purification, characterization, and primary structure of *Escherichia coli* protease VII with specificity for paired basic residues: Identity of protease VII and OmpT. *J. Bacteriol.* 170:5625–32

138. Swamy, K. H. S., Goldberg, A. L. 1982. Subcellular distribution of various proteases in *Escherichia coli. J. Bacteriol.* 149:1027–33

139. Takano, T. 1971. Bacterial mutants defective in plasmid formation: requirement for the *lon*⁺ allele. *Proc. Natl. Acad. Sci. USA* 68:1469–73

140. Tilly, K., McKittrick, N., Zylica, M., Georgopoulos, C. 1983. The dnaK protein modulates the heat shock response of *Escherichia coli. Cell* 34:641–46

141. Tilly, K., Spence, J., Georgopoulos, C. 1989. Modulation of stability of the *Escherichia coli* heat shock regulatory factor sigma 32. *J. Bacteriol.* 171:1585–89

142. Torres-Cabassa, A. S., Gottesman, S. 1987. Capsule synthesis in *Escherichia coli* K-12 is regulated by proteolysis. *J. Bacteriol.* 169:981–89

143. Torres-Cabassa, A., Gottesman, S., Frederick, R. D., Dolph, P. J., Coplin, D. L. 1987. Control of extracellular polysaccharide synthesis in *Erwinia stewartii* and *Escherichia coli* K-12: a common regulatory function. *J. Bacteriol.* 169:4525–31

144. Trempy, J. E., Gottesman, S. 1989. Alp: A suppressor of Lon protease mutants in *Escherichia coli. J. Bacteriol.* 171:3348–53

145. Trisler, P., Gottesman, S. 1984. *lon* transcriptional regulation of genes necessary for capsular polysaccharide synthesis in *Escherichia coli* K-12. *J. Bacteriol.* 160:184–91

146. Truitt, C. L., Haldenwang, W. G., Walker, J. R. 1976. Interaction of host and viral regulatory mechanisms: effect of the *lon* cell division defect on regulation of repression by bacteriophage lambda. *J. Mol. Biol.* 105:231–44

147. Walker, G. C. 1987. The SOS Response of *Escherichia coli.* See Ref. 6, pp. 1346–57.

148. Walker, J. R., Ussery, C. L., Allen, T. S. 1973. Bacterial cell division regula-

tion: lysogenization of conditional cell division mutants of *Escherichia coli* by bacteriophage lambda. *J. Bacteriol.* 111:1326–32

149. Ward, J. E. J., Lutkenhaus, J. 1985. Overproduction of FtsZ induces minicell formation in *E. coli. Cell* 42:941–49

150. Waxman, L., Goldberg, A. L. 1985. Protease La, the *lon* gene product, cleaves specific fluorogenic peptides in an ATP-dependent reaction. *J. Biol. Chem.* 260:12022–28

151. Waxman, L., Goldberg, A. L. 1986. Selectivity of intracellular proteolysis: protein substrates activate the ATP-dependent protease (La). *Science* 232: 500–3

152. Witkin, E. M. 1947. Genetics of resistance to radiation in *Escherichia coli. Genetics* 32:221–48

153. Woo, K. M., Chung, W. J., Ha, D. B., Goldberg, A. L., Chung, C. H. 1989. Protease Ti from *Escherichia coli* requires ATP hydrolysis for protein breakdown but not for hydrolysis of small peptides. *J. Biol. Chem.* 264:2088–91

154. Zehnbauer, B. A., Foley, E. C., Henderson, G. W., Markovitz, A. 1981. Identification and purification of the *lon*[+] (*capR*[+]) gene product, a DNA-binding protein. *Proc. Natl. Acad. Sci. USA* 78:2043–47

155. Zehnbauer, B. A., Markovitz, A. 1980. Cloning of gene *lon(capR)* of *Escherichia coli* K-12 and identification of polypeptides specified by the cloned deoxyribonucleic acid fragment. *J. Bacteriol.* 143:852–63

156. Zipser, D., Bhavsar, P. 1976. Missense mutations in the *lacZ* gene that result in degradation of β-galactosidase structural protein. *J. Bacteriol.* 127:1538–42

Annu. Rev. Genet. 1989. 23:199–225

HOMOLOGOUS RECOMBINATION IN MAMMALIAN CELLS

Roni J. Bollag, Alan S. Waldman[1], and R. Michael Liskay

Departments of Therapeutic Radiology and Human Genetics, Yale University School of Medicine, New Haven, Connecticut 06510

CONTENTS

[1]Current address: Walther Oncology Center and
                 Department of Biochemistry
                 Indiana University School of Medicine
                 Indianapolis, IN 46202

0066-4197/89/1215-0199$02.00

INTRODUCTION

Investigations of homologous recombination in cultured mammalian cells are providing insight into mechanisms for genetic rearrangement, as well as providing techniques for precisely manipulating the mammalian genome. This review focuses on three approaches used to study recombination mediated by cellular functions: (*a*) extrachromosomal recombination between transfected DNA molecules; (*b*) chromosomal recombination between repeated genes stably incorporated in the genome; and (*c*) targeted recombination between introduced DNA molecules and homologous sequences in the chromosome. Biochemical studies of homologous recombination in mammalian cells have been reviewed recently (22, 49, 51). Other classes of recombination events supported by mammalian cells also have been investigated, but are beyond the scope of this review. These recombination events include nonhomologous recombination (83), immunoglobulin gene rearrangement (27), and integrative recombination of retroviruses (97). In addition, the reader is referred to other reviews on in vivo studies of homologous recombination in mammalian cells (32, 50, 54, 108, 115).

EXTRACHROMOSOMAL RECOMBINATION

Extrachromosomal recombination (ECR) in mammalian cells has been studied extensively over the past decade. In contrast with studies of chromosomal recombination (see below), ECR systems allow results to be obtained rapidly, which in turn has generated a large body of data about these processes (50, 108). ECR can occur at high frequencies in mammalian cells, permitting the study of many independent recombination events. Another important asset of ECR systems is that the recombination substrates can easily be manipulated. Finally, ECR experiments elucidate how mammalian cells metabolize transfected DNA molecules, an issue important to the study of targeted recombination (see later section of this review).

There are, of course, some limitations to studies of ECR. Such work almost invariably involves the transfection of DNA into mammalian cells and, since observations are usually made within a few days of transfection, may be influenced by the transfection itself. In addition, some ECR studies are intended to shed light on the recombinational behavior of chromosomes. What is true for extrachromosomal sequences is not always true for chromosomal sequences, as is discussed below.

Recombination Schemes and Recombination Frequencies

ECR studies require the introduction of recombination substrates into the cultured cells. This is accomplished by any one of the common transfection procedures such as calcium phosphate/DNA coprecipitation (36), protoplast fusion (80), electroporation (76), or direct microinjection of DNA (17). To what extent each transfection procedure may influence ECR is not presently clear.

Early studies involved plaque assays as a way of monitoring the regeneration of lytic virus due to recombination between defective viral genomes (50, 108). Currently, the scheme most commonly employed involves monitoring the reconstruction, via recombination between two defective sequences, of a gene encoding a selectable marker. ECR between two defective Herpes simplex virus thymidine kinase *(tk)* gene sequences to produce a functional *tk* gene has been studied in mouse L*tk*⁻ cells (12, 13, 56, 93, 98). Recombination between defective *neo* gene sequences to produce a functional gene that confers resistance to the neomycin analog G418 has been extensively studied (5, 7, 30, 52, 79, 86, 87, 107). ECR schemes involving either nonreplicating or autonomously replicating DNA molecules have been reported. Both intramolecular and intermolecular recombination events have been monitored.

Systems using nonreplicating molecules commonly involve direct selection for expression of the reconstructed marker about 24 hours following transfection [although transient expression assays are sometimes used (129, 130)]. The appearance of colonies can take up to two weeks. Colony growth is dependent not only upon the initial recombination event that reconstructs the marker gene, but also upon subsequent stable integration and expression of the gene in the mammalian genome. ECR frequency is calculated as the number of colonies obtained due to recombination between two defective sequences divided by the number of colonies obtained when the cells are transfected with the wild-type marker gene. ECR frequencies ranging from 1–20% have been reported (30, 50, 52, 93, 98, 107, 108, 120). According to one report (30), when appropriate corrections are made for target sizes, it can be calculated that every transfected DNA molecule may undergo ECR.

When autonomously replicating plasmids are used, a more rapid screen for ECR is possible. The substrate is engineered so that the reconstructed marker is expressible and selectable in bacteria. In addition, the plasmid vector contains a second marker that is selectable in bacteria. Mammalian cells are transfected with the recombination substrate and low molecular weight DNA is isolated a few days later. This DNA is used to transform recombination-deficient *Escherichia coli* and selection for either the reconstructed marker gene or the second marker is applied. This type of ECR assay is not dependent upon integration of the reconstructed gene into the mammalian genome. ECR frequency is calculated as the number of bacterial colonies obtained following selection for the reconstructed marker divided by the number of colonies obtained following selection for the recombination-independent marker. Using such systems, recombination frequencies range from about 10^{-2}–10^{-4} for both intramolecular (86) and intermolecular events (5, 86). In this approach, care must be taken to demonstrate that recombination had actually occurred in the mammalian cells and not in the bacteria (5, 86, 87). It is difficult to demonstrate this unequivocally.

Direct Southern blotting analysis of DNA passaged through mammalian cells has also been used to discriminate between nonrecombined and recombined substrate (86). In these studies, the apparent intermolecular recombination frequency was reported to be $\sim 10^{-2}$. Other recombination strategies not involving genetic selection have been described (14).

It is noteworthy that ECR systems that require integration of the reconstructed marker into the mammalian genome yield recombination frequencies equal to or greater than those obtained with systems that do not require such integration. Wong & Capecchi (129) have suggested that ECR is coupled to the subsequent integration of the recombinant molecule. This suggestion was based upon the observation that as many as 50% of Rat-20 cells that had undergone an ECR event to reconstruct an *aprt* gene went on to stably integrate the gene (129). In contrast, integration of a transfected wild-type *aprt* gene occurred in only 5% of the cells that were *aprt*-positive in a transient expression assay. If ECR of non replicating molecules is indeed coupled to integration, then such ECR systems would likely yield an overestimation of ECR frequency.

Types of Recombination Events

A complex picture has emerged regarding the types of ECR events observed. Some groups have obtained results consistent with nonreciprocal events (gene conversions) as the predominant event (30), others have observed products consistent with both reciprocal and nonreciprocal events (7, 52, 86, 101, 120), while still others have made observations consistent with nonconservative events in which sequence information was lost during recombination (3, 19, 56, 91). Needless to say, individual cell clones, the precise DNA sequence used, transfection method, and the state of the substrate may all influence the outcome of any experiment. Unfortunately, there are no comparative reports that systematically assess the effect(s) of such variables. As discussed below, the various observations have suggested different models for recombination.

Timing of ECR

Using nonreplicating substrates, Folger et al (30) demonstrated that when two DNA molecules were microinjected separately into the nucleus of a mouse L cell, recombination between the two molecules could be detected only if the two injections were made within one hour of each other. These studies strongly suggest that ECR occurs rapidly. These studies also imply that at approximately one hour after entry into the nucleus, DNA molecules enter a state in which they are refractory to recombination. This state might involve interaction with the genome, or assembly into chromatin. Such notions are consistent with the observation that recombination between two chromosomal sequences proceeds at a rate several orders of magnitude lower than does ECR (see below). Also of interest is the observation that ECR is much more efficient than recombination between a transfected molecule and homologous chromosomal sequences (i.e. gene targeting, see below).

Another way of asking the question, "When does ECR occur?" is to determine whether such recombination is cell-cycle dependent. Reconstruction of a functional *aprt* gene via ECR of linear molecules in *Rat-2* cells was shown to peak in early to mid-S phase, at a frequency ($\sim 2 \times 10^{-2}$) about 15-fold higher than occurs in early G_1 (130). Interestingly, other work by the same group (128) has shown that random integration of transfected linear DNA into the genome of Rat-2 cells is not cell-cycle dependent. This suggests that the cellular mechanisms for ECR and random integration do not completely overlap.

Homology Requirements of ECR

The homology requirements of ECR were initially addressed by measuring recombination frequency as a function of the length of homology shared by two sequences. Rubnitz & Subramani (85) measured ECR rates by monitoring the production of a lytic SV40 genome generated by recombination between homologous sequences of varying length contained within a single replicating plasmid substrate. Recombination decreased linearly between 5243 bp and 214 bp of homology, dropped about tenfold between 214 and 163 bp, and then decreased linearly from 163 bp to 14 bp. Ayares et al (5) also observed such a biphasic relationship between intermolecular ECR frequency and homology length when nonreplicating substrates were used, although the point of inflection was determined to be between 400 and 300 bp. Interestingly, Ayares et al observed a strictly linear relationship between ECR frequency and homology length when the recombination substrates were allowed to replicate (5). In both cases, Ayares et al observed measurable recombination with as little as 25 bp of homology.

Biphasic kinetics is suggestive of two recombination mechanisms, one that operates on long homologies, another that operates on shorter homologies. The point of inflection is often considered the minimal amount of homology required for "efficient" recombination. Based on the work described above, this value appears to lie somewhere between 164 and 400 bp for ECR in mammals, considerably larger than analogous values in procaryotes (95, 96, 124).

An alternative way to address the homology requirements of recombination is to assess the effect that base-pair mismatch has on recombination rates. Using nonreplicating substrates, Waldman & Liskay (120) have demonstrated that evenly dispersed 19% nucleotide mismatch reduced both intra- and intermolecular ECR in mouse L cells about 3- to 15-fold relative to the rate of ECR between perfectly homologous sequences. These results were obtained when the DNA substrates were introduced into the cells by either calcium phosphate coprecipitation or by microinjection into nuclei, indicating that the mode of transfection did not profoundly influence these ECR studies. Curiously, 19% nucleotide mismatch was capable of reducing intrachromosomal recombination between closely linked sequences by a factor of greater than 1000 (120) (see below), suggesting that there are at least some mechanistic differences between extra- and intrachromosomal recombination. The observation that ECR is much less affected by heterology than is intrachromosomal recombination could reflect either distinct recombination pathways and/or the constraints that chromatin structure might impose upon recombination.

Double-Strand Breaks and Recombination Models

Based on yeast as a paradigm (74), the effect of double-strand breaks on ECR has been extensively studied. The general finding has been that the introduction of a double-strand break (by cleavage with a restriction enzyme) within, or very near, one of the homologous recombining sequences enhances the rate of ECR about tenfold (3, 7, 12, 13, 30, 50, 52, 56, 84, 86, 88, 91, 99, 101, 108, 120, 129). For intermolecular events, the placement of a break in both of the homologous recombining sequences has a synergistic effect and can increase recombination as much as 100-fold (3, 101, 119, 129). When a break is placed in only one homologous sequence, that sequence acts predominantly as the recipient of information in nonreciprocal exchanges, in those systems where such events are observable (101). In several studies (12, 13, 101, 129), breaks placed outside of the homologous recombining sequences (within vector sequences) had a substantially lesser effect on ECR.

In their simplest forms, single-strand invasion recombination models such as the Holliday and Meselson-Radding models (39, 67), appear inconsistent with the broken sequence acting as recipient in nonreciprocal exchanges. For such reasons, it has been argued (12, 13, 101, 129) that the effect of double-strand breaks suggests a gap-repair mechanism as originally described for yeast (109). In this model, a double-strand gap is initiated at the site of the double-strand break by the action of exonucleases, and then the gap is repaired by using a strand from the unbroken recombination partner as a template for DNA synthesis. Although distinct from the Holliday and Meselson-Radding models in the initiating steps, this model does invoke the formation and resolution of "Holliday intermediates" (39, 73, 109). Repair of a double-strand gap in transfected sequences via recombination with cotransfected sequences has been directly observed in cultured cells (7, 12, 13, 52, 88, 101).

Interestingly, some investigators have observed that breaks appropriately placed within vector sequences can have a significant stimulatory or inhibitory effect on ECR (3, 56, 58) and have suggested a mechanism involving "terminal pairing" (3, 56, 58, 91). In this model, the recombination substrate is digested in opposite directions by a $5'$-(or $3'$)-specific exonuclease acting at the site of the double-strand break. Exonucleolytic digestion continues bidirectionally until complementary strands of homologous sequences are exposed. The complementary strands then anneal to form a substrate for a nuclease that cuts DNA at the junction between paired and unpaired regions. The event is completed by a gap-filling process, or by replication. This type of process does not involve a true synapsis step (78), does not proceed via the classic Holliday intermediate (39), does not allow for gene conversions, and occurs with the obligatory loss of genetic information.

Consistent with such a nonconservative, destructive ECR mechanism was the observation by Chakrabarti & Seidman (19) that only one of two predicted products of an intramolecular ECR event between homologous sequences on a circular DNA molecule was ever recovered in monkey CV-1 cells. Seidman also rarely observed the formation of recombinant dimers resulting from intermolecular ECR between two circular molecules (91). Interestingly, Ayares et al (7) have reported on the rather efficient recovery of dimer molecules resulting from apparent conservative intermolecular ECR between replicating plasmids in COS cells. Perhaps the safest conclusion is that multiple ECR pathways operate in mammalian cells and that the multitude of variables from one study to the next favor different pathways. The isolation and characterization of mammalian recombination mutants would unquestionably aid in dissecting this multitude of pathways.

Single-Stranded Recombination Substrates

Rauth et al (79) presented evidence that a single-stranded DNA molecule containing a fragment of the *neo* gene could recombine extrachromosomally with a double-stranded autonomously replicating deletion derivative of the *neo* gene in a variety of mammalian cells. These authors also presented evidence that single-stranded DNA can directly integrate into a mammalian genome without first being converted to a double-stranded form (79). Whether the mechanisms responsible for recombination of single-stranded substrates are the same as those governing recombination between double-stranded substrates remains to be seen.

Related Extrachromosomal Metabolism of Transfected DNA

In addition to recombination, transfected DNA molecules are subject to other processes. It has been reported (53) that DNA molecules transfected by either calcium phosphate or DEAE-dextran suffer damage such as point mutations, deletions, and even insertion of host-cell sequences. Such damage has been observed in a variety of cell types and appears to occur independently of replication of the transfected molecule, shortly after the DNA enters the nucleus. Another report (118) indicates that the biologically active subpopulation of DNA molecules transfected by the DEAE-dextran method suffers about one double-strand break per 5–15 kb. This latter study also indicated that mammalian cells are very proficient at catalyzing DNA end-joining reactions. The efficiency of end-joining might approach 100%, does not require homology, and can occur between noncognate ends following processing of such ends by the addition or deletion of small numbers of nucleotides (83). Mammalian cells have also been shown to be proficient at correcting mismatched sequences when preformed heteroduplex DNA is introduced into cells (1, 6, 15, 31, 35, 38, 68, 125, 126).

ECR must occur in conjunction with, in competition with, or in spite of processes such as those mentioned above. It is clear that transfected DNA can be actively metabolized by mammalian cells in a variety of ways; what is not clear is the manner in which these complex processes intermesh to yield the final products of extrachromosomal recombination.

CHROMOSOMAL RECOMBINATION

Interchromosomal Recombination

Mitotic recombination between homologous chromosomes, here referred to as interchromosomal recombination, was first observed in *Drosophila* (104) and has been studied extensively in fungi (73). Initial studies designed to detect interchromosomal recombination in Chinese hamster ovary (CHO) cell hybrids (16, 82) as well as in human lymphocytes (34) were unsuccessful. Although linked heterozygous loci were found to segregate at detectable frequency, most segregations were due to chromosome breakage or loss, gene inactivation, or nondisjunction with chromosome duplication, rather than mitotic recombination. Likewise, no intragenic recombination at the X-linked *hprt* locus could be detected in CHO-cell hybrids (82, 110). However, Wasmuth & Vock Hall (123) were able to observe rare CHO-hybrid segregants that were consistent with mitotic recombination having occurred between the centromere and a marker locus. More recently, evidence for mitotic recombination between homologs in cultured murine lymphocytes has been reported (4, 46, 77). A role for recombination in exposing recessive alleles in

humans has been suggested by experiments tracing markers linked to retinoblastoma in tumor tissue from affected individuals (18).

Intrachromosomal Recombination

Intrachromosomal recombination (ICR) refers to genetic rearrangements
resulting from interactions between linked homologous sequences within a
chromosome. Extensive ICR studies in yeast (reviewed in 48) have provided a
paradigm for mammalian systems. Recombination can involve two modes of
exchange: reciprocal exchange (crossing over) and nonreciprocal exchange
(gene conversion). Gene conversion refers to a non-Mendelian segregation of
markers involving a unidirectional transfer of genetic information from one
molecule to its homologous partner. In fungi all products of meiotic exchange
can often be analyzed to allow definitive demonstration of nonreciprocality.
One hallmark of fungal gene conversion is a frequent association with reciprocal exchange of flanking markers (29, 73); this observation has led to
models for recombination in which nonreciprocal and reciprocal exchanges
are alternative outcomes to a single process (39, 67, 109). The frequency of
mitotic recombination in mammalian cells is sufficiently low that recombinants must be selected. The inability to recover all products of recombination makes rigorous proof of the mode of recombination impossible. In this
review, we have considered products that result in disruption of linkage of
flanking markers to involve reciprocal exchange, and those in which no
flanking marker exchange occurs to represent gene conversion. It must be
emphasized that .double reciprocal exchange between sister chromatids
following unequal pairing cannot be distinguished unequivocally from gene
conversion.

ICR has been suggested as a mechanism for generation of novel MHC class
I alleles $H\text{-}2k^{bm6}$ and $H\text{-}2k^{bm9}$ by mitotic gene conversion in germ cells of
inbred mouse strains (33). Unequal sister-chromatid exchanges have also
been suggested to result in rearrangements involving immunoglobulin gene
loci (44). However, experimental systems for ICR studies at natural loci have
not been reported. The advent of efficient means for stably integrating foreign
DNA into mammalian cells has allowed the study of ICR focusing on artificial
substrates that contain repeated selectable markers. These substrates are
designed such that reconstruction of a wild-type gene from two linked defective markers (recombination reporter genes *tk* or *neo;* see Table 1) is taken
to indicate recombination. In addition, most recombination substrates incorporate a separate dominant marker to facilitate selection for cells that have
integrated the substrate. These include xanthine guanine phosphoribosyl
transferase *(gpt), neo, tk,* and SV40 T antigen (Tag), as listed in Table 1.

The general assumption that the ICR substrate is integrated into the

Table 1 Rates of intrachromosomal recombination

Cell line	Reporter gene	Integration selection	Copy #	Rate ($\times 10^6$)	Types of products[d]	Orientation[e]	Reference
Ltk^-	tk	gpt	NR[a]	12–120[b]	all	d	62
Ltk^-	tk	neo	1	2.6–4.8	all	d	54, 64
Ltk^-	tk	neo	1	0.5–1.6	all	i	11
Ltk^-	tk	tk	NR	2–2.5[b]	all	d	59
3T6	neo	gpt	1–2	5–30[b]	all	d	99
3T6	neo	gpt	NR	0.13–1	all	d	107
3T6	neo	gpt	NR	0.023–0.43	all	i	107
CHO	tk	neo	1–2[c]	1.6–44.1	all	d	unpublished
Rat 3	tk	Tag	NR	5	SCE	d	105
Ltk^-	tk	neo	1–2[c]	0.8–1.7	GC	d	63
3T6	neo	gpt	2–5	0.66–2.7	GC	d,i	87

[a] Not reported
[b] Numbers represent frequencies (recombinations/cell) rather than rates (recombinations/cell/generation)
[c] Rates corrected for copy number
[d] All implies no restriction on recoverable products; GC, gene conversion; SCE unequal sister chromatid exchange
[e] These represent; d-direct; i-inverted; di-both.

chromosome is based upon several observations. For instance, stably inherited transferred genes have been shown experimentally to reside in chromosomes (75). These introduced sequences segregate faithfully to daughter cells in the absence of selection, as appropriate for chromosomal sequences. Furthermore, endogenous sequences flanking input DNA can be traced as junction fragments and these remain, in general, constant among subclones and recombinants.

Recombination Rates

Table 1 lists frequencies or rates of recombination from several ICR studies. Rates of recombination typically fall in the range of 10^{-6}–10^{-5} events/cell/generation. Direct quantitative comparison of studies is complicated by many factors. For instance, numbers of recombination substrate pairs vary from line to line. The relationship between copy number and recombination frequency is unclear, although for low copy numbers, a linear relationship appears to exist (55, 63). When corrected for copy number, studies with *tk* alleles in Ltk^- cells have shown no line-to-line variation, although similar studies in CHO cells revealed 20-fold rate variation among lines (R. J. Bollag, R. M. Liskay, unpublished observations). Significant line-to-line variation also has been reported for other cells (99, 107). Generally termed "position effects," this variation could reflect properties specific to the chromosomal region in which the substrates reside, such as chromatin structure and transcriptional activity, that has been shown to affect immunoglobulin gene recombination (10) and homologous recombination in yeast (111, 117). Alternatively, these effects could reflect damage to the substrate incurred during transfection (53, 92, 118), and/or perhaps clonal variation. Furthermore, as discussed below, the extent of homology shared by the interacting genes affects their ability to recombine. In addition, the distance between genetic markers varies from study to study, but the effect of this variation has not been systematically addressed. Ideally, a rate of recombination (recombinations per cell generation) is determined. However, many studies report frequencies (recombinations per cell), which may be misleading in the case of "jackpots," representing progeny of recombinants arising early during growth of a culture (65).

Substrates have been designed that allow recovery of the products of both reciprocal and nonreciprocal types of exchange (See Table 1: 11, 59, 62, 64, 99, 107), or that restrict viable products to those produced by either just reciprocal (105) or just nonreciprocal exchange (55, 61, 63, 87). Studies addressing the proportions of products generated by reciprocal and nonreciprocal exchanges have generally found a preponderance of gene conversions (11, 64, 87). In the following sections we discuss separately what is known about reciprocal and nonreciprocal exchanges in ICR.

Reciprocal Exchange

Reciprocal intrachromosomal interactions in somatic cells can be of two types: intrachromatid (between genes on the same DNA molecule) or sister chromatid (between genes on replicated daughter strands). Reciprocal exchange between direct repeats can result in deletion or duplication of one of the interacting genes, although only one product is generally recoverable (see Figure 1a). Recombinations generating deletions can involve either intrachromatid or sister-chromatid exchange (Figure 1a, pathways ii and iii) and account for 15–20% of all events between *tk* alleles in mouse L*tk*⁻ cells (54, 64). Crossovers leading to gene duplications most likely involve sister chromatids (Figure 1a, pathway iii) and have been observed in studies with duplicated *neo* genes (99), and with *tk* alleles (105; unpublished results).

For inverted repeats (Figure 1b), intrachromatid flanking marker exchange should result in inversion of the intervening sequence and no change in gene-copy number (11, 107). Recombination generating inversions have provided evidence for gene conversion accompanying reciprocal exchange, indicating that these modes of recombination are mechanistically related (11). Such inversions must involve intrachromatid interactions (Figure 1b), while reciprocal exchanges leading to increases in gene-copy number (such as duplications) must involve unequal sister-chromatid pairing. In recent experiments designed to determine the relative contributions of these types of interactions, sister-chromatid exchanges appear to predominate in mouse cells

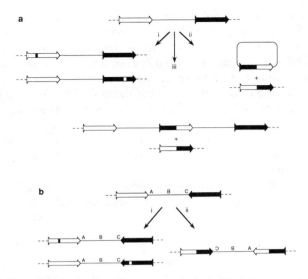

Figure 1   (a) Recombination between direct repeats. Pathway i leads to gene conversion either of gene (□) or of gene (■), generating recombinants in which one gene has received information from the other, but the donor gene remains unchanged (or segregates to the sister cell in the case of a sister chromatid interaction). Pathway ii represents intrachromatid reciprocal exchange, generating a circular product (which is generally lost) and a chromatid with a single hybrid gene. Exchange between sister chromatids following unequal pairing (pathway iii) can generate the identical single gene chromatid, as well as a chromatid with three gene products, the central gene being hybrid. In selective systems such as those reported in the text, which products of reciprocal exchange (pathways ii and iii) are recovered depends on the type of interaction and on the orientations of mutations in the genes. Products of ii and iii may also be generated by sister chromatid conversion as described (47).

(b) Recombination between inverted repeats. Pathway i is analogous to i in Figure 1a. Pathway ii represents intrachromatid reciprocal exchange, generating a chromatid in which the sequence between the points of exchange is inverted. Exchange between sister chromatids following unequal pairing leads to aberrant chromatids that are inviable (see 11).

(R. J. Bollag, R. M. Liskay, unpublished results). Interestingly, this is in contrast to studies in yeast, where intrachromatid products predominate (47).

Gene Conversion

Recombination in the absence of flanking marker exchange is consistent with a nonreciprocal mode of recombination. Such exchanges, termed gene conversions in this review, appear unidirectional, as the donor gene remains unchanged (Figure 1a and b, pathway i). Intrachromosomal gene conversion appears to involve the transfer of contiguous tracts of information (63, 103) of varying lengths (63, 121) from one sequence to another. For gene duplications of 1–2 kb conversion tracts are often less than 100 bp, but can cover as much as 358 bp (63). Essentially all studies involving full-length genes have suggested a significant role for gene conversion (11, 59, 62, 64, 99, 107), including *tk* gene studies, in which greater than 80% of recombinants involved apparent gene conversion (11, 64).

Although formal proof for gene conversion in mammalian cells remains elusive, several lines of evidence are strongly supportive. Since single reciprocal exchanges are less frequent than apparent gene conversions, it is unlikely that double reciprocal exchanges between sister chromatids would occur with sufficient frequency to explain the majority class of products. If double exchanges were predominant, then intrachromatid double exchanges should be common, since single intrachromatid exchanges occur at frequencies comparable to other types of reciprocal exchange (11, 107). However, recombinants resembling intrachromatid double exchanges are extremely rare, comprising at most 1% of all recombinations (11, 55). As mentioned above, gene conversion can accompany reciprocal recombination during intrachromatid exchange, an interesting effect since a close association between gene conversion and reciprocal exchange exists for fungal meiotic recombination (29, 73).

Homology Requirements of ICR

The results of a systematic evaluation of homology requirements revealed a linear decrease in recombination rate with decreasing length of shared homology from 2 kb down to 295 bp (61). At 200 bp, rates decreased by more than sevenfold and no recombinants were detected with only 95 bp of shared homology. Thus, recombination efficiency shows a marked reduction between 200–295 bp of shared homology. ECR studies revealed a similar dropoff at about 200 bp of homology, but recombination was detectable even with much shorter stretches of homology (5, 85).

Studies by Waldman & Liskay (120) demonstrated that the 19% of base-pair heterology between *tk* genes from HSV strains 1 and 2 reduced ICR at least 1000-fold relative to crosses with two strain 1 sequences. As discussed previously, this dramatic reduction was not observed during ECR (120). In a subsequent study, it was found that perfect homology adjacent to the heterologous sequences could substantially relieve the recombination block (121). These results were interpreted to mean that the heterology affected recombination initiation, and that, once initiated, recombination events could propagate through considerable heterology. Further related studies showed that two single base-pair mismatches interrupting the longest stretch of shared homology reduced ICR about 20-fold, whereas four mismatches clustered in an adjacent shorter region of homology had little effect (121). These findings led to the conclusion, first reached for bacteria (95, 124), that perfect stretches of homology rather than overall homology govern the efficiency of recombination. Furthermore, based on these studies the minimal length of perfect homology required for efficient recombination in mammalian cells appears to be between 134–232 bp (121).

Effect of Mutation Size on Recombination

The types of mutations used in recombination substrates include linker insertions, truncations, and internal deletions. In a systematic study, Letsou & Liskay (55) found that mutations of 1 bp, 8 bp or 16 bp (including base substitutions, deletions or insertions), were converted at a similar rate, about 10^{-6}. For larger lesions, rates were roughly inversely proportional to the size of the mutation; 100 bp and 1.5 kb insertions exhibited reductions of 10-fold and 100-fold, respectively, compared to recombination rates for smaller lesions. A single base-pair insertion was converted with 3- to 4-fold greater efficiency than a single base-pair deletion at the same site, suggesting a role for heteroduplex repair (55). Weiss & Wilson (125) found that preformed heteroduplex molecules with varying size deletions (2–200 bp) were all repaired at similar rates. Taken together with the results of Letsou & Liskay (55), this observation suggests that larger mutations affect the establishment of heteroduplex DNA intermediates rather than actual correction. In a study involving *neo* gene constructs, no statistically significant rate differences for deletions up to 167 bp were detected, possibly due to large variance in individual determinations (88).

Accuracy of Recombination

Studies of gene targeting (25, 112; see following section) and of extrachromosomal heteroduplex correction (126) have provided evidence that recombination-related processes are not always accurate. However, there is no parallel evidence for inaccuracy in ICR. As a first approximation, molecular hybridization analysis of hundreds of ICR recombinants has yielded data consistent with fidelity of ICR, in that wild-type restriction patterns are regenerated (11, 55, 59, 61–64, 87, 88, 99, 103, 105, 107, 121, 122). However, selection per se may prevent recovery of products of events that proceed with error. In a more refined analysis, DNA sequencing of 20 convertants revealed no mutations in over 5000 total bases sequenced, including over 2400 bases of intron sequence that should tolerate numerous mutations (103). Thus, even upon closer scrutiny, intrachromosomal gene conversion appears to proceed with fidelity.

An issue related to fidelity of recombination is whether the process is conservative. In certain ECR studies, the inability to recover one of the two recombining genes was interpreted as evidence for a nonconservative mechanism (19, 91). In contrast, ICR appears to be primarily conservative, since the ability to recover reciprocal recombinants between inverted chromosomal repeats requires that the process be conservative (11, 107).

Other Aspects of Intrachromosomal Recombination

Although limited genetic evidence for the involvement of heteroduplex DNA in ICR exists (55), formal proof is lacking. In physical studies of density-labeled cell DNA, evidence for a small fraction of hybrid DNA was found (69). Furthermore, a number of studies have demonstrated the ability of mammalian cells to correct a variety of mismatches extrachromosomally (1, 6, 15, 31, 35, 38, 68, 125, 126).

Using the ICR systems and lines described above, several studies have evaluated the ability of DNA-damaging agents to induce or alter recombination properties. Among the agents tested and shown to have a possible enchancing effect on recombination were: mitomycin C (59, 110, 122); UV light (122), and BPDE (a metabolite of benzo(a)pyrene) (122). UV light was shown to stimulate intermolecular extrachromosomal recombination between copies of Herpes simplex strains (23) and between *tk* alleles (71). Ionizing radiation was not found to have a detectable effect on ICR (122). These approaches may prove useful in evaluating the types of DNA modifications that play a role in initiating recombination. In addition to such physical and chemical agents, it may be of value to explore how the expression of gene products involved in DNA metabolism, such as polymerases, recombination enzymes, and nucleases that produce more specific DNA damage, might influence recombination.

Mutants

To date no known mammalian mutants exist that display an effect on homologous recombination between chromosomal sequences. Direct selection for mutants defective for recombination is difficult, due to the low overall frequency of recombination. Among the more promising approaches, based primarily upon analogy to yeast, bacteria, and flies, is the screening of mammalian cell mutants that were initially selected for hypersensitivity to DNA-damaging agents for a recombination phenotype (recently reviewed by Thompson, 115). Among the interesting candidates are the human syndromes in which affected individuals display enhanced sensitivity to radiation, induced chromosome breakage or rearrangement, and/or predisposition to cancer. Bloom's syndrome, for instance, has been of interest because of the observation of elevated levels of sister-chromatid exchanges (115), and a recent finding of defects in DNA ligase I in these cells (20, 127). Other diseases with relevance to recombination may include xeroderma pigmentosum, ataxia telangiectasia, Fanconi anemia, and Cockayne syndrome. In addition to these human-cell mutants, there are increasing numbers of mammalian-cell mutants selected for enhanced sensitivity to UV light, X-irradiation, chemical agents and various inhibitors of DNA metabolism. Detectable effects on ECR have been observed in CHO lines EM9 (40) and *xrs* mutants (37, 70), although there have been no reports of effects on ICR.

GENE TARGETING

The modification of chromosomal loci by homologous recombination is commonly referred to as gene targeting. The types of genetic manipulations in fungi made possible by targeted recombination have inspired investigators working with mammalian systems to pursue the development of gene disruption and correction technology. Gene-targeting procedures offer the potential to: (*a*) enhance studies of gene structure and function; (*b*) provide animal models for human genetic disease; and (*c*) provide an approach for human gene therapy. Despite the ability of mammalian cells to support and catalyze efficient homologous recombination between extrachromosomal DNA or between repeated chromosomal sequences, incoming DNA is subject most often to random integrations. The remarkable ability of mammalian cells to support nonhomologous recombination of incoming DNA has been discussed by Roth & Wilson (83). The aim of this section is to describe the development of gene-targeting strategies.

Targeted Modification of Artificial Loci

Most of the initial demonstrations of homologous recombination between transfected DNA and genomic sequences involved selection for reconstruction of introduced defective targets (57, 99, 102, 114). Frequencies of targeting to an integrated target sequence were 10^{-3}–10^{-7} of transfected cells. The approximate ratio of homologous to nonhomologous recombination products was 10^{-2}–10^{-5}. A common theme in targeting studies has been the use of input DNA linearized in a region of homology with the target as a means to enhance targeting (2, 24–26, 41, 43, 57, 90, 100, 102, 113, 114, 116, 131). This strategy is based upon findings in fungi that such linearization can increase targeting up to 1000-fold (28, 73, 74) and on studies of ECR in mammalian cells showing that such breaks stimulate recombination (see previous section).

Line-to-line variation in targeting efficiency, generally attributed to genomic position effect, has been observed in certain studies (57, 102) but not in others (99, 114). In one study with human cells, the ratio of homologous to nonhomologous events varied from 1 : 500 to 1 : 76, depending on the cell line used (102). These differences could result from variation between cell types, the properties of the recombination substrates and/or the method of introducing the DNA. Indeed, line-to-line variation in intrachromosomal recombination is seen in CHO cells but not in mouse L cells with identical recombination substrates (54; and our unpublished observations).

Evidence for three classes of products of targeting have been observed: (*a*) correction of the target by double crossover or by gene conversion (see Figure 2a), (*b*) integration of the plasmid to form a duplication (see Figure 2b), and

(c) correction of the input sequence followed by random integration elsewhere. A noteworthy point from most targeting studies (2, 24–26, 41, 43, 57, 90, 99, 100, 102, 113, 114, 116) is the general lack of nonhomologous integrations in the targeted cells.

Curiously, a high frequency of homologous interactions involving a particular amber mutation of *neo* resulted in unexpected second-site suppressor mutations in the target (60, 112, 114). Because these error-prone events appeared to require interaction between mismatched sequences, an error-prone heteroduplex repair process was proposed (112). In a separate targeting study with *neo*, errors were not observed (102).

Most studies of homologous recombination between an exogeneous sequence and its genomic counterpart have emphasized alteration of the genomic sequences. Several systems have been designed to specifically detect information flow from the chromosome onto an extrachromosomal element (42, 94, 106). In these studies the common theme was rescue of a defective incoming virus via recombination with viral sequences residing in the genome. Although the determination of frequencies was not as straightforward as in the "chromosome-target" systems, estimates ranged from 1% to as high as 25% of the transfected cells. Such systems are useful not only for rescuing mutations out of the genome but also should help to determine better means for altering the genome by homologous recombination.

Targeted Modification of Natural Loci

The first report of targeting to a natural locus (β-globin) was by Smithies et al (100). The system was a screening method designed to detect targeting at "nonselectable" loci. Insertion at the β-globin locus occurred in about 1 in 1000 transformed cells. Modification at the β-globin locus was successful in both β-globin expressing and nonexpressing cells. The insertions appeared to be simple one-copy insertions and no nonhomologous insertions of plasmid sequences were seen at other genomic sites. Later studies on these targeted lines indicated normal expression of β-globin at the targeted sites (72).

Targeted modification at natural loci using selection in established cell lines has been reported for the *aprt* gene (2), and for immunoglobulin genes (8, 9). The ratios of targeting to nonhomologous integrations were approximately 1 : 1000.

Further Refinements in Gene Targeting

Gene targeting in mouse embryonic stem (ES) cells offers a powerful genetic tool because these cells can be introduced into blastocysts and can contribute to the germ line (32). The first locus to be modified in ES cells, the X-linked *hprt* locus, is a good test system for several reasons: (*a*) it is present in only one active copy in these diploid cells; (*b*) there are strong positive and negative selections for expression; (*c*) *hprt* deficiency in humans causes severe mental retardation and has been a prime candidate for gene therapy.

Doetschman et al were able to correct a deletion mutation of *hprt* in ES cells at an average frequency of 1.4×10^{-6} per transfected cell (24). The incoming plasmid, containing 2.5–5 kb of *hprt* sequence, was cut at a unique site in the *hprt* sequence, transfected into *hprt*$^-$ ES cells, and selection for *hprt*$^+$ was imposed. In two cases the corrections resulted from a reciprocal exchange (insertion) and in three cases the products were consistent with crossover plus a gene-conversion event (24). No extraneous vector insertions were observed in five cases of targeting. Results of a related study demonstrated correction of the same deletion mutant at frequencies of approximately 10^{-7} (116). Twenty-eight of thirty lines selected for *hprt* correction contained the expected insertion at the *hprt* locus. Two lines had small deletions of the vector sequence. Significantly, in this study (116) the ability of the targeted ES cells to populate the germ line of chimeric mice was demonstrated.

Disruption of *hprt* in ES cells has been reported by two groups (25, 113). Thomas & Capecchi (113) disrupted the locus equally efficiently with two classes of vectors: (*a*) replacement vectors that result in placement of the *neo* gene into the eighth exon of *hprt* (see Figure 2a), or (*b*) insertion vectors that create a partial duplication of defective *hprt* sequences (see Figure 2b). The ratio of homologous to nonhomologous events was approximately 1:1000. A strong homology-dependence of targeting was observed. Doubling the size of the *hprt* segment (5.4kb vs 9.0kb) resulted in a 10- to 20-fold greater targeting frequency. No additional sequence changes were observed in the region of the disruptions. Because the *neo* based vector of Thomas & Capecchi (113) was optimized for transformation, efficient expression at a large number of genomic sites was likely. As one means for enriching for targeting,

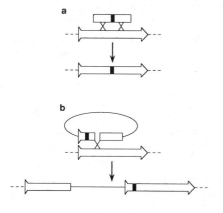

Figure 2 Two general targeting schemes: (a) Replacement—A portion of the input sequence interacts with the target via a double crossover or gene conversion. Information is transferred to the target without other changes in the genome. (b) Insertion—A plasmid carrying a portion of the target, cleaved by a restriction enzyme in a region of homology with the target, recombines via the double-strand break resulting in insertion of the plasmid and a partial duplication of the target. In (a) and (b) (■) represents a region of heterology between input and target sequence.

these authors suggested the use of a crippled *neo* gene that should function only after a targeting event.

Targeted disruption at *hprt* using selection with a promoterless *neo* gene has been reported for ES cells (25). Six of nine thioguanine-resistant lines (*hrpt⁻*) represented targeting events giving a frequency of approximately 2.5 $\times$ 10^{-5} per surviving cell. In two of six targeted lines, deletions of 14 and 37 base pairs were seen adjacent to the targeted insertion. Although the amount of *hprt* homology used by Doetschman et al (25) was less than in the Thomas & Capecchi study (113), the targeting frequency in the former study was higher. Possibly factors other than homology, such as the different targeted regions of *hprt,* or culture conditions, were responsible for the different targeting efficiencies.

Jasin & Berg (41) used an enhancerless *gpt* gene to obtain an estimated 100-fold enrichment for targeted integrations without direct selection at the target locus. Homologous integrations accounted for 44% of *gpt⁺* cells. A stimulation of targeting was achieved by linearizing the input DNA in a region of homology. An enrichment scheme by Sedivy & Sharp (90) used a *neo* gene fusion that required targeted recombination to restore reading frame and *neo* function. The enrichment factor for insertion and replacement vectors was estimated at 100-fold. The ratio of targeted to random events was between 1:2500 and 1:100. The more favorable ratio was achieved using a gel-purified homologous DNA fragment.

A different method for reducing the background of nonhomologous events, termed "positive/negative selection" (66), involves the use of two active selection elements: (*a*) a positive one, the *neo* gene, that can be expressed following either random integration or targeting, and (*b*) a negative element, the HSV *tk* gene, that can only be expressed following random integration. The original vector contained a portion of the *hprt* gene disrupted by *neo*, plus the HSV *tk* gene positioned in the vector so that it should integrate only by nonhomologous recombination. A nucleoside analog, gancyclovir, which is preferentially toxic to cells expressing HSV *tk,* was used to reduce the background of nonhomologous recombination. The *hprt* gene was disrupted in the predicted manner in 19 out of 24 ES cells coselected to be G418ʳ and gancyclovirʳ. Importantly, there was *no* direct selection against *hprt* expression. Without the gancyclovir-negative selection, the frequency of *hprt* targeting amongst G418ʳ colonies was only 1 per 1000 (113).

The *int-2* proto-oncogene also was disrupted using positive/negative selection (66). Four of eighty-one G418ʳ, gancyclovirʳ lines represented disruption of one autosomal copy of *int-2*. The absolute targeting frequency for *int-2* was approximately 20-fold lower than for *hprt,* possibly the result of a lower level of expression of *int-2*. *Int-2* expression, based on mRNA levels determined by Northern analysis, was 20- to 50-fold lower than *hprt* expression in the ES cell line used (M. Capecchi, personal communication). Capecchi and coworkers also successfully disrupted two homeobox loci in ES cells (*hox* 1.2 and 1.3) at frequencies similar to *hprt* (personal communication). In accordance with the notion that more actively transcribed sequences might be more amenable to targeting, the levels of transcription of hox 1.2 and 1.3, again based on Northern analysis, were similar to that of the *hprt* gene (M. R. Capecchi, personal communication). In yeast, a correlation between recombination rates and transcriptional activity has been reported (111, 117). Furthermore, there is evidence in mammalian cells that actively transcribed immunoglobulin-gene sequences are more likely to undergo rearrangement (10).

An approach for detecting rare targeting events at nonselectable loci uses the "polymerase chain reaction" (PCR), a method for specifically amplifying regions of DNA (89). Expeditious positioning of oligonucleotides allows amplification of unique recombinant junction fragments expected in targeted cells. Various means of screening for the targeted cells are subsequently employed. According to one study, it should be possible to detect targeting events in as few as 1 cell in 10,000 (45). Without the use of any selective procedures, Zimmer & Gruss (131) have used a PCR-based protocol to detect disruptions of the homeobox 1.1 gene in ES cells. Approximately 1 in 150 microinjected ES cells appeared to represent targeted disruptions. These investigators estimated a remarkably high ratio of homologous to nonhomologous events of 1 in 30. In these studies the size of the heterologous insert was only 20 base pairs, which is substantially less than the $\geqslant$1kb represented by the *neo* gene insert used in other studies (25, 43, 113). Perhaps differences in size of the disruption element can affect targeting (32, 113, 131). Indeed, systematic studies of intrachromosomal recombination in mouse cells indicate that large mutations are more difficult to correct (55). Another possibility is that microinjection per se accounts for more efficient targeting. It should be emphasized that the procedure of Zimmer & Gruss (131) has the advantage of not requiring expression of a selectable marker, e.g. *neo*, at the target locus.

The *en-2* gene locus of ES cells has been disrupted by a strategy that used the selectable marker *neo* and PCR (43). One in approximately 300 *neo*-expressing cells were disrupted at *en-2*. The absolute frequency of targeting was 10^{-7}. A related strategy involving a promoterless *neo* gene in conjunction with PCR has been used to target to an SV40 viral gene array in a human/mouse-hybrid cell line (26).

SUMMARY AND FUTURE PERSPECTIVES

Most attempts to overcome the propensity of mammalian cells to nonhomologously integrate introduced DNA have resorted to eliminating the nontargeted cells from analysis. Another approach involves the development of strategies to specifically inhibit nonhomologous events. Wilson and coworkers have conducted a series of studies designed to elucidate the nature of nonhomologous recombination between transfected DNA (for a review see 83). Of particular note, Chang & Wilson (21) found that the addition of dideoxy nucleotides to the recessed termini of DNA molecules could enhance homologous recombination 6- to 7-fold relative to nonhomologous events. These studies measured end-joining of extrachromosomal DNA molecules. The relevance of these end-joining reactions to the integration of DNA into the genome has been discussed (83). It is hoped that such modifications might inhibit nonhomologous integrations of input DNA, and thereby channel these molecules into homologous recombination (21, 83).

Evidence from certain studies casts some doubt on the idea that the inhibition of nonhomologous recombination should increase the "pool" of molecules available for homologous recombination and thereby raise the frequency of targeting events. Two different groups have demonstrated that the frequency of homologous interactions between the input DNA and its chromosomal copy is independent of the number of molecules microinjected (81, 114). In the simplest interpretation, these findings suggest that the chance encounter of incoming DNA with its chromosomal target is not rate-limiting.

In the schemes discussed above, the input DNA is available for targeting only transiently. The use of replicating vectors offers the potential to expand the interaction period that could contribute to an increase in the targeting efficiency. A replicating vector derived from the bovine papilloma virus has been used to detect targeting in mouse 3T6 fibroblasts (D. Strehlow & P. Berg, personal communication).

Other avenues for increasing targeting efficiencies could include further modification of the input DNA. Based on the finding that a modest increase in homology can produce a substantial increase in targeting efficiency (113), the use of larger pieces of genomic DNA as input, e. g. $\geqslant$30 kb, seems reasonable. The use of recombination activities from a mammalian source in conjunction with the input sequences could be of value. Finally, manipulation of the target-cell population might be beneficial. Synchronization of cells so as to maximize homologous recombination (128, 130) is one potential approach.

At present, nonrandom integrations of incoming DNA in mammalian cells are the rule in all test systems reported. However, homologous recombination at "nonselectable" loci is now possible, albeit at low absolute frequency, 10^{-3}–10^{-7}, necessitating enrichment or screening procedures. In general, accompanying nonrandom events do not occur in the targeted cell. Generally speaking, the frequency of targeting is dependent upon the extent of homology between the input and the target, linearization of the input DNA, the location of the target within the genome, and possibly the region within the gene. Most targeting events occur without detectable error, although some local rearrangements or mutations are observed. It seems that current targeting schemes in mouse embryonic stem cells will allow the generation of mutant mouse lines for many cloned genes. On the other hand, the general application of targeting to human gene therapy would appear to require higher absolute frequencies.

ACKNOWLEDGMENTS

The authors thank Dr. Merl Hoekstra for critical comments on this manuscript, and the many colleagues who communicated recent results. Likewise, special thanks are due to Celeste Soli for typing the manuscript. R. M. L. is a Leukemia Society of America Scholar. A. S. W. is a recipient of a Swebilius Cancer Research award. R. J. B. was supported by training grant T2 GM07499-12. This work was also supported by Public Health Service grants GM32741 to R. M. L. and a program project grant P01 CA39238 to Dr. William C. Summers from the National Institutes of Health.

Literature Cited

1. Abastado, J.-P., Cami, B., Dinh, T. H., Igolen, J., Kourilsky, P. 1984. Processing of complex heteroduplexes in *Escherichia coli* and *Cos-1* monkey cells. *Proc. Natl. Acad. Sci. USA* 81:5792–96

2. Adair, G. M., Nairn, R. S., Wilson, J. H., Seidman, M. M., Brotherman, K. A., et al. 1989. Targeted homologous recombination at the endogenous adenine phosphoribosyltransferase locus in Chinese hamster cells. *Proc. Natl. Acad. Sci. USA* 86:4574–78

3. Anderson, R. A., Eliason, S. L. 1986. Recombination of homologous DNA fragments transfected into mammalian cells occurs predominantly by terminal pairing. *Mol. Cell. Biol.* 6:3246–52

4. Applegate, M. L., Moore, M. M., Broder, C. B., Burrell, A., Juhn, G., et al. 1989. Molecular dissection of mutations at the heterozygous thymidine kinase locus in mouse lymphoma cells. *Proc. Natl. Acad. Sci. USA* In press

5. Ayares, D., Chekuri, L., Song, K.-Y., Kucherlapati, R. 1986. Sequence homology requirements for intermolecular recombination in mammalian cells. *Proc. Natl. Acad. Sci. USA* 83:5199–203

6. Ayares, D., Ganea, D., Chekuri, L., Campbell, C. R., Kucherlapati, R. 1987. Repair of single-stranded DNA nicks, gaps, and loops in mammalian cells. *Mol. Cell. Biol.* 7:1656–62

7. Ayares, D., Spencer, J., Schwartz, F., Morse, B., Kucherlapati, R. 1985. Homologous recombination between autonomously replicating plasmids in mammalian cells. *Genetics* 111:375–88

8. Baker, M. D., Pennell, N., Bosnoyan, L., Shulman, M. J. 1988. Homologous recombination can restore normal immunoglobulin production in a mutant hybridoma cell line. *Proc. Natl. Acad. Sci. USA* 85:6432–36

9. Baker, M. D., Shulman, M. J. 1988. Homologous recombination between transferred and chromosomal immunoglobulin k genes. *Mol. Cell. Biol.* 8:4041–47

10. Blackwell, T. K., Moore, M. W., Yancopoulos, G. D., Suh, H., Lutzker, S., et al. 1986. Recombination between immunoglobulin variable region gene segments is enhanced by transcription. *Nature* 324:585–89

11. Bollag, R. J., Liskay, R. M. 1988. Conservative intrachromosomal recombination between inverted repeats in mouse cells: association between reciprocal exchange and gene conversion. *Genetics* 119:161–69

12. Brenner, D. A., Smigocki, A. C., Camerini-Otero, R. D. 1985. Effect of insertions, deletions, and double-strand breaks on homologous recombination in mouse L cells. *Mol. Cell. Biol.* 5:684–91

13. Brenner, D. A., Smigocki, A. C., Camerini-Otero, R. D. 1986. Double-

strand gap repair results in homologous recombination in mouse L cells. *Proc. Natl. Acad. Sci. USA* 83:1762–66

14. Brouillette, S., Chartrand, P. 1987. Intermolecular recombination assay for mammalian cells that produces recombinants carrying both homologous and nonhomologous junctions. *Mol. Cell. Biol.* 7:2248–55

15. Brown, T. C., Jiricny, J. 1988. Different base/base mispairs are corrected with different efficiencies and specificities in monkey kidney cells. *Cell* 54:705–11

16. Campbell, C. E., Worton, R. G. 1981. Segregation of recessive phenotypes in somatic cell hybrids: role of mitotic recombination, gene inactivation, and chromosome nondisjunction. *Mol. Cell. Biol.* 1:336–46

17. Capecchi, M. R. 1980. High efficiency transformation by direct microinjection of DNA into cultured mammalian cells. *Cell* 22:479–88

18. Cavenee, W. K., Dryja, T. P., Phillips, R. A., Benedict, W. F., Godbout, R., et al. 1983. Expression of recessive alleles by chromosomal mechanisms in retinoblastoma. *Nature* 305:779–84

19. Chakrabarti, S., Seidman, M. M. 1986. Intramolecular recombination between transfected repeated sequences in mammalian cells is nonconservative. *Mol. Cell. Biol.* 6:2520–26

20. Chan, J. Y., Becker, F. F., German, J., Ray, J. H. 1987. Altered DNA ligase I activity in Bloom's syndrome cells. *Nature* 325:357–59

21. Chang, X.-B., Wilson, J. H. 1987. Modification of DNA ends can decrease end joining relative to homologous recombination in mammalian cells. *Proc. Natl. Acad. Sci. USA* 84:4959–63

22. Cox, M. M., Lehman, I. R. 1987. Enzymes of general recombination. *Annu. Rev. Biochem.* 56:229–62

23. Dasgupta, U. B., Summers, W. C. 1980. Genetic recombination of Herpes simplex virus, the role of the host cell and UV-irradiation of the virus. *Mol. Gen. Genet.* 178:617–23

24. Doetschman, T., Gregg, R. G., Maeda, N., Hooper, M. L., Melton, D. W., et al. 1987. Targetted correction of a mutant HPRT gene in mouse embryonic stem cells. *Nature* 330:576–78

25. Doetschman, T., Maeda, N., Smithies, O. 1988. Targeted mutation of the Hprt gene in mouse embryonic stem cells. *Proc. Natl. Acad. Sci. USA* 85: 8583–87

26. Dorin, J. R., Inglis, J. D., Porteous, D. J. 1989. Selection for precise chromosomal targeting of a dominant marker by

homologous recombination. *Science* 243:1357–60

27. Engler, P., Storb, U. 1988. Immunoglobulin gene rearrangement. In *Genetic Recombination*, ed. R. Kucherlapati, G. R. Smith, pp. 667–700. Washington, DC: Am. Soc. Microbiol.

28. Fincham, J. R. S. 1989. Transformation in fungi. *Microbiol. Rev.* 53:148–70

29. Fogel, S., Mortimer, R., Lusnak, K., Tavares, F. 1979. Meiotic gene conversion: A signal of the basic recombination event in yeast. *Cold Spring Harbor Symp. Quant. Biol.* 43:1325–41

30. Folger, K. R., Thomas, K., Capecchi, M. R. 1985. Nonreciprocal exchanges of information between DNA duplexes coinjected into mammalian cell nuclei. *Mol. Cell. Biol.* 5:59–69

31. Folger, K. R., Thomas, K., Capecchi, M. R. 1985. Efficient correction of mismatched bases in plasmid heteroduplexes injected into cultured mammalian cell nuclei. *Mol. Cell. Biol.* 5:70–74

32. Frohman, M. A., Martin, G. R. 1989. Cut, paste, and save: New approaches to altering specific genes in mice. *Cell* 56:145–47

33. Geliebter, J., Zeff, R. A., Melvold, R. W., Nathenson, S. G. 1986. Mitotic recombination in germ cells generated two major histocompatibility complex mutant genes shown to be identical by RNA sequence analysis: K^{bm9} and K^{bm6}. *Proc. Natl. Acad. Sci. USA* 83:3371–75

34. Gladstone, P., Fueresz, L., Pious, D. 1982. Gene dosage and gene expression in the *HLA* region: Evidence from deletion variants. *Proc. Natl. Acad. Sci. USA* 79:1235–39

35. Glazer, P. M., Sarkar, S. N., Chisholm, G. E., Summers, W. C. 1987. DNA mismatch repair detected in human cell extracts. *Mol. Cell. Biol.* 7:218–24

36. Graham, F. L., Van der Eb, A. J. 1973. A new technique for the assay of infectivity of human adenovirus 5 DNA. *Virology* 52:456–67

37. Hamilton, A. A., Thacker, J. 1987. Gene recombination in X-ray-sensitive hamster cells. *Mol. Cell. Biol.* 7:1409–14

38. Hare, J. T., Taylor, J. H. 1985. One role for DNA methylation in vertebrate cells is strand discrimination in mismatch repair. *Proc. Natl. Acad. Sci. USA* 82:7350–54

39. Holliday, R. 1964. A mechanism for gene conversion in fungi. *Genet. Res.* 5:282–304

40. Hoy, C. A., Fuscoe, J. C., Thompson, L. H. 1987. Recombination and ligation

of transfected DNA in CHO mutant EM9, which has high levels of sister chromatid exchange. *Mol. Cell. Biol.* 7:2007–11

41. Jasin, M., Berg, P. 1988. Homologous integration in mammalian cells without target gene selection. *Genes Dev.* 2:1353–63

42. Jasin, M., de Villiers, J., Weber, F., Schaffner, W. 1985. High frequency of homologous recombination in mammalian cells between endogenous and introduced SV40 genomes. *Cell* 43:695–703

43. Joyner, A. L., Skarnes, W. C., Rossant, J. 1989. Production of a mutation in mouse En-2 gene by homologous recombination in embryonic stem cells. *Nature* 338:153–56

44. Katzenberg, D. R., Tilley, S. A., Birshtein, B. K. 1989. Nucleotide sequence of an unequal sister chromatid exchange site in a mouse myeloma cell line. *Mol. Cell. Biol.* 9:1324–26

45. Kim, H.-S., Smithies, O. 1988. Recombinant fragment assay for gene targetting based on the polymerase chain reaction. *Nucleic Acids Res.* 16:8887–903

46. Kipps, T. J., Herzenberg, L. A. 1986. Homologous chromosome recombination generating immunoglobulin allotype and isotype switch variants. *EMBO J.* 5:263–68

47. Klein, H. L. 1988. Different types of recombination events are controlled by the *RAD1* and *RAD52* genes of *Saccharomyces cerevisiae*. *Genetics* 120:367–77

48. Klein, H. L. 1988. Recombination between repeated yeast genes. In *The Recombination of Genetic Material,* ed. K. B. Low, pp. 385–421. San Diego: Academic

49. Kolodner, R. D. 1990. Biochemistry of genetic recombination. *Annu. Rev. Biochem.* In preparation

50. Kucherlapati, R. 1986. Homologous recombination in mammalian somatic cells. In *Gene Transfer,* ed. R. Kucherlapati, pp. 363–81. New York: Plenum

51. Kucherlapati, R., Moore, P. D. 1988. Biochemical aspects of homologous recombination in mammalian somatic cells. In *Genetic Recombination,* ed. R. Kucherlapati, G. R. Smith, pp. 575–95. Washington, DC: Am. Soc. Micro.

52. Kucherlapati, R. S., Eves, E. M., Song, K.-Y., Morse, B. S., Smithies, O. 1984. Homologous recombination between plasmids in mammalian cells can be enhanced by treatment of input DNA. *Proc. Natl. Acad. Sci. USA* 81:3153–57

53. Lebkowski, J. S., DuBridge, R. B., Antell, E. A., Greisen, K. S., Calos, M. P. 1984. Transfected DNA is mutated in

monkey, mouse, and human cells. *Mol. Cell. Biol.* 4:1951–60

54. Letsou, A., Liskay, R. M. 1986. Intrachromosomal recombination in mammalian cells. See Ref. 50, pp. 383–409

55. Letsou, A., Liskay, R. M. 1987. Effect of the molecular nature of mutation on the efficiency of intrachromosomal gene conversion in mouse cells. *Genetics* 117:759–69

56. Lin, F.-L., Sperle, K., Sternberg, N. 1984. Model for homologous recombination during transfer of DNA into mouse L cells: Role for DNA ends in the recombination process. *Mol. Cell. Biol.* 4:1020–34

57. Lin, F.-L., Sperle, K., Sternberg, N. 1985. Recombination in mouse L cells between DNA introduced into cells and homologous chromosomal sequences. *Proc. Natl. Acad. Sci. USA* 82:1391–95

58. Lin, F.-L., Sperle, K. M., and Sternberg, N. L. 1987. Extrachromosomal recombination in mammalian cells as studied with single- and double-stranded DNA substrates. *Mol. Cell. Biol.* 7:129–40

59. Lin, F.-L., Sternberg, N. 1984. Homologous recombination between overlapping thymidine kinase gene fragments stably inserted into a mouse cell genome. *Mol. Cell. Biol.* 4:852–61

60. Liskay, R. M. 1986. Manipulation just off target. *Nature* 324:13

61. Liskay, R. M., Letsou, A., Stachelek, J. L. 1987. Homology requirement for efficient gene conversion between duplicated chromosomal sequences in mammalian cells. *Genetics* 115:161–67

62. Liskay, R. M., Stachelek, J. L. 1983. Evidence for intrachromosomal gene conversion in cultured mouse cells. *Cell* 35:157–65

63. Liskay, R. M., Stachelek, J. L. 1986. Information transfer between duplicated chromosomal sequences in mammalian cells involves contiguous regions of DNA. *Proc. Natl. Acad. Sci. USA* 83:1802–6

64. Liskay, R. M., Stachelek, J. L., Letsou, A. 1984. Homologous recombination between repeated chromosomal sequences in mouse cells. *Cold Spring Harbor Symp. Quant. Biol.* 49:183–89

65. Luria, S. E., Delbruck, M. 1943. Mutations of bacteria from virus sensitivity to virus resistance. *Genetics* 28:491–511

66. Mansour, S. L., Thomas, K. R., Capecchi, M. R. 1988. Disruption of the proto-oncogene int-2 in mouse embryo-derived stem cells: a general strategy for targeting mutations to non-selectable genes. *Nature* 336:348–52

67. Meselson, M. S., Radding, C. M. 1975. A general model for genetic recombination. *Proc. Natl. Acad. Sci. USA* 72: 358–61

68. Miller, L. K., Cooke, B. E., Fried, M. 1976. Fate of mismatched base-pair regions in polyoma heteroduplex DNA during infection of mouse cells. *Proc. Natl. Acad. Sci. USA* 73:3073–77

69. Moore, P. D., Holliday, R. 1976. Evidence for the formation of hybrid DNA during mitotic recombination in Chinese hamster cells. *Cell* 8:573–79

70. Moore, P. D., Song, K.-Y., Chekuri, L., Wallace, L., Kucherlapati, R. S. 1986. Homologous recombination in a Chinese hamster X-ray-sensitive mutant. *Mutat. Res.* 160:149–55

71. Nairn, R. S., Humphrey, R. M., Adair, G. M. 1988. Transformation depending on intermolecular homologous recombination is stimulated by UV damage in transfected DNA. *Mutat. Res.* 208:137–41

72. Nandi, A. K., Roginski, R. S., Gregg, R. G., Smithies, O., Skoultchi, A. I. 1988. Regulated expression of genes inserted at the human chromosomal B-globin locus by homologous recombination. *Proc. Natl. Acad. Sci. USA* 85: 3845–49

73. Orr-Weaver, T. L., Szostak, J. W. 1985. Fungal recombination. *Microbiol. Rev.* 49:33–58

74. Orr-Weaver, T. L., Szostak, J. W., Rothstein, R. J. 1981. Yeast transformation: a model system for the study of recombination. *Proc. Natl. Acad. Sci. USA* 78:6354–58

75. Pellicer, A., Robins, D., Wold, B., Sweet, R., Jackson, J., et al. 1980. Altering genotype and phenotype by DNA-mediated gene transfer. *Science* 209:1414–22

76. Potter, H., Weir, L., Leder, P. 1984. Enhancer-dependent expression of human K immunoglobulin genes introduced into mouse Pre-B lymphocytes by electroporation. *Proc. Natl. Acad. Sci. USA* 81:7161–65

77. Potter, T. A., Zeff, R. A., Frankel, W., Rajan, T. V. 1987. Mitotic recombination between homologous chromosomes generates H-2 somatic cell variants in vitro. *Proc. Natl. Acad. Sci. USA* 84: 1634–37

78. Radding, C. M. 1982. Homologous pairing and strand exchange in genetic recombination. *Annu. Rev. Genet.* 16: 405–37

79. Rauth, S., Song, K.-Y., Ayares, D., Wallace, L., Moore, P. D., Kucherlapati, R. 1986. Transfection and homologous recombination involving single-stranded DNA substrates in mammalian cells and nuclear extracts. *Proc. Natl. Acad. Sci. USA* 83:5587–91

80. Robert de Saint Vincent, B., Wahl, G. M. 1983. Homologous recombination in mammalian cells mediates formation of a functional gene from two overlapping gene fragments. *Proc. Natl. Acad. Sci. USA* 80:2002–6

81. Rommerskirch, W., Graeber, I., Grassmann, M., Grassman, A. 1988. Homologous recombination of SV40 DNA in COS7 cells occurs with high frequency in a gene dose independent fashion. *Nucleic Acids. Res.* 16:941–52

82. Rosenstraus, M. J., Chasin, L. A. 1978. Separation of linked markers in Chinese hamster cell hybrids: Mitotic recombination is not involved. *Genetics* 90:735–60

83. Roth, D., Wilson, J. 1988. Illegitimate recombination in mammalian cells. See Ref. 27, pp. 621–53

84. Roth, D. B., Porter, T. N., Wilson, J. H. 1985. Mechanisms of nonhomologous recombination in mammalian cells. *Mol. Cell. Biol.* 5:2599–607

85. Rubnitz, J., Subramani, S. 1984. The minimum amount of homology required for homologous recombination in mammalian cells. *Mol. Cell. Biol.* 4:2253–58

86. Rubnitz, J., Subramani, S. 1985. Rapid assay for extrachromosomal homologous recombination in monkey cells. *Mol. Cell. Biol.* 5:529–37

87. Rubnitz, J., Subramani, S. 1986. Extrachromosomal and chromosomal gene conversion in mammalian cells. *Mol. Cell. Biol.* 6:1608–14

88. Rubnitz, J., Subramani, S. 1987. Correction of deletions in mammalian cells by gene conversion. *Somatic Cell Mol. Genet.* 13:183–90

89. Saiki, R. K., Gelfand, D. H., Stoffel, S., Scharf, S. J., Higuchi, R., et al. 1988. Primer-directed enzymatic amplification of DNA with a thermostable DNA polymerase. *Science* 239:487–91

90. Sedivy, J. M., Sharp, P. A. 1989. Positive genetic selection for gene disruption in mammalian cells by homologous recombination. *Proc. Natl. Acad. Sci. USA* 86:227–31

91. Seidman, M. M. 1987. Intermolecular homologous recombination between transfected sequences in mammalian cells is primarily nonconservative. *Mol. Cell. Biol.* 7:3561–65

92. Seidman, M. M., Bredberg, A., Seetharam, S., Kraemer, K. H. 1987. Multi-

ple point mutations in a shuttle vector propagated in human cells: Evidence for an error-prone DNA polymerase activity. *Proc. Natl. Acad. Sci. USA* 84:4944–48

93. Shapira, G., Stachelek, J. L., Letsou, A., Soodak, L. K., Liskay, R. M. 1983. Novel use of synthetic oligonucleotide insertion mutants for the study of homologous recombination in mammalian cells. *Proc. Natl. Acad. Sci. USA* 80:4827–31

94. Shaul, Y., Laub, O., Walker, M. D., Rutter, W. J. 1985. Homologous recombination between a defective virus and a chromosomal sequence in mammalian cells. *Proc. Natl. Acad. Sci. USA* 82:3781–84

95. Shen, P., Huang, H. V. 1986. Homologous recombination in *Escherichia coli:* Dependence on substrate length and homology. *Genetics* 112:441–57

96. Singer, B. S., Gold, L., Gauss, P., Dogherty, D. H. 1982. Determination of the amount of homology required for recombination in bacteriophage T4. *Cell* 31:25–33

97. Skalka, A. M. 1988. Integrative recombination of retroviral DNA. See Ref. 27, pp. 701–24

98. Small, J., Scangos, G. 1983. Recombination during gene transfer into mouse cells can restore the function of deleted genes. *Science* 219:174–176

99. Smith, A. J. H., Berg, P. 1984. Homologous recombination between defective neo genes in mouse 3T6 cells. *Cold Spring Harbor Symp. Quant. Biol.* 49:171–181

100. Smithies, O., Gregg, R. G., Boggs, S. S., Koralewski, M. A., Kucherlapati, R. S. 1985. Insertion of DNA sequences into the human chromosomal β-globin locus by homologous recombination. *Nature* 317:230–234

101. Song, K.-Y., Chekuri, L., Rauth, S., Ehrlich, S., Kucherlapati, R. 1985. Effect of double-strand breaks on homologous recombination in mammalian cells and extracts. *Mol. Cell Biol.* 5:3331–36

102. Song, K.-Y., Schwartz, F., Maeda, N., Smithies, O., Kucherlapati, R. 1987. Accurate modification of a chromosomal plasmid by homologous recombination in human cells. *Proc. Natl. Acad. Sci. USA* 84:6820–24

103. Stachelek, J. L., Liskay, R. M. 1988. Accuracy of intrachromosomal gene conversion in mouse cells. *Nucleic Acids Res.* 16:4069–76

104. Stern, C. 1936. Somatic crossing over and segregation in *Drosophila mela-*

nogaster. Genetics 21:625–730

105. Stringer, J. R., Kuhn, R. M., Newman, J. L., Meade, J. C. 1985. Unequal homologous recombination between tandemly arranged sequences stably incorporated into cultured rat cells. *Mol. Cell. Biol.* 5:2613–22

106. Subramani, S. 1986. Rescue of chromosomal T-antigen sequences onto extrachromosomally replicating, defective simian virus 40 DNA by homologous recombination. *Mol. Cell. Biol.* 6:1320–25

107. Subramani, S., Rubnitz, J. 1985. Recombination events after transient infection and stable integration of DNA into mouse cells. *Mol. Cell. Biol.* 5:659–66

108. Subramani, S., Seaton, B. L. 1988. Homologous recombination in mitotically dividing mammalian cells. See Ref. 27, pp. 549–73

109. Szostak, J. W., Orr-Weaver, T. L., Rothstein, R. J., Stahl, F. W. 1983. The double-strand-break repair model for recombination. *Cell* 33:25–35

110. Tarrant, G. M., Holliday, R. 1977. A search for allelic recombination in Chinese hamster cell hybrids. *Mol. Gen. Genet.* 156:273–79

111. Thomas, B. J., Rothstein, R. 1989. Elevated recombination rates in transcriptionally active DNA. *Cell* 56:619–630

112. Thomas, K. R., Capecchi, M. R. 1986. Introduction of homologous DNA sequences into mammalian cells induces mutations in the cognate gene. *Nature* 324:34–38

113. Thomas, K. R., Capecchi, M. R. 1987. Site-directed mutagenesis by gene targeting in mouse-embryo derived stem cells. *Cell* 51:503–512

114. Thomas, K. R., Folger, K. R., Capecchi, M. R. 1986. High frequency targeting of genes to specific sites in the mammalian genome. *Cell* 44:419–28

115. Thompson, L. 1988. Mammalian cell mutations affecting recombination. See Ref. 27, pp. 597–620.

116. Thompson, S., Clarke, A. R., Pow, A. M., Hooper, M. L., Melton, D. W. 1989. Germ line transmission and expression of a corrected HPRT gene produced by gene targeting in embryonic stem cells. *Cell* 56:313–21

117. Voelkel-Meiman, K., Keil, R. L., Roeder, G. S. 1987. Recombination-stimulating sequences in yeast ribosomal DNA correspond to sequences regulating transcription by RNA polymerase I. *Cell* 48:1071–79

118. Wake, C. T., Gudewicz, T., Porter, T., White, A., Wilson, J. H. 1984. How

damaged is the biologically active sub-population of transfected DNA?. *Mol. Cell. Biol.* 4:387–98

119. Wake, C. T., Vernaleone, F., Wilson, J. H. 1985. Topological requirements for homologous recombination among DNA molecules tranfected into mammalian cells. *Mol. Cell. Biol.* 5:2080–89

120. Waldman, A. S., Liskay, R. M. 1987. Differential effects of base-pair mismatch on intrachromosomal versus extrachromosomal recombination in mouse cells. *Proc. Natl. Acad. Sci. USA* 84:5340–44

121. Waldman, A. S., Liskay, R. M. 1988. Dependence of intrachromosomal recombination in mammalian cells on uninterrupted homology. *Mol. Cell. Biol.* 8:5350–57

122. Wang, Y., Maher, V. M., Liskay, R. M., McCormick, J. J. 1988. Carcinogens can induce homologous recombination between duplicated chromosomal sequences in mouse L cells. *Mol. Cell. Biol.* 8:196–202

123. Wasmuth, J. J., Vock Hall, L. 1984. Genetic demonstration of mitotic recombination in cultured Chinese hamster cell hybrids. *Cell* 36:697–707

124. Watt, V. M., Ingles, C.J., Urdea, M. S., Rutter, W. J. 1985. Homology requirements for recombination in *Escherichia coli. Proc. Natl. Acad. Sci. USA* 82:4768–72

125. Weiss, U., Wilson, J. 1987. Repair of single-stranded loops in heteroduplex DNA transfected into mammalian cells. *Proc. Natl. Acad. Sci. USA* 84:1619–23

126. Weiss, U., Wilson, J. 1988. Heteroduplex-induced mutagenesis in mammalian cells. *Nucleic Acids Res.* 16:2313–22

127. Willis, A. E., Lindahl, T. 1987. DNA ligase I deficiency in Bloom's syndrome. *Nature* 325:355–57

128. Wong, E. A., Capecchi, M. R. 1985. Effect of cell cycle position on transformation by microinjection. *Somatic Cell Mol. Genet.* 11:43–51

129. Wong, E. A., Capecchi, M. R. 1986. Analysis of homologous recombination in cultured mammalian cells in transient expression and stable transformation assays. *Somatic Cell Mol. Genet.* 12:63–72

130. Wong, E. A., Capecchi, M. R. 1987. Homologous recombination between coinjected DNA sequences peaks in early to mid-S phase. *Mol. Cell. Biol.* 7:2294–95

131. Zimmer, A., Gruss, P. 1989. Production of chimaeric mice containing embryonic stem (ES) cells carrying a homoeobox Hox 1.1 allele mutated by homologous recombination. *Nature* 338:150–53

Annu. Rev. Genet. 1989. 23:227–50

MULTIPARTITE GENETIC CONTROL ELEMENTS: COMMUNICATION BY DNA LOOP

Sankar Adhya

Laboratory of Molecular Biology, National Cancer Institute, National Institutes of Health, Bethesda, Maryland 20892

CONTENTS

INTRODUCTION

Pardee et al (53) and Englesberg et al (14) provided a conceptual breakthrough in studying the molecular basis of gene regulation by proposing key steps for negative and positive control in transcription initiation. The concepts have evolved into the following summary view: The rate of transcription initiation by RNA polymerase from a promoter is determined by regulatory protein(s) that affect the activity of RNA polymerase, by interacting with specific DNA control sites. Regulatory proteins are of two types, a repressor (mediating negative control) or an activator (mediating positive control) of RNA polymerase function. Both types of regulatory proteins have two states, active or inactive, which are determined by environmental conditions. The change from one state to another could be a ligand-induced allosteric alteration or an enzyme-catalyzed covalent modification. For a repressor, the active form is able to bind a target DNA control site (operator) and block transcription, whereas the inactive form is unable to bind to the operator, thus allowing transcription. An activator protein in its active state binds to its target DNA site and facilitates or stimulates transcription. An inactive form of the activator fails to do so. Some regulatory proteins act as an activator in one state and as a repressor in another.

Biochemically, a repressor is commonly believed to act by sterically hindering RNA-polymerase binding to the promoter, whereas an activator is perceived to function by directly influencing either the RNA polymerase or the promoter DNA. Such molecular models of repressor and activator action directly at the level of promoter or RNA polymerase were accepted in concept because of the following basic premise: The DNA control elements to which the regulatory proteins bind to govern transcription initiation are adjacent to or overlap with the promoter. Thus, a direct influence of the DNA-bound regulatory proteins upon promoter or RNA polymerase is readily conceivable. However, more recent discoveries in some regulatory systems have revealed that the DNA control elements are multipartite in nature and are located at sites distant from the promoter: (*a*) In the *gal* operon of *Escherichia coli,* repression requires interaction of Gal repressor with two operators, one located contiguous to the two promoters of the operon and the other previously unknown, more than 100 bp away (33); (*b*) For repression of the two divergent promoters in the *ara* system of *E. coli,* the regulatory protein AraC must bind to an additional newly discovered operator locted more than 100 bp away from either promoter (11). To account for the requirement of such multipartite control sites and to explain how a regulatory protein influences the promoter-RNA polymerase function from a distant DNA site, our concept of the role of the regulatory proteins in transcription initiation has been extended by introducing the model of the DNA loop (11, 33, 36, 43). In this model, protein molecules bound to distal sites operate by direct protein-protein contact, as if they were adjacent, but in the process cause the intervening DNA segment to be looped out (Figures 1b and c). The generation of such DNA loop has potential implications on gene regulation: (*a*) It can affect the structure of the DNA within the loop that is now a topologically

a.

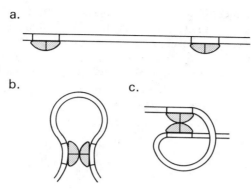

b.   c.

Figure 1   (*a*) Protein molecules (stippled) bound to two DNA sites that are spatially separated from each other. A dimeric protein is assumed to bind to each site. (*b*) The two DNA-bound protein dimers occupying the same face of the DNA helix interact with each other and generate a loop in the intervening DNA segment. (*c*) The two DNA-bound dimers occupying opposite faces of the DNA helix, given enough DNA flexibility, have the option of interacting with each other and producing a topologically different kind of DNA loop.

independent domain; (*b*) The two interacting proteins can influence the conformation of each other.

Based on the model of DNA looping and other new developments in *E. coli*, we discuss here the reformulation of how regulatory proteins and DNA sites interact with each other and thereby govern gene expression. We describe a few examples of control systems of *E. coli*, mostly negative control, which involve distant and multipartite sites. An excellent review of positive control systems has appeared recently (57). The genetic systems discussed here and the references cited are mainly to provide an example or to make a point and are not intended to be a comprehensive survey.

REGULATORY SYSTEMS WITH DNA LOOPING

Bipartite Operators in gal

The regulatory circuit of the *gal* operon in *E. coli* is shown in Figure 2. Two partially overlapping promoters control the expression of the operon. Cyclic AMP and its receptor protein (CRP) acting as a complex stimulate transcription initiation from *P1* and inhibit that of *P2* (1, 51). Gal-repressor protein, the product of an unlinked gene, represses both promoters. D-galactose or D-fucose induces the operon by binding to the repressor and making it inactive. Gal repressor acts by binding to two operators, O_E and O_I (33, 43); mutation at either site causes derepression. O_E and O_I, which are separated by 114 bp[1], are similar 16-bp sequences, each with a hyphenated dyad symmetry

[1]In this article, the distance in bp between the two DNA control elements, which are dyad symmetries are cited from center of symmetry to center of symmetry.

that binds a Gal-repressor dimer (44). Note that the promoters are located in between but contiguous to O_E. O_I is in fact located within the first structural gene. Although several models have been proposed that explain repression of the *gal* operon from two operator sites, the key experiments summarized below demonstrate the model originally proposed, in which repression requires the formation of a DNA loop by interaction of dimeric repressor molecule at O_E and O_I.

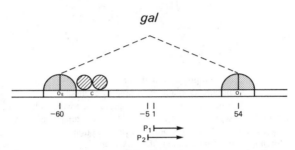

Figure 2   The regulatory segment of the *gal* operon of *Escherichia coli*. The location of the two operators (O_E and O_I), the cyclic AMP receptor protein (CRP) binding site *(C)*, and the start sites of transcription initiation from the two promoters (*P1* and *P2*) are shown. The drawings are not to scale. The stippled figures represent Gal repressor molecules and the cross-hatched circles represent a CRP monomer. The broken lines indicate a loop formation by interaction of the indicated DNA-bound proteins. The numbers represent distance in bp from the *P1* transcription initiation site.

OPERATOR CONVERSIONS   The functional relationship between the two *gal* operators (O_E^G and O_I^G) has been studied by converting one of them (by site-directed mutagenesis) to a *lac* operator, to yield a genotype of O_E^G–O_I^L or O_E^L–O_I^G (22) (Figure 3). In both cases, the operon is not repressed even in the simultaneous presence of Gal and Lac repressors. This defect in repression does not appear to be caused by weaker affinity of either operator for the corresponding repressor, because repression is not restored even when both repressors were highly overproduced. These results clearly show that mere occupation of the two operators is not sufficient for repression of the *gal* genes. In contrast, conversion of both operators to *lac* operators restored normal repression in the presence of the Lac repressor and made the operon inducible with IPTG. Thus, the *gal* operon can be repressed if both operators are occupied by similar repressors, either Lac repressor for O_E^L–O_I^L or Gal repressor for wildtype. The most likely explanation is that repression requires interaction between the two DNA-bound repressor dimers and that interaction is not possible between heterologous repressors.

COOPERATIVE BINDING   The affinities of O_E and O_I for Gal repressor have been measured using an in vivo titration assay (22). In this method, a

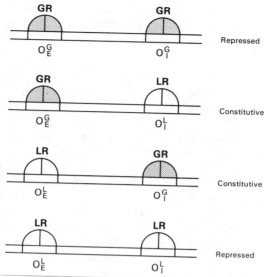

Figure 3 Repression of wild-type and "converted" *gal* operons by Gal and Lac repressors (22). Gal repressor molecules are marked by GR and Lac repressor by LR (see text for details).

wild-type bacterial strain, i.e. containing the relevant genes in single copy, is transformed with multicopy plasmids that carry various combinations of cloned *gal* and *lac* operators sequences at O_E and O_I. In the absence of inducer, depression of a chromosomal *gal* or *lac* operon reflects the sequestering of cellular repressors by operators on multicopy plasmids. The relative level of chromosomal derepression is a function of the affinities of the plasmid-borne operators for repressors. These results show that the affinity of Gal repressor for O_E^G–O_I^G is higher than the sum of the affinities for O_E^G and O_I^G. The same is true for O_E^L and O_I^L elements. In other words, the λ repressor binds to O_E and O_I cooperatively. Cooperative binding of phage λ repressor to two adjacent operator sites is due to protein-protein interaction between the two DNA-bound λ repressor molecules (29). The observation that Gal and Lac repressor bind to spatially separated O_E and O_I sites cooperatively implies an interaction between O_E- and O_I-bound repressor molecules and, as a result, the formation of a DNA loop. Cooperative binding of Lac repressor to two spatially separated *lac* operators has also been shown in vitro (5). However, cooperative binding of Gal repressor has not yet been observed in vitro, presumably because the appropriate conditions have not been identified.

ELECTRON MICROSCOPY   Electron microscopic observations provide the direct evidence that repressor binding to O_E and O_I in *gal* can generate a DNA loop. When linear *gal* DNA fragments with O_E^L–O_I^L genotype are mixed with

Lac repressor, about 30% of the DNA molecules contained a loop of the expected size (~94 bp) at the correct location on the DNA segment as seen by electron microscope (N. Mandal et al, in preparation).

REPRESSOR DEFECTIVE IN DNA LOOPING   A Lac repressor mutant that fails to derepress the *gal* operon of the O_E^L–O_I^L genotype has been characterized biochemically (N. Mandal et al, in preparation). While wild-type Lac repressor is a tetramer (50), the purified mutant repressor is a stable dimer in size and binds to a *lac*-operator DNA sequence with the same affinity as wild-type repressor. The alteration is caused by a frameshift mutation at the carboxy terminal end of the Lac repressor gene (N. Mandal et al, in preparation). When the O_E^L–O_I^L DNA fragment bound to the mutant repressor was investigated by electron microscopy, none was found with a loop (<1%), presumably because repressor tetramerization is needed for loop formation. The behavior of this mutant Lac repressor convincingly demonstrates that repression of the *gal* operon requires DNA loop formation by a repressor tetramer and two operators. The mechanisms by which a DNA loop may cause repression in *gal* and other systems is discussed later.

The Tripartite ara Operators

Another example of gene regulation involving DNA-loop formation is the *ara* system in *E. coli*. The *ara* regulatory circuit, diagrammed in Figure 4, was studied by Englesberg and coworkers who originally formulated the concept of positive control of gene regulation. It has two divergent promoters, P_C and P_B, subject to both negative and positive control by AraC protein (for review see 15). The gene encoding AraC is under the control of P_C. Both P_C and P_B are repressed by AraC and activated by cAMP and CRP; the repression of P_C gives rise to autoregulation and maintains a constant level of AraC in the cell. When AraC binds to inducer arabinose, it becomes an activator of P_B while continuing to repress P_C. Three operators of AraC have been identified: I, O_1 and O_2^{b2} (11, 42, 52). O_1 and I are located between P_C and P_B, while O_2 is located more than 100 bp apart from I and O_1 and is within the transcribed region of *araC*. CRP activates both P_C and P_B by binding to a single site, located between O_1 and I. The repressor and activator forms of AraC act by binding to different combinations of I, O_1 and O_2 (23, 41, 46). The model explaining most of the results in the *ara* system is as follows: Occupation of O_2 and I sites and simultaneous protein-protein contact between the two DNA-bound AraC molecules generate a DNA loop that keeps both P_C and P_B

---

[2]Although AraC protein acts both as repressor and activator, its binding sites are called operators for historical reasons.

repressed. When liganded to Arabinose, AraC binds to all three sites, I, O_1 and O_2. Binding of liganded AraC to O_1 and O_2 generates a DNA loop that keeps P_C repressed, while binding of the complex to an expanded version of I activates P_B. Formulation of this model, which is different from various models proposed earlier (42, 52), is based on the observations described below.

CO-OPERATIVE BINDING   AraC binding to the operators in vivo has been studied by a variety of DNA protection experiments before and after selective inactivation of the binding sites (23, 46). These experiments have shown that while binding to O_2 alone is low under both induced and repressed conditions, it is enhanced by the presence of the others, indicating pairwise cooperativity between I–O_2 and O_1–O_2. In wild-type, unliganded AraC binds to I and O_2 cooperatively (a diagnostic of DNA looping) and does not bind to O_1, and liganded AraC binds to O_1 and O_2 cooperatively, as well as to I independently.

BINDING TO I FOR ACTIVATION   Liganded AraC protein shows a solo high-affinity binding to an extended I site (41), and may be attributable to occupation of this site by two molecules of liganded AraC. It is only observed when liganded AraC activates P_B.

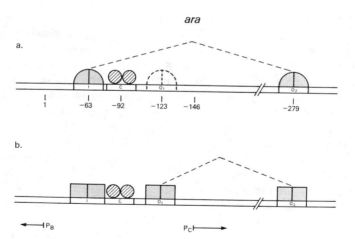

Figure 4   The regulatory region of the *ara* system in *Escherichia coli*. The binding sites of AraC protein are shown as I, O_1 and O_2, the CRP binding site as C, AraC molecules by stippled figures, and CRP by cross-hatched circles. (a) Binding of AraC without arabinose and (b) Binding of AraC when liganded with arabinose. The broken lines indicate the two DNA loops I–O_2 and O_1–O_2 discussed in the text.

PERIODICITY IN REPRESSION   In the absence of arabinose, the repression of the P_B promoter and presumably of the P_C promoter by AraC is a somewhat periodic property of the distance between O_2 and I. The distance between O_2 and I is 211 bp. A preliminary study has shown that by changing this distance, repression occurs only when the ratio of the distance to the number of bp per B-DNA helical turn is an integer (11). As discussed below changing the angular orientation of the two binding sites by a half-integral number of helical turns, to place them on opposite sides of DNA, is expected to interfere with looping and thus cause derepression.

SWITCH IN LOOP   When arabinose is added, cooperative binding switches from O_2–I to O_2–O_1 and transient derepression of P_C occurs during the transition of O_2–I to O_2–O_1 loop (32, 42). The latter has been attributed to the amount of time necessary to build up an effective concentration of liganded AraC for O_2–O_1 loop formation.

Tripartite Operators in Phage λ

We now describe the bacteriophage λ repressor-operator system first as an elegant example of how a single protein uses a multipartite, but adjacent, operator system to achieve negative and positive controls simultaneously and then describe how this has been used as an artificial system for studying the feasibility of DNA-loop formation.

ADJACENT OPERATORS   Figure 5 depicts aspects of the λ repressor-operator system crucial to our discussion (for review, see 56). Embedded between the two divergent λ promoters, P_{RM} and P_R, are three tandem operators, O_{R1}, O_{R2} and O_{R3} with 3- and 7-bp spaces between them, respectively). The operators are similar, but not identical, 17-bp sequences with an approximate dyad symmetry. λ repressor is a dimer, each subunit having two domains. Each amino domain contacts one-half of a dyad symmetry. P_{RM} is the promoter of gene cI, which encodes the repressor itself, while P_R is one of the early promoters needed for lytic development of λ. Genetic and biochemical studies have established that λ repressor simultaneously activates P_{RM} and represses P_R in one physiological condition and allows transcription from P_R but not from P_{RM} in the other (56, 58). The following results explain how the same protein achieves positive and negative control simultaneously by means of selective co-operative binding to adjacent multipartite operators.

1. In vitro repressor binding studies with DNA bearing one, two, or all three operators intact have established that repressor dimer binds to the three DNA sites with alternate pairwise cooperativity (34). Whereas the intrinsic relative affinities of repressor for O_{R1}, O_{R2} and O_{R3} are 2, 25, and 25 respectively, the corresponding observed affinities are 1, 1, and 25 in wild-

λO_R

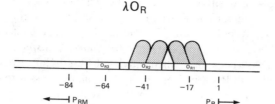

Figure 5 Diagrammatic representation of cooperative binding of λ repressor to the O_R segment of phage DNA. The three tandem operators are shown as O_{R1}, O_{R2} and O_{R3}, the rightward lytic promoter as P_R and leftward promoter for cI as P_{RM}. λ repressor, shown in stippled figures, binds to O_{R1} and O_{R2} cooperatively by protein-protein contact. O_{R3} remains vacant at low repressor concentration (see text for details).

type DNA and 0, 5, and 5 in an O_{R1} mutant DNA. Thus, repressor by cooperative binding occupies O_{R1} and O_{R2} in the presence of O_{R3} and to O_{R2} and O_{R3} in the absence of O_{R1}. The pairwise cooperativity originates from the fact that only two repressor dimers can cooperate at one time. However, repressor binding to O_{R1} and O_{R3} is not cooperative in an O_{R2} mutant. The reason for this non-cooperativity is discussed below.

2. All the evidence indicates that the cooperativity is mediated by interaction between the carboxyl domains of adjacent repressor dimers (34). When separated from the carboxyl domains, the amino domains bind to the operators non-cooperatively, following the relative intrinsic affinities of each site for intact repressor.

3. It has been shown that occupation of O_{R1}, and thus O_{R2}, turns off the lytic promoter P_R (58). At the same time, occupation of O_{R2}, without occupation of O_{R3}, activates P_{RM}, the maintenance promoter for repressor synthesis in a lysogen (58). Occupation of O_{R3}, which happens at a very high repressor concentration, represses P_{RM} (58). Thus, λ repressor autoregulates at two levels: occupation of O_{R2} (low repressor concentration) stimulates P_{RM} and of O_{R3} (high repressor concentration) represses P_{RM}. The former situation represents the normal physiological repressor concentration in a lysogenic cell in which repressor concentration is not high enough to bind to O_{R3}. Note that this way, i.e. by occupation of O_{R1} and O_{R2}, repressor acts both as a negative and positive autoregulatory protein to maintain its constant level in a repressed lysogen.

Mechanistically, the negative control of P_{RM} and P_R by λ repressor has been ascribed to steric interference of RNA polymerase binding to the two promoters (58). This model is accepted in these two cases because of the complete overlap of O_{R1} and O_{R2} with P_R and of O_{R3} with P_{RM}, and because of the biochemical demonstration that λ repressor sterically hinders RNA polymerase binding (see 58). The activation of P_{RM} by repressor is a well-studied example of positive control in gene regulation (24, 57). It has been suggested that repressor at O_{R_2} stimulates transcription from P_{RM} by a direct contact with RNA polymerase. The idea of a direct contact is strongly supported by the following experiments: (a) *Shared phosphate*. A phosphate group in the DNA backbone when ethylated by ethylnitrosourea interferes with binding of both RNA polymerase to P_{RM} and repressor to O_{R2} (28); (b) *Positive control mutants*. Repressor mutants (called *pc*) that are defective in positive control have been isolated and characterized (20, 24, 28). These repressor mutants bind normally to O_{R2} but fail to activate transcription. In these mutants, the position of the altered amino acid in the repressor, when occupying O_{R2} is on a patch of the protein surface that should according to structural predictions approach (touch) RNA polymerase closely.

SPATIAL SEPARATION   Because λ repressor binds cooperatively to adjacent operator sites with 3- or 7-bp spacing by protein-protein contact, in principle it generates a microloop in DNA. It thus provides a simple system for studying the formation of larger DNA loops. When two λ operators (one weak and one strong) are placed 40 to 70 bp apart, instead of 3 or 7 bp, λ repressor can still bind cooperatively to the separated operators in some of these cases (29). As with adjacent sites, cooperative binding from separated sites involves the same protein-protein contact. This is inferred because a mutant repressor that has lost its ability to show cooperative binding to adjacent sites has also lost its ability to bind cooperatively to separated sites (30). The following experiments confirmed the hypothesis that cooperativity of λ repressor binding to spatially separated sites generates a loop in the intervening DNA.

1. Since λ repressor binds mostly to one face of an operator DNA helix, repressor molecules bound to separated operators must be located on the same face of the entire helix for protein-protein contact (Figure 6a; 27, 29). Consistently, cooperative binding of λ repressor was observed only when the center-to-center distance of the two operators constituted an integral number of turns of B-DNA helix (5, 6, or 7 turns). Repressor binding was non-cooperative for operators n + ½ (4.6, 5.5 or 6.4) helical turns apart (Figure 6b). Apparently, in this in vitro system, DNA curving for face-to-face protein-protein contact is energetically feasible, but DNA twisting (to bring repressors separated by non-integral helical turns together) is not. We also note here that the λ operator O_{R1} does not cooperate with O_{R3} in repressor binding in their natural location if the O_{R2} site is mutated (34). Since O_{R1} and O_{R3} are separated in this context by 4.3 helical turns, the absence of cooperative binding likely results from the DNA twisting problem. In support of the idea that nonintegral DNA turns prevent looping because of torsional stress in twisting, it has been shown that cooperative binding to two operators separated by non-integral helical turns is restored if a short single-stranded gap is

introduced in the intervening DNA segment, thereby removing the torsional stress needed for free twisting (29).

2. In forming a DNA loop of 5 or 6 helical turns, the intervening DNA must curve. That this segment of DNA in a loop is curved was evident from the fact that cooperative binding was always accompanied by the appearance of an alternating set of enhanced and diminished DNase-sensitive phosphodiester bonds between the two operator segments (29). The enhanced and diminished DNase-sensitive bonds are spaced about 5 bp (½ B-DNA helix) apart. In addition, the enhanced DNase-sensitive sites lie to the outside and diminished sensitive sites lie to the inside of the DNA curvature, as expected.

3. Bent DNA molecules with a small protein complex inside the bend (presumed to be a diagnostic of a DNA fragment containing a short loop of about 50 bp) have been observed using electron microscopy only when λ repressor was bound cooperatively to the DNA fragment with two spatially separated operators (19).

DNA looping was not observed in vitro with spatially separated λ operators when the spacing was large, e.g. 20 B-DNA helical turns (27). We discuss this aspect and the length-dependent energetics of DNA curving and twisting for loop formation in a later section.

Pseudo-Operators in lac

Secondary operator sites, known as pseudo-operators, have even been found in the *lac* operon, one of the primary systems that brought about the concept of an operator. They too seem to contribute to repression. The *lac* operon contains a strong repressor binding site O_1 (the originally identified operator) and two weak binding sites, O_2 (within the *lacZ* structural gene) and O_3 (upstream), located 402 bp and 93 bp from O_1, respectively (18, 60; see Figure 7). The intrinsic affinities of O_2 and O_3 are much weaker than that of O_1 (5, 16, 17, 54). Because additional operators contribute to repression in other systems, the *lac* operon is now being reinvestigated. Several observations suggested a role of DNA looping involving O_1 and the secondary

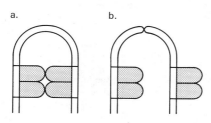

a.   b.

Figure 6 (a) Diagrammatic representation of DNA looping as a consequence of two λ repressor dimers binding to two O_R operators, separated artificially by six B-DNA helical turns (29). (b) Failure of loop formation by λ repressor when the two operators are separated by 6.5 helical turns. In the former case, the two repressor molecules occupy the same face of the DNA helix, while in the latter they occupy different faces.

operators in repression: (*a*) O^C mutations in O_1, which reduce its affinity for repressor 1000-fold in vitro, nonetheless cause only 20% or less constitutivity in vivo (48, 61). (*b*) A single O_1 sequence fused to a different promoter, e.g. *trp* promoter, does not repress the promoter fully (25). (*c*) Placing two *lac* operators, one upstream and one downstream of a *trp* or *lac* promoter and spatially separated by greater than 100 bp, results in a high level of repression (4, 25). (*d*) A second efficient intact operator placed 185 bp or 283 bp upstream of a mutant O_1 restores repression (49). (*e*) Mutation of O_2 causes a low but significant constitutive expression of the *lac* operon in vivo (13). (*f*) Two suitably spaced *lac* operators form a DNA loop in vitro, as seen by electron microscopy as well as by its characteristic mobility shift during gel electrophoresis under appropriate repressor concentrations (39). At very high repressor concentrations, two sites are independently occupied by repressor molecules. Since *lac* repressor is a tetramer with two operator-binding regions, one in each dimer, the loop formation at low repressor concentration is caused by a tetramer contacting two intramolecular DNA sites. Formation of a DNA loop is also favored by correct phasing of the two operators. (*g*) A repressor tetramer binds to O_1 and O_3 in a supercoiled DNA in vitro with similar and high affinities. Such binding makes a segment of DNA between the two operators hypersensitive to dimethylsulfate and potassium permanganate, indicating DNA distortion (5). A simple interpretation of these results is the formation of a DNA loop. (*h*) Repressor binding to O_1 and O_2 is cooperative in vivo (16). Binding to O_2 strengthens repressor binding to O_1 threefold, whereas binding to O_1 strengthens binding to O_2 12-fold.

Tripartite Operators in the deo Operon

Initiation of transcription of the operon coding the nucleoside and deoxynucleoside catabolizing enzymes in *E. coli* (the *deo* operon) occurs from two promoters, *P*1 and *P*2, which are located 599 bp apart (7, 72). The

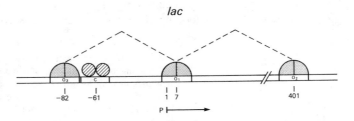

Figure 7 The regulatory sites of the *lac* operon of *Escherichia coli*. The three operators are marked as O_1, O_2 and O_3 and the CRP binding site as *C*. The CRP-dependent *lac* promoter is *P*. The broken lines indicate the two potential DNA loops discussed in the text. Lac repressors are shown as stippled figures and CRP as cross-hatched circles.

structure of the regulatory region of the *deo* operon is shown in Figure 8. *P*1 and *P*2 are negatively controlled by two repressors, DeoR and CytR (72). Whereas *P*1 is repressible by DeoR, *P*2 is negatively controlled both by DeoR and CytR and is positively controlled by cyclic AMP and CRP. Here we briefly discuss the negative control of *P*2 by DeoR that seems to involve DNA looping. By analyzing *deo-lacZ* genetic fusions, which carry different parts of the *deo* regulatory region coupled to a *lacZ* reporter gene, three operator sites for DeoR binding have been shown to exist; two of these, O_1 and O_2, are 599 bp apart and completely overlap *P*1 and *P*2, respectively, whereas the third site O_E is present 279 bp upstream of O_1 (73). Although DeoR-mediated repression of *P*1 and *P*2 is small (about twofold) in the presence of only its cognate operator (O_1 or O_2) and repression of *P*1 or *P*2 is enhanced 20- to 30-fold by the presence of both O_1 and O_2, full repression requires all three sites. Further investigations have shown that efficient repression can also be obtained by varying the distance between O_1 and O_2 within a range of 24–997 bp (7, 73). Significant repression of *P*2 can also be achieved by having the cognate O_2 operator fixed and moving the O_1 operator even up to 1–5 kilo bp downstream of the O_2 in a manner independent of orientation (8). However, the degree of repression is reduced somewhat with increasing distance. Note that there is no periodicity of repression in this distance range. The involvement of a DNA loop for repression in the *deo* operon would explain these results. In this model, a DNA loop between O_1 and O_2 is required for repression of *P*1 and *P*2, but this loop may be unstable and the third operator O_E helps repression by forming an insurance loop with O_1 or O_2.

FEASIBILITY OF DNA LOOPING

Therodynamic Considerations

In order to form a loop, protein bound to two DNA sites must face each other in a given geometry. The energy cost and probability of two distant DNA-

deo

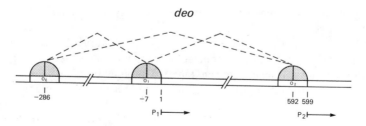

Figure 8 The regulatory region of the *deo* operon of *Escherichia coli*. The three DeoR repressor binding site are O_E, O_1 and O_2. The two promoters are P_1 and P_2. DeoR molecules are stippled. The broken lines indicate the DNA loops proposed in the text.

control elements communicating with each other by protein molecules bound to each site may be considered in terms of known properties of DNA in solution. A DNA segment of 200 bp or less may be stiff and energy would be required for (a) curving the DNA to bring the two protein binding sites together and (b) twisting and/or writhing the DNA for proper face-to-face contact of the two bound proteins.

In the absence of a direct energy supply like ATP hydrolysis, we expect that the available sources are free energy of protein-DNA and protein-protein interactions, as well as free energy released by any DNA supercoiling in the formation of a loop structure. These theoretical expectations have been tested experimentally with the DNA loop formation in the λ system in which two repressor binding sites are separated by different DNA helical turns, as already discussed (29). The energy generated by two DNA-bound repressor dimers in forming a tetramer is about 8.5 kcal/mole (R. Sauer, quoted in 27). Since it is more than the estimated cost for example, of 6.5 kcal/mole for curving of a DNA segment with 6.0 B-DNA helical turns, looping is thermodynamically permissible (27). But if the two operator sites are separated by 5.5 B-DNA helical turns, the DNA additionally requires twisting for a face-to-face encounter of the DNA-bound proteins and the extimated extra cost is 11–12 kcal/mole. DNA looping in this case has not been observed, presumably because a significant energy source for this step is lacking.

For longer pieces of DNA, the helix flexibility is expected to make the energy requirement for both curving and twisting insignificant. However, cooperative binding of λ repressor to sites separated by more than 200 bp, i.e. 20 turns, was not detectable in vitro (27). It has been pointed out that for large DNA pieces, the limiting factor in looping is the probability of the two sites on the DNA molecule coming close to each other, i.e. the concentration of one being near the other (67). Thus according to these expectations, DNA looping is unlikely with large DNA segments between two operators.

The Question of Periodicity

We have discussed above the relative inflexibility and torsional stiffness of short DNA segments and unavailability of large amounts of energy required for DNA twisting. Under these conditions, one would predict that the formation of a loop by face-to-face contact between proteins should be a periodic function of the distance between the two protein binding sites. When λ repressor was studied in vitro for its ability to induce DNA looping, the two operator sites on DNA were spaced apart by 48, 52, 57, 63, 67, and 73 bp. Of these only the species that were 52, 63, and 73 bp apart showed the cooperative binding diagnostic of looping (29). Assuming a 10.4 bp helix size in solution, 52, 63, and 73 represent 5, 6 and 7 integral number of turns. The species separated by 48, 57 and 67 bp that show non-cooperative binding have

~n + ½ helical turns (4.6, 5.5, and 6.4, respectively) between them. In a much more elaborate study, a similar periodicity was observed when Lac repressor was used to bind to DNA fragments carrying two *lac* operators separated from each other by 63 to 535 bp (39). Formation of a DNA loop, as judged by electron microscopy, gel electrophoretic mobility measurements, and DNase I hypersensitivity assays was facilitated when the two operators were correctly phased. Stable loop structures were detected only when the two operators were separated by an integral number of B-DNA helical turns from 6 to 21. When the angular orientation was distorted, for example, from 15 turns (158 bp) by deleting or adding 2, 3, 4, 5, or 6 bp, the loop gradually became unstable in both directions. Unlike λ repressor, Lac repressor was able to generate a loop with a spacing of as large as 220 bp.

Although the above results of in vitro studies of looping by λ and Lac repressors match theoretical expectations well, it must be pointed out that other factors (known or unknown) may influence DNA loop formation and thus make it hard to predict the probability of periodicity of DNA looping in a given system in vivo: (*a*) naturally curved or bent DNA sequences, if present in the intervening segment, may help or hinder DNA looping, depending upon the relative direction of the curve or bend with respect to the direction of DNA looping. (*b*) The cost of twisting DNA may be reduced somewhat by flexibility of the DNA binding protein, which can overcome differences in the angular orientation of the two binding sites on DNA. Some flexibility in λ repressor is indicated by the observation that it shows cooperative binding to two sites with center-to-center distance varying from 1.9–2.3 turns of B-DNA helix (34). (c) For long DNA spacings, flexibility may make the energy requirement for both bending and twisting insignificant. As mentioned before, DNA seems to loop with operator separations of kilobases in the *deo* operon (8), as opposed to expectations from in vitro experiments. We also noted that the expected periodicity of repression was not observed at this range of DNA length, presumably because DNA flexibility allows loop formation by contacts between proteins lying on different faces of DNA, as depicted in Figure 1c. (*d*) The state of DNA supercoiling is known to aid cooperative binding, presumably by releasing energy (5). At elevated negative DNA supercoiling in vitro, the stability of loops increased as the spacing requirement for loop formation diminished (38). (*e*) Other proteins, binding to the intervening segment, may bend DNA in the same direction or opposite to looping, thereby either helping or hindering looping. Protein-induced DNA bending greatly contributes to increasing the concentration of one end of a small DNA fragment around the other end (10, 37). (*f*) Helix size of the DNA region may be different from the assumed value of 10.4 bp/turn observed in vitro.

A new in vivo study in the *ara* system has indicated a periodic behavior of the proposed DNA looping, as studied by AraC-mediated repression of P_B, centered around a helix size of 11.1bp (40) as opposed to a helix size of 10.5 reported in an earlier study (11).

MECHANISM OF NEGATIVE CONTROL

During the early studies with Lac and λ repressors, it was commonly believed that repressor binding to an overlapping operator inhibits transcription initiation by hindering the binding of RNA polymerase to the promoter. Because of (a) the multipartite nature of the operators in several systems, (b) their location at sites far away from the promoters and (c) the actual demonstration in some systems that repressor does not inhibit RNA-polymerase binding, we must consider repressor action at other levels of RNA-polymerase action as well. Based on in vitro studies, transcription initiation and subsequent events by RNA polymerase can be divided into several steps (6, 47). In principle a repressor could act at any of these steps to block transcription.

Block of RNA polymerase binding (steric hindrance)   RNA polymerase binds to a promoter initially to form what is known as the closed complex. If the operator at least partially overlaps with the site of closed complex formation, then repressor would sterically hinder RNA polymerase access to the site, either by free or linear diffusion.

It must be pointed out that, for weak promoters, activator proteins are needed to stimulate RNA-polymerase activity. For these promoters, instead of blocking RNA-polymerase function, the repressor may inhibit transcription either by blocking the binding of the activator (steric hindance) or contacting the activator to prevent the activator's own contact with RNA polymerase.

Block of isomerization (competition for stable binding)   A closed complex of RNA polymerase isomerizes to a more stable form, called the open complex, in which the helix is opened and there is a change in the conformation of RNA polymerase as well as in the size of the DNA segment contacted by RNA polymerase (B. Krummell & M. Chamberlin, personal communication). Repressor could block isomerization by occupying part of the DNA segment needed for a stable complex formation with RNA polymerase or by distorting the DNA in such a way that RNA polymerase fails to open the helix for open complex formation. The repressor Arc may repress the *P*ant promoter of bacteriophage P22 of *Salmonella typhimurium* by affecting isomerization (75).

Inhibition of initiation   After open complex formation, RNA polymerase forms phosphodiester bonds without leaving the promoter. This step may have several transient intermediate stages, producing discrete sizes of short RNA oligonucleotides (B. Krummel & M. Chamberlin, personal communication). Repressor bound to an operator may directly contact and freeze RNA polymerage at the open complex state and prevent the latter from initiating phosphodiester bond synthesis. It has been shown that Gal repressor does not block open complex formation at the *gal* promoters (A. Majumdar & S. Adhya, unpublished results). This repressor very likely inhibits RNA polymerase at this stage or at the next stage of promoter clearance.

Block of promoter clearance   RNA polymerase leaves the promoter region as a transcription bubble and elongates RNA. Bound repressor may contact RNA polymerase and prevent it from clearing the promoter. RNA polymerase would continue to make short RNA oligomer.

Transcription termination   If an operator is placed at a DNA site along the transcription path after promoter clearance, then the repressor may act to provoke transcription termination. Such a property of the repressor has been demonstrated with Lac repressor at an operator site (O_3) located at the end of its own gene *(lacI)* (64) or by inserting a synthetic *lac* operator in synthetic operons (9).

THE ROLE OF DNA LOOPING IN REPRESSION   We have discussed the fact that multipartite operators function in gene regulation by forming a DNA loop through protein-protein contact. To generalize, we can assume that such contacts between proteins bound to adjacent multipartite sites, as for the adjacent operators in phage λ, generate a microloop. The main unanswered question remaining is how a DNA loop contributes to repression. Its actual role in inhibiting transcription at one of the above steps may vary from system to system and could be either passive or active. (*a*) In a passive model, only one operator located contiguous to the promoter in a bipartite system plays a direct role. Repressor binding at this site causes repression by any one of the mechanisms discussed above. The second operator contributes to repression indirectly. Repressor binding at this site because of cooperative binding via a DNA loop enhances the stability of the first repressor-operator complex. (*b*) Alternatively, DNA-loop formation involving two operators directly contributes to repression either by changing the DNA structure or by altering RNA-polymerase conformation. A looped DNA would topologically be an independent domain capable of assuming a different conformation. The structure of such a domain profoundly affects site-specific recombination within this domain (62). A promoter altered in this way may be refractory to isomerization by RNA polymerase. A DNA loop may facilitate additional protein-protein contacts that help repression. For example, one or both of the repressor molecules in a loop may also contact RNA polymerase directly and inhibit its action. Such contact between repressor(s) and RNA polymerase (or a DNA-bound activator protein) generates a DNA-multiprotein complex of a higher order structure that topographically contains more than one DNA loop. A repressor-RNA polymerase contact made this way may block isomerization, initiation, or promoter clearance by the polymerase. From the properties of the repressor proteins that we have described, we discuss the role of a loop in the effective repression in the respective systems.

Lac repressor   It is very likely that repressor loops O_1-O_2 or O_1-O_3 form a repression apparatus that blocks transcription initiation (Figure 7). It was originally believed that repressor binding to O_1 hinders the formation of a closed complex. Recently, however, *lac* DNA molecules carrying both RNA polymerase and Lac repressor have been found (69). Whether an O_1-O_2 or O_1-O_3 loop helps repression by increasing repressor binding to O_1 or contributes to a later stage remains to be answered.

DeoR repressor O_1 and O_2 are the actual operators for $P1$ and $P2$ promoters, respectively. DeoR binding to O_1 and O_2 most likely sterically hinders binding of RNA polymerase to the respective promoters (Figure 8). Normally, O_1-O_2 loop may stabilize binding to both, thus helping repression passively. The alternating loops O_E-O_1 and O_E-O_2 may contribute to enhanced repressor binding to O_1 and O_2 respectively, if the O_1-O_2 loop fails to materialize.

GalR repressor   Since Gal repressor does not block binding of RNA polymerase to the promoter, the O_E-O_I loop (Figure 2) brings about repression by an active role, either by changing the promoter structure or by allowing a repressor-RNA polymerase contact to form a higher-order multiloop structure. Thus, a loop very likely contributes to repression in *gal* by blocking isomerization, initiation, or promoter clearance.

AraC   AraC exerts its many controls through different stages of cooperative interaction with DNA-forming alternate loops (Figure 3). Although it is simple to perceive that the $I-O_2$ and the O_1-O_2 loops cause repression by excluding closed-complex formation at P_B and P_C, respectively, the mechanism of repression of P_C by the O_2-I loop is most likely, as for *gal*, at a post RNA-polymerase binding level. However, the mechanism by which liganded AraC binding to I activates P_C is completely unknown. Any model explaining the activation mechanism must also take into account the fact that CRP binding to a site between O_1 and I is additionally required for P_B and P_C activation. Note that P_C is not activated by liganded AraC.

In the discussion above, we have pointed out the various biochemical stages at which a repressor could act to freeze transcription and have noted that repressors from different systems may function at different stages. Furthermore, it is also likely that repressors may have evolved to function at more than one stage. At least some of the repressor molecules possess the flexibility to freeze transcription, given the opportunity to do so, at more than one stage. In support of this concept, we refer to the experiments showing that Lac repressor, in addition to blocking RNA polymerase binding, can inhibit transcription at the stage of isomerization or initiation (22) as well as elongation (9, 64), as discussed in previous sections.

POSITIVE CONTROL ELEMENTS AT DISTANT LOCATIONS

The gene regulatory proteins that we have discussed exemplify the participation of a DNA loop in the regulatory processes. The examples are mainly repressors, some of which also act as activators. There are cases of pure activator proteins, whose action depends also on DNA control sites located far from the promoter they control (57). To illustrate the role of a DNA loop in such systems, we cite only one example here. The GlnG (also called NtrC) protein of enteric bacteria binds to distant sites to activate transcription from a promoter of the gene *glnA*, which encodes glutamine synthetase (reviewed in 21, 59). GlnG has two binding sites located at -140- and -108-bp with respect to the *glnA* promoter. A phosphorylated form of GlnG protein (GlnG $\sim$P) has been shown to bind to these sites and catalyze the isomerization of the closed complex of RNA polymerase 6^{54}-holoenzyme and *glnA* promoter to an open complex in the presence of ATP (26, 31, 55). If this reaction requires a physical contact between DNA-bound GlnG$\sim$P and RNA polymerase in the closed complex, a contact is easily envisioned if the intervening DNA sequence is looped out. Such a model also explains how the GlnG binding site can be moved kilobases away from the promoter, either in the downstream or upstream direction, without loss of promoter activation (59). Interestingly, for transcription activation, the GlnG binding site cannot be adjacent to the *glnA* promoter and must be at least 70 bp away from it (59). Below this minimal distance, energy constraints may prohibit the required DNA curving or twisting for GlnG-RNA polymerase contact by DNA looping. It is likely that a loop for positive activation of RNA polymerase, unlike the repression loop in *gal* and λ, discussed previously, may be transient in nature; GlnG-RNA polymerase conduct should have a higher turnover rate.

CURRENT AFFAIRS

The examples of gene regulation studied in the past (in prokaryotes) were thought to have one operator or one activator binding site. In both cases, these sites were also found to be located close enough to the promoter to so that the regulatory proteins could physically affect the binding or activity of RNA polymerase. Such simple systems of gene regulation may be an exception rather than the rule. Several new facts and concepts have emerged from the progress made during the last few years: (*a*) DNA control elements are frequently multipartite; (*b*) The control elements can be located far away from the promoters they control; (*c*) DNA looping permits the formation of a DNA multiprotein complex of a necessarily higher order structure, involving multipartite dispersed DNA control elements. Such higher order structures may be aided by another frequently observed property of regulatory proteins, i.e. DNA bending. Specific protein-induced DNA bending is an important element in the formation of higher order structures of biological consequence (12, 71). How DNA binding, looping, and higher order structure translate into gene regulation remain to be elucidated.

Regulation of transcription initiation by distant DNA elements, called enhancers or silencers, now seems to be a common phenomenon in eukaryotes (35, 65, 68). Such elements very likely act by forming a DNA loop. A frequent observation among Drosophila developmental genes is the occurrence of extended regions of DNA control elements upstream. The regulatory protein Ubx encoded by the *ubx* gene (ultrabithorax) mediates in vitro the formation of a DNA loop involving such DNA sites (in its own promoter) and provides a mechanistic basis for the regulatory effects of Ubx (2; P. A. Beachy, J. P. Varkey, D. von Kessler, personal communication). Proper stereospecific alignment between an enhancer element and some 21-bp repeat sequences by protein-protein interaction has been shown to be a requirement in gene expression in simian virus 40 (70). Protein-protein interaction giving rise to DNA loops is also found when steroid hormone (e.g. progesterone) receptor binds to specific DNA control elements located at a distance from each other (reviewed in ref. 3). In yeast, genes are silenced (repressed) when they are located at HML or HMR locus in the chromosome. Repression requires flanking *cis*-acting silencer elements (*E* and *I*) that are located kilobases away from the promoter (*P*) (30a). At the HML locus, DNA loop formation has been observed through electron microscope involving either the silencers (*E-I* loop) or a silencer and the promoter (*E-P* or *I-P* loop). Loop formation requires a nuclear protein called RAP-1. It has been suggested that an *E-P* or *I-P* looping brings about repression by excluding the binding of RNA polymerase or an essential transcription factor to DNA. Even micro-loops, in some form, may seem to play an important role in regulation of eukaryotic genes. It has been shown that the yeast repressor protein $\alpha2$ is a flexible dimer that binds to physically separated operator half-symmetry sites (63). The two half sites are spaced $2 + \frac{1}{2}$ DNA helical turns apart, thus positioning them to opposite faces of the helix. The structure of $\alpha2$ allows a dimer to reach across its operator such that it occupies two half sites and allows binding of another protein in between. For a review of genetic action of distant DNA elements in eukaryotes, we refer to other recent excellent articles (45, 76).

If looping out intervening DNA serves the purpose of bringing two DNA-bound proteins (homologous or heterologous) together for the regulation of transcription, then it may very likely be part of other complex multiprotein reactions, replication and recombination. It is routine thinking now to ask if DNA looping plays a role in regulating DNA replication. It has been demonstrated that DNA looping helps several dispersed DNA-bound proteins for making additional protein-protein or DNA-protein contacts, which are otherwise prohibitory, in the formation of a productive *intasome* complex to carry out site-specific recombination in bacteriophage λ (12, 71).

EPILOGUE

We have seen during the last decade an explosion in our knowledge of genetic regulatory mechanisms, both in prokaryotes and eukaryotes. They have revealed fascinating features not envisioned before and made the subject perhaps the most understood area of molecular biology. We end this exposition about our efforts to reformulate the early concepts of genetic regulatory mechanisms to incorporate new facts and thoughts with a quote from an earlier author.

> "My salad days,
> When I was green in judgment, Cold in blood,
> To say as I said then" (66).

ACKNOWLEDGMENTS

I appreciate the discussions with my co-workers in the laboratory and thank Althea Gaddis for skillful typing of this manuscript.

Literature Cited

1. Adhya, S., Miller, W. 1979. Modulation of the two promoters of the galactose operon of *Escherichia coli*. *Nature* 279:492–94
2. Beachy, P. A., Krosnow, M. A., Gagis, E. R., Hogness, D. S. 1988. An *ultrabithorax* protein binds sequences near its own and the *antennapedia* PI promoters. *Cell* 55:1069–81
3. Beato, M. 1989. Gene regulation by steroid hormones. *Cell* 56:335–44
4. Besse, M. von Wilcken-Bergmann, B., Muller-Hill, B. 1986. Synthetic *lac* operator mediates repression through *lac* repressor when introduced upstream and downstream from *lac* promoter. *EMBO J.* 5:1377–81
5. Borowiec, J. A., Zhang, L., Sasse-Dwight, S., Gralla, J. D. 1987. DNA supercoiling promotes formation of a bent repression loop in *lac* DNA. *J. Mol. Biol.* 196:101–11
6. Chamberlin, M. 1974. The selectivity of transcription. *Annu. Rev. Biochem.* 43:721–75
7. Dandanell, G., Hammer, K. 1985. Two operator sites separated by 599 base pairs are required for *deoR* repression of the *deo* operon of *Escherichia coli*. *EMBO J.* 4:3333–38
8. Dandanell, G., Valentin-Hansen, P., Love-Larsen, J. E., Hammer, K. 1987. Long-range cooperativity between gene regulatory sequences in a prokaryote. *Nature* 325:823–26
9. Deuschle, U., Gentz, R., Bujard, H. 1986. *Lac* repressor blocks transcribing RNA polymerase and terminates transcription. *Proc. Natl. Acad. Sci. USA* 83:4134–37
10. Dripps, D., Wartell, R. M. 1987. DNA bending induced by the catabolite activator protein allows ring formation of 144 bp DNA. *J. Biomol. Struct. Dyn.* 5:1–13
11. Dunn, T. M., Haber, S., Ogden, S., Schleif, R. F. 1984. An operator at −280 base pairs that is required for repression of *araBAD* operon promoter: addition of DNA helical turns between the operator and promoter cyclically hinders repression. *Proc. Natl. Acad. Sci. USA* 81:5017–20
12. Echols, H. 1986. Multiple DNA-protein interactions governing high precision DNA transactions. *Science* 133:1050–56
13. Eismann, E., von Wilcken-Bergman, B., Muller-Hill, B. 1987. Specific destruction of the second *lac* operator decreases repression of the *lac* operon in *Escherichia coli* fivefold. *J. Mol. Biol.* 195:949–52

14. Englesberg, E., Irr, J., Power, J., Lee, N. 1965. Positive control of enzyme synthesis by gene C in the L-arabinose system. *J. Bacteriol.* 90:946–57
15. Englesberg, E., Wilcox, G. 1974. Regulation: positive control. *Annu. Rev. Genet.* 8:219–42
16. Flashner, Y., Gralla, J. D. 1988. Dual mechanism of repression at a distance in the *lac* operon. *Proc. Natl. Acad. Sci. USA* 85:8968–72
17. Fried, M., Crothers, D. M. 1981. Equilibrium kinetics of the *lac* repressor-operator interactions by polyacrylamide gel electrophoresis. *Nucleic Acids. Res.* 9:6505–25
18. Gilbert, S., Gralla, J., Majors, J., Maxam, A. 1975. Lactose operator sequences and the action of *lac* repressor. In *Protein Ligand Interactions*, ed. H. Sund, G. Blauer, pp. 193–210. Berlin: de Gruyter
19. Griffith, J., Hochschild, A., Ptashne, M. 1986. DNA loops induced by cooperative binding of λ repressor. *Nature* 322:750–52
20. Guarente, L., Nye, J. S., Hochschild, A., Ptashne, M. 1982. A mutant λ repressor with a specific defect in its positive control function. *Proc. Natl. Acad. Sci. USA* 79:2236–39
21. Gussin, G. N., Ronson, C. W., Ausubel, F. M. 1986. Regulation of nitrogen fixation genes. *Annu. Rev. Genet.* 20:567–91
22. Haber, R., Adhya, S. 1988. Interaction of spatially separated protein-DNA complexes for control of gene expression: operator conversions. *Proc. Natl. Acad. Sci. USA* 86:9683–87
23. Hamilton, E. P., Lee, N. 1988. Three binding sites for AraC protein are required for autoregulation of *araC* in *Escherichia coli*. *Proc. Natl. Acad. Sci. USA* 85:1749–53
24. Hawley, D. R., McClure, W. R. 1983. The effect of a λ repressor mutation on the activation of transcription initiation from λ P_{RM} promoter. *Cell* 32:327–33
25. Herrin, J. L. Jr., Bennett, G. N. 1984. Role of DNA regions flanking the tryptophan promoter of *Escherichia coli* II. Insertion of *lac* operator fragments. *Gene* 32:349–56
26. Hirschman, J., Wong, P. K., Sei, K., Keener, J., Kustu, S. 1985. Products of nitrogen regulatory genes *ntrA* and *ntrC* of enteric bacteria activate *glnA* transcription in vitro: evidence that the *ntrA* product is a sigma factor. *Proc. Natl. Acad. Sci. USA* 82:7525–29

27. Hochschild, A. 1989. Protein-protein interaction and DNA loop formation. *Biological Effects of DNA Topology*, ed. J. Wang, N. Cozzarelli. New York: Cold Spring Harbor Lab. In press

28. Hochschild, A., Irwin, N., Ptashne, M. 1983. Repressor structure and the mechanism of positive control. *Cell* 32:319–25

29. Hochschild, A., Ptashne, M. 1986. Cooperative binding of λ repressors to sites separated by integral turns of the DNA helix. *Cell* 44:681–87

30. Hochschild, A., Ptashne, M. 1988. Interaction at a distance between λ repressors disrupts gene activation. *Nature* 336:353–57

30a. Hofmann, J. F.-X., Laroche, T., Barnd, A. H., Gasser, S. 1989. RAP-1 factor is necessary for DNA loop formation in vitro at the silent mating type locus *HML. Cell* 57:725–37

31. Hunt, J. P., Magasanik, B. 1985. Transcription of *glnA* by purified *Escherichia coli* components: Core RNA polymerase and the products of *glnF, glnG* and glnL. *Proc. Natl. Acad. Sci. USA* 82:8453–57

32. Huo, L., Martin, K. J., Schleif, R. 1988. Alternative DNA loops regulate the arabinose operon in *Escherichia coli. Proc. Natl. Acad. Sci. USA* 85:5444–48

33. Irani, M., Orosz, L., Adhya, S. 1983. A control element within a structural gene: the *gal* operon of *Escherichia coli. Cell* 32:783–88

34. Johnson, A. D., Meyer, B. J., Ptashne, M. 1979. Interactions between DNA-bound repressors govern regulation by the λ phage repressor. *Proc. Natl. Acad. Sci. USA* 76:5061–65

35. Johnson, P. F., Landschultz, W. H., Graves, B. J., McKnight, S. L. 1987. Identification of a rat liver nuclear protein that binds to the enhancer core element of three animal viruses. *Genes Dev.* 1:133–46

36. Kolata, G. 1983. Control element found within structural gene. *Science* 220:294

37. Kotlarz, D., Fritsch, A., Buc, H. 1986. Variations of intramolecular ligation rates allow the detection of protein-induced bends in DNA. *EMBO J.* 5: 799–803

38. Kramer, H., Amouyal, M., Nordheim, A., Muller-Hill, B. 1988. DNA supercoiling changes the spacing requirement of two *lac* operators for DNA loop formation with *lac* repressor. *EMBO J.* 4:547–56

39. Kramer, H., Niemoller, M., Amouyal, M., Revet, B., von Wilcken-Bergmann, B., et al. 1987. *lac* repressor forms loops with linear DNA carrying two suitably spaced *lac* operators. *EMBO J.* 6:1481–91

40. Lee, D.-H., Schleif, R. F. 1989. *In vivo* DNA loops in *araCBAD:* size limits and helical repeat. *Proc. Natl. Acad. Sci. USA* 86:476–80

41. Lee, N., Francklyn, C., Hamilton, E. P. 1987. Arabinose induced binding of AraC protein to *araI$_2$* activates the *araBAD* operon promoter. *Proc. Natl. Acad. Sci. USA* 84:8814–18

42. Lee, N. L., Gielow, W. O., Wallace, R. G. 1981. Mechanism of *araC* autoregulation and the domains of two overlapping promoters, P_C and P_{BAD} in the L-arabinose regulatory region of *Escherichia coli. Proc. Natl. Acad. Sci. USA* 78:752–56

43. Majumdar, A., Adhya, S. 1984. Demonstration of two operator elements in *gal: in vitro* repressor binding studies. *Proc. Natl. Acad. Sci. USA* 81:6100–4

44. Majumdar, A., Adhya, S. 1986. Probing the structure of Gal operator-repressor complexes. *J. Biol. Chem.* 262:13258–62

45. Maniatis, T., Goodbourn, S., Fischer, J. A. 1987. Regulation of inducible and tissue-specific gene expression. *Science* 236:1237–45

46. Martin, K., Huo, L., Schleif, R. 1986. The DNA loop model for *ara* repression: AraC protein occupies the proposed loop sites *in vivo* and repression-negative mutations lie in these same sites. *Proc. Natl. Acad. Sci. USA* 83:3654–58

47. McClure, W. R. 1985. Mechanism and control of transcription initiation in prokaryotes. *Annu. Rev. Biochem.* 54:171–204

48. Mossing, M. C., Record, M. T. Jr. 1985. Thermodynamic origins of specificity in the *lac* repressor-operator interaction: adaptability in the recognition of mutant operator sites. *J. Mol. Biol.* 186:296–305

49. Mossing, M. C., Record, M. T. Jr. 1986. Upstream operators enhance repression of the *lac* promoter. *Science* 233:889–92

50. Muller-Hill, B., Beyreuther, K., Gilbert, W. 1971. Lac repressor from *Escherichia coli. Methods Enzymol.* 21D: 483–90

51. Musso, R., DiLauro, R., Adhya, S., deCrombrugghe, B. 1977. Dual control for transcription of the galactose operon by cyclic AMP and its receptor protein at two interspersed promoters. *Cell* 12: 847–54

52. Ogden, S., Haggerty, D., Stonner, C. M., Kolodrubetz, D., Schleif, R. 1980.

The *Escherichia coli* L-arabinose operon: binding sites of the regulatory proteins and a mechanism of positive and negative regulation. *Proc. Natl. Acad. Sci. USA* 77:3346–50

53. Pardee, A. B., Jacob, F., Monod, J. 1959. The genetic control and cytoplasmic expression of "inducibility" in the synthesis of β-galactosidase. *J. Mol. Biol.* 1:165–78

54. Pfahl, M., Gulde, V., Bourgeois, S. 1979. "Second" and "third operator" of the *lac* operon: an investigation of their role in the regulatory mechanisms. *J. Mol. Biol.* 127:339–44

55. Popham, D. L., Szeto, D., Keener, J., Kustu, S. 1989. Function of a bacterial activator protein that binds to transcriptional enhancers. *Science* 243:629–35

56. Ptashne, M. 1986. *A Genetic Switch: Gene Control and Phage* λ. Cell Press/Blackwell Sci.

57. Ptashne, M. 1989. How gene activators work. *Sci. Am.* 260:40–47

58. Ptashne, M., Jeffrey, A., Johnson, A. D., Maurer, R., Meyer, B. J., et al. 1980. How the λ repressor and Cro work. *Cell* 19:1–11

59. Reitzer, L. J., Magasanik, B. 1986. Transcription of *glnA* in *Escherichia coli* is stimulated by activator bound to sites far from the promoter. *Cell* 45:785–92

60. Reznikoff, W. S., Winter, R. B., Hurley, C. K. 1974. The location of the repressor binding sites in the *lac* operon. *Proc. Natl. Acad. Sci. USA* 71:2314–18

61. Sadler, J. R., Smith, T. F. 1971. The nature of lactose operator constitutive mutations. *J. Mol. Biol.* 59:273–305

62. Saldanha, R., Flanagan, P., Fennewald, M. 1987. Recombination by resolvase is inhibited by *lac* repressor simultaneously binding operators between res sites. *J. Mol. Biol.* 196:505–16

63. Sauer, R. T., Smith, D. L., Johnson, A. D. 1988. Flexibility of the yeast α2 repressor enables it to occupy the ends of its operator, leaving the center free. *Genes Dev.* 2:807–16

64. Sellitti, M. A., Pavco, P. A., Steege, D. A. 1987. Lac repressor blocks in vivo transcription of *lac* control region DNA. *Proc. Natl. Acad. Sci. USA* 84:3199–203

65. Serfling, E., Jasin, M., Schaffner, W. 1985. Enhancers and eukaryotic gene transcription. *Trends Genet.* 1:224–30

66. Shakespeare, W. 1960. *The Tragedy of Antony and Cleopatra,* ed. M. Mack. Baltimore: Penguin

67. Shore, D., Langowski, J., Baldwin, R. L. 1981. DNA flexibility studied by covalent closure of short fragments into circles. *Proc. Natl. Acad. Sci. USA* 78:4833–37

68. Sippel, A. E., Borgmeyer, U., Puschel, A. W., Rupp, R. A. W., Stief, A., et al. 1987. Multiple nonhistone protein-DNA complexes in chromatin regulate the cell- and stage-specific activity of an eukaryotic gene. In *Structure and Function of Eukaryotic Chromosomes,* ed. W. Henning, pp. 255–68. Berlin/Heidelberg: Springer, Verlag

69. Straney, S. B., Crothers, D. M. 1987. Lac repressor is a transient gene-activating protein. *Cell* 51:699–707

70. Takahashi, K., Vigneron, M., Matthes, H., Wildeman, A., Zenke, M., et al. 1986. Requirement of sterospecific alignments for initiation from simian virus 40 early promoter. *Nature* 319:121–26

71. Thompson, J. F., Landy, A. 1988. Empirical estimation of protein-induced DNA bending angles: applications to λ site-specific recombination complexes. *Nucleic Acids Res.* 16:9687–705

72. Valentin-Hansen, P. 1985. DNA sequences involved in expression and regulation of *doeR, cytR* and cAMP/CRP-controlled genes in *Escherichia coli.* In *Gene Manipulation and Expression,* ed. R. E. Glass, J. Spizek, pp. 273–88. London: Croom Helm

73. Valentin-Hansen, P., Aiba, H., Schumperli, D. 1982. The structure of tandem regulatory regions in the *deo* operon of *Escherichia coli* K12. *EMBO J.* 1:317–22

74. Valentin-Hansen, P., Albrechtsen, B., Love-Larsen, J. E. 1986. DNA-protein recognition: demonstration of three genetically separated operator elements that are required for repression of the *Escherichia coli deoCABD* promoters by DeoR repressor. *EMBO J.* 5:2015–21

75. Vershon, A. K., Liao, S.-M., McClure, W. R., Sauer, R. T. 1987. Interaction of the bacteriophage P22 Arc repressor with operator DNA. *J. Mol. Biol.* 195:323–31

76. Wang, J. C., Giaver, G. N. 1988. Action at a distance along DNA. *Science* 240:300–4

Annu. Rev. Genet. 1989. 23:251–87

THE POPULATION GENETICS OF *DROSOPHILA* TRANSPOSABLE ELEMENTS*

Brian Charlesworth

Department of Ecology and Evolution, University of Chicago, 1103 East 57th St, Chicago, Illinois 60637

Charles H. Langley

Laboratory of Molecular Genetics, NIEHS, P.O. Box 12233, Research Triangle Park, North Carolina 27709

CONTENTS

INTRODUCTION

The discovery that the genomes of all carefully investigated species contain dispersed, repeated DNA elements, capable of inserting replicas of themselves into novel genomic locations, is one of the most exciting recent developments in molecular genetics (7, 24, 58, 124). This discovery has occasioned vigorous debate about the nature of the forces that affect the distribution of such transposable elements (TEs) in populations of their host organisms (24, 37, 63, 103, 109, 136). There has been significant recent progress in both empirical and theoretical population studies of TEs (18, 25, 27, 28, 45, 46, 61).

In this paper, we review both population data from the study of *Drosophila* TEs, and theory pertaining to the interpretation of such data. The data on *Drosophila* populations are by far the most abundant, and they provide the most critical tests of theory. We do not consider in detail information on *P* elements, except when relevant to general points that we wish to make, since these have recently been the subject of excellent reviews (16, 45, 46). We also omit any consideration of the taxonomic distribution of families of TEs, except when directly relevant to population dynamics. Nor do we review the many important theoretical investigations of the population biology of TEs that are not closely related to the interpretation of empirical studies (e.g. 55, 63, 104, 105, 108, 138). The literature on dispersed repetitive sequences in other taxa is not covered here, except where it can be usefully compared with the results for *Drosophila*.

The theoretical and empirical studies reviewed here yield the conclusion that *Drosophila* TEs are maintained in populations as a result of transpositional increase in copy number, and that their spread is checked by one or more opposing forces. In other words, the concept that TEs are essentially intragenomic parasites (37, 109) is supported. We also conclude that the genetic and evolutionary mechanisms influencing the distribution of TEs in populations significantly affect their distribution among different regions of the genome, and thus may explain certain aspects of chromosome organization.

DROSOPHILA TRANSPOSABLE ELEMENTS

Basic Structure and Properties

Several recent reviews cover the basic properties of *Drosophila* transposable elements (7, 50, 118). A few important features are worth mentioning here as a basis for further discussion. The diversity of structure of *Drosophila* TEs approaches that known for all organisms. Structural differences between the 50 or so families of *D. melanogaster* TEs (50, 118) are thought to reflect important differences in their biology and mechanism of transposition. A typical element is between 1–9 kb in length, and often has direct or inverted repeats at each end. Little is known about the structural characteristics of TEs in other *Drosophila* species, except that species related to *D. melanogaster* share many elements showing substantial sequence homology with those in *D. melanogaster*, whereas increasingly distant species show less similarity (19, 22, 39, 46, 62, 66, 93, 126; J. Hey & W. F. Eanes, in preparation). The majority of *D. melanogaster* TEs are probably either retroviruses or retroposons (7, 50, 146). They are assumed to "retrotranspose" via an RNA intermediate that is reverse-transcribed into DNA and integrated at a new site in the genome of the host. The primary distinction between retroviruses and retroposons is that the former are capable of infecting a different cell or even host, while the latter can replicate only through insertion into new chromosomal sites. No *Drosophila* TE has been demonstrated to have an extracellular phase to its life history, despite their possession of many structural properties similar to those of retroviruses, including direct long terminal repeats (LTRs). Yet other *Drosophila* elements *(I, F)* resemble mammalian retroposons like *L1* in lacking LTRs, but contain the genes needed for replication via reverse transcription. Other *Drosophila* TEs, such as the *P* and *hobo* elements with inverted terminal repeats, and the *FB* element with multiple internal repeats, are thought to transpose by direct DNA replication. Except for the elements involved in hybrid dysgenesis *(P, I* and, probably, *hobo)*, which mobilize at a high rate in the F1 progeny of matings between flies lacking elements and flies that possess them (14, 16, 45, 46, 148), few of these distinctions among *Drosophila* TEs are known to be important in their population biology.

Most *Drosophila* TEs appear to be distributed throughout the genome, as dispersed middle repetitive DNA, although many families tend to accumulate differentially in the proximal euchromatin and heterochromatin (see below). If there is site-specificity for insertions, the number of sites available for occupation by a given family appears to be very large (50, 118). An exception to this is provided by the Type I and Type II ribosomal insertion sequences of *Drosophila* and other insects. These show strong specificity for the ribosomal cistrons, but occasionally are found inserted elsewhere (72, 116, 145, 146). They are retroposons without LTRs, and have considerable sequence similarity with the other *Drosophila* elements of this class (146).

Rates of Movement of Elements

Most spontaneous morphological mutations used in *Drosophila* genetics are apparently caused by the insertions of TEs (50, 58, 118). The low rate at which such mutations occur and the low rate of occurrence of lethal mutations, despite the fact that approximately 10% of the DNA of the fly consists of middle repetitive (putatively transposable) sequences, indicates that such insertion events are infrequent, except for the mobilization of *P, I* or *hobo* elements in hybrid dysgenesis. Similarly, the great stability of the majority of mutations caused by TE insertions indicates that the spontaneous loss of elements from their chromosomal locations is usually very rare or imprecise (118, 143). For this reason, relatively few experiments have been performed that are expressly designed to measure the rates of insertion of new elements into new chromosomal sites (transposition) or loss of elements from old sites (excision).

From the population point of view, one would ideally like to have estimates of the probability u that an element at a given site produces a new copy and the probability v that it is excised (see Table 1). The experiments reviewed below almost certainly overestimate the true rate of excision, in the sense of precise removal of elements (including their terminal repeats) from their chromosomal locations, since detailed molecular characterizations of the events have not been carried out. Imprecise excision, in which a substantial part of the TE sequence remains inserted, may lead to reversion or alteration in a mutant phenotype associated with the insertion (46, 50). Homologous recombination between the direct LTRs of a retrovirus-like element, generating an extra-chromosomal circle and leaving one LTR behind, mimics excision (50). Crossing over between homologous elements located at different chromosomal sites (ectopic exchange) leads to rearrangements, such as deletions or inversions, and could cause the loss of much of the DNA of one or both of the elements concerned. There is experimental evidence for ectopic exchange between TEs and the concomitant production of rearrangements in *D. melanogaster,* apparently at a much higher rate than true excision (34, 56). Such exchange may be confined to individuals heterozygous for the presence of TEs at the chromosomal sites involved in the exchange event (34, 78; B. H. Judd, personal communication).

Movement of the composite transposable element *TE,* which consists of host sequences flanked by *FB* sequences (111), has been studied by Ising & Block (67), using special marker stocks to detect transposition and excision. Different forms of *TE,* associated with different host sequences, appear to move at different rates; values of u range from 10^{-5} to 2.5×10^{-4}. Excision of one version *(TE30)* takes place at a relatively high rate (3×10^{-4}), but others are much more stable. The rate of transposition of *Dm225* has been estimated in similar experiments to be approximately 3×10^{-4} (115). Estimates of v for several elements (from the rate of disappearance of the mutant phenotypes for which they are responsible) range from a high value of 5.1×10^{-3} for the *FB* element associated with the w^c mutation, to zero for most other elements (143). The overall result for mutants due to *gypsy, copia, roo, F, BEL,* and *hobo,* together with four other uncharacterized insertional mutations, is 3 excision events out of 465,726 flies ($v = 6.4 \times 10^{-6}$), pooled over both *P-M* dysgenic and nondysgenic crosses. The rates for *P* elements are 10^{-3} for dysgenic crosses and 10^{-4} for nondysgenic crosses (143).

Eggleston et al (43) studied the rate of appearance and loss of nineteen families of elements in the euchromatic section of the X chromosome of *D. melanogaster* males in both *P-M* dysgenic and nondysgenic crosses, by in situ hybridization to polytene salivary chromosomes. The values of u for these elements can be calculated by multiplying the published estimates of rates of insertion by 9, assuming that the probability of insertion of a new copy into the X is 1/9 (the proportion of the euchromatin represented by the X in a male). For elements other than P, the overall values of u are 4.0×10^{-4} for dysgenic crosses and 1.0×10^{-5} for nondysgenic crosses (pooled value: 2.9×10^{-4}). The corresponding u values for P elements are 0.26 and 5.1×10^{-3}. The values of v are 5.8×10^{-4} and 0 for elements other than P in dysgenic and nondysgenic crosses respectively (pooled value: 3.6×10^{-5}), compared with corresponding values of 5.1×10^{-4} and 0 for P elements. A similar analysis of an in situ hybridization study (K. Harada, K. Yukuhiro & T. Mukai, in preparation) of insertions and losses of *I, hobo, copia, 412,* and *17.6* (in recessive lethal second chromosomes maintained by crossing to a balancer stock) yields mean estimates of u and v of 5.7×10^{-5} and 7.7×10^{-6} respectively for the three latter families. In contrast, *I* and *hobo* show much higher rates of movement ($u = 2.2 \times 10^{-3}$ and 6.0×10^{-3} respectively; v unknown and 1.5×10^{-3} respectively). The high rates of movement of *I* and *hobo* in these stocks may reflect hybrid dysgenesis.

Although these rate estimates are subject to considerable error, particularly as the calculations assume random insertions of elements into the euchromatin and ignore the heterochromatin, they suggest that transposition rates are typically rather low (of the order of 10^{-4}). Excision rates are probably even lower, possibly by an order of magnitude or more. There is no statistically significant evidence for mobilization of elements other than P elements in *P-M* dysgenic crosses, despite earlier claims to the contrary (43, 143). There are several reports in the literature of unusually high rates of movement of elements. Unfortunately, in some cases the stocks concerned lacked genetic markers or were not known for certain to be initially homozygous at all relevant sites (9, 10, 59, 84, 112), so that contamination or Mendelian segregation of elements cannot be excluded as causes of apparent movement. In others, movement of TEs is correlated with loss or alteration of the mutant allele characterizing the stock (53, 54), so that contamination is again a possibility. Nonetheless, it is clear that there may be variation between elements belonging to a given family in their rates of movement of TEs. For example, Gerasimova's ct^{MR2} allele (53, 54) and the mutator factor of Lim (87, 88; B. H. Judd personal communication) provide examples in which the normally quiescent *gypsy* element appears to be unusually mobile. The *tom* element of *D. anannassae* seems also sometimes to transpose at an unusually high rate (125). These cases may reflect the origin by mutation of an element that escapes the normal process of control of transposition (31).

REGULATION OF TRANSPOSITION   There is extensive evidence in prokaryotic systems for regulation of the rate of transposition, such that u is a decreasing function of the number of elements of the family in question in the same cell (7, 31, 75). The low rate at which *Drosophila* TEs typically transpose and excise has precluded the detection and characterization of any regulation of either process in *Drosophila,* with the exception of the elements involved in hybrid dysgenesis, that seem to have the capacity to repress their own transposition (14, 16, 45, 46, 148). However, recent molecular studies of *P* elements indicate that repressor activity appears to be a property of defective elements (46), which are widespread in natural populations with the *P* cytotype (15). The molecular basis of repression in the case of the *I* and *hobo* elements is unknown at present, although it is known that various kinds of defective *hobo* elements are often present in the genome (14, 15). As far as retroviral-like elements are concerned, the only evidence suggestive of regulation is their greatly increased copy number in cultured cells (114, 137). This can, however, equally well be accounted for by relaxed selection against elements under these artificial conditions.

POPULATION MODELS OF TRANSPOSABLE ELEMENTS

Population Statics

The first kind of prediction from population models concerns the distribution of elements belonging to a given family between different individuals within a population (30, 77). The parameters used to describe this distribution are listed in the upper portion of Table 1. The state of a population of size N at a given chromosomal site i can be described by the frequency x_i with which the $2N$ copies of that site are occupied by an element. If the element frequencies at different sites are independent (no linkage disequilibrium), a list of values of x_i for the entire array of sites ($i = 1, 2 \ldots m$) provides a complete description of the state of the population.

The mean copy number $\bar{n}$ is equal to $2 \Sigma_i x_i$. The variance of copy number is given by the following expression (30):

$$V_n = \bar{n} (1 - \bar{x}) - 2m\sigma_x^2 + 4 \sum_{i<j} D_{ij} \qquad 1.$$

If σ_x^2 and the sum of the D_{ij}s are sufficiently small, only the first term need be considered; this corresponds to a binomial distribution of n across individuals. If $\bar{x} \ll 1$, as frequently seems to be the case (see below), then there is a Poisson distribution of n, with $V_n \approx \bar{n}$. Variation between sites in element frequencies reduces the variance below binomial expectation; if there is an overall tendency for the frequency of occupation of one site to be higher if a

Table 1   Parameters for describing transposable element population biology

Static parameters	
m	number of occupable sites in a haploid genome
n	number of copies of a given family of elements in a given individual
$\bar{n}$	mean number of copies of a given family of elements per individual in a population
V_n	variance in copy number between individuals within a population
N	number of breeding individuals in a population
x_i	frequency of elements at the ith occupable site in a population
$\bar{x}$	mean of x_i over all sites ($\bar{x} = \bar{n}/2m$).
σ_x^2	variance of x_i between sites
D_{ij}	coefficient of linkage disequilibrium in element frequency between the ith and jth occupable sites.

Dynamic parameters	
u_n	the germ-line probability of transposition per generation of an element belonging to a given family, in a host individual carrying n elements of that family. (The functional dependence of u on n, denoted by the subscript n, allows for possible regulation of the rate of transposition in response to copy number.)
v	the germ-line probability of excision per generation of an element of a given family
w_n	the fitness of a host individual carrying n members of a given family, relative to a value of one for an element-free individual
$\bar{w}$	the mean of w_n over all individuals in the population

neighboring site is occupied, such that $\Sigma D_{ij} > 0$ (net positive linkage disequilibrium), the variance will be increased; the opposite tendency (net negative linkage disequilibrium) reduces the variance.

Element Dynamics in an Infinite Population

A model of the transmission of a family of elements from generation to generation needs to incorporate the following features, summarized in the lower part of Table 1:

TRANSPOSITION AND EXCISION   If transposition is not automatically accompanied by excision of the parental elements, there is clearly an expected gain of copy number of nu_n in the germ-line of an individual with n elements.

Excision of elements at rate v in an individual with n elements reduces copy number by nv. The net change in mean copy number for the population is approximately $\bar{n}(u_{\bar{n}}-v)$ (30). If there is no force other than excision opposing transposition, and no regulation of transposition rate (so that u is independent of n), transposition leads to fixation of elements at each site if $u>v$. If transposition is regulated, and if the rate of transposition at low copy number (u_{o}) exceeds v, mean copy number will increase until $\bar{n} \approx \hat{n}$ (such that $u_{\hat{n}}=v$), or until all available sites have filled up with elements (30). The evidence reviewed above on the values of u and v suggests that *Drosophila* elements do indeed generally have the capacity to increase in copy number as a result of an excess of transposition events over excision. Experiments on the introduction of P elements into stocks that lack them provide further evidence for this (46). Unless regulation of transposition is so strong that there is no transposition at all above a certain threshold copy number, maintenance of an equilibrium with element frequency polymorphism by transposition regulation alone requires a non-zero excision rate, since otherwise copy number will increase indefinitely.

NATURAL SELECTION   Several models have been proposed for the effects of natural selection on the abundance of elements within populations (27, 30, 69, 78, 99). The simplest model assumes that the fitness of a host individual with copy number n for a given family is a decreasing function of n, w_n. Assuming no linkage disequilibrium between sites, the following equation is obtained for the change in copy number per generation (27, 30):

$$\Delta \bar{n} \approx \bar{n}\,(\bar{n} - \bar{x})\,\frac{\partial \ln \overline{w}}{\partial \bar{n}} + \bar{n}\,(u_{\bar{n}} - v) \qquad\qquad 2.$$

The approximate equilibrium value of $\bar{n}$, $\hat{n}$, is given by setting this expression to zero. It is relatively easy to find functional forms that result in an equilibrium with low element frequencies at each site; for example, with $w_n = \exp-\tfrac{1}{2}\,tn^2$, we have $\hat{n} \approx (u_{\hat{n}}-v)/t$. An equilibrium is possible even in the absence of excision ($v = 0$). In the absence of regulated transposition, the stability of this equilibrium requires that the logarithm of fitness decline more steeply than linearly with increasing n (27, 30). Given that low frequencies of transposition seem to be the norm for *Drosophila* elements, this shows that even a weak pressure of selection (as measured by t, which is the slope of the relation between the logarithm of fitness and copy number at a copy number of 1) is capable of maintaining a balance with transpositional increase in copy number. For example, a mean copy number of 50 elements per diploid genome would be maintained if t is one-fiftieth of the excess of the rate of transposition over excision. With rates of movement of the order of 10^{-4}, this implies that selection coefficients of the order of 10^{-5} against individual insertions would be needed to maintain a balance in this case. The mean fitness of the population at equilibrium, relative to the fitness of an element-free individual, is equal to exp $- \hat{n}(u_{\hat{n}} - v)$ regardless of the form of the selection function (27). There are approximately 50 families of elements in D. *melanogaster,* and the mean copy for most of them is probably of the order of 10 per individual or less (33, 50; J. Hey & W. F. Eanes, in preparation). With $\hat{n} = 500$ and $u_{\hat{n}} - v = 10^{-4}$, $\bar{w} = 0.95$, implying that the mean fitness of a D. *melanogaster* population is barely affected by the presence of TEs.

POSSIBLE MODES OF SELECTION   This section considers modes of selection that could lead to the stabilization of element copy numbers in the face of transpositional increase in copy numbers. First, it is well-established that the insertion of elements into or near genes frequently alters their expression (124). Studies of *Drosophila* mutations affecting viability have shown that the most spontaneous mutations with detectable effects have relatively small, detrimental effects (127). P element insertions in D. *melanogaster* are frequently accompanied by such detrimental mutations (42, 52, 92, 150). But the above proof that a very weak intensity of selection is sufficient to check the spread of elements casts doubt on the possibility that selection against insertional mutations is the main factor in controlling element frequencies, in the absence of self-regulated transposition or other forces (29). The studies of Mukai and colleagues on viability mutations indicate a mean selection coefficient of the order of 2% against homozygous detrimental mutations, and 0.7% against heterozygotes (127). If all insertions resulted in a mutation with this magnitude of effect on fitness, elements would never be able to spread in the face of selection, unless transposition rates are higher than seems realistic for most elements, or if regulation of transposition is such that transposition rates are very high for low copy numbers. The latter condition might well have prevailed during the initial spread of P elements through populations of D. *melanogaster* (45, 46).

A solution to this dilemma would at first sight seem to be the possibility that a large fraction of insertions have little or no effect on fitness, because they involve sites that are sufficiently remote from active genes. The average impact on fitness of an element insertion is then the product of the probability that it involves a selectively significant site, and the selection coefficient against a mutation induced in such a site. Numerical analysis of a model with a mixture of neutral and selected sites shows, however, that the majority of elements in an equilibrium population are found at the neutral sites, with element frequencies of 50% or more, reaching fixation if there is no excision (B. Charlesworth, in preparation). Mean copy number may become very large in this case, if there is a large number of occupable neutral sites. This is inconsistent with most observations on *Drosophila* elements (see below).

A modification to the two-class model that could in principle explain the population data is to assume that there are two classes of selected sites; one class being composed of the sites subject to relatively strong selection of the sort considered above, and the other subject to much weaker selection against insertions. These might correspond to insertions into transcribed sequences and nontranscribed flanking regions, for example. In this model, if weakly selected sites are sufficiently common, the dynamics of elements are essentially controlled by them, and copy numbers stabilize at values close to those predicted by Eq. 2, with $\bar{n}$ and $\partial \ln \bar{w}/\partial \bar{n}$ corresponding to values for the weakly selected sites, and the abundance of elements in the strongly selected sites being effectively negligible. Comparisons of DNA sequences among species suggest that the rate of nucleotide substitution in the 5' and 3' flanking regions of genes is higher than that for nonsynonymous changes in coding regions, but considerably slower than the rate for pseudogenes (86). This suggests that there may indeed be selection against insertions of elements into nontranscribed regions, but it is unclear whether neutral sites are so rare that the problems discussed above can be ignored. At all events, the models lead to the conclusion that equilibrium element frequencies will be negligibly low in sectors of the genome, such as coding sequences, where insertions have fitness effects comparable to those of typical spontaneous mutations. Measurable frequencies in large populations will be observed only at sites where insertions have little or no negative direct effects on fitness. This is, in fact, observed in restriction map surveys of segments of the *Drosophila* genome (see below).

These considerations lead to the conclusion that, although selection against insertional mutations may be a factor in stabilizing element frequencies in natural populations, it may not be the only force involved. One possibility is that regulation of transposition rates may act in addition to selection, but the evidence for such regulation is equivocal in *Drosophila* (see above). Another possible mechanism that could contain TE copy number is ectopic exchange, leading to the production of deleterious chromosome rearrangements (78, 99), and reducing the fitness of individuals as a function of the number of elements that they carry (see above). Population models have been developed that take into account heterogeneity in rates of ectopic recombination over different parts of the genome (78). If ectopic exchange events between element pairs occur with a frequency that is proportional to the rates of normal crossing-over in their neighborhood, the equilibrium density of elements in a genomic region is inversely proportional to the frequency of exchange in that region. If ectopic exchange is confined to elements at nearby locations, as is suggested by the *Drosophila* experiments, abundances are inversely proportional to the square root of the frequency of exchange.

THE EVOLUTION OF TRANSPOSITION RATES   If elements are maintained in populations as a result of a balance between transpositional increase in copy number and the counter-effects of selection, there is clearly an advantage to the host to reduce copy number by decreasing the rate of transposition and increasing the rate of excision. It might seem at first sight that there would be a selective advantage to the elements themselves to act similarly, and preserve the existence of their host, as has been frequently argued by students of conventional host-parasite relationships. But, just as in the case of host-parasite relations (94), this selection pressure conflicts with the fact that the representation of the parasite in the next generation is increased by a greater reproductive rate, corresponding in the case of TEs to $u-v$. The outcome of this conflict is not obvious intuitively.

This question has been studied theoretically by Charlesworth & Langley (31). They concluded that, if the spread of TEs in a given family is contained by the deleterious fitness effects of insertional mutations, then there is an effective selection pressure on elements that repress transposition of all elements of that family in the same cell only if the frequency of recombination in the genome is very low. This is because the selective advantage to repression arises from a reduction in the numbers of other elements in the neighborhood of the repressing element. If there is relatively free recombination, this number is effectively controlled by the input of elements due to exchange with meiotic partners each generation, rather than by transposition. Recombination is far too frequent in an organism such as *Drosophila* to permit a selective advantage to repression, although there can be an advantage in low recombination frequency organisms such as bacteria or highly self-fertilizing plants (31). This conclusion also holds if the generation of chromosome rearrangements by ectopic exchange is the mechanism of containment of elements, since the models depend only on the assumption of a decreasing relation between fitness and copy number.

We also investigated the possibility that the induction of dominant lethal or sterile chromosome rearrangements by transposition events could favor repression in a diploid organism, since here the disadvantage of transposition is immediate and cannot be diluted by recombination (31). Provided that the product of mean copy number and the probability of a dominant lethal event per transposition event is sufficiently high, repression can be advantageous. The germ-cell death and zygotic lethality associated with high rates of transposition of *P, I* or *hobo* elements in hybrid dysgenic crosses, presumably due to chromosome breaks (14, 16, 45, 46, 148), act in a similar way (31). Since hybrid dysgenesis is the only evidence in *Drosophila* for repression of transposition, it is tempting to speculate that the driving force for the evolution of such repression is based on this kind of mechanism, and this appears to be quantitatively plausible (31). But, given the fact that repression in the *P-M* system appears to be due to defective elements (see above), and that their spread can be accounted for by an intranuclear transpositional advantage (see below), it is not clear whether it is necessary to invoke an additional factor to explain repressor elements. (It is, of course, possible that defective elements with repressor activity might have an advantage over ones without, but this has not been analyzed quantitatively.)

Finally, there is always an advantage to repression of transposition in the somatic tissues, since such transposition does not enhance the representation of elements in the next generation and any somatic mutations or chromosome breaks will tend to lower the fitness of the host (31). Most *Drosophila* elements appear to be somatically stable, except for the *mariner* element of *D. mauritiana* (20), which undergoes excision at a significant frequency in somatic cells. Precise excision in somatic tissues could indeed be selectively advantageous, because it lowers the number of potentially deleterious elements (31). The molecular basis of somatic repression of transposition is partially understood for the *P* element (46): the 2–3 intron of the ORF of a complete *P* element is not spliced out in somatic cells. The drastic effect on fitness of a failure to regulate splicing has recently been demonstrated by the use of a construct (Δ–2–3) with this intron deleted, and that is therefore mobilized in somatic cells in dysgenic crosses (47).

Element Dynamics in Finite Populations

Since the deterministic forces acting on TEs are so weak, a full interpretation of statistics on element frequencies in natural populations requires that effects of genetic drift be taken into account.

THE PROBABILITY DISTRIBUTION OF ELEMENT FREQUENCIES   Extensions to finite populations of the models described above yield the form of the probability distribution of element frequencies per site for a given family, attained when the forces of transposition, excision, and selection come into statistical equilibrium with random changes in element frequencies (the stationary distribution of element frequencies) (27, 30, 70, 77).

The following parameters provide a complete but approximate description of the form of $\phi(x)$, the stationary probability density for element frequency x at a site: N_e is the effective size of a local population; $\hat{n}$ is the equilibrium value of $\bar{n}$ for a large population; $\alpha = 4 N_e \hat{n}/(2m - \hat{n}u)$; $\beta = 4 N_e(s+v)$, where s is the value of $- \partial \ln \bar{w}/ \partial \bar{n}$ at $\bar{n} = \hat{n}$. α measures the effect of drift, and the effect of transposition in causing insertions into a given site; β measures the joint effects of drift, excision and selection. ϕ is approximated by a beta distribution (30):

$$\phi(x) \approx \frac{\Gamma(\alpha+\beta)}{\Gamma(\alpha)\Gamma(\beta)} \, x^{\alpha-1}(1-x)^{\beta-1} \qquad\qquad 3.$$

This is the formula for a closed population; if $\hat{n} \ll 2m$, the effect of migration can be included by adding $4N_eM$ to β, where M is the frequency of immigrants into a local population. If m is sufficiently large compared with $\hat{n}$, α can be neglected in Eq. 3. This yields a simple formula for the expected number of sites per haploid genome with frequency x, $\Phi(x) = \frac{1}{2} \hat{n}x^{-1}(1-x)^{\beta-1}$ (70). As will be seen below, Eq. 3 can be applied to population data in order to obtain estimates of the parameters α and β.

STOCHASTIC LOSS OF ELEMENTS   The process of the chance loss of elements from finite populations, suggested as an explanation for the absence of the elements responsible for hybrid dysgenesis from old laboratory stocks of *D. melanogaster* (44), has been modelled (27, 71). In a very small population, the expected time to loss of all elements is of the order of at least $1/v$, unless the initial mean copy number per individual is very small; the time is substantially larger than $1/v$ in larger populations (27). This is difficult to reconcile with the fact that laboratory strains lacking active *P* elements may be only 30 years old or less, corresponding to 750 generations at 25° (16, 45, 46). A high rate of excision of *P* elements from their chromosomal sites, and a very small mean number of active elements per individual in the populations from which the laboratory strains in question were collected, would be required to rescue the stochastic loss hypothesis for *P* elements. Since the mean haploid copy number is of the order of 30 for natural populations (117), the latter condition is unlikely to be satisfied. Other data relevant to the question of stochastic loss of *P* elements are reviewed by Engels (45, 46). Kidwell's recent invasion hypothesis (73) is most probably valid, particularly in view of the evidence for the presence of *P* elements in unrelated species and their absence from the closest relatives of *D. melanogaster* (46). It is conceivable that the stochastic loss hypothesis is valid for the *I* element system of hybrid dysgenesis, since these have a lower mean copy number per individual than *P* elements (117) and the laboratory strains that lack active *I* elements are over 60 years old (16). They are also present in the relatives of *D. melanogaster* (22). Whether or not excision rates for *I* elements are sufficiently high to account for their loss over 1500 generations is unclear; there is currently no evidence for precise excision of these elements (50), although *I-R* dysgenic crosses produce unstable mutations that reflect the presence of *I* element insertions (21, 113). Changes from the *I* to *R* state have not, however, been observed in laboratory stocks maintained for over ten years (16).

A striking feature of *P* and *hobo* elements is that a substantial fraction of elements are defective and unable to produce the enzymes needed for their own transposition (14, 15, 46, 91). They can, however, transpose in the presence of complete elements. Kaplan et al (71) have constructed a model in which defective elements are generated by mutation from functional elements. The equilibrium state of a large population is such that a large fraction of elements are defective, provided that the defective elements suffer no replicative disadvantage in host cells that contain a sufficient number of complete elements to supply them with transposase functions. In addition, defective elements might enjoy a selective advantage, due to faster transposition, and hence tend to displace complete ones (130). In a finite population, stochastic loss of complete elements may occur relatively fast (in terms of evolutionary time), because of their low equilibrium numbers. This could lead to the evolution of an element family, all of whose members are incapable of transposition. The ultimate fate of such a family would be mutational degeneration or extinction, the latter of course requiring a non-zero rate of removal by excision, selection, or ectopic exchange.

MULLER'S RATCHET   This is a process by which insertions of elements into genomic regions that lack crossing over can cause a gradual increase in copy number over time, as a result of the effects of genetic drift. The theory of Muller's ratchet has been worked out for the case of conventional deleterious mutations (49, 60, 110), based on an idea of Muller's (102), but essentially the same results apply to TEs provided that excision rates are sufficiently small (27). Consider a segment of chromosome in an infinitely large population. If there is some rate of insertion of elements into this segment, then (as we have seen) an equilibrium in the copy-number distribution between individuals will be established under counter-selection or regulated transposition. If the rate of insertion is high in relation to the pressure of selection, the frequency of chromosomes lacking elements in the given segment is low. In a finite population, this means that this class of chromosomes is vulnerable to loss by drift. If there is crossing over in the segment in question, it is possible for the zero-element class to be regenerated by crossing over, and so the form of the distribution of element numbers will remain similar to that for an infinite population. If there is no crossing over, however, the zero class will be permanently lost in the absence of excision. Once it is lost, the class with one element in the region in question will be vulnerable to loss, and so on, such that the distribution of copy number moves steadily to the right in a ratchet-like process. The speed of this process depends on the inverse of the product of the species population size and the initial frequency of the zero class (60, 110), hence it will probably proceed very slowly in natural populations of *Drosophila*.

The effect of the ratchet in producing a build-up of elements in regions where crossing over is suppressed is distinguished from the similar effects of ectopic exchange and transposition discussed above by the fact that the ratchet produces an unlimited build-up of elements, given sufficient time, whereas the latter process will only produce a limited increase, unless exchange is totally suppressed and there is no regulation of copy numbers. The ratchet also fails to operate in the presence of even low frequencies of recombination or excision (110), and will thus only produce an excess of elements that excise at low frequency in genomic regions almost totally lacking crossing over, except in very small populations.

SEQUENCE SIMILARITY BETWEEN MEMBERS OF THE SAME FAMILY   The expected degree of DNA-sequence similarity between members of the same family of elements has been studied intensively (17, 28, 65, 106, 107, 129). The models assume that elements differentiate as a result of neutral mutations occurring at a rate ν per element per generation; sequence similarity is enhanced by the effects of genetic drift (causing fixation of nucleotide sequence variants), and unbiased gene conversion between homologous elements within the same host nucleus (at a rate λ per element per generation). Provided that $4N_e u \gg 1$, as is suggested by most of the available *Drosophila* data (see below), the equilibrium probability of sequence identity between a randomly chosen pair of elements from different chromosomal sites in a population of effective size N_e is:

$$f \approx \frac{1}{2N_e \hat{n} \nu + 1} \qquad\qquad 4.$$

This is identical to the standard formula for a single-copy gene (74), except that the product of effective population size and haploid equilibrium mean copy number replaces the effective population size. Surprisingly, as Slatkin first pointed out (129), there is no dependence of the equilibrium probability of sequence identity on the rate of gene conversion, unless λ and $u \leq 1/2N_e$.

Unfortunately, there is a dearth of information from detailed sequence studies on the degree of genetic divergence between *Drosophila* elements. The available data from restriction mapping and DNA/DNA hybridization analysis of *Drosophila* retrovirus-like elements (51, 140) does not permit any distinction between coding and non-coding substitutions in the element sequences, and so it is not clear whether or not the high degree of similarity (3–7% divergence) fits the neutral model (17).

POPULATION DATA FROM IN SITU HYBRIDIZATION STUDIES

In *Drosophila,* the technique of in situ hybridization of labelled probes to the polytene salivary gland chromosomes permits identification of the sites of homologous DNA elements. A survey of stocks containing chromosomes isolated from natural populations, and made isogenic by standard genetic procedures, provides a picture of the haploid genomes present in the population from which the chromosomes were sampled. The resolution of this technique is somewhat coarse compared with restriction mapping of limited portions of the genome (see below), since at best the locations of elements can only be determined down to the level of the salivary chromosome bands shown in the Lefevre photographic map (81). Nonetheless, they provide a useful picture of the general properties of the distribution of elements in populations (3, 33, 42, 83, 99, 100, 147). Surveys of isofemale lines or inbred stocks derived from a natural population but not subjected to isogenization, or of larvae sampled directly from a population, provide less interpretable data, since sites that are heterozygous and homozygous for the presence of an element cannot be distinguished (117, 147). Samples from laboratory populations, especially those subjected to artificial selection, may also provide a distorted picture of element frequencies, because of increased homozygosity due to inbreeding (8, 11–13). Such studies can, at best, provide only qualitative information concerning the population properties of elements in nature. Accordingly, in the following sections we emphasize the conclusions from studies of isogenic chromosomes sampled from natural populations.

Copy Number Distributions between Homologous Chromosomes

As shown above, if element frequencies for a given family are low and similar at individual chromosomal sites, and if linkage disequilibrium is negligible, a Poisson distribution of copy number is expected between different individuals, with the variance in copy number approximating the mean. This result applies equally well to the haploid copy numbers determined for isogenic chromosomes or inbred stocks. Elements present in the centromeric heterochromatin cannot be detected by in situ hybridization, owing to the underreplication of the heterochromatic DNA during polytenization (68), and so the data come exclusively from the counts of numbers of elements in the euchromatic divisions of the salivary chromosomes. In addition, the tendency for elements to accumulate in the proximal regions of the euchromatin (see below) means that tests of fit to the Poisson distribution must be conducted using counts for the distal 18 divisions of each salivary chromosome arm only, since heterogeneity in element abundances between chromosomal regions will cause deviations from Poisson (33).

Published data on such copy numbers for sets of isogenic chromosomes show generally good agreement with the Poisson distribution for twelve element families (3, 33, 42, 83, 99, 100). Most of the data are for the X chromosome, apart from the data of Montgomery et al (99) on *roo, 412,* and *297,* and unpublished data of B. Charlesworth & A. Lapid on ten families (including these three) for the second and third chromosomes. The one striking exception is the case of *roo* on the X chromosome in one population, which has a mean copy number of 11.4 and a variance of 4.4, and deviates from Poisson with $p < 0.01$ (33). Detailed analysis suggests that this is a chance result, largely reflecting an effect of nonsignificant linkage disequilibrium terms (33). Excess variances over Poisson expectation for total genomic numbers of *P, 297*, and *copia* sampled from two Japanese populations have been reported (147), but these are hard to interpret since one of the populations was studied using isofemale lines, and the X was apparently not made isogenic in the other.

Early studies of the distribution of *Drosophila* elements between inbred laboratory strains gave the impression of uniformity of copy number for a given family, in contrast to the wide variation in the identity of occupied sites. This led to the suggestion that there is selection for an optimal copy number (135, 149). However, reanalysis of these data shows that the variance in copy number is approximately the same as the mean, consistent with the Poisson expectation (29).

Distribution of Elements between Nonhomologous Chromosomes

The hypothesis that TE frequencies are stabilized by the deleterious effects of recessive insertional mutations predicts that the mean equilibrium element frequencies for X chromosomal sites will be lower than for for autosomal sites, since X chromosomal mutational effects are expressed in the hemizygous state in males, compared with the predominantly heterozygous state of rare autosomal mutations. A quantitative model of the relative equilibrium mean copy numbers for the X chromosomes and autosomes of *D. melanogaster* has been developed both for the hypothesis of insertional mutation effects, and for the null hypothesis that X-linked and autosomal elements are eliminated at the same rate (78, 99). The results of scoring the numbers of copies of the retrovirus-like elements *copia, roo* and *412* on sets of 20 X, 2nd and 3rd chromosomes from a natural population shows no significant difference for *copia* and *roo* between the observed relative abundances of elements on the X and autosomes and the expectation on the null hypothesis, whereas *412* shows a significant ($p < 0.001$) deviation from the null expectation, but agrees with the expectation on the insertional mutation hypothesis (78, 99). There is a similar lack of evidence for a deficiency of *P* elements on the X (42). These results suggest that, while there may be some fitness effects of insertional mutations, they are not necessarily the chief factor involved in containing the spread of TEs. This is in accord with the theoretical conclusions discussed above. It is unclear why the above differences between families should exist; further data on a wider variety of families are clearly desirable.

Element Frequencies at Individual Chromosomal Sites

A convenient summary of the results of comparisons of the salivary chromosome band locations of members of a given element family among independently isolated, homologous chromosomes is provided by the occupancy profile, which gives the numbers of sites at which hybridization is detected at that site in 1, 2, 3 . . . *S* separate chromosomes of a sample of size *S* (100). Data from *D. melanogaster* population surveys of isogenic chromosome lines show clearly that, for the twelve families that have been surveyed (3, 33, 83, 100), element frequencies are low for sites in the distal 18 divisions of the salivary chromosomes. Again, most of the data are for the X chromosome, but the available data sets for the major autosomal arms suggest that the profiles are similar to those for the X (33; B. Charlesworth & A. Lapid, unpublished data). The data show that most of the time a site is occupied only once by members of a given family in a sample of size 10–20. With high-copy number elements, such as *roo,* occupancies of up to seven have been observed; low-copy number families rarely show occupancies higher than two. Qualitatively similar results appear to hold for the *mdg-1* element (8, 11–13), but the data were not obtained from isogenic stocks and hence cannot be treated quantitatively.

A very different result has been reported for four uncharacterized families of repetitive DNA from *D. algonquin* and *D. affinis,* studied in a small number of inbred stocks (62). Copy numbers per haploid genome are much higher than those typical of *D. melanogaster* (in the hundreds), and many bands show element frequencies of 50% or more. A similar result appears also to be true for an element of the Hawaiian species *D. heteroneura* and *D. silvestris* (66). The presence of the *FB* element has also been reported (126) at identical sites in several species of the *melanogaster* species group, consistent with fixation of elements at these sites, although it is not clear whether or not the hybridization reflects flanking sequences and/or accumulation of elements at the base of the chromosomes. Detailed population surveys have not been carried out for species other than *D. melanogaster,* and these isolated reports underline the importance of such work in the future.

Sampling theory based on Eq. 3 can be applied to the analysis of in situ hybridization data on element frequencies (30, 69, 70, 77, 100). One model assumes $\alpha = 0$, and uses an extension of Ewens' sampling theory for neutral alleles (48) to estimate β (70, 77, 100). An alternative model assumes equal frequencies of $1/S$ at each site (70). Multiple occupancy is then due purely to the chance occupation at different sites within the same salivary chromosome band in different chromosomes of the sample. A similar model, with equal population frequencies of elements at each band, was suggested by Charlesworth & Charlesworth (30), who applied a maximum likelihood method (85) to the estimation of the number of occupable sites per chromosome (*m*) and the element frequency per occupable band (x). A general model (30) involves joint estimation of α and β by a minimum χ^2 method of fitting the observed occupancy profile to that predicted from Eq. 3. This method also provides estimates of m and the expected element frequency per occupable band, $\hat{x} = \alpha/(\alpha+\beta)$.

Table 2 displays the estimates of α, β, $\hat{x}$ and m obtained for a number of elements of *D. melanogaster*. It should be stressed that the fits obtained with the model fitting α and β jointly are usually not significantly better than those obtained with the assumption that $\alpha = 0$, or that element frequencies are equal at each site. Estimates of α and β are subject to considerable uncertainty, except for the high-copy number elements such as *roo, 2161* and *297*. Nevertheless, expected element frequencies per occupable site are at most a few per cent, and the generally large values of β obtained are highly significant when compared with small values, reflecting the low frequencies of multiple occupation. *P* elements have not been included in this analysis, since these elements are probably far from equilibrium (46, 73), but low *P* element frequencies are indeed observed (3, 117).

These findings effectively rule out the possibility that element frequencies are controlled by regulated transposition in the absence of excision and/or

Table 2 Estimates of the parameters of the probability distribution of element frequencies for *Drosophila melanogaster* elements from natural populations

Element	α	β	$\hat{x}$	m	Element	α	β	$\hat{x}$	m
roo[2]	0.8	12.5	0.060	191	2210[2]	25	550	0.044	15
[3]	3.4	32	0.096	128					
[4]	2.5	28	0.082	132					
2156[2]	0	∞	0	∞	2217[2]	0.4	40	0.009	204
2158[2]	0	∞	0	∞	297[1]	0.05	16.5	0.003	1340
					[2]	0	5.5	0	∞
2161[2]	2.5	35	0.067	58	412[1]	0	30	0	∞
					[2]	17	380	0.043	55
2181[2]	∞	∞	0.028	46	*copia*[1]	∞	∞	0.020	82
					[2]	∞	∞	0.019	63
I[5]	∞	∞	0.028	95					

[1] data for the X (100)
[2] data for the X (33)
[3,4] data for *roo* for 3L and 3R respectively (33)
[5] data for the X (83)
Note: $4N_eM$ is equal to 8.8 for autosomal loci and 6.5 for X-linked loci (126). The contributions of selection and excision to β can be found by deducting these from the appropriate entries in the table.

selection, such that transposition rates are zero in an equilibrium population (see above). An estimate of the transposition rate u can be obtained as follows: if the expected number of elements per individual is unchanging over time, then the condition for equilibrium derived from Eq. 2 implies that $\beta \approx 4N_e u$. Estimates of N_e for East Coast $D.$ *melanogaster* populations from the frequencies of allelism between recessive lethals suggest values of the order of 2×10^4 (101). With a β value of 20, this yields an estimate of 2.5×10^{-4} for u. The estimates of N_e are subject to considerable uncertainty, so that this value should not be taken too seriously, but it is in good agreement with the values of u estimated directly (see above). For the more abundant elements, the number of occupable bands is either infinite or of similar magnitude to the total number of bands on the photographic maps of the salivary chromosomes (33). This suggests that elements do not have a high degree of site-specificity of requirements for insertions, at least at the level of salivary chromosome bands.

The differences among elements in the values of β suggest that there may be real differences between elements in rates of transposition. Unfortunately, there are at present no reliable quantitative estimates of the rates of transposition for individual elements that could be used to compare with these estimates. The magnitude of β in these data is generally so high (10 or more) that drift must be rather ineffective in relation to the deterministic forces affecting element frequencies; for element frequencies to be predominantly controlled by drift, β values of the order of one would be required (144). Values of this magnitude were, however, found for elements of $D.$ *algonquin* and $D.$ *affinis* (62).

Linkage Disequilibrium and Related Phenomena

There is no evidence for nonrandom associations in frequency between elements at different sites, either between or within element families, in the data set of Charlesworth & Lapid (33) on ten element families in a sample of fourteen X chromosomes. This holds even at the tip and base of the chromosome, where crossing over is rare (89) and conditions are more favorable for the maintenance of linkage disequilibrium. This finding is consistent with the general agreement of the copy-number distributions with the Poisson distribution (see above). In addition, there is no evidence for a correlation in copy number per chromosome between different element families. There is an apparently random distribution of elements along the chromosomes, with respect to the identity of sites occupied at least once in the sample, if the basal region of the chromosome (subdivisions 18D1–20A3) is excluded. The only exception is *2161*, which shows evidence of accumulating in divisions 15–18. This suggests that there is little tendency for transposition events to produce insertions into chromosomal sites close to that occupied by the parent element, as has been reported for the maize element *Ac* (123).

There are, however, correlations between certain element families with respect to the identity of sites occupied at least once in the sample, but no evidence for a significant correlation within chromosomes of the kind earlier reported by Montgomery & Langley (100). The absence of a within-chromosome effect suggests that the correlations reflect either some degree of shared site-specificity for insertion, or protection of certain regions against removal by selection or excision. There is no evidence for correlations between element families with respect to sites at which elements reach high frequencies, suggesting that the high-element frequencies observed at some sites for certain families do not reflect an unusual affinity of these sites for element insertions, and are thus probably due to genetic drift or hitch-hiking (33).

Accumulation of Elements in Regions of Restricted Crossing Over

ECTOPIC EXCHANGE   A test for selection against chromosome rearrangements produced by ectopic exchange between elements is provided by asking whether or not the abundance of elements tends to be higher in regions where exchange is reduced in frequency. *D. melanogaster* again provides useful material for carrying out such a test, since there is pronounced suppression of meiotic crossing over in the telomeric and centromeric regions of the euchromatin of the chromosome arms (89). It is straightforward to use in situ hybridization to determine the distribution of the numbers of elements over the salivary chromosome maps in chromosomes sampled from natural populations, and to compare this with the distribution expected on a null hypothesis of no variation in element abundance with respect to the rate of meiotic crossing over (33, 78).

The results of a study of *roo* show clear evidence for an excess of elements at the base of the X chromosome, of the order of threefold over the number of elements that would be expected if elements were inserted in proportion to the physical size of the region (78). The picture is less clear for the other major chromosomes, with a weak tendency for elements to accumulate at the base of the euchromatin. There is no evidence for an accumulation of elements at the tips; rather, there is a slight tendency for *roo* to be less frequent there than expected. A more recent study of ten elements, including *roo,* on a sample of X chromosomes (33) provides highly significant evidence for an accumulation of elements at the base of the euchromatin (only *2217* and *2210* failed to show a strong effect). Again, there is no evidence for an excess of elements at the tip. Unpublished data of B. Charlesworth & A. Lapid indicate that there is also a tendency for many of these element families to accumulate at the bases of the autosomes. There is evidence for accumulation of *P* elements at high frequency at the extreme tip of the X chromosome in some natural populations (3), but this may be due to site-specificity of insertion or excision rather than an effect of exchange frequency, in view of the lack of evidence for such an effect with other elements. There is no evidence for accumulation of *P* elements at the bases of chromosomes (3, 117).

These observations should not be taken as providing watertight evidence in favor of the ectopic-exchange hypothesis, since an alternative mechanism exists for creating a build-up of elements in regions where crossing over is suppressed (see discussion of Muller's ratchet above). Furthermore, the failure to detect an excess near the chromosome tips presents problems of interpretation for either model. One possible solution is that crossing over near the tip is induced by an interchromosomal effect on crossing over of inversion heterozygosity for the inversions present in natural populations (78, 90). Other evidence for an accumulation of elements in regions of restricted crossing over of the *D. melanogaster* genome is provided by laboratory cultures of balanced lethal inversions (99), and unpublished data of W. F. Eanes & C. Wesley showing an excess of *P* elements in naturally occurring inversions.

An extreme example of the differential concentration of elements in genomic regions where crossing over is suppressed is provided by the numerous *Drosophila* elements that appear to be disproportionately abundant in the β-heterochromatin that forms the boundary between the euchromatin (which gives a regular banding pattern in the polytene chromosomes) and the α-heterochromatin (which is not polytenized, and consists mainly of highly repeated satellite sequences). Many families of elements hybridize only infrequently to the euchromatic regions of the polytene chromosomes, but show strong, diffuse staining of the β-heterochromatin (33, 50, 118, 149). It is presently not clear whether or not TEs are abundant in the α-heterochromatin as well as β-heterochromatin. It is possible that the mechanisms just discussed are responsible for this concentration, since crossing over is virtually absent in this part of the genome (121). Another possibility is that the relative rarity of functional genes in heterochromatin (64, 121) means that insertional mutations do not effectively oppose the spread of elements in these regions. The two-class models discussed above suggest that a very large concentration of elements should be found in the heterochromatin if this were the case. Clones containing material that hybridizes to the β-heterochromatin or to the euchromatin/heterochromatin junction often contain many sequences homologous to known TEs or which are present as dispersed copies in the euchromatin and hence represent presumptive TEs (72, 96, 97, 141, 142, B. Wakimoto, personal communication). Detailed characterization of some of these clones has revealed the presence of multiple insertions of TEs, sometimes with several different types of element inserted in an overlapping fashion into the same location, presumably as a result of a sequence of independent insertion events (35, 36, 116). It is interesting to note that certain of these clones also show hybridization to chromosome 4, another region where crossing over is restricted, although it is not entirely clear whether this reflects the presence of TEs or of tandemly repeated arrays that are also abundant in these regions (96, 97). But for at least some element families (*copia, 412* and *297*) a disproportionality in abundance between the euchromatin and heterochromatin is not observed (114, 135). In these cases there may be a relative dearth of sites for insertion within the heterochromatin, which outweighs the lack of exchange and effects of insertions on fitness in this region.

Furthermore, TEs may have difficulty in transposing once inserted into the α-heterochromatin, due to the lack of transcription in this region (68, 95). If elements cannot mobilize once inserted, there will be no selection to prevent their decay under the pressures of mutation and drift. A gradual build-up of degenerate elements in these sections of the chromosome might thus be expected. The defective forms of the *I* element, responsible for the *R* cytotype characteristic of *D. melanogaster* lines that lack complete *I* elements, are also concentrated in the β-heterochromatin (21), and the same appears to be true for the defective *P* elements found in species of *Drosophila* other than *D. melanogaster* (46).

MULLER'S RATCHET   It is doubtful whether Muller's ratchet (see above) could account for the excess of elements in the proximal euchromatin of *D. melanogaster* chromosomes described above. In the models of selection against deleterious mutations normally used to analyse Muller's ratchet, the process is very sensitive to the relative values of the per genome mutation rate and the strength of selection against the mutations concerned (60, 110). In the context of TEs inserting into the basal euchromatin of the X, the relevant "mutation" rate U is the product of the per genome transposition rate (the mean copy number per haploid genome times u), and the probability of insertion into the base. If elements are distributed approximately randomly across the chromosomes, as seems to be the case in *D. melanogaster* (33, 78, 99), the "mutation" rate is thus equal to the product of u, the haploid copy number for the X and the proportion of the X represented by the base. In the case of *roo*, the most abundant element studied, this is equal to $1.7u$ (33). The values for other elements are smaller. If the population is approximately at equilibrium, and excision is absent (as is necessary for the ratchet to work), we have $u \approx s$. Hence, $U \approx 1.7 s$ for *roo*. This is close to the boundary value for which the ratchet can operate in the total absence of recombination, even in very small populations (60, 110). *D. melanogaster* is known to have reasonably high effective local population sizes and rates of migration between populations (101, 128), and there are significant frequencies of crossing over in the basal euchromatin (89). It is therefore theoretically rather implausible that a ratchet mechanism could have been responsible for the build-up of elements observed here. It is possible, though, that the concentration of *Drosophila* elements in the centric heterochromatin (see above) is due to Muller's ratchet, since exchange is virtually absent here (121). Similarly, the excess of middle repetitive DNA in the neo-Y chromosome of *D. miranda* (132), could be due to the operation of the ratchet. This chromosome is the result of a centric fusion between an autosome and the original Y chromosome, and so has been maintained strictly through the male line, without crossing over, since the establishment of the fusion. Finally, the small size of laboratory populations means that the ratchet will operate much more rapidly than in large natural populations. The apparent build-up of elements near the breakpoints of balanced lethal inversions in laboratory stocks could thus also be due to the ratchet (99).

POPULATION DATA FROM RESTRICTION MAP SURVEYS

General Findings

Much finer resolution concerning the identity of the sites occupied by TEs in a sample of chromosomes from a natural population is provided by restriction-mapping surveys of defined genomic regions. Several such surveys have now been published for various different *Drosophila* species (1, 2, 4–6, 40, 41, 76, 79, 80, 82, 98, 119, 120, 134). Tables 3 and 4 summarize these and unpublished findings from *D. melanogaster* and from other *Drosophila* species, respectively. Variation due to the presence of large insertions (greater than 400 bases long) is readily distinguishable from nucleotide-site polymorphism and variation due to small deletions and insertions. In two surveys of specific gene regions in *D. melanogaster,* the large insertions found segregating in natural populations were cloned and characterized (4, 82). These insertions were all similar in sequence to previously cloned and characterized middle repetitive sequences. Based on these results, the large insertions detected in the other studies listed in Tables 3 and 4 are assumed to be TEs.

Overall, there is remarkable concordance between the results of these surveys and the results from in situ hybridization studies. The general conclusion is that each particular TE insertion is very rare at a particular chromosomal site in a natural population, while the density of TEs is sufficiently high that one or more element insertions are commonly found in genomic regions more than 10 kb in length, if 20 or so independent chromosomes are sampled.

The data on the *Adh* region in *D. melanogaster* also show that TE insertions occur only in haplotypes located toward the tips of the phylogenetic tree connecting the set of chromosomes sampled, suggesting that a deterministic force such as selection or regulation of transposition rate is keeping frequencies low (57).

Results of Surveys of D. melanogaster Populations

RELATIONSHIP OF INSERTION SITES TO TRANSCRIPTIONAL UNITS   Element insertions are rarely found in transcriptional units in *Drosophila* genes sampled from nature. Among the surveys listed in Table 3, only one case of a TE was unambiguously mapped within a transcriptional unit (in the intron of *Ddc*, with no associated phenotype). Since such large insertions are usually in or very near the transcriptional unit in most spontaneous mutations used in *Drosophila* genetics (118), it is likely (as discussed above) that such insertions are deleterious and rapidly eliminated from the population. Naturally occurring insertions outside the transcriptional unit have only rarely been associated with reduced gene expression (*copia* in *Adh* [4], and several large insertions in *Zw* [N. Miyashita, personal communication]). Occasionally, large insertions found in these restriction-map surveys appear to cluster in a particular region flanking the transcription unit(s), e.g in *Adh* (4), and in *vermilion* and *Om* in *D. ananassae* (134, 134a). It is not clear whether this reflects inherent clustering at the stage of insertion, reduced deleterious phenotypic effects of insertions at such locations, or less efficient genetic mechanisms for the removal of insertions at these sites.

X CHROMOSOMES VERSUS AUTOSOMES   If selection on slightly deleterious effects of TE insertions in the flanking regions of genes were responsible for keeping the frequencies of these insertions rare, then (as discussed above) a substantial difference between the densities of TEs on the *X* and autosomes would arise (78, 99). As with the in situ hybridization work, restriction-map surveys provide little evidence to support such a hypothesis. In Table 3, the average density of large insertions on the *X* chromosome is not significantly lower than in autosomal regions. Neither is the observed proportion of homozygous insertions lower for the *X*, as might be expected under more intense purifying selection.

REGIONS OF REDUCED CROSSING OVER   There are reasons that we have already noted to expect TEs to accumulate in the distal and proximal regions of chromosomes, where meiotic crossing over is reduced in frequency. Table 3 gives the results of surveys of restriction map variation in the *y-ac-sc* region at the tip of the *X* chromosome of *D. melanogaster*. The density of large insertions in this region (0.004) is the same as in the central portion of the X

Table 3 Summary of insertional variation found in natural populations of *Drosophila melanogaster*

Region	Size of region (kb)	Sample size	Density per kb	Proportion of homozygotes	Reference or source
X Chromosome:					
forked	25	64	0.002	0.000	C. H. Langley, N. Miyashita (in preparation)
Notch	37	60	0.002	0.028	120
per	52	78	0.002	0.000	D. Stern, W. Noon, E. Kindahl, C. F. Aquadro (in preparation)
vermilion	24	64	0.002	0.000	C. H. Langley, N. Miyashita (in preparation)
white	45	64	0.011	0.067	98
white	45	38	0.010	0.046	76
zeste-tko	34	64	0.001	0.000	1
Total or mean[2]	217	—	0.004	0.017	
Reduced crossing over:					
su(f)	24	64	0.004[†]	0.011	C. H. Langley, N. Miyashita (in preparation)
y-ac-sc	106	64	0.006	0.085	2
y-ac-sc	120	49	0.003	0.056	6
y-ac-sc	31	109	0.003	0.000	41
Zw	13	64	0.029	0.063	N. Miyashita (in preparation)
Zw	13	127	0.015	0.041	40
Total or mean[2]	143	—	0.006	0.041	

Autosomes:

Amy	15	85	0.009	0.009	80	
Adh	13	48	0.018	0.000*	4	
Ddc	65	46	0.003	0.000		C. F. Aquadro et al (in preparation)
rosy	100	60	0.002	0.030	5	C. F. Aquadro, V. Bansal, W. Noon (in preparation)
Hsp70	25	29	0.006	0.000	82	
Total or mean[2]	218	—	0.004	0.014		
Grand total or mean[2]	578	—	0.005	0.022		

* Subsequent sequencing of apparently identical insertions demonstrated that the TE involved (*2161*) is actually inserted at different sites in each chromosome (W. Quattlebaum, C. H. Langley, C. F. Aquadro, unpublished data). Each TE in the *Adh* region is thus unique.

† The *su(f)* transcriptional unit is surrounded by repetitive sequences that may be TEs fixed in the population (A. Mitchelson & K. O'Hare personal communication). In that case, density of TEs per kb would be greater than 0.5, and the proportion of homozygous TEs would approach 1.0.

[1] The proportion of homozygous large insertions is the probability that a large insertion found in the sample would be homozygous in a zygote formed by two random (but distinct) chromosomes from the sample.

[2] The means were calculated by weighting by the size of the region surveyed (kilobases). Where the same region was surveyed more than once, the unweighted average of these surveys was calculated and then weighted by the size of the region.

chromosome, consistent with the in situ hybridization surveys discussed above.

The restriction-map survey of the proximal *su(f)* region of the X chromosome cannot be compared to the in situ surveys directly, since it is located at 20F4, well into the β-heterochromatic region of the X where discrete sites of hybridization are difficult to identify. As would be expected from its location in the β-heterochromatin (96, 97), the *su(f)* transcriptional unit is surrounded with repetitive DNA (A. Mitchelson & K. O'Hare, personal communication). A restriction-map survey indicates that all sampled chromosomes characterized for the *su(f)* region are surrounded by identical repetitive sequences that may well be TEs (C. H. Langley & N. Miyashita, in preparation). Out of 64 chromosomes sampled, six have an additional TE-like insertion (three of the sampled haplotypes have identical insertions or derivatives thereof). These results indicate an accumulation of elements in this region. The *Zw* region, located more distally in the basal euchromatin of the X, also shows a higher than average density of TEs. Table 3 also shows that the average proportion of homozygous insertions appears somewhat higher for loci in regions of restricted recombination (even when the uniform repetitive region of *su(f)* is excluded). This reflects a tendency for large insertions in these regions to be found at higher frequencies. Thus, the limited data on insertional variation at the base of the X chromosome are consistent with the in situ hybridization results. However, the nature of repetitive sequences in the basal regions of the chromosomes is so different from the rest of the chromatin (96, 97) that further surveys of slightly more distal regions and of the proximal regions of the autosomes will be required for an adequate picture of the distribution of TEs in these regions of reduced crossing over.

Results for Species other than D. melanogaster

D. ANANASSAE   Three regions of the proximal part of the metacentric X chromosome of *D. ananassae* have been surveyed (134, 134a). As Table 4 shows, the *vermilion* region has two unique insertions and one highly polymorphic insertion, while no large insertions were observed in the *forked* region. Many insertions (large and small) are present in the *Om* region. The frequencies of these are generally low. There is not yet enough data to determine if the densities and distribution of TE-size insertions in *D. ananassae* are different from those found in *D. melanogaster*. The *Om* and *forked* loci are in regions of relatively high rates of crossing over, whereas *vermilion* is in a region of reduced crossing over (134, 134a). This is of interest in view of the appreciable element frequencies observed there.

D. PSEUDOOBSCURA   The initial two surveys of restriction-map variation in this species have yielded very little evidence of large-insertion polymorphism (119; C. F. Aquadro et al, in preparation).

Table 4 Summary of insertional variation found in natural populations of *Drosophila* species other than *D. melanogaster*

Region	Size of region (kb)	Sample size	Density per kb	Proportion of homozygotes	Reference or source
D. ananassae					
vermilion	15	60	0.049	0.368	134
forked	15	60	0.002	0.000	134
Om (1D)	37	60	0.014	0.054	134a
Total or mean	67	—	0.019	0.141	
D. simulans					
Adh	13	38	0.000	—	C. F. Aquadro, K. Sykes, P. Nelson (in preparation)
per	52	38	0.000	—	D. Stern, W. Noon, E. Kindahl, C. F. Aquadro (in preparation)
rosy	100	30	0.001	0.000	5, C. F. Aquadro, V. Bansal, W. Noon (in preparation)
Total or Mean	165	—	0.001	0.000	
D. pseudoobscura					
Adh	32	20	0.002	0.00	119
Amy	25	121	0.000	—	C. F. Aquadro, A. Weaver, K. Rice, S. W. Schaeffer, W. W. Anderson (in preparation)
Total or Mean	57	—	0.001	0.000	

D. SIMULANS   The *rosy, per* and *Adh* regions of the *D. simulans* have been surveyed (5; C. F. Aquadro et al, in preparation; D. Stern et al, in preparation). Consistent with reports based on in situ hybridization (39; J. Hey & W. F. Eanes, in preparation), the density of large insertions (0.001) is dramatically lower in this species than in *D. melanogaster* (Table 4). The few large insertions found are unique. The reasons for this difference, which contrasts with the lower level of nucleotide-site variation in *D. melanogaster* compared with *D. simulans* (5), are obscure.

FUTURE PROSPECTS   Unfortunately there have not yet been any restriction-mapping surveys of *D. affinis, algonquin, heteroneura* or *silvestris,* where in situ hybridization surveys have suggested much higher abundances and element frequencies, at least for some types of middle repetitive DNA families (62, 66). If the initial in situ surveys of these species are borne out by further, more detailed studies, then the range of variation in the distribution of TEs among *Drosophila* species may be large, and afford opportunities to investigate situations quite different from that in *D. melanogaster*.

CONCLUSIONS

The main conclusion from the models and data presented above is that they are consistent with the view that TEs are maintained in *Drosophila* populations as a result of transpositional increase in copy number, balanced by some opposing force or forces. This view convincingly explains the fact that elements in *D. melanogaster* (and several other species) are usually present at low frequencies at individual chromosomal sites into which they can insert. Similar results have been obtained in yeast (23) and bacteria (61). In addition, restriction fragment length polymorphism (RFLP) in mice may be due largely to variation in the sites of insertion of L1 elements (25). These data are almost impossible to reconcile with the hypothesis that elements persist as a result of favorable mutations associated with their transpositional activities (26, 103, 136). It is possible, however, that certain classes of mutation, especially chromosome rearrangements, could be generated as a result of imprecise excision of elements (46), recombination between LTRs (50), or ectopic exchange between elements at different locations (34, 56, 78), and could become fixed in populations by drift or selection without an increase in element frequency at the sites in question. Thus, TEs could act indirectly as a source of mutational variation for evolutionary change, but this would not have much influence on their distribution within populations.

   D. melanogaster and related species have much less repetitive dispersed DNA than typical vertebrates and many plants (131). This difference may well reflect the low frequencies of TE insertions in most *Drosophila* populations that have been studied. Despite the great premium placed on the discovery of RFLPs in human genetics and the abundance of repetitive DNA in the human genome, it is remarkable that so few human DNA polymorphisms are known to be attributable to the presence versus absence of a TE. Virtually all middle repetitive dispersed sequences appear to be fixed in the human population (122). At present, we can only speculate about the causes of these differences between taxa (78).

We do not yet know the nature of the forces opposing the spread of TEs in *Drosophila;* regulation of rate of transposition in relation to number of element copies per individual, selection against insertional mutations, and the induction of highly deleterious chromosome rearrangements by ectopic exchange may all be implicated. In view of the weak intensities of forces affecting element frequencies in natural populations, direct experimental tests to discriminate between the various possibilities discussed above will clearly be difficult. The careful collection and analysis of quantitative population data should, however, continue to provide means of examining these questions.

The other major conclusion that emerges from these studies is that there are theoretical reasons to expect TEs to accumulate in regions where ectopic exchange and/or regular meiotic exchange are suppressed. The data reviewed above suggest that there is indeed such an accumulation in the basal euchromatin and β-heterochromatin of *Drosophila* chromosomes, although not in the distal sections of the chromosomes where meiotic exchange is also suppressed. The population-genetic reasons for this accumulation are very different from those proposed for the accumulation of nontransposable, highly repeated sequences in regions of reduced recombination (32, 133, 139), which are abundant in the heterochromatin (68, 95). It is clear that the population dynamics of repeated DNA sequences must have a considerable evolutionary influence on the structure of the chromosome. Despite the lack of sympathy for mathematical modelling of population processes expressed in the writings of many molecular biologists (38, 68), the interplay between experimental and theoretical investigations of these population processes will shed light on the old problem of the euchromatin/heterochromatin distinction and the newer problems concerning the distributions of repetitive DNA sequences.

ACKNOWLEDGMENTS

This work was supported by NSF grant BSR-8-16629 and PHS grant 1-R01-GM 36405-01, and a grant to B. Charlesworth from the Louis Block fund of The University of Chicago. We thank numerous colleagues for providing us with copies of their unpublished work, and Deborah Charlesworth, Walter Eanes, Gail Simmons and an anonymous reviewer for their comments on the manuscript.

Literature Cited

1. Aguadé, M., Miyashita, N., Langley, C. H. 1989. Restriction-map variation at the *zeste-tko* region in natural populations of *Drosophila melanogaster*. *Mol. Biol. Evol.* 6:123–30
2. Aguadé, M., Miyashita, N., Langley, C. H. 1989. Reduced variation in the *yellow-achaete-scute* region in natural populations of *Drosophila melanogaster*. *Genetics*. In press
3. Ajioka, J. W., Eanes, W. F. 1989. The accumulation of P-elements on the tip of the X chromosome in populations of *Drosophila melanogaster*. *Genet. Res.* 53:1–6
4. Aquadro, C. F., Deese, S. F., Bland, M. M., Langley, C. H., Laurie-Ahlberg, C. C. 1986. Molecular population genetics of the alcohol dehydrogenase gene region of *Drosophila melanogaster*. *Genetics* 114:1165–90
5. Aquadro, C. F., Lado, K. M., Noon, W. A. 1988. The *rosy* region of *Drosophila melanogaster* and *Drosophila simulans*. I. Contrasting levels of naturally occurring DNA restriction map variation and divergence. *Genetics* 119:875–88
6. Beech, R. N., Leigh Brown, A. J. 1989. Insertion-deletion variation at the *yellow-achaete-scute*-region in two natural populations of *Drosophila melanogaster*. *Genet. Res.* 53:7–16
7. Berg, D. E., Howe, M. M., eds. 1989. *Mobile DNA*. Washington, DC: Am. Soc. Microbiol.
8. Bièmont, C. 1986. Polymorphism of the *mdg-1* and I mobile elements in *Drosophila melanogaster*. *Chromosoma* 93:393–97
9. Bièmont, C., Auoar, A. 1987. Copy-number dependent transpositions and excisions of the *mdg-1* mobile element in inbred lines of *Drosophila melanogaster*. *Heredity* 58:39–47
10. Bièmont, C. Auoar, A., Arnault, C. 1987. Genome reshuffling of the *copia* element in an inbred line of *Drosophila melanogaster*. *Nature* 329:742–43
11. Bièmont, C., Belyaeva, E. G., Pasyukova, E. G., Kogan, G. 1985. Mobile gene localization and viability in *Drosophila*. *Experientia* 41:1474–76
12. Bièmont, C., Gautier, C. 1987. *Mdg-1* mobile element heterozygosity in *Drosophila melanogaster*. *Heredity* 58:167–72
13. Bièmont, C., Terzian, C. 1988. *Mdg-1* mobile element polymorphism in selected *Drosophila melanogaster* populations. *Genetica* 76:7–14
14. Blackman, R. K., Grimaila, R., Koehler, M. M. D., Gelbart, W. M. 1987. Mobilization of hobo elements residing within the decapentaplegic gene complex: suggestion of a new hybrid dysgenesis system in *Drosophila melanogaster*. *Cell* 49:497–505
15. Boussy, I. A., Healy, M. J., Oakeshott, J. G., Kidwell, M. G. 1988. Molecular analysis of the P-M gonadal dysgenesis cline in eastern Australian *Drosophila melanogaster*. *Genetics* 119:889–902
16. Bregliano, J.-C., Kidwell, M. G. 1983. Hybrid dysgenesis determinants. In *Mobile Genetic Elements*, ed. J. A. Shapiro, pp. 363–410. New York: Academic
17. Brookfield, J. F. Y. 1986. A model for DNA sequence evolution within transposable elements. *Genetics* 112:393–407
18. Brookfield, J. F. Y. 1986. Population biology of transposable elements. *Philos. Trans. R. Soc. London. Ser. B* 312:217–26
19. Brookfield, J. F. Y., Montgomery, E. A., Langley, C. H. 1984. Apparent absence of transposable elements related to the P elements of *D. melanogaster* in other species of *Drosophila*. *Nature* 310:330–32
20. Bryan, G. J., Jacobson, J. W., Hartl, D. L. 1987. Heritable somatic excision of a *Drosophila* transposable element. *Science* 235:1636–38
21. Bucheton, A., Paro, R., Sang, H. M., Pelisson, A., Finnegan, D. J. 1984. The molecular basis of I-*R* hybrid dysgenesis: identification, cloning and properties of the I factor. *Cell* 40:327–38
22. Bucheton, A., Simonelig, M., Vaury, C., Crozatier, M. 1986. Sequences similar to the I transposable element involved in IR hybrid dysgenesis occur in other *Drosophila* species. *Nature* 322:650–52
23. Cameron, J. R., Loh, E. Y., Davis, R. W. 1979. Evidence for transposition of dispersed repetitive DNA families in yeast. *Cell* 16:739–51
24. Campbell, A. 1983. Transposons and their evolutionary significance. In *Evolution of Genes and Proteins*, ed. M. Nei, R. K. Koehn, pp. 258–79. Sunderland, Mass: Sinauer
25. Casavant, N. C., Hardies, S. C., Funk, F. D., Comer, M. B., Edgell, M. H., Hutchison, C. A. 1988. Extensive movement of LINES ONE sequences in β-globin loci of *Mus caroli* and *Mus domesticus*. *Mol. Cell. Biol.* 8:4669–74

26. Chao, L., Vargas, C., Spear, B. B., Cox, E. C. 1983. Transposable elements as mutator genes in evolution. *Nature* 303:633–35

27. Charlesworth, B. 1985. The population genetics of transposable elements. In *Population Genetics and Molecular Evolution,* ed. T. Ohta, K. Aoki, pp. 213–32. Berlin: Springer-Verlag

28. Charlesworth, B. 1986. Genetic divergence between transposable elements. *Genet. Res.* 48:111–18

29. Charlesworth, B. 1988. The maintenance of transposable elements in natural populations. In *Plant Transposable Elements,* ed. O. J. Nelson, pp. 189–212. New York: Plenum

30. Charlesworth, B., Charlesworth, D. 1983 The population dynamics of transposable elements. *Genet. Res.* 42:1–27

31. Charlesworth, B., Langley, C. H. 1986. The evolution of self-regulated transposition of transposable elements. *Genetics* 112:359–83

32. Charlesworth, B., Langley, C. H., Stephan, W. 1986. The evolution of restricted recombination and the accumulation of repeated DNA. *Genetics* 112:947–62

33. Charlesworth, B., Lapid, A. 1989. A study of ten transposable elements on X chromosomes from a population of *Drosophila melanogaster. Genet. Res.* In press

34. Davis, P. S., Shen, M. W., Judd, B. H. 1987. Asymmetrical pairings of transposons in and proximal to the white locus of *Drosophila* account for four classes of regularly occurring exchange products. *Proc. Natl. Acad. Sci. USA* 84:174–78

35. Dawid, I. B., Olong, E. O., Di Nocera, P. P., Pardue, M. L. 1981. Ribosomal insertion-like elements in *Drosophila melanogaster* are interspersed with mobile sequences. *Cell* 25:399–408

36. Di Nocera, P. P., Dawid, I. B. 1983. Interdigitated arrangement of two oligo(a)-terminated DNA sequences in *Drosophila. Nucleic Acids Res.* 11:5475–82

37. Doolittle, W. F., Sapienza, C. 1980. Selfish genes, the phenotype paradigm and genome evolution. *Nature* 284:601–7

38. Dover, G. A. 1989. The age of innocence is over. *Trends Genet.* 5:30–31

39. Dowsett, A. P., Young, M. W. 1982. Differing levels of dispersed repetitive DNA among closely related species of *Drosophila. Proc. Natl. Acad. Sci. USA* 79: 4570–74

40. Eanes, W. F., Ajioka, J. W., Hey, J., Wesley, C. 1989. Restriction map variation associated with the G6PD polymorphism in natural populations of *Drosophila melanogaster. Mol. Biol. Evol.* In press

41. Eanes, W. F., Labate, J., Ajioka, J. W. 1989. Restriction map variation associated with the *yellow-achaete-scute* region in four populations of *Drosophila melanogaster. Mol. Biol. Evol.* In press

42. Eanes, W. F., Wesley, C., Hey, J., Houle, D. 1988. The fitness consequences of *P* element insertion in *Drosophila melanogaster. Genet. Res.* 52:17–26

43. Eggleston, W. B., Johnson-Schlitz, D. M., Engels, W. R. 1988. P-M hybrid dysgenesis does not mobilize other transposable element families in *Drosophila melanogaster. Nature* 331:368–70

44. Engels, W. R. 1981. Hybrid dysgenesis in *Drosophila* and the stochastic loss hypothesis. *Cold Spring Harbor Symp. Quant. Biol.* 45:561–66

45. Engels, W. R. 1986. On the evolution and population genetics of hybrid-dysgenesis causing transposable elements in *Drosophila. Philos. Trans. R. Soc. London Ser. B* 312:205–15

46. Engels, W. R. 1989. P elements in *Drosophila.* See Ref. 7

47. Engels, W. R., Benz, W. K., Preston, C. R., Graham, P. L., Phillis, R. W., Robertson, H. M. 1987. Somatic effects of P element activity in *Drosophila melanogaster:* Pupal lethality. *Genetics* 117:745–57

48. Ewens, W. J. 1972. The sampling theory of selectively neutral alleles. *Theor. Popul. Biol.* 3:87–112

49. Felsenstein, J., Yokoyama, S. 1976. The evolutionary advantage of recombination. II. Individual selection for recombination. *Genetics* 83:845–59

50. Finnegan, D. J., Fawcett, D. H. 1986. Transposable elements in *Drosophila melanogaster. Oxford Surv. Eukaryot. Genes* 3:1–62

51. Finnegan, D. J., Rubin, G. M., Young, M. W., Hogness, D. S. 1978. Repeated gene families in *Drosophila melanogaster. Cold Spring Harbor Symp. Quant. Biol.* 42:1053–63

52. Fitzpatrick, B. J., Sved, J. A. 1986. High level of fitness modifiers induced by hybrid dysgenesis in *Drosophila melanogaster. Genet. Res.* 48:89–94

53. Gerasimova, T. I., Matyunina, L. V., Ilyin, Y. V., Georgiev, G. P. 1984. Simultaneous transposition of different mobile elements. Relation to multiple

mutagenesis in *Drosophila melanogaster. Mol. Gen. Genet.* 194:517–22

54. Gerasimova, T. I., Mizrokhi, L. J., Georgiev, G. P. 1984. Transposition bursts in genetically unstable *Drosophila melanogaster. Nature* 309:714–15

55. Ginzburg, L. R., Bingham, P. M., Yoo, S. 1984. On the theory of speciation induced by transposable elements. *Genetics* 107:331–41

56. Goldberg, M. L., Shen, J.-Y., Gehring, W. J., Green, M. M. 1983. Unequal crossing-over associated with asymmetrical synapsis between nomadic elements of the *Drosophila* genome. *Proc. Natl. Acad. Sci. USA* 80:5017–21

57. Golding, G. B., Aquadro, C. F., Langley, C. H. 1986. Sequence evolution within populations under multiple types of mutation. *Proc. Natl. Acad. Sci. USA* 83:427–31

58. Green, M. M. 1980. Transposable elements in *Drosophila* and other Diptera. *Annu. Rev. Genet.* 14:109–20

59. Gvozdev, V. A., Belyaeva, E. S., Ilyin, Y. V., Amosova, I. S., Kaidanov, L. Z. 1981. Selection and transposition of mobile dispersed genes in *Drosophila melanogaster. Cold Spring Harbor Symp. Quant. Biol.* 45:673–85

60. Haigh, J. 1978. The accumulation of deleterious genes in a population- Muller's ratchet. *Theor. Popul. Biol.* 14: 251–67

61. Hartl, D. L., Medhora, M., Green, L., Dykhuizen, D. E. 1986. The evolution of DNA sequences in *Escherichia coli. Philos. Trans. R. Soc. London Ser. B* 312:191–204

62. Hey, J. 1989. The transposable portion of the genome of *Drosophila algonquin* is very different from that in *D. melanogaster. Mol. Biol. Evol.* 6:66–102

63. Hickey, D. A. 1982. Selfish DNA: a sexually transmitted parasite. *Genetics* 101:519–31

64. Hilliker, A. J., Appels, R., Schalet, A. 1980. The genetic analysis of *D. melanogaster* heterochromatin. *Cell* 21:607–19

65. Hudson, R. R., Kaplan, N. L. 1986. On the divergence of members of a transposable element family. *J. Math. Biol.* 24:207–15

66. Hunt, J. A., Bishop, J. G., Carson, H. L. 1984. Chromosomal mapping of a middle-repetitive DNA sequence in a cluster of five species of Hawaiian *Drosophila. Proc. Natl. Acad. Sci. USA* 81:7146–50

67. Ising, B., Block, K. 1981. Derivation-dependent distribution of insertion sites for a *Drosophila* transposon. *Cold Spring Harbor Symp. Quant. Biol.* 45: 527–44

68. John, B., Miklos, G. L. G. 1988. *The Eukaryote Genome in Development and Evolution.* London: Allen & Unwin

69. Kaplan, N. L., Brookfield, J. F. Y. 1983. The effect on homozygosity of selective differences between sites of transposable elements. *Theor. Popul. Biol.* 23:273–80

70. Kaplan, N. L., Brookfield, J. F. Y. 1983. Transposable elements in Mendelian populations. III. Statistical results. *Genetics* 104:485–95

71. Kaplan, N. L., Darden, T., Langley, C. H. 1985. Evolution and extinction of transposable elements in Mendelian populations. *Genetics* 109:459–80

72. Kidd, S. J., Glover, D. M. 1980. A DNA segment from *D. melanogaster* which contains five tandemly repeating units homologous to the major rDNA insertion. *Cell* 19:103–19

73. Kidwell, M. G. 1983. Evolution of hybrid dysgenesis determinants in *Drosophila melanogaster. Proc. Natl. Acad. Sci. USA* 79:341–404

74. Kimura, M., Crow, J. F. 1964. The number of alleles that can be maintained in a finite population. *Genetics* 49:725–38

75. Kleckner, N. 1981. Transposable elements in prokaryotes. *Annu. Rev. Genet.* 15:341–404

76. Langley, C. H., Aquadro, C. F. 1987. Restriction map variation in natural populations of *Drosophila melanogaster: white* locus region. *Mol. Biol. Evol.* 4:651–63

77. Langley, C. H., Brookfield, J. F. Y., Kaplan, N. L. 1983. Transposable elements in Mendelian populations. I. A theory. *Genetics* 104:457–72

78. Langley, C. H., Montgomery, E. A., Hudson, R. H., Kaplan, N. L., Charlesworth, B. 1988. On the role of unequal exchange in the containment of transposable element copy number. *Genet. Res.* 52:223–35

79. Langley, C. H., Quattlebaum, W. F. 1982. Restriction map variation in the *Adh* region of *Drosophila. Proc. Natl. Acad. Sci. USA* 79:5631–25

80. Langley, C. H., Shrimpton, A. E., Yamazaki, T., Miyashita, N., Matsuo, Y., Aquadro, C. F. 1988. Naturally occurring variation in the restriction map of the *Amy* region of *Drosophila melanogaster. Genetics* 119:619–29

81. Lefevre, G. 1976. A photographic representation of the polytene chromosomes of *Drosophila melanogaster* salivary glands. In *The Genetics and Biology of*

Drosophila, ed. M. Ashburner, E. Novitski, 1A:31–36. Orlando, Fla: Academic

82. Leigh Brown, A. J. 1983. Variation at the 87A heat shock locus in *Drosophila melanogaster. Proc. Natl. Acad. Sci. USA* 80:5350–54

83. Leigh Brown, A. J., Moss, J. E. 1987. Transposition of the I element and *copia* in natural populations of *Drosophila melanogaster. Genet. Res.* 49:121–28

84. Lewis, A. P., Brookfield, J. F. Y. 1987. Movement of *Drosophila melanogaster* transposable elements other than *P* elements in a *P-M* hybrid dysgenic cross. *Mol. Gen. Genet.* 208:506–10

85. Lewontin, R. C., Prout, T. 1956. Estimation of the number of different classes in a population. *Biometrics* 12:211–23

86. Li, W.-H., Luo, C.-C., Wu, C.-I. 1985. Evolution of DNA sequences. In *Molecular Evolutionary Genetics,* ed. R. J. MacIntyre, pp. 1–94. New York: Plenum

87. Lim, J. K. 1979. Site-specific instability in *Drosophila melanogaster:* the origin of mutation and the cytogenetic evidence for site-specificity. *Genetics* 93:681–701

88. Lim, J. K., Simmons, M. J., Raymond, J. D., Cox, N. M., Doll, R. F., Culbert, T. P. 1983. Homologue destabilization by a putative transposable element in *Drosophila melanogaster. Proc. Natl. Acad. Sci. USA* 80:6624–27

89. Lindsley, D. L. Sandler, L. 1977. The genetic analysis of meiosis in female *Drosophila. Philos. Trans. R. Soc. London Ser. B* 277:295–312

90. Lucchesi, J. C. 1976. Inter-chromosomal effects. See Ref. 81, pp. 315–30

91. McGinnis, W., Shermoen, A. W., Beckendorf, S. K. 1983. A transposable element inserted just 5' to a Drosophila glue protein gene alters gene expression and chromatin structure. *Cell* 34:75–84

92. Mackay, T. F. C. 1986. Transposable element-induced fitness mutations in *Drosophila melanogaster. Genet. Res.* 48:77–87

93. Martin, G., Wiernasz, D., Schedl, P. 1983. Evolution of *Drosophila* repetitive-dispersed DNA. *J. Mol. Evol.* 19:203–13

94. May, R. M., Anderson, R. M. 1983. Epidemiology and genetics in the coevolution of parasites and hosts. *Proc. R. Soc. London Ser. B* 219:281–313

95. Miklos, G. L. G. 1985. Localized highly repetitive DNA sequences in vertebrate and invertebrate genomes. See Ref. 86, pp. 240–321

96. Miklos, G. L. G., Healy, M. J., Pain, P., Howells, A. J., Russell, R. J. 1984. Molecular genetic studies on the euchromatin-heterochromatin junction in the X chromosome of *Drosophila melanogaster.* I. A cloned entry point near to the uncoordinated *(unc)* locus. *Chromosoma* 89:218–27

97. Miklos, G. L. G., Yamamoto, M., Davies, J., Pirotta, V. 1988. Microcloning reveals a high frequency of repetitive sequences characteristic of chromosome 4 and the β-heterochromatin of *Drosophila melanogaster. Proc. Natl. Acad. Sci. USA* 85:2051–55

98. Miyashita, N., Langley, C. H. 1988. Molecular and phenotypic variation of the *white* locus region in *Drosophila melanogaster. Genetics* 120:199–212

99. Montgomery, E. A., Charlesworth, B., Langley, C. H. 1987. A test for the role of natural selection in the stabilization of transposable element copy number in a population of *Drosophila melanogaster. Genet. Res.* 49:31–41

100. Montgomery, E. A., Langley, C. H. 1983. Transposable elements in Mendelian populations. II. Distribution of three *copia*-like elements in a natural population. *Genetics* 104:473–83.

101. Mukai, T., Yamaguchi, O. 1974. The genetic structure of natural populations of *Drosophila.* XI. Genetic variability in a local population. *Genetics* 76:339–66

102. Muller, H. J. 1964. The relation of recombination to mutational advance. *Mutat. Res.* 1:2–9

103. Nevers, P., Saedler, H. 1977. Transposable genetic elements as agents of instability and chromosomal rearrangements. *Nature* 268:109–15

104. Ohta, T. 1981. Population genetics of selfish DNA. *Nature* 292:648–49

105. Ohta, T. 1983. Theoretical study on the accumulation of selfish DNA. *Genet. Res.* 41:1–16

106. Ohta, T. 1985. A model of duplicative transposition and gene conversion of repetitive DNA families. *Genetics* 110:145–58

107. Ohta, T. 1986. Population genetics of an expanding family of transposable elements. *Genetics* 113:145–59

108. Ohta, T., Kimura, M. 1981. Some calculations on the amount of selfish DNA. *Proc. Natl. Acad. Sci. USA* 78:1129–32

109. Orgel, L. E., Crick, F. H. C. 1980. Selfish DNA: the ultimate parasite. *Nature* 284:604–7

110. Pamilo, P., Nei, M., Li, W-H. 1987. Accumulation of mutations in sexual and asexual populations. *Genet. Res.* 49: 135–46

111. Paro, C., Goldberg, M. L., Corces, V. G., Gehring, W. J. 1985. Molecular analysis of large transposable elements carrying the *white* locus of *Drosophila melanogaster*. *EMBO J.* 2:853–60

112. Pasyukova, E. G., Belyaeva, E. S., Kogan, G. L., Kaidanov, L. Z., Gvozdev, V. A. 1986. Concerted transpositions of mobile genetic elements coupled with fitness changes in *Drosophila melanogaster*. *Mol. Biol. Evol.* 3:299–312

113. Pelisson, A. 1981. The *I-R* system of hybrid dysgenesis in *Drosophila melanogaster:* are *I* factor insertions responsible for the mutagenic effect of the *I-R* interaction? *Mol. Gen. Genet.* 183:123–29

114. Potter, S. S., Brorein, W. J., Dunsmuir, P., Rubin, G. M. 1979. Transposition of elements of the *412, copia,* and *297* dispersed repeated gene families in *Drosophila. Cell* 17:415–27

115. Rasmuson, B., Westerberg, B. M., Rasmuson, A., Gvozdev, V. A., Belyaeva, E. S., Ilyin, Y. V. 1981. Transpositions, mutable genes, and the dispersed gene family *Dm225* in *Drosophila melanogaster. Cold Spring Harbor Symp. Quant. Biol.* 45:545–51

116. Roiha, H., Miller, J. R., Woods, L. C., Glover, D. M. 1981. Arrangements and rearrangements of sequences flanking the two types of rDNA insertion in *Drosophila melanogaster. Nature* 290:749–53

117. Ronsseray, S., Anxolabehère, D. 1986. Chromosomal distribution of *P* and *I* transposable elements in a natural population of *Drosophila melanogaster. Chromosoma* 94:433–40

118. Rubin, G. M. 1983. Dispersed repetitive DNA sequences in *Drosophila*. In *Mobile Genetic Elements*, ed. J. A. Shapiro, pp. 329–61. Orlando Fla: Academic

119. Schaeffer, S. W., Aquadro, C. F., Anderson, W. W. 1987. Restriction-map variation in the alcohol dehydrogenase region of *Drosophila pseudoobscura. Mol. Biol. Evol.* 4:254–65

120. Schaeffer, S. W., Aquadro, C. F., Langley, C. H. 1988. Restriction-map variation in the *Notch* region of *Drosophila melanogaster. Mol. Biol. Evol.* 5:30–40

121. Schalet, A., Lefevre, G. 1976. The proximal region of the X chromosome. See Ref. 81, pp. 848–928

122. Schmid, C. W., Shen, C.-K. J. 1985. The evolution of interspersed repetitive DNA sequences in mammals and other vertebrates. See Ref. 86, pp. 323–58

123. Schwartz, D., 1989. Patterns of *Ac* transposition in maize. *Genetics* 121: 125–28

124. Shapiro, J. A., ed. 1983. *Mobile Genetic Elements*. New York: Academic

125. Shrimpton, A. E., Montgomery, E. A., Langley, C. H. 1986. *Om* mutations in *Drosophila ananassae* are linked to insertions of a transposable element. *Genetics* 114:125–35

126. Silber, J., Bazin, C., Lemeunier, F., Aulard, S., Volovitch, M. 1989. Distribution and conservation of the foldback transposable element in *Drosophila. J. Mol. Evol.* 28:220–24

127. Simmons, M. J., Crow, J. F. 1977. Mutations affecting fitness in *Drosophila* populations. *Annu. Rev. Genet.* 11:49–78

128. Singh, R. S., Rhomberg, L. R. 1987. A comprehensive study of genic variation in natural populations of *Drosophila melanogaster*. I. Estimates of gene flow from rare alleles. *Genetics* 115:313–22

129. Slatkin, M. 1985. Genetic differentiation of transposable elements under mutation and unbiased gene conversion. *Genetics* 110:145–58

130. Spradling, A. C. 1986. P-element mediated transformation. In *Drosophila: A Practical Approach*, ed. D. B. Roberts, pp. 175–97. Oxford: IRL Press

131. Spradling, A. C., Rubin, G. M. 1981. *Drosophila* genome organization: conserved and dynamic aspects. *Annu. Rev. Genet.* 15:219–64

132. Steinemann, M. 1982. Multiple sex chromosomes in *Drosophila miranda:* a system to study the degeneration of a chromosome. *Chromosoma* 89:59–76

133. Stephan, W. 1986. Recombination and the evolution of satellite DNA. *Genet. Res.* 47:167–74

134. Stephan, W., Langley, C. H. 1989. Molecular variation in the centromeric region of the X chromosome in three *Drosophila ananassae* populations. I. Contrasts between the *vermilion* and *forked* loci. *Genetics* 121:89–99

134a. Stephan, W. 1989. Molecular variation in the centromeric region of the X chromosome in three *Drosophila ananassae* populations. II. The *OM(1D)* locus. *Mol. Biol. Evol.* In press

135. Strobel, E., Dunsmuir, P., Rubin, G. M. 1979. Polymorphism in the locations of elements of the *412, copia* and *297* dispersed repeated gene families in *Drosophila. Cell* 17:429–39

136. Syvanen, M. 1984. The evolutionary implication of mobile genetic elements. *Annu. Rev. Genet.* 18:271–93

137. Tchurikov, N. A., Ilyin, Y. V.,

Skryabin, K. G., Ananiev, E. V., Krayev, A. S., et al. 1981. General properties of mobile dispersed genetic elements in *Drosophila melanogaster*. *Cold Spring Harbor Symp. Quant. Biol.* 45:655–65

138. Uyenoyama, M. K. 1985. Quantitative models of hybrid dysgenesis: rapid evolution under transposition, extrachromosomal inheritance and fertility selection. *Theor. Popul. Biol.* 27:176–201

139. Walsh, J. B. 1987. Persistence of tandem arrays: implications for satellite and simple sequence DNAs. *Genetics* 115:553–67

140. Wensink, P. W. 1978. Sequence homology with families of *Drosophila melanogaster* middle repetitive DNA. *Cold Spring Harbor Symp. Quant. Biol.* 62:1033–39

141. Wensink, P. W., Finnegan, D. J., Donelson, J. E., Hogness, D. S. 1974. A system for mapping DNA sequences in the chromosomes of *Drosophila melanogaster*. *Cell* 3:315–25

142. Wensink, P. W., Tabata, S., Pachi, C. 1979. The clustered and scrambled arrangement of moderately repetitive elements in *Drosophila* DNA. *Cell* 18:1231–46

143. Woodruff, R. C., Blount, J. L., Thompson, J. N. 1987. Hybrid dysgenesis is not a general release mechanism for DNA transpositions. *Science* 237:1206–7

144. Wright, S. 1931. Evolution in Mendelian populations. *Genetics* 16:97–159

145. Xiong, Y., Burke, W. D., Jakubczak, J. L., Eickbush, T. H. 1988. Ribosomal DNA insertions R1Bm and R2Bm can transpose in a sequence specific manner to locations outside the 28S genes. *Nucleic Acids Res.* 16:10561–73

146. Xiong, Y., Eickbush, T. H. 1988. Similarity of reverse-transcriptase-like sequences of viruses, transposable elements and mitochondrial introns. *Mol. Biol. Evol.* 5:675–90

147. Yamaguchi, O., Yamazaki, T., Saigo, K., Mukai, T., Robertson, A. 1987. Distributions of three transposable elements, *P, 297* and *copia,* in natural populations of *Drosophila melanogaster*. *Jpn. J. Genet.* 62:205–16

148. Yannopoulos, G., Stamatis, N., Monastirioti, M., Hatzopoulos, P., Louis, C. 1987. Hobo is responsible for induction of hybrid dysgenesis by strains of *Drosophila melanogaster* bearing the male recombination factor *23.5MRF*. *Cell* 49:487–95

149. Young, M. 1979. Middle repetitive DNA: a fluid component of the *Drosophila genome*. *Proc. Natl. Acad. Sci. USA* 76:6274–78

150. Yukuhiro, K., Harada, K., Mukai, T. 1985. Viability mutations induced by the *P* element in *Drosophila melanogaster*. *Jpn. J. Genet.* 60:531–37

Annu. Rev. Genet. 1989. 23:289–310

GENETIC ANALYSIS OF PROTEIN STABILITY AND FUNCTION

Andrew A. Pakula

Division of Biology, California Institute of Technology, Pasadena, California 91125

Robert T. Sauer

Department of Biology, Massachusetts Institute of Technology, Cambridge, Massachusetts 02139

CONTENTS

INTRODUCTION

There is currently a great deal of interest in understanding the amino-acid sequence determinants of protein stability and function. This is important not

0066-4197/89/1215-0289$02.00

only for ongoing studies aimed at dissecting the structure and activities of biologically important proteins, but also for the realization of longer term goals such as the prediction of protein structure from sequence and the design of proteins with novel activities. Detailed genetic and biophysical studies of proteins are beginning to improve our overall understanding of protein structure-function relationships and should allow considerably more progress in the near future.

Genetic studies of protein structure and activity generally center on the properties of proteins altered by deletions or point mutations. Two basic strategies are commonly used. In the first, one creates a specific alteration in the coding sequence and asks, "What is the effect of this alteration?" In the second, one creates pools of randomly altered genes, applies a screen or selection to identify those encoding proteins with a specific phenotype, and then asks, "What kinds of sequence alterations can cause this effect?" The directed approach is most useful when there is already enough information about the structure or activity of the protein to formulate specific questions about the roles of particular residues. The random approach is particularly useful for identifying important residues in an unbiased way in the absence of detailed information from other sources or studies.

MAKING AND MAPPING MUTATIONS

Traditionally, mutations have been generated by treating cells with agents such as nitrosoguanidine, EMS, and UV light. These mutations are then located by genetic mapping. For the study of protein function and stability, this approach is rapidly being replaced by methods involving manipulations of cloned genes. Mutations may be generated by directed mutagenesis, rapidly localized to specific restriction fragments using recombination in vitro, and then analyzed by DNA sequencing.

Numerous methods are available for the random mutagenesis of cloned genes. In general, these permit a broader, less biased, mutagenic specificity than has been possible with traditional techniques. Furthermore, several strategies are available for limiting random mutagenesis to portions of a DNA molecule. Thus, a specific gene, or only selected regions of a gene, can be mutagenized without creating changes in the rest of the cloning vector. Specific mutations can be constructed in cloned genes using oligonucleotide-directed mutagenesis. This technique permits the creation of proteins with one or more defined amino acid change(s). Such changes can also be created by synthesis of double-stranded DNA cassettes that are then returned to the gene in vitro. Cassette mutagenesis can also be used as a powerful technique for localized random mutagenesis when some or all of the base positions in the cassette are synthesized with a mixture of wild-type and mutant nucleotides.

Thus, single nucleotide pairs, single codons, or blocks of codons may be randomly mutagenized with extremely high efficiency. Many such methods are described in several reviews (7, 66, 79) and are not discussed here. Instead, we concentrate on general mutagenic strategies and discuss what these methods have taught us regarding the sequence determinants of protein structure, stability, and activity.

The domain is the basic unit of protein structure and function. For proteins with multiple domains, deletion analysis can often rapidly identify large portions of a protein sequence that are not required for a particular activity. For example, in *Escherichia coli* alanine-tRNA synthetase (875 residues), the COOH-terminal 415 residues of the protein can be deleted and the truncated protein still retains amino acylation activity (27). In like fashion, the site-specific DNA-binding activities of the GAL4 (881 residues) and GCN4 (281 residues) transcriptional regulatory proteins of *Saccharomyces cerevisiae* reside in independent domains of 60–100 residues (26, 31). In such cases, it clearly makes sense to use deletions to identify structural domains and thus restrict the problem being investigated to the greatest possible extent. However, the effects of deletions (or insertions) within structural domains are generally too drastic to provide very much useful information for structure-function studies. At this more detailed level, missense mutations provide the major tool for further dissection of structure and activity.

Studies of mutant proteins can be roughly divided into two classes: Some focus on the identification of residues that are directly involved in binding or enzymatic activities. Others concentrate on the importance of specific residues and interactions in the folding of proteins, and in the stability of protein structures. Although these two types of studies have clearly different goals, they are intimately related in the sense that protein folding and the maintenance of a stably folded structure are almost always prerequisites for activity. Thus, putative active-site mutations must be shown to be free of severe effects on structure and stability, and putative stability mutations must be distinguished from those that disrupt function but not structure.

FOLDED AND UNFOLDED PROTEIN STRUCTURES

As a rule, proteins fold and unfold spontaneously in a reaction that can be described in terms of a simple, two-state equilibrium. The unfolding of a monomeric protein can be modeled as

$$\text{folded} \rightleftharpoons \text{unfolded} \qquad K_u = \frac{[\text{unfolded}]}{[\text{folded}]}$$

where the equilibrium constant, K_u, is a measure of the ratio of unfolded to folded protein molecules. The free energy change upon unfolding (ΔG_u) can be calculated from K_u by

$$\Delta G_u = -RT \ln (K_u)$$

where ΔG_u represents the difference between the free energies of the folded and the unfolded states, R is a constant (1.98 cal/mol-°K), and T is the temperature in °K. The conversion from terms of the equilibrium constant to terms of free energy is useful, because this permits the net stability of the folded protein to be directly compared to the energetic contributions of specific interactions. At 37°C, a 1 kcal/mol decrease in ΔG_u corresponds to a fivefold increase in K_u. Values of ΔG_u for protein unfolding range from about 3–15 kcal/mol under physiological conditions of temperature and pH (49). Single destabilizing mutations can decrease the stability of some proteins to the point where most molecules are unfolded (see discussion on Destabilizing Mutations). If ΔG_u for a protein is 3 kcal/mol at 37°C, for example, then the fraction of unfolded protein would be 0.7%. A mutation that decreased the stability of this protein by 4 kcal/mol would increase the fraction of unfolded protein to 80%. Hence, a fivefold loss in activity would be expected simply on the basis of the decreased concentration of folded, active molecules. In reality, the activity loss could be considerably greater if, for example, the destabilizing mutation also affected the specific activity of the folded protein. Moreover, in the cell, processes such as aggregation or proteolysis that rapidly and irreversibly remove unfolded protein, may magnify the phenotypic effects of destabilization.

Protein structures contain an impressive array of stabilizing interactions; these include hydrophobic and packing interactions, hydrogen bonds, and salt bridges. As a result, it is often difficult to imagine that changing a single side chain could result in a serious perturbation of structure or stability. However, although the forces favoring protein folding contribute a large amount of energy and involve a large number and variety of interactions, they are nearly offset by the entropic cost of folding. This entropic penalty is due to the enormous loss of conformational freedom that occurs as the protein goes from a denatured state with many possible conformations to a native state with only one or a few conformations. Thus, a net stability of 5 kcal/mol may arise as the difference between a favorable energy of 300 kcal/mol and an unfavorable energy of 295 kcal/mol. Clearly, in such a case, small fractional changes in the energies favoring and opposing folding can shift the balance and lead to unfolding. Such changes can occur as a consequence of alterations in temperature, pH, and the concentration of denaturants, as well as by the introduction of mutations.

Measuring Protein Stability

The folded and unfolded forms of a protein almost always have different spectral or hydrodynamic properties. As a result, the fraction of unfolded protein molecules can generally be determined by monitoring an appropriate physical property as a function of changes in temperature, urea concentration, or guanidinium-HCl concentration (49). Susceptibility to proteolysis provides another means of determining protein stability, because most native proteins are relatively resistant to cleavage, whereas denatured proteins are exquisitely sensitive (50). Thus, the rate at which a purified protein is degraded by a protease will depend on the fraction of molecules that are unfolded, and proteolysis in vitro can be used to compare the stabilities of a mutant protein and its wild-type counterpart (22).

Most single-domain proteins unfold in a cooperative fashion, i.e. a given molecule is either folded or unfolded. Figure 1 shows thermal denaturation experiments for a wild-type protein and a mutant that displays reduced stability. Although the fraction of molecules that are unfolded varies as a function of temperature for both proteins, the unfolding transition for the mutant occurs over a lower temperature range than that for the wild-type protein. At any given temperature in the transition zone, K_u and ΔG_u values can be calculated for both molecules. The difference in stability can then be expressed as $\Delta\Delta G_u$, which is defined as ΔG_u(wild-type) $- \Delta G_u$(mutant). We refer to $\Delta\Delta G_u$ values when we say, for example, that an Ile→Val substitution destabilizes a protein by 1 kcal/mol. In many cases, it is also convenient to refer to the temperature at which half of the protein molecules are unfolded, T_m, as a rough measure of the stability of a protein.

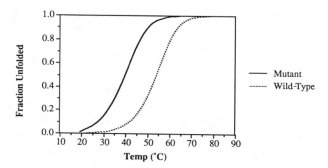

Figure 1 Thermal denaturation of a hypothetical protein and a mutant derivative with reduced stability. The stabilities of the two proteins can be directly compared at temperatures where their transition zones overlap (ca. 35–50°C in this case). Stabilities at temperatures outside of the transition zone can be calculated if ΔH and ΔC_p for unfolding are known (for discussion, see references 6, 49).

TOLERANCE OF RESIDUE POSITIONS TO SUBSTITUTIONS

An initial question concerning any protein is how many of its amino acids are really critical for structure or function? Can the protein be inactivated by substitutions anywhere in the sequence or are only a few key residues really important? In the sections that follow, we first discuss studies of defective mutations for several proteins with known three-dimensional structures and for which deleterious mutations have been isolated and identified by random mutagenesis. Subsequently, we discuss the phenotypically neutral mutations that have been studied in several proteins.

Mutations Causing Reduction or Loss of Activity

Many different missense mutations, each causing a defective phenotype, have been isolated in the genes encoding staphylococcal nuclease (67), phage T4 lysozyme (2), the N-terminal domain of λ repressor (20), λ Cro (52), and yeast iso-1-cytochrome *c* (18, 19). The severity of the mutant phenotypes varies for the different proteins, and often varies among the collection of missense alleles for a given protein. For example, the T4 lysozyme mutants were each isolated on the basis of a temperature-sensitive phenotype, and thus known to be able to fold and function at the permissive temperature. By contrast, many of the mutant forms of the other proteins showed no activity at any temperature. Nevertheless, in each case, defective mutations can clearly occur at many positions. For example, residue substitutions at 32 of the 66 positions in λ Cro, and 55 of the 149 positions in staphylococcal nuclease are known to result in diminished activity or loss of activity. Moreover, the sites of these mutations are not obviously clustered in the protein sequences or within the crystal structures of any of the five proteins.

What kinds of mutations result in a defective phenotype? The striking observation is that most mutant substitutions appear to affect activity indirectly via effects on protein structure or stability. This conclusion is supported by several findings. First, these mutations occur at positions for which no evidence exists for a direct functional role; that is, they are found at sites distant from the active site/binding regions of the proteins. Second, several mutant proteins of this class have been purified and shown to be less stable than wild-type for each of the five proteins (6, 22, 51, 55, 62, 68). Finally, most of these mutations affect side chains that would be expected to play important structural roles. These include side chains that are buried in the protein structure, side chains involved in hydrogen bonds or electrostatic interactions, and side chains with special properties, such as glycine and proline. We return later to a discussion of each of these types of mutations.

The degree to which a side chain is buried in the native protein is usually defined by computer calculation of its fractional accessibility to water (57); low solvent accessibilities indicate that residues are buried, whereas high accessibilities indicate that residues are exposed on the protein surface. In Figure 2 the likelihood of isolating a destabilizing substitution is plotted as a function of the fractional accessibility of the wild-type side chain for staphylococcal nuclease, T4 lysozyme, λ repressor, λ Cro, and yeast iso-1-cytochrome c. Buried or core residues are obviously the most common sites of destabilizing mutations for each of the five proteins, suggesting that these residues are extremely important for the maintenance of protein structure and stability. However, certain exposed or partially exposed side chains must also be structurally important, as some destabilizing mutations also occur at these positions.

As might be expected, at least some of the mutations that disrupt protein activity do so in a direct fashion. In staphylococcal nuclease, λ repressor and λ Cro, approximately one quarter to one third of the defective mutations alter residues that are directly involved in function. In staphylococcal nuclease, these include substitutions at twelve positions within the active site or polynucleotide binding region. In λ repressor and λ Cro, mutations occur at about ten positions that form a significant portion of the DNA-binding surfaces of each protein. Active site mutations are not represented among the T4-lysozyme mutants, but are not expected as these temperature-sensitive mutants have wild-type or near wild-type activities at low temperatures. None

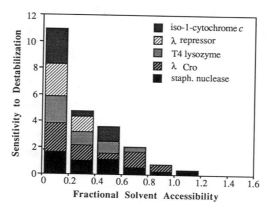

Figure 2 Normalized probability of isolating destabilizing mutations as a function of fractional residue solvent accessibility. Data are compiled for the N-terminal domain of λ repressor (20), phage T4 lysozyme (2), λ Cro (52), staphylococcal nuclease (67), and yeast iso-1-cytochrome c (19). For each accessibility class, the number of positions at which destabilizing mutations occur was divided by the total number of positions in that class. This value was then normalized by dividing by the frequency of such mutations for the entire protein. Thus, a value of 1 indicates residues with an average susceptibility to destabilizing mutations.

of the mutations in iso-1-cytochrome c abolishes electron transfer function while retaining normal stability. The absence of this class of mutations may simply reflect the sampling of potential mutants or could indicate that functional residues in this protein play a dual role and are involved both in function and in structural stability.

The generalization that most missense mutations act by affecting protein stability does not hold in all cases. For example, missense mutations causing a null phenotype in EcoR1 primarily affect residues located at the protein-DNA interface or at the protein-protein dimer interface (77). Only a few mutations apparently alter the stability of the monomers in this case. It is possible that the EcoR1 monomer is more stable than the proteins discussed above. In this case, mutations that are sufficiently destabilizing to cause a complete loss of activity might be rare. Alternatively, the clustering of mutations in the EcoR1 case may reflect the stringency of the mutant selection, which demands elimination of even trace amounts of enzyme activity.

Neutral Mutations

The studies discussed above show that defective mutations can generally occur at many positions throughout a protein sequence. What, however, can be concluded about the sites where mutations were not isolated? Are amino acid changes at these positions silent or is the catalog of defective mutations simply incomplete? A general way of addressing this question is to ask whether amino acid substitutions can be functionally tolerated at any given residue position. Two strategies have been used for the efficient generation of mutations that can then be scored for a neutral phenotype. The first is tRNA-mediated suppression of *amber* (UAG) mutations. Strains have now been isolated or constructed that allow the efficient insertion of Ala, Cys, Gln, Gly, His, Leu, Lys, Phe, Pro, Ser, and Tyr at *amber* codons (for review, see ref. 47). The second involves codon randomization via cassette mutagenesis (56). Here a double-stranded DNA cassette is chemically synthesized with one or more codons randomized by the inclusion of all four bases during synthesis. The cassette is then recloned into the gene and introduced into cells by transformation. Genes encoding active proteins can then be identified by a selection or screen, and sequenced.

Miller and colleagues have provided the most extensive view of neutral mutations in their work on suppressed nonsense mutations at 142 of the 360 codon positions in the *lac* repressor gene (35, 44). In these studies, the phenotypes of some 1500 single residue changes were scored and approximately half of these changes were found to be phenotypically silent. At 28 residue positions, all substitutions tested were tolerated. At an additional 54 sites, at least half of all substitutions were tolerated and, in many of these cases, only proline was not tolerated. These results indicate that the identity of the side chain is not a critical determinant of either structure or activity for a significant number of positions in a protein.

Figure 3 shows the nearly exhaustive set of neutral mutations that have been isolated in part of the N-terminal domain of λ repressor following cassette mutagenesis (56; J. Reidhaar-Olson, unpublished data). All the residues in the mutagenized region are distant from the DNA in the crystal structure of the protein-DNA complex (28), and thus any effects on activity caused by substitutions in this region must be mediated indirectly via protein structure or stability. As shown in the figure, the neutral mutations isolated at six positions included only the wild-type residue or a single conservative substitute. The method of mutagenesis used in these experiments ensures that all residue substitutions are represented in the population prior to selection. Hence, recovery of only a few neutral substitutions indicates that most other substitutions are not functional and have been selected against. Five of these positions are buried in the active dimeric form of N-terminal domain. At most surface positions, however, a large number of chemically different side chains are allowed, including those that are charged, uncharged, large, small, hydrophilic, and hydrophobic. Clearly, residue positions that display this degree of tolerance do not play essential roles in protein structure or stability. By contrast, the finding that allowed substitutions are highly restricted for buried residues suggests that these side chains carry fundamentally important information for protein folding and stability.

DESTABILIZING MUTATIONS

Substitutions Affecting the Hydrophobic Core

It should be evident from the preceding sections that residues buried within the core appear to be extremely important determinants of protein structure and stability. In proteins of known structure, the cores are composed chiefly of hydrophobic residues and, more rarely, of polar residues that can satisfy

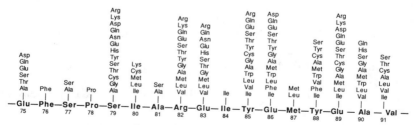

Figure 3 Neutral residue substitutions in the N-terminal domain of λ repressor. [Data for residues 84–91 are from ref. (56). Data for residues 75–83 are from the unpublished work of John Reidhaar-Olson.]

their hydrogen-bonding potential by forming hydrogen bonds with the protein backbone or other side chains (13). These core residues pack together efficiently to fill the protein interior (57). The characteristics of close packing and hydrophobicity are presumably important for two reasons. First, these internal packing interactions must, in some sense, determine the overall shape of the protein. Second, because of the hydrophobic effect, the shielding of nonpolar-core side chains from water contributes to the stabilization of the native protein structure (30).

In an unfolded protein, water is thought to be organized in cage-like structures around the hydrophobic side chains. When these side chains are transferred to the nonpolar environment of the core, the structured water is released. This increases the entropy of the solvent and thereby helps to stabilize the native protein. The magnitude of the hydrophobic effect may be estimated from free energies calculated for the transfer of amino acid side chains from water to nonpolar solvents such as ethanol or octanol (14, 48). These calculations suggest that the hydrophobic effect provides the largest free energy contribution to the stability of folded proteins (13, 30).

The hydrophobic contribution of individual buried side chains to stabilization has been examined in a number of different proteins (33, 40, 41, 53, 78). For each, the effects on protein stability of several different substitutions at a single site have been determined. In T4 lysozyme, for example, the effects of 13 different substitutions for Ile3, a residue that is about 80% buried in the structure, have been determined (40). The most deleterious mutations involve replacing Ile3 with larger side chains such as Trp and Tyr, or polar or charged residues such as Ser, Thr, and Asp. Each of these substitutions decreases stability by 1.7 to 3.2 kcal/mol. By contrast, substitution of Ile3 with the smaller but nonpolar Ala was found to decrease stability only by 0.7 kcal/mol.

It is difficult, however, to draw any general conclusions regarding the expected magnitudes of stability changes resulting from changes at buried or partially buried positions. For example, in the small ribonuclease barnase, substitution of the buried Ile196 side chain by Ala results in a destabilization of 4.0 kcal/mol (33). Although this Ile→Ala change is chemically identical to the Ile→Ala change in lysozyme, the observed destabilization is 5–6 times larger for barnase than for lysozyme. A similarly large destabilization has been observed for the Leu57→Ala mutation in λ repressor (53). This mutation reduces stability by 4–5 kcal/mol and reduces the T_m of the protein from 54°C to 20°C.

Why are the destabilizing effects of Ile→Ala or Leu→Ala mutations so different in the different cases? One possible factor is the degree to which the side chain being studied is truly buried within the protein. For the T4-lysozyme studies, Ile3 is only partially buried and is quite near the protein surface. Hence amino acid substitutions can probably be accommodated by local adjustments in packing interactions, and polar side-chain atoms can probably satisfy their hydrogen-bonding needs by extending out into solvent. This in fact happens for the Ile3→Tyr change (40). By contrast, the Ile and Leu residues replaced in barnase and λ repressor are completely buried in the hydrophobic core. Replacing these residues with a smaller side chain like Ala may require leaving an energetically unfavorable hole in the hydrophobic core. Here, the reduced stability will result both from the loss of hydrophobic interactions and from the cost of having a cavity in the protein interior (33). Cavities are presumed to be energetically expensive because some van der Waals interactions between the protein and water in the unfolded state will not be replaced by energetically comparable interactions within the folded protein. As a result, there will be a net decrease in stability due to these lost packing interactions.

The reader should not be left with the impression that the most deleterious mutations in the hydrophobic core will only decrease protein stability by 4–5 kcal/mol. In fact, for completely buried positions such as those discussed above, replacing hydrophobic residues such as Leu and Ile with extremely polar or charged residues seems likely to destabilize the protein to a considerably greater extent. Such large changes in ΔG_u are less likely to be measured than more moderate ones owing to the technical difficulties involved in purifying and studying extremely unstable proteins.

Glycine and Proline Substitutions

Glycine lacks a β-carbon and can therefore assume many backbone dihedral angles that are energetically unfavorable for other amino acids (13). This property is extremely important because it allows glycine to be used in certain types of reverse turns where positive dihedral angles are required (59). Replacing glycines in such turns with any other residue would be expected to be destabilizing unless the protein could form an alternative type of turn. In λ Cro, the destabilizing Gly48→Ala and Gly15→Glu substitutions affect glycines with positive dihedral angles in turns, and thus presumably act by this mechanism (52).

The pyrrolidine ring of proline constrains its Φ dihedral angle to values near $-60°$. Thus, proline should be destabilizing at positions where significantly different backbone torsional angles are required. An example occurs in staphylococcal nuclease, where replacing His121 ($\Phi = -170°$) with Pro results in complete loss of activity (67). In addition, proline is not found in the middle or at the C-terminal ends of most α-helices (12, 58). The exclusion of proline from helices is thought to be a consequence of steric clashes between the pyrrolidine side chain and the β-carbon of the previous residue (64) and/or because one of the α-helical hydrogen bonds is lost as a result of proline not having a peptide-NH group. Destabilizing mutations in which α-helical residues are replaced by prolines are reasonably common. For example, the Leu12→Pro and Ser35→Pro defective mutations in λ repressor (20) both affect surface residues in α-helices. Clearly, the effects of these mutations are caused by insertion of the proline and not by the loss of the wild-type side chain, because other substitutions such as Leu12→-Gln or Ser35→Leu are fully functional at both positions (21; J. Reidhaar-Olson, unpublished data).

Substitutions of the type Xaa → Gly or Pro → Xaa (where Xaa represents any other amino acid) may cause destabilization by increasing the entropy of unfolding (43). These entropy increases would occur because the backbone of glycine has more accessible conformations in the unfolded state than other residues, whereas the backbone of proline has fewer accessible conformations. The Pro35→Leu and Pro76→Leu mutations in yeast iso-1-cytochrome c may destabilize the protein, in part, by this mechanism (19, 55). However, both of these substitutions affect residues that are inaccessible to solvent and thus also alter packing and hydrophobic interactions in the core. Surface mutations of the reverse type, Gly → Xaa and Xaa → Pro result in stabilization of T4 lysozyme (43) and λ repressor (23) but only by about 0.4–0.8 kcal/mol. These free energy changes may represent the degree of destabilization that results solely from the conformational entropy changes that occur upon replacing proline or introducing glycine.

Substitutions Affecting Hydrogen Bonds and Electrostatic Interactions

Several uncertainties make it is difficult a priori to assess the importance of hydrogen bonds or salt bridges in protein structures. First, any hydrogen bond or electrostatic interaction that is made in the folded protein is formed at the cost of breaking similar bonds with solvent in the unfolded form. Second, the strength of electrostatic interactions depends on the extent to which they are shielded by solvent, and it can be difficult to assess these shielding terms for interactions at protein surfaces. Nevertheless, in model systems involving enzyme-substrate binding, the energetic contributions of hydrogen bonds that do not involve charged residues range from 0.5 to 1.5 kcal/mol, whereas hydrogen bonds involving charged residues may contribute as much as 4 kcal/mol (5, 15, 71). The experimental studies described below suggest that hydrogen bonding and electrostatic interactions can contribute modestly to protein stabilization.

A significant number of the destabilizing mutations in staphylococcal nuclease, λ Cro, λ repressor, and T4 lysozyme affect residues whose side chains participate in hydrogen bonds. For example, the Thr157→Ile mutation in T4 lysozyme disrupts a network of hydrogen bonds mediated via the threonine hydroxyl group (17). Studies of the stabilities and structures of a set of mutants at position 157 suggest that loss of this side chain hydrogen bond decreases the overall stability of T4 lysozyme by about 1.2 kcal/mol (3). A similar degree of destabilization has been measured for the Thr113→Val mutation in dihydrofolate reductase, which disrupts several hydrogen bonds mediated by the wild-type side chain (54).

In proteins, salt bridges can occur between positively and negatively charged side chains. Surface salt bridges in bovine pancreatic trypsin inhibitor (10) and dihydrofolate reductase (54) seem to contribute about 1 kcal/mol to overall stability, although hydrogen bonding may also play a role in the latter case. Electrostatic interactions between charged side chains and the ends of α-helices are also possible; because of the alignment of the peptide dipoles, α-helices bear a partial positive charge at their N-terminal ends and a partial negative charge at their C-terminal ends (25). Stabilizing interactions of this type in T4 lysozyme (46) and barnase (60) appear to contribute from 0.8 to 2 kcal/mol to protein stability.

Substitutions Affecting the Denatured State

As we have seen, it is often possible to rationalize the effects of destabilizing mutations in terms of the folded structure of a protein. Matthews (42) has argued that this suggests that most destabilizing substitutions exert their effects primarily on the folded state of a protein. However, Shortle and his colleagues have found that several mutations in staphylococcal nuclease alter the physical properties of the unfolded state (68–70). Because the overall stability of a protein depends on the free energies of both the folded and unfolded states, it is not unreasonable that a mutation could exert its effect primarily via the unfolded state. However, at present, it is not clear how to partition the effects of the staphylococcal nuclease mutations between perturbation of the energies of the unfolded and folded states.

We have already mentioned substitutions involving proline and glycine that may affect protein stability by altering the conformational entropy of the unfolded state. Disulfide linkages between cysteine residues are also thought to stabilize folded proteins by reducing the number of conformations accessible to the unfolded protein and thus reducing the entropy of unfolding (30). Although disulfide bonds are extremely rare in intracellular proteins, they are common in secreted proteins and provide potential targets for destabilizing mutations. The introduction of new disulfide bonds has been a common strategy for attempting to increase protein stability through rational design. However, the stabilization afforded by such covalent cross-links is highly dependent upon structural context and position. Some new disulfides do stabilize the protein, while others have no effect, or may actually destabilize the protein (73, 74).

CONFORMATIONAL CHANGES IN MUTANT PROTEINS

Until now, we have been discussing amino-acid sequence changes in terms of their effects on the equilibrium between the folded and unfolded conformations of a protein. It is also worth asking if single point mutations can cause significant changes in activity by altering the conformation of the folded protein. There are a few cases where mutations have been shown to cause propagated conformational changes. For example, replacing Pro86 on the surface of T4 lysozyme causes a conformational change by allowing extension of an α-helix (1). Although the observed structural change is modest (residues 81–83 shift positions by no more than 1.4 Å), some of these changes occur 11 Å from the site of the mutant substitution. Another case in which a single substitution causes nonlocal conformational changes occurs in staphylococcal nuclease. Here, a Glu43→Asp substitution at a partially buried position in the enzyme active site results in detectable changes at residues as far as 30 Å away (75). In both the lysozyme and nuclease cases, however, the observed changes are small in terms of the overall structure. Moreover, because proteins are somewhat flexible, it is not obvious a priori that small conformational changes would cause large reductions in activity. The cases discussed do not resolve this issue. In the T4-lysozyme case, the observed changes are distant from the active site, and the mutant enzyme has normal stability and activity. In the staphylococcal nuclease case, the mutation alters an active site residue, and thus it is difficult to determine the extent to which the loss of activity is caused by the conformational change.

Overall, misfolding appears to be rare. Most mutant proteins that have been studied thus far have conformations that are extremely similar to wild type. For example, Matthews and his colleagues have solved the crystal structures of more than 50 mutant forms of T4 lysozyme and found that in almost all cases the mutant and wild-type structures are extremely similar, with structural differences occurring only at or near the site of the mutant substitution (3, 17, 40, 42, 43, 46). This is true even for mutant proteins that are significantly less stable than wild-type.

IDENTIFYING RESIDUES IMPORTANT FOR FUNCTION

Many genetic analyses of proteins are directed towards answering functional questions rather than those concerning protein structure or stability per se. Which are the active site residues? Which residues mediate binding and specificity. These questions have been approached both by studies of defective mutants and by studies of neutral mutations.

As we have seen, mutations affecting active site residues are usually present in collections of defective mutations, but so are mutations that affect structure and stability. Hence to conclude that a defective mutation affects a functionally important residue, it must first be shown that it does not affect structure or stability. This has been established in some cases by purifying the mutant proteins and determining their stabilities. For example, in studies of defective λ repressor mutants, it was shown that a subset of the mutant proteins had thermal stabilities almost identical to wild-type and yet had operator binding affinities reduced by 100-fold or more (20, 22, 45). The conclusion that these "activity" mutations identify residues in or near the DNA-recognition surface of the protein has been directly supported by the crystal structure of the protein-DNA complex (28). Similar identification of DNA-binding residues by biochemical characterization of purified mutant proteins has been reported for EcoR1 (77) and P22 Arc repressor (72).

It is sometimes possible to infer that mutant proteins are stably folded without purification and subsequent study. For proteins that are active only as oligomers, stably folded but inactive proteins may have a transdominant negative phenotype because mixed oligomers containing wild-type and mutant subunits have dramatically reduced activities. For example, most dominant-negative mutations in the Trp repressor affect side chains in or near the DNA-binding surface of the protein (32, 63).

With current methods of cassette mutagenesis, functionally important residues can also be identified by studies of neutral mutations. For example, a cassette method was used to mutagenize regions of about 30 base-pairs in the *arc* repressor gene such that most cassettes contained from two to four mutations (8). Following an activity selection, functionally neutral residue substitutions were identified at 24 of the 53 positions of Arc. In a separate screening experiment, mutant Arc sequences that could still fold into a stable structure were isolated, and substitutions, some conservative and some non-conservative, were identified at 41 positions. Comparison of these two sets of neutral mutations revealed that the N-terminal residues of Arc could tolerate substitutions when formation of a stable structure was required but not when function was required, suggesting that these residues form part of the operator-binding surface of the protein. The identification of this region of Arc as the likely DNA-binding region has also been supported by studies of defective mutant proteins (72) and chimeric proteins with Arc-binding specificity (37).

There can clearly be problems in interpretation for any of the experiments discussed above. Some stably folded mutant proteins might have subtly altered conformations that are responsible for their decreased activity; a mutant substitution may exert its main effect directly on activity but also cause a modest decrease in stability. Nevertheless, with appropriate caution, functionally important residues can usually be identified. It is generally easiest to do this when dealing with mutations that cause significant reductions in activity. There are presumably a large number of ways, many of them subtle, to reduce protein activity by a factor of two. By contrast, there are relatively few ways to reduce activity by a factor of 100-fold or more, and most of these will involve large and easily detectable changes in protein stability or the alteration of functionally important residues.

MUTATIONS THAT ENHANCE STABILITY AND ACTIVITY.

Different genetic strategies have been used to identify mutations that enhance protein stability and/or activity. Most use some means of reducing the activity of the protein of interest, followed by a selection or screen to detect variants with increased activity. For example, the activity of a protein might be reduced by mutation (21, 24, 45, 51), by decreasing its intracellular level, by increasing the temperature (11, 38, 39), or by decreasing the concentration of required cofactors (32). The parental gene is then mutagenized and strains with increased activity can be isolated and analyzed.

In studies in which one starts with a gene bearing a loss of activity mutation, pseudo-revertants can arise at the site of the original mutation or at second-sites within the gene or in other genes (76). The most common types of second-site suppressor mutations are those that act globally to overcome the original defect by increasing protein stability, activity, or level. For example, if a defective mutation destabilizes a protein by 2 kcal/mol, then a second-site substitution might act by increasing stability by a comparable amount. In such a case, an otherwise wild-type protein bearing the suppressor mutation should be more stable than wild-type. Mutations that increase protein stability have been identified in this way for staphylococcal nuclease (65, 67) and λ Cro (51). Enhanced stability mutations have also been identified in kanamycin nucleotidyltransferase and subtilisin by selecting or screening for activity at elevated temperatures (11, 38, 39).

Some second-site suppressor mutations act by increasing activity directly. For example, amino acid substitutions in λ repressor that increase operator-binding affinity as much as 600-fold have been identified by their ability to suppress both stability and activity mutations (21, 45). A similar class of mutations has been identified in Trp repressor, but by direct selection for activity at low concentrations of the co-repressor, tryptophan (32, 36).

PROTEOLYTIC SENSITIVITY OF MUTANT PROTEINS

Since unfolded proteins are usually better substrates for proteolytic digestion than their folded counterparts, intracellular proteolysis of unstable proteins can play an important role in mutant phenotypes. For example, the Ile30→Leu mutation in λ Cro affects a residue in the hydrophobic core and reduces the T_m of the protein from 40°C to 35°C (51). This decrease in stability alone would only be expected to cause a modest decrease in activity by reducing the concentration of folded, active Cro. However, whereas wild-type Cro has an intracellular half-life of 60 min, the mutant half-life is reduced to 11 min. Hence, proteolysis amplifies the effect of this destabilizing mutation by reducing the steady-state level of the mutant protein.

Several findings suggest that the stability of a folded protein is an important determinant of its rate of degradation. First, proteins that contain amino acid analogs or are prematurely terminated are often degraded rapidly in the cell (16). Second, good correlations exist between the measured or inferred thermal stabilities of specific mutant proteins and the rates at which they are degraded in *E. coli* (52, 53). Finally, second-site suppressor mutations that increase the thermodynamic stability of unstable mutant proteins have also been shown to increase resistance to intracellular proteolysis (51).

The rate of intracellular proteolysis of mutant proteins can also be influenced by determinants other than the stability of the native structure. For example, the N-terminal residues of some proteins appear to be important in determining their susceptibility to ubiquitin-mediated degradation in the yeast *S. cerevisiae* (4). In *E. coli,* the identity of residues at the C-terminal ends of some proteins influences their rates of intracellular degradation (9, 53). For example, frameshift mutations near the C-terminus of the Arc repressor result in the addition of extra C-terminal residues that suppress the proteolytic instability of unstable Arc mutants without affecting the thermal stability or activity of the protein (9). In addition to sequence determinants, the solubility of mutant proteins can also affect their proteolytic resistance. Some proteins aggregate to form inclusion bodies, presumably because they are unfolded or incompletely folded, and thus escape proteolytic attack (29). Because of these factors, increased susceptibility to intracellular degradation does not by itself provide sufficient evidence to conclude that a mutant is thermodynamically unstable. In similar fashion, a mutant protein could be resistant to intracellular proteolysis and yet not be stably folded. Nevertheless, susceptibility to degradation can be a convenient indicator of thermodynamic stability for some proteins.

SUMMARY

There is tremendous variability in the importance of individual amino acids in protein sequences. On the one hand, nonconservative residue substitutions can be tolerated with no loss of activity at many residue positions, especially those exposed on the protein surface. On the other hand, destabilizing mutations can occur at a large number of different sites in a protein, and for many proteins such mutations account for more than half of the randomly isolated missense mutations that confer a defective phenotype. At sites that are key determinants of stability or activity, even residue substitutions that are generally considered to be conservative (e.g., Glu↔Asp, Asn↔Asp, Ile↔Leu, Lys↔Arg and Ala↔Gly) can have severe phenotypic effects. Unfortunately, this means that there is no simple way to infer the likely effect of an amino acid substitution on the basis of sequence information alone. A nonconservative Gly→Arg substitution could be phenotypically silent at one position while a conservative Asn→Asp change could lead to complete loss of activity at another position.

For proteins whose structures are known, it is often possible to predict whether particular residue substitutions will be destabilizing, as long as detailed estimates of the destabilization energy are not required. Substitutions that introduce polar groups, large cavities, or overly large side chains into the hydrophobic core are potentially the most destabilizing. Substitutions that disrupt hydrogen bonding or electrostatic interactions can also have significant effects, although the destabilization caused by these substitutions is smaller than that caused by severe core mutations. Destabilizing substitutions that involve replacing glycines in turns, or introducing prolines into α-helices and other disallowed positions are also reasonably common. Finally, most solvent exposed residues can apparently be freely substituted without serious effects on protein stability. Although exceptions may occur, these generalizations serve to summarize a large body of information and can be rationalized in physical and chemical terms.

It is an especially encouraging result that proteins appear to tolerate most substitutions, even those that are destabilizing, without significant changes in the native structure. For proteins whose structures are known, this means that it is reasonable to interpret mutant phenotypes in terms of the wild-type structure. For proteins whose structures are not known, it is reasonable to infer that mutations that reduce activity without affecting stability are directly involved in function. Detailed studies of the structure of the mutant proteins are still needed, but, because induced conformational changes are rare, such efforts are usually worthwhile.

Because proteins are so diverse, it is always dangerous to extrapolate too far. We note that most of the studies described here concern small, globular, single-domain proteins whose folded and unfolded structures are in dynamic equilibrium. Fibrous proteins, proteins that are extremely thermostable, or proteins that contain multiple interacting domains may face special problems in folding (34). Moreover, indirect effects of mutations mediated via protein conformation are much more likely to be common for allosteric proteins, which can exist in distinctly different quaternary structures (61). Nevertheless, the basic principles of protein structure and activity established in the simpler and more readily studied systems should still form the groundwork for studies on more complicated proteins.

ACKNOWLEDGMENTS

We thank Jim Bowie and Dawn Parsell for comments on the manuscript, John Reidhaar-Olson for communication of unpublished data, Jim Bowie and Neil Clarke for assistance with solvent accessibility calculations, and Lynn Kleina and Jeffrey Miller for a preprint of their manuscript (35). A.A.P is a postdoctoral fellow of the American Cancer Society, California Division. Work in the author's laboratory was supported by NIH grants AI-15706 and AI-16892 to R.T.S.

Literature Cited

1. Alber, T., Bell, J. A., Sun, D.-P., Nicholson, H., Wozniak, J. A., et al. 1988. Replacements of Pro86 in phage T4 lysozyme extend an alpha-helix but do not alter protein stability. *Science* 239:631–35

2. Alber, T., Sun, D.-P., Nye, J. A., Muchmore, D. C., Matthews, B. W. 1987. Temperature-sensitive mutations of bacteriophage T4 lysozyme occur at sites with low mobility and low solvent accessibility in the folded protein. *Biochemistry* 26:3754–58

3. Alber, T., Sun, D.-P., Wilson, K., Wozniak, J. A., Cook, S. P., Matthews, B. W. 1987. Contributions of hydrogen bonds of Thr157 to the thermodynamic stability of phage T4 lysozyme. *Nature* 330:41–46

4. Bachmair, A., Finley, D., Varshavsky, A. 1986. In vivo half-life of a protein is a function of its amino-terminal residue. *Science* 234:179–86

5. Bartlett, P. A., Marlowe, C. K. 1987. Evaluation of intrinsic binding energy from a hydrogen bonding group in an enzyme inhibitor. *Science* 235:569–71

6. Becktel, W. J., Schellman, J. A. 1987. Protein stability curves. *Biopolymers* 26:1859–77

7. Botstein, D., Shortle, D. 1985. Strategies and applications of in vitro mutagenesis. *Science* 229:1193–201

8. Bowie, J. U., Sauer, R. T. 1989. Identifying determinants of folding and activity for a protein of unknown structure. *Proc. Natl. Acad. Sci. USA* 86: 2152–56

9. Bowie, J. U., Sauer, R. T. 1989. Identification of C-terminal extensions that protect proteins from intracellular proteolysis. *J. Biol. Chem.* 264:7596–602

10. Brown, L. R., DeMarco, A., Richarz, R., Wagner, G., Wuthrich, K. 1978. The influence of a single salt bridge on static and dynamic features of the globular solution conformation of the basic pancreatic trypsin inhibitor. *Eur. J. Biochem.* 88:87–95

11. Bryan, P. N., Rollence, M. L., Pantoliano, M. W., Wood, J., Finzel, B. C., et al. 1986. Proteases of enhanced stability: Characterization of a thermostable variant of subtilisin. *Proteins* 1: 326–34

12. Chou, P. Y., Fasman, G. D. 1978. Empirical predictions of protein conformation. *Annu. Rev. Biochem.* 47: 251–76

13. Creighton, T. E. 1984. *Proteins. Structures and Molecular Properties.* New York: Freeman

14. Fauchere, J.-L., Pliska, V. 1983. Hydrophobic parameters π of amino-acid side chains from the partitioning of N-acetyl-amino-acid amides. *Eur. J. Med. Chem. Chim. Ther.* 18:369–75

15. Fersht, A. R., Shi, J. P., Knill-Jones, L., Lowe, D. M., Wilkinson, A. J., et al. 1985. Hydrogen bonding and biological specificity analyzed by protein engineering. *Nature* 314:235–38

16. Goldberg, A. L., St. John, A. C. 1976. Intracellular protein degradation in mammalian and bacterial cells: Part 2. *Annu. Rev. Biochem.* 45:747–803

17. Grütter, M. G., Gray, T. M., Weaver, L. H., Alber, T., Wilson, K., Matthews, B. W. 1987. Structural studies of mutants of the lysozyme of bacteriophage T4: The temperature-sensitive mutant protein Thr157→Ile. *J. Mol. Biol.* 197:315–29

18. Hampsey, D. M., Das, G., Sherman, F. 1986. Amino acid replacements in yeast iso-1-cytochrome *c:* comparison with the phylogenetic series and the tertiary structure of related cytochromes-c. *J. Biol. Chem.* 261:3259–71

19. Hampsey, D. M., Das, G., Sherman, F.

1988. Yeast iso-1-cytochrome *c*: genetic analysis of structural requirements. *FEBS Lett.* 231:275–83

20. Hecht, M. H., Nelson, H. C. M., Sauer, R. T. 1983. Mutations in lambda repressor's amino-terminal domain: Implications for protein stability and DNA binding. *Proc. Natl. Acad. Sci. USA* 80:2676–80

21. Hecht, M. H., Sauer, R. T. 1985. Phage lambda repressor revertants: Amino acid substitutions that restore activity to mutant proteins. *J. Mol. Biol.* 186:53–63

22. Hecht, M. H., Sturtevant, J. M., Sauer, R. T. 1984. Effect of single amino acid replacements on the thermal stability of the NH₂-terminal domain of phage lambda repressor. *Proc. Natl. Acad. Sci. USA* 81:5685–89

23. Hecht, M. H., Sturtevant, J. M., Sauer, R. T. 1986. Stabilization of lambda repressor against thermal denaturation by site-directed Gly→Ala changes in alpha-helix 3. *Proteins* 1:43–46

24. Ho, C., Jasin, M., Schimmel, P. 1985. Amino acid replacements that compensate for a large polypeptide deletion in an enzyme. *Science* 229:389–93

25. Hol, W. G., Halie, L. M., Sander, C. 1981. Dipoles of the alpha-helix and beta-sheet and their role in protein folding. *Nature* 294:532–36

26. Hope, I. A., Struhl, K. 1986. Functional dissection of a eukaryotic transcriptional activator protein, GCN4 of Yeast. *Cell* 46:885–94

27. Jasin, M., Regan, L., Schimmel, P. 1983. Modular arrangement of functional domains along the sequence of an aminoacyl tRNA synthetase. *Nature* 306:441–47

28. Jordan, S. R., Pabo, C. O. 1988. Structure of the lambda complex at 2.5 Å resolution: Details of the repressor-operator interactions. *Science* 242:893–99

29. Kane, J. F., Hartley, D. L. 1988. Formation of recombinant protein inclusion bodies in Escherichia coli. *Tibtech* 6:95–101

30. Kauzmann, W. 1959. Some factors in the interpretation of protein denaturation. *Adv. Protein Chem.* 14:1–63

31. Keegan, L., Gill, G., Ptashne, M. 1986. Separation of DNA binding from the transcription-activating function of a eukaryotic regulatory protein. *Science* 231:699–704

32. Kelley, R. L., Yanofsky, C. 1985. Mutational studies with the trp repressor of *Escherichia coli* support the helix-turn-helix model of repressor recogni-

tion of operator DNA. *Proc. Natl. Acad. Sci. USA* 82:483–87

33. Kellis, J. T. J., Nyberg, K., Sali, D., Fersht, A. R. 1988. Contribution of hydrophobic interactions to protein stability. *Nature* 333:784–86

34. King, J. 1986. Genetic analysis of protein folding pathways. *Bio/Technology* 4:297–303

35. Kleina, L. G., Miller, J. H. 1989. Genetic studies of the *lac* repressor: XIII. Extensive amino acid replacements generated by the use of natural and synthetic nonsense suppressors. *J. Mol. Biol.* Submitted

36. Klig, L., Yanofsky, C. 1988. Increased binding of operator DNA by trp super-repressor EK49. *J. Biol. Chem.* 263:243–46

37. Knight, K. L., Sauer, R. T. 1989. DNA binding specificity of the Arc and Mnt repressors is determined by a short region of N-terminal residues. *Proc. Natl. Acad. Sci. USA* 86:797–801

38. Liao, H., McKenzie, T., Hagerman, R. 1986. Isolation of a thermostable enzyme variant by cloning and selection in a thermophile. *Proc. Natl. Acad. Sci. USA* 83:576–80

39. Matsumura, M., Aiba, S. 1985. Screening for thermostable mutants of kanamycin nucleotidyltransferase by the use of a transformation system for a thermophile, *Bacillus stearothermophilus. J. Biol. Chem.* 260:15298–303

40. Matsumura, M., Becktel, W. J., Matthews, B. W. 1988. Hydrophobic stabilization in T4 lysozyme determined directly by multiple substitutions of Ile 3. *Nature* 334:406–10

41. Matsumura, M., Yahanda, S., Yasumura, S., Yutani, K., Aiba, S. 1988. Role of tyrosine-80 in the stability of kanamycin nucleotidyltransferase analyzed by site-directed mutagenesis. *Eur. J. Biochem.* 171:715–20

42. Matthews, B. W. 1987. Genetic and structural analysis of the protein stability problem. *Biochemistry* 26:6885–88

43. Matthews, B. W., Nicholson, H., Becktel, W. J. 1987. Enhanced protein thermostability from site-directed mutations that decrease the entropy of unfolding. *Proc. Natl. Acad. Sci. USA* 84:6663–67

44. Miller, J. H., Coulondre, C., Hofer, M., Schmeissner, U., Sommer, H., et al. 1979. Genetic studies of the *lac* repressor: ix. generation of altered proteins by the suppression of nonsense mutations. *J. Mol. Biol.* 131:191–222

45. Nelson, H. C., Sauer, R. T. 1985. Lambda repressor mutations that in-

crease the affinity and specificity of operator binding. *Cell* 42:549–58

46. Nicholson, H., Becktel, W. J., Matthews, B. W. 1988. Enhanced protein thermostability from designed mutations that interact with α-helix dipoles. *Nature* 336:651–56

47. Normanly, J., Abelson, J. 1989. tRNA identity. *Annu. Rev. Biochem.* 58:1029–49

48. Nozaki, Y., Tanford, C. 1971. The solubility of amino acids and two glycine peptides in aqueous ethanol and dioxane solutions. *J. Biol. Chem.* 246: 2211–17

49. Pace, C. N. 1975. The stability of globular proteins. *CRC Crit. Rev. Biochem.* 3:1–43

50. Pace, C. N., Barrett, A. J. 1984. Kinetics of tryptic hydrolysis of the arginine-valine bond in folded and unfolded ribonuclease T1. *Biochem. J.* 219:411–17

51. Pakula, A. A., Sauer, R. T. 1989. Amino acid substitutions that increase the thermal stability of the λ Cro protein. *Proteins* 5:202–10

52. Pakula, A. A., Young, V. B., Sauer, R. T. 1986. Bacteriophage lambda cro mutations: Effects on activity and intracellular degradation. *Proc. Natl. Acad. Sci. USA* 83:8829–33

53. Parsell, D. A., Sauer, R. T. 1989. The structural stability of a protein is an important determinant of its proteolytic susceptibility in *E. coli. J. Biol. Chem.* 264:7590–95

54. Perry, K. M., Onuffer, J. J., Touchette, N. A., Herndon, C. S., Gittelman, M. S., et al. 1987. Effect of single amino acid replacements on the folding and stability of dihydrofolate reductase from *Escherichia coli. Biochemistry* 26:2674–82

55. Ramdas, L., Sherman, F., Nall, B. T. 1986. Guanidine hydrochloride induced unfolding of mutant forms of iso-1-cytochrome *c* with replacement of proline-71. *Biochemistry* 25:6952–58

56. Reidhaar-Olson, J. F., Sauer, R. T. 1988. Combinatorial cassette mutagenesis as a probe of the informational content of protein sequences. *Science* 241: 53–57

57. Richards, F. M. 1977. Areas, volumes, packing and protein structure. *Annu. Rev. Biophys. Biophys. Chem.* 6:151–76

58. Richardson, J. S., Richardson, D. C. 1988. Amino acid preferences for specific locations at the ends of alpha helices. *Science* 240:1648–52

59. Rose, G. D., Gierasch, L. M., Smith, J. A. 1985. Turns in peptides and proteins. *Adv. Protein Chem.* 37:1–109

60. Sali, D., Bycroft, M., Fersht, A. R. 1988. Stabilization of protein structure by interaction of alpha-helix dipole with a charged side chain. *Nature* 335:740–43

61. Schachman, H. K. 1988. Can a simple model account for the allosteric transition of aspartate transcarbomoylase? *J. Biol. Chem.* 263:18583–86

62. Schellman, J. A., Lindorfer, M., Hawkes, R., Grutter, M. 1981. Mutations and protein stability. *Biopolymers* 20:1989–99

63. Schevitz, R. W., Otwinowski, Z., Joachimiak, A., Lawson, C. L., Sigler, P. B. 1985. The three-dimensional structure of *trp* repressor. *Nature* 317: 782–86

64. Schimmel, P. R., Flory, P. J. 1968. Conformational energies and configurational statistics of copolypeptides containing ʟ-proline. *J. Mol. Biol.* 34: 105–20

65. Shortle, D. 1986. Guanidine hydrochloride denaturation studies of mutant forms of staphylococcal nuclease. *J. Cell. Biochem.* 30:281–89

66. Shortle, D., DiMaio, D., Nathans, D. 1981. Directed mutagenesis. *Annu. Rev. Genet.* 15:265–94

67. Shortle, D., Lin, B. 1985. Genetic analysis of staphylococcal nuclease: Identification of three intragenic "global" suppressors of nuclease-minus mutations. *Genetics* 110:539–55

68. Shortle, D., Meeker, A. K. 1986. Mutant forms of staphylococcal nuclease with altered patterns of guanidine hydrochloride and urea denaturation. *Proteins* 1:81–89

69. Shortle, D., Meeker, A. K. 1989. Residual structure in large fragments of staphylococcal nuclease: Effects of amino acid substitutions. *Biochemistry* 28: 936–44

70. Shortle, D., Meeker, A. K., Freire, E. 1988. Stability mutants of staphylococcal nuclease: Large compensating enthalpy-entropy changes for the reversible denaturation reaction. *Biochemistry* 27: 4761–68

71. Tronrud, D. E., Holden, H. M., Matthews, B. W. 1987. Structures of two thermolysin-inhibitor complexes that differ by a single hydrogen bond. *Science* 235:571–74

72. Vershon, A. K., Bowie, J. U., Karplus, T. M., Sauer, R. T. 1986. Isolation and analysis of arc repressor mutants: evidence for an unusual mechanism of DNA binding. *Proteins* 1:302–11

73. Wells, J. A., Powers, D. B. 1986. In vivo formation and stability of engi-

neered disulfide bonds in subtilisin. *J. Biol. Chem.* 261:6564–70

74. Wetzel, R., Perry, L. J., Baase, W. A., Becktel, W. J. 1988. Disulfide bonds and thermal stability in T4 lysozyme. *Proc. Natl. Acad. Sci. USA* 85:401–5

75. Wilde, J. A., Bolton, P. H., Dell' Acqua, M., Hibler, D. W. Pourmotabbed, T., Gerlt, J. A. 1988. Identification of residues involved in a conformational change accompanying substitutions for glutamate-43 in staphylococcal nuclease. *Biochemistry* 27:4127–32

76. Yanofsky, C. 1971. Tryptophan biosynthesis in *Escherichia coli*. Genetic determination of the proteins involved. *J. Am. Med. Assoc.* 218:1026–35

77. Yanofsky, S. D., Love, R., McClarin, J. A., Rosenberg, J. M., Boyer, H. W., Greene, P. J. 1987. Clustering of null mutations in the EcoRI endonuclease. *Proteins* 2:273–82

78. Yutani, K., Ogasahara, K., Tsujita, T., Sugino, Y. 1987. Dependence of conformational stability on hydrophobicity of the amino acid residue in a series of variant proteins substituted at a unique position of tryptophan synthase alpha subunit. *Proc. Natl. Acad. Sci. USA* 84: 4441–44

79. Zoller, M. J., Smith, M. 1987. Oligonucleotide-directed mutagenesis: a simple method using two oligonucleotide primers and a single-stranded DNA template. *Methods Enzymol.* 154:329–50

Annu. Rev. Genet. 1989. 23:311–36

PROKARYOTIC SIGNAL TRANSDUCTION MEDIATED BY SENSOR AND REGULATOR PROTEIN PAIRS

Lisa M. Albright, Eva Huala, and Frederick M. Ausubel

Department of Genetics, Harvard Medical School and Department of Molecular Biology, Massachusetts General Hospital, Boston, Massachusetts 02114

CONTENTS

311

0066-4197/89/1215-0311$02.00

INTRODUCTION

In bacteria, signal transduction in response to a wide variety of environmental stimuli is mediated by pairs of proteins that communicate with each other by a conserved mechanism involving protein phosphorylation. Initial genetic and biochemical analyses of the gene pairs *envZ/ompR* (33), *ntrB/ntrC* (11, 82), and *phoR/phoB* (119) suggested that one member of each pair codes for a sensor (modulator) protein (EnvZ, NtrB, PhoR) that modifies the activity of its partner, a regulator (effector) protein (OmpR, NtrC, PhoB). Subsequently, DNA sequence analysis revealed that these three sensor/regulator (modulator/effector) pairs belong to a large superfamily of signal-transducing regulatory pairs, characterized by the feature that the sensor proteins are all similar in their C-domains while their regulator partners share homology in their N-domains (3, 14, 19, 35, 55, 86, 92, 110, 113, 126, 127, 131).

The finding that NtrB phosphorylates its partner NtrC (82) and the similarity in domain organization among other homologous regulatory pairs suggested the following general model for signal transduction for these systems (Figure 1; 86, 94, 127): The N-domain of the sensor protein receives an environmental signal, transducing that signal to its C-domain. Utilizing a mechanism conserved among the homologous regulatory pairs, the C-domain of the sensor phosphorylates the N-terminal domain of the regulator. Phosphorylation of the regulator alters the activity of its C-domain, which ultimately carries out the appropriate response. Thus, diverse environmental signals are transduced through a common mechanism to regulate a wide range of processes in bacteria, including response to nutrient deprivation, osmolarity changes, chemotaxis, sporulation, symbiosis, and bacterial pathogenesis (Table 1).

The term "sensor" as used here applies only to the function of the protein within a regulatory pair. Because some of these regulatory pairs are located in

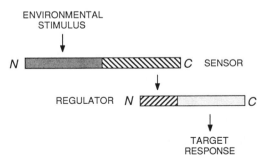

Figure 1 Model for signal transduction in homologous regulatory pairs. Diagonal shading indicates conserved domains.

```
                                              1
                                        _____
KpNtrB  116  RLSQEQLQHAQQIAARDLVRGLAHEIKNPLGGLRGAAQLLSKALPDPA....LMEYTKVI
RpNtrB  126  KMDRQLTHRSAARSVIALAAMLAHEIKNPLSGIRGAAAQLLEQQASSED.....RLLTRLI
RlDctB  392  AVQQDLVQANRLAILGQVAAGVAHEINQPVATIRAYADNARTFLDRGQ.TAPAGENLESI
StPgtB  359  AMQDELIQTAKLAVVGQTMTTLAHEINQPLNALSMYLFTAGRALEQGQ.SGQARNTLRKA
EcCpxA  225  FNQMVTALERMMTSQQRLLSDISHELRTPLTRLQLGTALLRRRSGESK......ELERT
EcEnvZ  220  FNHMAAGVKQLADDRTLLMAGVSHDLRTPLTRIRLRLAEMMSEQDGYLA.........ESI
EcPhoR  190  VARDVTQMHQLEGARRNFFANVSHELRTPLTVLQGYLEMMNEQPLEGA....VREKALHTM
BsPhoR  337  VFHDMTETKKLEQMRKDFVANVSHELKTPITSIKGFNETLLDGAMEDK..EALSEFLSII
EcPhoM  242  ALESMRVKLEGKNYIEQYYYALTHELKSPLAAIRGAAAEILREGPPPEV...VARFTDNI
AtWHVirA 451 VLARRLEHAQRLEAVGTLAGGIAHEFNNILGSILGHAELAQNSVSRTS...VTRRYIDYI
RmFixL  221  QIQAELARLARLRLNEMGEMASTLAHELNQPLSAIANYSHGCTRILRDMD..DAVATRIREA
EcUhpB  290  NQHLAERLLETEESVRRDVARELHDDIGQTIIAIRTQAGIVQRLAADR..RQREAERAAH
BsDegS  166  KQDFGLRITEAQEEERKRVSREJHDGPAQMLANVMMRSELIIERIFRDRGAEDGFQEIKNL
StCheA  370  LTLVGSSTELDKSLIIERIIDPLTHLVRNSLDHGI..........
StCheA-N 25  QHLLDLVPESPDAEQLNAIFRAAHSIKGGAGTFGFT.........
```

```
KpNtrB  172  IEQADRLRNLVDRLL..GPQHPGMHVIT.ESIHKVAERVVKLVSMEL....PDNVKLVRDY
RpNtrB  181  CDEADRIVITLVDRMEVFGDDRPVARGP.VNIHSVLDHVKRLAQSGF....ARNVRFIEDY
RlDctB  451  AALTERIGSIITEELKTFARKGRGSAEP.TGLKDVIEGAVMLLRSRF....AGRMDTLDID
StPgtB  418  EGLINRIDAIIRSLREFYRRACLETPLYPVDARQTFVAAWELLAMR.....HQSRQGALS
EcCpxA  278  ETEAQRLDSMINDLLVMSRNQQKNALVSETIKANQLWSEVLDNAAF..EAEQMGRSLTVN
EcEnvZ  271  NKDIEECNAIIEQFIDYLRTGQEMPMEMADLNAVLGEVIAAES.......GYERETEATAL
EcPhoR  247  REQTQRMEGLVKQLLTLSKIEAAPTHLLNEKVDVPMMLRVVEREAQ...TLSQKKQTFTF
BsPhoR  395  LKESERLQSLVDQLLDLSKIEQQNFTLSIETFEPAKMLGEIETLLK.HKADEKGISLHLN
EcPhoM  298  LTQNARMQALVETLLRQARLENRQEVVLTAVDVAALFRRVSEARTVQ...LAEKKITLHV
AtWHVirA 508 ISSGDRAMLIIDQILTLSRKQERMIKPFSVSELVTEIAP.LLRMAL....PPNIELSRF
RmFixL  279  LEEVASQSLRAGQIIKHLREFVTKGETEKAPEDIRKLVEESAALALVGSREQGVRTVFEY
EcUhpB  348  RTTIAGRLRVRRLLGRLRPRQLDDLITLEQAIRSLMREMELEGRGIV....SHLEWRIDE
BsDegS  216  RQNVRNALYEVRRIIYDLRPMALDDLGLIPTLRKYLYTTEEYNGKVKI..HFQCIGETED
StCheA  .........................................................
StCheA-N 61  .ILQETTHLMENLLDEARRGEMQLNIDIINLFLETKDIMQE...
```

```
                                         2                                    3
                                      _____                               ___
KpNtrB  225  DPSLPE.LPHDPDQIEQVLLNIVRNALQALGPEGGEITLRTRTAFQ......LTLHGVRY
RpNtrB  236  DPSLPP.VLANQDQLTQVFLNLVKNAAEAVADLGTDAEIQLTTAFRPGVRLSVPGKKSRV
RlDctB  506  LPPDELQVMGNRIRLEQVLINLLQNALEAVAPKAGEGR...........VEIRTSTD
StPgtB  473  LPTDTVWNSGEDVRIQQVLVNVLANALDACSHDAV...............IAVTWQTQ
EcCpxA  336  FPPGPWPLYGNPNALESALENIVRNALRYSHTK................IEVGFAVD
EcEnvZ  324  YPPGSIE.VKMHPLSIKRAVANMVVNAARYGNGW................IEVGFAVD
EcPhoR  304  EIDNGLKVSGNEDQLRSAISNLVYNAVNHTPEGTH...............IKVSSGTE
BsPhoR  454  VPKDPQYVSGDPYRLKQVFLNLVMNALHYTPTEGG...............ITVRWQVR
EcPhoM  355  TPTEVN.VAAEPALLEQALGNLLDNAIKYSGSGIPIT..............VAINVKPR
AtWHVirA 563 DQMQSV...ESPLELQQVLINICKNASQAMTANGQIDIIISQAFLPVKKI..LAHGVMPP
RmFixL  339  LPGAEM.VLVDRIQVQQVLLNLVRNAIEAMRHVDRREL.............TIRTMPAD
EcUhpB  404  SALSENQRVITLFRVQEGLNNIVKHADASA..................VITLQGWQQ
BsDegS  274  QRLAPQFEVALERLAQEAVSNALKHSESEE.................IIVKVEIT
StCheA  404  ........EMPEKRLEAGKNVVGN.......................LILSAEHQ
```

```
                       3                                              4
                    ___                                            ____
KpNtrB  278  RLAARIDVEDNGPGIPSHLQDT.........................LFYPMVSG......REG
RpNtrB  295  SLPLEFCVKDNGSGVPEDLLPN.........................LFDPFVTT......KQT
RlDctB  552  AGMVTVTVADNGPGIPTEIRKG.........................LFTPFNTS......KES
StPgtB  516  GEALEVYIADNGPGMPVALLPS.........................LLKPFTTS......KAV
EcCpxA  377  KDGTITITVDDDGPGVSPEDREQ........................IFRPFYRTDEARDRESG
EcEnvZ  364  PNRAWFQVEDNGPGIAPEQRKH........................LFQPFVRGDSAR..TIS
EcPhoR  347  PHGAEFSVEDNGPGIAPEHIPR.........................LFERFYRVDKARSRQTG
BsPhoR  497  EKDIQTEVADSGIGIQKEEIPR.........................LTERFYRVDKDRSRNSG
EcPhoM  397  QEHVTLKVLDTGSGIPDYALSR........................IFERFYSLPRAN..GQK
AtWHVirA 620 GDYVLLSISDNGGGIPEAVLPH.........................IFEPFFSTRA...RNG
RmFixL  384  PGEVAVVVEDTGGGIPEEVAGQ.........................LFKPFVIT......KAS
EcUhpB  442  DERLMLVIEDDGSGLPAGSGNKV.......................LASPECAI.......RN
BsDegS  312  KDFVIILMIKDNGKGFDLKEAKEK......................................KNK
StCheA  428  QGNICIEVTDDGAGLNRERILAKAMSQGMAVNENMTDDEVGMLIFAPGFSTAEQVT.DVS
```

```
            5
            _____
KpNtrB  311  GTGLGLSIARSLIDQHSGKIEFT.SWPGHTEFSVYLPIRK*
RpNtrB  328  GSGLGLALVAKIVGDHGGIIECE.SQPRKTTFRVLDADVQ...  11 aa *
RlDctB  585  GLGLGLVISKDIVGDYGGRMDVA.SDSGGIRFVIRLP*
StPgtB  549  GLGIGLSISVSLMAQMKGDLRLAS.TLTRNACVMLQESV...   7 aa *
EcCpxA  416  GTGLGLAIVETAIQQHRGWVKAEDSPLGGLRVLVMLPHVS...  2 aa *
EcEnvZ  401  GTGLGLAIVQRIVDNHNGMLELGTSERGGLSIRAWLPVPV...  10 aa *
EcPhoR  386  GSGLGLAIVKHAVNNHHESRLNIESTVGKGTRFSEVIPERL... 6 aa *
BsPhoR  536  GTGLGLAIVKHHLIEAHEGKIDVTSELGRGTIVFTVTLKRAA... 4 aa *
EcPhoM  434  SSGLGLAMVSEVARLFMGEVILRNVQEGGVLASLRLHRHF...  1 aa *
AtWHVirA 655 GTGLGLASVHGHISAFAGYIDVSSTVGHGTRFDIYLLPPSS... 134 aa *
RmFixL  417  CMGIGLSISKRLIVEAHACQFES....KNEAGGATFRFTLPAYL... 8 aa *
EcUhpB  475  GACWHLIIHFLSARHACQRFS.....TSTLCLRFDDVAVS... 10 aa *
BsDegS  338  SFGLLGMKERVDLLEGTMTID...SKIGLGTFIMIKVPLSL*
StCheA  487  GRGVGMDVVKRNIQEMCGHVEIQSKQGSGTTIRILLPLTL... 145 aa *
```

Figure 2 Homology among sensor components. Boxes surround amino acids with conservative replacements according to the following groupings: ILVM, C, PAGST, HKR, QNED. References for sequences are in Table 1. WHVirA: wide host range virA; CheA-N: N-terminal region of CheA.

Table 1 Superfamily of signal-transducing regulatory pairs

Organism[a]	Environmental signal	Sensor[g]	Regulator[h]	Target	References
NtrC Regulator Subclass (utilizes σ^{54})[d]:					
Ec, Kp, Rm, Rl, Bj, Bsp	Nitrogen status	NtrB[b,j]	NtrC[k]	Nitrogen assimilation genes	19, 53, 65, 86
Rl	C$_4$-dicarboxylates	DctB[bc]	DctD	dctA	92
St	Phosphoglycerate	PgtB[bc]	PgtA	pgtP	34, 49, 115a, 131, 132
OmpR Regulator Subclass[e]:					
Ec, St	Medium osmolarity	EnvZ[bcj]	OmpR[k]	ompC, ompF	26, 60
Bs, Ec	Phosphate status	PhoR[bcj]	PhoB[k]	phoA, phoE, psi	69, 69a, 96, 97, 99, 119
Ec	Carbon status?	PhoM[bc]	ORF2	?; can be phoA	3, 122
At	Acetosyringone	VirA[c]	VirG	vir genes	58, 106, 127
Ec	Sexual state?	CpxA[c]	SfrA[i]	Conjugation genes	2, 20, 100
Ec	Redox state?	ArcB	SfrA[i]	Aerobic metabolism genes	46–48, Iuchi & Lin, personal communication
St	pH?, C & N status?	PhoQ[bc]	PhoP	pagC, other vir genes	74

FixJ Regulator Subclass[f]:

Rm	O$_2$ status	FixL[bc]	FixJ	*nifA, fixK*	14, 17, 118
Ec	Glucose-6-phosphate	UhpB[bc]	UhpA	*uhpT*	27, 125, 126
Bs	?	Degs[b]	DegU	Exoenzyme genes	35, 55
Ec	?	?	UvrC-23Kd	?	34, 98, 115a
Bp	?	BvgC[bc]	BvgA	Virulence genes	7, 73, 109

CheB Regulator Subclass (methylesterase):

Ec, St	attractants, etc.	CheA[j]	CheB[k]	Chemoreceptors (MCPs)	79, 89, 107, 110

CheY Regulator Subclass (regulator consists solely of conserved N-terminal domain):

Ec, St	attractants, etc.	CheA[j]	CheY[k]	Flagellar motor	79, 89, 107, 110
Bs	nutrient status?	?	SpoOF	Sporulation genes	62, 117

Miscellaneous Regulator:

Bs	nutrient status?	?	SpoOA	Sporulation genes	21, 62

[a] At, *Agrobacterium tumefaciens*; Bj, *Bradyrhizobium japonicum*; Bsp, *Bradyrhizobium* species; Bp, *Bordetella pertussis*; Bs, *Bacillus subtilis*; Ec, *Escherichia coli*; Kp, *Klebsiella pneumoniae*; Rl, *Rhizobium leguminosarum*; Rm, *Rhizobium meliloti*; St, *Salmonella typhimurium*.

[b] Sensor and regulator genes are located in the same operon

[c] Sensor predicted to be membrane-bound

[d] Regulator central domain also homologous to NifA (12, 19, 92), TyrR (13, 34), and HrpS (32)

[e] Regulator C-domain also homologous to ToxR (75)

[f] Regulator C-domain also homologous to UvrC-ORF1 (35, 55, 98), LuxR (34, 35, 55), MalT (34, 35, 55), GerE (34), and RcsA (34)

[g] Sensor component has also been called "modulator" (66) and its conserved domain has been called a "transmitter motif" (54)

[h] Regulator component has also been called "effector" (66) and its conserved domain has been called a "receiver motif" (54)

[i] Also known as ArcA, Dye, FexA, Msp, Seg

[j] Sensor shown to be a protein kinase

[k] Regulator shown to be a substrate for sensor protein kinase

the middle of a signal-transduction pathway, the sensor is not always the component utilized by the cell to perceive the primary environmental stimulus. The terms "modulator" and "effector" have also been used to refer to the sensor protein kinase and its regulator substrate, respectively (66).

HOMOLOGY BETWEEN SIGNAL TRANSDUCING REGULATORY PAIRS

Table I summarizes several features of gene products that have been identified as of March 1989 to be homologous to the superfamily of signal-transducing regulatory pairs. Many of the sensor components are predicted to be transmembrane proteins with a substantial periplasmic domain (Table 1). The regulator component is often, though not exclusively, a regulator of gene expression. Most of the transcriptional regulators can be further subclassified by homology between their C-domains (Table 1). The sensor and regulator genes are often organized in an operon (Table 1); in at least one case (*envZ/ompR*) (60), translational coupling has been shown.

The Sensors

The conserved domain of the sensor class extends over approximately 250 amino acids and is usually located at the C-terminus of the protein (Figure 2). Although most members show significant homology extending throughout the entire domain, five regions showing the strongest conservation can be delineated; four of these surround at least one invariant amino acid. The alignments between different sensors and regulators have been evaluated statistically (14, 86, 126) using the computer program PIRALIGN. In most cases scores are greater than 10 standard deviation units above the average score of alignments of the same statistically jumbled sequences (16). For UhpB, CheA and DegS, alignment values $> +5$ standard deviation units can be achieved with only one or two other members of the set. On the other hand, each of these sensor proteins has been shown genetically or biochemically to function with their respective regulator partner(s) (35, 38, 105, 110, 126). Because of the low alignment scores, CheA, UhpB and DegS have been manually aligned in Figure 2 to the most highly conserved regions.

The Regulators

The conserved domain in the regulator class extends over approximately 120 amino acids and in most cases is located at the N-terminus of the protein (Figure 3, see page 325). Most members show strong homology throughout the entire domain; however, four regions with the strongest conservation can be delineated. Regions 1 and 2 have an invariant aspartic acid residue that may be important since NtrC is phosphorylated at an aspartic acid residue within its N-terminal domain (52, 124).

Identification of New Signal Transducing Regulatory Pairs

Using a new algorithm to search for homology to the most highly conserved amino acids of region 4 (Figure 3), Henikoff et al. (34) have identified three new potential members of the regulator class; PgtA from *Salmonella typhimurium* (Table 1), "TrpO" in the *Pseudomonas aeruginosa trpAB* region, and "PetO" in the *Rhodopseudomonas capsulata petABC* region. Kofoid & Parkinson (54) took a different approach to look for other pairs of proteins that retain fundamental structural features similar to the conserved domains of the signal-transducing regulatory pairs. Using relaxed matching criteria to search for what they called "transmitter" and "receiver" motifs, they identified a number of other pairs of proteins, such as PtsG/Crr, which interact in the phosphoenolpyruvate-dependent glucose phosphotransferase uptake system (PTF) (112) and MalE/MalF, which probably interact in maltose transport (4). Interestingly, by Kofoid & Parkinson's criteria, CheA and NtrB appear to carry both transmitter and receiver motifs. Both of these sensor components monitor the environment indirectly through interaction with other proteins (53, 89, 110).

CONSERVED INTERACTION BETWEEN SENSOR AND REGULATOR

The conservation of protein domains in the signal-transducing regulatory pairs suggests that the interaction between each pair involves a mechanism conserved in evolution. In each of the regulatory pairs discussed below, the sensor component is a protein kinase that phosphorylates its regulator partner. In this section we summarize the evidence that phosphorylation of the regulator by the sensor represents a conserved mechanism for signal transduction between partners.

The NtrB/NtrC System

The NtrB/NtrC pair regulates the expression of several genes required for nitrogen assimilation. The most well-characterized target gene is *glnA*, which codes for glutamine synthetase, the major enzyme in enteric bacteria for assimilating low concentrations of NH_3. Transcription of *glnA* occurs from tandem promoters *glnA*p1, which is repressed by NtrC, and *glnA*p2, which is activated by NtrC (91). Activation mediated by NtrC requires the sigma factor NtrA (also called σ^{54} or RpoN) (53). Under conditions of nitrogen starvation, high levels of transcription from the *glnA*p2 promoter are induced by NtrC and NtrA-polymerase (53, 65–67, 90).

The protein kinase activity of the sensor NtrB (also called GlnL or NR_{II}) consists first of an autophosphorylation at a histidine residue using ATP as a phosphate donor and subsequent transfer of that phosphate to the regulator NtrC (also called GlnG or NR_I) (52, 82, 124). Phosphorylation of NtrC correlates with its ability to act as a transcriptional activator in vitro; efficient transcription from *glnAp2* requires NtrB and ATP in addition to NtrC and NtrA-RNA polymerase (53, 82, 85). The continuous presence of NtrB is probably required because phospho-NtrC is unstable and undergoes auto-dephosphorylation (52, 124). Phospho-NtrC is also rapidly dephosphorylated in a regulated manner in a reaction requiring NtrB, ATP, and another regulatory protein called P_{II} (product of *glnB*) (52, 82), resulting in reduced transcriptional activity of NtrC (53, 82). This is consistent with the effect of P_{II} in vivo, which is to antagonize NtrB action in response to nitrogen sufficiency (53, 65). Information about the nitrogen status of the cell is transmitted to P_{II} by a uridylyltransferase/uridylyl-removing enzyme (UTase-UR; product of *glnD*) which uridylylates P_{II}, probably in response to a low ratio of intracellular glutamine to 2-ketoglutarate. P_{II} in its uridylylated form appears to have no effect on NtrB activity, but when unmodified promotes NtrB-mediated dephosphorylation (i.e. inactivation) of Phospho-NtrC.

The following regulatory cascade therefore operates in the NtrB/NtrC system: The uridylyl-removing activity of UTase-UR is stimulated by a signal, possibly a high intracellular glutamine to 2-ketoglutarate ratio, indicative of nitrogen sufficiency. This increases the concentration of unmodified P_{II} in the cell, which in turn increases the function of NtrB in stimulating removal of phosphate from NtrC, leading to inactivation of NtrC as a transcriptional activator and turn off of *glnA* expression. Conversely, nitrogen starvation favors NtrB-catalyzed phosphorylation of NtrC, which is then active in transcription of *glnAp2* (66).

The CheA/CheY + CheB System

Chemotactic behavior is achieved by biasing the direction of flagellar rotation in response to changes in external stimuli over time (89, 107). External stimuli are transduced across the inner membrane by a family of methyl-accepting chemotaxis proteins (MCPs; 5). Wild-type cells alternate between smooth swimming and tumbling. "Gutted" chemotactic mutants, which lack all of the known signal-transducing regulatory gene products, show smooth swimming behavior because of counterclockwise rotation of the flagella. Addition of the sensor CheA, the regulators CheY and CheB, and two other proteins, CheW and CheZ, is sufficient to restore signalling to gutted mutants (129).

CheW is thought to couple a primary signal received by the MCPs to the sensor CheA (10) that in turn signals to both CheY and CheB. The regulator CheY in its activated form biases the flagellar motor toward clockwise (CW) rotation, resulting in tumbling (15, 56, 129). The regulator CheB is a methylesterase which methylates MCPs to regulate their signalling capability, permitting adaptation of the signalling machinery to an environmental stimulus. Addition of repellent or removal of attractant increases the activity of the CheB methylesterase (107, 108).

In vitro, CheA undergoes autophosphorylation at His48 using ATP as a donor (37, 38, 39, 130). Phosphate is transferred from phospho-CheA to CheY or CheB (38, 130). Phospho-CheY and phospho-CheB undergo a relatively slow hydrolysis (38, 130). CheZ accelerates hydrolysis of P_i from CheY (38, 88), consistent with its in vivo role as an antagonist of CheY action (56, 129).

Although the physiological role of the protein kinase activity of CheA has not been shown directly, mutagenesis of the *cheA* gene has shown a strong correlation between mutants defective in chemotaxis and those altered either in autophosphorylation or in transfer of phosphate to CheB and CheY (37, 88). In addition, ATP depletion antagonizes the ability of CheY to generate a tumble signal (103), consistent with the model that phospho-CheY promotes clockwise rotation of the flagellum, resulting in tumbling (although the ability of CheY to bind ATP may indicate a more direct role (103)). Likewise, the CheB-methylesterase activity is inhibited by ATP depletion (103), suggesting that phospho-CheB has increased methylesterase activity, resulting in net removal of methyl groups from MCPs. Thus, phospho-CheY and phospho-CheB appear to mediate the response to negative chemotactic stimuli, and CheA appears to act as an integrator of signals from the MCPs (37, 89).

The CheA/CheB, CheY system is an example of a signalling network which is slightly more complex than the NtrB/NtrC system, in which one sensor and one regulator transmit a single environmental signal. In the CheA/CheB, CheY system, a single sensor (CheA) signals to two different regulators (CheY and CheB). Conversely, the sensors CpxA and ArcB each appear to be capable of transmitting a unique environmental signal through the same regulator, SfrA (46, 47).

The EnvZ/OmpR System

The sensor EnvZ and the regulator OmpR control expression of the outer membrane porins OmpC and OmpF (33) which are regulated in a reciprocal manner by medium osmolarity: low osmolarity favors OmpF synthesis while high osmolarity favors OmpC synthesis (26). The regulator OmpR is absolutely required for expression of both genes, while *envZ* null mutants show significantly reduced expression of both genes (29, 77). The interaction of OmpR with the *ompF* and *ompC* promoters is likely to be complex given that the binding sites for the two promoters are different (51, 76, 87). In addition, there appears to be post-transcriptional regulation of *ompF* through a small antisense RNA called *micF* (26). EnvZ and OmpR are also required for normal expression of the tripeptide permease gene, *tppB*, but this regulation is independent of the medium osmolarity (30).

The biochemistry of the sensor EnvZ has been difficult because of its membrane location. Autophosphorylation of wild-type EnvZ (41) and of EnvZ containing N-terminal deletions that remove up to 179 amino acids (1, 24, 41, 42) has been demonstrated. This phosphate linkage exhibits a pH-dependent stability similar to that of phospho-CheA and phospho-NtrB, suggesting a similar phospho-histidine linkage (24, 42). In vitro, EnvZ containing a 38 amino acid N-terminal deletion phosphorylates OmpR, which stimulates the ability of OmpR to activate transcription from the *ompF* promoter (41). Phospho-OmpR is dephosphorylated in a reaction mediated by EnvZ and requiring ATP (1a).

The PhoR/PhoB System

The PhoR/PhoB pair regulate expression of the phosphate regulon, which includes *phoA* and other *p*hosphate *s*tarvation-*i*nducible *(psi)* genes. In vitro, PhoR protein lacking the N-terminal 83 amino acids autophosphorylates and phospho-PhoR transfers phosphate to PhoB. Phosphorylation of PhoB correlates with an increase in its ability to bind DNA upstream of the *pstS* promoter and to activate transcription at that promoter (K. Makino, personal communication).

Cross-Talk

If phosphorylation of the regulator by the sensor represents a conserved mode of signal transduction, unless each sensor is tuned perfectly to its cognate regulator, it seems likely that a particular sensor might "cross-talk" with (i.e. phosphorylate or dephosphorylate) regulators from other systems. This occurs in both directions for the CheA/CheY, CheB and NtrB/NtrC systems. NtrC is phosphorlyated in vitro by CheA, and the phospho-NtrC produced in this manner stimulates in vitro transcription from *glnAp2* (84). In the other direction, phospho-NtrB transfers phosphate in vitro to CheY (84). In vivo, overproduction of a mutant allele of *ntrB* (NtrB2302), which phosphorylates but does not dephosphorylate NtrC, suppresses the smooth-swimming phenotype of a *cheA* mutant. Unless NtrB has some other CheA-like activity, this strongly suggests that phospho-CheY generates a tumbling signal (84). In one other example of cross-talk in vitro, NtrB has been shown to phosphorylate SpoOA (52).

In vivo, cross-talk between signal-transducing regulatory pairs provides a means for integrating many types of environmental stimuli. If cross-talk contributed significantly to target activity, one might expect that a null sensor mutant in one system could have a pleiotropic effect in other systems. Although no such dramatic evidence exists, there is some genetic evidence that cross-talk can occur in vivo under special circumstances.

PhoM and the PhoR/PhoB system provide the most compelling genetic evidence for cross-talk. The PhoR/PhoB pair regulate expression of *phoA* and other phosphate *starvation-inducible (psi)* genes in response to phosphate level. In the absence of the PhoR sensor, phosphate-independent expression of *phoA* (coding for alkaline phosphatase) can occur. This expression depends on the function of an unlinked gene, *phoM* (70, 121), whose product is homologous to the sensor class. PhoM-dependent expression of *phoA* still requires the regulator PhoB, consistent with phosphorylation of PhoB by PhoM. The physiological relevance of PhoM-dependent expression of *phoA* is unclear, because *phoM* mutations have no effect on *phoA* expression in a *phoR*$^+$ background (99, 119). In fact, the ORF directly upstream of *phoM* in the *phoM* operon (ORF2) is homologous to the regulator class (3). The normal function of the putative PhoM/ORF2 pair therefore seems likely to be the regulation of different class of genes. Recent evidence that PhoM responds to catabolite repression suggests that the target of the PhoM/ORF2 pair may be a class of catabolite-responsive genes (120, 122).

One other hint that cross-talk can occur in vivo is also suggested by the fact that expression or overproduction of the regulator proteins NtrC (65), OmpR (25, 29, 30, 77, 87, 102), DctD (L. Albright & C. Ronson, unpublished data), CheY (15, 56, 103, 129), UhpA (125, 126), PgtA (49, 132), FixJ (36), and PhoB (68, 96) in the absence of their sensor partners elicits a target response that, in most cases, is constitutive. These observations might be explained by low-level cross-talk from the sensors of other systems. Alternatively, the regulator protein may have some inherent activity in the absence of sensor modification/phosphorylation.

The general absence of pleiotropic phenotypes for null mutants in sensor proteins and the suggestion by the PhoR-PhoM example that cross-talk is not seen until normal interactions are interrupted indicates that a high degree of specificity within each pair has been achieved, probably in large part by mutational optimization. The genetic organization of many of the pairs into operons (see Table 1), in some cases with translational coupling (*envZ/ompR;* 60), may also contribute to the maintenance of a specific interaction.

FUNCTION OF THE SENSOR

The concentration of conserved amino-acid residues in the C-domain of the sensors (Figure 2) and the large variation in the structure of the N-domains suggests that the N-domain has the capacity to receive a particular environmental signal and that the conserved C-domain transmits the signal through phosphorylation of the regulator. Several experiments described in this section address the question of whether the receiving and transmitting functions of the sensor are carried out by two distinct domains.

Signal Detection

The signal-transducing regulatory pairs appear to be using one of two modes of signal detection: either the sensor receives an environmental stimulus directly, or the environmental stimulus is first perceived and transduced by a primary environmental sensor into an intracellular (or intramembrane) signal that is then detected by the sensor component of the two-component system. The latter is clearly the case for CheA and NtrB, which are cytoplasmic proteins and function as part of an intracellular relay mechanism. In the chemotaxis system, the detection of external ligands by transmembrane receptor proteins called MCPs (89) is coupled to the sensor protein CheA through CheW (10, 89). In the NtrB/NtrC system, GlnA (glutamine synthetase) can be thought of as the primary sensor of environmental ammonia (66, 67). Glutamine synthetase generates an intracellular signal (glutamine) that is relayed to the NtrB sensor protein by the UTase-UR and P_{II} proteins as described in an earlier section. DegS is another example of a sensor component thought to be a cytoplasmic protein. In the DegS/DegU system, two additional genes encoding polypeptides of 46 and 60 amino acid are implicated in regulation of degradative exoenzymes in *Bacillus subtilis,* although the signal to which this regulatory pair responds remains unknown (35, 55).

In some cases where the sensor component of a regulatory pair is a membrane protein, there is also evidence for the involvement of proteins other than the sensor component in signal perception. In the case of the DctB/DctD pair, the regulator DctD activates expression of the *dctA* dicarboxylate-transport gene in the presence of external C4-dicarboxylic acids. Insertions in *dctA* result in constitutive synthesis of a *dctA-lacZ* fusion (92). It is possible that the transport protein DctA and the putative protein kinase DctB cooperate in detecting the presence of C4-dicarboxylic acids outside the cell (92). A similar phenomenon is seen in the PhoR/PhoB system that activates *phoA,* the gene coding for alkaline phosphatase. Mutations in the *pst-phoU* locus, which, with the exception of *phoU,* codes for the high-affinity inorganic phosphate transport system, lead to constitutive expression of *phoA.* This suggests that phosphate repression is achieved by transport of inorganic phosphate by the *pst* system, with PhoU functioning to transduce that signal to PhoR (80, 116, 119).

Many other sensor components predicted by hydropathy profiles to have a disposition across the inner membrane may serve as the primary sensor of an environmental stimulus. These latter sensors appear to be similar in structure to the chemoreceptors (MCPs), which have a periplasmic ligand-binding domain and a cytoplasmic domain that interacts with the chemotactic signalling apparatus (Table I; 6). For the most part, the presumptive cytoplasmic domain of these transmembrane sensors corresponds to the conserved sensor domain postulated to interact with the regulator protein. The sensors EnvZ

(23, 59), VirA (58), UhpB (125) and PgtB (131) have been localized to membrane fractions, and VirA appears to span the inner membrane in a manner similar to that of MCPs (128). These sensor proteins may therefore bind specific external ligands and have a function similar to that of MCPs.

Domain Structure of Sensors

For those sensors that appear to have a transmembrane location, the separation of signal-detection and signal-transmission functions into separate periplasmic and cytoplasmic domains, respectively, makes sense both from a functional and an evolutionary point of view. Experimental evidence supports this concept. For example, deletions of 38, 55, and 179 amino acids at the N-terminus of the EnvZ protein are predicted to remove EnvZ from its membrane location. The presence of EnvZ carrying a 38 or 55 amino acid N-terminal deletion partially complements an *envZ* null mutant but does not restore normal regulation of OmpF and OmpC (42, 77). Significantly, EnvZ carrying a 38 or 179 amino acid deletion retains autophosphorylation and OmpR kinase activity (24, 41, 42), consistent with the separation of the signal-reception and signal-transmission functions into two distinct domains.

In contrast to EnvZ, there is both biochemical and genetic evidence in the case of CheA that some determinants for interaction with CheY and CheB occur in the N-domain. Although the strongest homology between CheA and other sensors lies near its C-terminus (Figure 2), autophosphorylation of CheA occurs near the N-terminus of the protein at His48 (37). At a genetic level, Oosawa et al (88) performed in vitro mutagenesis of the *cheA* gene, screening for chemotaxis-defective mutants, and mapped each mutation to one of five regions defined by restriction sites. Five mutations mapping to region I (amino acids 1–170) gave rise to CheA proteins capable of undergoing autophosphorylation but defective in transfer of phosphate to CheY and CheB. Four mutations produced full-size CheA protein that could not undergo autophosphorylation. One of these mapped in region I and the other three mapped in region IV, roughly amino acids 420–540, corresponding to regions 3, 4 and 5 of Figure 2. These results indicate that determinants at the N-terminus and in conserved region IV are necessary for autophosphorylation, but that interaction with CheB and CheY is determined by the N-terminus alone (88). This is consistent with the ability of an N-terminal 18-kd CheA peptide to transfer phosphate to CheB and CheY, but not to undergo autophosphorylation (37). CheA has two translational starts 90 amino acids apart, giving rise to $CheA_L$ and $CheA_S$ (104). Only $CheA_L$ undergoes autophosphorylation, indicating that at least some of the determinants for autophosphorylation are located within the first 90 amino acids of the protein (37, 38).

One way to reconcile the CheA data with the overall model of sensor/ regulator interactions (Figure 1) is to postulate that the conserved sensor domain in CheA$_L$ has been split into two parts. One part would consist of amino acids 1–170 of CheA, corresponding to region 1 of Figure 2, and the second part would consist of amino acids 290–420 of CheA (region IV as defined by Oosawa et al (88)), corresponding to regions 2–5 of Figure 2. Although the sites of autophosphorylation in NtrB or EnvZ have not yet been mapped, it is tempting to speculate, as others have, that phosphorylation of all of the sensor components takes place at the invariant histidine in region 1 of Figure 2 (124). This interpretation is consistent with the findings that NtrB autophosphorylation occurs at a histidine (124) and that the pH stability of phospho-EnvZ and phospho-PhoR is consistent with phospho-histidine (24, 42; A. Makino, personal communication). An alignment of the region surrounding His48 of CheA with region 1 of the sensors is shown in Figure 2. This alignment shows no statistical significance by *PIRALIGN* criteria (alignment values <2). Nevertheless, because of the data cited above, it appears possible that determinants for recognition and interaction of the sensor with the regulator lie in region 1 and adjacent portions of the conserved sensor domain, while determinants for autophosphorylation are contained in both parts of the domain. It is intriguing that the portion of the conserved sensor domain that includes region 1 is in general more weakly conserved than the rest of the domain, which might be expected if determinants for specific recognition of the regulator lie in this region.

Because CheA is an unusual sensor in its overall structure (54) and in its ability to communicate with two regulators, it may have unique features. Therefore caution must be used in generalizing the results obtained with CheA to other sensor proteins. In any case, because CheA is an intracellular sensor, there are no architectural constraints that would necessarily prevent separation of sensor subdomains to the N- and C-terminal positions of the protein. More data on the location of autophosphorylation of other sensor components are needed before a general model for the function of sensor proteins can be formulated.

FUNCTION OF THE REGULATOR

The presence of a conserved N-domain is the distinguishing feature of the regulators (Figure 3). The conservation of the N-domain points to this portion of the regulator participating in the conserved interaction with a sensor protein. This is borne out by experiments with NtrC discussed below. In addition, good evidence exists for the division of signal reception and response functions into two separate domains within regulators containing a second domain. Most regulators that function as transcriptional activators can be grouped into subclasses based on homology within their C-domains (Table 1). These structural features of the regulators suggest that a common mechanism of transcriptional regulation exists for each subclass and that the regulators evolved by the "mixing and matching" of domains. In support of this latter idea, the C-domain of most of the subclasses is shared by at least one other protein that does not function as a member of a signal-transducing regulatory pair (Table 1). The relationships between the regulator subclasses has been elegantly illustrated in a recent paper on methods for searching nucleic acid databases (34).

NtrC Subclass

NtrC belongs to a group of transcriptional activators requiring the alternate sigma factor NtrA (RpoN; σ^{54}) (31, 57, 90, 93). In addition to NtrC, members of this group include DctD, PgtA, and NifA (Table 1). Of these, PgtA is the only one for which an NtrA requirement has not been shown. These four regulators (NtrC, DctD, NifA, and PgtA) share a strongly conserved block of 238 amino acids in the central region and a less well conserved block of 45 amino acids at the C-terminal end that contains a

Figure 3   Homology among regulator components. See *Figure 2* for other details.

helix-turn-helix DNA-binding motif (19, 92). However, NifA, which activates structural genes required for nitrogen fixation in a variety of Gram-negative bacteria (12), does not contain the conserved N-terminal domain shared by the other regulators considered in this review.

There is good evidence that both the N-terminal and central domains of the NtrC subclass can function autonomously. In the case of the N-terminal domain, NtrC is phosphorylated at an aspartic acid residue in the amino terminal 12.5 kd of the protein (52, 124). This peptide alone, which essentially constitutes the conserved domain shared by all regulators, is an efficient substrate for the NtrB kinase, and undergoes autodephosphorylation and regulated dephosphorylation with kinetics comparable to that of the entire protein (52). This indicates that the N-domain of the regulator component contains sufficient determinants for interaction with and modification by the sensor component.

The presence of a large highly conserved block of amino acids in the central domain of NtrC, DctD, and NifA has led to the prediction that the central domain interacts directly with NtrA-RNA polymerase holoenzyme to stimulate transcription from NtrA-dependent promoters (12, 19, 92). Evidence in support of this prediction comes from experiments with all three activators. For example, the *Rhizobium leguminosarum* DctD regulator can drive low-level constitutive expression of its target *dctA* promoter in the absence of the sensor DctB (L. Albright & C. Ronson, unpublished data). A modified DctD protein that deletes the conserved N-domain yields high constitutive levels of *dctA* expression. This not only indicates that the N-domain of DctD is unnecessary for transcriptional activation but also that the N-terminal domain has a regulatory role (J. Stigter, L. Albright & F. Ausubel, unpublished data; B. T. Nixon, personal communication). Similarly, deletion analysis of the NifA proteins from *Klebsiella pneumoniae* (78), *Rhizobium meliloti* (8, 40), and *Bradyrhizobium japonicum* (22) has shown that neither the N-domain of NifA nor its C-terminal helix-turn-helix domain is required for transcriptional activity, and that the central domain of NifA shared by NtrC, DctD, and PgtA is sufficient to activate transcription (40). In addition, several point mutations in NtrC that result in a specific defect in the ability to activate transcription map to highly conserved amino acid residues in the central domain (57).

The central domain of the NtrC class is also found in HrpS, a regulatory protein in *Pseudomonas syringae* pv. *phaseolicola* which is required for pathogenesis of host plants and induction of a hypersensitive defense response on nonhost plants. HrpS consists of the central domain and a possible helix-turn-helix motif at its C-terminus (32). The central domain is additionally found in TyrR, a repressor of general and specific aromatic amino acid transport and general aromatic amino acid biosynthesis in *E. coli* (13, 34). The homology of TyrR to the activators of the NtrC subclass is particularly intriguing in that it appears to function mainly as a repressor, although a positive role in regulation of *tyrP* may be indicated (50; C. Stephens, personal communication).

Finally, NtrA has been implicated in the regulation of the *Pseudomonas putida xylABC* (xylenes and toluene metabolism) (44), carboxypeptidase G2 (18), and pilin genes (45), of the *Caulobacter crescentus flbG* and *flaN* flagellar genes (83), and of the *E. coli hydB* and *fdhF* genes required for formate metabolism (9, 95). The previous domain homologies predict that the regulatory proteins for these genes will have homology to the central domain of the NtrC subfamily (already shown for XylR (44)).

OmpR Subclass

OmpR, PhoB, PhoM-ORF2, VirG and SfrA share a common C-domain (127) that is also found as an N-domain in ToxR, a transcriptional activator for cholera toxin, pilus, and outer membrane proteins in *Vibrio cholerae* (75). OmpR (51, 76, 87), PhoB (68), and ToxR (75) each bind DNA upstream of the transcriptional start site of their target promoters; each of the binding sites contains short direct repeats. A 16-kd proteolytic fragment of the OmpR protein corresponding to the C-domain binds to the same sequence upstream of the *ompF* promoter as does the full-length protein, although the fragment is about tenfold less efficient in binding (114). Experiments using a protein fusion between ToxR and PhoA (alkaline phosphatase) have demonstrated that the homologous N-domain of ToxR is directly responsible for DNA-binding and transcriptional activation (75). The C-domain of the ToxR is thought to be localized in the periplasm on the basis of *toxR*::Tn*phoA* fusions and is implicated in the mediation by ToxR of a response to NaCl but not to pH and amino acids (73, 75).

CheB Subclass

In the case of the CheB regulator, the N-domain of inactive CheB appears able to compete with wild-type CheB and CheY for phospho-CheA (108). When expressed at high levels, a CheB-LacZ fusion protein that deletes the C-terminal half of CheB inhibits the swarming ability of wild-type cells as well as the ability of the wild-type methylesterase activity to increase in response to negative stimuli. The same phenomenon is seen with three different mutant CheB proteins that are catalytically inactive due to missense mutations that map in the C-terminal half of CheB.

A number of observations indicate that the catalytic esterase activity of CheB resides in its C-domain, and that its N-domain regulates the esterase activity in response to negative stimuli (101, 108). A 21-kd proteolytic fragment of the CheB methylesterase, representing the C-terminal 60% of the protein, has fifteenfold higher methylesterase activity in vitro than the entire protein (101). In addition, N-terminal deletions of up to one half of CheB retain methylesterase activity in vivo. The normal increase of methylesterase activity in response to negative stimuli is impaired in these deletions although the decrease in activity in response to positive stimuli is not (108). The same work showed that the response to both negative and positive stimuli depends on *cheA* function, although others (105) have suggested that only the response to negative stimuli requires CheA.

CheY Subclass

The structure of the CheY and SpoOF regulators, which consists only of the conserved N-domain found in other regulators, clearly does not allow for a separation of their signal reception and response functions into separate domains. The three-dimensional structure of CheY was recently reported (111).

GENETIC ANALYSIS OF SIGNAL TRANSDUCTION IN REGULATORY PAIRS

The model in Figure 1 predicts that each component receives an input in its N-domain, and that the N- and C-domains interact in an intramolecular reaction to generate the appropriate output through the C-domain. Input and output mutants should therefore cluster in the N- and C-domains, respectively, of each component. Mutations defective in intramolecular transduction could reside in either domain; moreover, it should be possible to find allele-specific intragenic suppressors within individual sensor and regulator genes.

In previous genetic analyses, the most intensively studied mutations have been those resulting in constitutive activation of promoters normally regulated by a signal-transducing pair. These mutations most likely fall into the intramolecular transduction category and, as expected, map throughout both the sensor and regulator genes. In the NtrB/NtrC system, mutants selected for high constitutive expression of *glnA* map in either *ntrB* or *ntrC* (63, 64, 91). NtrB prepared from one such mutant (*ntrB*2302) does not mediate in vitro dephosphorylation of NtrC in the presence of P_{II} (82). The *ntrC*610 allele, which in vivo results in elevated levels of *glnA* expression in the absence of NtrB, maps to the central domain of NtrC but other NtrC constitutive mutants map to the N-terminal domain (D. Popham & S. Kustu, personal communication). A similar allele of *ntrC* that also maps in the central domain was recently reported (123). Significantly, the NtrC610 protein cannot undergo regulated dephosphorylation (52). This indicates that determinants in the central region of NtrC mediate interaction of its N-domain with NtrB and P_{II}. In the DegS/DegU system, mutations resulting in overproduction of a number of degradative enzymes have been mapped to the conserved domain in each component. Mutations in DegS map at amino acids 236 (V→M) and 218

(G→E). Those in DegU map at amino acids 12 (H→L) and 107 (E→K) (35). In the EnvZ/OmpR system, an *ompR* constitutive mutant with a phenotype of OmpF^c OmpC$^-$ (the *ompR2* class) maps to the C-domain (V203→M) (81) and its gene product binds to the *ompF* promoter but not the *ompC* promoter (76). Conversely, an OmpR mutant with a OmpF$^-$ OmpC^c phenotype (OmpR3) maps to the N-domain (81; R15→C) and its gene product binds to both the *ompF* and *ompC* promoters, as does wild-type OmpR protein. The same OmpF$^-$ OmpC^c phenotype is seen with *envZ* alleles 160, 11, and 473 (71, 115). *envZ160* maps in the N-domain (L35→Q) (71) and *envZ11* maps in the C-domain (T247→R) (71). Other unmapped mutations of this class are the R1b mutants of *phoR*, which give rise to *phoM*-independent constitutive expression of *phoA* (119), and the *pho510* allele of the *phoM* locus, which produces cAMP-independent expression of *phoA* in a *phoR*-mutant background (120).

The *envZ11* mutation is particularly interesting as it may provide an example of a mutation in a sensor for which an allele-specific suppressor mutation in the regulator has been obtained. *ompR77* (L16→Q) partially suppresses the OmpF$^-$ and the OmpC^c phenotypes of the *envZ11* mutation in an allele-specific manner, consistent with a direct interaction between EnvZ and OmpR (71). However, a simple interpretation of these mutants is difficult because both *envZ11* and *envZ473* show pleiotropic repression of *phoA*, *phoE*, *malE*, and *lamB* (115), a phenotype not seen in *envZ* null alleles (29, 77). The pleiotropic repression by the *envZ11* mutation is also relieved by *ompR77* (71). In addition, suppression of the *envZ11* LamB$^-$ and OmpF$^-$ phenotypes can occur by mutations in the *rpoα* operon, coding for the α subunit of RNA polymerase and several ribosomal proteins (28), and a mutation in *rpoα* can interfere with suppression of *envZ11* by *ompR77* (72). This may indicate that interaction between OmpR and RNA polymerase can "feed back" to affect the interaction between EnvZ and OmpR or that EnvZ modification of OmpR affects the interaction between OmpR and RNA polymerase.

Recently, it has been shown that EnvZ11 and OmpR3 are each defective in EnvZ-mediated dephosphorylation of OmpR. In addition, OmpR77 is defective in phosphotransfer from EnvZ11 but not from EnvZ (1a).

SUMMARY AND PROSPECTS

The evidence is now quite convincing that a common mechanism of interaction involving phospho-transfer has been conserved among the regulatory pairs discussed in this review. In many of the systems, a more detailed understanding of the interaction of the regulator with its target and the role of phosphorylation in that interaction is needed. In addition, the mechanism(s) of intramolecular signal transduction within each component remains to be determined. Nevertheless, the phospho-transfer reaction, which must still be demonstrated in many of the signal-transducing regulatory pairs, has provided a powerful assay for intermediate steps in the signal transduction from sensor to regulator component. Modern mutagenesis schemes aimed at identifying mutations and suppressors within each regulatory pair, coupled with the phosphotransfer assay, should prove extremely useful in dissecting determinants involved in signal reception, transduction, and output in each component (88).

It should also be possible to design mutagenesis schemes based on the domain structure of the sensor component to uncover other components (if they exist) necessary for signalling in systems where the nature of the signals being perceived is not understood. As has already been pointed out (86, 94), the predictive value of the signal-transducing regulatory pair homology also points the way in cases where regulator components have been found (such as Spo0A and Spo0F) but where the sensor component(s) has not been identified by conventional means.

ACKNOWLEDGMENTS

We thank our many colleagues who provided reprints and unpublished information; Boris Magasanik, Sydney Kustu, John S. Parkinson, and Tom Silhavy for critically reading the manuscript, Clive Ronson for initiating the effort to write this chapter and for literature searches, and Rachel Hyde for help in literature searches and manuscript preparation.

Literature Cited

1. Aiba, H., Mizuno, T., Mizushima, S. 1989. Transfer of phosphoryl group between two regulatory proteins involved in osmoregulatory expression of the *ompF* and *ompC* genes in *Escherichia coli*. *J. Biol. Chem.* 264:8563–67

1a. Aiba, H., Nakasai, F., Mizushima, S., Mizuno, T. 1989. Evidence for the physiological importance of the phosphotransfer between the two regulatory components, EnvZ and OmpR, in osmoregulation in *Escherichia coli*. *J. Biol. Chem.*, In press

2. Albin, R., Weber, R., Silverman, P. M. 1986. The Cpx proteins of *Escherichia coli* K12. *J. Biol. Chem.* 261:4698–705

3. Amemura, M., Makino, K., Shinagawa, H., Nakata, A. 1986. Nucleotide sequence of the *phoM* region of *Escherichia coli:* four open reading frames may constitute an operon. *J. Bacteriol.* 168:294–302

4. Ames, G. F.-L. 1986. Bacterial periplasmic transport systems: structure, mechanism and evolution. *Annu. Rev. Biochem.* 55:397–425

5. Ames, P., Chen, J., Wolff, C., Parkinson, J. S. 1989. Structure-Function studies of bacterial chemosensors. *Cold Spring Harbor Symp. Quant. Biol.* 53:59–65

6. Ames, P., Parkinson, J. S. 1988. Transmembrane signalling by bacterial chemoreceptors: *Escherichia coli* transducers with locked signal output. *Cell* 55:817–26

7. Arico, B., Stibitz, S., Royo, C., Miller, J., Falkow, S, et al. 1989. Gene products required for expression of *Bordetella pertussis* virulence factors share homology with prokaryotic signal transduction proteins., *Proc. Natl. Acad. Sci. USA*, In press

8. Beynon, J. L., Williams, M. K., Cannon, F. C. 1988. Expression and functional analysis of the *Rhizobium meliloti nifA* gene. *EMBO J.* 7:7–14

9. Birkmann, A., Sawers, R. G., Bock, A.

1987. Involvement of the *ntrA* gene product in the anaerobic metabolism of *Escherichia coli*. *Mol. Gen. Genet.* 210:535–42

10. Borkovich, K. A., Kaplan, N., Hess, J. F., Simon, M. I. 1989. Transmembrane signal transduction in bacterial chemotaxis involves ligand-dependent activation of phosphate group transfer. *Proc. Natl. Acad. Sci. USA* 86:1208–12

11. Bueno, R., Pahel, G., Magasanik, B. 1985. Role of *glnB* and *glnD* gene products in regulation of the *glnALG* operon of *Escherichia coli*. *J. Bacteriol.* 164:816–22

12. Buikema, W. J., Szeto, W. W., Lemley, P. V., Orme-Johnson, W. H., Ausubel, F. M. 1985. Nitrogen fixation specific regulatory genes of *Klebsiella pneumoniae* and *Rhizobium meliloti* share homology with the general nitrogen regulatory gene *ntrC* of *K. pneumoniae*. *Nucleic Acids Res.* 13:4539–55

13. Chye, M.-l., Pittard, J. 1987. Transcription control of the *aroP* gene in *Escherichia coli* K-12: analysis of operator mutants. *J. Bacteriol.* 169:386–93

14. David, M., Daveran, M.-L., Batut, J., Dedieu, A., Domergue, O., et al. 1988. Cascade regulation of *nif* gene expression in *Rhizobium meliloti*. *Cell* 54:671–83

15. David, S., Matsumura, P., Eisenbach, M. 1986. Restoration of flagellar clockwise rotation in bacterial envelopes by insertion of the chemotaxis protein CheY. *Proc. Natl. Acad. Sci. USA* 83:7157–61

16. Dayhoff, M. O., Barker, W. C., Hunt, L. T. 1983. Establishing homologies in DNA sequences. *Methods Enzymol.* 91:524–45

17. Ditta, G., Virts, E., Palomares, A., Kim, C.-k. 1987. The *nifA* gene of *Rhizobium meliloti* is oxygen regulated. *J. Bacteriol.* 169:3217–23

18. Dixon, R. 1986. The *xylABC* promoter from the *Pseudomonas putida* TOL plasmid is activated by nitrogen regulatory genes in *Escherichia coli*. *Mol. Gen. Genet.* 203:129–36

19. Drummond, M., Whitty, P., Wootton, J. 1986. Sequence and domain relationships of *ntrC* and *nifA* from *Klebsiella pneumoniae*: homologies to other regulatory proteins. *EMBO J.* 5:441–47

20. Drury, L. S., Buxton, R. S. 1985. DNA sequence analysis of the *dye* gene of *Escherichia coli* reveals amino acid homology between the Dye and OmpR proteins. *J. Biol. Chem.* 260:4236–42

21. Ferrari, F. A., Trach, K., Lecoq, D.,

Spence, J., Ferrari, E., et al. 1985. Characterization of the *spo0A* locus and its deduced product. *Proc. Natl. Acad. Sci. USA* 82:2647–51

22. Fischer, H.-M., Bruderer, T., Hennecke, H. 1988. Essential and nonessential domains in the *Bradyrhizobium japonicum* NifA protein: identification of indispensable cysteine residues potentially involved in redox reactivity and/or metal binding. *Nucleic Acids Res.* 16:2207–24

23. Forst, S., Comeau, D., Norioka, S., Inouye, M. 1987. Localization and membrane topology of EnvZ, a protein involved in osmoregulation of OmpF and OmpC in *Escherichia coli*. *J. Biol. Chem.* 262:16433–38

24. Forst, S., Delgado, J., Inouye, M. 1989. Phosphorylation of OmpR by the osmosensor EnvZ modulates the expression of the *ompF* and *ompC* genes in *Escherichia coli*. *Proc. Natl. Acad. Sci. USA*, In press

25. Forst, S., Delgado, J., Ramakrishnan, G., Inouye, M. 1988. Regulation of *ompC* and *ompF* expression in *Escherichia coli* in the absence of *envZ*. *J. Bacteriol.* 170:5080–85

26. Forst, S., Inouye, M. 1988. Environmentally regulated gene expression for membrane proteins in *Escherichia coli*. *Annu. Rev. Cell Biol.* 4:21–42

27. Friedrich, M. J., Kadner, R. J. 1987. Nucleotide sequence of the *uhp* region of *Escherichia coli*. *J. Bacteriol.* 169:3556–63

28. Garrett, S., Silhavy, T. J. 1987. Isolation of mutations in the α operon of *Escherichia coli* that suppress the transcriptional defect conferred by a mutation in the porin regulatory gene *envZ*. *J. Bacteriol.* 169:1379–85

29. Garrett, S., Taylor, R. K., Silhavy, T. J., Berman, M. L. 1985. Isolation and characterization of ΔompB strains of *Escherichia coli* by a general method based on gene fusions. *J. Bacteriol.* 162:840–44

30. Gibson, M. M., Ellis, E. M., Graeme-Cook, K. A., Higgins, C. F. 1987. OmpR and EnvZ are pleiotropic regulatory proteins: positive regulation of the tripeptide permease (*tppB*) of *Salmonella typhimurium*. *Mol. Gen. Genet.* 207:120–29

31. Gussin, G. N., Ronson, C. W., Ausubel, F. M. 1986. Regulation of nitrogen fixation genes. *Annu. Rev. Genet.* 20:567–91

32. Grimm, C., Panopoulos, N. J. 1989. The predicted protein product of a pathogenicity locus from *Pseudomonas*

syringae is homologous to a highly conserved domain of several prokaryotic regulatory proteins. *J. Bacteriol.*, In press

33. Hall, M. N., Silhavy, T. J. 1981. Genetic analysis of the *ompB* locus in *Escherichia coli* K-12. *J. Mol. Biol.* 151:1–15

34. Henikoff, S., Wallace, J. C., Brown, J. P. 1989. Finding protein similarities with nucleotide sequence databases. *Methods Enzymol.*, In press

35. Henner, D. J., Yang, M., Gerrari, E. 1988. Localization of *Bacillus subtilis* *sacU* (Hy) mutations to two linked genes with similarities to the conserved procaryotic family of two-component signalling systems. *J. Bacteriol.* 170:5102–09

36. Hertig, C., Ya Li, R., Louarn, A.-M., Garnerone, A.-M., David, M., et al. 1989. *Rhizobium meliloti* regulatory gene *fixJ* activates transcription of *R. meliloti* *nifA* and *fixK* genes in *Escherichia coli*. *J. Bacteriol.* 171:1736–38

37. Hess, J. F., Bourret, R. B., Simon, M. I. 1988. Histidine phosphorylation and phosphoryl group transfer in bacterial chemotaxis. *Nature* 336:139–43

38. Hess, J. F., Oosawa, K., Kaplan, N., Simon, M. I. 1988. Phosphorylation of three proteins in the signalling pathway of bacterial chemotaxis. *Cell* 53:79–87

39. Hess, J. F., Oosawa, K., Matsumura, P., Simon, M. I. 1987. Protein phosphorylation is involved in bacterial chemotaxis. *Proc. Natl. Acad. Sci. USA* 84:7609–13

40. Huala, E., Ausubel, F. M. The central domain of *Rhizobium meliloti* NifA is sufficient to activate transcription from the *R. meliloti* *nifH* promoter. *J. Bacteriol.* 171:3354–65

41. Igo, M. M., Ninfa, A. J., Silhavy, T. J. 1989. A bacterial environmental sensor that functions as a protein kinase and stimulates transcriptional activation. *Genes Dev.* 3:598–605

42. Igo, M. M., Silhavy, T. J. 1988. EnvZ, a transmembrane environmental sensor of *Escherichia coli* K-12, is phosphorylated in vitro. *J. Bacteriol.* 170:5971–73

43. Ikenaka, K., Tsung, K., Comeau, D. E., Inouye, M. 1988. A dominant mutation in *Escherichia coli* OmpR lies within a domain which is highly conserved in a large family of bacterial regulatory proteins. *Mol. Gen. Genet.* 211:538–40

44. Inouye, S., Nakazawa, A., Nakazawa, T. 1988. Nucleotide sequence of the regulatory gene *xylR* of the *TOL* plasmid from *Pseudomonas putida*. *Gene* 66:301–6

45. Ishimoto, K., Lory, S. 1989. Formation of pilin in *Pseudomonas aeruginosa* requires the alternative σ factor (RpoN) of RNA polymerase. *Proc. Natl. Acad. Sci. USA* 86:1954–57

46. Iuchi, S., Cameron, D. C., Lin, E. C. C. 1989. A second global regulator gene *(arcB)* mediating repression of enzymes in aerobic pathways of *Escherichia coli*. *J. Bacteriol.* 171:868–73

47. Iuchi, S., Furlong, D., Lin, E. C. C. 1989. Differentiation of *arcA*, *arcB*, and *cpxA* mutant phenotypes of *Escherichia coli* by sex pilus formation and enzyme regulation. *J. Bacteriol.* 171:2889–93

48. Iuchi, S., Lin, E. C. C. 1988. *arcA (dye)*, a global regulatory gene in *Escherichia coli* mediating repression of enzymes in aerobic pathways. *Proc. Natl. Acad. Sci. USA* 85:1888–92

49. Jiang, S.-Q., Yu, G.-Q., Li, Z.-G., Hong, J.-S. 1988. Genetic evidence for modulation of the activator by two regulatory proteins involved in the exogenous induction of phosphoglycerate transport in *Salmonella typhimurium*. *J. Bacteriol.* 170:4304–08

50. Kasian, P. A., Davidson, B. E., Pittard, J. 1986. Molecular analysis of the promoter operator region of the *Escherichia coli* K-12 *tyrP* gene. *J. Bacteriol.* 167:556–61

51. Kato, M., Aiba, H., Mizuno, T. 1989. Molecular analysis by deletion and site-directed mutagenesis of the cis-acting upstream sequence involved in activation of the *ompF* promoter in *Escherichia coli*. *J. Biochem. (Tokyo)* 105:341–47

52. Keener, J., Kustu, S. 1988. Protein kinase and phosphoprotein phosphatase activities of nitrogen regulatory proteins NTRB and NTRC of enteric bacteria: roles of the conserved amino-terminal domain of NTRC. *Proc. Natl. Acad. Sci. USA* 85:4976–80

53. Keener, J., Wong, P., Popham, D., Wallis, J., Kustu, S. 1987. A sigma factor and auxiliary proteins required for nitrogen-regulated transcription in enteric bacteria. In *RNA Polymerase and the Regulation of Transcription*, ed. W. S. Reznikoff, R. R. Burgess, J. E. Dahlberg, C. A. Gross, M. T. Record, Jr., M. P. Wickens. pp. 159–75. New York: Elsevier

54. Kofoid, E. C., Parkinson, J. S. 1988. Transmitter and receiver modules in bacterial signaling proteins. *Proc. Natl. Acad. Sci. USA* 85:4981–85

55. Kunst, F., Debarbouille, M., Msadek, T., Young, M., Mauel, C., et al. 1988. Deduced polypeptides encoded by the *Bacillus subtilis sacU* locus share homology with two-component sensor-regulator systems. *J. Bacteriol.* 170: 5093–101

56. Kuo, S. C., Koshland, D. E., Jr. 1987. Roles of the *cheY* and *cheZ* gene products in controlling flagellar rotation in bacterial chemotaxis of *Escherichia coli*. *J. Bacteriol.* 169:1307–14

57. Kustu, S., Santero, E., Popham, D., Keener, J. 1989. Expression of sigma^{54} (ntrA)-dependent genes is probably united by a common mechanism. *Microbiol. Rev.* In press

58. Leroux, B., Yanofsky, M. F., Winans, S. C., Ward, J. E., Zeigler, S. F., et al. 1987. Characterization of the *virA* locus of *Agrobacterium tumefaciens:* a transcriptional regulator and host range determinant. *EMBO J.* 6:849–856

59. Liljestrom, P. 1986. The EnvZ protein of *Salmonella typhimurium* LT-2 and *Escherichia coli* K-12 is located in the cytoplasmic membrane. *FEMS Microbiol. Lett.* 36:145–50

60. Liljestrom, P., Laamanen, I., Palva, E. T. 1988. Structure and expression of the ompB operon, the regulatory locus for the outer membrane porin regulon in *Salmonella typhimurium* LT-2. *J. Mol. Biol.* 201:663–73

61. Liljestrom, P., Luokkamaki, M., Palva, E. T. 1987. Isolation and characterization of a substitution mutation in the *ompR* gene of *Salmonella typhimurium* LT2. *J. Bacteriol.* 169:438–41

62. Losick, R., Youngman, P., Piggot, P. J. 1986. Genetics of endospore formation in *Bacillus subtilus*. *Annu. Rev. Genet.* 20:625–69

63. MacFarlane, S. A., Merrick, M. J. 1987. Analysis of the *Klebsiella pneumoniae ntrB* gene by site-directed in vitro mutagensis. *Mol. Microbiol.* 1:133–42

64. MacNeil, T., MacNeil, D., Tyler, B. 1982. Fine-structure deletion map and complementation analysis of the *glnA-glnL-glnG* region in *Escherichia coli*. *J. Bacteriol.* 150:1302–13

65. Magasanik, B. 1982. Genetic control of nitrogen assimilation in bacteria. *Annu. Rev. Genet.* 16:135–68

66. Magasanik, B. 1988. Reversible phosphorylation of an enhancer binding protein regulates the transcription of bacterial nitrogen utilization genes. *Trends Biochem. Sci.* 13:475–79

67. Magasanik, B., Neidhardt, F. C. 1987. Regulation of carbon and nitrogen utilization. In *Escherichia coli and Salmonella typhimurium: Cellular and Molecular Biology*, ed. F. C. Neidhardt, J. Ingraham, K. B. Low, B. Magasanik, M. Schaechter, H. E. Umbarger, pp. 1318–25. Washington, DC: Am. Soc. Microbiol

68. Makino, K., Shinagawa, H., Amemura, M., Kimura, S., Nakata, A. 1988. Regulation of the phosphate regulon of *Escherichia coli:* activation of *pstS* transcription by PhoB protein in vitro. *J. Mol. Biol.* 203:85–95

69. Makino, K., Shinagawa, H., Amemura, M., Nakata, A. 1986. Nucleotide sequence of the *phoR* gene, a regulatory gene for the phosphate regulon of *Escherichia coli*. *J. Mol. Biol.* 192:549–56

69a. Makino, K., Shinagawa, H., Amemura, M., Nakata, A. 1986. Nucleotide sequence of the *phoB* gene, the positive regulatory gene for the phosphate regulon of *Escherichia coli* K-12. *J. Mol. Biol.* 190:37–44

70. Makino, K., Shinagawa, H., Nakata, A. 1984. Cloning and characterization of the alkaline phosphatase positive regulatory gene *(phoM)* of *Escherichia coli*. *Mol. Gen. Genet.* 195:381–90

71. Matsuyama, S.-I., Mizuno, T., Mizushima, S. 1986. Interaction between two regulatory proteins in osmoregulatory expression of *ompF* and *ompC* genes in *Escherichia coli:* a novel *ompR* mutation suppresses pleiotropic defects caused by an *envZ* mutation. *J. Bacteriol.* 168:1309–14

72. Matsuyama, S.-I., Mizushima, S. 1987. Novel *rpoA* mutation that interferes with the function of OmpR and EnvZ, positive regulators of the *ompF* and *ompC* genes that code for outer-membrane proteins in *Escherichia coli* K12. *J. Mol. Biol.* 195:847–53

73. Miller, J. F., Mekalanos, J. J, Falkow, S. 1989. Coordinate regulation and sensory transduction in the control of bacterial virulence. *Science* 243:916–22

74. Miller, S. I., Kukral, A. M., Mekalanos, J. J. 1989. A two-component regulatory system *(phoP-phoQ)* controls *Salmonella typhimurium* virulence. *Proc. Natl. Acad. Sci. USA* 86:5054–58

75. Miller, V. L., Taylor, R. K., Mekalanos, J. J. 1987. Cholera toxin transcriptional activator ToxR is a transmembrane DNA binding protein. *Cell* 48:271–79

76. Mizuno, T., Kato, M., Jo, Y.-L., Mizushima, S. 1988. Interaction of OmpR, a positive regulator, with the osmo-

regulated *ompC* and *ompF* genes of *Escherichia coli*. Studies with wild-type and mutant OmpR proteins. *J. Biol. Chem.* 263:1008–12

77. Mizuno, T., Mizushima, S. 1987. Isolation and characterization of deletion mutants of *ompR* and *envZ*, regulatory genes for expression of the outer membrane proteins OmpC and OmpF in *Escherichia coli*. *J. Biochem.* 101:387–96

78. Morett, E., Cannon, W., Buck, M. 1988. The DNA-binding domain of the transcriptional activator protein NifA resides in its carboxy terminus, recognises the upstream activator sequences of *nif* promoters and can be separated from the positive control function of NifA. *Nucleic Acids Res.* 16:11469–88

79. Mutoh, N., Simon, M. I. 1986. Nucleotide sequence corresponding to five chemotaxis genes in *Escherichia coli*. *J. Bacteriol.* 165:161–66

80. Nakata A., Amemura, M., Makino, K., Shinagawa, H. 1987. Genetic and biochemical analysis of the phosphate-specific transport system in *Escherichia coli*. In *Phosphate Metabolism and Cellular Regulation in Microorganisms,* ed. A. Torriani-Gorini, F. G. Rothman, S. Silver, A. Wright, E. Yagil, pp. 150–55. Washington, DC: Am. Soc. Microbiol.

81. Nara, F., Matsuyama, S.-I., Mizuno, T., Mizushima, S. 1986. Molecular analysis of mutant *ompR* genes exhibiting different phenotypes as to osmoregulation of the ompF and ompC genes of *Escherichia coli*. *Mol. Gen. Genet.* 202:194–99

82. Ninfa, A. J., Magasanik, B. 1986. Covalent modification of the *glnG* product, NR_I, by the *glnL* product, NR_{II}, regulates the transcription of the *glnALG* operon in *Escherichia coli*. *Proc. Natl. Acad. Sci. USA* 83:5909–13

83. Ninfa, A. J., Mullin, D. A., Ramakrishnan, G., Newton, A. 1989. *Escherichia coli* σ^{54} RNA polymerase recognizes *Caulobacter crescentus flbG* and *flaN* flagellar gene promoters in vitro. *J. Bacteriol.* 171:383–91

84. Ninfa, A. J., Ninfa, E. G., Lupas, A. N., Stock, A., Magasanik, B., et al. 1988. Crosstalk between bacterial chemotaxis signal transduction proteins and regulators of transcription of the Ntr regulon: evidence that nitrogen assimilation and chemotaxis are controlled by a common phosphotransfer mechanism. *Proc. Natl. Acad. Sci. USA* 85:5492–96

85. Ninfa, A. J., Reitzer, L. J., Magasanik, B. 1987. Initiation of transcription at the bacterial *glnA*p2 promoter by purified *E.*

coli components is facilitated by enhancers. *Cell* 50:1039–46

86. Nixon, B. T., Ronson, C. W., Ausubel, F. M. 1986. Two-component regulatory systems responsive to environmental stimuli share strongly conserved domains with the nitrogen assimilation regulatory genes *ntrB* and *ntrC*. *Proc. Natl. Acad. Sci. USA* 83:7850–54

87. Norioka, S., Ramakrishnan, G., Ikenaka, K., Inouye, M. 1986. Interaction of a transcriptional activator, OmpR, with reciprocally osmoregulated genes, *ompF* and *ompC*, of *Escherichia coli*. *J. Biol. Chem.* 261:17113–19

88. Oosawa, K., Hess, J. F., Simon, M. I. 1988. Mutants defective in bacterial chemotaxis show modified protein phosphorylation. *Cell* 53:89–96

89. Parkinson, J. S. 1988. Protein phosphorylation in bacterial chemotaxis. *Cell* 53:1–2

90. Popham, D. L., Szeto, D., Keener, J., Kustu, S. 1989. Function of a bacterial activator protein that binds to transcriptional enhancers. *Science* 243:629–35

91. Reitzer, L. J., Magasanik, B. 1985. Expression of *glnA* in *Escherichia coli* is regulated at tandem promoters. *Proc. Natl. Acad. Sci. USA* 82:1979–83

92. Ronson, C. W., Astwood, P. M., Nixon, B. T., Ausubel, F. M. 1987. Deduced products of C4-dicarboxylate transport regulatory genes of *Rhizobium leguminosarum* are homologous to nitrogen regulatory gene products. *Nucleic Acids Res.* 15:7921–34

93. Ronson, C. W., Nixon, B. T., Albright, L. M., Ausubel, F. M. 1987. *Rhizobium meliloti ntrA (rpoN)* gene is required for diverse metabolic functions. *J. Bacteriol.* 169:2424–31

94. Ronson, C. W., Nixon, B. T., Ausubel, F. M. 1987. Conserved domains in bacterial regulatory proteins that respond to environmental stimuli. *Cell* 49:579–81

95. Sankar, P., Shanmugam, K. T. 1988. Biochemical and genetic analysis of hydrogen metabolism in *Escherichia coli:* the *hydB* gene. *J. Bacteriol.* 170:5433–39

96. Seki, T., Yoshikawa, H., Takahashi, H., Saito, H. 1987. Cloning and nucleotide sequence of *phoP*, the regulatory gene for alkaline phosphatase and phosphodiesterase in *Bacillus subtilis*. *J. Bacteriol.* 169:2913–16

97. Seki, T., Yoshikawa, H., Takahashi, H., Saito, H. 1988. Nucleotide sequence of the *Bacillus subtilis phoR* gene. *J. Bacteriol.* 170:5935–38

98. Sharma, S., Stark, T. F., Beattie, W.

G., Moses, R. E. 1986. Multiple control elements for the *uvrC* gene unit of *Escherichia coli. Nucleic Acids Res.* 14: 2301–18

99. Shinagawa, H., Makino, K., Amemura, M., Nakata, A. 1987. Structure and function of the regulatory genes for the phosphate regulon in *Escherichia coli.* See Ref. 80, pp. 20–25

100. Silverman, P. M. 1986. Host cell-plasmid interactions in the expression of DNA donor activity by F^+ strains of *Escherichia coli* K-12. *Bioessays* 2:254–59

101. Simms, S. A., Keane, M. G., Stock, J. 1985. Multiple forms of the CheB methylesterase in bacterial sensing. *J. Biol. Chem.* 260:10161–68

102. Slauch, J. M., Garrett, S., Jackson, D. E., Silhavy, T. J. 1988. EnvZ functions through OmpR to control porin gene expression in *Escherichia coli* K-12. *J. Bacteriol.* 170:439–41

103. Smith, J. M., Rowsell, E. H., Shioi, J., Taylor, B. L. 1988. Identification of a site of ATP requirement for signal processing in bacterial chemotaxis. *J. Bacteriol.* 170:2698–704

104. Smith, R. A., Parkinson, J. S. 1980. Overlapping genes at the *cheA* locus of *Escherichia coli. Proc. Natl. Acad. Sci. USA* 77:5370–74

105. Springer, M. S., Zanolari, B. 1984. Sensory transduction in *Escherichia coli:* regulation of the demethylation by the CheA protein. *Proc. Natl. Acad. Sci. USA* 81:5061–65

106. Stachel, S. E., Zambryski, P. C. 1986. *virA* and *virG* control the plant-induced activation of the T-DNA transfer process of *A. tumefaciens. Cell* 46:325–33

107. Stewart, R. C., Dahlquist, F. W. 1987. Molecular components of bacterial chemotaxis. *Chem. Rev.* 87:997–1025

108. Stewart, R. C., Dahlquist, F. W. 1988. N-terminal half of CheB is involved in methylesterase response to negative chemotactic stimuli in *Escherichia coli. J. Bacteriol.* 170:5728–38

109. Stibitz, S., Aaronson, W., Monack, D., Falkow, S. 1989. Phase variation in *Bordetella pertussis* by frameshift mutation in a gene for a novel two-component system. *Nature* 338:266–69

110. Stock, A., Chen, T., Welsh, D., Stock, J. 1988. CheA protein, a central regulator of bacterial chemotaxis, belongs to a family of proteins that control gene expression in response to changing environmental conditions. *Proc. Natl. Acad. Sci. USA* 85:1403–7

111. Stock, A. M., Mottonen, J. M., Stock, J. B., Schutt, C. E. 1989. Three-dimensional structure of CheY, the response regulator of bacterial chemotaxis. *Nature* 337:745–49

112. Stock, J. B., Waygood, E. B., Meadow, N. D., Postma, P. W., Roseman, S. 1982. Sugar transport by the bacterial phosphotransferase system. *J. Biol. Chem.* 257:14543–52

113. Tanaka, T., Kawata, M. 1988. Cloning and characterization of *Bacillus subtilis iep,* which has positive and negative effects on production of extracellular proteases. *J. Bacteriol.* 170:3593–600

114. Tate, S.-I., Kato, M., Nishimura, Y., Arata, Y., Mizuno, T. 1988. Location of DNA-binding segment of a positive regulator, OmpR, involved in activation of the *ompF* and *ompC* genes of *Escherichia coli. FEBS Lett.* 242:27–30

115. Taylor, R. K., Hall, M. N., Silhavy, T. J. 1983. Isolation and characterization of mutations altering expression of the major outer membrane porin proteins using the local anaesthetic procaine. *J. Mol. Biol.* 166:273–82

115a. Timme, T. L., Lawrence, C. B., Moses, R. E. 1989. Two new members of the OmpR superfamily detected by homology to a sensor-binding core domain. *J. Mol. Evol.* 28:545–52

116. Torriani-Gorini. 1987. The birth and growth of the *pho* regulon. See Ref 80, pp. 3–11

117. Trach, K., Chapman, J. W., Piggot, P. J., Hoch, J. A. 1985. Deduced product of the stage 0 sporulation gene SpoOF shares homology with the SpoOA, OmpR, and SfrA proteins. *Proc. Natl. Acad. Sci. USA* 82:7260–64

118. Virts, E. L., Stanfield, S. W., Helinski, D. R., Ditta, G. S. 1988. Common regulatory elements control symbiotic and microaerobic induction of *nifA* in *Rhizobium meliloti. Proc. Natl. Acad. Sci. USA* 85:3063–65

119. Wanner, B. L. 1987. Phosphate regulation of gene expression in *Escherichia coli.* See Ref. 67, pp. 1326–33

120. Wanner, B. L. 1987. Control of *phoR*-dependent bacterial alkaline phosphatase clonal variation by the *phoM* region. *J. Bacteriol.* 169:900–3

121. Wanner, B. L., Wilmes, M. R., Hunter, E. 1988. Molecular cloning of the wild-type *phoM* operon in *Escherichia coli* K-12. *J. Bacteriol.* 170:279–88

122. Wanner, B. L., Wilmes, M. R., Young, D. C. 1988. Control of bacterial alkaline phosphatase synthesis and variation in an *Escherichia coli* K-12 *phoR* mutant by adenyl cyclase, the cyclic AMP receptor protein, and the *phoM* operon. *J. Bacteriol.* 170:1092–102

123. Weglenski, P., Ninfa, A., Ueno-Hishio, S., Magasanik, B. 1989. Mutations in the *glnG* gene of *Escherichia coli* that result in increased activity of nitrogen regulator I. *J. Bacteriol.* 171:4479–85

124. Weiss, V., Magasanik, B. 1988. Phosphorylation of nitrogen regulator I (NR$_I$) of *Escherichia coli*. *Proc. Natl. Acad. Sci. USA* 85:8919–23

125. Weston, L. A., Kadner, R. J. 1987. Identification of Uhp polypeptides and evidence for their role in exogenous induction of the sugar phosphate transport system of *Escherichia coli* K-12. *J. Bacteriol.* 169:3546–55

126. Weston, L. A., Kadner, R. J. 1988. Role of *uhp* genes in expression of the *Escherichia coli* sugar-phosphate transport system. *J. Bacteriol.* 170:3375–83

127. Winans, S. C., Ebert, P. R., Stachel, S. E., Gordon, M. P., Nester, E. W. 1986. A gene essential for *Agrobacterium* virulence is homologous to a family of positive regulatory loci. *Proc. Natl. Acad. Sci. USA* 83:8278–82

128. Winans, S. C., Kerstetter, R. A., Ward, J. E., Nester, E. W. 1989. A protein required for transcriptional regulation of *Agrobacterium* virulence genes spans the cytoplasmic membrane. *J. Bacteriol.* 171:1616–22

129. Wolfe, A. J., Conley, M. P., Kramer, T. J., Berg, H. B. 1987. Reconstitution of signaling in bacterial chemotaxis. *J. Bacteriol.* 169:1878–85

130. Wylie, D., Stock, A., Wong, C.-Y., Stock, J. 1988. Sensory transduction in bacterial chemotaxis involves phosphotransfer between *che* proteins. *Biochem. Biophys. Res. Comm.* 151:891–96

131. Yang, Y.-L., Goldrick, D., Hong, J.-S. 1988. Identification of the products and nucleotide sequences of two regulatory genes involved in the exogenous induction of phosphoglycerate transport in *Salmonella typhimurium*. *J. Bacteriol.* 170:4299–303

132. Yu, G.-Q., Hong, J.-S. 1986. Identification and nucleotide sequence of the activator gene of the externally induced phosphoglycerate transport system of *Salmonella typhimurium*. *Gene* 45:51–57

Annu. Rev. Genet. 1989. 23:337–70

EVOLUTIONARY QUANTITATIVE GENETICS: HOW LITTLE DO WE KNOW?

N. H. Barton

Department of Genetics and Biometry, University College London, 4 Stephenson Way, London NW1 2HE, U.K.

Michael Turelli

Department of Genetics, University of California, Davis, California 95616

CONTENTS

0066-4197/89/1215-0337$02.00

INTRODUCTION

The centenary of Galton's (84) "Natural Inheritance" emphasizes the continuity between the ideas of the early pioneers and the modern renaissance of evolutionary quantitative genetics (142). Many questions posed by Galton, Weldon, and Pearson remain only partially answered. Weldon (248) asserted that "the problem of animal evolution is essentially a statistical problem," and argued that legitimate conjectures on macroevolution must be based on understanding the factors shaping microevolution: variation, selection, and heredity. As he noted, this demands statistical analysis of phenotypic variation (including both variances and correlations), of fitness as a function of phenotype, and of the similarities between parents and offspring. His call, in 1893, for additional empirical studies of selection in the wild was repeated by Endler in 1986. In his first statistical analyses of multiple characters, Weldon (247, 248) asked whether phenotypic correlations and variances from one population were applicable to others. Like Lofsvold (156) and Kohn & Atchley (129), he found that correlations were relatively constant, but variances were not. This question of relative constancy of quantitative genetic parameters remains central to evaluations of macroevolutionary analyses that assume constant genetic parameters (170, 237).

Galton found that useful predictions could be made without understanding the mechanism of inheritance. Statistical predictions also dominated the work of Galton's disciple Karl Pearson, who felt that scientific advance required replacing hopes of understanding causes with empirical studies based on correlation (183). Statistical reasoning still dominates most quantitative genetics theory. Despite Fisher's reconciliation of Mendelism with resemblances among relatives (79), the standard equations for the response of phenotypic means to selection are essentially pre-Mendelian. They rest on the properties of the multivariate normal (Gaussian) distribution developed by Pearson (181), among others (225). As expected from their origin in statistics rather than genetics, these equations are only approximately consistent with multilocus Mendelian inheritance (32; 238a).

Galton himself doubted "whether any organic laws are so strictly in accordance with that of the error function [normal distribution] as to admit of building vast mathematical edifices on that insecure foundation" (186). Nevertheless, many recent analyses treat the Gaussian selection equations as laws for the behavior of quantitative traits on which precise predictions, possibly involving selection over thousands of generations, may be based. Here we examine whether evolutionary inferences are robust to the departures from the usual Gaussian approximations that must arise from Mendelian inheritance.

Lewontin's 1967 review of population genetics did not mention quantitative characters (151). However, after ten years of looking at allozyme variation, many evolutionists became disillusioned with trying to measure selection on allozymes (146, 153). Lande's elegant analyses of phenotypic evolution (132, 134) and of the maintenance of polygenic variation (131, 133, 136), and Lande & Arnold's analysis of selection on multiple traits (143) led to a new research program. There is an emerging awareness, however, that some of these analyses involve questionable assumptions. Despite the relative ease of documenting selection on quantitative characters, understanding the nature of variation and selection may not be so much easier for quantitative traits than for allozymes.

Evolutionary quantitative genetics covers a large literature. For instance, two recent reviews (173, 203) summarize heritability estimates from 270 studies of outbred, nondomestic animal populations (for other reviews see 11, 18, 33, 72, 78, 124, 141, 142, 170, 196, 254). Our emphasis is on the genetic basis of quantitative variation, and its effects on the evolution of quantitative characters: long-term dynamics depend on the genetics and cannot be understood from phenotypic analyses. Yet our review indicates that few experiments approach basic issues such as the number of loci that contribute to within-population variation, the rate of polygenic mutation, the extent of pleiotropy, the mechanisms that maintain additive variance, and the reasons for reduced fitness of extreme phenotypes. These questions are accessible: one aim of this review is to suggest research that complements the primarily descriptive data gathered over the last decade.

THE GENETICS OF QUANTITATIVE VARIATION

Most continuously varying characters in most populations will respond to directional selection (25a, 153, 254). There are few exceptions: the best-known involve selection for directional asymmetry (56, 167). Moreover, artificial selection can often move a population's mean phenotype well outside its initial range. Changes of four or more phenotypic standard deviations are common (75). Some selection experiments have altered means by more than ten times the initial phenotypic standard deviation within 100 generations (e.g. 71, 73, 255). These rates far exceed any seen in the fossil record (41, 90, 103, 130). Despite this evidence for pervasive additive genetic variance, the genetic basis of this variation remains obscure. How many loci are involved, what are their mutation rates, and how do they affect the phenotype?

How Many Loci?

Around 1910 Johannsen, East, and Nilsson-Ehle established that continuous variation is often due to environmental effects and the segregation of several

Mendelian loci (191, 254). However, we still do not know whether the number of loci responsible for most genetic variation is small (5 to 20) or large (100 or more). If few are involved, it will be feasible to map and manipulate the loci. If there are many, mapping may identify chromosome regions with large effects, but the genetics of quantitative variation will be difficult to unravel. The fact that phenotypic distributions are often approximately Gaussian does not tell us much, because even three segregating loci can produce a nearly normal distribution of phenotypes with high heritability (232).

The number of loci determines the applicability of statistical descriptions of polygenic evolution. The "infinitesimal model," originated by Fisher (79) and extended by Bulmer (28), assumes that contributions to the genetic variance are additive, relatively small, and come from enough loci to give an approximately Gaussian distribution of breeding values. If this is so, the short-term dynamics of the population mean and additive genetic variance can be predicted from phenotypic observations that do not require Mendelian analysis. However, departures from the predictions of the infinitesimal model are expected after a time proportional to a "variance effective" number of loci (16, 149, 238).

The number of loci underlying quantitative variation can be estimated either using biometrical techniques, which rely on means and variances from a cross between two populations, or using genetic markers to map chromosome regions that contribute to differences. Both have severe limitations. Although interest centers on within-population variation, both methods count only the number of loci that contribute to differences between populations. Because the differences are often produced by artificial selection, there is a bias toward finding alleles of major effect (139).

The widely used Wright-Castle formula (38, 254) estimates an "effective number" of loci (denoted n_E) by comparing the variance arising from segregation in the F_2 with the difference between the parental means. If there is no epistasis and the divergent populations are homozygous at the relevant loci, n_E underestimates n, the actual number of loci. If there is neither dominance nor epistasis, $n_E \leq n$ even if the loci are polymorphic in the parental lines (51, 137). Because n_E may exceed n if these polymorphic loci exhibit dominance or epistasis, it is important to test for nonadditivity (the tests in (164, 165) do not require homozygous parental stocks if mating is random).

Estimates of n_E are typically between 5 and 20 (137, 254), but n may in fact be much larger. Linkage, which must enter as n increases, sets an upper bound on estimates from F_2s (66): these cannot exceed three or four times the haploid chromosome number (cf. 137, 234). Estimates increase if crossing is continued over many generations. For example, the increase in genetic variance at the center of a natural hybrid zone gives an estimate of n_E, provided that linkage disequilibria in the zone can be independently estimated using discrete markers (see 15).

Visible markers have long been used to identify chromosome regions containing loci affecting quantitative traits (180, 231). Shrimpton & Robertson (216) analyzed the contribution of the third chromosome to the difference in bristle number between selected lines of *Drosophila melanogaster*. They found 17 disjunct regions, each contributing at least 0.3 phenotypic standard deviations to the line differences. Their data suggest a highly leptokurtic distribution of effects, with relatively few regions producing most of the difference. This may be because the differences were produced by strong artificial selection (139). The abundance of molecular markers (e.g. 179, 227) and efficient statistical methods based on likelihood (144) aid the location of polygenes. However, these new methods share many of the same assumptions (and hence, limitations) as the biometrical methods. Mapping works best when relatively few regions ($n_E < 5$) are responsible for most of the differences. Regions can be further subdivided, but very large samples are needed to detect small effects.

Such studies support the view that most quantitative changes are polygenic, but they do not determine the actual number of loci involved, nor do they exclude the possible importance of a few leading factors (216). Molecular studies show that one "locus" may be further subdivided: for example, variation in Adh activity in *D. melanogaster* may be due to sequence variation over many kilobases surrounding the coding region, as well as to a protein polymorphism (4). The observation of large variances in effect across regions of similar size (144, 216) sets an upper bound on the number of loci, but this has not been statistically analyzed.

Part of the difficulty in counting quantitative trait loci is that different numbers are relevant to different questions (cf. 232). Breeders will be content to identify a few loci with large effects. In contrast, the dynamics of additive variance under selection depends on a "variance-effective" number that accounts for variation within a population; and this has not yet been related to the effective number n_E that accounts for differences between populations produced by selection. Ideally, we would like to know the joint distribution of frequencies and effects (on both morphology and fitness) of alleles segregating in natural populations.

The Paradox of Per-Character versus Per-Locus Mutation Rates: What are Polygenes?

Lynch (158) reviews many estimates of V_m, the additive genetic variance introduced by mutation each generation (50). This quantity is important for long-term selection response (110), neutral phenotypic evolution (160, 239), and some models of polygenic mutation-selection balance (33). In contrast, only three sets of experiments estimate the *number* of new mutations per generation affecting quantitative characters ($\Sigma\mu$; see 234, 238). These suggest a total mutation rate of $\Sigma\mu \approx 10^{-2}$ per character, a value difficult to reconcile with typical per-locus mutation rate estimates of about 10^{-6}–10^{-5}. Little progress has been made in resolving this paradox: the per-character estimates may be experimental artifacts; or hundreds of loci with pleiotropic effects may contribute; or the mutation rates at loci that contribute to quantitative variation may be qualitatively higher than those estimated in studies of lethals, visible mutants, and allozymes. We do not even know what sort of allelic differences are involved: are they coding changes that alter the proteins, or noncoding differences that lead to different amounts of gene product (cf. 64, 82, 145, 175)?

Comparative studies may help to resolve the nature of polygenic differences. Restriction-enzyme analysis reveals more nucleotide variation in *D. simulans* than in *D. melanogaster,* but much less insertion-deletion variation, a reduction proportional to the lower abundance of transposable elements in *D. simulans* (5; C. F. Aquadro, personal communication). Thus transposable elements may cause most insertion-deletion variation. However, there are similar amounts of intrapopulation protein variation in each species, and comparable amounts of phenotypic and additive genetic variation for wing and thorax length within sympatric populations (228, 229; P. Capy, personal communication). In contrast, Cohan & Hoffmann (53) found lower rates of selection response for ethanol tolerance in *D. simulans* than in *D. melanogaster.* Thus, no simple patterns have yet emerged.

Gene Action: Additivity, Dominance, Epistasis, and Pleiotropy

What is the relation between genotype and phenotype (206)? Many quantitative characters follow an approximately Gaussian distribution on an appropriate scale (75, 84, 254). The traditional explanation is that the genetic component is the sum of effects from many independently segregating loci (79), though approximate normality for phenotypes is consistent with few segregating loci (232). Fisher's additive gene action model is the basis for most quantitative genetic theory (32, 75), and yet it is hard to know how often it applies.

Powers (187) showed that in any one population he could always find a scale on which the distribution of fruit weight per locule in tomatoes was approximately Gaussian, but no single transformation could approximate normality and eliminate epistasis in all stocks and all environments. Although most genetic variation can often be described statistically by the additive component of variance, this does not preclude strong epistasis or complete dominance (62, 18). Biometric analysis of means from crosses between divergent lines can reveal more about departures from additive gene action than analysis of variance. Some data are consistent with simple additivity, but many are not (see 117, 122, 137, 164, 200, 254). Thus, the robustness of conclusions based on additive gene action models should be investigated.

The approximate normality of phenotypes is rather puzzling, given the likelihood of epistasis. Wright (254) reconciled approximate additivity with the known interactions among coat-color alleles in guinea pigs by postulating that composite characters like height and weight are controlled by independent sets of loci, within which there is interaction, but between which there is additivity. Alternatively, if alleles have sufficiently small effects, then a Taylor's series expansion shows that any continuous relation between gene and phenotype leads to approximate additivity (cf. 200). However, this requires many loci and that every allele has a small effect compared with the changes in phenotype that can be produced by its locus.

Interest in nonadditive effects has been renewed by recent experiments and theories. Population bottlenecks can increase the additive variance (26, 27), possibly because random drift can raise the frequency of rare recessives so that they contribute additive variance; epistasis can have similar effects (26, 92, 159). It is not clear whether this would be important in adaptation following a founder event: if recessives are rare because they have deleterious effects, the additive variance would be expected to decrease again as they are selected out (13).

It is widely believed that pleiotropy is ubiquitous (37, 254): this has important implications for the maintenance of variation and the limits to selection. However, many experiments that claim to show pervasive pleiotropy are of dubious value, because they do not contain adequate controls (e.g. 95), or they ignore the effects of major mutations on body size, which may cause allometric effects on many characters, exaggerating the extent of pleiotropy (e.g. 70). Further experiments are needed to survey the effects of spontaneous mutations on multivariate morphology and physiology.

Wright (254) argued that alleles at one locus often vary in their range of pleiotropic effects, but his claim is based on only a few coat-color mutations. Alternative alleles at complex loci in *D. melanogaster,* such as *Notch* (8) and *daughterless* (39), show very different pleiotropic effects. But these loci act early in development and effect many aspects of physiology, and so may be quite different from most loci affecting quantitative traits. Such complexity, if pervasive, contradicts Wagner's simple model of pleiotropy (244), which assumes that alleles differ on a linear scale that maps onto all characters affected.

More data are also needed on dominance relations of pleiotropic loci. Keightley & Kacser (120) claim that alleles tend to display similar dominance patterns for all the characters they affect. In contrast, if antagonistic pleiot-

ropy is to maintain variation (204), then alleles must have different dominance relations for different fitness components such that the favorable allele
is always dominant (cf. 111).

PATTERNS OF VARIATION

Much of quantitative genetics relies on estimates of components of phenotypic variance and covariance. Evolutionists have used these to infer constraints
on long-term evolution. This is feasible only if the estimates are reasonably
accurate and are constant across populations and environments. Here, we
review empirical patterns of variance and covariance; the effects of selection
are considered in a later section.

There are several summaries of standard estimation procedures for genetic
variances and covariances (17, 75), and of recent developments, emphasizing
likelihood and resampling techniques (107, 170, 214). Sampling errors tend
to be extremely large: very large samples are needed to give a good chance of
detecting low heritabilities, and the necessary sample size must be roughly
quadrupled to detect between-population heritability differences of comparable magnitude (127, 128). Sufficiently large samples are rarely obtained from
nature.

Differences between Populations

The best-known comparison of genetic covariances is Arnold's laboratory
analysis of garter snake populations (6). It illustrates apparent constancy,
though with small samples. Berven (20) also used rather small samples to
compare heritabilities (h^2) and genetic correlations for life history traits in two
frog populations. Developmental rates showed similar heritabilities, but estimates for body size at metamorphosis were quite different ($h^2 = 0.66 \pm 0.31$
versus 0.07 ± 0.13). Estimates of additive and environmental correlations
between developmental rates and body size were strikingly different
($r_A = 0.65 \pm 0.37$ versus -0.86 ± 0.08; $r_E = -0.62$ versus 0.25), but the first
estimate of r_A is probably unreliable because one of the heritabilities is not
significant. In contrast, estimates of phenotypic correlations were fairly similar ($r_P = -0.42$ versus -0.26).

Three formal statistical comparisons of genetic covariance matrices have all
found significant differences (21, 129, 156, 237). Billington et al's results
(21) are puzzling. They studied differences in means and genetic and
phenotypic covariances in adjacent grass populations occupying ecologically
different agricultural habitats. Their large-sample likelihood tests indicated
significant genetic differentiation, and four of the 10 traits examined showed
significantly different additive genetic covariances, of opposite signs, between adjacent populations. However, samples were small, distributions non-

normal, and their results seem to indicate more statistical power than expected (127).

Genetic covariances can also be compared via correlated responses to directional selection. Replicate selection lines must be taken from each population to indicate the precision of the estimates. This technique has been used to compare migratory (Iowa) and nonmigratory (Puerto Rico) milkweed bugs. These populations show similar phenotypic and genetic correlations for some traits, but apparently different correlations for others (68, 69). Cohan & Hoffmann (53) summarize other examples of differences (and similarities) in correlated responses, using both intra- and inter-specific comparisons. For instance, selection for ethanol tolerance in replicate lines from five populations of *D. melanogaster* shows consistent differences in the physiological changes underlying the response (52).

Effects of the Environment

A general feature of quantitative traits is that alternative genotypes respond differently to different environments (54. 57, 100, 177, 193, 206, 209). Parsons (178) proposes that additive variance, at least for characters related to fitness, tends to increase under conditions that reduce longevity and/or fecundity (i.e. under "stress"); he argues that "stress" accentuates differences among genotypes. However, he concentrates on direct measures of genetic and phenotypic variance; if one considers heritabilities, which include both genetic and environmental variance, the data are more ambiguous. The agricultural literature provides several examples of decreased heritabilities under suboptimal conditions (e.g. 147, 177). Tantawy et al (229) found that for both *D. melanogaster* and *D. simulans,* the phenotypic coefficient of variation for wing length was highest at extreme temperatures, whereas heritability was highest at an intermediate temperature.

Shaw (213) found that "broad sense" genetic correlations between leaf length and number in *Salvia lyrata* were lower at denser plant spacings. In *D melanogaster,* Service & Rose (211) observed increased genetic correlations between fitness components in a "novel" environment. A simple theoretical analysis of this phenomenon is presented by Bell & Koufopanou (18), who summarize data showing lower phenotypic correlations between fitness components under natural versus laboratory conditions. Environmental extremes, which are likely to engender strong natural selection, are thus also likely to alter the genetic and phenotypic parameters of quantitative traits.

Almost all heritability estimates are obtained under relatively controlled conditions. Fewer than 20 of 270 heritability studies cited by Roff & Mousseau (173, 203) involve nondomestic environments. Some studies compare wild-caught parents with their lab-reared offspring. Prout (190) found that despite significant heritability for wing length in laboratory populations of *D.*

melanogaster, the regression of lab-reared offspring on their wild-caught fathers was not significant. He conjectured that laboratory studies may inflate h^2 estimates relative to the values in nature, where environmental variance might be higher. In contrast, Coyne & Beecham (57) used Prout's design (190) and found significant heritabilities for both bristle number and wing length in another natural population of *D. melanogaster* (cf. 108, 240). Lower bounds for h^2 in nature can usually be obtained by combining laboratory estimates of h^2 with regressions of lab-reared offspring on their wild-caught parents (57; 196a).

In wild-bird populations, in which relatives can be identified, many parent-offspring studies have demonstrated significant heritability in the field (e.g. 23, 241; for comparable work with plants see 169). Cross-fostering (e.g. 22, 67) shows that common environments may not significantly alter parent-offspring regressions (although they may contribute to resemblances between sibs; 1). These results reinforce the view that most morphological traits exhibit substantial heritable variation in nature, and hence will respond to directional selection. It remains to be determined whether this extends to physiological and behavioral traits (see 25a).

Relation between Heritability and Effect on Fitness

Characters most strongly associated with net fitness tend to exhibit lower heritabilities. In a natural population of collared flycatchers, there is a negative correlation between the estimated heritability of a trait and the square of its correlation with lifetime reproductive success (101). In a review of 1500 heritability estimates, Roff & Mousseau (173, 203) found that "life history" traits (i.e. traits such as fecundity, viability, and development rate that are ineluctably tied to net fitness) have lower heritabilities than morphological and physiological traits (cf. 210). From 130 Drosophila studies, they calculated median heritabilities of about 0.10 for the former and 0.35 for the latter (203). Although Fisher's "Fundamental Theorem" is often invoked to explain this pattern, the connection is tenuous, because the "Theorem" only suggests that fitness itself should have little additive variance at equilibrium (74, 80). The observed pattern may, in fact, result because traits more closely related to fitness experience greater environmental variance rather than having lower additive variance.

Genetic and Environmental Correlations

Under the usual Gaussian approximation, multivariate selection response is governed by additive genetic covariances and selection gradients. However, genetic covariances are difficult to estimate, so one might hope that they are revealed by phenotypic covariances. This will only be so if genetic and environmental correlations are similar and all the traits have similar heritabilities. Correlations may be similar if genetic and nongenetic perturbations have similar effects on development (105). Similarities have been found in some surveys (46, 75), but not others (18, 203; J. A. Coyne, personal communication). Nevertheless, agreement between phenotypic and genetic correlations does not help us understand multivariate selection response, because both heritabilities and variances vary considerably among morphological characters (173, 208).

Roff & Mousseau (203) found only positive phenotypic and genetic correlations between morphological traits and between morphological traits and life history traits, presumably because all the traits examined are correlated with body size. Riska (195) and Slatkin (221) use simple developmental models to describe the creation and implications of such correlations. Additional data on patterns of covariation are best motivated as tests of causal hypotheses based on mathematical or physiological models (47). For instance, Hoffmann & Parsons (112) note that resistance to various forms of physiological stress is associated with decreased metabolic rates. The prediction of positive genetic correlations in resistance to various other stresses was supported by correlated responses to selection for desiccation tolerance in *D. melanogaster*.

If there is little additive variance for net fitness, there should be negative genetic correlations between fitness components (249). Direct evidence for this is weak (18, 42, 203). However, indirect estimates, based on correlated selection responses, are less error-prone and do generally indicate negative genetic correlations (18).

Overall, the data on genetic variances and covariances provide little support for the simplifying assumption that they will remain relatively constant in natural populations. These estimates should be viewed as local analyses that apply only to the specific environment and population in which they are made (152). This fact is recited more often in the classroom than it is acknowledged in the literature. Genetic variances and covariances cannot be expected a priori to reveal long-term genetic potential or constraints any more than the linear terms of a Taylor's series can serve as an accurate guide to the global behavior of a nonlinear function.

RESPONSE TO SELECTION

Darwin's explanation of evolution by natural selection (65) did not depend on how traits are passed on. However, to understand the rate and direction of evolution, and the limits to selection, we need a quantititative theory of the response to selection. The fundamental difficulty is that the phenotypic distribution of the few characters we observe depends on genotype frequencies over many loci, a problem analogous to that in thermodynamics, where bulk properties reflect the hidden motions of many molecules. Fortunately, the symmetries of physics cause complex systems to evolve in ways independent of molecular details; in contrast, Mendelian genetics does not allow polygenic systems to be described by a tractable number of variables. The basic conclusion of the theoretical work reviewed below is that while the immediate response to selection is predictable, long-term changes depend on genetic parameters about which we know very little.

Changes over a single generation depend only on the statistical relation between parent and offspring. Suppose that the regression of offspring phenotype (z_O) on the midparent value ($m = (z_{P_1} + z_{P_2})/2$ is linear, with residual fluctuations ζ caused by nongenetic factors and by segregation within families: $z_O = h^2(m - \bar{z}_P)/2 + \bar{z}_P + \zeta$ (overbars denote averages; 218, 222). This phenotypic model leads directly to the standard equation $R = h^2 S$ ($R = \bar{z}_O - \bar{z}_P$ is the "selection response" caused by the "selection differential" $S = \bar{m} - \bar{z}_P$). This model is supported by the fact that regressions between parents and offspring are often linear (84). The response equation also follows from the more stringent condition that the joint distribution of parent and offspring phenotypes and breeding values is Gaussian (32, 181). This can be reconciled with Mendelian inheritance if genes act additively (79); with a very large number of unlinked genes, and Gaussian environmental effects, ζ will follow a normal distribution, independent of parental values, and the regression coefficient h^2 is the fraction of the phenotypic variance attributable to additive genetic effects (32, 77).

There are several alternative descriptions of selection. Analyses of natural selection usually start with fitness as a function of the character, $W(z)$, rather than with the selection differential. Robertson (199) and Price (188) showed that the selection differential is the covariance between relative fitness and the character (63, 175a). This representation can be useful where fitness depends on interactions with relatives (e.g. 96) or in hierarchical selection analyses (e.g. 106). Another representation is based on the "adaptive landscape," a graph of mean fitness against the character mean (217, 251). If selection simply shifts the distribution, then $R = V_A \partial \log \bar{W}/\partial \bar{z}$ (132; V_A is the additive genetic variance). However, with a limited number of loci, with epistasis, or with linkage disequilibria generated by selection, the shape of the distribution will change, and $\log \bar{W}$ will depend on the variance and higher moments as well as just the mean. Terms involving higher-order selection gradients ($\partial \log \bar{W}/\partial V$, etc) must then be added to describe selection response (16, 238a). Similarly, the standard formula $R = h^2 S$ would only extend to nonadditive gene action and linkage disequilibria if the distribution of parents and offspring remains Gaussian (32, 238a). With genotype-environment interactions, frequency-dependence, or maternal effects, the response to selection may actually be in the opposite direction to that expected (45, 89, 91,

102, 125). With epistasis and linkage disequilibrium, recombination can change the mean even without selection (32, 99).

The response equations are easily extended to selection on multiple characters. Selection gradients are especially useful, because they give an intuitive picture of progress towards local peaks on a multidimensional adaptive landscape (134, 143). The rate of change is proportional to the matrix of additive genetic variances and covariances, and so is constrained by this matrix. This idea has been applied to various problems (e.g. 34, 134, 242, 243, 257), with a simple conclusion: though the rate of change is affected by genetic covariances, the population will generally reach some local optimum (but see 189a). The final outcome is only affected if there are alternative adaptive peaks, or if the covariance matrix is singular, implying a lack of heritable variance in one or more directions (43, 168, 176). This simple picture of increasing adaptation can be distorted by frequency-dependent selection (see the discussions of sexual selection in 124, 141).

Seen thus, patterns of genetic variation are unlikely to impose strong constraints on evolution of mean phenotypes. However, the "adaptive landscape" can mislead (192). When very many characters are involved, directions of possible evolutionary change may be limited, even though one observes heritable variation for most arbitrarily chosen character combinations. The total number of genetic degrees of freedom cannot be greater than the number of segregating alleles, and would be smaller (roughly the number of loci) if each locus acts in a particular direction. This number must be smaller than the (effectively infinite) number of characters (126; see 192). Indeed some studies have found genetic correlations near 1 for morphological traits (e.g. 212).

The evolution of the mean can be extrapolated over many generations if the parent-offspring regression stays constant, and if the offspring phenotype depends on only its parents, and not on more distant ancestors (cf. 125, 194). This is guaranteed only with an infinite number of additive loci, exponential directional selection (i.e. $W_{(z)} = e^{sz}$, which changes the mean without changing the additive variance), no maternal effects or environmental correlations, and no interactions between gene effects and a changing environment. Because both genetic and phenotypic variances and covariances differ between relaxed taxa and depend on the environment (see above), and because predictions for laboratory populations are inaccurate over more than a few generations (49, 83), we can have little faith in quantitative extrapolations over millennia (237). Departures from the idealized model may have little effect in the short term, but will dominate over long periods. A further difficulty is that the realized selection differential depends on pleiotropic effects, as well as selection on the observed characters; thus, it can change unpredictably as the system evolves.

How does selection alter the shape of the distribution of breeding values? With asexual reproduction or selfing, the distribution would directly reflect relative fitnesses ($\Psi^*(z) \approx (W/\bar{W})\Psi(z)$). But sexual reproduction and recombination strongly constrain the distribution: with many loci and additive effects, it will approach a Gaussian without selection. If a strongly skewed or bimodal distribution is favored, the population can only produce it if selection is strong enough to overcome recombination and segregation, or if one or a few genes have major effects (as may happen with disruptive selection on laboratory (230) or natural (215, 223) populations). The genetic variance and the shape of the distribution can change in two ways. First, selection may build up associations ("linkage disequilibria") between loci; these will be broken down by recombination and so will only change the variance temporarily. Second, allele frequencies can change. However, for polygenic traits, these allele frequency changes will be slow. Suppose that additive variance, V_A, is due to n loci, each of effect α; then α scales as $n^{-1/2}$. Selection on the mean ($\partial\log\bar{W}/\partial\bar{z}$) produces a selection coefficient of $s \approx \alpha$ on each locus, and the mean responds at a rate $\approx n\alpha s \approx V_A$. Selection on the variance ($\partial\log\bar{W}/\partial V$) only produces a selection coefficient $s \approx \alpha^2$, and a response $\approx n\alpha^2 s \approx V_A^2/n$ (16, 32). Similarly, selection on the k'th moment of the distribution only produces a response due to changes at individual loci at a rate $\approx V_A^k/n^{k-1}$. Permanent changes in the variance and higher moments of the distribution will be slow, will depend on the number of loci and their effects, and will be significantly influenced by weak pleiotropic selection on individual alleles.

A major simplification occurs if we consider a very large number of loci, each with small additive effects; there may be dominance but not epistasis. This "infinitesimal model" (28, 32, 226) is equivalent (if loci are unlinked) to "phenotypic" models with a fixed within-family variance ($V_E + V_{A,LE}/2$), where $V_{A,LE}$ denotes the additive variance that the current allele frequencies would produce in the absence of linkage disequilibrium (77, 116). Changes in the expressed additive variance, V_A, are solely due to linkage disequilibria, and so V_A will converge rapidly to $V_{A,LE}$ after selection is relaxed. Selection will generally distort the distribution away from a Gaussian (32, 238a). These transient effects become much more important with selfing, because disequilibria then take longer to decay, and because there can be persistent deviations from Hardy-Weinberg proportions (250). With epistasis, they affect the mean as well as the variance: the selection response will consist of a permanent component proportional to the additive genetic variance, and transient components proportional to additive $\times$ additive and higher-order additive interactions (32, 99).

The infinitesimal model is independent of the distribution of effects of individual loci, and is therefore useful for short-term predictions (e.g. 10, 224). In contrast, evolutionists have tended to apply Lande's Gaussian allelic model (131), which is motivated by questionable assumptions about mutation-selection balance (238) and is much more restrictive than Fisher's assumption that *breeding values* are approximately Gaussian. Despite its shortcomings, the infinitesimal model provides an alternative framework for analyzing problems ranging from the coevolution of female preferences with male traits (cf. 141) to the increase in quantitative variance in clines caused by linkage disequilibria (cf. 219; J. M. Szymura & N. H. Barton, unpublished data). However, because the variance contributed by individual loci, $V_{A,LE}$, is constant in the "infinitesimal" limit, this model cannot explain the maintenance of variation.

Unfortunately, the long-term effects of selection on genetic variance depend on the distribution of effects of individual loci. For example, stabilizing selection of the form $W_{(z)} = \exp[-z^2/(2V_S)]$ reduces variance at a rate $\Sigma(C_4 - C_2^2)/V_S$ (C_2 and C_4 are the variance and the fourth central moment of the distribution of allelic effects at individual loci (16); linkage disequilibria and interactions between alleles and loci are ignored). If variation is contributed by rare alleles, the distribution will be highly leptokurtic ($C_4 >> C_2^2$), and stabilizing selection will eliminate variance much more rapidly than if the distribution of allelic effects is Gaussian ($C_4 = 3C_2^2$). A set of uncorrelated characters will evolve independently only if the underlying distribution of allelic effects is Gaussian: with rare alleles and pleiotropic effects, stabilizing selection on one character will reduce the variance of other characters (238).

WHAT MAINTAINS HERITABLE VARIATION FOR MORPHOLOGICAL TRAITS?

The central paradox is that we see abundant polygenic variation, together with stabilizing selection that is expected to eliminate that variation. Genetic variation is manifest in correlations between relatives and in sustained responses to directional selection. Evidence for stabilizing selection comes from the reduced fitness of extreme phenotypes (see below), and the constancy of form over time (166).

Following Robertson (200), we contrast two kinds of explanation for this apparent contradiction. Variation might be maintained by forces acting directly on the character of interest. This has received most attention, because it can explain variation in terms of measurable parameters. However, an organism's development and behavior can be described by an indefinitely large number of continuous characters; these depend in an obscure way on variation at a much smaller number of genes, each of which may have pleiotropic effects on many characters. Variation in any one trait might therefore simply be a side-effect of polymorphisms maintained by forces independent of the observed character.

Direct Explanations

MUTATION-SELECTION BALANCE   The idea that mutation may suffice to balance stabilizing selection was popularized by Lande (131). It was supported by the realization that mutation generates enough quantitative variability to account for sustained responses to artificial selection (82, 109). The models have been extensively reviewed (33, 236, 238), so we only outline the basic ideas.

There are two key difficulties in applying the theory: first, the rate at which stabilizing selection eliminates variance depends on the distribution of allelic effects at individual loci; and second, it is difficult to determine the relevant intensity of selection. Single-character analyses implicitly assume that the relevant selection is that observed on the character considered; this may not be valid (235). The genetics of variation is often described by "continuum of alleles" models, in which mutations at each locus have a continuous range of effects, drawn from a normal distribution with variance α^2 (61). If the effects are much smaller than the standing variance at a single locus ($\alpha^2 << C_2$), the distribution of allelic effects will be approximately Gaussian ($C_4 = 3C_2^2$; 121, 234). This is the basis of Lande's Gaussian allelic models (131, 133, 136), which assume that the joint distribution of effects of all loci is multivariate Gaussian. Selection then erodes variation at a rate $(C_4 - C_2^2)/(2V_S) = C_2^2/V_S$ per locus; this is balanced by mutational input $V_m/2n$ per locus (with $V_m = 2n\mu\alpha^2$), giving $V_g = \sqrt{2nV_mV_S}$.

If new mutations have effects much larger than the standing genetic variance, the distribution of allelic effects will become highly leptokurtic ($C_4 \approx \alpha^2C_2 >> C_2^2$): most variation will be contributed by rare alleles. The equilibrium genetic variance is then $V_g = 4n\mu V_S$ (16). This is proportional to the net mutation rate ($n\mu$) and is independent of the effects of individual mutants (α). This is because, if mutations have large effects, each must be eliminated by the death of a diploid individual: the loss in fitness is therefore twice the total mutation rate $[V_g/(2V_S) = 2n\mu]$ (61). For the Gaussian allelic approximation to apply, alleles must have such small effects that most are transformed by new mutations before they cause a selective death.

The "rare alleles" result can also be derived from models with two alleles at each locus (29, 148, 253). A complication is that many stable equilibria are possible, involving different combinations of + and − alleles. For some, the mean is very close to the optimum and the variance is close to the rare-alleles prediction; for others, the mean deviates from the optimum, giving a much larger variance, close to the Gaussian allelic prediction (12). However, this may be an artifact of a highly symmetric model: drift, linkage disequilibrium, and variation in allelic effects across loci all tend to bring the variance down to the rare-alleles value (13; N. H. Barton, unpublished data). The distinction between the Gaussian and rare-alleles predictions lies in the shape of the distribution of allelic effects and not in the number of alleles: using five alleles, Slatkin (221a) developed an approximation that bridges both extremes. Several analyses show that the effect of linkage disequilibrium is generally weak under both directional and stabilizing selection (28, 32, 35, 36, 104, 119, 201, 238a).

Sampling drift reduces additive variance by a factor $[1 - 1/(2N_e)]$ per generation (N_e denotes the effective population size); with no selection, this balances mutation to give an equilibrium variance $V_g = 2NV_m$ (50, 161). Because $V_m/V_g \approx 10^{-3}$ (131, 158), drift will only deplete heritabilities significantly in populations of less than $\approx 1,000$ (135). Drift can also establish different combinations of alleles, even if stabilizing selection maintains the same overall phenotype: this could lead to reproductive isolation. If there is a continuum of allelic effects, the contribution of each locus can drift freely, subject only to the constraint on the overall mean (131). One measure of the degree of isolation is the increase in genetic variance in a population formed by hybridization between two diverging lines. It will rise indefinitely, reaching $(1 + t/4N_e)V_g$ after t generations of divergence. (Linkage disequilibria in the hybrid population are transient, and so are ignored.) With a finite number of rare alleles, divergence is slower, because the population must jump between distinct stable equilibria. If selection on each locus is much stronger than drift ($\alpha^2/V_S > 10/N_e$, say), shifts are essentially impossible, while if selection is weaker than drift ($\alpha^2/V_S < 1N_e$), substitutions will occur at roughly the neutral rate, and divergence will occur over a time scale of $\approx 1/\mu$ generations (13, 81).

Because we know so little about the alleles responsible for quantitative variation, we have no direct evidence of the distribution of allelic effects. However, Turelli (234) argues that if variation is indeed maintained by mutation-selection balance, this distribution is likely to be dominated by rare alleles. This is because estimates of $V_m \approx 10^{-3}V_g$ can be reconciled with the assumption of the Gaussian allelic analysis that mutant effects are small (i.e. $\alpha^2 < C_2$) only if per-locus mutation rates are extraordinarily high ($\mu > V_m/V_g \approx 10^{-3}$). So, if variation is maintained by a mutation-selection balance and the relevant per-locus mutation rates are no higher than 10^{-4}, the variation must be based on rare alleles with appreciable effects ($\alpha >> \sqrt{V_g/n}$), and the mutation rate per character must be high ($n\mu \approx 10^{-2}$). In general, mutation can account for high heritabilities only if most genes affect most characters (so that $n >> 100$), or if loci affecting quantitative traits are highly mutable (see above; 234). This is possible: mutation is a plausible explanation of quantitative genetic variation if one ignores the consequences of pleiotropy. When it is taken into account, the distinction between Gaussian and "rare alleles" distributions of allelic effects becomes critical (see 238 and below).

OVERDOMINANCE INDUCED BY GENOTYPE × ENVIRONMENT INTERAC-
TIONS   Can direct selection on quantitative traits maintain polymorphisms at
the underlying loci? Under the additive model, with constant allelic effects
and a constant environment, strong stabilizing selection maintains at most one
polymorphism: either the single genotype nearest the optimum is fixed, or the
pair of alleles giving genotypes closest to the optimum are maintained by
heterozygote advantage (252, cf. 175b). Rose's model of "antagonistic pleiotropy" between fitness components (204) maintains polymorphism by inducing overdominance; however, it requires that favorable effects are always
dominant (see above). Much more variation may be maintained with epistasis
(89a). Fluctuations in phenotypic fitnesses will not easily maintain variation.
In the Gaussian allelic model, changes in the optimum do not affect the
variance at all (133), while with rare alleles, there is only a slight increase
above the constant-optimum mutation-selection equilibrium (238). With a
continuum of alleles, as in Via & Lande's Gaussian allelic model (242),
variation in the effects of alleles across environments cannot maintain variation, because a single homozygous genotype can be fittest in all environments
(88).

In contrast, temporal and spatial fluctuations in the effects of discrete
alleles can maintain variation (88): if alleles have different (but additive)
effects in different environments, then heterozygotes will tend to have lower
phenotypic variance than homozygotes, and higher mean fitness under
stabilizing selection. This is a special case of Gillespie's SAS-CFF model
(86), in which fitness is a concave function of some randomly fluctuating
additive variable. If the polymorphism is to be stable, temporal fluctuations
must be rapid ($\tau \ll V_S/\alpha^2$); the scaled variance of allelic effects must be
large relative to the number of alleles [$(\text{var}(\alpha)/\bar{\alpha}^2) > k$]; and the alleles must
have similar means and variances of effects (N. H. Barton, unpublished data).
However, because selection will sieve out those alleles satisfying these
constraints, substantial variation can be maintained.

Because fitnesses and allelic effects in any one environment will not be
representative of the average across microenvironments, this hypothesis
seems especially hard to test, and it seems more plausible as an explanation
for the maintenance of variation in nature than in long-established laboratory
populations that experience relatively constant environments. Mukai (174)
suggests that genotype-environment interactions may explain the excess additive variance in viability seen in some *D. melanogaster* populations. Fluctuating allelic effects may also contribute to preserving more additive variance in
populations subject to fluctuating versus constant laboratory environments
(e.g. 162).

FREQUENCY-DEPENDENT SELECTION ON THE CHARACTER   If different
phenotypes exploit separately regulated resources, variation can be main-

tained by frequency-dependent selection (31, 32, 48, 205, 220). This mechanism is suggested by, for example, the different seeds used best by birds with different beak shapes (97, 98, 223). However, it presents two difficulties. First, if it is to explain abundant variation in many traits, there must be many separately regulated resources. Second, frequency dependence is not obviously consistent with the general observation of stabilizing selection. If populations could match their phenotypic distribution to the availability of resources, fitnesses would be uniform at equilibrium. In fact, with many freely recombining and additive loci, the distribution is constrained to be roughly normal. Thus, only the mean and variance can adjust to the fitness surface (220); and there will be variance in fitness at equilibrium. Nevertheless, it is not clear that extreme phenotypes will necessarily be less fit.

SPATIAL VARIATION   There is surprisingly little theory on spatial variation in continuous traits (32, 76, 138, 219). Slatkin (219) showed that the population mean can track variations in the optimum, provided that these occur over a distance larger than $\approx \sqrt{V_S/V_g}$ times the dispersal distance (defined as the standard deviation of the distance between parents and offspring). As with most results concerning the mean, this does not depend on genetic details. The effect of spatial heterogeneity on the genetic variance is more complicated. The mixing of different populations generates linkage disequilibria, and also increases heterozygosity at individual loci: both increase the variance. Suppose there is stabilizing selection, with an optimum that changes in an infinite linear gradient by β per dispersal distance. Under the Gaussian allelic model, mixing introduces variance β^2 per generation at each locus, in a manner similar to the variance V_m/n introduced by mutation. This maintains genetic variance $2n\beta\sqrt{V_S}$ in the face of stabilizing selection V_S. With many loci, large amounts of variation can be maintained (see eq. 33 in 219). However, gene flow is much less effective if loci are close to fixation. Then, the mean will track a changing optimum through a succession of clines at individual loci. With two alleles of effect $\pm\alpha$ at each locus, gene flow maintains genetic variance of $\beta\sqrt{V_S/2}$ (derived by adding a diffusion term to Latter's model (148)). With $V_S = 20V_e$, a heritability of 50% could only be maintained by a steep cline, of one phenotypic standard deviation in three dispersal distances. So, if alleles are rare, spatial variation is not a plausible source of heritable variance.

Pleiotropic Models

Instead of considering only the trait(s) of interest, an alternative approach is suggested by the view that most genes affect many traits, and that each trait is affected by many genes. Ubiquitous pleiotropy does not *necessarily* mean that selection on one trait will affect variation of others: in Lande's model (136), enough alleles segregate at each locus to give a multivariate Gaussian distribution of effects, allowing independent responses to selection on indefinitely many characters. However, the limited number of alleles and the limited physiological effects of each locus constrain the response. Even with only two characters about 100 alleles are needed to approach the Gaussian approximation (235; see also 244). There can be constraints even with a continuous and Gaussian distribution of allelic effects, if the effects at a locus are limited to a smaller number of degrees of freedom than the number of characters (244). To illustrate the problem, consider a rare allele, with effect α_i on the i'th character; this character is subject to independent stabilizing selection of intensity V_{Si}. The allele reduces fitness by $\Sigma_i[\alpha_i^2/(2V_{Si})]$; mutation will raise it to frequency $2\mu/\Sigma_i(\alpha_i^2/V_{Si})$, and it will contribute variance $4\mu\alpha_j^2/\Sigma_i(\alpha_i^2/V_{Si})$ to character j. Thus, stabilizing selection on character j does not act only on this character, but instead reduces the variance of all pleiotropically related characters, even without genetic correlation (cf. 238). Does this rather counter-intuitive prediction accord with the facts? Strong stabilizing selection in the laboratory can reduce both the genetic and environmental variance (163, 230); however, these studies are not decisive, because they involve strong selection on single traits. It would be interesting to know whether strong stabilizing selection on one character reduces the genetic variance of others, and whether traits under stronger stabilizing selection show lower genetic variance in natural populations.

The models discussed above describe pleiotropy in terms of stabilizing selection on many characters. A simpler approach is to consider the net effects of alleles on fitness and on the character of interest (110). Quantitative variation is now maintained as a side-effect of polymorphisms maintained for other reasons. This has the advantage that it can describe variation in fitness components that are under directional selection, as well as morphological traits under stabilizing selection. First, consider polymorphisms maintained by deleterious mutation. Mutations that reduce heterozygote fitness by s, and have additive effect α on a trait, will contribute a variance (summed over n loci) of $2n\mu\alpha^2/s = V_m/s$. Because $V_m \approx 10^{-3}V_e$, any such mutations must be very mildly deleterious ($s \approx 10^{-3}$) if appreciable heritability is to be maintained. If selection is mediated by stabilizing selection of $V_S \approx 20V_e$ on many traits (cf. 235), then $s = \Sigma\alpha^2/2V_S \approx 10^{-3}$, and so the total effect of alleles on all traits can be no more than $\Sigma\alpha^2 \approx 2V_S10^{-3} \approx 0.04V_e$. Thus, if high heritabilities are to be explained by any kind of mutation-selection balance, they must be based on alleles with only slightly deleterious effects, and with very small net effects on all traits under selection. This is hard to reconcile with data on Drosophila viability: estimates of selection against spontaneous mutations are much higher ($\approx 2\%$; 64).

Balancing selection can more easily maintain quantitative genetic variation. Overdominance can account for high heritabilities (30, 87, 197, 252); however, highly heritable variation in asexuals and predominantly selfing plants seems to rule out any general role for overdominance, whether direct or indirect (e.g. 114, 157).

Extreme phenotypes are often associated with reduced fitness (72). This could be because of direct selection on the trait; selection on correlated characters (143, 182); or pleiotropic fitness effects of the underlying genes that are not associated with morphology (e.g. recessive lethals; 200). If variation is maintained as a side-effect of deleterious mutations or overdominance, there will be apparent stabilizing selection with $V_S \approx 2/s$, where s is the selection on each polymorphic locus (197; N. H. Barton, unpublished data). Because we have seen that individual polymorphisms must be weakly selected, this is unlikely to be significant: observations of strong stabilizing selection must be explained either by direct effects of the character, or by correlations with selected traits.

Though it is hard to be definite without more direct evidence, our analysis suggests that quantitative genetic variation for most morphological traits may be maintained by the pleiotropic effects of many diverse polymorphisms. Pleiotropy is more plausible if many loci contribute to variation, present observations on pairs of selected lines need to be extended to natural populations. An experimental test for pleiotropy might be to look for correlated responses to selection on the variance: stabilizing selection on one character should reduce the variance of other characters that share the same genes, even if they are uncorrelated.

Most recent attention has centered on whether quantitative genetic variation is maintained by mutation-selection balance. We need better estimates of per-locus and per-character mutation rates ($\Sigma \mu$), determination of the nature of the relevant alleles, and a survey of the joint effects of spontaneous mutations on fitness and multivariate morphology. This would be laborious, but feasible using balancer chromosomes in Drosophila.

LIMITS TO SELECTION, NEW MUTATIONS, AND ALLELES OF LARGE EFFECT

Most discussions of selection limits have aimed at maximizing the response to artificial selection. However, a principal motivation for studying quantitative genetic variation in nature is to understand how populations adapt in the long term, and in particular, how genetic constraints affect morphological change. Response to selection may cease either because additive genetic variation has been exhausted, or because of counterbalancing selection (see 189a). Robertson (198) showed that in small populations, variation will mainly be lost by drift, and the final response will be at most $2N_e$ times the response in the first generation. Because a plateau is often reached before this predicted limit, even with lines of a few tens of individuals, and because the mean often returns towards its original value when selection is relaxed, the consensus has been that selection limits usually result from opposing natural selection (75, 83, 200, 258). At first sight, this conflicts with the observation that a larger final response is reached with a larger population (see 245). The explanation may be that natural selection usually acts not against the character itself, but rather on the pleiotropic effects of the selected alleles. At equilibrium, variation can be maintained in a balance between artificial selection on the character and natural selection against pleiotropic effects. This is supported by the fact that the response to strong artificial selection is often caused by recessive lethals (e.g. 83, 113, 256). In a large population, there is a greater chance that alleles will be found that increase the character without producing large deleterious side-effects.

In many cases, response to artificial selection has continued over many generations, even when the lines were started from a limited base (e.g. 71, 82, 245, 255). Such continued responses are likely to be due to new muta-tions, rather than to preexisting variation. This is consistent with the observed production of polygenic variation (V_m; 109) and, in some cases, has been confirmed by showing that alleles responsible for selection response could not be found in the base population (82, 113, 256). The same conclusion may apply to natural populations. Although these will usually be larger than artificially selected populations, larger population size increases the chance of accumulating new mutations at least as much as it increases the chance of picking up preexisting variants. The uncertainty in the comparison is that the response in nature, where selection may generally be weaker and less fo-cussed on single traits than in typical artificial selection experiments, is likely to be based on alleles with smaller deleterious side-effects than in the labora-tory. Because the eventual response to directional selection depends on the selected alleles' effects on net fitness, not just on the character considered, and because these alleles may arise de novo, observed patterns of additive variances and covariances may tell us little about the constraints on evolution-ary change. This view is reinforced by experiments selecting for faster developmental rate in Drosophila, in which little progress is made despite evidence for genetic variation (see 48b).

The relative importance to phenotypic evolution of quantitative versus qualitative variants has been debated since the 1890s (191), and as recently as 1984, Gottlieb (93) proposed that qualitative differences between plant spe-cies may be caused by few loci (but see 58, 94). The usual argument against the importance of "macromutations" in evolution is that they are likely to have harmful pleiotropic effects that prevent fixation in natural populations, even though they may be overcome by intense artificial selection (43, 139). In spite of the strength of this argument, ecologically significant morphological polymorphisms (such as wing dimorphisms in insects) may indeed be governed by single loci (202). Natural selection is, at least occasionally, extremely strong (24, 72). This could, in principle, lead to the fixation of alleles with deleterious pleiotropic effects, and some experiments suggest that further evolution could ameliorate these deleterious effects (e.g. 48a, 150). Genetic analysis of population and species differences are essential to resolve this uncertainty (e.g. 59).

ESTIMATION OF FITNESSES AND APPLICATION OF MODELS TO MICROEVOLUTION AND MACROEVOLUTION

Lande & Arnold (143) extended methods developed by Pearson (182) to analyze multivariate selection. Their analyses attempt to disentangle changes in the means and variances that are caused by directional and stabilizing selection acting directly on each character, from changes caused by selection on correlated traits included in the analyses. Their technique involves using linear and quadratic regression to fit the observed estimates of fitness components to the character values. It has been generalized to incorporate multiple components of fitness, including sexual selection (7), and it has been widely applied (142). These regression analyses are subject to two sorts of errors (143): statistical problems associated with the application of regression, and the confounding influences of "hidden characters" not included in the multivariate analysis.

Mitchell-Olds & Shaw (171) review both classes of problems, emphasizing the fact that analyzing more characters, in an attempt to uncover the "targets of selection," increases the likelihood that subsets will be highly correlated. This "multicollinearity" reduces the reliability of the parameter estimates, and so the effort expended on measuring more and more characters may be better spent on manipulative experiments. To circumvent the normality assumptions required by most significance tests, they recommend resampling techniques. Finally, they point out that the definitions of stabilizing and disruptive selection proposed in Lande & Arnold (143) are inconsistent with standard usage, and suggest alternative statistical criteria.

Schluter (207) clarifies the assumptions implicit in Lande & Arnold's definition of stabilizing selection and presents a univariate, nonparametric analysis of selection. A key motivation for quadratic selection analysis is that under the Gaussian infinitesimal model only the linear and quadratic terms contribute to evolutionary changes in the mean and additive variance. However, as noted above, departures from this model, especially with respect to the dynamics of the additive variance, are expected. Moreover, this motivation is relevant only if selection response is predicted and tested; but no analysis based on these multivariate methods has yet attempted this. Evolutionary predictions are feasible only if *net fitness* of phenotypes, not just its components, is estimated; but most studies have concentrated on individual components of fitness or individual episodes of selection. For such descriptive purposes, it may be more informative to describe fitness surfaces nonparametrically.

"Hidden characters" that contribute to fitness differences pose a second problem. Lande & Arnold (143) stressed that their multivariate analysis can only determine the relative importance of selection *on characters included in the study*. An incomplete disentanglement is better than none at all, but hidden selected characters can cause serious complications (see 200). One approach to this problem is suggested by Crespi & Bookstein (60), who apply path analysis under the assumption that measured characters are only indicators of the underlying variables "size" and "shape" that are the true targets of selection. However, the rationale for this a priori assumption is unclear, and it is not obvious that their statistical procedure is more likely to uncover the causes of fitness differences.

These causes can be investigated in several ways. In some cases, experimental manipulations or functional arguments can show that a trait directly affects fitness: for example, widowbirds with artificially lengthened tails get more mates (3), and bill shapes in Galapagos and African finches correspond to their use of different seeds (97, 98, 223). Robertson (200) suggested an ingenious genetic manipulation, in which different homozygous backgrounds are used to shift a character mean; stabilizing selection on the trait is indicated if selection on genetically variable chromosomes then shifts the character back to its original value. Linney et al (154) examined the larval viability of flies with different adult bristle counts in two populations: a large base population and a collection of four nearly homozygous lines with the same phenotypic distribution. They found that in both, extreme phenotypes had reduced viability; this rules out overdominance as a cause of the observed stabilizing selection. Because larvae do not have bristles, selection must be due to pleiotropic effects of alleles determining bristle number.

Empirical studies of selection in nature lead to two major conclusions. First, natural selection can be extremely intense, approaching the intensity of selection applied in artificial selection studies (e.g. 24, 72). Hence, Darwin's pessimism over the chance of measuring the selection pressures that produce species differences was probably ill-founded (65). Second, selection pressures often vary dramatically across populations, years, and parts of the life cycle (e.g. 85, 97, 115, 189).

Despite the increasing number of field studies of quantitative characters, some of which include estimates of both h^2 and S, the prediction $R = h^2S$ and its multivariate extensions have rarely been tested in nature (Boag (22) is an exception). The lack of fit found in many lab experiments suggests that significant departures may be expected in nature, where allelic effects may fluctuate with environmental conditions (88), and where selection may occur during inaccessible portions of the life cycle (see 55).

Interestingly, an apparently successful, and perhaps unique, attempt to predict both selection differentials and the corresponding dynamics of the mean in a natural population involves a fundamental misapplication of quantitative genetics theory (172). Morris (172) regressed offspring phenotypes directly on parental phenotypes, without considering deviations from population means. His results support Pearson's view that regression models can be useful predictive tools, at least in the short term, without understanding mechanisms. Despite its flaws, this analysis of diapause microevolution in a natural population should challenge evolutionary quantitative geneticists to attempt predictive rather than merely descriptive analyses.

Qualitatively different rates of change are seen in artificial selection experiments, in the responses to natural selection in novel environments and in the fossil record (e.g. 2, 90, 103). The apparent slowness of macroevolutionary change may be because relatively rapid fluctuations in the direction of selection average out (19, 40, 85, 90, 140). If only end points are compared, evolutionary rates are so slow that extraordinarily weak selection suffices to explain them (41, 132), and genetic drift can rarely be ruled out (103, 132, 160, 239). The macroevolutionary pattern of apparent stasis followed by rapid change can be readily explained using the paradigms of the neo-Darwinian synthesis (e.g. 43, 123, 140, 184).

These conclusions, based primarily on univariate analyses, are robust to the departures from standard selection theory we have discussed. For instance, Lande's analysis of the minimum selection necessary to account for observed evolutionary rates (132) depends only on average parameter values and does not require constant heritabilities or the exact validity of $R = h^2 S$. In contrast, because the rates are so slow, predictions about the *direction* of the very weak multivariate selection necessary to produce an observed change in a suite of characters will not be robust to the assumption that the genetic covariances are constant (237). Genetic covariances are relevant to long-term selection response if the adaptive landscape presents several peaks, if there is a complete lack of genetic variation in some direction, or if one of the characters is neutral. This last may explain some examples of "preadaptation," i.e. production of character states that are not currently selected. Despite the descriptive value of quantitative genetic models, their use for evolutionary predictions remains uncertain.

CONCLUSIONS

What has been accomplished by the application of quantitative genetics theory to evolution? Given that there is additive variance for most traits, we know they can respond to selection. Even very weak selection can produce rates of phenotypic evolution that are extremely fast on a geological time scale. Thus, there is no basis for claims that developmental constraints (in the accepted sense of genetic correlations; 168, 176) cause stasis and no support for the view that population bottlenecks are necessary for rapid morphological evolution (14, 43, 44, 140). Quantitative genetics theory provides a language in which many important evolutionary questions can be described and possible explanations explored. However, the standard selection equations have been taken too literally; and genetic assumptions with little empirical support have gained undue credibility.

The alternative explanations for the maintenance of variation and the causes of fitness differences can be distinguished, but only with additional information on mutation rates; numbers, effects and frequencies of alleles; and correlations of genetic variances and covariances with population structure and functional constraint. Because of the complexity of the dynamics of additive genetic variances and covariances, it may be impossible to use selection equations to uncover the forces that produce differences between taxa. Laboratory experiments will be needed to complement additional descriptive and manipulative field studies, and so help us to move beyond descriptions of variance components and selection.

ACKNOWLEDGMENTS

We thank G. E. Bradford, B. Charlesworth, J. A. Coyne, D. Currie, H. Dingle, J. H. Gillespie, A. Gimelfarb, W. G. Hill, A. A. Hoffmann, K. S. Jackson, J. S. Jones, A. Liebowiz, T. Mitchell-Olds, T. A. Mousseau, J. Peck, T. Prout, B. Riska, D. A. Roff, H. B. Shaffer, and M. Slatkin for helpful discussions and comments on earlier drafts. Our research is supported in part by grants from the Science and Engineering Research Council, the National Science Foundation (BSR-8866548), and the Institute of Theoretical Dynamics at University of California, Davis.

Literature Cited

1. Alatalo, R. V., Lundberg, A. 1986. Heritability and selection on tarsus length in the pied flycatcher *(Ficedula hypoleuca)*. *Evolution* 40:574–83
2. Anderson, W. W. 1973. Genetic divergence in body size among experimental populations of *Drosophila pseudoobscura* kept at different temperatures. *Evolution* 27:278–84
3. Andersson, M. B. 1982. Female choice selects for extreme tail length in a widowbird. *Nature* 199:818–20
4. Aquadro, C. F., Desse, S. F., Bland, M. M., Langley, C. H., Laurie-

Ahlberg, C. C. 1986. Molecular population genetics of the alcohol dehydrogenase gene region of *Drosophila melanogaster*. *Genetics* 114:1165–90
5. Aquadro, C. F., Lado, K. M., Noon, W. A. 1988. The *rosy* region of *Drosophila melanogaster* and *Drosophila simulans*. I. Contrasting levels of naturally occurring DNA restriction map variation and divergence. *Genetics* 119: 875–88
6. Arnold, S. J. 1981. Behavioral variation in natural populations. I. Phenotypic, genetic and environmental correlations

between chemoreceptive responses to prey in the garter snake, *Thamnophis elegans. Evolution* 35:489–509

7. Arnold, S. J., Wade, M. J. 1984. On measurement of natural and sexual selection: theory. *Evolution* 38:709–19

8. Artavanis-Tsakonas, S. 1988. The molecular biology of the *Notch* locus and the fine tuning of differentiation in Drosophila. *Trends Genet.* 4:95–100

9. Ashburner, M., Carson, H. L., Thompson, J. N. Jr., eds. 1983. *The Genetics and Biology of Drosophila, Vol. 3c.* New York: Academic

10. Atkins, K. D., Thompson, R. 1986. Predicted and realized responses to selection for an index of bone length and body weight in Scottish Blackface sheep. *Anim. Prod.* 43:421–35

11. Barker, J. S. F., Thomas, R. H. 1987. A quantitative genetic perspective on adaptive evolution. See Ref. 155, pp. 3–23

12. Barton, N. H. 1986. The maintenance of polygenic variation through a balance between mutation and stabilizing selection. *Genet. Res.* 47:209–16

13. Barton, N. H. 1989. The rate of divergence under stabilising selection, mutation and drift. *Genet. Res.* In press

14. Barton, N. H., Charlesworth, B. 1984. Genetic revolutions, founder effects, and speciation. *Annu. Rev. Ecol. Syst.* 15:133–64

15. Barton, N. H., Hewitt, G. M. 1985. Analysis of hybrid zones. *Annu. Rev. Ecol. Syst.* 16:113–48

16. Barton, N. H., Turelli, M. 1987. Adaptive landscapes, genetic distance and the evolution of quantitative characters. *Genet. Res.* 49:157–73

17. Becker, W. A. 1984. *Manual of Quantitative Genetics.* Pullman, WA: Academic Enterprises. 4th ed.

18. Bell, G., Koufopanou, V. 1986. The cost of reproduction. *Oxford Surv. Evol. Biol.* 3:83–131

19. Bell, M. A. 1985. Patterns of temporal change in single morphological characters of a Miocene stickleback fish. *Paleobiology* 11:258–71

20. Berven, K. A. 1987. The heritable basis of variation in larval developmental patterns within populations of the wood frog *(Rana sylvatica). Evolution* 41:1088–97

21. Billington, H. L., Mortimer, A. M., McNeilly, T. 1988. Divergence and genetic structure of adjacent grass populations. I. Quantitative genetics. *Evolution* 42:1267–77

22. Boag, P. T. 1983. The heritability of external morphology in Darwin's ground finches *(Geospiza)* on Isla Daphne Major, Galapagos. *Evolution* 37:877–94

23. Boag, P. T., Grant, P. R. 1978. Heritability of external morphology in Darwin's finches. *Nature* 274:793–94

24. Boag, P. T., Grant, P. R. 1981. Intense natural selection in a population of Darwin's Finches (Geospizinae) in the Gálapagos. *Science* 214:82–85

25. Bradbury, J. W., Andersson, M. B., eds. 1987. *Sexual Selection: Testing the Alternatives.* New York: Wiley

25a. Bradshaw, A. D. 1984. The importance of evolutionary ideas in ecology—and vice versa. In *Evolutionary Ecology,* ed. B. Shorrocks, pp. 1–25. Oxford: Blackwell

26. Bryant, E. H., McCommas, S. A., Combs, L. M. 1986. The effect of an experimental bottleneck upon quantitative genetic variation in the housefly. *Genetics* 114:1191–1211

27. Bryant, E. H., Meffert, L. M. 1988. Effect of an experimental bottleneck on morphological integration in the housefly. *Evolution* 42:698–707

28. Bulmer, M. G. 1971. The effect of selection on genetic variability. *Am. Nat.* 105:201–11

29. Bulmer, M. G. 1972. The genetic variability of polygenic characters under optimizing selection, mutation, and drift. *Genet. Res.* 19:17–25

30. Bulmer, M. G. 1973. The maintenance of the genetic variability of quantitative characters by heterozygote advantage. *Genet. Res.* 22:9–12

31. Bulmer, M. G. 1974. Density-dependent selection and character displacement. *Am. Nat.* 108:45–58

32. Bulmer, M. G. 1980. *The Mathematical Theory of Quantitative Genetics.* Oxford: Clarendon

33. Bulmer, M. G. 1989. Maintenance of genetic variability by mutation-selection balance: a child's guide through the jungle. *Genome* 31:

34. Bürger, R. 1986. Evolutionary dynamics of functionally constrained phenotypic characters. *IMA J. Math. Appl. Med. Biol.* 3:265–87

35. Bürger, R. 1989. Linkage and the maintenance of heritable variation by mutation-selection balance. *Genetics* 121:175–84

36. Bürger, R., Wagner, G. P., Stettinger, F. 1989. How much heritable variation can be maintained in finite populations by a mutation selection balance? *Evolution.* In press

37. Caspari, E. 1952. Pleiotropic gene action. *Evolution* 6:1–18

38. Castle, W. E. 1921. An improved

method of estimating the number of genetic factors concerned in cases of blending inheritance. *Science* 54:223

39. Caudy, M. 1988. *daughterless*, a Drosophila gene essential for both neurogenesis and sex determination, has sequence similarities to *myc* and the *achaete-scute* complex. *Cell* 55:1061–67

40. Charlesworth, B. 1984. Some quantitative methods for studying evolutionary patterns in single characters. *Paleobiology* 10:308–18

41. Charlesworth, B. 1984. The cost of phenotypic evolution. *Paleobiology* 10:319–27

42. Charlesworth, B. 1987. The heritability of fitness. See Ref. 25, pp. 21–40

43. Charlesworth, B., Lande, R., Slatkin, M. 1982. A neo-Darwinian commentary on macroevolution. *Evolution* 36:474–98

44. Charlesworth, B., Rouhani, S. 1988. The probability of peak shifts in a founder population. II. An additive polygenic trait. *Evolution* 42:1129–45

45. Cheverud, J. M. 1984. Evolution by kin selection: A quantitative genetic model illustrated by maternal performance in mice. *Evolution* 38:766–77

46. Cheverud, J. M. 1988. A comparison of genetic and phenotypic correlations. *Evolution* 42:958–68

47. Clark, A. G. 1987. Senescence and the genetic-correlation hang-up. *Am. Nat.* 129:932–40

48. Clarke, B. C., Shelton , P. R., Mani. G. S. 1988. Frequency dependent selection, metrical characters and molecular evolution. *Philos. Trans. R. Soc. London, Ser. B* 319:634–40

48a. Clarke, G. M., McKenzie, J. A. 1987. Developmental stability of insecticide resistant phenotypes in the blowfly; a result of canalizing natural selection. *Nature* 325:345–46

48b. Clarke, J. M., Maynard Smith, J., Sondhi, K. C. 1961. Asymmetrical response to selection for rate of development in *Drosophila subobscura*. *Genet Res.* 3:70–81

49. Clayton, G. A., Knight, G. R., Morris, J. A., Robertson, A. 1957. An experimental check on quantitative genetic theory. III. Correlated responses. *J. Genet.* 55:171–80

50. Clayton, G. A., Robertson, A. 1955. Mutation and quantitative variation. *Am. Nat.* 89:151–58

51. Cockerham, C. C. 1986. Modifications in estimating the number of genes for a quantitative character. *Genetics* 114:659–64

52. Cohan, F. M., Hoffmann, A. A. 1986. Genetic divergence under uniform selection. II. Different responses to selection for knockdown resistance to ethanol among *Drosophila melanogaster* populations and their replicate lines. *Genetics* 114:145–63

53. Cohan, F. M., Hoffmann, A. A. 1989. Uniform selection as a diversifying force in evolution: Evidence from Drosophila. *Am. Nat.* In press

54. Comstock, R. E., Moll, R. H. 1963. Genotype-environment interactions. In *Statistical Genetics in Plant Breeding*, ed. W. D. Hanson, H. F. Robinson, pp. 164–96. Washington: NAS-NRC Publ. 982

55. Conner, J. 1988. Field measurements of natural and sexual selection in the fungus beetle, *Bolitotherus cornutus*. *Evolution* 42:736–49

56. Coyne, J. A. 1987. Lack of response to selection for directional asymmetry in *Drosophila melanogaster*. *J. Hered.* 78:119

57. Coyne, J. A., Beecham, E. 1987. Heritability of two morphological characters within and among natural populations of *Drosophila melanogaster*. *Genetics* 117:727–37

58. Coyne, J. A., Lande, R. 1985. The genetic basis of species differences in plants. *Am. Nat.* 126:141–45

59. Coyne, J. A., Orr, H. A. 1989. Two rules of speciation. In *Speciation and Its Consequences*, ed. D. Otte, J. Endler, pp. 180–207. Sunderland, MA: Sinauer

60. Crespi, B. J., Bookstein, F. L. 1989. A path-analytic model for the measurement of selection on morphology. *Evolution* 43:18–28

61. Crow, J. F., Kimura, M. 1964. The theory of genetic loads. *Proc. 11th Int. Congr. Genet.* 2:495–505

62. Crow, J. F., Kimura, M. 1970. *An Introduction to Population Genetics Theory*. New York: Harper & Row

63. Crow, J. F., Nagylaki, T. 1976. The rate of change of a character correlated with fitness. *Am. Nat.* 110:207–13

64. Crow, J. F., Simmons, M. J. 1983. The mutation load in *Drosophila*. See Ref. 9, pp. 2–35

65. Darwin, C. 1859. *The Origin of Species*. London, UK: Murray

66. Dempster, E. R., Snyder, L. A. 1950. A correction for linkage in the computation of number of gene differences. *Science* 111:283–85

67. Dhondt, A. A. 1982. Heritability of blue tit tarsus length from normal and cross-fostered broods. *Evolution* 36:418–19

68. Dingle, H., Evans, K. E. 1987. Re-

sponses in flight to selection on wing length in non-migratory milkweed bugs, *Oncopeltus fasciatus. Entomol. Exp. Appl.* 45:289–96

69. Dingle, H., Evans, K. E., Palmer, J. O. 1988. Responses to selection among life history traits in a nonmigratory population of milkweed bugs *(Oncopeltus fasciatus). Evolution* 42:79–92

70. Dobzhansky, Th., Holz, A. M. 1943. A re-examination of the problem of manifold effects of genes in *Drosophila melanogaster. Genetics* 28:295–303

71. Dudley, J. W. 1977. 76 generations of selection for oil and protein percentage in maize. See Ref. 185, pp. 459–73

72. Endler, J. A. 1986. *Natural Selection in the Wild.* Princeton, NJ: Princeton Univ. Press

73. Enfield, F. D. 1977. Selection experiments in tribolium designed to look at gene action issues. See Ref. 185, pp. 177–90

74. Ewens, W. J., Thomson, G. 1977. Properties of equilibria in multi-locus genetic systems. *Genetics* 87:807–19

75. Falconer, D. S. 1981. *Introduction to Quantitative Genetics.* London: Longman. 2nd ed.

76. Felsenstein, J. 1977. Multivariate normal genetic models with a finite number of loci. See Ref. 185, pp. 227–46

77. Felsenstein, J. 1981. Continuous genotype models and assortative mating. *Theor. Popul. Biol.* 19:341–57

78. Felsenstein, J. 1988. Phylogenies and quantitative characters. *Annu. Rev. Ecol. Syst.* 19:445–71

79. Fisher, R. A. 1918. The correlation between relatives under the supposition of Mendelian inheritance. *Trans. R. Soc. Edinburgh* 52:399–433

80. Fisher, R. A. 1958. *The Genetical Theory of Natural Selection.* New York: Dover. 2nd ed.

81. Foley, P. 1987. Molecular clock rates at loci under stabilizing selection. *Proc. Natl. Acad. Sci. USA* 84:7996–8000

82. Frankham, R. 1988. Exchanges in the rDNA multigene family as a source of genetic variation. See Ref. 246, pp. 236–42

83. Frankham, R., Jones, L. P., Barker, J. S. F. 1968. The effects of population size and selection intensity in selection for a quantitative character in Drosophila. III. Analysis of the lines. *Genet. Res.* 12:267–83

84. Galton, F. 1889. *Natural Inheritance.* London: Macmillan

85. Gibbs, H. L., Grant, P. R. 1987. Oscillating selection on Darwin's finches. *Nature* 327:511–12

86. Gillespie, J. H. 1978. A general model to account for enzyme variation in natural populations. V. The SAS-CFF model. *Theor. Popul. Biol.* 14:1–45

87. Gillespie, J. H. 1984. Pleiotropic overdominance and the maintenance of genetic variation in polygenic traits. *Genetics* 107:321–30

88. Gillespie, J. H., Turelli, M. 1989. Genotype-environment interactions and the maintenance of polygenic variation. *Genetics* 137:129–38

89. Gimelfarb, A. 1986. Multiplicative genotype-environment interaction as a cause of reversed response to directional selection. *Genetics* 114:333–43

89a. Gimelfarb, A. 1989. Genotypic variation for a quantitative character maintained under stabilizing selection without mutations: Epistasis. *Genetics,* In press

90. Gingerich, P. D. 1983. Rates of evolution: Effects of time and temporal scaling. *Science* 222:159–61

91. Goodnight, C. J. 1985. The influence of environmental variation on group and individual selection in a cress. *Evolution* 39:545–58

92. Goodnight, C. J. 1988. Epistasis and the effect of founder events on the additive genetic variance. *Evolution* 42:441–54

93. Gottlieb, L. D. 1984. Genetics and morphological evolution in plants. *Am. Nat.* 123:681–709

94. Gottlieb, L. D. 1985. Reply to Coyne and Lande. *Am. Nat.* 126:141–45.

95. Gottlieb, L. D., de Vienne, D. 1988. Assessment of pleiotropic effects of a gene substitution in pea by two-dimensional polyacrilamide gel electrophoresis. *Genetics* 119:705–10

96. Grafen, A. 1985. A geometric view of relatedness. *Oxford Surv. Evol. Biol.* 2:28–89

97. Grant, B. R., Grant, P. R. 1989. Natural selection in a population of Darwin's finches. *Am. Nat.* 133:377–93

98. Grant, P. R. 1986. *Ecology and Evolution of Darwin's Finches.* Princeton, NJ: Princeton Univ. Press

99. Griffing, B. 1960. Theoretical consequences of truncation selection based on individual phenotype. *Aust. J. Biol. Sci.* 13:307–43

100. Gupta, A. P., Lewontin, R. C. 1982. A study of reaction norms in natural populations of *Drosophila pseudoobscura Evolution* 36:934–48

101. Gustafsson, L. 1986. Lifetime reproductive success and heritability: Empirical support for Fisher's fundamental theorem. *Am. Nat.* 128:761–64

102. Haldane, J. B. S. 1931. A mathematical

theory of natural and artificial selection. VII. Selection intensity as a function of mortality rate. *Proc. Cambridge Philos. Soc.* 27:137–42

103. Haldane, J. B. S. 1949. Suggestions as to quantitative measurement of rates of evolution. *Evolution* 3:51–56

104. Hastings, A. 1988. Disequilibrium in two-locus mutation-selection balance models. *Genetics* 118:543–47

105. Hegmann, J. P., DeFries, J. C. 1970. Are genetic correlations and environmental correlations correlated? *Nature* 226:284–86

106. Heisler, I. L., Damuth, J. 1987. A method for analyzing selection in hierarchically structured populations. *Am. Nat.* 130:582–602

107. Henderson, C. R. 1988. Progress in statistical methods applied to quantitative genetics since 1976. See Ref. 246, pp. 85–90

108. Highton, R. 1960. Heritability of geographic variation in trunk segmentation in the red-backed salamander, *Plethodon cinereus. Evolution* 14:351–60

109. Hill, W. G. 1982. Predictions of response to artificial selection from new mutations. *Genet. Res.* 40:255–78

110. Hill, W. G., Keightley, P. D. 1988. Interrelations of mutation, population size, artificial and natural selection. See Ref. 246, pp. 57–70

111. Hoekstra, R. F., Bijlsma, R., Dolman, A. J. 1985. Polymorphism from environmental heterogeneity: models are only robust if the heterozygote is close in fitness to the favored homozygote in each environment. *Genet. Res.* 45:299–314

112. Hoffmann, A. A., Parsons, P. A. 1989. Selection for increased resistance in *Drosophila melanogaster:* Additive genetic control and correlated responses for other stresses. *Genetics.* 122:837–45

113. Hollingdale, B. 1971. Analysis of some genes from abdominal bristle selection lines in *Drosophila melanogaster. Theor. Appl. Genet.* 41:292–301

114. Imam, A. G., Allard, R. W. 1965. Population studies of predominantly self-pollinated species. VI. Genetic variability between and within natural populations of wild oats from differing habitats in California. *Genetics* 51:49–62

115. Kalisz, S. 1986. Variable selection on the timing of germination in *Collinsia verna* (Scrophulariaceae). *Evolution* 40:479–91

116. Karlin, S. 1988. Non-Gaussian models of quantitative traits. See Ref. 246, pp. 123–44

117. Kearsey, M. J., Kojima, K. 1967. The genetic architecture of body weight and egg hatchability in *Drosophila melanogaster. Genetics* 56:23–37

118. Keightley, P. D. 1989. Models of quantitative variation of flux in metabolic pathways. *Genetics* 121:869–76

119. Keightley, P. D., Hill, W. G. 1983. Effects of linkage on response to directional selection from new mutations. *Genet. Res.* 42:193–206

120. Keightley, P. D., Kacser, H. 1987. Dominance, pleiotropy and metabolic structure. *Genetics* 117:331–41

121. Kimura, M. 1965. A stochastic model concerning the maintenance of genetic variability in quantitative characters. *Proc. Natl. Acad. Sci. USA* 54:731–36

122. Kinghorn, B. P. 1987. The nature of 2-locus epistatic interactions in animals: evidence from Sewall Wright's guinea pig data. *Theor. Appl. Genet.* 73:595–604

123. Kirkpatrick, M. 1982. Quantum evolution and punctuated equilibria in continuous genetic characters. *Am. Nat.* 119:833–48

124. Kirkpatrick, M. 1987. Sexual selection by female choice in polygynous animals. *Annu. Rev. Ecol. Syst.* 18:43–70

125. Kirpatrick, M., Lande, R. 1989. The evolution of maternal characters. *Evolution.* 43:485–503

126. Kirkpatrick, M., Lofsvold, D. 1989. The evolution of complex quantitative characters. *Genome* 31:

127. Klein, T. W. 1974. Heritability and genetic correlation: Statistical power, population comparisons, and sample size. *Behav. Genet.* 4:171–89

128. Klein, T. W., DeFries, J. C., Finkbeiner, C. T. 1973. Heritability and genetic correlation: Standard errors of estimates and sample size. *Behav. Genet.* 3:355–64

129. Kohn, L. A. P., Atchley, W. R. 1988. How similar are genetic correlation structures? *Evolution* 42:467–81

130. Kurtén, B. 1959. Rates of evolution in fossil mammals. *Cold Spring Harbor Symp. Quant. Biol.* 24:205–14

131. Lande, R. 1975. The maintenance of genetic variability by mutation in a polygenic character with linked loci. *Genet. Res.* 26:221–35

132. Lande, R. 1976. Natural selection and random genetic drift in phenotypic evolution. *Evolution* 30:314–34

133. Lande, R. 1977. The influence of the mating system on the maintenance of genetic variability in polygenic characters. *Genetics* 86:485–98

134. Lande, R. 1979. Quantitative genetic

analysis of multivariate evolution applied to brain:body size allometry. *Evolution* 34:402–16

135. Lande, R. 1980. Genetic variation and phenotypic evolution during allopatric speciation. *Am. Nat.* 116:463–79

136. Lande, R. 1980. The genetic covariance between characters maintained by pleiotropic mutations. *Genetics* 94:203–15

137. Lande, R. 1981. The minimum number of genes contributing to quantitative variation between and within populations. *Genetics* 99:541–53

138. Lande, R. 1982. Rapid evolution of sexual isolation and character divergence in a cline. *Evolution* 36:213–23

139. Lande, R. 1983. The response to selection on major and minor mutations affecting a metrical trait. *Heredity* 50:47–65

140. Lande, R. 1986. The dynamics of peak shifts and the pattern of morphological evolution. *Paleobiology* 12:343–54

141. Lande, R. 1987. Genetic correlations between the sexes in the evolution of sexual dimorphism and mating preferences. See Ref. 25, pp. 83–94

142. Lande, R. 1988. Quantitative genetics and evolutionary theory. See Ref. 246, pp. 71–84

143. Lande, R., Arnold, S. J. 1983. The measurement of selection on correlated characters. *Evolution* 37:1210–26

144. Lander, E. S., Botstein, D. 1989. Mapping Mendelian factors underlying quantitative traits using RFLP linkage maps. *Genetics* 121:185–99

145. Langley, C. H., Montgomery, E., Quattlebaum, W. F. 1982. Restriction map variation in the Adh region of Drosophila. *Proc. Natl. Acad. Sci. USA* 79:5631–35

146. Langley, C. H., Voelker, R. A., Leigh Brown, A. J., Ohnishi, S., Dickson, B., Montgomery, E. 1981. Null allele frequencies at allozyme loci in natural populations of *Drosophila melanogaster*. *Genetics* 99:151–56

147. Lasslo, L. L., Bradford, G. E., Torell, D. T., Kennedy, B. W. 1985. Selection for weaning weight in Targhee sheep in two environments. I. Direct response. *J. Anim. Sci.* 61:376–86

148. Latter, B. D. H. 1960. Natural selection for an intermediate optimum. *Aust. J. Biol. Sci.* 13:30–35

149. Latter, B. D. H. 1965. The response to artificial selection due to autosomal genes of large effect. I. Changes in gene frequency at an additive locus. *Aust. J. Biol. Sci.* 18:585–98

150. Lenski, R. E. 1988. Experimental studies of pleiotropy and epistasis in *Escherichia coli*. II. Compensation for maladaptive effects associated with resistance to virus T4. *Evolution* 42:433–40

151. Lewontin, R. C. 1967. Population genetics. *Annu. Rev. Genet.* 1:37–70

152. Lewontin, R. C. 1974. The analysis of variance and the analysis of causes. *Am. J. Hum. Genet.* 26:400–11

153. Lewontin, R. C. 1974. *The Genetic Basis of Evolutionary Change.* New York: Columbia Univ. Press

154. Linney, R., Barnes, B. W., Kearsey, M. J. 1971. Variation for metrical characters in Drosophila populations. III. The nature of selection. *Heredity* 27:163–74

155. Loeschcke, V., ed. 1987. *Genetic Constraints on Adaptive Evolution.* Berlin: Springer-Verlag.

156. Lofsvold, D. 1986. Quantitative genetics of morphological differentiation in Peromyscus. I. Tests of the homogeneity of genetic covariance structure among species and subspecies. *Evolution* 40:559–73

157. Lynch, M. 1984. The limits to life history evolution in Daphnia. *Evolution* 38:465–82

158. Lynch, M. 1988. Design and analysis of experiments on random drift and inbreeding depression. *Genetics* 120:791–807

159. Lynch, M. 1988. The rate of polygenic mutation. *Genet. Res.* 51:137–48

160. Lynch, M. 1989. The rate of morphological evolution in mammals from the standpoint of the neutral expectation. *Evolution* 43:1–17

161. Lynch, M., Hill, W. G. 1986. Phenotypic evolution by neutral mutation. *Evolution* 40:915–35

162. Mackay, T. F. C. 1981. Genetic variation in varying environments. *Genet. Res.* 37:79–93

163. Mather, K. 1983. Response to selection. See Ref. 9, pp. 155–221

164. Mather, K., Jinks, J. L. 1971. *Biometrical Genetics.* Ithaca, NY: Cornell Univ. Press. 2nd ed.

165. Mather, K., Jinks, J. L. 1977. *Introduction to Biometrical Genetics.* Ithaca, NY: Cornell Univ. Press

166. Maynard Smith, J. 1983. The genetics of stasis and punctuation. *Annu. Rev. Genet.* 17:11–25

167. Maynard Smith, J., Sondhi, K. C. 1960. The genetics of a pattern. *Genetics* 45:1039–50

168. Maynard Smith, J. M., Burian, R., Kauffman, S., Alberch, P., Campbell, J., et al. 1985. Developmental con-

straints and evolution. *Quart. Rev. Biol.* 60:265–87

169. Mitchell-Olds, T. 1986. Quantitative genetics of survival and growth in *Impatiens capensis. Evolution* 40:107–16

170. Mitchell-Olds, T., Rutledge, J. J. 1986. Quantitative genetics in natural plant populations. A review of the theory. *Am. Nat.* 127:379–402

171. Mitchell-Olds, T., Shaw, R. G. 1987. Regression analysis of natural selection: Statistical inference and biological interpretation. *Evolution* 41:1149–61

172. Morris, R. F. 1971. Observed and simulated changes in genetic quality in natural populations of *Hyphantria cunea. Can. Entomol.* 103:893–906

173. Mousseau, T. A., Roff, D. A. 1987. Natural selection and the heritability of fitness components. *Heredity* 69:181–97

174. Mukai, T. 1988. Genotype-environment interaction in relation to the maintenance of genetic variability in populations of *Drosophila melanogaster.* See Ref. 246, pp. 21–31

175. Mukai, T., Cockerham, C. C. 1977. Spontaneous mutation rates at enzyme loci in *Drosophila melanogaster. Proc. Natl. Acad. Sci. USA* 74:2514–17

175a. Nagylaki, T. 1989. Rate of evolution of a character without epistasis. *Proc. Natl. Acad. Sci. USA* 86:1910–13

175b. Nagylaki, T. 1989. The maintenance of genetic variability in two-locus models of stabilizing selection. *Genetics* 122:235–48

176. Oster, G. F., Shubin, N., Murray, J. D., Alberch, P. 1988. Evolution and morphogenetic rules: the shape of the vertebrate limb in phylogeny and ontogeny. *Evolution* 42:862–84

177. Pani, S. N., Lasley, J. F. 1972. Genotype × environment interactions in animals. *Res. Bull. 992, Univ. Missouri—Columbia.*

178. Parsons, P. A. 1987. Evolutionary rates under environmental stress. *Evol. Biol.* 21:311–47

179. Paterson, A. H., Lander, E. S., Hewitt, J. D., Peterson, S., Lincoln, S. E., Tanksley, S. D. 1988. Resolution of quantitative traits into Mendelian factors by using a complete RFLP linkage map. *Nature* 335:721–26

180. Payne, F. 1918. An experiment to test the nature of the variation on which selection acts. *Indiana Univ. Stud.* 5, No. 36

181. Pearson, K. 1896. Mathematical contributions to the theory of evolution. III. Regression, heredity and panmixia. *Philos. Trans. R. Soc. London Ser. A* 187:253–318

182. Pearson, K. 1903. Mathematical contributions to the theory of evolution XI. On the influence of natural selection on the variability and correlation of organs. *Philos. Trans. R. Soc. London. Ser. A* 200:1–66

183. Pearson, K. 1930. *Life, Letters and Labours of Francis Galton, Vol. IIIA.* London: Cambridge Univ. Press

184. Petry, D. 1984. The pattern of phyletic evolution. *Paleobiology* 8:56–66

185. Pollack, E., Kempthorne, O., Bailey, T. B. Jr., eds. 1977. *Proceedings of the International Conference on Quantitive Genetics.* Ames, IA: Iowa State Univ. Press

186. Porter, T. M. 1986. *The Rise of Statistical Thinking 1820–1900,* p. 300. Princeton, NJ: Princeton Univ. Press

187. Powers, L. 1950. Determining scales and the use of transformations in studies of weight per locule of tomato fruit. *Biometrics* 6:145–63

188. Price, G. R. 1970. Selection and covariance. *Nature* 227:520–21

189. Price, T. D. 1984. Sexual selection on body size, territory and plumage variables in a population of Darwin's finches. *Evolution* 38:327–41

189a. Price, T., Kirkpatrick, M., Arnold, S. J. 1988. Directional selection and the evolution of breeding dates in birds. *Science* 240:798–99

190. Prout, T. 1958. A possible difference in genetic variance between wild and laboratory populations. *Dros. Inf. Serv.* 32: 148–49

191. Provine, W. B. 1971. *The Origins of Theoretical Population Genetics.* Chicago: Univ. Chicago Press

192. Provine, W. B. 1986. *Sewall Wright and Evolutionary Biology.* Chicago: Univ. Chicago Press

193. Rendel, J. M. 1979. Canalization and selection. See Ref. 233, pp. 139–56

194. Reznick, D. 1981. "Grandfather effects": the genetics of interpopulational differences in offspring size in the mosquito fish. *Evolution* 35:941–53

195. Riska, B. 1986. Some models for development, growth, and morphometric correlation. *Evolution* 40:1303–11

196. Riska, B. 1989. Composite traits, selection response, and evolution. *Evolution.* In press

196a. Riska, B., Prout, T., Turelli, M. 1989. Laboratory estimates of heritabilities and genetic correlations in nature. *Genetics.* In press

197. Robertson, A. 1956. The effect of selection against extreme deviants based on deviation or on homozygosis. *J. Genet.* 54:236–48

198. Robertson, A. 1960. A theory of limits in artificial selection. *Proc. R. Soc. London, Ser. B.* 153:234–49

199. Robertson, A. 1966. A mathematical model of the culling process in dairy cattle. *Anim. Prod.* 8:95–108

200. Robertson, A. 1967. The nature of quantitative genetic variation. In *Heritage from Mendel*, ed. R. A. Brink, pp. 265–80. Madison, WI: Univ. Wis. Press

201. Robertson, A., Hill, W. G. Population and quantitative genetics of many linked loci in finite populations. *Proc. R. Soc. London, Ser. B* 219:253–64

202. Roff, D. A. 1986. The evolution of wing dimorphism in insects. *Evolution* 40:1009–20

203. Roff, D. A., Mousseau, T. A. 1987. Quantitative genetics and fitness: lessons from Drosophila. *Heredity* 58:103–18

204. Rose, M. R., Service, P. M., Hutchinson, E. W. 1987. Three approaches to trade-offs in life-history evolution. See Ref. 155, pp. 91–105

205. Roughgarden, J. 1972. Evolution of niche width. *Am. Nat.* 106:683–718

206. Scharloo, W. 1987. Constraints in selection response. In *Genetic Constraints on Adaptive Evolution*, ed. V. Loeschcke, pp. 125–49. Berlin: Springer

207. Schluter, D. 1988. Estimating the form of natural selection on a quantitative trait. *Evolution* 1988:849–61

208. Schluter, D., Smith, J. N. M. 1986. Genetic and phenotypic correlations in a natural population of song sparrows. *Biol. J. Linn. Soc.* 29:23–36

209. Schnee, F. B., Thompson, J. N. Jr. 1984. Conditional polygenic effects in the sternopleural bristle system of *Drosophila melanogaster*. *Genetics* 108:409–24

210. Schwaegerle, K. E., Garbutt, K., Bazzaz, F. A. 1986. Differentiation among nine populations of Phlox. I. Electrophoretic and quantitative variation. *Evolution* 40:506–17

211. Service, P. M., Rose, M. R. 1985. Genetic covariation among life-history components: The effect of novel environments. *Evolution* 39:943–45

212. Shaffer, H. B. 1986. Utility of quantitative genetic parameters in character weighting. *Syst. Zool.* 35:124–34

213. Shaw, R. G. 1986. Response to density in a wild population of the perennial herb *Salvia lyrata:* Variation among families. *Evolution* 40:492–505

214. Shaw, R. G. 1987. Maximum-likelihood approaches applied to quantitative genetics of natural populations. *Evolution* 41:812–26

215. Sheppard, P. M. 1958. *Natural Selection and Heredity.* London: Hutchinson

216. Shrimpton, A. E., Robertson, A. 1988. The isolation of polygenic factors controlling bristle score in *Drosophila melanogaster*. II. Distribution of third chromosome bristle effects within chromosome sections. *Genetics* 118:445–59

217. Simpson, G. G. 1944. *Tempo and Mode in Evolution*. New York: Columbia Univ. Press

218. Slatkin, M. 1970. Selection and polygenic characters. *Proc. Natl. Acad. Sci. USA* 66:87–93

219. Slatkin, M. 1978. Spatial patterns in the distributions of polygenic characters. *J. Theor. Biol.* 70:213–28

220. Slatkin, M. 1979. Frequency- and density-dependent selection on a quantitative character. *Genetics* 93:755–71

221. Slatkin, M. 1987. Quantitative genetics of heterochrony. *Evolution* 41:799–811

221a. Slatkin, M. 1987. Heritable variation and heterozygosity under a balance between mutations and stabilizing selection. *Genet. Res.* 50:53–62

222. Slatkin, M., Lande, R. 1976. Niche width in a fluctuating environment-density independent model. *Am. Nat.* 110:31–55

223. Smith, T. B. 1987. Bill size polymorphism and intraspecific niche utilization in an African finch. *Nature* 329:717–19

224. Sorensen, D. A., Hill, W. G. 1983. Effects of disruptive selection on genetic variance. *Theor. Appl. Genet.* 65:173–80

225. Stigler, S. M. 1986. *The History of Statistics*. Cambridge, MA: Belknap

226. Tallis, G. M. 1987. Ancestral covariance and the Bulmer effect. *Theor. Appl. Genet.* 73:815–20

227. Tanksley, S. D., Medina-Filho, H., Rick, C. M. 1982. Use of naturally-occurring enzyme variation to detect and map genes controlling quantitative traits in an interspecific backcross of tomato. *Heredity* 49:11–25

228. Tantawy, A. O., Mallah, G. S. 1961. Studies of natural populations of Drosophila. I. Heat resistance and geographical variation in *Drosophila melanogaster* and *D. simulans*. *Evolution* 15:1–14

229. Tantawy, A. O., Mallah, G. S., Tewfik, H. R. 1964. Studies of natural populations of Drosophila. II. Heritability and response to selection for wing length in *Drosophila melanogaster* and *D. simulans* at different temperatures. *Genetics* 49:935–48

230. Thoday, J. M. 1972. Disruptive selection. *Proc. R. Soc. London, Ser. B.* 182:109–43

231. Thoday, J. M. 1979. Polygene mapping: Uses and limitations. See Ref. 233, pp. 219–33

232. Thoday, J. M., Thompson, J. N. Jr. 1976. The number of segregating genes implied by continuous variation. *Genetica* 46:335–44

233. Thompson, J. N. Jr., Thoday, J. M., eds. 1979. *Quantitative Genetic Variation*. New York: Academic

234. Turelli, M. 1984. Heritable genetic variation via mutation-selection balance: Lerch's zeta meets the abdominal bristle. *Theor. Popul. Biol.* 25:138–93

235. Turelli, M. 1985. Effects of pleiotropy on predictions concerning mutation-selection balance for polygenic traits. *Genetics* 111:165–95

236. Turelli, M. 1986. Gaussian versus non-Gaussian genetic analyses of polygenic mutation-selection balance. In *Evolutionary Processes and Theory*, ed. S. Karlin, E. Nevo, pp. 607–28. New York: Academic

237. Turelli, M. 1988. Phenotypic evolution, constant covariances and the maintenance of additive variance. *Evolution* 43:1342–47

238. Turelli, M. 1988. Population genetic models for polygenic variation and evolution. See. Ref. 246, pp. 601–18

238a. Turelli, M., Barton, N. H. 1989. Dynamics of polygenic characters under selection. *Theor. Popul. Biol.* In press

239. Turelli, M., Gillespie, J. H., Lande, R. 1988. Rates tests for selection on quantitative characters during macroevolution and microevolution. *Evolution* 42:1085–89

240. Underhill, D. K. 1969. Heritability of some linear body measurements and their ratios in the leopard frog *Rana pipiens. Evolution* 23:268–75

241. van Noordwijk, A. J., Balen, J. H., van, Scharloo, W. 1980. Heritability of ecologically important traits in the great tit, *Parus major. Ardea* 68:193–203

242. Via, S., Lande, R. 1987. Evolution of genetic variability in a spatially heterogeneous environment: effects of genotype-environment interaction. *Genet. Res.* 49:147–56

243. Wagner, G. P. 1984. Coevolution of functionally constrained characters: prerequisites for adaptive versatility. *Biosystems* 17:51–55

244. Wagner, G. P. 1989. Multivariate mutation-selection balance with constrained pleiotropic effects. *Genetics* 122:223–34

245. Weber, K. E. 1986. *The effect of population size on response to selection*. PhD thesis. Harvard Univ.

246. Weir, B. S., Eisen, E. J., Goodman, M. M., Namkoong, G., eds. 1988. *Proceedings of the Second International Conference on Quantitative Genetics*. Sunderland, MA: Sinauer

247. Weldon, W. F. R. 1892. Certain correlated variations in *Cragnon vulgaris. Proc. R. Soc. London* 51:2–21

248. Weldon, W. F. R. 1893. On certain correlated variations in *Carcinus moenas. Proc. R. Soc. London* 54:318–29

249. Williams, G. C. 1957. Pleiotropy, natural selection and the evolution of senescence. *Evolution* 11:398–411

250. Wright, A. J. 1988. Some applications of the covariances of relatives with inbreeding. See Ref. 246, pp. 10–20

251. Wright, S. 1932. The roles of mutation, inbreeding, crossbreeding and selection in evolution. *Proc. 6th Int. Congr. Genet.* 1:356–66

252. Wright, S. 1935. Evolution of populations in approximate equilibrium. *J. Genet.* 30:257–66

253. Wright, S. 1935. The analysis of variance and the correlation between relatives with respect to deviations from an optimum. *J. Genet.* 30:243–56

254. Wright, S. 1968–78. *Evolution and the Genetics of Populations, Vols. 1–4*. Chicago: Univ. Chicago Press

255. Yoo, B. H. 1980. Long-term selection for a quantitative character in large replicate populations of *Drosophila melanogaster*. I. Response to selection. *Genet. Res.* 35:1–17

256. Yoo, B. H. 1980. Long-term selection for a quantitative character in large replicate populations of *Drosophila melanogaster*. II. Lethals and visibles with large effects. *Genet. Res.* 35:19–31

257. Zeng, Z.-B. 1988. Long-term correlated response, interpopulation covariation, and interspecific allometry. *Evolution* 42:363–74

258. Zeng, Z.-B., Hill, W. G. 1986. The selection limit due to the conflict between truncation selection and stabilizing selection with mutation. *Genetics* 114:1313–28

Annu. Rev. Genet. 1989. 23:371–93

THE MOLECULAR GENETICS OF 21-HYDROXYLASE DEFICIENCY

Walter L. Miller and Yves Morel*

Department of Pediatrics and The Metabolic Research Unit, University of California, San Francisco, CA 94143-0978

CONTENTS

INTRODUCTION

Steroid 21-hydroxylase is one of several enzymes needed to convert cholesterol to cortisol, the principal human glucocorticoid. The pathways leading to cortisol and other human adrenal steroids are well understood, and the molecular biology of the various enzymes has been studied in detail (for review, see 50). The steroidogenic pathway (Figure 1) shows that deficient activity of several enzymes might impair cortisol synthesis. Diminished concentrations of plasma cortisol are detected by the pituitary, which in turn increases synthesis of proopiomelanocortin (POMC) and the release of various POMC-derived peptides, including adrenocorticotrophic hormone (ACTH). ACTH, and possibly other POMC-derived peptides, stimulate adrenal steroidogenesis and promote considerable adrenal growth. As a result, deficient activity of any of the enzymes required for cortisol synthesis results

*Current address: INSERM U34, 29 rue Soeur Bouvier, 69322 Lyon Cedex 05, France

371

0066-4197/89/1215-0371$02.00

in congenital adrenal hyperplasia (CAH) (for review, see 52). Depending on the clinical criteria used for its diagnosis, the population examined and the type of survey procedure used, CAH has a reported incidence between 1 in 5000 and 1 in 20,000 persons (see below). Deficient 21-hydroxylase activity accounts for over 90% of CAH, and hence the terms "congenital adrenal hyperplasia" and "21-hydroxylase deficiency" are often used interchangeably. Because forms of CAH other than 21-hydroxylase deficiency are rare, and because very little is known about them at the molecular level (28, 45, 86), this review deals only with 21-hydroxylase deficiency.

BIOLOGY OF 21-HYDROXYLASE

The adrenal produces three principal categories of steroid hormones. *Mineralocorticoids,* principally aldosterone, instruct the kidney to retain Na^+ ions; therefore, they are crucial for regulating intravascular volume and blood pressure. *Glucocorticoids,* principally cortisol, regulate carbohydrate metabolism and a wide variety of other physiologic processes. *Adrenal androgens,* principally androstenedione and testosterone, are responsible for

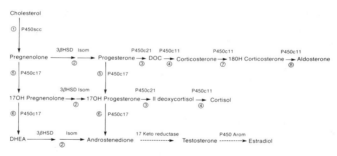

Figure 1   Principal pathways of human adrenal steroidogenesis. Other quantitatively and physiologically minor steroids are also produced. The chemical identities of the enzymes are shown by each reaction, and the traditional names of the enzymatic activities correspond to the circled numbers. Reaction 1: Mitochondrial cytochrome P450scc mediates 20α-hydroxylation, 22-hydroxylation, and scission of the C20–22 carbon bond to convert cholesterol to pregnenolone. Reaction 2: A non-P450 enzyme bound to the endoplasmic reticulum mediates 3β-hydroxysteroid dehydrogenase and isomerase activities. Reaction 3: P450c21 in the endoplasmic reticulum mediates 21-hydroxylation of both progesterone and 17-hydroxyprogesterone. Reactions 4, 7 and 8: Mitochondrial P450c11 exerts three clearly distinguishable activities: 11β-hydroxylation (4), 18-hydroxylation (7), and 18-methyl oxidase activity (8). Reactions 5 and 6: P450c17 in the endoplasmic reticulum mediates both 17α-hydroxylase (5) and 17,20 lyase (6) activities. Reactions 9 and 10 are found principally in the testes and ovaries; 17-ketosteroid reductase, a non-P450 enzyme of the endoplasmic reticulum, produces testosterone (9), which may then be converted to estradiol by P450aro (10), another P450 enzyme of the endoplasmic reticulum mediating the aromatization of the A ring of the steroid nucleus.

pubic and axillary hair in females, but have no essential role. However, the same enzymes that produce adrenal androgens are needed to produce testosterone and estrogen in the gonads, which are required for reproduction.

Steroid 21-hydroxylation occurs in two of these adrenal steroidogenic pathways. Progesterone is 21-hydroxylated to yield deoxycorticosterone (DOC) in the mineralocorticoid pathway and 17-hydroxyprogesterone (17OHP) is 21-hydroxylated to yield 11-deoxycortisol in the glucocorticoid pathway. Inferences from clinical observations of patients with 21-hydroxylase deficiency led to the widely held belief that separate 21-hydroxylase enzymes mediated these two reactions. However, isolation and purification to homogeneity for a single enzyme having both activities (41) disproved this.

Steroid 21-hydroxylation is mediated by a specific form of cytochrome P450, termed P450c21. Like most other P450 enzymes, P450c21 is bound to the endoplasmic reticulum and receives electrons from NADPH via a generic flavoprotein, P450 reductase, which acts as an electron-transport intermediate for all microsomal P450 enzymes. Some other P450 enzymes, including the steroidogenic enzymes P450scc and P450c11, are found in mitochondria and use a different electron-transfer scheme (for review, see 50).

Some steroid 21-hydroxylation occurs in nonadrenal tissues, especially in the fetus and during pregnancy (10, 11). This extra-adrenal 21-hydroxylation persists in patients with deficient adrenal 21-hydroxylase, and mRNA for P450c21 is not found in human or rat extra-adrenal tissues having 21-hydroxylase activity (47a, 80). In the rabbit, the 21-hydroxylase activity found in liver is clearly mediated by P450-1 (79), which has no significant amino acid similarity to P450c21. By contrast, authentic P450c21 mRNA can be found in mouse liver (2) and bovine testis (15); thus, the tissue distribution of P450c21 varies among different mammals. 21-hydroxylation by enzymes other than P450c21 confounds the examination of hormonal phenotypes in persons having disorders of P450c21. Thus, some conclusions about 21-hydroxylase "genetics" based on examining serum cortisol levels are, at best, problematic.

PHENOTYPIC MANIFESTATIONS OF 21-HYDROXYLASE DEFICIENCY

Physiology of 21-Hydroxylase Deficiency

For patients having complete deficiency of P450c21 (e.g. in a patient with homozygous gene deletion), the resulting phenotype can be deduced correctly by inspection of Figure 1. Inability to convert progesterone to DOC results in aldosterone deficiency. In the absence of aldosterone, the kidney cannot retain Na^+ at normal concentrations (135–145 mM) in blood and extracellular fluid; Na^+ concentrations fall to the low 100s and the kidney inappropriately retains K^+ and H^+. This results in hypotension, shock, cardiovascular collapse, and death. As the control of fluids and electrolytes in the fetus can be maintained by the mother's kidneys, the "salt-losing crisis" described above develops following birth, usually during the second week of life.

Inability to convert 17OHP to 11-deoxycortisol results in glucocorticoid deficiency. In addition to impairing postnatal carbohydrate metabolism and other processes, cortisol deficiency is also manifested prenatally. Low concentrations of fetal cortisol result in increased production and secretion of ACTH from the fetal pituitary. In addition to stimulating adrenal hyperplasia, ACTH, which works via cAMP as an intracellular second messenger, stimulates transcription of the genes for all of the steroidogenic enzymes (37). It particularly stimulates transcription of P450scc (36; C.C.D. Moore , W. L. Miller, unpublished data), the rate-limiting step in steroidogenesis. This increased gene transcription results in increased enzyme production and activity, with consequent accumulation of non 21-hydroxylated steroids, especially 17OHP. As the pathways in Figure 1 indicate, these steroids are converted to testosterone.

In the male fetus, the testes produce high concentrations of testosterone in early to mid-gestation. This testosterone differentiates external male genitalia from the pluripotent embryonic precursor structures. In the female fetus, the ovaries are steroidogenically quiescent. No sex steroids or other factors are needed for differentiation of the female external genitalia (i.e. the default state of the human fetus is a female phenotype). In the male fetus with 21-hydroxylase deficiency, the additional testosterone produced in the adrenals has little if any demonstrable phenotypic effect. In the female fetus with 21-hydroxylase deficiency, the testosterone inappropriately produced by the adrenals causes varying degrees of virilization (masculinization) of the external genitalia. At birth, these female infants, who retain normal ovaries, fallopian tubes and a uterus, may have "ambiguous" genitalia or may be sufficiently virilized so that they appear to be male, resulting in errors of sex assignment at birth. This physiology leads to the other common terms for 21-hydroxylase deficiency: "virilizing adrenal hyperplasia" or "adreno-genital syndrome" (for reviews, see 52, 71).

Clinical Forms of 21-Hydroxylase Deficiency

A wide range of phenotypes exists in CAH, some of which are less severe. While these are generally described as different diseases or different forms of 21-hydroxylase deficiency, we emphasize that there is a continuous spectrum of manifestations of this disease, ranging from the severe "salt-wasting form" described above to clinically inapparent forms that may be normal variants. Thus, the diagnostic entities listed below are mainly a clinical convenience.

SALT-WASTING CAH   The phenotype described above is that of a patient with salt-wasting 21-hydroxylase deficiency. Females with this disorder are frequently diagnosed at birth because of ambiguous development of the external genitalia. When this happens, the absent mineralocorticoid and glucocorticoid hormones can be administered as drugs given by mouth, and the ambiguous genitalia can be corrected by a series of plastic surgical procedures. The steroid drug management is difficult, and these patients frequently have decreased fertility, short stature and other problems (57). Males with this disorder generally go undiagnosed at birth and either come to medical attention during the salt-wasting crisis that follows 7–15 days later, or die.

SIMPLE VIRILIZING CAH   Females with virilized, ambiguous genitalia and increased concentrations of 17OHP but who do not suffer a salt-losing crisis have long been recognized as having the "simple virilizing" form of CAH. The existence of this clinical variant first led to the incorrect hypothesis that there were two adrenal 21-hydroxylases. Males with this disorder often escape diagnosis until age 4–7 years, when they come to medical attention because of inappropriate virilization (pubic, axillary and facial hair, etc). Although these males are tall for age when diagnosed, they invariably become short adults because of androgen-induced premature termination of bone growth (epiphyseal fusion). Because the adrenal normally produces 100 times as much cortisol as aldosterone, mild defects (point mutations) in P450c21 are much less likely to affect mineralocorticoid secretion than cortisol secretion. Thus, patients with simple virilizing CAH simply have a less severe disorder of P450c21. This is reflected physiologically by the increased plasma renin activity in these patients, which reflects hyperstimulation of the mineralocorticoid pathway.

NONCLASSICAL CAH   Many people have very mild forms of 21-hydroxylase deficiency. These may be evidenced by mild to moderate hirsutism, virilism, menstrual irregularities and decreased fertility in adult women (so-called "late onset CAH") (13, 40, 48) or there may be no phenotypic manifestations at all, other than an elevated response of plasma 17OHP in response to administration of ACTH (so-called "cryptic CAH") (43). Despite the minimal manifestations of this disorder, these individuals also have hormonal evidence of a mild impairment in mineralocorticoid secretion (22), as predicted from the existence of a single adrenal 21-hydroxylase.

There has been considerable medical debate about how to classify individuals into these categories. This difficulty is due in very large part to the fact that each diagnostic category represents a typical patient along a continuous spectrum of disease phenotypes. As shown below, this is due to a broad spectrum of molecular genetic lesions in the 21-hydroxylase gene.

Incidence

The reported incidence of 21-hydroxylase deficiency varies widely. Perinatal screening for elevated concentrations of serum 17OHP in several countries yields an incidence of 1 in 14,000 for "classical" CAH (i.e. salt-wasting and simple virilizing CAH) and 1 in 60 for heterozygous carriers (63). However, this calculation excluded two groups—the Yupik Eskimos of Alaska, a genetic isolate where salt-losing CAH occurs in 1 in 280 persons, and the population of Réunion, a French island in the Indian Ocean east of Madagascar, where the incidence is 1 in 2,000 (63).

Nonclassical CAH is clearly much more common, but these data are also variable. One group reports very high frequencies: 1 in 27 for Ashkenazi Jews, 1 in 50–60 for Hispanics and Yugoslavs, 1 in 330 for Italians and 1 in 1,000 for other Caucasians (73, 75); other studies show lower frequencies (12, 42). Diagnosis of nonclassical CAH requires family studies, as the hormonal data (17OHP responses to ACTH) in these individuals may be indistinguishable from unaffected heterozygous carriers of more severe forms. The high incidence and lack of decreased fertility in most individuals with nonclassical CAH indicates that this may be a variant of normal and not a disease.

GENETICS OF THE 21-HYDROXYLASE LOCUS

HLA Linkage

Diagnosis and genetic tracking of 21-hydroxylase deficiency is greatly facilitated by its linkage to the human major histocompatibility complex (MHC) (20). This complex, which lies on band 21.3 of the short arm of chromosome 6, contains genes coding for three major classes of proteins: HLA-A, -B, and -C (class I); HLA-DP, -DQ, and -DR (class II); serum complement components C2 and C4, properdin factor Bf, and tumor necrosis factors α and β (class III) (Figure 2). HLA typing is now widely used for prenatal diagnosis and for identifying heterozygous family members. Recombination frequency estimates map the 21-hydroxylase locus very close to C4 (24, 61) and to HLA-B and -DR (3). Examination of DNA by pulsed-field gel electrophoresis indicates the P450c21 locus lies about 600 kb from HLA-B and about 400 kb from HLA-DR (19, 77).

Genetic linkage disequilibrium is well established between CAH and certain specific HLA types. Salt-losing CAH is associated with HLA-B60, and the rare HLA type Bw47 is very strongly associated with salt-losing CAH (21, 24). Most individuals carrying the supratype HLA-A3, Bw47, DR7, C2C,

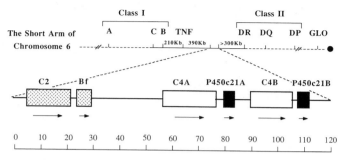

Figure 2 Map of the short arm of chromosome 6. The upper line shows the organization of the Class I, III and II regions of the major histocompatibility locus (MHC). The MHC map is roughly to scale with the dotted lines based on recombinational distances and the solid line based on mapping by pulsed field gel electrophoresis. The distances from the MHC to the telomere (left) and centromere (right) are not to scale. The second line shows about 120 kb of the class III region. C2: Complement factor 2, Bf: Properdin factor Bf, C4A and C4B: Non-allelic genes for complement factor 4, P450c21A: The pseudogene for 21-hydroxylase, P450c21B: The functional gene for 21-hydroxylase. The arrows indicate transcriptional orientation, and the scale indicates distances in kilobases.

BfF, C4A91, C4BQ0 have salt-losing CAH, but the similar supratype HLA-A3, Bw47, DR7, BfS, C4A3, C4B1 is not associated with CAH (18). HLA Bw51 is often associated with simple virilizing CAH in some populations (34), and 30–50% of haplotypes for nonclassical CAH carry HLA-B14 (67). HLA-B14 is often associated with a duplication of C4B, especially in the two supratypes HLA-Aw33, B14, DR1, C2C, BfS, C4A2, B1B2, and HLA-Aw28, B14, C2C, BfS, C3A2, B1B2 (53, 76). By contrast, all HLA-B alleles can be found linked to CAH.

21-Hydroxylase Genes

21-hydroxylase is encoded by the P450c21B gene, also termed P450XXIA2 (58) or CYP21A2 (59). Two P450c21 loci, P450c21A and B, are duplicated in tandem with the C4A and C4B loci encoding the fourth component of serum complement (8, 82) (Figure 2). Another pair of genes, XA and XB, encoded on the opposite strand of DNA from the P450c21 and C4 genes, and having unknown function, partially overlaps exon 10 of P450c21 (54). Sequencing studies (33, 69, 84) show that the P450c21A locus is an inactive pseudogene. It has an 8-bp deletion in exon 3 and a single-base insertion in exon 7, each resulting in a downstream, in-phase translational termination codon; in addition, it has a C→T transition in exon 8 causing a third stop codon. Thus, only the P450c21B gene encodes 21-hydroxylase. The P450c21 genes consist of 10 exons, are about 3.4 kb long, and differ in only 87 or 88 of these bases (33, 69). This very great degree of sequence similarity indicates these two genes are evolving in tandem (concerted evolution) (35, 44, 51). The P450c21 genes of mice (2) and cattle (14, 74) are also duplicated and linked to leukocyte antigen loci. However, while only the P450c21B gene functions in human beings, only the P450c21A gene functions in mice (64), and both genes appear to function in cattle (50). Thus, the duplication of the C4/P450c21 locus appears to predate mammalian speciation, but the inactivation of the human P450c21A gene postdates this speciation that occurred about 85 million years ago.

C4 Genes

The tandemly duplicated C4A and C4B loci are both functional and their encoded serum proteins can be distinguished functionally and immunologically: C4A has a higher binding efficiency than C4B, but C4B has greater hemolytic activity. At least 34 alleles of C4 have been described (6, 30).

The C4A gene is 22 kb long and encodes a 5-kb mRNA. The C4B gene exists in a long (22 kb) form and a short (16 kb) form due to the presence or absence of a 6.8-kb intron at the 5' end of the gene (7, 88). As a result of this polymorphism, the 5' end of the C4B gene is characterized by 6.0- or 5.4-kb *Taq*I fragments, corresponding to the long and short C4B genes, respectively, but 5' end of the C4A gene can be distinguished from either of these, as it lies on a 7.0-kb *Taq* Ifragment (5, 72).

SOUTHERN BLOTTING STUDIES IN CAH

Mapping of the P450c21 Genes in Normals and in CAH

P450C21 GENES IN NORMAL PERSONS   Some restriction enzymes can distinguish the two P450c21 genes in normal persons. The P450c21A pseudogene is characterized by 3.2- and 2.4-kb *Taq*I, 12-kb *Bgl*II, 4.0-kb *Kpn*I and 2.0-kb *Pvu*II fragments, whereas the functional P450c21B gene is characterized by 3.7- and 2.5-kb *Taq*I, 11-kb *Bgl*II, 2.9-kb *Kpn*I and 1.8-kb *Pvu*II fragments. *Eco*RI digests are difficult to interpret because the 3' sites are polymorphic, and the enzymatic digestion is somewhat unpredictable, apparently due to variable methylation of these *Eco*RI sites. *Taq*I and *Bgl*II digestions are especially powerful for analyzing this complex locus if a genomic 3.1-kb *Eco*RI/*Bam*HI probe is used (53). Two identical *Taq*I sites are located 3.7 kb apart on the two P450c21 genes (Figure 3). An additional *Taq*I site is found in the 5' flanking DNA of the P450c21A pseudogene generating a 3.2-kb *Taq*I fragment, thus differentiating it from the functional gene (3.7-kb *Taq*I fragment). *Bgl*II defines a longer region extending 1.6 kb upstream from the initiation codons of the two P450c21 genes to 6–7 kb downstream from the polyadenylation sites. This latter *Bgl*II site is polymorphic, yielding a 12-kb fragment for the P450c21A pseudogene and an 11-kb fragment for the P450c21 gene (Figure 3).

P450c21 GENES IN CAH   Southern blotting studies of DNA from patients with CAH have been performed by numerous groups, detecting a wide variety of RFLP patterns in various groups of patients but finding grossly normal P450c21 genes in most patients. White et al (83) initially reported deletion of the P450c21B gene evidenced by an absent 3.7-kb *Taq* I fragment and an absent *Eco*RI fragment in a patient homozygous for HLA-A3, -B47, -DR7, C4A91, C4BQ0 and having severe salt-losing CAH. Consistent with this finding, absence of the 3.2-kb *Taq*I fragment suggested deletion of the P450c21A pseudogene in some hormonally normal persons, including homozygotes for HLA A1, -B8, -DR3, C4AQ0, C4B1 (9, 83). These initial observations created considerable enthusiasm for relying heavily on data from DNA cut with *Taq*I. One study using *Taq*I digests alone reported that the P450c21B and C4B genes were deleted in 10 of the 16 chromosomes from eight patients who carried a C4BQ0 allele (i.e. undetectable C4B protein) (72). However, two of these chromosomes lacking the 3.7-kb *Taq*I fragment retained both the C4A and C4B genes, and hence probably had gene con-

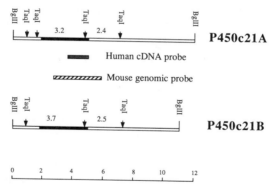

Figure 3   Map of the P450c21A and B genes. 5′ is to the left and 3′ is to the right, as in Figure 2. The P450c21A pseudogene is shown lying on a 12-kb *Bgl*II fragment while the functional P450c21B gene lies on an 11-kb *Bgl*II fragment. An additional *Taq*I site is found just 5′ to the P450c21A pseudogene so that it lies on a 3.2-kb *Taq*I fragment, while the P450c21B gene lies on a 3.7-kb *Taq*I fragment. These *Taq*I and *Bgl* II fragments can be distinguished readily by probing Southern blots with P450c21 cDNA. The DNA bearing the P450c21A and B genes can also be distinguished by examining the 3′ flanking DNA with a genomic probe that extends 3′ to the P450c21 genes. The 3′ flanking DNA of P450c21A is characterized by a 2.4-kb *Taq*I fragment, while the 3′ flanking DNA of the functional P450c21B gene is characterized by a 2.5-kb *Taq*I fragment. This distinction has provided a powerful tool for distinguishing gene deletions from gene conversions (53, 55).

versions. Other reports have similarly shown deleted C4 and P450c21 genes in patients selected for certain HLA and complement types (9, 25, 81). One of these studies excluded patients with the HLA-A3, -B47, -DR7, C4A91, C4BQ0 haplotype which is closely associated with P450c21B gene deletions, but still found seven deletions among 30 CAH-bearing chromosomes, evidenced by absent 3.7-kb *Taq*I and 12-kb *Bgl*II fragments (81). A similar study found nine deletions among 40 such chromosomes but showed that the absence of 12-kb *Bgl*II fragment is due to a large deletion extending from the 3' end of P450c21B (70). Thus, absent 3.7-kb *Taq*I fragments were initially regarded as diagnostic of P450c21B deletions, and these *Taq*I digests suggested that 25–40% of patients with salt-losing CAH had gene deletions.

Gene Conversions

However, patients lacking the 3.7-kb *Taq*I fragment who nonetheless retained the P450c21B gene were also reported early (17, 65). This phenomenon is generally termed "gene conversion," as the *Taq*I structure of the P450c21B gene is converted to that of the P450c21A pseudogene (17, 31, 32, 38, 46, 49, 53, 55). The term "gene conversion" originally referred to a phenomenon in yeast genetics, wherein a gene on one chromosome had its structure and function "converted" to that of another gene on an homologous chromosome (68). Gene conversion thus describes a change that appears to have occurred by a crossover event, but the reciprocal crossover product is not found. Because it is not possible to search for reciprocal crossover products in human genetics, the term is used somewhat differently, and probably less precisely, referring to a genetic change in which a segment of one gene replaces the corresponding segment of a closely related gene (4). Thus, a hallmark of gene conversion is that the number of closely related genes remains constant while their diversity decreases. Numerous mechanisms have been proposed for such events (see 49 for references), but present evidence in yeast favors the creation of double-stranded breaks that yield a double-stranded gap followed by repair by copying both strands of the undamaged homolog (23, 60, 78). Gene conversion-like events have also been reported with a broad array of human genes. When gene conversion causes 21-hydroxylase deficiency, the *Taq*I digestion pattern of the functional P450c21B gene is converted to that of the P450c21A pseudogene. This gene conversion changes the 5' end of the P450c21B gene to the 5' end of the P450c21A gene, as detected by Southern blotting studies (17, 53, 55), by oligonucleotide probing (31), or by sequencing (29).

The relative frequency of gene conversion versus gene deletion in CAH has been controversial. Jospe et al (38) reported nine deletions and three gene conversions among 28 chromosomes that cause salt-losing CAH (38). Matteson et al (46) found absent 3.7-kb *Taq*I fragments on 8 out of 20 chromosomes that cause salt-losing CAH but found that gene-deletion models were inconsistent with the results with four other restriction enzymes; they proposed that those eight apparent deletions represented gene conversions, unequal crossovers or polymorphisms, but not physical loss of the P450c21B gene. We have reanalyzed the Southern blots from the families bearing these eight chromosomes that lack the 3.7-kb *Taq*I fragment (46). Alleles *a* and *c* in family 8 have classic gene conversions, and allele *c* in family 1 has a deletion of P450c21B, but the lesions in the other five chromosomes still remain unresolved. A study of Japanese patients examined Southern blots probed with a series of oligonucleotides hybridizing specifically and exclusively to either the P450c21A or B gene (31). Among the 22 chromosomes examined, seven lacked the 3.7-kb *Taq*I fragment, but all retained the 3' half of the P450c21B gene; this suggests these absent 3.7-kb *Taq*I bands were due to gene conversion. In other Japanese CAH patients, six of 21 unrelated, affected chromosomes lacked the 3.7-kb *Taq*I fragment; of these six, three were clearly due to gene conversion, and gene conversion or deletion could not be distinguished in the other three (29). Gene conversion was confirmed for one chromosome by DNA sequence analysis (29). By contrast, the New York group recently showed that in their patient population, 13 of 15 CAH chromosomes lacking the 3.7-kb *Taq*I fragment had bona fide gene deletions (85). Thus, the frequency of gene deletion versus gene conversion varies greatly among published studies of small series of patients.

Genetic Alterations in the C4 Genes Associated with CAH

Deletions, duplications and conversions of the C4 genes are frequently associated with CAH. The C4AQ0 "allele," which corresponds to absent C4A activity and no electrophoretically detectable protein, is found in both normal and CAH populations. This "allele" can, in fact, be due to several different structural changes, including deletion of the C4A gene (9, 25, 72), nonexpression of a mutated C4A gene, and expression of C4B allotypes at both the C4A and C4B loci due to conversion of the C4A gene to a C4B gene (26, 47, 62, 89).

Deletions of the P450c21A pseudogene are generally associated with deletions of the either C4A or C4B (but not both) but are not associated with CAH. In our survey of a normal Western European population, 14% of normal chromosomes bore P450c21A gene deletions, while only 4% of CAH chromosomes had a deletion of this pseudogene (53). This is consistent with HLA linkage data that show a negative linkage disequilibrium between CAH and HLA-B8 or C4AQ0, which are associated with deletions of C4A and P450c21A.

Duplication of the P450c21A pseudogene is closely inked with HLA B14 (56, 81); these two markers are found in about 1/3 to 1/2 of individuals with nonclassical CAH. However, duplicated P450c21A pseudogenes and HLA B14 can also be found in simple virilizing CAH and on normal chromosomes (53). Since both deletions and duplications of the P450c21A pseudogene can be found in normals, it is clear that structural changes in the P450c21A pseudogene are not causally involved in nonclassical CAH but are merely statistically associated with it.

Frequency of Various Genetic Lesions in CAH

To determine the frequency of genetic defects causing CAH, we studied 57 French families with 68 patients affected with CAH and 165 unaffected family members (53). For this study we used a new DNA probing tactic employing a P450c21 genomic probe, rather than a cDNA probe. This genomic probe (Figure 3) can distinguish the *Taq*I-digested P450c21 A and B

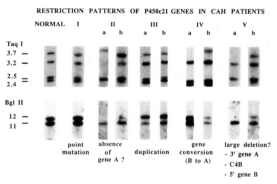

Figure 4   Representative Southern blots of genomic DNA from patients with severe salt-wasting 21-hydroxylase deficiency. DNA samples from 10 representative individuals were cleaved with *Taq*I (above) and *Bgl*II (below), blotted and probed with the murine P450c21 genomic probe diagrammed in Figure 3. DNA from a normal person is shown at left; the banding pattern is identical in most patients (haplotype I), indicating point mutation (or micro gene conversions) in both alleles of P450c21B. In haplotypes II-V, *a* represents a patient homozygous for the haplotype and *b* represents a patient exhibiting compound heterozygosity with haplotype I. In haplotype II the P450c21A gene is deleted and in haplotype III this pseudogene is duplicated. Haplotypes II and III can also be found in 14.3 and 1.2%, respectively, of normal chromosomes that do not carry CAH (53). Therefore, the grossly intact P450c21B genes in these haplotypes must carry the same kinds of lesions as the P450c21B genes in haplotype I. Haplotype IV shows a conversion of the 5' end of the P450c21B gene to a P450c21A pseudogene. As a result, there is no 3.7-kb band, but the presence of the P450c21B gene is shown by the persistence of the 2.5-kb *Taq*I band and the 11-kb *Bgl*II band. Haplotype V shows the results of deletion of about 30 kb of DNA extending from the middle of the P450c21A pseudogene to an apparently precisely homologous point in the P450c21B gene. Thus, the 2.4-kb *Taq* I band representing the 3' flanking DNA of P450c21A and the 3.7-kb band representing P450c21B are absent. This deletion creates a nonfunctional hybrid P450c21A/B gene. Therefore, the 5' end has the characteristics of P450c21A, evidenced by the 3.2-kb *Taq*I band, and the 3' end has the characteristics of the P450c21B gene, evidenced by the 2.5-kb *Taq*I and 11-kb *Bgl*II bands.

genes by the 3.7-kb and 3.2-kb *Taq*I fragments and by 2.5 and 2.4 kb *Taq*I fragments lying downstream (3') from these. This greatly simplified the analysis of this complex locus and permitted the assignment of 114 of 116 affected chromosomes to five haplotypes (Figure 4). The two remaining chromosomes presented a confusing array of restriction endonuclease patterns akin to those described previously (16, 46). Such restriction endonuclease patterns might arise from compound heterozygosity of different lesions or uncharacterized gene rearrangements. Among these 116 chromosomes, we found that 88 (76%) had grossly intact P450c21B genes and hence probably carried point mutations. Chromosomes bearing gene deletions and gene conversions were equally abundant, 13 each, representing 11% of chromosomes carrying CAH (53).

Table I summarizes the literature on Southern blotting studies in CAH. If these results are combined, one finds 84 deletions (19%) and 41 conversions (9%) among 453 informative chromosomes bearing CAH. However, many of these studies examined only patients with salt-losing CAH or specifically chose (or excluded) certain HLA types, biasing the results toward gene

Table 1 Summary of published Southern blotting studies showing the number of chromosomes carrying each type of P450c21 gene lesion

References	Phenotype	Point Mutations*			Gene conversion	30 kb Deletion
		Alone	With deleted Pseudogene	With duplicated Pseudogene		
38	SW	15	0	1	3	9 (6 with B47)
72 (with C4 null)	SW	9	1	0	2	8 (5 with B47)
70	SW + SV	31	0	0	0	9
53	SW	53	2	2	11	12 (6 with B47)
	SV	15	0	3	2	1
	NC	8	0	3	0	0
81 (without B47)	SW	15	1	0	0	#
	SV	7	1	3	0	#
	NC	2	0	7	0	#
85	SW + SV	55	—	—	2	13 (6 with B47)
46	SW	10	2	0	2 + 5**	1
33	SW	6	1	0	7	0
	SV	4	0	2	0	0
	NC	2	0	0	0	0
29	SW + SV	15	—	—	3 + (0–3)***	(0–3)***
15a	SW + SV	24	5	4	6	19 (10 with B47)
64a	SW + SV	18	0	1	3	12 (2 with B47)

Abbreviations: SW = salt-wasting; SV = simple virilizing; NC = nonclassical
*Includes genes with micro-conversions that do not affect RFLP patterns.
**The genetic lesion in 5 chromosomes cannot be determined.
***At least 3 chromosomes had gene conversion; another 3 either gene conversion or gene deletion.
#These patients (6 SW, 1 SV, 1 NC) are included in the 13 deletions studied by White et al (85).
—Not reported.

deletions. Even when studies do not have such a design bias, the increased likelihood that patients with salt-losing CAH will be diagnosed, and the low ascertainment of persons with nonclassical CAH, will inherently bias the results. Thus, the available data indicate that about 75% of patients with salt-losing CAH have point mutations in their P450c21B genes, and that the remaining 25% are about equally divided into gene deletions and large gene conversions.

Nature of the Gene Deletion causing CAH

Among the 84 chromosomes bearing deleted P450c21B genes (Table 1), only one kind of deletion has been reported. This deletion encompasses about 30 kb and extends from the middle of the P450c21A pseudogene, through the C4B gene, and includes the 5' portion of the P450c21B gene. Deletion of this DNA thus results in a locus containing only an intact, functional C4A gene and a nonfunctional hybrid P450c21A/B recombinant pseudogene. Gene-specific oligonucleotide probing (31, 39, 85), careful RFLP analysis with multiple restriction endonucleases (53, 70, 85), and DNA sequencing all confirm that these P450c21A/B recombinants contain the 8-bp deletion found in exon 3 of P450c21A. Attempts to identify the precise junction site between the P450c21A and B segments suggest several possible junction sites downstream from the 8-bp deletion in exon 3 and upstream from the C→T transition in exon 8 of P450c21A.

However, all such P450c21A/B recombinant pseudogenes appear to be composed of A and B gene fragments broken at precisely homologous loci. All such A/B recombinants are characterized by 3.2-kb *Taq* I fragments, indicating that the A/B recombinant is the same size as the P450c21A pseudogene. Furthermore, Harada's sequencing data (29) show no other bases added or deleted. Thus, while the 30-kb deletions commonly occurring in this region may have slightly different ends, they always begin and end at precisely corresponding bases in the P450c21A and B loci. Deletion of 30 kb of DNA was confirmed recently by pulsed field electrophoresis of DNA digested with *Bss*HII. With this enzyme, both P450c21 and both C4 genes lie on a single 110-kb fragment, but 19 of 58 chromosomes causing CAH had only an 80-kb fragment, showing deletion of 30 kb of DNA (15a).

Such P450c21A/B recombinants resulting from gene deletion cannot be distinguished from the "converted" P450c21B genes in patients having large gene conversions. Thus, chromosomes bearing gene conversions are characterized by retaining two different C4 genes, retaining two indistinguishable P450c21A or A/B hybrids, and retaining the characteristic 2.4- and 2.5-kb *Taq*I fragments lying immediately 3' to the P450c21A and B genes, respectively. However, P450c21B genes "converted" to P450c21A genes cannot be distinguished from the P450c21A/B recombinants resulting from the 30-kb deletion(s).

POINT MUTATIONS

Polymorphism in the P450c21B Gene

Most chromosomes (75%) encoding CAH do not carry alterations in the P450c21B gene detectable by Southern blotting using either cDNA or genomic probes; hence, these genes appear to carry point mutations. Determining the genetic lesions in these genes requires cloning and sequencing the affected DNA. Specific amplification of affected sequences by polymerase chain-reaction technology (87) has not been effective because the 98% sequence identity in the P450c21A and B genes prohibits synthesis of suitably specific primers. Furthermore, many of the single-base lesions in the P450c21B gene that cause CAH are the same as base changes found in the P450c21A pseudogene, further confounding PCR amplification. The three published sequences for the normal P450c21B gene (33, 69, 84) show 76 differences within the 3400-base coding region of the P450c21B gene, suggesting considerable polymorphism in this functional gene. Most of these differences are base changes also found in introns of the P450c21A pseudogene, but 15 of these polymorphic differences occur in the exons of the functional P450c21B gene, and five of these different bases change codons. An insertion of three bases (CTG, leucine) between codons 9 and 10 occurs in the P450c21A pseudogene and in one reported functional P450c21B gene (69), resulting in P450c21 proteins of 494 or 495 amino acids. Codon 101 can be AGG (Arg) or it can be AAG (Lys), as it is in the P450c21A pseudogene, and codon 426 can be either CGC· (Arg) or CCG (Pro), with either also occurring in the pseudogene (33, 69, 84). Thus, the P450c21B genes are polymorphic and may contain sequences "characteristic" of the P450c21A pseudogene. Therefore, great care must be taken in interpreting nucleotide sequence differences in mutated P450c21B genes.

Most P450c21B Point Mutations are Actually Small Gene Conversions

Nine mutant P450c21B genes causing CAH have been sequenced (Table 2). These contain a total of 54 base changes, of which 44 are also found in the P450c21A pseudogene. These base changes are not localized to a specific region but are scattered throughout the P450c21B gene. These observations suggest that most of these apparent point mutations are actually gene conversions (49).

Four lines of evidence indicate that small, localized gene conversion events, which do not change the RFLP patterns of the P450c21 genes, are responsible for most apparent point mutations. First, the very great sequence similarity in the P450c21A and B loci indicates considerable intragenic exchange of sequences. Second, most of the phenotypically silent polymorphic base changes in normal P450c21B genes are also found in the P450c21A

Table 2 Summary of published sequencing studies identifying lesions and characterizing enzymatic activity encoded by P450c21 genes having apparent point mutations

Gene	Mutation(s)[+]	Phenotype	P450c21 mRNA*	21OH activity*	Present in P450c21A	Gene conversion	References
1	172 Ile (ATC)→Asn (AAC)	SW	Present	ND	Yes	Probable**	1
2	Aberrant splicing in intron 2	SW	ND	ND	Yes	Probable	69
	258 Ser (AGC)→Thr (ACC)				No	No	
	493 Asn (AAC)→Ser (AGC)				No	No	
3	281 Val (GTG)→Leu (TTG)	SW	Decreased	ND	Yes	Probable	27
	318 Stop codon (TAG)				Yes	Probable	
4	318 Stop codon (TAG)	SW	ND	ND	Yes	Probable	79a
	493 Asn (AAC)→Ser (AGC)				No	No	
5	236 Ile (ATC)→Asn (AAC)	SW	Present	Absent	Yes	Yes	32
	237 Val (GTG)→Glu (GAG)				Yes	Yes	
	239 Met (ATG)→Lys (AAG)				Yes	Yes	
6	?Regulation upstream from −71	SW	Present	Present	—	—	32
7	Aberrant splicing in intron 2	SV	Present	Absent	Yes	Probable	32
8	Aberrant splicing in intron 2	SV	Present	Absent	Yes	Probable	32
9	211 Val (GTG)→Leu (CTG)	NC	ND	ND	No	No	76
	281 Val (GTG)→Leu (TTG)				Yes	Yes	

Abbreviations: SW, salt-wasting; SV, simple virilizing; NC, nonclassical; ND, not determined

[+] Mutation is listed if this base change is not found in the published sequences of the normal P450c21B gene.

*From transfection studies.

***"Probable" gene conversions refer to base changes in the B gene that also exist in the A gene but are found in a pattern that would require more than one genetic crossover event. For example, Amor et al (1) report a mixed P450c21 B/A hybrid having the structure BABAB with residue 172 lying in the first A region.

pseudogene. Third, 44 of the 54 base changes found in nine mutated P450c21B genes that cause CAH are also found in P450c21A genes. Fourth, 11 of the 15 phenotypically expressed changes in these genes that cause CAH are also found in the P450c21A pseudogene (Table 2).

Small, localized gene conversions may occur more than once in a single P450c21 gene. Several mutant P450c21B genes have been described having structures suggesting multiple crossover events, i.e. having alternating segments derived from the B and A genes. Genes having an apparent BABAB structure (1) and a BABABABAB structure (27) have been described. A summary of the principal classes of genetic lesions associated with 21-hydroxylase deficiency is presented in Figure 5.

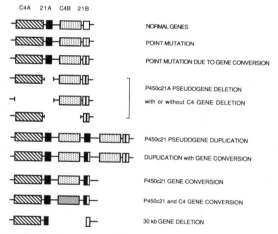

Figure 5   Diagram of the various types of changes causing severe, salt-losing 21-hydroxylase deficiency. The top diagram shows the normal arrangement of the genes for C4A (broad diagonal lines), P450c21A (solid black), C4B (dotted) and P450c21B (open). Individual point mutations, indicated by a fine line showing a single base change (second line), appear to be less common than micro gene conversions, indicated by a broad line showing conversion of a whole region of 21B sequences to 21A sequences (third line). These two types of lesions constitute haplotype I (Figure 4). The fourth through sixth lines show deletion of the P450c21A pseudogene, either without or with deletion of one of the two C4 genes (but not both). These three types of lesions constitute haplotype II. The seventh and eighth lines show duplication of the P450c21A pseudogene associated with a micro conversion or point mutation in P450c21B and showing the absence or presence of a conversion of the 3' end of one of the 21A pseudogenes to the sequences of a 21B gene. These lesions constitute haplotype III. P450c21B conversions, where the 5' end of the 21B gene acquires 21A sequences, can occur with or without conversion of the C4B gene, as shown in the ninth and tenth lines, respectively. These lesions constitute haplotype IV. Deletion of about 30 kb of DNA extending from the middle of the 21A pseudogene to the homologous point in the 21B gene (last line) is responsible for haplotype V.

Effects of Point Mutations on P450c21 mRNA and Enzymatic Activity

Because only a few mutant genes have been sequenced to date, it is not yet possible to draw consistent conclusions about the structure/function relationships of various amino acid replacements. One gene causing nonclassical CAH of the typical form linked to HLA B14,DR1 has been sequenced showing several nucleotide changes. These include the insertion of a CTG (Leu) codon between codon 9 and 10, a Leu to Val change in codon 211, and a silent change in Leu codon 248, all as seen in the P450c21A pseudogene. This gene also contained a Leu to Val change in codon 281, which the authors speculate was the cause of the nonclassical CAH, because this change is not seen in other mammals, whereas the Leu to Val change at codon 211 is. Oligonucleotide probing indicated this change at codon 281 was also found in 8 of 8 patients with nonclassical CAH and HLA B14, DR1, but was not found in another patient with a different HLA type (76).

Adrenal tissue from individuals with CAH is very rarely available; hence, direct studies of the effects of various point mutations have been confined to the transfer of mutated genes into cultured cells. Some of these mutations diminish P450c21 activity (as measured by conversion of 17OHP added to the culture medium to 11-deoxycortisol), while P450c21 mRNA remains present, but other mutations diminish both the mRNA and the enzymatic activity. In one gene, three clustered mutations change the sequence Ile-Val-Glu-Met in exon 6 (codons 236–239) to Asn-Glu-Glu-Lys, resulting in severe salt-losing congenital adrenal hyperplasia in the patient and absent 21-hydroxylase activity in transfected COS cells (32). Despite this, normal amounts of P450c21 mRNA were detected in the COS cells, suggesting that an abnormal P450c21 protein was made that lacked 21-hydroxylase activity (32). Two mutant P450c21B genes have been sequenced having an A→G change in the second intron, creating a new 3' splice acceptor sequence (CAG) 12 bases upstream from this intron's normal 3' splice acceptor site (32). Although this would suggest that an aberrant protein would result having four additional amino acids encoded by these 12 bases, S_1 nuclease protection experiments showed that the mutant mRNA was not spliced at this predicted site but was spliced in 3 locations 19–32 bases upstream from the normal 3' splice acceptor site. Such aberrant splicing will cause frame shifts that lead to premature downstream stop codons or to an abnormal protein containing 11 extra amino acids. Consistent with this, COS cells transfected with this P450c21B mutant have no 21-hydroxylase activity, despite having P450c21 mRNA (32). Although this model and the absent 21-hydroxylase activity in the COS cells would suggest a lesion that causes severe salt-losing CAH, the patient homozygous for this lesion only had simple virilizing CAH. It is possible that the abnormal protein containing 11 extra residues had some P450c21 activity or that a small number of RNA transcripts were spliced correctly in vivo, producing small amounts of normal P450c21 mRNA and protein, which would thus permit 21-hydroxylation of some progesterone so that aldosterone deficiency was not manifested physiologically. However, because the forced expression of this mutant gene by an SV-40 promotor in COS cells is clearly nonphysiologic, proof of this can only come from direct examination of adrenal RNA. This example clearly shows that it is not always possible to predict the phenotype expressed in vivo from a gene's behavior when expressed in a heterologous system.

Most mutations described to date are in the amino terminal half of the P450c21 protein. By contrast, the principal functional regions of the protein are in the carboxy-terminal half of the sequence: the heme binding site of P450c21 is at amino acids 421–439, and the presumed steroid binding site is at amino acids 342–358 (66). This suggests that the activity of P450c21 is very sensitive to minor alterations throughout its structure. Furthermore, finding most mutations in the amino-terminal half of the protein is consistent with finding all gene conversion and gene deletion events in exons 1–8 of P450c21. Changes in exons 9 and 10 appear to be rare (31, Table 2). This may be due to evolutionary pressure on this region to retain the 3' untranslated region and 3' flanking DNA of the P450c21B gene, as this DNA also contains the 3' end of a newly discovered gene transcribed from the opposite DNA strand and overlapping exon 10 of P450c21B (54).

ACKNOWLEDGMENTS

We thank Carol Dahlstrom for typing the manuscript. This research was supported by NIH Grant DK37922 and March of Dimes Grant 6-396 to WLM, and by NIH Fogarty International Fellowship TWO3935 to YM.

Literature Cited

1. Amor, M., Parker, K. L., Globerman, H., New, M. I., White, P. C. 1988. Mutation in the CYP21B gene (Ile-172→Asn) causes steroid 21-hydroxylase deficiency. *Proc. Natl. Acad. Sci. USA* 85:1600–4

2. Amor, M., Tosi, M., Duponchel, C., Steinmetz, M., Meo, T. 1985. Liver cDNA probes disclose two cytochrome P450 genes duplicated in tandem with the complement C4 loci of the mouse H-2S region. *Proc. Natl. Acad. Sci. USA* 82:4453–57

3. Aston, C. E., Sherman, S. L., Morton, N. E., Speiser, P. W., New, M. I. 1988. Genetic mapping of the 21-hydroxylase locus: estimation of small recombination frequencies. *Am. J. Hum. Genet.* 43: 304–10

4. Baltimore, D. 1981. Gene conversion: Some implications for immunoglobulin genes. *Cell* 24:592–94

5. Belt, K. T., Carroll, M. C., Porter, R. R. 1984. The structural basis of the multiple forms of human complement component C4. *Cell* 36:907–14

6. Carroll, M. C., Belt, K. T., Palsdottir, A., Yu, Y. 1985. Molecular genetics of the fourth component of human complement and steroid 21-hydroxylase. *Immunol. Rev.* 87:39–60

7. Carroll, M. C., Campbell, R. D., Bentley, D. R., Porter, R. R. 1984. A molecular map of the human major histocompatibility class III region linking complement genes C4, C2 and factor B. *Nature* 307:237–41

8. Carroll, M. C., Campbell, R. D., Port-

er, R. R. 1985. Mapping of steroid 21-hydroxylase genes adjacent to complement component C4 genes in HLA, the major histocompatibility complex in man. *Proc. Natl. Acad. Sci. USA* 82: 521–25

9. Carroll, M. C., Pasldottir, A., Belt, K. T., Porter, R. R. 1985. Deletion of complement C4 and steroid 21-hydroxylase genes in the HLA class III region. *EMBO J.* 4:2547–52

10. Casey, M. L., MacDonald, P. C. 1982. Extraadrenal formation of a mineralocorticoid: Deoxycorticosterone and deoxycorticosterone sulfate biosynthesis and metabolism. *Endocr. Rev.* 3:396–403

11. Casey, M. L., Winkel, C. A., MacDonald, P. C. 1983. Conversion of progesterone to deoxycorticosterone in the human fetus: Steroid 21-hydroxylase activity in fetal tissues. *J. Steroid Biochem.* 18:449–52

12. Chetkowski, R. J., DeFazio, J., Shamonki, I., Judd, H. L., Chang, R. J. 1984. The incidence of late-onset congenital adrenal hyperplasia due to 21-hydroxylase deficiency among hirsute women. *J. Clin. Endocrinol. Metab.* 58:595–98

13. Chrousos, G. P., Loriaux, D. I., Mann, D. L., Cutler, G. B. 1982. Late-onset 21-hydroxylase deficiency mimicking idiopathic hirsutism or polycystic ovarian disease: An allelic variant of congenital virilizing adrenal hyperplasia with a milder enzymatic defect. *Ann. Intern. Med.* 96:143–48

14. Chung, B., Matteson, K. J., Miller, W. L. 1985. Cloning and characterization of the bovine gene for steroid 21-hydroxylase (P-450c21). *DNA* 4:211–19

15. Chung, B., Matteson, K. J., Miller, W. L. 1986. Structure of a bovine gene for P450c21 (steroid 21-hydroxylase) defines a novel cytochrome P450 gene family. *Proc. Natl. Acad. Sci. USA* 83: 4243–47 ·

15a. Collier, S., Sinnott, P. J., Dyer, P. A., Price, D. A., Harris, R., Strachan, T. 1989. Pulsed field gel electrophoresis identifies a high degree of variability in the number of tandem 21-hydroxylase and complement C4 gene repeats in 21-hydroxylase deficiency haplotypes. *EMBO J.* 8:1393–1402

16. Dawkins, R. L., Martin, E., Kay, P. H., Garlepp, M. J., Wilton, A. N., Stuckey, M. S. 1987. Heterogeneity of steroid 21-hydroxylase genes in classical congenital adrenal hyperplasia. *J. Immunogenet.* 14:89–98

17. Donohoue, P. A., Van Dop, C., McLean, R. H., White, P. C., Jospe, N., Migeon, C. J. 1986. Gene conversion in salt-losing congenital adrenal hyperplasia with absent complement C4B protein. *J. Clin. Endocrinol. Metab.* 62:995–1002

18. Donohoue, P. A., Van Dop, C., Migeon, C. J., McLean, R. H., Bias, W. B. 1987. Coupling of HLA-A3,Cw6, Bw47, DR7 and a normal CA21HB steroid 21-hydroxylase gene in the old order Amish. *J. Clin. Endocrinol. Metab.* 65:980–86

19. Dunham, I., Sargent, C. A., Trowsdale, J., Campbell, R. D. 1987. Molecular mapping of the human major histocompatibility complex by pulsed-field gel electrophoresis. *Proc. Natl. Acad. Sci. USA* 84:7237–41

20. Dupont, B., Oberfield, S. E., Smithwick, E. R., Lee, T. D., Levine, L. S. 1977. Close genetic linkage between HLA and congenital adrenal hyperplasia (21-hydroxylase deficiency). *Lancet* 2:1309–12

21. Dupont, B., Pollack, M. S., Levine, L. S., O'Neill, G. J., Hawkins, B. R., New, M. I. 1981. Congenital adrenal hyperplasia: Joint report from the eight international histocompatibility workshop. In *Histocompatibility Testing 1980*, ed. P. I. Terasaki, pp. 693–706. Berlin: Springer-Verlag

22. Fiet, J., Gueux, B., Raux-Demay, M. C., Kuttenn, F., Vexiau, P., et al. 1989. Increased plasma 21-deoxycorticosterone levels in late-onset adrenal 21-hydroxylase deficiency suggest a mild defect of the mineralocorticoid pathway. *J. Clin. Endocrinol. Metab.* 68:542–47

23. Fincham, J. R. S., Oliver, P. 1989. Initiation of recombination. *Nature* 338: 14–15

24. Fleischnick, E., Awdeh, Z. L., Raum, D., Granados, J., Alosco, S. M., et al. 1983. Extended MHC haplotypes in 21-hydroxylase deficiency congenital adrenal hyperplasia: Shared genotypes in unrelated patients. *Lancet* 1:152–56

25. Garlepp, M. J., Wilton, A. N., Dawkins, R. L., White, P. C. 1986. Rearrangement of 21-hydroxylase genes in disease-associated MHC supratypes. *Immunogenetics* 23:100–5

26. Giles, C. M., Uring-Lambert, B., Boksch, W., Braun, M., Goetz, J., et al. 1987. The study of a French family with two duplicated C4A haplotypes. *Hum. Genet.* 77:359–65

27. Globerman, H., Amor, M., Parker, K. L., New, M. I., White, P. C. 1988. Nonsense mutation causing steroid 21-

hydroxylase deficiency. *J. Clin. Invest.* 82:139–44

28. Globerman, H., Rosler, A., Theodor, R., New, M. I., White, P. C. 1988. An inherited defect in aldosterone biosynthesis caused by a mutation in or near the gene for steroid 11-hydroxylase. *New Engl. J. Med.* 319:1193–97

29. Harada, F., Kimura, A., Iwanaga, T., Shimozawa, K., Yata, J., Sasazuki, T. 1987. Gene conversion-like events cause steroid 21-hydroxylase deficiency in congenital adrenal hyperplasia. *Proc. Natl. Acad. Sci. USA* 84:8091–94

30. Hauptmann, G., Tappeiner, G., Schifferli, J. A. 1988. Inherited deficiency of the fourth component of human complement. *Immunodefic. Rev.* 1:3–22

31. Higashi, Y., Tanae, A., Inoue, H., Fujii-Kuriyama, Y. 1988. Evidence for frequent gene conversions in the steroid 21-hydroxylase (P-450c21) gene: implications for steroid 21-hydroxylase deficiency. *Am. J. Hum. Genet.* 42:17–25

32. Higashi, Y., Tanae, A., Inoue, H., Hiromasa, T., Fujii-Kuriyama, Y. 1988. Aberrant splicing and missense mutations cause steroid 21-hydroxylase [P-450(C21)] deficiency in humans: Possible gene conversion products. *Proc. Natl. Acad. Sci. USA* 85:7486–90

33. Higashi, Y., Yoshioka, H., Yamane, M., Gotoh, O., Fujii-Kuriyama, Y. 1986. Complete nucleotide sequence of two steroid 21-hydroxylase genes tandemly arranged in human chromosome: a pseudogene and a genuine gene. *Proc. Natl. Acad. Sci. USA* 83:2841–45

34. Holler, W., Scholz, S., Knorr, D., Bidlingmaier, F., Keller, E., Albert, E. D. 1985. Genetic differences between the salt-wasting, simple virilizing and nonclassical types of congenital adrenal hyperplasia. *J. Clin. Endocrinol. Metab.* 60:757–63

35. Hood, L., Campbell, J H., Elgin, S. C. R. 1975. The organization expression and evolution of antibody genes and other multigene families. *Annu. Rev. Genet.* 9:305–53

36. Inoue, H., Higashi, Y., Morohashi, K., Fujii-Kuriyama, Y. 1988. The 5' flanking region of the human P-450(scc) gene shows responsiveness to cAMP-dependent regulation in a transient gene-expression system in Y-1 adrenal tumor cells. *Eur. J. Biochem.* 171:435–40

37. John, M. E., John, M. C., Boggaram, V., Simpson, E. R., Waterman, M. R. 1986. Transcriptional regulation of steroid hydroxylase genes by corticotropin. *Proc. Natl. Acad. Sci. USA* 83:4715–19

38. Jospe, N., Donohoue, P. A., Van Dop, C., McLean, R. H., Bias, W., Migeon, C. J. 1987. Prevalence of polymorphic 21-hydroxylase gene (CA21HB) mutations in salt-losing congenital adrenal hyperplasia. *Biochem. Biophys. Res. Commun.* 142:798–804

39. Jospe, N., Donohoue, P. A., Van Dop, C., Migeon, C. J. 1988. Crossing-over sites within the 21-hydroxylase (CYP21) genes causing salt-losing congenital adrenal hyperplasia (CAH). *Pediatr. Res.* 23:279A (Abstr.)

40. Kohn, B., Levine, L. S., Pollack, M. S., Pang, S., Lorenzen, F., et al. 1982. Late-onset steroid 21-hydroxylase deficiency: a variant of classical congenital adrenal hyperplasia. *Clin. Endocrinol. Metab.* 51:817–27

41. Kominami, S., Ochi, H., Kobayashi, Y., Takemori, S. 1980. Studies on the steroid hydroxylation system in adrenal cortex microsomes: Purification and characterization of cytochrome P450 specific for steroid C21 hydroxylation. *J. Biol. Chem.* 255:3386–94

42. Kuttenn, F., Couillin, P., Girard, F., Billaud, L., Vincens, M., et al. 1985. Late-onset adrenal hyperplasia in hirsutism. *New Engl. J. Med.* 313:224–31

43. Levine, L. S., Dupont, B., Lorenzen, F., Pang, S., Pollack, M. S., et al. 1981. Genetic and hormonal characterization of the cryptic 21-hydroxylase deficiency. *J. Clin. Endocrinol. Metab.* 53:1193–98

44. Liebhaber, S. A., Goossens, M., Kan, Y. W. 1981. Homology and concerted evolution of the α-1 and α-2 loci of human α-globin. *Nature* 290:26–29

45. Matteson, K. J., Chung, B., Urdea, M. S., Miller, W. L. 1986. Study of cholesterol side-chain cleavage (20, 22 desmolase) deficiency causing congenital lipoid adrenal hyperplasia, using bovine-sequence P450scc oligodeoxyribonucleotide probes. *Endocrinology* 118:1296–305

46. Matteson, K. J., Phillips, J. A. III, Miller, W. L., Chung, B., Orlando, P. J., et al. 1987. P450XXI (steroid 21-hydroxylase) gene deletions are not found in family studies of congenital hyperplasia. *Proc. Natl. Acad. Sci. USA* 84:5858–62

47. McLean, R. H., Donohoue, P. A., Jospe, N., Bias, W. B., Van Dop, C., Migeon, C. J. 1988. Restriction fragment analysis of duplication of the fourth component of complement (C4A). *Genomics* 2:76–85

47a. Mellon, S. H., Miller, W. L. 1989. Extra-adrenal steroid 21-hydroxylation is not mediated by P450c21. *J. Clin. Invest.* In press

48. Migeon, C. J., Rosenwask, Z., Lee, P. A., Urban, M. D., Bias, W. B. 1980. The attenuated form of congenital adrenal hyperplasia as an allelic form of 21-hydroxylase deficiency. *J. Clin. Endocrinol. Metab.* 51:647–49

49. Miller, W. L. 1988. Gene conversions, deletions, and polymorphisms in congenital adrenal hyperplasia. *Am. J. Hum. Genet.* 42:4–7

50. Miller, W. L. 1988. Molecular biology of steroid hormone synthesis. *Endocr. Rev.* 9:295–318

51. Miller, W. L., Eberhardt, N. L. 1983. Structure and evolution of the growth hormone gene family. *Endocr. Rev.* 4:97–130

52. Miller, W. L., Levine, L. S. 1987. Molecular and clinical advances in congenital adrenal hyperplasia. *J. Pediatr.* 111:1–17

53. Morel, Y., Andre, J., Uring-Lambert, B., Hauptmann, G., Betuel, H., Tosi, M., Forest, M. G., David, M., Bertrand, J., Miller, W. L. 1989. Rearrangements and point mutations of P450c21 genes are distinguished by five restriction endonuclease haplotypes identified by a new probing strategy in 57 families with congenital adrenal hyperplasia. *J. Clin. Invest.* 83:527–36

54. Morel, Y., Bristow, J., Gitelman, S. E., Miller, W. L. 1989. Transcript encoded on the opposite strand of the human 21-hydroxylase/C4 gene locus. *Proc. Natl. Acad. Sci. USA.* 86:6582-86

55. Morel, Y., David, M., Forest, M. G., Betuel, H., Hauptmann, G., Andre, J., Bertrand, J., Miller, W. L. 1989. Gene conversions and rearrangements cause discordance between inheritance of forms of 21-hydroxylase deficiency and HLA types. *J. Clin. Endocrinol. Metab.* 68:592–99

56. Mornet, E., Couillin, P., Kutten, F., Raux, M. C., White, P. C., et al. 1986. Associations between restriction fragment length polymorphisms detected with a probe for human 21-hydroxylase (21-OH) and two clinical forms of 21-OH deficiency. *Hum. Genet.* 74:402–8

57. Mulaikal, R. M., Migeon, C. J., Rock, J. A. 1987. Fertility rates in female patients with congenital adrenal hyperplasia due to 21-hydroxylase deficiency. *New Engl. J. Med.* 316:178–82

58. Nebert, D. W., Adesnik, M., Coon, M. J., Estabrook, R. W., Gonzalez, F. J., et al. 1987. The P450 gene superfamily: Recommended nomenclature. *DNA* 6:1–11

59. Nebert, D. W., Nelson, D. R., Adesnik, M., Coon, M. J., Estabrook, R. W., et al. 1989. The P450 superfamily: Updated listing of all genes and recommended nomenclature for the chromosomal loci. *DNA* 8:1–13

60. Nicolas, A., Treco, D., Schultes, N. P., Szostak, J. W. 1989. An initiation site for meiotic gene conversion in the yeast *Saccharomyces cerevisiae. Nature* 338:35–39

61. O'Neill, G. J., Dupont, B., Pollack, M. S., Levine, L. S., New, M. I. 1982. Complement C4 allotypes in congenital adrenal hyperplasia due to 21-hydroxylase deficiency: Further evidence for different allelic variants at the 21-hydroxylase locus. *Clin. Immun. Immunopathol.* 23:312–22

62. Palsdottir, A., Arnason, A., Fossdal, R., Jensson, O. 1987. Gene organization of haplotypes expressing two different C4A allotypes. *Hum. Genet.* 76:220–24

63. Pang, S., Wallace, M. A., Hofman, L., Thuline, H. C., Dorche, C., et al. 1988. Worldwide experience in newborn screening for classical congenital adrenal hyperplasia due to 21-hydroxylase deficiency. *Pediatrics* 81:866–74

64. Parker, K. L., Chaplin, D. D., Wong, M., Seidman, J. G., Smith, J. A., Schimmer, B. P. 1985. Expression of murine 21-hydroxylase in mouse adrenal glands and in transfected Y-1 adrenocortical tumor cells. *Proc. Natl. Acad. Sci. USA* 82:7860–64

64a. Partanen, J., Koskimies, S., Sipila, I., Lisanen, V., 1989. Major histocompatibility complex gene markers and restriction fragment analysis of steroid 21-hydroxylase (CYP 21) and complement C4 genes in classical congenital adrenal hyperplasia patients in a single population. *Am. J. Hum. Genet.* 44:660–70

65. Phillips, J. A. III, Burr, I. M., Orlando, P., Matteson, K. J., Chung, B., Miller, W. L. 1985. DNA analysis of human steroid 21-hydroxylase genes in congenital adrenal hyperplasia and controls using a bovine genomic clone. *Am. J. Hum. Genet.* 37:A171 (Abstr. 505)

66. Picado-Leonard, J., Miller, W. L. 1988. Homologous sequences in steroidogenic enzymes, steroid receptors, and a steroid-binding protein suggest a consensus steroid binding sequence. *Mol. Endocrinol.* 2:1145–50

67. Pollack, M. S., Levine, L. S., O'Neill, G. L. 1981. HLA linkage and B14, DR1, BfS haplotype association with the genes for late onset and cryptic 21-hydroxylase deficiency. *Am. J. Hum. Genet.* 33:540–50

68. Radding, C. M. 1978. Genetic recombination: Strand transfer and mismatch repair. *Annu. Rev. Biochem.* 47: 847–80

69. Rodrigues, N. R., Dunham, I., Yu, C. Y., Carroll, M. C., Porter, R. R., Campbell, R. D. 1987. Molecular characterization of the HLA-linked steroid 21-hydroxylase B-gene from an individual with congenital adrenal hyperplasia. *EMBO J.* 6:1653–61

70. Rumsby, G., Carroll, M. C., Porter, R. R., Grant, D. B., Hjelm, M. 1986. Deletion of the steroid 21-hydroxylase and complement C4 genes in congenital adrenal hyperplasia. *J. Med. Genet.* 23: 204–9

71. Saenger, P. 1984. Abnormal sex differentiation. *J. Pediatr.* 104:1–17

72. Schneider, P. M., Carroll, M. C., Alper, C. A., Rittner, C., Whitehead, A. S., et al. 1986. Polymorphism of the human complement C4 and steroid 21-hydroxylase genes. *J. Clin. Invest.* 78: 650–57

73. Sherman, S. L., Aston, C. E., Morton, N. E., Speiser, P. W., New, M. I. 1988. A segregation and linkage study of classical and nonclassical 21-hydroxylase deficiency. *Am. J. Hum. Genet.* 42:830–38

74. Skow, L. C., Womack, J. E., Petresh, J. M., Miller, W. L. 1988. Synteny mapping of the genes for steroid 21-hydroxylase, alpha-A-crystallin, and class I bovine leukocyte antigen in cattle. *DNA* 7:143–49

75. Speiser, P. W., Dupont, B., Rubenstein, P., Piazza, A., Kastelan, A., New, M. I. 1985. High frequency of nonclassical steroid 21-hydroxylase deficiency. *Am. J. Hum. Genet.* 37:650–67

76. Speiser, P. W., New, M. I., White, P. C. 1988. Molecular genetic analysis of nonclassical steroid 21-hydroxylase deficiency associated with HLA-B14,DR1. *New Engl. J. Med.* 319:19–23

77. Spies, T., Blanck, G., Bresnarhan, M., Sands, J., Srominger, J. L. 1989. A new cluster of genes within the human major histocompatibility complex. *Science* 243:214–17

78. Szostak, J. W., Orr-Weaver, T. L., Rothstein, R. J. 1983. The double-strand break repair model for recombination. *Cell* 33:25–35

79. Tukey, R. H., Okono, S., Barnes, H., Griffin, K. J., Johnson, E. F. 1985. Multiple gene-like sequences related to the rabbit hepatic progesterone 21-hydroxylase cytochrome P450-1. *J. Biol. Chem.* 260:13347–54

79a. Urabe, K., Kimura, K., Harada, F., Iwanaga, T., Sasazuki, T. 1989. Gene conversion in steroid 21-hydroxylase genes. *Am. J. Hum. Genet.* In press

80. Voutilainen, R., Miller, W. L. 1986. Developmental expression of genes for the steroidogenic enzymes P450scc (20, 22 desmolase), P450c17 (17α-hydroxylase/17,20 lyase) and P450c21 (21-hydroxylase) in the human fetus. *J. Clin. Endocrinol. Metab.* 63:1145–50

81. Werkmeister, J. W., New, M. I., Dupont, B., White, P. C. 1986. Frequent deletion and duplication of the steroid 21-hydroxylase genes. *Am. J. Hum. Genet.* 39:461–69

82. White, P. C., Grossberger, D., Onufer, B. J., Chaplin, D. D., New, M. I., et al. 1985. Two genes encoding steroid 21-hydroxylase are located near the genes encoding the fourth component of complement in man. *Proc. Natl. Acad. Sci. USA* 82:1089–93

83. White, P. C., New, M. I., Dupont, B. 1984. HLA-linked congenital adrenal hyperplasia results from a defective gene encoding a cytochrome P450 specific for steroid 21-hydroxylation. *Proc. Natl. Acad. Sci. USA* 81:7505–9

84. White, P. C., New, M. I., Dupont, B. 1986. Structure of the human steroid 21-hydroxylase genes. *Proc. Natl. Acad. Sci. USA* 83:5111–15

85. White, P. C., Vitek, A., Dupont, B., New, M. I. 1988. Characterization of frequent deletions causing steroid 21-hydroxylase deficiency. *Proc. Natl. Acad. Sci. USA* 85:4436–40

86. Winter, J. S. D., Couch, R. M., Muller, J., Perry, Y. S., Ferreira, P., et al. 1989. Combined 17-hydroxylase and 17,20 desmolase deficiencies: Evidence for synthesis of defective cytochrome P450c17. *J. Clin. Endocrinol. Metab.* 68:309–16

87. Wong, C., Dowling, C. E., Saiki, R. K., Higuchi, R. G., Erlich, H. A., Kazazian, H. H. 1987. Characterization of β-thalassemia mutations using direct genomic sequencing of amplified single copy DNA. *Nature* 330:384–86

88. Yu, Y. C., Belt, K. T., Giles, C. M., Campbell, R. D., Porter, R. R. 1986. Structural basis of the polymorphism of the human complement components C4A and C4B: Gene size, reactivity and antigenicity. *EMBO J.* 5:2873–81

89. Yu, Y. C., Campbell, R. D. 1987. Definitive RFLPs to distinguish between the human complement C4A/C4B isotopes and the major rodgers/chido determinants: application to the study of C4 null alleles. *Immunogenetics* 25:383–90

Annu. Rev. Genet. 1989. 23:395–408

HABITUATION: Heritable Variation in the Requirement of Cultured Plant Cells for Hormones

Frederick Meins, Jr.

Friedrich Miescher-Institut, Postfach 2543, CH-1002 Basel, Switzerland

CONTENTS

INTRODUCTION

Plant tissues usually require sources of the growth hormones auxin and cytokinin for continuous proliferation in culture on an otherwise complete medium (19). In 1942, Gautheret reported that certain strains of carrot tissue can gradually lose their requirement for exogenous auxin. He called this phenomenon *accoutumance à l'auxine* and later *anergie à l'auxine*, which is usually translated as auxin habituation (17, 95). It was soon recognized that similar variation can occur for cytokinins and, more rarely, for certain

0066-4197/89/1215-0395$02.00

vitamins (18, 88). Habituation is now defined more generally as a stable heritable loss in the requirement of cultured plant cells for growth factors (8).

Habituation for growth hormones is of interest for several reasons. First, the classical experiments of Skoog & Miller (84) established that the relative concentrations of auxin and cytokinin in the culture medium regulate the formation of root and shoot meristems from unorganized tissues. It has been proposed that auxin and cytokinin are primary morphogens responsible for the basic body plan of plants (76). Habituation provides an experimental system for studying stable changes in the metabolism of these hormones under precisely controlled conditions in culture. Second, plant development is highly regulative. Cell lineages are not, in general, fixed and highly specialized cells remain totipotent (67, 75). This has led to the view that determination in plants is primarily at the supracellular level (25). Nevertheless, studies of habituation have shown that some determined states can be inherited by individual cells and that these cells undergo a process similar to transdetermination (58). Finally, habituation poses the fundamental problem of how totipotent cells in the same organism can transmit different phenotypes to their mitotic progeny.

The aim of this contribution is to review the basic features of habituation for auxin and cytokinin. I summarize the evidence that habituation results from reversible modifications in cell heredity, known as epigenetic changes, and discuss possible mechanisms for these changes.

THE NATURE OF THE HERITABLE CHANGE

Mitotic Transmission

By comparing the growth response of tissues on media with and without auxin, Gautheret (17) found that during serial propagation, tissues often increase in degree of habituation. He argued that habituation was some type of "enzymic adaptation" resulting in gradual changes in phenotype, which are transmitted to daughter cells. Direct evidence that habituation is inherited by individual cells has come from experiments in which numerous single cells were isolated and the cloned lines assayed for their growth factor requirement. These studies show that the cytokinin- and auxin-autotrophic states of cultured tobacco tissues persist in cloned lines for many cell generations over a period of years (7, 40, 78, 89).

Cloned lines isolated from cytokinin-habituated tissues of tobacco differ widely in their relative growth rate on media with and without added cytokinin (7, 89). These differences can persist when cells are subcloned, indicating that individual cells differ in degree of habituation (50). When slightly habituated clones are serially propagated for long periods of time on cytokinin-free medium, the distribution of subclones obtained from the cloned lines increases in degree of habituation (50). Therefore cytokinin habituation is a gradual rather than all-or-nothing change in cellular heredity.

Mutation versus Epigenetic Change

In principle, habituation could result from mutations, epigenetic changes, or a combination of both processes. Mutations are usually rare, spontaneous changes in genetic constitution that are essentially irreversible. The concept of epigenetic, i.e. developmental, change was proposed to account for the somatic transmission of patterns of gene expression (24, 66). These changes differ from most mutations in several important ways (47). First, the cell-heritable alteration is directed, i.e. it occurs regularly in response to specific inducers. Under inductive conditions the rate of epigenetic change is greater than 10^{-3} per cell generation, whereas well-characterized gametic mutations in plants usually exhibit spontaneous rates of less than 10^{-5} (86). Second, the variant phenotype is potentially reversible, and the reversal process is directed and occurs at higher rates. Third, the range of phenotypes generated by epigenetic changes is limited by the genetic potentiality of the organism. Finally, by definition, epigenetic changes are not transmitted meiotically.

In the case of totipotent plant cells, epigenetic changes can be distinguished from most mutations by regenerating plants from variant clones and assaying tissues from the regenerated plants and their sexual progeny for the variant trait (47). If the variant phenotype persists in tissues of the regenerated plant and is inherited in sexual crosses according to well-established laws of inheritance, it is reasonable to conclude that the variation does not have an epigenetic basis. Failure to detect the variant trait in the regenerated plants is more difficult to interpret. Because certain phenotypes are only expressed in culture or in a tissue-specific fashion, it is essential to assay cultured tissues derived from the same type as the tissue from which the variant was derived. Finally, it is possible that variant cells are no longer organogenetic and that the regenerated plants arise from cells generated by low rates of back mutation. To distinguish between back mutation and directed reversal, the rate at which buds arise must be measured on a per cell generation basis. If the regeneration rate is high, e.g. greater than 10^{-3} per cell generation and the variant phenotype also arises at high rates, then the variants probably result from epigenetic changes.

There have been several attempts to apply such tests to the problem of habituation. Plant regeneration and reversal of habituation were obtained with cloned lines of "White Burley" tobacco habituated for both auxin and cytokinin (40, 41). Sacristán and Wendt-Gallitelli (79) were able to regenerate *Crepis capillaris* plants from nonhabituated tissues and tissues habituated for auxin and cytokinin. They showed that tissues cultured from the regenerants exhibited the nonhabituated phenotype of comparable tissues from seed-grown plants. Although the plants were not regenerated from cloned materials, habituated tissues and the regenerated plants carried a specific chromosomal abnormality absent in the original nonhabituated line, making it likely that the habituated line was of single-cell origin.

Convincing evidence that cytokinin habituation has an epigenetic basis has come from studies of pith tissue from "Havana 425" tobacco, which habituates rapidly under inductive conditions in culture (51). Primary explants of pith tissue habituate for cytokinin when incubated on auxin-containing medium at 35°C, ca. 10°C above the standard culture temperature or in the presence of trace amounts of the cytokinin kinetin (51). Under these conditions, some of these explants form dense, white, hyperplastic nodules, which are cytokinin-habituated. These nodules provide a marker for habituation events. Assuming that such events occur randomly, then the proportion of explants without nodules can be used to estimate the rate of conversion of cytokinin-requiring (C^-) cells to cytokinin-habituated (C^+) cells. This rate was roughly 5×10^{-3} per cell generation (57), which is 100–1000 higher than the rates reported for somatic mutation in *Nicotiana* species (9, 80).

Cytokinin-habituated (C^+) clones regularly form fertile, normal-appearing tobacco plants (7). Pith tissues from these plants exhibit the C^- phenotype and in this regard are indistinguishable from seed-grown control plants regenerated from nonhabituated (C^-) pith cells. Therefore, at some point in the plant regeneration process at least some mitotic progeny of C^+ cells revert to the C^- phenotype. This reversion is not the result of rare back mutations. Recently isolated C^- and C^+ clones form buds, the first step in plant regeneration, at the same rate (52). Moreover, the rate at which C^- buds are initiated from C^- and C^+ cells is high, $>4 \times 10^{-3}$ initiations per cell generation. Therefore, plants either develop directly from progeny or C^+ cells or from revertant cells that arise in cloned tissues at the high rates expected for epigenetic changes.

REGULATION OF HABITUATION

Developmental Factors

The developmental state of cells strongly influences their tendency to habituate. There is a gradient along the stem of the tobacco plant in the incidence of cytokinin habituation of pith explants—high near the top of the plant and low near the bottom (57, 91). Different tissue types also differ in competence for cytokinin habituation (56). Pith tissue of tobacco consists of inducible C^- cells that habituate at high rates when treated with cytokinins and noninducible C^- cells that do not habituate under these conditions. On the other hand, tissue cultured from the lamina of the leaf appears to consist almost exclusively of noninducible C⁻ cells, which are indistinguishable in their cytokinin requirement from noninducible C⁻ cells of pith origin. Finally, tissues cultured from the cortex of the stem do not require cytokinin for proliferation in culture; these tissues consist almost exclusively of constitutive C⁺ cells.

When the different cortex-, leaf-, and pith-derived cells are cloned, the majority of the clones, >95%, exhibit the tissue-specific cytokinin requirement (56). Tissues of plants regenerated from these clones exhibit the cytokinin phenotype expected for comparable tissues of seed-grown plants, which indicates that the different mitotically inherited states are "reset" at some point during plant regeneration (53). Therefore, changes in competence for habituation, like the habituation process itself, have an epigenetic basis.

The cytokinin phenotypes and changes in phenotype of tobacco cells established by cell cloning and plant regeneration experiments are summarized in Figure 1. Three distinct, cell-heritable states of cytokinin habituation have been identified: a stable "on" state, corresponding to constitutive C⁺ cells of stem-cortex origin; a stable "off" state corresponding to noninducible C⁻ cells of pith and leaf origin; and a "bistable" state corresponding to inducible C⁻ cells of pith that can shift back and forth between the C⁻ and C⁺ phenotypes.

Genetic Factors

Ultimately, the variety and stability of epigenetic states is under genetic regulation. Tissues cultured from different plant species, and even different varieties of the same species, often differ in their tendency to habituate (17). This is particularly striking in interspecific genetic hybrids of *Nicotiana* that are tumor-prone (4, 31). When placed in culture, tissues from these hybrids habituate readily for auxin and cytokinin, whereas tissues from the parent species, which are not tumor-prone, require at least one of the two hormones (6, 81). Studies of tumorigenesis-defective mutants and complementation

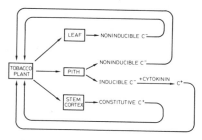

Figure 1 Epigenetic states of cytokinin requirement of cultured tobacco cells. C⁻, cytokinin-requiring phenotype; C⁺, cytokinin-autotrophic phenotype (From Meins (49)).

tests with cells transformed with Ti plasmids defective at loci encoding production of hormones have shown that cytokinin- and auxin-autotrophy are not additive traits (64). Inheritance of the hormone- autotrophic states is complex. One parental species contributes one or several genetic factors controlling tumor initiation, while the other parental species contributes a larger number of genetic factors influencing the degree of growth-factor autotrophy and tumor expression (2, 85).

In the case of intra- and interspecific genetic hybrids of *Phaseolus*, which are not tumor-prone, cytokinin autotrophy is regulated at a single locus (59). There is evidence that the C^- and C^+ phenotypes result from differences in the metabolism of cytokinins with saturated and unsaturated N^6-side chains (39, 60).

Two unlinked genes, *Habituated leaf-1 (H1-1) and Habituated leaf-2 (Hl-2)*, that regulate the cytokinin requirement of cultured tobacco tissues have been identified (54, 55). Plants carrying the partially dominant *Hl-1* allele or the dominant *Hl-2* allele are normal in appearance. They differ from wild-type plants in that tissues cultured from pith and leaf explants consistently exhibit a C^+ phenotype. Thus, the *Hl* alleles convert cells in the "off" and "bistable" states to cells in the "on" state characteristic of cultured stem-cortex tissue.

Cloned lines of C^- leaf cells rapidly give rise to C^+ variants when subcultured on media containing successively lower concentrations of cytokinin (54). Plants regenerated from cloned lines of these variants exhibit an habituated-leaf phenotype that is inherited as a dominate monogenic trait at the *Hl-2* locus. The discovery of this mutation is of particular interest because it demonstrates that cytokinin habituation can result from genetic as well as epigenetic changes.

Estimates based on cell-population models show that C^+ variants arise from cultured C^- cells at high rates, 10^{-3} to 10^{-2} per cell generation, in the range observed for induction and reversion of epigenetic forms of cytokinin habituation (48). Although the original *Hl-2* mutants were recovered from these C^+ variants, too few have been characterized to ascertain whether or not rapid phenotypic variation always results from genetic changes.

The function of the *Hl-2* locus is not known. There is evidence, however, that the *Hl-1* locus has a role in plant tumorigenesis (20, 21). Neoplastic growth of cells transformed with the Ti plasmid depends on the expression of genes in the T-DNA region of the plasmid that is integrated into the genome of the host cell (28). The most important of these are the *tmr* locus, which encodes isopentenyl transferase, a key enzyme in cytokinin biosynthesis, and the *tms1* and *tms2* loci, which encode enzymes for the biosynthesis of the auxin, indole-3-acetic acid (63). Although wild-type tobacco cells transformed with Ti plasmids defective at the *tmr* locus are not tumorigenic, the tumor phenotype is expressed in *Hl-1/Hl-1* cells transformed with these plasmids. Thus, the *Hl-1* allele can compensate for a defective *tmr* gene in the plasmid. Nevertheless, *Hl-1* is not a cellular form of *tmr*; it does not replace the function of *tmr* in tumor inception; it requires expression of T-DNA genes in addition to *tms1*, and *tms2* to give the tumor phenotype (1); and, unlike *tmr*, it does not result in tumor cells with high levels of cytokinins (22).

THE CELLULAR BASIS FOR HABITUATION

Physiological Mechanisms

HORMONE AUTOTROPHY AND HORMONE METABOLISM   Most physiological studies of habituation have focused on the relationship between the autotrophic phenotype and the capacity of cells to produce hormones. Precursor feeding experiments and direct measurements of hormone contents show that habituated cells have the capacity to produce the hormones for which they are habituated (37, 46).

More detailed experiments aimed at measuring differences in the hormonal physiology of habituated and nonhabituated cells have produced conflicting results. There are reports that auxin and cytokinin are not present in either cell type (22, 23, 65, 96); that the hormones are present in both cell types at roughly the same concentration (35, 36); and that hormone concentrations are elevated in habituated cells but not nonhabituated cells (15, 33, 73). These discrepancies reflect the more general problem of relating hormone concentration in a tissue to hormone action (90). The primary active form of the many types of auxins and cytokinins produced by plant cells has not been identified. There is striking clonal and physiological variation in the hormone content of cultured tissues. Sister clones can differ by up to ten-fold in cytokinin and auxin content (22, 72). The levels of these hormones vary dramatically with time after transferring tissues (20, 22, 23, 72), and even change with phases of the cell cycle in partially synchronized cell suspensions (68, 69). Finally, auxin and cytokinin interact in regulating hormone accumulation, making it difficult to assay habituated and nonhabituated cells under comparable physiological conditions. For example, auxin-promoted metabolic degradation of cytokinins (72) appears to be the explanation for the failure to detect the major cytokinin, zeatin riboside, in either C^- or C^+ tobacco cells cultured on auxin-containing medium (22, 23). When these cells are incubated on auxin-free medium, zeatin riboside can be detected in C^+ cells but not in C^- cells; this suggests that C^+ cells have a greater capacity than C^- cells to accumulate the cytokinin.

To summarize, although the issue is not settled, there is evidece that habituation results from an enhanced accumulation by cells of the hormone for which they are habituated. It is not known whether this is due to increased synthesis, decreased degradation, or a combination of both mechanisms.

A BIOCHEMICAL SWITCH MODEL   Two types of mechanisms have been proposed for the mitotic inheritance of habituation: self-perpetuating regulatory circuits and conservative genetic modifications (51). Delbrück (13) suggested that epigenetic variation in protozoa could be accounted for by regulatory circuits that can exist in alternative stable states. Later, this concept was extended to multicellular organisms by Monod & Jacob (62). The simplest bistable system is one in which a metabolite acts autocatalytically to promote its own biosynthesis. Provided there is a mechanism for degrading the metabolite, this positive-feedback loop can exist in an "off" state, in which the steady-state level of the metabolite is low, or in an "on" state in which the steady-state level of the metabolite is high. Provided the feedback loop is not broken when cells divide, it is possible for two genetically identical cells to inherit the "on" and "off" states.

This form of inheritance is well documented in prokaryotic organisms, e.g. "pseudomutation" of the *lac* operon in *E. coli* (70) and the maintenance of the lysogenic state of bacteriophage λ by the cI gene product (74). In the case of higher plants, the hormones ethylene and auxin exhibit autocatalytic physiological regulation that has the potential for generating alternative heritable states. For example, flowers of the Japanese morning glory produce ethylene when treated with ethylene (32). Cells cultured from tumor-prone *Nicotiana glauca* x *langsdorffii* hybrids, sycamore, and tobacco produce the auxin indole-3-acetic acid in response to auxins added to the culture medium (11, 61, 69).

A biochemical switch model for cytokinin habituation has been proposed that specifies that the C^+ state is maintained by a positive-feedback loop in which cytokinins, or related cell-division promoting factors, either induce their own biosynthesis or inhibit their own degradation (49, 51). Evidence for this hypothesis is indirect. First, cytokinins induce cells to shift to the C^+ state. Explants of tobacco-pith tissue rapidly habituate when treated with the cytokinin kinetin at concentrations far below the concentration optimum for proliferation of C^- cells. This induction shows a sharp concentration threshold, as expected for an autocatalytic system (57). Second, if the putative positive-feedback loop is broken by blocking cytokinin production, then C^+ cells should shift to the C^- inducible state. C^+ pith tissue reverts to the C^- inducible phenotype when incubated under nonpermissive conditions thought to inhibit cytokinin accumulation (51). Finally, if cytokinins act autocatalytically to induce their own production, then cytokinins should induce cells to accumulate cytokinins. Provided tissues are incubated on auxin-free medium, treatment with the synthetic cytokinin, kinetin, induces the accumulation of the naturally occurring cytokinin, *trans*-zeatin riboside, by C^- tissues, but does not further elevate the levels of this cytokinin in C^+ tissues (22).

GENETIC MECHANISMS   In principle, habituation could also result from reversible genetic alterations such as the methylation of cytosines in DNA (27), the transposition of controlling elements (14), gene amplification (87), or site-specific recombination (29, 30, 83, 92). Cell-heritable regulation of gene expression by DNA methylation or gene transposition is well documented in higher plants. Plant tissues differ in their degree of DNA methylation, which can be altered by treating seedlings or cultured tissues with hormones (71, 94). Genes introduced into plant cells by Ti-plasmid-mediated transformation can undergo methylation with a concomitant loss of expression (3, 26, 44, 93). In some cases, reversible changes in expression and patterns of methylation of these genes are transmitted meiotically as well as mitotically (44). Methylation is also thought to be the mechanism for heritable inactivation of the transposable elements *Mu* (5, 10), *Ac* (12, 82) and *Spm* (16).

Transposable controlling elements in the *Suppressor-mutator* (Spm) family discovered in maize (45), which are also active when introduced into tobacco cells (42), are of particular interest. These elements can turn on and off the expression of genes with which they are associated. They exhibit reversible, heritable changes in state and can lead to graded alterations in the level of gene expression (16, 43).

Although the formal similarities between habituation and variation caused by DNA methylation and transposable elements are striking, there is at present no evidence that these genetic alterations are involved in the habituation process. What is needed is well-characterized, cloned, marker genes that can be used to screen for differences in DNA sequences in cells expressing different states of habituation.

CONCLUSION

The few cases studied in detail show that habituation, at least for cytokinins, results from gradual, progressive changes in cellular heredity. Habituation occurs at high rates relative to classical mutations and is readily reversible. The capacity to habituate is influenced by the developmental and physiological state of cells. Different states of competence for habituation arise in the plant and persist when cells are cultured. These states are reset at some point in plant regeneration by a process similar to transdetermination.

Cytokinin habituation can result from epigenetic changes, which are reset during plant regeneration, and by genetic changes, which persist through meiosis and are inherited at specific genetic loci. Leaf cells of tobacco bearing the *Hl-2* allele mimic cortex cells in their cytokinin phenotype; this suggests that epigenetic changes affect expression of the *Hl-2* gene or that this gene modifies the resetting of epigenetic states of cytokinin requirement. There is evidence that the *Hl-2* gene can change state at the same high rates as epigenetic changes such as the resetting of leaf and cortex states, habituation of pith cells, and organ initiation in culture (52). This raises the possibility that both forms of habituation result from reversible genetic alterations.

The molecular basis for habituation is not known, nor has it been established that different heritable changes in growth-factor autotrophy have a common underlying mechanism. There is indirect evidence for a biochemical switch model, which can account for cell-heritable changes in the cytokinin requirement of tobacco pith cells. This simple model does not, however, provide an adequate explanation for different degrees of habituation, the progressive nature of habituation, or cell-heritable variation in the competence of cells to habituate.

Whatever its exact mechanism, there are hints that habituation may have a causal role in morphogenesis. Auxin and cytokinin can act as primary morphogens to induce the formation of shoot and root initials from parenchyma cells (84). This involves a stable change of cells from an auxin- and cytokinin-dependent phenotype to an hormone-independent phenotype. Self-perpetuating states of hormone production have been proposed as a basic mechanism for generating the bipolar structure of the plant body (76). The induction of hormone-producing cells by hormones as suggested by the biochemical switch model is also consistent with the "infectious" character of several hormone-dependent developmental processes in plants (38). Thus, it is plausible that gradients of hormones either induce hormone production by competent cells or select for cells differing in hormone requirement and that this results in the formation of specific organized structures (76, 77).

ACKNOWLEDGMENTS

I thank Jean-Pierre Jost, Yoshi Nagamine and Tsvi Sachs for stimulating discussions and Sue Thomas for help in preparing this manuscript.

Literature Cited

1. Aebi, R. 1988. *Eigenschaften eines Pflanzenhormongenes mit potentiell onkogenen Funktionen: Komplementationsstudien mit einem klonierten Hormongen.* Diploma Diss. Univ. Basel, Basel, Switzerland
2. Ahuja, M. R. 1968. An hypothesis and evidence concerning the genetic components controlling tumor formation in *Nicotiana. Mol. Gen. Genet.* 103:176–84
3. Amasino, R. M., Powell, A. L. G., Gordon, M. P. 1984. Changes in T-DNA methylation and expression are associated with phenotypic variation and plant regeneration in a crown gall tumor line. *Mol. Gen. Genet.* 197:437–46
4. Bayer, M. H. 1983. Genetic tumors: physiological aspects of tumor formation in interspecies hybrids. In *Molecular Biology of Plant Tumors,* ed. G. Kahl, J. S. Schell, pp. 33–67. New York: Academic
5. Bennetzen, J. L. 1985. The regulation of *Mutator* function and *Mu* transposition. *UCLA Symp. Mol. Cell. Biol. (NS)* 83:343–53
6. Bennici, A., Buiatti, M., Tognoni, F., Rosellini, D., Giorgi, L. 1972. Habituation in *Nicotiana bigelovii* tissue cul-

tures: Different behavior of two varieties. *Plant Cell Physiol.* 13:1–6
7. Binns, A., Meins, F. Jr. 1973. Evidence that habituation of tobacco pith cells for cell division-promoting factors is heritable and potentially reversible. *Proc. Natl. Acad. Sci. USA* 70:2660–62
8. Butcher, D. N. 1977. Plant tumour cells. In *Plant Tissue and Cell Culture*, ed. H. E. Street, pp. 429–61. Oxford: Blackwell
9. Carlson, P. S. 1974. Mitotic crossing-over in a higher plant. *Genet. Res.* 24:109–12
10. Chandler, V. L., Walbot, V. 1986. DNA modification of a maize transposable element correlates with loss of activity. *Proc. Natl. Acad. Sci. USA* 83:1767–71
11. Cheng, T.-Y. 1972. Induction of indoleacetic acid synthetases in tobacco pith explants. *Plant Physiol.* 50:723–27
12. Chomet, P. S., Wessler, S., Dellaporta, S. L. 1987. Inactivation of the maize transposable element *Activator(Ac)* is associated with DNA modification. *EMBO J.* 6:295–302
13. Delbrück, M. 1949. In the discussion following a paper by T. M. Sonneborn, G. H. Beale. *Colloq. Int. CNRS* 7:25
14. Döring, H.-P., Starlinger, P. 1986. Molecular genetics of transposable elements in plants. *Annu. Rev. Genet.* 20:175–200
15. Dyson, W. H., Hall, R. H. 1972. N^6-(Δ^2-isopentenyl)adenosine: Its occurrence as a free nucleoside in an autonomous strain of tobacco tissue. *Plant Physiol.* 59:45–47
16. Fedoroff, N. V., Banks, J. A., Masson, P. 1989. Developmental determination of *Spm* expression. *UCLA Symp. Mol. Cell. Biol. (NS)* 92:51–63
17. Gautheret, R. J. 1955. Sur la variabilité des proportiétés physiologiques des cultures de tissus végétaux. *Rev. Gen. Bot.* 62:1–106
18. Gautheret, R. J. 1955. The nutrition of plant tissue cultures. *Annu. Rev. Plant Physiol.* 6:433–84
19. Gresshoff, P. M. 1979. Phytohormones and growth and differentiation of cells and tissues cultured *in vitro*. In *Phytohormones and Related Compounds—A Comprehensive Treatise*, ed. D. S. Letham, P. B. Goodwin, T. J. V. Higgins, 2:1–29. Amsterdam: Elsevier/North-Holland
20. Hansen, C. E. 1986. *Studies on the variation of cytokinin content in tobacco tumor cells, and the involvement of host genes for expression of the tumor phenotype.* PhD Diss., Univ. Basel, Basel, Switzerland
21. Hansen, C. E., Meins, F. Jr. 1986. Evidence for a cellular gene with potential oncogenic activity in plants. *Proc. Natl. Acad. Sci. USA* 83:2492–95
22. Hansen, C. E., Meins, F. Jr., Aebi, R. 1987. Hormonal regulation of zeatin-riboside accumulation by cultured cells. *Planta* 172:520–25
23. Hansen, C. E., Meins, F. Jr., Milani, A. 1985. Clonal and physiological variation in the cytokinin content of tobacco-cell lines differing in cytokinin requirement and capacity for neoplastic growth. *Differentiation* 29:1–6
24. Harris, M. 1964. *Cell Culture and Somatic Variation*, New York: Holt, Rinehard & Winston
25. Henshaw, G. G., O'Hara, J. F., Webb, K. J. 1982. Morphogenetic studies in plant tissue cultures. *Symp. Br. Soc. Biol.* 4:231–51
26. Hepburn, A. G., Clarke, L. E., Pearson, L., White, J. 1983. The role of cytosine methylation in the control of nopaline synthase gene expression in a plant tumor. *J. Mol. Appl. Genet.* 2:315–29
27. Holliday, R. 1987. The inheritance of epigenetic defects. *Science* 238:163–70
28. Hooykaas, P. J. J., Schilperoort, R. A. 1984. The molecular genetics of crown gall tumorigenesis. *Adv. Genet.* 22:210–83
29. Iida, S., Meyer, J., Kennedy, K. E., Arber, W. 1982. A site specific conservative recombination system carried by bacteriophage P-1 mapping the recombinase gene *cin* and the crossover sites *cix* for the inversion of the C segment. *EMBO J.* 1:1445–54
30. Kamp, D., Kardas, E., Ritthaler, W., Sandulache, R., Schmucker, R., Stern, B. 1984. Comparative analysis of invertible DNA in phage genomes. *Cold Spring Harbor Symp. Quant. Biol.* 49:301–11
31. Kehr, A. E. 1965. The growth and development of spontaneous plant tumors. *Encycl. Plant Physiol.* 15(2):184–96
32. Kende, H., Baumgartner, B. 1974. Regulation of ageing in flowers of *Ipomoea tricolor* by ethylene. *Planta* 116:279–89
33. Kerbauy, G. B., Monteiro, W. R., Kraus, J. E., Hell, K. G. 1988. Some physiological and structural aspects of cytokinin-autonomy in the callus of tobacco (*Nicotiana tabacum* L.). *J. Plant Physiol.* 132:218–22
34. Kerbauy, G. B., Peters, J. A., Hell, K. G. 1986. Cytokinin autotrophy and differentiation in tissue cultures of

haploid *Nicotiana tabacum* L. *Plant Sci.* 45:125–32

35. Keves, C., Coumans, M., De Greef, W., Hofinger, M., Gaspar, T. 1981. Habituation in sugarbeet callus: auxin content, auxin protectors, peroxydase pattern and inhibitors. *Physiol. Plant.* 51:281–86

36. Köves, E., Szabó, M. 1987. Ethylene production in habituated and auxin-requiring tobacco callus cultures. Does ethylene play a role in the habituation? *Physiol. Plant.* 69:351–55

37. Laloue, M., Pethe-Terrine, C., Guern, J. 1981. Uptake and metabolism of cytokinins in tobacco cells: Studies in relation to the expression of their biological activities. In *Metabolism and Molecular Activities of Cytokinins*, ed. J. Guern, C., Péaud-Lenoël, pp. 80–96. Heidelberg:Springer-Verlag

38. Lang, A. 1965. Progressiveness and contagiousness in plant differentiation and development. *Encyl. Plant Physiol.* 15(1):409–23

39. Lee, Y.-H., Mok, M. C., Mok, D. W. S., Griffin, D A., Shaw, G. 1985. Cytokinin metabolism in *Phaseolus* embryos. Genetic difference and the occurrence of novel zeatin metabolites. *Plant Physiol.* 77:635–41

40. Lutz, A. 1971. Aptitudes morphogénétiques des cultures de tissus d'origine unicellulaire. *Colloq. Int. CNRS* 193:163–68

41. Lutz, A., Belin, C. 1974. Analyse des aptitudes organogénétiques de clones d'origine unicellulaire issus d'une souche anergiée de tabac, en fonction de concentrations variées en acide indolylacétique et en kinétine. *C. R. Acad. Sci. Ser. D* 279:1531–33

42. Masson, P., Fedoroff, N. V. 1989. Mobility of the maize suppressor-mutator element in transgenic tobacco cells. *Proc. Natl. Acad. Sci. USA* 86:2219–23

43. Masson, P., Surosky, R., Kingsbury, J. A., Fedoroff, N. V. 1987. Genetic and molecular analysis of the *Spm-dependent a-m2* alleles of the maize *a* locus. *Genetics* 177:117–37

44. Matzke, M. A., Primig, M., Trnovsky, J., Matzke, A. J. M. 1989. Reversible methylation and inactivation of marker genes in sequentially transformed tobacco plants. *EMBO J.* 8:643–49

45. McClintock, B. 1956. Controlling elements and the gene. *Cold Spring Harbor Symp. Quant. Biol.* 21:197–216

46. Meins, F. Jr. 1982. Habituation of cultured plant cells. See Ref. 4, pp. 3–31

47. Meins, F. Jr. 1983. Heritable variation

in plant cell culture. *Annu. Rev. Plant Physiol.* 34:327–46

48. Meins, F. Jr. 1985. Cell heritable changes during development. In *Plant Genetics*, ed. M. Freeling, pp. 45–59. New York: Liss

49. Meins, F. Jr. 1989. A biochemical switch model for cell-heritable variation in cytokinin requirement. In *The Molecular Basis of Plant Development*, ed. R. Goldberg, pp. 13–24. New York: Liss

50. Meins, F. Jr., Binns, A. 1977. Epigenetic variation of cultured somatic cells: Evidence for gradual changes in the requirement for factors promoting cell division. *Proc. Natl. Acad. Sci. USA* 74:2928–32

51. Meins, F. Jr., Binns, A. N. 1978. Epigenetic clonal variation in the requirement of plant cells for cytokinins. In *The Clonal Basis of Development*, ed. S. Subtelny, I. M. Sussex, pp. 185–201. New York: Academic

52. Meins, F. Jr., Binns, A. N. 1982. Rapid reversion of cell-division-factor habituated cells in culture. *Differentiation* 23:10–12

53. Meins, F. Jr., Foster, R. 1986. Transdetermination of plant cells. *Differentiation* 30:188–89

54. Meins, F. Jr., Foster, R. 1986. A cytokinin mutant derived from cultured tobacco cells. *Dev. Genet.* 7:159–65

55. Meins, F. Jr., Foster, R., Lutz, J. 1983. Evidence for a Mendelian factor controlling the cytokinin requirement of cultured cells. *Dev. Genet.* 4:129–41

56. Meins, F. Jr., Lutz, J. 1979. Tissue-specific variation in the cytokinin habituation of cultured tobacco cells. *Differentiation* 15:1–6

57. Meins, F. Jr., Lutz, J. 1980. The induction of cytokinin habituation in primary pith explants of tobacco. *Planta* 149:402–7

58. Meins, F. Jr., Wenzler, H. 1986. Stability of the determined state. *Symp. Soc. Exp. Biol.* 40:155–70

59. Mok, M. C., Mok, D. W. S., Armstrong, D. J., Rabakoarihanta, A., Kim, S.-G. 1980. Cytokinin autonomy in tissue cultures of Phaseolus: A genotypic-specific and heritable trait. *Genetics* 94:675–86

60. Mok, M. C., Mok, D. W. S., Dixon, S. C., Armstrong, D. J., Shaw, G. 1982. Cytokinin structure-activity relationships and the metabolism of N^6-(Δ^2-isopentenyl)adenosine-8-^{14}C in *Phaseolus* callus tissues. *Plant Physiol.* 70:173–78

61. Moloney, M. M., Hall, J. F., Robinson,

G. M., Elliott, M. C. 1983. Auxin requirements of sycamore cells in suspension culture. *Plant Physiol.* 71:927–31

62. Monod, J., Jacob, F. 1961. General conclusions: Telenomic mechanisms in cellular metabolism, growth, and differentiation. *Cold Spring Harbor Symp. Quant. Biol.* 26:389–401

63. Morris, R. O. 1986. Genes specifying auxin and cytokinin biosynthesis in phytopathogens. *Annu. Rev. Plant Physiol.* 37:509–38

64. Nacmias, B., Ugolini, S., Ricci, M. D., Pellegrini, M. G., Bogani, P., et al. 1987. Tumor formation and morphogenesis on different *Nicotiana* sp and hybrids induced by *Agrobacterium tumefaciens* T-DNA mutants, *Dev. Genet.* 8:61–72

65. Nakajima, H., Yokota, T., Matsumoto, T., Noguchi, M., Takahashi, N. 1979. Relationship between hormone content and autonomy in various autonomous tobacco cells cultured in suspension. *Plant Cell Physiol.* 20:1489–500

66. Nanney, D. L. 1958. Epigenetic control systems. *Proc. Natl. Acad. Sci. USA* 44:712–17

67. Narayanaswamy, S. 1977. Regeneration of plants from tissue cultures. In *Plant Cell, Tissue, and Organ Culture,* ed. J. Reinert, Y. P. S. Bajaj, pp. 179–248. Berlin: Springer-Verlag

68. Nishinari, N., Syono, K. 1980. Identification of cytokinins associated with mitosis in synchronously cultured tobacco cells. *Plant Cell Physiol.* 21:383–93

69. Nishinari, N., Yamaki, T. 1976. Relationship between cell division and endogenous auxin in synchronously cultured tobacco cells. *Bot. Mag.* 89:73–81

70. Novick, A., Weiner, M. 1957. Enzyme induction as an all-or-none phenomenon. *Proc. Natl. Acad. Sci. USA* 43:553–66

71. Palmgren, G., Mattsson, O., Okkels, F. T. 1988. Tissue specific levels of DNA-methylation in plants. *15th Congr. Scand. Soc. Plant Physiol.,* Turku, Finland, July 31–Aug. 5, 1988. *Physiol. Plant J.* 73(2):11A

72. Palni, L. M. S., Burch, L., Horgan, R. 1988. The effect of auxin concentration on cytokinin stability and metabolism. *Planta* 174:231–34

73. Pengelly, W. L., Meins, F. Jr. 1983. Growth, auxin requirement, and indole-3-acetic acid content of cultured crown-gall and habituated tissues of tobacco. *Differentiation* 25:101–5

74. Pirrotta, V. 1976. The λ repressor and its action. *Curr. Top. Microbiol. Immunol.* 74:21–54

75. Poethig, R. S. 1987. Clonal analysis of cell lineage patterns in plant development. *Am. J. Bot.* 74:581–94

76. Sachs, T. 1986. Cellular interaction in tissue and organ development. *Symp. Soc. Exp. Biol.* 40:181–210

77. Sachs, T. 1988. Epigenetic selection: An alternative mechanism of pattern formation. *J. Theor. Biol.* 134:547–59

78. Sacristán, M. D., Melchers, G. 1977. Regeneration of plants from "habituated" and "*Agrobacterium*-transformed" single-cell colonies of tobacco. *Mol. Gen. Genet.* 152:111–17

79. Sacristán, M. D., Wendt-Gallitelli, M. F. 1971. Transformation to auxinautotrophy and its reversibility in a mutant line of *Crepis capillaris* callus culture. *Mol. Gen. Genet.* 110:355–60

80. Sand, S. A., Sparrow, A. H., Smith, H. H. 1960. Chronic gamma irradiation effects on the mutable V and stable R loci of *Nicotiana. Genetics* 45:289–308

81. Schaeffer, G. W., Smith, H. H. 1963. Auxin-kinetin interaction in tissue cultures of Nicotiana species and tumor-conditioned hybrids. *Plant Physiol.* 38:291–97

82. Schwartz, D., Dennis, E. 1986. Transposable activity of the *Ac* controlling element is regulated by its degree of methylation. *Mol. Gen. Genet.* 205: 476–82

83. Silverman, M., Zieg, J., Mandel, G., Simon, M. 1980. Analysis of the functional components of the phage variation system. *Cold Spring Harbor Symp. Quant. Biol.* 45:17–26

84. Skoog, F., Miller, C. O. 1957. Chemical regulation of growth and organ formation in plant tissues cultivated *in vitro. Symp. Soc. Exp. Biol.* 11:118–31

85. Smith, H. H. 1988. The inheritance of genetic tumors in *Nicotiana* hybrids. *J. Hered.* 79:277–83

86. Sparrow, A. H., Sparrow, R. C. 1976. Spontaneous somatic mutation frequencies for flower color in several *Tradescantia* species and hybrids. *Environ. Exp. Bot.* 16:23–43

87. Stark, G. R., Wahl, G. M. 1984. Gene amplification. *Annu. Rev. Biochem.* 53:447–91

88. Street, H. E. 1969. Growth in organized and unorganized systems—knowledge gained by culture of organs and tissue explants. In *Plant Physiology,* ed. F. C. Steward, 5B:2–224. New York: Academic

89. Tandeau de Marsac, N., Jouanneau, J.-P. 1972. Variation de l'exigence en cytokinine de lignées clonales de cellules de tabac. *Physiol. Vég.* 10:369–80

90. Trewavas, A. J., Cleland, R. E. 1983. Is plant development regulated by changes in the concentration of growth substances or by changes in the sensitivity to growth substances? *Trends Biochem. Sci.* 8:354–57

91. Turgeon, R. 1982. Cytokinesis, cell expansion, and the potential for cytokinin-autonomous growth in tobacco pith. *Plant Physiol.* 70:1071–4

92. van de Putte, P., Plasterk, R., Kuijpers, A. 1984. A phage μ *gin* complementing function and an invertible DNA region in *Escherichia coli* K-12 are situated on the genetic element E-14. *J. Bacteriol.* 158:517–22

93. van Slogteren, G. M. S., Hooykaas, P. J. J., Schilperoort, R. A. 1984. Silent T-DNA genes in plant lines transformed by *Agrobacterium tumefaciens* are activated by grafting and 5-azacytidine treatment. *Plant Mol. Biol.* 3:333–36

94. Vanyushin, B. F., Bashkite, E. A., Fridrich, A., Chvoika, L. A. 1981. Methylation of DNA in wheat seedlings and the influence of phytohormones. *Biokhimiya* 46:47–54

95. White, P. R. 1951. Neoplastic growth in plants. *Q. Rev. Biol.* 26:1–16

96. Yokota, T., Takahashi, N. 1981. Hormonal autonomy of tobacco cells. *Int. Assoc. Plant Tissue Culture Newsl.* 34:2–4

Annu. Rev. Genet. 1989. 23:409–23

MOLECULAR STRUCTURE OF HUMAN CHROMOSOME 21

G. D. Stewart, M. L. Van Keuren, J. Galt, S. Kurachi, M. J. Buraczynska

Howard Hughes Medical Institute, Department of Pediatrics, University of Michigan Medical Center, Ann Arbor, Michigan 48109-0650

D. M. Kurnit

Howard Hughes Medical Institute, Departments of Pediatrics and Human Genetics, University of Michigan Medical Center, Ann Arbor, Michigan 48109-0650

CONTENTS

INTRODUCTION

Chromosome 21 is of interest because it is the smallest human autosome, accounting for just under 2% of the genome, yet it is also significant clinically: when present in three copies it causes Down syndrome (35, 52), and

409

0066-4197/89/1215-0409$02.00

anomalies of a gene that maps to chromosome 21 cause a form of familial Alzheimer's disease (21, 88, 99). Its small size coupled with its clinical relevance ensure that this bellwether human chromosome has been and will continue to be the subject of intense scientific study. In particular, we define the organization of this chromosome by examining:

1. Repeated sequences on the short arm of the chromosome.
2. Genes on the long arm.
3. A genetic linkage map of the chromosome.
4. Future gene mapping studies: long-range restriction fragments, sequencing, and isolation of genes expressed in different fetal tissues.

REPEATED SEQUENCES ON 21p

There are three major types of repeated sequences on the short arm of chromosome 21: the ribosomal, alphoid, and 724 sequences.

Ribosomal Sequences

These are found in the stalks of the short arms of all five human acrocentric chromosome pairs (23), i.e chromosomes 13, 14, 15, 21, and 22. There are several hundred copies per genome, consisting of 28S, 18S, and spacer DNAs (17, 31), averaging about 30 tandemly repeated copies per acrocentric chromosome (80). Although numerical variation would be tolerated, major sequence changes of this structural gene would be lethal. Thus quantitative but not qualitative changes in the DNA are the rule for this gene (17, 31, 81). Unequal exchange will tend to homogenize these gene arrays. If a mutant copy becomes amplified, then natural selection will tend to eliminate them.

Alphoid Sequences

These centromeric repeats of a nontranscribed canonical 171-bp repeat are found on all human autosomes (48, 58, 107). At increased stringencies of hybridization, it is possible to distinguish members of this family that have been conserved through evolution. A subset of this family has been identified that localizes to the human acrocentrics (38). Although crossovers between different acrocentric chromosomes (likely to be mitotic (37, 47)) are infrequent per generation, they are frequent enough to be observed en passant (57, 69), and they likely explain why an alphoid family exists that is specific to human acrocentrics in general, but not to an individual acrocentric chromosome. The unfortunate aspect of this inter-acrocentric exchange is that there is unlikely to be an alphoid sequence specific to a given acrocentric chromosome (i.e. chromosome 21). On chromosome 21 (and probably the other acrocentrics) the bulk of the alphoid sequences appear to be located on the short arm (M. L. Van Keuren, unpublished data). Thus, when alphoid probes are hybridized to somatic cell hybrids that contain deletions of the short arm, less hybridization occurs as less alphoid sequence is present. Although the alphoid sequences surround the centromere, the exact relationship between the centromere and alphoid sequences remains unclear.

724 Family

This family of repeated sequences localizes primarily to the human acrocentric chromosomes (49, 50). A rodent-human somatic-cell mapping panel establishes particular EcoRI fragments of this interspersed repetitive sequence on both the proximal and on the distal portions of 21p (50). Unfortunately, again there are no bands that are specific to human chromosome 21, so that this sequence cannot be used in man to diagnose the inheritance of a particular human acrocentric homolog. In fact, the only chromosome 21 short arm sequences useful for this purpose may be ones like pPW235D *(D21S5)* that although they map to more than one chromosome, particular RFLPs may map to a specific chromosome (104).

A related family of 724 sequences, the 7D family, differs from the larger 724 family in that the fragments detected by p21-7D are less numerous and do not cross-hybridize detectably to rodent DNAs (50; D. M. Kurnit, G. D. Stewart, M. L. Van Keuren, unpublished observations). The 7D family extends beyond the 724 sequences to the pericentromeric region of chromosome 21, onto the very proximal region of 21p and perhaps proximal 21q. This means that the centromere of chromosome 21 is likely to be adjacent to, or even sandwiched between 7D sequences.

Taken together, the data for these repetitive families (ribosomal, alphoid, 724, and 7D) along with occasional low-order repetitive sequences on 21p, such as pPW235D *(D21S5),* that are also present on other chromosomes (104) demonstrate that detectable sequence variation on 21p is in the main quantitative rather than qualitative, due to the numerology rather than the sequence-specificity of repetitive human families. Smith (85) suggests that such arrays of tandemly repeating stretches of DNA can maintain their homology by means of recombination between sister chromatids. He also suggests that this mechanism would explain how large variations are found in the numbers of repeats. Thus, the highly conserved sequence of the ribosomal genes can be maintained by unequal crossing over: any new variant that occurs in a tandem array will have a finite probability of increasing and becoming the predominant, and ultimately the only, sequence in the array of natural selection did not act to eliminate it. In the case of ribosomal genes, unequal crossing over will act in conjunction with natural selection to homogenize the array while maintaining a functional intact gene (18). This means that quantification of numerical variation detected as chromosome heteromorphisms using a microscope (34), rather than analysis of sequence changes using the Southern blot (87), is useful to detect heteromorphisms on 21p that define individual number 21 chromosomes. Figure 1 details the pattern of nonribosomal sequence organization on chromosome 21p and the pericentromeric region of that chromosome.

SINGLE-COPY AND REPEATED SEQUENCES ON 21q

In contrast to the sequence organization on 21p discussed above, the DNA organization of 21q is typical of the remainder of the genome, i.e. it consists of single-copy sequences interspersed with moderately repetitive sequences (7). Presumably, the transcriptional pattern of chromosome 21 mirrors the pattern of gene localization on this chromosome, as indicated both by R-banding and clinical criteria (4, 25, 27, 32, 35, 36, 41, 52–54, 62, 64, 68, 72–74, 77, 108). Thus, the finding that the mental retardation and many of the clinical disabilities of Down syndrome is due to three copies of the distal portion of chromosome 21 relates well to the findings that this portion of chromosome 21 is early-replicating, bands darkly following the R-banding procedure, and is relatively gene-rich compared to the remainder of the chromosome (32, 45).

Figure 2 details the pattern of genes that map to chromosome 21q. It can be seen from this figure that the majority of genes mapped so far are assigned to the distal portion of the long arm. The genes were mapped to chromosome 21q using a variety of methods. In the case of susceptibility to Alzheimer's disease, the presumed locus maps to the proximal, apparently pericentromeric region of chromosome 21 (21, 88, 99). Here, the actual locus has not been cloned; rather, its position has been inferred from linkage studies. It is noteworthy that although the β-amyloid locus *(APP)* maps to 21q21 (22, 71, 79, 95), this position is distal to the putative Alzheimer's gene (97, 100).

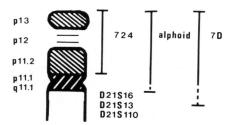

Figure 1 Non-ribosomal DNA sequences on the short arm and in the pericentromeric region of chromosome 21. The short arm appears to consist almost entirely of repeated sequences that may spread over onto the long arm. The exact distance from the centromere of the three single-copy loci shown *(D21S16, D21S13, D21S110)* is not known. They are, however, the loci closest to the centromere of chromosome 21 yet identified.

Thus, the nature of the gene on chromosome 21 responsible for the Alzheimer's phenotype remains in question. Superoxide dismutase-1 *(SOD-1)* has been mapped to 21q22.1 (84) and cloned by Groner and colleagues (56). Although the major phenotype of Down syndrome maps distal to this gene (44), transgenic mice that overexpress *SOD-1* purportedly show some of the phenotypic aspects of Down syndrome such as abnormal tongue morphology (5, 16). Other genes mapping to chromosome 21 that have been cloned include: an estrogen-induced gene in breast cancer *(BCEI)* (11), phosphofructokinase liver type *(PFKL)* (55, 102), α-crystallin *(CRYA1)* (30, 75), leucocyte adhesion molecule-1 *(CD18)* (1, 42, 59, 92), the oncogene *ETS2* (105) and its related sequence *ERG* (76), collagen type VI α1 and α2 *(COL6A1, COL6A2)* (106), and the *S100B* molecule (2). Other genes that have not yet been cloned but that map to chromosome 21 include three enzymes of the purine de novo biosynthesis pathway, phosphoribosyl glycinamide synthetase, phosphoribosyl aminoimidazole synthetase, phosphoribosyl glycinamide formyltransferase *(PRGS, PAIS,* and *PGFT,* respectively) thought to be contained in a multi-functional complex (9, 28, 63, 70), the heat-shock 70 kD protein *(HSP70)* (29), and the α and β interferon receptors *(IFNAR, IFNBR)* (78, 93). The regional mapping of cystathionine β-synthase *(CBS)* was accomplished using a rat cDNA probe (64). In addition to these genes a number of anonymous expressed sequences (11H8, cM21, cE8, 7U, 5L and 3L) have also been cloned and subregionally mapped (67). Of all the mapped genes, only 11H8 and *ETS2* appear to map within the Down syndrome region, based on the work of Korenberg et al (44).

For almost all the genes mapping to chromosome 21, it is possible to find some way of tying each gene in with the Down syndrome phenotype. As mentioned above, *SOD-1* may affect tongue morphology (5, 16). *ETS2* is close to or at the breakpoint found in the M2 subgroup of acute myelogenous

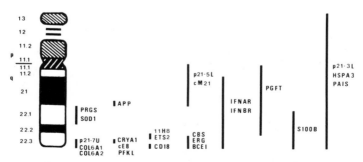

Figure 2   Genes currently mapped to chromosome 21 include cloned genes, enzyme activites, and anonymous cDNAs. All have been mapped using somatic cell hybrids. Reference to the cloning and/or mapping is given in the text.

leukemia (15, 26). The *CD18* gene is implicated in leukocyte adhesion deficiency disease (3), and other immunological problems are found in Down syndrome (98) that could implicate *IFNAR* and *IFNBR* as well as *CD18*. *CRYA1* and *CO16A1* and *CO16A2* map near the Down syndrome region and both gene products are found in the eye. There are eye problems found in Down syndrome patients, including neonatal cataracts (74). Collagen and *CD18* could also play a role in cell migration (109), aggregation and/or structural formation of tissue such as the heart. Finally, *PFKL, CBS,* and the purine biosynthetic genes could have pathogenic consequences during development. Thus, although no gene is directly implicated in Down syndrome pathology as yet, few if any of these genes can be ruled out from playing a role in the syndrome. On the other hand, since the majority of genes on chromosome 21 have yet to be cloned, phenotype-genotype correlations cannot yet be made for most symptoms in Down syndrome. The most likely way such work will be done is by cloning genes within this region by a deliberate strategy rather than by simply mapping genes to this region.

PHYSICAL MAP OF CHROMOSOME 21

Rodent/human somatic cell hybrids that carry different pieces of chromosome 21 have been utilized to detail the mapping pattern of cloned single-copy sequences on chromosome 21q (Figure 3) (101). The map constructed

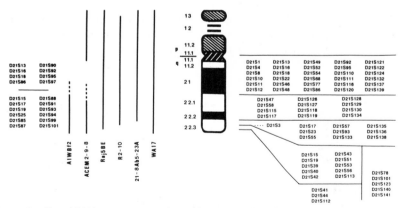

Figure 3   Cloned DNA segments mapping to chromosome 21. The position of each locus (D21S . . .) on chromosome 21 as defined by somatic cell hybrids is listed. The solid lines represent the regions of chromosome 21 present in each hybrid. Broken lines represent ambiguity. Some probes have been mapped against hybrid A1WBf2 and are placed on the left. This breakpoint has not been positioned exactly with respect to the other breakpoints. Thus, most probes mapped using A1WBf2 (61) are also included in the list at the right hand side defined by the remaining 5 hybrids (101). D21S3 is at the breakpoint of the R2-10 hybrid. The loci listed (13, 14, 19, 43, 61, 65, 89, 104) will all appear in the Human Gene Mapping library database.

using these hybrids is concordant with the linkage map of chromosome 21 (96) and with other methodologies. One of these other methodologies is the use of irradiation-reduced hybrids that allows the mapping of fragments of chromosome 21 DNA (24). Unlike 21p, 21q has a normal host of single-copy sequences, and within these there is a normal range of polymorphism. The isolation of polymorphic DNA sequences that span chromosome 21q in turn can be used to create a linkage map (see below), and to determine the parental origin of nondisjunction (13, 89–91) and whether nondisjunction is associated with crossing over. In contrast with an earlier report that nondisjunction was associated with achiasmate meiosis for chromosome 21 (103), it is now apparent that crossing over accompanies a significant plurality of (and perhaps even all) nondisjunction events involving chromosome 21 (8).

The extreme proximal and distal portions of chromosome 21q remain two primary areas of future inquiry. Investigations in the very proximal region of 21q indicate that the centromere is likely to be sandwiched between probe pGSE9 *(D21S16)* and p21-7D (Figure 1). Isolation of probes in the pericentric region will focus on the isolation of sequences at and just distal to the centromere of chromosome 21. Investigations at the distal end of 21q indicate that the most distal polymorphic loci (using rodent-human somatic cell hybrids) are the gene CD18 and the variable nucleotide tandem repeat (VNTR) probe at *D21S112* (82; G. D. Stewart, unpublished data). Walking from these loci will be useful to isolate the most distal polymorphic loci on 21q that can be used to determine whether crossing over on 21q is associated with nondisjunction. Additionally, it should be possible to isolate chromosome 21 VNTRs by screening a chromosome 21-specific cosmid library using the method of Nakamura et al (66). During the past 8 years a large number of probes have been isolated and mapped to chromosome 21 (39). The use of mapping panel hybrids has allowed precise regional mapping of these probes on the long arm of chromosome 21. The large number of probes available has allowed precise characterization of the breakpoints in hybrids used for the mapping, and in a complementary fashion the numerous hybrids available have permitted the mapping of these probes. Although the D21 numbers assigned by Human Gene Mapping 10[1] will exceed 149, several probes have been dropped, have turned out not to be single copy or have not been regionally mapped. Those that have been mapped are shown in Figure 3. The sources of each of the probes for these loci can be found in Human Gene Mapping 9.5 (39). The probes are well spread over chromosome 21, and many identify RFLPs (40).

[1]The Tenth International Workshop on Human Gene Mapping was held June 11–17, 1989. The Human Gene Mapping Library in New Haven, CT maintains an up-to-date listing of all assignments, polymorphisms, etc. At the time of writing, this database contained listings up to D21S149. Updates are made on a regular basis.

LINKAGE MAP OF CHROMOSOME 21

Tanzi et al (96) have constructed a linkage map of chromosome 21 (Figure 4). This map orders a number of single-copy probes on 21q that detect polymorphisms. The linkage map thus produced is in agreement with the map produced by use of somatic cell hybrids. This endeavor has also enabled the determination of where crossovers occur on this smallest human autosome. The answer is concordant with results found in other species: the terminalization of chiasma is associated with more crossovers distally on chromosome 21 (33). This result underscores the importance of having distal polymorphic loci on 21q (such as CD18 and D21S112 (Figure 3)) to detect crossovers between the centromere and the distal end of chromosome 21q.

PROSPECTS FOR FUTURE GENE MAPPING ON CHROMOSOME 21

Irradiation Hybrids

A new method of producing irradiation hybrids has been developed (12). A hamster hybrid cell line containing human chromosome 21 only is heavily irradiated and fused to an unirradiated hamster cell line. Human chromosome 21 fragments are then identified by hybridizing chromosome 21-specific DNA probes to DNA from the cell ines. The distance between the probes appears to

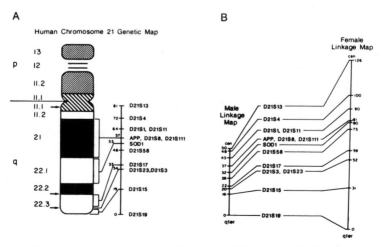

Figure 4   Sex-averaged and male/female linkage maps of chromosome 21. Includes polymorphic loci for cloned genes and random DNA segments. Distances between loci are in centimorgans. Reproduced with permission from Tanzi et al (96).

be in proportion to the coincident occurrence of probes in each line, i.e. if the probes always occur together they are likely to be on the same fragment and map closely together. Analysis of large numbers of cell lines allows construction of a map. This technique may prove very useful in bridging the gap between long-range (linkage and somatic cell hybrids) and short-range (pulsed field gel electrophoresis) methods of analysis.

Long-Range Restriction Digests

A number of investigators (20) have embarked on using long-range restriction digests to map chromosome 21. Coupling the physical methodologies mentioned above to methods designed to slice up chromosome 21 (which contains about 50×10^7 base pairs (60)) into pieces ranging from several hundred kilobases to several megabases will yield even finer maps of chromosome 21. Although tentative at this stage, these long-range restriction maps will clearly become standard and requisite in the very near future.

Sequence of Chromosome 21

The preparation of long-range restriction maps is prerequisite for another project, the sequencing of chromosome 21. Upon preparation of long-range restriction maps of chromosome 21 (some of which will be complementary, but this is reasonable since complementarity will involve the repetitiveness necessary to verify such maps), the sequencing of chromosome 21 can proceed in an ordered fashion. Although it remains unclear whether fluorescent sequencing (86), multiplex sequencing (10), or a combination will be most efficient, the technology for rapid genomic sequencing will soon be brought to bear on chromosome 21. This technology, coupled with the mapping methodologies discussed above, promises to yield strategic insights into the organization of chromosome 21.

Isolation of Genes Encoded by Chromosome 21

Although successful completion (which will require 5–10 years) of the above protocols will yield the determination of open reading frames on chromosome 21, such analyses still will not reveal the tissues, timing, or level of expression of various genes encoded by chromosome 21. For these purposes, analysis of trisomy 21 requires isolation of these expressed sequences on chromosome 21. Of the 10^4–10^5 sequences encoded by the human genome (6), 21q by size and by banding should encode 100–300 genes. Although ca. 24 genes have been mapped to chromosome 21 to date, no obvious phenotype of Down syndrome has been explained unequivocally by the overexpression of a particular gene. Presently, the correlation between increased expression of α-crystallin and congenital cataracts represents the best example of a relationship between overexpression and mapping on chromosome 21.

However, the major aspects of the phenotype (mental retardation, short stature and congenital heart defects) of Down syndrome (74) remain unexplained. Explanation of these phenotypic effects will require analyses of other genes.

To achieve analyses of genes on chromosome 21, our laboratory has adapted the recombination-based methodology of Seed (83) for the twin purposes of gene isolation and mapping, i.e. the reverse mapping of genes expressed by chromosome 21. In our modification, flow-sorted (51, 94) genomic DNA from chromosome 21 is cloned in a miniplasmid vector and allowed to recombine with genic DNA (cDNA) cloned in the specialized phasmid vector, Sumo 15A (46). In this manner, we propose to isolate expressed sequences from chromosome 21. By isolating all the genes in the "Down syndrome region" of chromosome 21, the "Down syndrome genes" will be identified. Our specialized method should allow us to achieve this goal. The success of this protocol in pilot studies demonstrates that coupling such an enterprise to the genomic initiative to sequence the human genome (starting with chromosome 21) should prove fruitful for deciphering the molecular basis for trisomy 21 and the successful construction of a transcriptional map of this chromosome.

ACKNOWLEDGMENTS

Supported by (NIH) 5-RO1-HL37703–02 and March of Dimes Birth Defects Foundation grant #6-501. D.M.K. is an Investigator of the Howard Hughes Medical Institute.

Literature Cited

1. Akao, Y. Utsumi, K. R., Naito, K. Ueda, R., Takahashi, T., et al. 1987. Chromosomal assignments of genes coding for human leucocyte common antigen, T-200 and lymphocyte function-associated antigen-1, LFA-1 subunit. *Somat. Cell Mol. Genet.* 13:273–78

2. Allore, R., O'Hanlon, D., Price, R., Neilson K., Willard, H. F., et al. 1988. Gene encoding the beta subunit of S100 protein is on chromosome 21: implications for Down syndrome. *Science* 239:1311–13

3. Anderson, D. C., Springer, T. A . 1987. Leukocyte adhesion deficiency: an inherited defect in the Mac-1, LFA-1, and p150,95 glycoproteins. *Annu. Rev. Med.* 38:175–94

4. Aula, P., Leisti, J., von Koskull, H. 1973. Partial trisomy 21. *Clin. Genet.* 4:241–51

5. Avraham, K. B., Schickler, M., Sapoznikov, D., Yarom, R., Groner, Y. 1988. Down's syndrome: Abnormal neuromuscular junction in tongue of transgenic mice with elevated levels of human Cu/Zn-superoxide dismutase. *Cell* 47: 823–29

6. Bishop, J. O. 1974. The gene numbers game. *Cell* 21:81–85

7. Britten, R. J., Kohne, D. E. 1968. Repeated sequences in DNA. Hundreds of thousands of copies of DNA sequences have been incorporated into the genomes of higher organisms. *Science* 161:529–40

8. Buraczynska, M., Stewart, G. D., Sherman, S., Freeman, V, Grantham, M., et al. 1988. Studying nondisjunction of chromosome 21 with cytogenetic markers on the short arm and DNA markers encompassing the long arm. In *Ann. NY Acad. Sci.* (In press)

9. Chadefaux, B., Allard, D., Réthoré, M. O., Raoul, O., Poissonnier, M., et al. 1984. Assignment of human phospho-

ribosylglycinamide synthetase locus to region 21q22.1. *Hum. Genet.* 66:190–92

10. Church, G. M., Keiffer-Higgins, S. 1988. Multiplex DNA sequencing. *Science* 240:185–88

11. Cohen-Haguenauer, O., Van Cong, N., Prud'homme, J. F., Jegou-Foubert, C., Gross, M. S., et al. 1985. A gene expressed in human breast cancer and regulated by estrogen in MCF-7 cells is located on chromosome 21. *8th Int. Workshop Hum. Gene Mapp. Cytogenet. Cell Genet.* 40:606

12. Cox, D. R., Price, E. R., Burmeister, M., Sheffield, V., Murray, E., et al. 1988. Fine structure genetic analysis of human chromosome 21 using radiation hybrid mapping. *Am. J. Hum. Genet.* 43(Suppl.):A140(0560)

13. Davies, K. E., Harper, K., Bonthron, D., Krumlauf, K., Polkey, A., et al. 1984. Use of a chromosome 21 cloned DNA probe for the analysis of nondisjunction in Down syndrome. *Hum. Genet.* 66:54–56

14. Donis-Keller, H., Green, P., Helms, C., Cartinhour, S., Weiffenbach, B., et al. 1987. A genetic linkage map of the human genome. *Cell* 51:319–37

15. Drabkin, H., Van Keuren, M., Sacchi, N., Papas, T., Patterson, D. 1986 Localization of the chromosome 21 Huets-2 gene and translocation in t(8:21) AML. *J. Cell Biochem* (Suppl.) 10A:18

16. Epstein, C. J., Avraham, K. B., Lovett, M., Smith, S., Elroy-Stein, O., et al. 1987. Transgenic mice with increased CuZn-superoxide dismutase activity: an animal model of dosage effects in Down's syndrome. *Proc. Natl. Acad. Sci. USA* 84:8044–48

17. Evans, H. J., Buckland, R. A., Pardue, M. L. 1974. Location of the genes coding for 18S and 28S ribosomal RNA in the human genome. *Chromosoma* 48:405–26

18. Frankham, R., Briscoe, D. A., Nurthen, R. K. 1978. Unequal crossing over at the rRNA locus as a source of quantitative variation. *Nature* 272:80–81

19. Galt, J., Boyd, E., Connor, J. M., Ferguson-Smith, M. A. 1989. Isolation of chromosome-21-specific DNA probes and their use in the analysis of nondisjunction in Down syndrome. *Hum. Genet.* 81:113–19

20. Gardiner, K., Watkins, P., Münke, M., Drabkin, H., Jones, C., et al 1988. Partial physical map of human chromosome 21. *Somat. Cell Mol. Genet.* 14:623–38

21. Goate, A. M., Haynes, A. R., Owen, M. J., Farrall, M., James, L. A., et al.

1989. Six families with early onset Alzheimer's disease show linkage between the disease and DNA markers on the long arm of chromosome 21: No evidence for non-allelic genetic heterogeneity. *Lancet* i:352–54

22. Goldgaber, D., Lerman, M. I., McBridge, O. W., Saffiotti, U., Gajdusek, D. C. 1987. Characterization and chromosomal localization of a cDNA encoding brain amyloid of Alzheimer's disease. *Science* 235:877–80

23. Goodpasture, C., Bloom., S., Hsu, T.C., Arright, F. E. 1976. Human nucleolus organizers: the satellites or the stalks. *Am. J. Hum. Genet.* 28:559–66

24. Graw, S., Davidson, J., Gusella, J., Watkins, P., Tanzi, R., et al. 1988. Irradiation-reduced human chromosome 21 hybrids. *Somat. Cell Mol. Genet.* 14:233–42

25. Habedank, M., Rodewald, A. 1982. Moderate Down's syndrome in three siblings having partial trisomy 21q22.2→qter and therefore no *SOD-1* excess. *Hum. Genet.* 60:74–77

26. Hagemeijer, A., Garson, O. M., Kondo, K. 1982. 4th Int. Workshop Chromosome Leuk. 1982 (1984): Translocation (8;21)(q22;q22) in acute nonlymphocytic leukemia. *Cancer Genet Cytogenet.* 11:284–87

27. Hagemeijer, A., Smit, E. M. E. 1977. Partial trisomy 21. Further evidence that trisomy of band 21q22 is essential for Down's phenotype. *Hum Genet.* 38:15–23

28. Hards, R. G., Benkovic, S. J., Van Keuren, M. L., Graw, S. L., Drabkin, H. A., et al. 1986. Assignment of a third purine biosynthetic gene (glycinamide ribonucleotide transformylase) to human chromosome 21. *Am. J. Hum. Genet.* 39:179–85

29. Harrison, G. S., Drabkin, H. A., Koa, F. T., Hart, J., Hart, I. M., et al. 1987. Chromosomal location of human genes encoding major heat-shock proteins. *Somat. Cell Mol. Genet.* 13:119–30

30. Hawkins, J. W., Van Keuren, M. L., Piatigorsky, J., Liao, L. M., Patterson, D., Kao, F. T. 1987. Confirmation of assignment of the human alpha crystallin gene (CRYA1) to chromosome 21 with regional localization to q22.3. *Hum. Genet.* 76:375–80

31. Henderson, A. S., Warburton, D., Atwood, K. C. 1972. Location of ribosomal DNA in the human chromosome complement. *Proc. Natl. Acad. Sci. USA* 69:3394–98

32. Hoehn, H. W. 1975. Functional im-

plications of differential chromosome banding. *Am. J. Hum. Genet.* 27:676–86

33. Hulten, M., Saadallah, N., Wallace, B. M. N., Cockburn, D. J. 1985. Meiotic investigations of aneuploidy in the human. *Basic Life Sci.* 36:75–90

34. Jacobs, P. A. 1977. Human chromosome heteromorphisms (variants). *Prog. Med. Genet. II:* 252–71

35. Jacobs, P. A., Baikie, A. G., Brown, W. M. C., Strong, J. A. 1959. The somatic chromosomes in mongolism. *Lancet* 1:710–11

36. Jeziorowska, A., Jakubowski, L., Lach, J., Kaluzewski, B. 1988. Regular trisomy 21 not accompanied by increased copper-zinc superoxide dismutase *(SOD1)* activity. *Clin. Genet.* 33:11–19

37. John, B., Miklos, G. L. G. 1979. Functional aspects of satellite DNA and heterochromatin. *Int. Rev. Cytol.* 58:1–109

38. Jørgensen, A. L., Bostock, C, J., Bak, A. L. 1987. Homologous subfamilies of human alphoid repetitive DNA on different nucleolus organizing chromosomes. *Proc. Natl. Acad. Sci. USA* 84:1075–79

39. Kaplan, J.-C., Emanuel, B. 1988. Report of the committee on the genetic constitution of chromosome 22. Human Gene Mapping 9.5: Update to the *9th Int. Workshop Hum. Gene Mapp. Cytogenet. Cell Genet.* 49:104–6

40. Kidd, K. K., Bowcock, A. M., Pearson, P. L., Schmidtke, J., Willard, H. F. et al. 1988. Report of the committee on human gene mapping by recombinant DNA techniques. Human Gene Mapping 9.5: Update to the *9th Int. Workshop Hum. Gene Mapp. Cytogenet. Cell Genet.* 49:132–218

41. Kirkilionis, A. J. 1986. Down syndrome with apparently normal chromosomes: An update. *J. Pediatr.* 108:793–94

42. Kishimoto, T. K., O'Connor, K., Lee, A., Roberts, T. M., Springer, T. A. 1987. Cloning of the β subunit of the leukocyte adhesion proteins: Homology to an extracellular matrix receptor defines a novel supergene family. *Cell* 48:681-90

43. Korenberg, J. R., Croyle, M. L., Cox, D. R. 1987. Isolation and regional mapping of DNA sequences unique to human chromosome 21. *Am. J. Hum. Genet.* 41:963–78

44. Korenberg, J. R., Pulst, S. M., Kawashima, H., Ikeuchi, T., Yamamoto, K., et al. 1988. Familial Down syndrome with normal karyotype: molecular definition of the region. *Am. J. Hum. Genet.* 43(Suppl.):A110 (0439)

45. Korenberg, J. R., Rykowski, M. C. 1988. Human genome organization: Alu, lines, and the molecular structure of metaphase chromosome bands. *Cell* 53:391–400

46. Kurachi, S., Baldori, N., Kurnit, D. M. 1989. Sumo 15A: A lambda phasmid that permits selection and counterselection for supF in the plasmid πVX recombination-based assay for nucleotide sequence homologies. *Gene.* In press

47. Kurnit, D. M. 1979. Satellite DNA and heterochromatin variants: The case for unequal mitotic crossing over. *Hum. Genet.* 47:169–86

48. Kurnit, D. M., Maio, J. J. 1974. Variable satellite DNA's in the African green monkey, *Cercopithecus aethips. Chromosoma* 45:387–400

49. Kurnit, D. M., Neve, R L., Morton, C. C., Bruns, G. A. P., Ma, N. S. F., et al. 1984. Recent evolution of DNA sequence homology in the pericentromeric region of human acrocentric chromosomes. *Cytogenet. Cell Genet.* 38:99–105

50. Kurnit, D. M., Roy, S., Stewart, G. D., Schwedock, J., Neve, R. L., et al. 1986. The 724 family of DNA sequences is interspersed about the pericentromeric regions of human acrocentric chromosomes. *Cytogenet. Cell Genet.* 43:109–16

51. Latt, S. A., Lalande, M., Kunkel, L. M., Schreck, R., Tantravahi, U. 1985. Applications of fluorescence spectroscopy to molecular cytogenetics. *Biopolymers* 24:77–95

52. Lejeune, J., Gauthier, M., Turpin, R. 1959. Etudes des chromosomes somatiques de neuf enfants mongoliens. *C. R. Acad. Sci.* 248:1721–22

53. Leonard, C., Gauthier, M., Sinet, P. M., Selva, J., Huret, J. L. 1986. Two Down syndrome patients with rec (21),dupq,inv(21)(p11;q2109) from a familial pericentric inversion. *Ann. Genet.* 29:181–83

54. Leschot, N. J., Slater, R. M., Joenje, H., Becker-Bloemkolk, M. J., de Nef, J. J. 1981. SOD-A and chromosome 21. *Hum. Genet.* 57:220–23

55. Levanon, D., Danciger, E., Dafni, N., Groner, Y. 1987. Construction of a cDNA clone containing the entire coding region of the human liver-type phosphofructokinase. *Biochem. Biophys. Res. Comm.* 147:1182–87

56. Lieman-Hurwitz, J., Dafni, N., Lavie, V., Groner, Y. 1982. Human cytoplasmic superoxide dismutase cDNA clone: a probe for studying the molecular biol-

ogy of Down's syndrome. *Proc. Natl. Acad. Sci. USA* 79:2808–11

57. Livingston, G. K., Lockey, J. E., Witt, K. S., Rogers, S. W. 1985. An unstable giant satellite associated with chromosomes 21 and 22 in the same individual. *Am. J. Hum. Genet.* 37:553–60

58. Maio, J. J. 1971. DNA strand reasociation and polyribonucleotide binding in the African green monkey, *Cercopithecus aethiops. J. Molec. Biol.* 56:579–95

59. Marlin, S. D., Morton, C. C., Anderson, D. C., Springer, T. A. 1986. LFA-1 Immunodeficiency disease. Definition of the genetic defect and chromosomal mapping of α and β subunits of the lymphocyte function-associated antigen 1 (LFA-1) by complementation in hybrid cells. *J. Exp. Med.* 164:855–67

60. Mendelsohn, M. L., Mayall, B. H., Bogart, E., Moore, D. H., Perry, B. H. 1973. DNA content and DNA-based centromeric index of the 24 human chromosomes. *Science* 179:1126–29

61. Millington-Ward, A., Wassenaar, A. L. M., Pearson, P. L. 1985. Restriction fragment length polymorphic probes in the analysis of Down's syndrome. *8th Int. Workshop Hum. Gene Mapp. Cytogenet. Cell Genet.* 40:699

62. Miyazaki, K., Yamanaka, T., Ogasawara, N. 1987. A boy with Down's syndrome having recombinant chromosome 21 but no SOD-1 excess. *Clin. Genet.* 32:383–87

63. Moore, E. E., Jones, C., Kao, F-T., Oates, D. C. 1977. Synteny between glycineamide ribonucleotide synthetase and superoxide dismutase (soluble). *Am. J. Hum. Genet.* 29:389–96

64. Münke, M., Kraus, J. P., Ohura, T., Francke, U. 1988. The gene for cystathionine β-synthase (CBS) maps to the subtelomeric region on human chromosome 21q and to proximal mouse chromosome 17. *Am. J. Hum. Genet.* 42:550–59

65. Nakamura, Y., Carlson, M., Krapscho, K., Ballard, L., Leppert, M., et al. 1988. Isolation and mapping of a polymorphic DNA sequence (pMCT15) on chromosome 21 (D21S113). *Nucl. Acids. Res.* 16:9882

66. Nakamura, Y., Carlson, M., Krapcho, K., Kanamori, M., White, R. 1988. New approach for isolation of VNTR markers. *Am. J. Hum. Genet.* 43:854–59

67. Neve, R. L., Stewart, G. D., Newcomb, P., Van Keuren, M. L., Patterson, D., et al. 1986. Human chromosome 21-encoded cDNA clones. *Gene* 49:361–69

68. Niebuhr, E. 1974. Down's syndrome: The possibility of a pathogenetic segment on chromosome no. 21. *Humangenetik* 21:99–101

69. Nielsen, J., Freidrich, U., Hreidarsson, A. B., Noel, B., Quack, B., et al. 1974. Brilliantly fluorescing enlarged short arms D or G. *Lancet* 1:1949–50

70. Patterson, D., Graw, S., Jones, C. 1981. Demonstration by somatic cell genetics of correidnate regulation of genes: two enzymes of purine synthesis assigned to human chromosome 21. *Proc. Natl. Acad. Sci. USA* 78:405–9

71. Patterson, D., Gardiner, K., Kao, F-T., Tanzi, R., Watkins, P., et al. 1988. Mapping of the gene encoding the β-amyloid precursor protein and its relationship to the Down syndrome region of chromosome 21. *Proc. Natl. Acad. Sci. USA* 85:8266–70

72. Pellissier, M. C., Laffage, M., Philip, N., Passage, E., Mattei, M.-G., Mattei, J.-F. 1988. Trisomy 21q223 and Down's phenotype correlation evidenced by in situ hybridization. *Hum. Genet.* 80:277–81

73. Pfeiffer, R. A., Kessel, E. K., Soer, K-H. 1977. Partial trisomies of chromosome 21 in man. Two new observations due to translocations 19;21 and 4;21. *Clin. Genet.* 11:207–13

74. Peuschel, S. M. 1982. Biomedical aspects in Down syndrome. In *Down Syndrome: Advances in Biomedicine and the Behavioral Sciences*, ed S. N. Pueschel, J. F. Rynders, pp. 383–89. Cambridge: Ware. 524 pp.

75. Quax-Jeuken, Y., Quax, W., Van Rens, G., Meera Kahn, P., Bloemendal, H., 1985. Complete structure of the αB-crystallin gene: conservation of the exon-intron distribution in the two nonlinked α-crystallin genes. *Proc. Natl. Acad. Sci. USA* 82:5819–23

76. Rao, V. N., Papas, T. S., Reddy E. S. P. 1987. *Erg*, a human ets-related gene on chromosome 21: alternative splicing, polyadenylation and translation. *Science* 237:635–39

77. Raoul, O., Carpentier, S., Dutrillaux, B., Mallet, R., Lejeune, J. 1976. Trisomies partielles du chromosome 21 par translocation maternelle 5(15;21) (q262;q2). *Ann. Genet.* 19:187–90

78. Raziuddin, A., Sarkar, F. H., Dutkowski, R., Schulman, L., Ruddle, F. H., et al. 1984. Receptors for human α and β interferon but not γ interferon are specified by human chromosome 21. *Proc. Natl. Acad. Sci. USA* 81:5504–8

79. Robakis, N. K., Ramakrishna, N., Wolfe, G., Wisniewski, H. M. 1987.

Molecular cloning and characterization of a cDNA encoding the cerebrovascular and the neuritic plaque amyloid peptides. *Proc. Natl. Acad. Sci. USA* 84:4190–94

80. Schmickel, R. D. 1973. Quantitation of human ribosomal DNA: hybridization of human DNA with ribosomal RNA for quantitation and fractionation. *Pediatr. Res.* 7:5–12

81. Schmickel, R. D., Knoller, M. 1977. Characterization and location of the human genes for ribosomal ribonucleic acid. *Pediatr. Res.* 11:929–35

82. Schumm, J. W., Knowlton, R. G., Braman, J. C., Barker, D. F., Botstein, D., et al. 1988. Identification of more than 500 RFLPs by screening random genomic clones. *Am. J. Hum. Genet* 42:143–59

83. Seed, B. 1983. Purification of genomic sequences from bacteriophage libraries by recombination and selection in vivo. *Nucl. Acids Res.* 11:2427–45

84. Sinet, P. M., Coururier, J., Dutrillaux, B., Poissonnier, M., Raoul, O., et al. 1976. Trisomie 21 et superoxide dismutase-1 (IPO-A): tentative de localisation sur la sous-bande 21q22.1. *Exp. Cell Res.* 97:47–55

85. Smith, G. P. 1976. Evolution of repeated DNA sequences by unequal crossover. *Science* 191:528–35

86. Smith, L. M., Sanders, J. Z., Kaiser, R. J., Hughes, P., Dodd, C., et al. 1986. Fluorescence detection in automated DNA sequence analysis. *Nature* 321:674–79

87. Southern, E. M. 1975. Detection of specific sequences among DNA fragments separated by gel electrophoresis. *J. Mol. Biol.* 98:503–17

88. St. George-Hyslop. P. H., Tanzi, R. E., Polinsky, R. J., Haines, J. L., Nee, L., et al. 1987. The genetic defect causing familial Alzheimer's disease maps on chromosome 21. *Science* 235:885–90

89. Stewart, G. D., Harris, P., Galt, J., Ferguson-Smith, M. A. 1985. Cloned DNA probes regionally mapped to human chromosome 21 and their use in determining the origin of nondisjunction. *Nucl. Acids. Res.* 13:4125–32

90. Stewart, G. D., Hassold, T. J., Berg, A., Watkins, P., Tanzi, R., Kurnit, D. M. 1988. Trisomy 21 (Down syndrome): Studying nondisjunction and meiotic recombination using cytogenetic and molecular polymorphisms that span chromosome 21. *Am. J. Hum. Genet.* 42:227–36

91. Stewart, G. D., Hassold, T. J., Kurnit, D. M. 1988. Trisomy 21: Molecular and cytogenetic studies of nondisjunction. *Adv. Hum. Genet.* 17:99–140

92. Stewart, G. D., Tanzi, R. E., Kishimoto, T. K., Buraczynska, M., Haines, J. L. et al. 1988. The CD18 gene maps to distal chromosome 21q22.3: RFLPs create a highly informative terminal haplotype. *Am. J. Hum. Genet.* 43(Suppl.):A160(0636)

93. Tan, Y. H., Tischfield, J., Ruddle, F. H. 1973. The linkage of genes for the human interferon-induced antiviral protein and indophenol oxidase-B traits to human chromosome G-21. *J. Exp. Med.* 137:317–30

94. Tantravahi, U., Stewart, G. D., Van Keuren, M., McNeil, G., Roy, S., et al. 1988. Isolation of DNA sequences on human chromosome 21 by application of a recombination-based assay to DNA from flow-sorted chromosomes. *Hum. Genet.* 79:196–202

95. Tanzi, R. E., Gusella, J. F., Watkins, P. C., Bruns, G. A. P., St. George-Hyslop, P., Van Keuren, M. L., et al. 1987. Amyloid β protein gene: cDNA, mRNA distribution, and genetic linkage near the Alzheimer locus. *Science* 235:880-84

96. Tanzi, R. E., Haines, J. L., Stewart, G. D., Watkins, P. C., Gibbons, K. T., et al. 1988. Genetic linkage map for the long arm of human chromosome 21. *Genomics* 3:129–36

97. Tanzi, R. E., St. George-Hyslop, P. H., Haines, J. L., Polinsky, R. J., Nee, L., et al. 1987. The genetic defect in familial Alzheimer's disease is not tightly linked to the amyloid β-protein gene. *Nature* 329:156–57

98. Thase, M. D. 1982. Longevity and mortality in Down's syndrome. *J. Ment. Defic. Res.* 26:177

99. Van Broeckhoven, C., Backhovens, H., Van Hul, W., Van Camp. G., Stinissen, P., et al. 1988 Genetic linkage analysis in early onset familial Alzheimer's dementia. *Neuropsychopharmacology*, ed. W. E. Bunney Jr., H. Hippius, G. Laakmann, G. M. Schmauss. *Proc. 16th Congr. Int Neur. Psychopharmacol.* Munich, Aug. Springer-Verlag. In press

100. Van Broeckhoven, C., Genthe, A. M., Vandenberghe, A., Horsthemke, B., Backhovens, H., et al. 1987. Failure of familial Alzheimer's disease to segregate with the A4-amyloid gene in several European families. *Nature* 329:153–55

101. Van Keuren, M. L., Watkins, P. C., Drabkin, H. A., Jabs, E. W., Gusella, J. F., et al. 1986. Regional localization of DNA sequences on chromosome 21 us-

ing somatic cell hybrids. *Am. J. Hum. Genet.* 38:793–804

102. Vora, S., Francke, U. 1981. Assignment of the human gene for liver-type 6-phosphofructokinase isozyme (PFKL) to chromosome 21 using somatic cell hybrids and anti-L antibody. *Proc. Natl. Acad. Sci. USA* 78:3738–42

103. Warren, A. C., Chakravarti, A., Wong, C., Slaugenhaupt, S. A., Holloran, S. L., et al. 1987. Evidence for reduced recombination of the nondisjoined chromosomes 21 in Down syndrome. *Science* 237:652–54

104. Watkins, P. C., Tanzi, R. E., Gibbons, K. T., Ricoli, J. V., Landes, G., et al. 1985. Isolation of polymorphic DNA segments from human chromosome 21. *Nucl. Acids Res.* 13:6075–88

105. Watson, D. K., McWilliams-Smith, M. J., Nunn, M. F., Duesberg, P. H., O'Brien, S. J., et al. 1985. The ets sequence from the transforming gene of avian erythroblastosis virus, E26, has unique domains on human chromosomes 11 and 21: Both loci are transcriptionally active. *Proc. Natl. Acad. Sci USA* 82:7294–98

106. Weil, D., Mattei, M-G., Passage, E., Cong, N. V., Pribula-Conway, D., et al. 1988. Cloning and chromosomal localization of human genes encoding the three chains of type VI collagen. *Am. J. Hum. Genet.* 42:435–45

107. Willard, H F., Waye, J. S. 1987. Hierarchical order in chromosome-specific human alpha satellite DNA. *Trends Genet.* 3:192–98

108. Williams, J. D., Summitt, R. L., Martens, P. R., Kimbrell, R. A. 1975. Familial Down syndrome due to t(10;21) translocation: Evidence that the Down phenotype is related to trisomy of specific segment of chromosome 21. *Am. J. Hum. Genet.* 27:478–85

109. Wright, T. C., Orkin, R. W., Destrempes, M., Kurnit, D. M. 1984. Increased adhesiveness of Down syndrome fetal fibroblasts *in vitro*. *Proc. Natl. Acad. Sci. USA* 81:2426–30

Annu. Rev. Genet. 1989. 23:425–53

POPULATION GENETICS AND EVOLUTION OF SPECIES RELATED TO *DROSOPHILA MELANOGASTER*

R. S. Singh

Department of Biology, McMaster University, Hamilton, Ontario, Canada L8S4K1

CONTENTS

0066-4197/89/1215-0425$02.00

INTRODUCTION

Drosophila melanogaster has played an important role in the development of experimental genetics and presently it is one of the most widely used higher organisms in molecular biology. The impact of recent discoveries in molecular biology on population genetics and the extensive use of *D. melanogaster* have made the group of species related to it a model system for studying molecular evolutionary genetics. Currently, considerable effort is going into the molecular population genetics and evolution of these species, but there is a wide literature on almost all aspects of its biology, ranging from molecules to morphology and from behavior to ecology. Some of these recent reviews are as follows: species phylogeny (57, 69), phenotypic and physiological variation (32, 57, 96), ecology and historical biogeography (57, 83, 96, 97), gene-enzyme systems (75, 89), behavioral genetics (95), chromosomal variation (69), fitness modifiers (84, 106), enzyme variation and biochemical adaptation (20, 125), DNA sequence variation (53, 73), and transposable elements (this issue).

The present review is focussed on the population aspects of genetic variation, i.e. the variation and evolution of genetic structure within and between species. Population survey data, by identifying probable cases of gene loci under selection, provide information for planning biochemical and physiological studies of adaptation on one hand, and molecular population genetic studies of gene structure on the other. But their usefulness is more than that. Population variation data, obtained from parallel studies of genetic, physiological, and phenotypic variation provide a general framework of the spatiotemporal pattern of variation that, in turn, helps to decide on the specific evolutionary hypotheses of variation to be tested. Further knowledge gained from detailed molecular, biochemical, physiological or ecological studies can then be readily incorporated into an integrated and explanatory picture of genetic structure in natural populations.

HISTORICAL BIOGEOGRAPHY

A comprehensive historical biogeography of the *D. melanogaster* species subgroup is given by Lachaise et al (57). At present, the *D. melanogaster* subgroup is known to contain eight species: *orena, erecta, yakuba, teissieri, melanogaster, simulans, mauritiana,* and *sechellia.* The subgroup is almost certainly Afrotropical in origin and all known facts are consistent with this conjecture. Six of the eight members of this subgroup are endemic to Africa and at least three of the eight species (*melanogaster, yakuba,* and *teissieri*) breed on a variety of native plants. Also, the two cosmopolitan and most widely studied members of the subgroup, *D. melanogaster* and *D. simulans,* show more gene-enzymes and mtDNA variation in Africa than anywhere else (40, 41, 111, 113).

D. erecta has been recorded from the south and mid-Ivory Coast. It is a rare species and appears to be closely associated with a few species of *Pandanus* (monocots: Pandanaceae). *D. orena* has been found only once (Mount Lefo, West Cameroon) and the single known strain to date comes from fewer than ten wild-caught males and only one female. The species appear to be strictly confined to submontane forest. Its rarity in the wild and the difficulties associated with its laboratory breeding suggest that it may be a specialist species.

D. teissieri and *D. yakuba* have strikingly similar geographic ranges, extending from eastern Guinea in the northeast to Zimbabwe in the southeast. There is also a clear east-west differentiation in their geographic ranges, with that of *D. teissieri* being more western and that of *D. yakuba* more eastern. *D. teissieri* is far more abundant than *D. yakuba* in the westernmost end of their geographic range (e.g. the Guinean mountains) and the reverse is true in the east (e.g. in the Kenya highlands). These two species also display ecological divergence; *D. teissieri* is chiefly a forest species, while *D. yakuba* is basically an open field species.

D. melanogaster and *D. simulans* are both cosmopolitan with exceptional colonizing ability. They are adapted to both temperate and tropical climates and are found on all the major continents. *D. melanogaster* is most abundant in the west Afrotropical region where it occurs in both wild and domesticated habitats. Exceedingly rare in this region and found only in domesticated habitats, *D. simulans* is more abundant in the easternmost mainland, Madagascar, Mascarene, Camoro, and Seychelles where *D. melanogaster* is rare and confined mainly to domestic coastal habitats. During its worldwide colonization, *D. melanogaster* appears to have gone through three distinct phases: the colonization of the Eurasian continent in prehistoric time, the colonization of the Far East in historic time and the more recent colonization of America and Australia (32, 57). In contrast, *D. simulans* appears to have started its worldwide colonization much more recently and is still expanding (57).

D. mauritiana and *D. sechellia* are endemic to *Mauritius* and Seychelles, respectively, in the Indian Ocean. *D. sechellia* is strictly associated for breeding with the fruits of the rubiaceous shrub *Morinda citrifolia,* which is widespread throughout the Indopacific area. *D. mauritiana,* although restricted to Mauritius, is an "abundant, broad-niched, opportunistic, and domestic species" (57). All of its identified resources are introduced fruits; its ability to ultilize banana (an unfamiliar resource) and its habit of entering human habitats is quite similar to *D. melanogaster* and *D. simulans.* Thus it appears to have all the qualities of a potential colonizer.

Of the 146 species of host plants identified as breeding sites for at least 1 drosophilid in the Afrotropical mainland, 63 have yielded at least 1 of the species of the *D. melanogaster* subgroup. These 63 species are distributed between 45 genera and 29 families! *D. melanogaster, D. teissieri* and *D. yakuba* have been reported from the largest number of plants and thus they can be considered generalist and ranked as *yakuba* > *melanogaster* > *teissieri*. Very little is known about the breeding site of *orena, simulans* and *mauritiana*. Although *erecta, orena, mauritiana,* and *sechellia* can be classified as specialist to some extent, on the basis of both native and current resource ultilization and their geographic distribution, one member of each of these species pair can probably be classified as having a broader (>) niche than the other: *D. erecta* > *D. orena, D. yakuba* > *D. teissieri, D. melanogaster* > *D. simulans,* and *D. mauritiana* > *D. sechellia.* These species-pairs provide excellent material for evolutionary genetic and ecological studies.

GENETIC POLYMORPHISMS AND GEOGRAPHIC DIFFERENTIATION

There is a large body of data on genetic variation in natural populations of *D. melanogaster* and much of this literature has been summarized (89, 112, 113). Relatively less data exists on geographic variation in *D. simulans* (21, 109) and even less on *D. mauritiana* and *D. sechellia* (16). In the following pages we discuss genetic variation of these four species. There is very little published data on population variation in the other four members of this subgroup, i.e. *D. orena, D. erecta, D. yakuba,* and *D. teissieri;* these species are presently being studied and should provide valuable genetic data to compare with ecological data, as did *D. melanogaster* and *D. simulans* (21, 109).

Gene-Enzyme Variation

Until recently a large body of data on genic variation in *D. melanogaster* has had more to do with ascertaining geographic patterns with prechosen polymorphic loci than with estimating heterozygosity or identifying different kinds of polymorphic loci (71). However, in one of the most extensive studies all these aspects of variation were pursued simultaneously (113). A total of 117 loci (coding for 79 enzymes and 38 abundant proteins) were studied in 15 geographic populations sampled from around the world. The overall picture of genic variation in *D. melanogaster* can be summarized as follows: (*a*) About half (52%) of the loci are polymorphic in one or more populations and the other half (48%) are uniformly monomorphic everywhere. (*b*) An average population is polymorphic for 43% of its gene loci and an average individual is heterozygous for 10%. (c) The average within-locality heterozygosity ($H_S = 1 - \Sigma \bar{p^2}$) for polymorphic loci is uniformly distributed over the range of heterozygosity observed; i.e. given that a locus has any local variation, it is nearly as likely to have a lot as a little. (d) The distribution of total heterozygosity ($H_T = 1 - \Sigma \bar{P}^2$) has a prominent bimodal distribution. The lower mode consists of loci with a single prominent allele and a few uncommon ones and the upper mode consists of clinally varying loci with a high fixation index, F_{ST} ($= H_T - H_S/H_T$) (e.g. Adh and G6-pd), loci with many alleles in high frequency (e.g. Ao and Xdh) and loci with two alleles in high frequency in all populations but with little interpopulational differentiation (e.g. Est-6 and α-Fuc). (e) The distribution of F_{ST} is strongly skewed, with a prominent mode at 8–10% and a long tail of high values reaching a maximum of 58%. Two-thirds of all loci fall within the bell-shaped distribution centered on an F_{ST} of 8–10%, a result compatible with the notion that they are experiencing a common tendency toward small interlocality differences owing to extensive gene flow among populations (55, 112, 124).

Previous genic variation data on *D. simulans* are less extensive and are summarized elsewhere (21, 75). In the most comprehensive study, 114 gene-structure loci were studied in four mainland (from Europe and Africa) and one island (Seychelles) populations of *D. simulans* (21) and a comparative picture of genic variation between the two sibling species can be summarized as follows: (a) *D. melanogaster* shows a significantly higher proportion of loci polymorphic (on 99% criterion) that *D. simulans* (52% vs 39%, p < 0.05), but on a more stringent criterion (95%) much of this difference disappears (35% vs 30%). Thus, in *D. melanogaster* many polymorphic loci maintain only low frequency (< 5%) alleles. (b) Since low frequency alleles have little effect on heterozygosity, both species have similar mean heterozygosity (10.2% vs 9.4%) and mean number of alleles per locus (1.48 vs 1.59). (c) The two species share some highly polymorphic loci but differ in the number of loci showing latitudinal variation; such loci are more common in *D. melanogaster* than in *D. simulans*. (d) Both in terms of F_{ST} and D statistics, *D. simulans* show significantly less geographic differentiation than *D. melanogaster*. (e) The differences in geographic divergence between the two species (F_{ST}) are limited to loci on the X and second chromosomes only; loci on the third chromosome show similar levels of geographic divergence in both species. These differences between chromosomes are not apparently related to the presence of inversion polymorphisms in *D. melanogaster*, as the third-chromosome inversions show an order of magnitude higher geographic divergence than the second-chromosome inversions; this is just the opposite of what we would expect if inversions were the main cause of high geographic differentiation of allozyme loci in *D. melanogaster* (113).

Most of the homologous loci studied in *D. melanogaster* (113) and *D.*

simulans (21) have also been recently studied in *D. mauritiana* and *D. sechellia* (R. S. Singh & M. Choudhary, unpublished data). Only one population (a mixture of isofemale lines) has been studied in each case so there is no picture of geographic variation in these insular species. However, the overall level of genic variation (in terms of polymorphic loci, heterozygosity, and number of alleles) is much less in *D. mauritiana* and *D. sechellia* than in *D. melanogaster* and *D. simulans*. *D. sechellia* shows the least variability of them all. Some of the reduction in genic variation may result from founder effect but some may also be attributable to their evolutionary history when the insular species may have experienced a narrower environmental variation than their cosmopolitan siblings.

Two-Dimensional Electrophoretic Protein Variation

Compared with older one-dimensional electrophoretic techniques, two-dimensional gel electrophoresis can resolve many more different polypeptides—usually several hundred—on a single gel, and thus permit the rapid survey of numerous loci for polymorphisms (90). Although this technique has been used in a number of organisms its application in population genetics is still limited (2, 24–26, 67, 92, 93, 119), partly because of technical difficulties associated with its routine use and partly because it is suspected of being inherently less sensitive and thus having less resolving power than the 1DE technique (79).

We have used the 2DE technique and have scored allelic variation in approximately 300 polypeptides of *D. melanogaster* and *D. simulans* adult male reproductive tracts (24–26). The overall percent loci polymorphic (P) and mean heterozygosity (H) are much lower with 2DE than with 1DE. Approximately 9% of *D. simulans* and *D. melanogaster* reproductive-tract polypeptides are polymorphic, with H = 1–3%. Thus, overall, soluble enzymes and abundant soluble proteins are three to six times more frequently polymorphic and are at least three times more heterozygous than the 2DE proteins. In contrast, if only polymorphic loci are considered, the differences in H between protein sets become much smaller or even disappear. The main source of difference in overall heterozygosity is the preponderance ($\sim 90\%$) of apparent monomorphism among the 2DE proteins.

One approach to assessing the relative sensitivity of 1DE and 2DE is direct "cross calibration" with unknown variants of "model" proteins (110). Some results (79) suggest that only 50–60% of variants detectable at a locus by 1DE are separable by 2DE. But other workers (119) maintain that 2DE detects 70–90% of the variants that 1DE does—and thus that they cannot explain two- to fivefold discrepancies in variation estimates in terms of differing physical sensitivities.

Another approach to the question relies on partial *correction,* with technical modifications, of putatively low sensitivity in 2DE. Our work is in this category: we used a polychromatic silver stain, and electrophoretic conditions were chosen to optimally resolve the proteins being studied. Despite this effort, we obtained overall estimates of variation very close to earlier 2DE estimates in *Drosophila* (67) and in vertebrates (2). The high proportion of monomorphic loci, the segregation of multiple alleles at polymorphic loci, and the similarity of mean heterozygosity at polymorphic loci all indicate that if our 2DE system is physically less sensitive than 1DE, then the deficiency is expressed mostly in terms of polymorphic loci falsely scored as monomorphic rather than in terms of hidden variation spread evenly between loci. 1DE studies, designed to reveal electrophoretically cryptic variation by altering experimental conditions, show a strong positive correlation between the number of alleles detected at a locus under "standard" conditions and the number of additional alleles revealed under altered conditions (27). There is also a marked tendency for loci initially monomorphic (or diallelic) to display no new variation on closer examination (22). In this light, the importance of 2DE data is that highly constrained structural loci may be more common than generally supposed on the basis of 1DE data.

Mitochondrial DNA Variation

D. melanogaster natural populations show a considerable mtDNA variation (40, 41, 103). Using 10 restriction enzymes, of which four showed polymorphism, a total of 23 haplotypes were distinguished among 144 isofemale lines examined in our lab from around the world (41; L. Hale & R. S. Singh, unpublished data). A small number of haplotypes are found in more than two populations, but most are unique to a single population.

D. melanogaster populations can be divided into three geographic regions, according to the amount and distribution of variants, a proposition that agrees with the conclusion based on the biogeographical data (57). These regions are the New World, Euro-Africa, and the Far East. The Euro-African populations are individually the most diverse, and the other two individually less diverse as most have only one high frequency haplotype. However, the Far East populations are more differentiated and so the total regional diversity there is as high as it is in the Euro-African region. By contrast, New World regional diversity is much lower, as virtually all populations there have the same high frequency variant.

Considering mtDNA alone, a species history for *D. melanogaster* can be inferred that places the ancestors of the present worldwide populations in Europe and Africa. Euro-African populations are individually the most diverse and collectively as diverse as anywhere, and the "ancestral" haplotype is found at high frequency (41). Colonization of Far East and New World populations would seem to be quite different, as allozymes and biogeography clearly indicate (32, 57, 111). North American populations show the effects of an apparently relatively recent colonization, in that only one derivative haplotype is found of those most common in this region.

The amount and pattern of mtDNA variation in *D. simulans* is strikingly different from that in *D. melanogaster* (8, 86, 115; L. Hale & R. S. Singh, unpublished data). Using the same ten restriction enzymes as with *D. melanogaster* on a total of 92 worldwide strains showed restriction site polymorphism for only one enzyme (Hinf I), giving rise merely 4 haplotypes in total. Although there is considerably less variation in *D. simulans*, all variants were observed in widely separated populations in each continent. This is in sharp contrast to *D. melanogaster* where 60% of the variants are unique to single populations.

In our study of *D. simulans*, we focussed on the mtDNAs of worldwide, continental populations, which have a nearly identical restriction type. Solignac, David and colleagues (8, 115) have shown that mtDNA from the whole of *D. simulans* is actually much more diverse, but this diversity is distinctly discontinuous. Three fairly divergent types of *D. simulans* mtDNA have been identified, with few or no polymorphisms within each (115). Two of these types, SiI and SiIII, are very restricted in distribution, the former confined to the Seychelles and Pacific islands, and the latter mostly confined to the islands of Madagascar and Réunion. The other type, SiII, is found exclusively in most cosmopolitan populations of *D. simulans* and is the type surveyed in our study (L. Hale & R. S. Singh, unpublished data). If all three types are included in calculating, the species nucleotide diversity index, π_T becomes 0.0197, about thirty times larger than for SiII alone ($\pi_T = 0.0006$). These values can be compared with $\pi_T = 0.0019$ in *D. melanogaster*.

The mtDNA in *D. melanogaster* is also highly variable in size (40). All size variants map to the A+T-rich region of the mtDNA molecule and have a size range of 18.2–19.9 kbp. Of the 136 lines for which size classes could be precisely ascribed, 112 lines were homoplasmic and showed a total of 13 different mtDNA size classes. The remaining 24 lines were heteroplasmic for size variants and a total of nine size classes was represented among them. Size variation also occurs within SiI and SiII types of *D. simulans*, which, as in *D. melanogaster*, is concentrated in the A+T-rich region. The size variation in *D. simulans* is due to both, tandem repeats (2–4 times) of a 470-bp unit and a domain of variable length that flanks the A+T-rich region (8). Although there have been no detailed molecular studies, the distribution of mtDNA size classes suggests that the same molecular mechanisms probably also hold for *D. melanogaster* (40).

Heteroplasmy is quite common in these species and virtually all of it involves size variants. *D. simulans* lines show a lower level of heteroplasmy (5.5%) than that reported in its closely related species: 18% in *D. melanogas-*

ter (40), 18% in *D. mauritiana,* and 12% in *D. sechellia* (8). The presence of the most common (which is also the smallest) variant in almost all heteroplasmic lines and the frequency distribution of size classes in *D. melanogaster* populations have been used to suggest a high rate of mutational occurrence of mtDNA size variants and some form of natural selection against them (40).

Gene-Regulation Variation

The most important enzyme-activity variation data have been produced by Laurie-Ahlberg and her associates (63–65). Several points emerge: first, the amount of enzyme activity varies widely, with estimates ranging from almost no variation for G3PD to nearly 100% variation for IDH and ADH on chromosome II. Second, some of the non-specific enzymes (e.g. ADH and AO) tend to show the highest observed activity variation. Finally, the activity variation due to chromosome II and III shows a moderate but significant correlation ($r = 0.475$, $p < 0.05$); this means that much of the enzyme activity variation is not linked to the locus and is due to genetic factors spread over both chromosomes. However, there are some enzymes (e.g. AO, ADH, ME) that show high activity variation because of the chromosome on which they are located; and the reverse is true for others (PGI and IDH). For at least some of the enzymes in the former category a major share of activity variation may be due to allozymes, i.e. due to activity differences between structural alleles. In fact, for ADH and AO, 86% and 82% respectively, of the enzyme activity differences were attributable to allozymes. Since the level of enzyme activity is a major component of gene-regulation variation and since enzyme activity can be affected by both structural and regulatory genetic changes, it is important to know if the two types of genetic variation (i.e. structural vs regulatory) are independent or dependent on each other. It is not known if structurally monomorphic loci are equally, more, or less, variable than the polymorphic loci with respect to activity variation controlled by modifiers. The relationship between the two kinds of variation has been investigated by correlation analyses between estimates of enzyme activity variation and genic heterozygosity for protein–coding loci in *D. melanogaster* (R. S. Singh, unpublished data). Overall, no correlation was found between activity variation and number of structural alleles or between activity variation and several measures of genic diversity if all loci (i.e. monomorphic and polymorphic) were considered.

However, at present, correlation studies on gene-structure and gene-regulation variation suffer from several conceptual problems; for example, the inclusion of *all* structurally monomorphic loci in the analysis would distort any positive correlations that may exist. This is because there are two types of monomorphic loci—those that are functionally constrained and generally show little variation and those that are monomorphic due to random genetic drift; the two groups of monomorphic loci may have very different levels of gene-regulation variation.

A second problem is that the estimates of genic heterozygosity are usually based on *all* observed alleles (majors and minors), although it may be more reasonable to expect that a long-term association between the two kinds of variation would involve only major alleles. A third problem is that while the genic diversity statistics are based on a *qualitative* (i.e. discrete allelic forms) description of genic variation, the activity variation statistics are based on the *quantitative* (i.e. amount of enzyme) effects of gene-regulation variation; the two measures are not equivalent. In future characterization of gene-regulation variation in terms of alleles DNA sequence changes rather than their quantitative effect on enzyme activity would solve this problem as it would then be possible to compare both gene-structure and gene-regulation variation in terms of sequence heterozygosity.

DNA Sequence Variation

DNA sequence variation studies have revolutionized population genetics; the effect is seen in several areas. Firstly, new phenomena and molecular processes have been discovered at the level of DNA that have important bearing on population genetics. For example, variation in codon usage and rates of silent replacement have an impact on the mechanisms of natural selection (74, 105), and multigene families, mobile genetic elements, and various types of mutations (e.g. insertion, deletion, transition, and transversion) are providing a much needed window to the mechanisms and role of mutation in evolution (For a recent review, see 85). Then, too, DNA sequence data have greatly extended the power of linkage-disequilibrium and gene-regulation studies by allowing fine-scale mapping of both *coding* and *noncoding* DNA (10). Finally, DNA sequence studies allow measurement of both silent and nonsilent changes in the gene (unlike the gene-enzyme variations that measure only nonsilent or replacement changes), as well as changes in the noncoding part of the gene, and thus provide a mechanism to detect varying intensities of selection in different parts of the same gene (54, 56, 88).

The overall impact of these findings has given a new dimension to population genetics. The fine-scale details of gene-structure variation (in terms of haplotypes) coupled with the slow rate of changes between generations (by mutation or intralocus recombination) enable one to differentiate between the identity of alleles by kind vs descent and to trace their history over evolutionary time. Another major impact is that DNA sequence studies allow the study of variation and the evolution of homologous sequences between species and thus they open up the possibility of comparing rates of evolution in much more precise terms than was possible before.

More data is available on DNA variation bearing on population genetics and evolution in *D. melanogaster* than any other species. In addition to the DNA sequence data for *Adh* gene (53), DNA restriction variation data are available for *Adh* (3, 55, 61), *Amylase* (62), notch (102), rosy (4), white (60, 81) and Hsp 87A (66) loci. Too few loci have been studied to generalize but already some interesting observations have emerged. These include (*a*) a considerable sequence variation in and around the gene, (*b*) different levels of variation in different parts of the same gene, (*c*) the presence of numerous low-frequency small insertion/deletion sequences, and some large sequences that are almost always transposable elements, and (*d*) varying levels of sequence variation, insertions/deletions and linkage disequilibrium among different chromosome regions.

Besides demonstrating the extensive role of purifying selection in the coding DNA (10, 23, 105), two aspects of DNA variation data are most useful for testing selective neutrality of alleles. The first is the relationship between sequence divergence (D) between populations and species, and the level of sequence variation within population and species (H) (referred to as "the D/H ratio"), and the second is the comparison of the level and pattern of DNA vs allozyme (e.g. silent vs nonsilent) variation between populations and species (4). Because both aspects of DNA variation data bear directly on genetic structure within and between species, we discuss them in more detail in a later section.

Quantitative Characters

The geographic pattern of quantitative variation in *D. melanogaster* and *D. simulans* is in some respects very similar to that seen for gene-enzyme variation in these species. Phenotypic characters appear to be much more sensitive to geographic differentiation than single enzyme loci. All phenotypic characters, except mating behavior (42), that have been studied in *D. melanogaster* show substantial variation within populations and strong geographic differentiation between populations (14, 29, 46, 101, 121; reviewed in 32, 69). These include biometrical traits such as adult weight, thorax length, wing size, sternopleural and abdominal chaetae numbers, and ovariole number; physiological traits such as reproductive capacity, tolerance to ethanol, temperature and desiccation; behavioral traits such as phototaxis and sexual activity; and morphological traits such as female abdomen and thoracic trident pigmentation. Many of these characters, e.g. adult weight, wing length, bristle number, ethanol tolerance, ovariole number, and thoracic trident pigmentation show strong latitudinal clines in *D. melanogaster*.

In contrast, although the extent of quantitative variation within local populations of *D. simulans* appears to be similar, the level of geographic differentiation is significantly lower than *D. melanogaster*. *D. simulans* also differs in characters showing clinal variation in *D. melanogaster: D. simulans* shows either no cline, e.g. for ethanol tolerance, or a weaker cline, e.g. for adult weight, ovariole number and thoracic trident pigmentation. The variation in the presence/absence as well as in the strength of phenotypic clines in these species appears in line with a host of ecologically relevant variables that differ between them, suggesting a differentiation of their ecological niches (96, 97). The character-specific patterns of latitudinal clines for phenotypic characters and the overall geographic uniformity of allozyme and mitochondrial DNA variation in *D. simulans* suggest that this species is undergoing diversifying selection but because of its recent colonization, the accumulated geographic differences are not as large as in *D. melanogaster*.

GENETIC VARIATION AND LIFE-CYCLE STAGES

Although different tissues and developmental stages have been often used as source material in gene-enzyme and abundant protein variation surveys, no systematic attempt has been made to see if estimates of genic heterozygosity vary between different life-history stages of the same organism. At present, we know of no systematic survey of protein variation in different developmental stages of *Drosophila,* but some indirect information regarding this type of variation can be obtained from our 2DE comparison of male reproductive-tract and larval wing-discs proteins in the *D. melanogaster* species complex (25).

The similarity index (F = % shared polypeptides) for imaginal disc proteins of different species is higher in every case than for reproductive tract proteins. That is, fewer polypeptides in imaginal discs were nonidentical between species of each pair and, of these nonidentical polypeptides, more could be matched with a homologous one in the other species than was possible with reproductive-tract polypeptides. This study was done with only one line of each species and so the variation within species was not taken into account. However, the range of sampling variation in F, from random allele sampling at loci with overlapping polymorphisms can be no greater than the total percentage of such overlapping polymorphic loci; thus the differences in F values become more statistically significant. For instance, overlapping polymorphisms occur at ~10% frequency between *D. melanogaster* and *D. simulans* (113); the difference between reproductive-tract and imaginal disc F values for these species is ~20% (25). Interestingly, Ohnishi et al (93), using proteins of adult whole-body homogenates on two-dimensional gels stained with Coomassie blue, obtained F values much closer to those for imaginal discs than for male reproductive tracts.

Assuming that genetic divergence between species is related to genetic variation within species, these results suggest that proteins expressing in the early stages of development may be much less polymorphic than proteins expressing in the later stages. More tissues and developmental stages need to be examined to separate the effect of variation between tissues and between developmental stages on the level of heterozygosity.

DETECTION OF SELECTION FROM POPULATION DATA

Genetic Structure of Sibling Species

The dual aspect of genic variation, i.e. equal within but different between populations, between D. melanogaster and D. simulans requires two sorts of explanations and various hypotheses have been considered in detail (21, 109). These include high mutation rate in D. melanogaster, population bottleneck in D. simulans, differences in species niche-width or in selection "strategies" and, of course, selective neutrality. After considering all the factors together, it was concluded (21) that, barring recent colonization, variation in niche-width and/or in genetic "strategies" of adaptation are the major contributing factors to the varying levels of geographic differentiation in these species.

Mitochondrial DNA is a better molecule for discriminating between the recent colonization and selection hypotheses as intraspecific sequence variants are held to be neutral and lack of recombination allows phylogenetic inference between the extant sequence variants (5). The restriction fragment length polymorphisms of mtDNA show that there is considerably less variation in D. simulans, but all variants are observed in widely separated populations. This differs sharply from D. melanogaster where 60% of the mtDNA variants are unique to populations (41). These results support the hypothesis that the greater geographic uniformity of allozyme variation in D. simulans is due to its relatively recent worldwide colonization (21; Hale & Singh, unpublished data). In other words, the geographic uniformity of D. simulans may stem from the shorter time its recent colonization has allowed to equilibrate with its physical and biotic environment (21). These results do not rule out the role of selection (see the section on phenotypic variation) but simply suggest that colonization history is an important factor in explaining the contrasting patterns of geographic differentiation between D. melanogaster and D. simulans.

In contrast to enzyme and mtDNA variation, the DNA sequence variation gives a rather different picture. Aquadro et al (4) have shown that DNA sequence variation around the rosy locus is about six times greater in D. simulans than D. melanogaster populations sampled from Raleigh, NC. They have hypothesized that differences in species' effective size are the major determinant of the contrasting levels and patterns of DNA sequence vs allozyme variation. A large effective population size (N) in D. simulans would lead to a loss of deleterious enzyme alleles (purifying selection due to large –Ns; 91) as well as geographic homogeneity among populations (due to large *Nm*). While more genes and populations, especially from Europe and Africa, need to be examined, an alternative explanation for the above result would be the abundance of segregating inversions, insertion/deletion sequences and transposable element in *D. melanogaster* and their notable absence in *D. simulans*. Directional selection coupled with linkage (e.g. due to inversions) and purifying selection caused by insertion/deletion sequences and transposable elements (82) would result in a greater loss of sequence variation in *D. melanogaster*. This may also explain why in *D. melanogaster* the levels of DNA sequence variation for X chromosome and autosomes are not different as we would expect (60, 102).

As for the differences in the levels of geographic differentiation, we emphasize the role of a relatively recent worldwide colonization, instead of differences in the level of gene flow (*Nm*). Laboratory measurement of dispersal (78) and indirect measurements of gene flow based on rare alleles (112) do not provide any evidence of higher gene flow in *D. simulans* than *D. melanogaster*.

Genetic Structure of Populations

The genetic structure of a species can be better understood by considering together data from different types of genetic markers. Although such markers may be under the influence of different types and degrees of selection pressure, they would be affected similarly by such factors as population size and gene flow. The parallel studies of gene-enzyme and mtDNA variation with the *same* populations make such comparisons possible in *D. melanogaster*.

In *D. melanogaster,* all population pairs within continents show rather low genetic distance (113) and intercontinental pairs high genetic distance with one important exception: Populations in the temperate zone show no relationship to distance. Populations from Hamilton, Ottawa, France, and Australia all temperate, cluster together, while Texas (subtropical) clusters with West Africa (tropical). Is the separate clustering of temperate and tropical populations from different continents an indication of macro climatic selection (temperate vs tropics) or common ancestry? This question can be answered by looking at the mtDNA variation of these populations. If the clustering of populations with respect to gene-enzyme variation is due to their common ancestry, then these populations should also cluster with respect to their mtDNA variation.

The mtDNA haplotypes of temperate (or tropical) populations do not cluster in the same manner as the allozymes (L. Hale, R. S. Singh, unpublished data). For example, all North American populations share one mtDNA haplotype, but with respect to allozyme variation, Hamilton and Ottawa cluster with France and Australia, and Texas with West Africa. France shares mtDNA variation with both Australia and West Africa, but with respect to allozymes it resembles Australia and not West Africa. These results suggest that macro climatic selection (i.e., temperate vs tropic) is an important determinant of geographic variation in *D. melanogaster*. However, this is really not a demonstration of "area effect" as the populations sampled are from the two extremes of the latitudinal gradient that is known to produce clines in *D. melanogaster* (see below). The only such observation in *D. melanogaster* that is not related to latitudinal variation is the presence of significant body size differences among populations from French West Indies, Africa, and the Far East. The Far East populations have significantly larger body size and appear to be an old race of *D. melanogaster* (31), and the French West Indies populations have smaller body size than the Afro-tropical populations (13). This regional variation in body size is independent on the latitudinal clinal variation in body size seen on the major continents.

The genetic structure within species can also be examined separately with different nuclear marker loci. Lewontin & Krakauer (70) asserted that, under the neutral theory, the F_{ST} for each locus ought to estimate a single parametric value dictated by population structure. Statistical heterogeneity among observed values of F_{ST} could then serve as a demonstration of the influence of selection on at least some loci. The F_{ST} in *D. melanogaster* populations not only show a very high variance, but there is a clear indication of inhomogeneity within the data (113). Two-thirds of all loci fall within the bell-shaped distribution centered on an F_{ST} of 8–10%, a result compatible with the notion that they are experiencing a common tendency for small interlocality differences to build up as a result of the species' population structure and high gene flow (112). The remaining third of variable loci, however, has markedly more interlocality differentiation and includes enzymes that exhibit clinal variation in the ratio of the two most abundant alleles [e.g. Adh and G-3pd (88)].

The use of heterogeneity among observed values of F_{ST} as a test of neutrality has been criticized on the ground that hierarchical grouping among sample localities and local abundance of newly arisen neutral alleles can inflate the variance in F_{ST} (100). Although there are few locally abundant alleles in *D. melanogaster* and therefore the F_{ST} test cannot be strictly used to single out individual loci under selection, the comparison of F_{ST} distribution among loci still provides a tool for identifying groups of loci affected similarly by selection vs drift.

Clinal Variation

Clinal variation is one of the strongest indications of selection in nature (35) and in *D. melanogaster* clinal variation has been shown for inversions (51, 52, 80, 122), for several morphological characters (14, 29, 32, 69), and for allozymes (1, 38, 48, 87, 88, 113, 118, 122). The major axis of geographic variation in *D. melanogaster* is with respect to latitude. At the present time, of the 117 loci studied in our lab, 18 show similar north-south variation in the frequency of the most common allele in three separate continental transects (Americas, Europe/Africa, Far East) (113). Clinal variation also occurs in the southern hemisphere (38). However, although the proposition is generally accepted that clinal variation for morphological characters and inversions is mostly due to natural selection, the allozyme clines are looked upon with suspicion. This is because latitudinal variation in allele frequencies alone may not indicate a real cline and even if a latitudinal gene cline existed, natural selection is not necessarily the cause (see 99 for a description of selective and non-selective clines). It can be argued for example that the latitudinal clines for allozymes are due to their association with inversions. While the parallel and complementary nature of the allozyme clines within and between the northern and southern hemispheres, respectively, suggests equilibrium clines, i.e. due to natural selection, the effects of selection at linked loci cannot be ruled out. Although only DNA sequencing can unravel the effect of tight linkage on association of polymorphisms between sites/loci, inversions in general do not seem to constrain the pattern of geographic variation at allozyme loci in *D. melanogaster* (113, 118). The allozyme loci studied are in general too loosely linked to be constrained in their variation pattern by linkage (59).

A test of the hypothesis that the latitudinal gene clines in *D. melanogaster* are steady state clines, i.e. due to gene flow between previously isolated populations (104), can be performed by examining the mtDNA variation. As shown above, all North American populations share the same mtDNA haplotype found only in Japan and Australia, a result we suspect of the introduction of *D. melanogaster* from North America (L. Hale & R. S. Singh, unpublished data). Similarly, Europe and Africa show unique mtDNA haplotypes, as do Korea and Vietnam. Thus, in spite of the differences in the mtDNA haplotypes, populations from the three transects (i.e. America, Europe/Africa, Far East) show parallel allozyme variation, strongly suggesting that gene flow alone cannot explain these clines. A similar argument has been used for latitudinal variation at the silent vs replacement sites of *Adh* in *D. melanogaster* (M. Krietman, personal communication).

It has recently been shown that seasonal cyclic variation in allele frequencies can generate latitudinal clines because most environmental factors that vary seasonally also vary latitudinally (99). Although temporal allele frequency variation has been shown in some studies and not in others (7, 18, 37), it is not known if seasonal temporal variation plays an important role in the latitudinal clines of *D. melanogaster*.

Statistical Tests of Neutrality: The D/H Ratio

The neutral theory (50), under which both standing variation within species and rate of divergence between species are increasing functions of neutral mutation rate, predicts a monotonic increasing relationship between the two levels of genetic variation. However, most tests of selective neutrality are based on some statistical properties of genetic variation within population or on the rate of genetic divergence between populations and species (36, 85), and the statistical relationship between the two levels of variation has not been used for testing selective neutrality (19, 114, 120). Skibinski & Ward (114) used the relationship between heterozygosity (H) and genetic distance (D) on pooled data from a variety of taxa and showed that the regression of D on H was positive and does not differ significantly from the expectation of the neutral theory.

A positive correlation between D and H may also be interpreted as a result of selection following the argument that highly heterozygous loci would evolve more rapidly. To analyze this we need to do a regression of D on H using single-locus heterozygosity between related species and ask whether the extent of interpopulation differentiation at a locus within a species is positively correlated with the degree of divergence between species at that locus. Such a positive correlation would be consistent with the notion that the same evolutionary process (drift or selection) acts as the major cause of genetic variation and change both *within* and *between* species.

Between *D. melanogaster* and *D. simulans* there is a clear absence of an overall correlation between Nei's genetic distance (D) and mean heterozygosity (H) for each locus. Surprisingly, we found a strong *negative* correlation between interspecies (D) and intraspecies (H) diversity (R. S. Singh, unpublished data). Another feature of the data is the absence of loci showing both high D and high H. What these empirical relationships show, therefore, is the absence of a positive correlation between present-day interspecies divergence at a locus, and the extent of present-day intraspecies variation at the locus. These data provide evidence for the importance of two kinds of selective constraint operating between species: a possible role for balancing selection in lowering the average rate of divergence of polymorphic loci and for founder effect or rapid substitution of alleles in increasing the average rate of divergence of monomorphic loci between species.

The D/H ratio has been used with DNA sequence variation of *Adh* and the role of both balancing and purifying selections has been suggested in different parts of this gene (44, 56).

GENETIC DIVERGENCE BETWEEN SPECIES

Levels and Patterns of Species Divergence

In most comparative genetic studies of species comparisons Nei's genetic distance (D) has been used as a measure of genetic divergence (6, 85). The advantage of using D is that it combines information on both allele numbers and their frequencies, and the changes can be inferred at the level of codons. The D is also linearly related to time in closely related organisms (i.e. with

D<1). Another useful measure, although less commonly used, would be the proportion of alleles that are *not* shared between species, especially between closely related taxa. Unique alleles can give qualitative information about isolation of populations and about their history. Still a third measure would be the proportion of loci that are *alternately* fixed for different alleles between two species. The three statistics together would be more useful for studying the evolutionary history of populations and species than D alone, as different evolutionary processes can affect D in a variety of ways and lead to similar overall D between pairs of taxa that differ in their evolutionary history.

The above three measures of genetic divergence for *D. melanogaster* and *D. simulans* show several interesting points (R. S. Singh, unpublished data): (*a*) *D. melanogaster* harbors nearly twice as many unique alleles as *D. simulans* (27% vs 16%), and chromosome X and 2, but not 3, contribute to this allelic diversity. All chromosomes show an equal proportion of shared alleles. (*b*) Chromosome 2 has significantly more alternatively fixed (monomorphic) loci than chromosome X or chromosome 3. (*c*) As a consequence of (*a*) and (*b*), mean genetic distance is nearly twice for chromosome 2 than for chromosome 3. (*d*) Ten percent of all enzyme loci examined have become fully diverged; none of the abundant protein loci shows complete divergence. The loci for which one species is monomorphic and the other polymorphic are also interesting. At most of these loci, one of the species apparently carries one of the alleles of the polymorphic set found in the other species. Two simple logical possibilities exist for the origin of this situation: either one species (perhaps through one or more population bottlenecks) has lost all but one of the alleles of an ancestral polymorphism, or one species has acquired one or more alleles in addition to an ancestral monomorphism. DNA sequence studies can distinguish between the two scenarios.

The mean genetic distance between *D. melanogaster* and *D. simulans* based on 112 homologous loci is 0.179, which give a divergence time, t = 0.09 Myr (R. S. Singh, unpublished data). This is much less than the average divergence time of 2–3.5 Myr based on DNA sequence data (116) and again suggests that functional (purifying selection) and/or selective (balancing selection) constraints have played an important role in the evolution of proteins.

Evolutionary Constraints on Species Divergence

As shown above, there are various types of selective and nonselective evolutionary factors that jointly shape the level and pattern of genetic divergence between *D. melanogaster* and *D. simulans*. These factors must be known before one can make a meaningful comparison of genetic divergence between any species (107). We summarize them here briefly:

Phylogenetic relatedness can cause the monomorphic or polymorphic state of a locus to be highly correlated between related species, which is the case between *D. melanogaster* and *D. simulans* (R. S. Singh, unpublished data). This type of positive correlation is expected for two species that arose relatively recently from the same ancestral gene pool. Another explanation, which need not exclude the historical one, invokes similar kinds of natural selection on homologous loci in the two species. As shown above, both factors seem to have played roles in the divergence of *D. melanogaster* and *D. simulans*.

Historical constraints operate as a result of genetic drift and population bottlenecks (affecting species' effective size), colonization and rapid expansion into new areas, and species hybridization and introgression (115). These are all essentially ecological and genetic "accidents" that can have profound effects on species divergence.

The effects of molecular *functional constraints* on the number of alleles and heterozygosity have been considered in the past (39), but their effect on genetic divergence between species is as yet unexplored. Recently, we have compared genetic identity and number of alleles (*unique* or *shared*) between *D. melanogaster* and *D. simulans* for various classes of proteins and enzymes classified according to some commonly used structural/functional criteria (R. S. Singh, unpublished data). Enzymes show higher divergence (or lower genetic identity) than abundant soluble proteins, and among enzymes those with externally generated substrates show higher divergence than those with internally generated substrates. *D. melanogaster* and *D. simulans* share fewer alleles at "null tolerant" loci than at "null nontolerant" loci, suggesting that essential enzymes are under strong purifying selection.

Selective constraint, as shown above, can affect species divergence in two opposite ways: it can retard divergence by maintaining balanced polymorphisms, or it can accelerate divergence by favoring different alleles in different species (adaptive divergence). The data on *D. melanogaster* and *D. simulans* show strong evidence for both purifying and balancing selection.

Developmental constraints take into account the developmental roles, as distinct from the "housekeeping" roles, of genes and proteins and their effects on the level of genetic variation. There are as yet very few data on this aspect of genetic variation. Our two-dimensional electrophoretic studies have shown that proteins that express during the early stages of development are more likely to be conserved than those that express late during the development. These results parallel the general observation that in closely related species phenotypic differences, if they occur, are usually found during the later stages of development; both of these observations are expected under the concept of "ontogeny recapitulates phylogeny."

PHYLOGENETIC RELATIONSHIPS

The phylogenetic relationship among the members of the *D. melanogaster* subgroup has been recently considered in detail using genetic information from all available sources and some revisions have been proposed (for review see 57). What was originally known as the *Yakuba* complex has now been split into two complexes: the *Yakuba* complex (*Yakuba* + *teissieri*), and the *erecta* complex (*erecta* + *orena*). Within the *D. melanogaster* complex, data from various sources support separation of *D. melanogaster* from the other three species, i.e. *D. simulans, D. mauritiana,* and *D. sechellia.* In contrast, the phylogenetic relationship of the last three species is controversial and depending upon the genetic system used, different relationships emerge among the three members. For example, consideration of reproductive relationships (58), cyst length (49) and mitochondrial DNA (115) show *D. simulans* to be closer to *D. mauritiana;* allozyme genetic distances (16) and *Adh* gene sequences (30) show *D. simulans* to be closer to *D. sechellia,* and metaphase chromosome banding pattern (68) and single-copy nuclear DNA-DNA hybridization data (12) show *D. mauritiana* to be closer to *D. sechellia.*

Since the purpose of phylogenetic relationship in closely related taxa is to shed light on the evolutionary relationships of populations during speciation, we cannot decide the issue simply by counting the number of pieces of evidence for or against any particular phylogeny; we need to make some a posteriori judgment as to which characters are more likely to provide an unambiguous answer. The phylogenetic relationships based on male genitalia and nuclear DNA-DNA hybridization data should be treated with caution because of the possible involvement of selection in the former (30) and unequal rates of evolution in different component of nuclear DNA in the latter (12, 76). The *Adh* gene sequence would also seem unreliable for constructing phylogeny, because of its unequal rates of evolution in the three species (23). DNA sequence data based on several genes or electrophoretic data based on many enzymes and abundant proteins may provide an unambiguous picture. Allozyme data (R. S. Singh, unpublished data) show *D. simulans* and *D. mauritiana* closer to each other than either is to *D. sechellia,* a relationship also supported by our 2DE genetic variation data for male reproductive-tract and larval wing-discs proteins (25).

A likely scenario of speciation in the three "*D. simulans*-like" species is that after separation from *D. melanogaster,* the prototype *D. simulans* was a predominant species on the east coast of Africa and independently gave rise to *D. sechellia* and *D. mauritiana,* probably in that order. Upon a posteriori weighting of characters, there is no strong evidence for branching of *D. simulans* prior to that of *D. sechellia* and *D. mauritiana* (57). Furthermore, while the latter two species have indeed diverged significantly from *D. simulans* and may have even gone through a certain amount of parallel evolution attributable to regional and climatic reasons, the patterns of change in the two insular species are not similar enough to be used as a basis for arguing for a common ancestor or convergent evolution (30).

SPECIES DIVERGENCE AND GENETIC THEORIES OF SPECIATION

Genetic models of species formation make predictions as to how many, or what proportion, of genes are involved in species formation. Mayr's "peripatric" model (77), an elaborate version of the basic allopatric model, suggested massive reorganization of the gene pool, with extensive allele substitution or fixation, in the course of achieving reproductive isolation. Early results from electrophoretic studies, although based on fewer loci, disproved this prediction in its most extreme interpretation (6), and our recent results based on a larger sample of loci reinforce this conclusion (108). However, the early studies usually failed to detect complete allelic divergence at *any* loci, and thus left open the question whether such divergence occurs during species formation, and if so, at what proportion of loci.

It now appears, from our results, that in a sufficiently large sample complete divergence between closely related species can be detected at a small percentage of loci. Since *D. melanogaster* and *D. simulans* have apparently been separated for at least 2 Myr, they are probably far from their earliest stages of genetic differentiation. Thus, our figure of about 10% completely diverged loci would seem likely to include many changes attributable to the side effects of population processes (e.g. random genetic drift) accompanying species formation, or to subsequent phyletic divergence, and not causally related to the establishment of reproductive isolation per se. If as few as, say, 10% of the completely divergent loci were actually involved in reproductive isolation, this would amount to a total of only about 50–100 genes in a genome the size of that carried by *Drosophila* [5,000–10,000 genes (98)]. Although this is still a large absolute number of genes, it is perhaps not surprising that complete interspecies allelic divergence has not been widely detected by electrophoresis, especially with species more closely related than are *D. melanogaster* and *D. simulans*.

Genetic models involving only small numbers of genes have periodically been proposed. The most important distinction between most of these models and the allopatric type is not, however, numbers of genes *per se,* but rather the postulation of a "privileged" class of genes that play a key role in species formation, while being constrained in their rates and/or modes of evolution within species (17, 117). Our data contain patterns that are at least consistent with the importance of a more or less distinct class of genes in the separation between *D. melanogaster* and *D. simulans*. These genes may differ quantitatively from the rest of the genome with respect to their rates of evolution, if not qualitatively in terms of their functions or modes of evolution.

Despite the relatively small proportion of diverged loci, we have found evidence for the importance of selective constraint operating within species from both purifying and balancing selection (R. S. Singh, M. Choudhary, unpublished data). This suggests that the major source of genic divergence between species may be substitutions at loci that are maintained almost monomorphic within species. Thus, one distinguishing feature of loci with high potential for interspecies divergence may be tight constraint by purifying selection, perhaps combined with the presence or appearance of rare alleles capable of fixation by directional selection. If such genetic conditions coincided with the ecological conditions favoring interpopulation divergence, allele substitutions at essentially monomorphic loci could play an important role in species formation (123).

With respect to the functional aspects of loci involved in species formation, we are intrigued by the seeming tendency for hybrid male sterility to arise in the early stages of species differentiation in *Drosophila* (11). The male reproductive tract proteins appear to be more divergent between species of the *D. melanogaster* complex than are proteins of certain "nonreproductive" tissues such as imaginal wing discs. Furthermore, differences in protein amount and presence/absence were very frequent among proteins of the male reproductive tract, and most frequent among proteins with organ-specific distributions within the reproductive tract. Preferential genetic involvement of male-reproductive-tract functions in postmating isolation, like the involvement of sexual behavior in premating isolation, is a possibility that needs to be investigated (108).

FUTURE PROSPECTS

Although much remains to be learned, *D. melanogaster* and *D. simulans* are at present the two best understood species with respect to their genetic variation. This has been achieved not simply by using a large numer of techniques with a large number of randomly selected populations, but by using all the techniques with the *same* populations in a sequential manner. No one technique can replace all the others as each has a different sensitivity and each provides a different perspective.

The multidisciplinary approach to the study of genetic variation has shown that both functional (purifying selection) and selection (balancing selection) constraints play an important role in the population dynamics of genetic variation within and between species. The 2DE studies suggest a far stronger functional constraint on protein variation than has previously been thought. The importance of these results lies not in the higher percentage of monomorphic loci per se but in raising the possibility that a smaller proportion of genes may contain "major" polymorphisms and most of these may be enzymes rather than abundant proteins. Furthermore, the 2DE results give a developmental perspective of genetic variation and raise the possibility that early acting genes may be less polymorphic than late acting genes. On the other hand, the presence of substantial gene-regulation variation for many structurally monomorphic loci (65) and the possibility of a "major" silent polymorphism at the *Adh* locus (45) raises the possibility that gene-regulation polymorphisms may be quite common. Genetic studies of "major" gene-regulation polymorphisms, and 2DE studies of functional constraint on early vs late-acting and/or tissue specific genes are important areas for future research.

Comparisons of genetic structure with appropriate data can discriminate between the role of selection and history in the evolution of variation within as well as between species. Spatially varying selection rather than heterosis emerges as the most important explanation of genic polymorphism in *D. melanogaster* and *D. simulans* (21, 72, 113). *D. melanogaster,* as it has spread from continent to continent, appears to have developed parallel or complementary latitudinal clines in Euro-Africa, the Far East, North America, Australia and most recently in Japan. The clinal variation appears to be more common and stronger for phenotypic characters than for allozymes. The older populations of *D. melanogaster* from Europe and the Far East also have accumulated much new genetic variation not found in North America, Australia, or Japan. The combined use of genic diversity and genetic distance statistics from gene-enzymes, mtDNA, and nuclear DNA sequence studies shows that while some diversifying selection on a few phenotypic characters and allozyme loci has occurred, recent colonization (or shared ancestry) is the dominant cause of reduced geographic differentiation in *D. simulans.*

Genetic studies of species differences with respect to "housekeeping" vs reproduction-related genes (behavioral or gametic) need to be pursued. These studies promise to identify genetic variation for species formation and thus are complementary to the approach involving species crosses pioneered by Dobzhansky (34) and more recently used by Coyne and associates (28, 94).

As exemplified by the *Adh* locus, the DNA sequence data allow a case-by-case analysis of the role of selection, mutation, and historical accidents in the evolution of a gene (73). These analyses essentially rely on molecular imprints left behind by evolutionary forces acting in the past: the presence of excess polymorphism in and around the gene (due to selection), the relationship between sequence variation within population and sequence divergence between populations, and retrospective behavior of haplotype frequencies over time (54, 73). It is equally important to push physiological studies of enzyme polymorphisms to see if the *apparent* physiological and environmental bases of selection (e.g. ethanol in ADH and starch in Amylase) are important (15, 33, 43, 47), or even essential (9), aspects of selection for maintaining polymorphism at the locus.

D. melanogaster and its related species would almost certainly continue to receive increased attention in population and evolutionary genetics. To turn this into a good evolutionary story, however, as pointed out in the section on historical biogeography, we need to learn more about quantitative and physiological variations and their relation to molecular variation and ecology in the various pairs of ecologically comparable species. Although we know something about physiological, quantitative, and behavioral differences between the two cosmopolitan members of this subgroup, we still know relatively little about their ecology, and very little indeed about the genetic structure of other species in this group. More effort is badly needed in this direction.

ACKNOWLEDGMENTS

This review was written during a sabbatical leave at the University of California, Davis. I wish to express my gratitude to Subodh Jain for his wonderful hospitality and for the pleasure of having many discussions about the needs and ways of increasing cross communication between population genetics and ecology. I thank Jean David and his colleagues at CNRS, Gif-Sur-Yvette, France, for sharing with me their research material as well as many pieces of unpublished information on the *D. melanogaster* species group. In appreciation of his leadership in providing research ideas, methodological breakthroughs, and training to graduate students in population genetics, his influence on biology and philosophy, and his outspokenness against the misuse and abuse of biology, this paper is dedicated to Professor R. C. Lewontin on the occasion of his sixtieth birthday.

Literature Cited

1. Anderson, P. R., Oakeshott, J. G. 1984. Parallel geographical patterns of allozyme variation in two sibling *Drosophila* species. *Nature* 308:729–31
2. Aquadro, C. F., Avise, J. C. 1981. Genetic divergence between rodent species assessed by using two-dimensional electrophoresis. *Proc. Natl. Acad. Sci. USA* 76:3784–88
3. Aquadro, C. F., Deese, S. F., Bland, M. M., Langley, C. H., Laurie-Ahlberg, C. C. 1986. Molecular population genetics of the alcohol dehydrogenase gene region of *Drosophila melanogaster*. *Genetics* 114:1165–90
4. Aquadro, C. F., Lado, K. M., Noon, W. A. 1988. The *rosy* region of *Drosophila melanogaster* and *Drosophila simulans*. I. Contrasting levels of naturally occurring DNA restriction map variation and divergence. *Genetics*, 119:875–88
5. Avise, J. C., Giblin-Davidson, C., Laerm, J., Patton, J. C., Lansman, R. A. 1979. Mitochondrial DNA clones and matriarchal phylogeny within and

among geographic populations of the pocket gopher. *Geomys. pinetics. Proc. Nat. Acad. Sci.* USA 76:6694–98

6. Ayala, F. J. 1975. Genetic differentiation during the speciation process. *Evol. Biol.* 8:1–78

7. Berger, E. M. 1970. A temporal survey of allelic variation in natural and laboratory populations of *D. melanogaster. Genetics* 67:121–36

8. Baba-Aissa, F., Solignac, M., Dennebouy, N., David, J. R. 1988. Mitochondrial DNA variability in *Drosophila simulans:* quasi absence of polymorphism within each of the three cytoplasmic race. *Heredity* 61:419–26

9. Bijlsma-Meeles, E., Bijlsma, R. 1988. The alcohol dyehdrogenase polymorphism in *Drosophila melanogaster:* Fitness measurements and predictions under conditions with no alcohol stress. *Genetics* 120:743–53

10. Bodmer, M., Ashburner, M. 1984. Conservation and change in the DNA sequences coding for alcohol dehydrogenase in sibling species of *Drosophila. Nature* 309:425–30

11. Bock, I. R. 1984. Interspecific hybridization in the genus *Drosophila. Evol Biol.* 18:41–70

12. Caccone, A., Amato, G. D., Powell, J. R. 1988. Rates and patterns of scnDNA and mtDNA divergence within the *Drosophila melanogaster* subgroup. *Genetics* 118:671–83

13. Capy, P., David, J. R., Allemand, R., Carton, Y., Febuay, G., et al. 1986. Genetic analysis of *Drosophila melanogaster* in the French West Indies and comparison with populations from other parts of the world. *Genetica* 69:167–76

14. Capy, P., David, J. R., Robertson, A. 1988. Thoracic trident pigmentation in natural populations of *Drosophila simulans:* a comparison with *D. melanogaster. Heredity* 61:263–68

15. Clark, A. G., Doane, W. W. 1984. Interactions between the *Amylase* and *Adipose* chromosomal regions of *Drosophila melanogaster. Evolution* 38(5):957–82

16. Cariou, M. L. 1987. Biochemical phylogeny of the eight species in the *Drosophila melanogaster* subgroup, including *D. sechellia* and *D. orena. Genet. Res.* 50:181–85

17. Carson, H. L. 1987. The genetic system, the deme and the origin of species. *Annu. Rev. Genet.* 21:405–23

18. Cavener, D. R, Clegg, M. T. 1981. Temporal stability of allozyme frequencies in a natural population of *Drosophila melanogaster. Genetics* 98:613–23

19. Chakraborty, R. 1984. Relationship between heterozygosity and genetic distance in the three major races of man. *Am. J. Phys. Anthropol.* 65:249–58

20. Chambers, G. K. 1988. The Drosophila alcohol dehydrogenase gene-enzyme system. *Adv. Genet.* 25:39–107

21. Choudhary, M., Singh, R. S. 1987. A comprehensive study of genic variation in natural populations of *Drosophila melanogaster.* III. Variation in genetic structure and their causes between *Drosophila melanogaster* and its sibling species *Drosophila simulans. Genetics* 117:697–710

22. Choudhary, M., Singh, R. S. 1987. Historical effective size and the level of genetic diversity in *Drosophila melanogaster* and *D. pseudoobscura. Biochem. Genet.* 25:41–51

23. Cohn, V. H., Moore, L. P. 1988. Organization and evolution of the alcohol dehydrogenase gene in *Drosophila. Mol. Biol. Evol.* 5:154–66

24. Coulthart, M. B., Singh, R. S. 1988. Low genic variation in male-reproductive tract proteins of *Drosophila melanogaster* and *D. simulans. Mol. Biol. Evol.* 5(2):167–81

25. Coulthart, M. B., Singh, R. S. 1988. High level of divergence of male-reproductive tract proteins, between *Drosophila melanogaster* and its sibling species, *D. simulans. Mol. Biol. Evol.* 5(2):182–91

26 . Coulthart, M. B., Singh, R. S. 1988. Differing amounts of genetic polymorphism in testes and male accessory glands of *Drosophila melanogaster* and *Drosophila simulans. Biochem. Genet.* 26:153–64

27. Coyne, J. A. 1982. Gel electrophoresis and cryptic protein variation. In *Isozymes: current typics in biological and medical research.* ed. M. C. Ratazzi, J. G. Scandalios, G. S. Whitt, 6:1–32. New York: Liss

28. Coyne, J. A. 1984. Genetic basis of male sterility in hybrids between two closely related species of *Drosophila. Proc. Nat. Acad. Sci.* 81:4444–47

29. Coyne, J. A., Beecham, E. 1987. Heritability of two morphological characters within and among natural populations of *Drosophila melanogaster. Genetics* 117:727–37

30. Coyne, J. A., Kreitman, M. 1986. Evolutionary genetics of two sibling species, *Drosophila simulans* and *D. sechellia. Evolution* 40:673–91

31. David, J. R., Boquet, C., Pla, E. 1976.

New results on the genetic characteristics of the Far East race of *Drosophila melanogaster*. *Genet. Res.* 28:253–60

32. David, J. R., Capy, P. 1988. Genetic variation of *Drosophila melanogaster* natural populations. *TIG* 4:106–11

33. DeJong, G., Scharloo, W. 1976. Environmental determination of selective significance or neutrality of amylase variants in *Drosophila melanogaster*. *Genetics* 84:77–94

34. Dobzhansky, Th. 1936. Studies on Hybrid Sterility II. Localization of sterility factors in *Drosophila pseudoobscura* hybrids. *Genetics* 21:113–3

35. Endler, J. 1988. *Selection in the Wild*. Princeton, N.J.: Princeton Univ. Press

36. Ewens, W. J. 1977. Population genetics theory in relation to the neutralist-selectionist controversy. *Adv. Hum. Genet.* 8:67–134

37. Franklin, I. R. 1981. An analysis of temporal variation at isozyme loci in *Drosophila melanogaster*. See ref. 38, pp. 217–36

38. Gibson, J. B., Oakeshott, J. G. (ed.) 1981. *Genetic Studies of Drosophila Populations*. Canberra: Australian Natl. Univ.

39. Gillespie, H. J., Kojima, K. 1968. The degree of polymorphism in enzymes involved in energy production compared to that in non-specific enzymes in two *D. ananassae* populations. *Proc. Natl. Acad. Sci. USA* 61:582–85

40. Hale, L. R., Singh, R. S. 1986. Extensive size variation and heteroplasmy in mitochondrial DNA among geographic populations of *D. melanogaster*. *Proc. Natl. Acad. Sci. USA* 83:8813–17

41. Hale, L. R., Singh, R. S. 1987. Mitochondrial DNA variation and genetic structure in populations of *Drosophila melanogaster*. *Mol. Biol. Evol.* 4(6):622–37

42. Henderson, N. R., Lambert, D. M. 1982. No significant deviation from random mating of worldwide populations of *Drosophila melanogaster*. *Nature* 300:437–40

43. Hickey, D. A. 1979. Selection on amylase allozymes in *Drosophila melanogaster:* selection experiments using several independently derived pairs of chromosomes. *Evolution* 33:1128–37

44. Hudson, R. R., Kreitman, M., Aguadé, M. 1987. A test of neutral molecular evolution based on nucleotide data. *Genetics* 116:153–59

45. Hudson, R. R., Kaplan, N. L. 1988. The coalescent process in models with selection and recombination. *Genetics* 120:831–40

46. Hyytia, P., Capy, P., David, J. R., Singh, R. S. 1985. Enzyme and quantitative variation in European and African populations of *D. simulans*. *Heredity* 54:209–17

47. Inoue, Y., Tobari, Y. N., Tsuno, K., Watanabe, T. K. 1984. Association of chromosome and enzyme polymorphisms in natural and cage populations of *Drosophila melanogaster*. *Genetics* 106:267–77

48. Johnson, F. M., Schaffer, H. E. 1973. Isozyme variability in species of the genus *Drosophila*. VII. Genotype-environment relationships in populations of *D. melanogaster* from the eastern U.S. *Biochem. Genet.* 10:149–63

49. Joly, D. 1987. Between-species divergence of cyst length distributions in the *Drosophila melanogaster* species complex. *Jpn. J. Genet.* 62:257–63

50. Kimura, M. 1983. *The Neutral Theory of Molecular Evolution*. Cambridge: Cambridge Univ. Press

51. Knibb, W. R. 1982. Chromosomal inversion polymorphisms in *Drosophila melanogaster*. II. Geographic clines and climatic associations in Australia, North America and Asia. *Genetica* 58:213–21

52. Knibb, W. R., Oakeshott, J. G., Gibson, J. B. 1981. Chromosome inversion polymorphisms in *D. melanogaster*. I. Latitudinal clines and association between inversions in Australian populations. *Genetics* 98:833–47

53. Kreitman, M. 1983. Nucleotide polymorphism at the alcohol dehydrogenase locus of *Drosophila melanogaster*. *Nature* (London) 304:412–17

54. Kreitman, M. 1987. Molecular population genetics. *Oxford Survey Evol. Biol.* 4:38–60

55. Kreitman, M., Aquadé, M. 1986. Genetic uniformity in two populations of *Drosophila melanogaster* as revealed by filter hybridization of four-nucleotide-recognizing restriction enzyme digests. *Proc. Natl. Acad. Sci. USA* 83:3562–66

56. Kreitman, M., Aquadé, M. 1986. Excess polymorphism at the *Adh* locus in *Drosophila melanogaster*. *Genetics* 114:93–110

57. Lachaise, D., Cariou, M. L., David, J. R., Lemeunier, F., Tsacas, L., et al. 1988. Historical biogeography of the *Drosophila melanogaster* species subgroup. *Evol. Biol.* 22:159–225

58. Lachaise, D., David, J. R., Lemeunier, F., Tsacas, L., Ashburner, M. 1986. The reproductive relationships of *Drosophila sechellia* with *D. mauritiana*, *D. simulans* and *D. melanogaster* from the

Afrotropical region. *Evolution* 40:262–71

59. Langley, C. H. 1977. Non-random association between allozymes in natural populations of *Drosophila melanogaster*. In *Measuring Selection in Natural Populations*, ed. F. B. Christiansen, T. M. Fenchel, pp. 265–73. Berlin: Springer-Verlag

60. Langley, C. H., Aquadro, C. F. 1987. Restriction-map variation in populations of *Drosophila melanogaster*: white-locus region. *Mol. Biol. Evol.* 4:651–63

61. Langley, C. H., Montgomery, E., Quattlebaum, W. F. 1982. Restriction map variation in the *Adh* region of *Drosophila*. *Proc. Nat. Acad. Sci. USA*, 79:5631–35

62. Langley, C. H., Shrimpton, A. E., Yamazaki, T., Miyashita, N., Matsuo, Y., et al. 1988. Naturally occurring variation in the restriction map of the *Amy* region of *Drosophila melanogaster*. *Genetics* 119:619–29

63. Laurie-Ahlberg, C. C., Maroni, G., Bewley, G. C., Lucchesi, J. C., Weir, B. S. 1980. Quantitative variation in enzyme activities in natural populations of *Drosophila melanogaster*. *Proc. Nat. Acad. Sci. USA* 77:1073–77

64. Laurie-Ahlberg, C. C., Williamson, J. H., Cochrane, B. J., Wilton, A. N., Chasalow, F. I. 1981. Autosomal factors with correlated effects on the activities of the glucose-6-phosphate and 6-glucophosphonate dehydrogenase in *Drosophila melanogaster*. *Genetics* 99:127–50

65. Laurie-Ahlberg, C. C., Wilton, A. H., Curtsinger, J. W., Emigh, T. H. 1982. Naturally occurring enzyme activity variation in *Drosophila melanogaster*. I. Sources of variation for 23 enzymes. *Genetics* 102:191–206

66. Leigh-Brown, A. J. 1983. Variation at the 87A heatshock locus in *Drosophila melanogaster*. *Proc. Nat. Acad. Sci. USA* 80:5350–54

67. Leigh Brown, A. J., Langley, C. H. 1979. Reevaluation of level of genic heterozygosity in natural population of *Drosophila melanogaster* by two-dimensional electrophoresis. *Proc. Nat. Acad. Sci. USA* 76:2381–84

68. Lemeunier, F., Ashburner, M. 1984. Relationships within the *melanogaster* species subgroup of the genus *Drosophila (Sophophora)*. IV. The chromosomes of two new species. *Chromosoma* 89:343

69. Lemeunier, F., David, J. R., Tsacas, L., Ashburner, M. 1985. The *Drosophila melanogaster* species group. In *The Genetics and Biology of Drosophila*, ed. M. Ashburner, H. L. Carson, J. N. Thompson, Jr., 3e:147–256. New York: Academic

70. Lewontin, R. C., Krakauer, J. 1973. Distribution of gene frequency as a test of the theory of the selective neutrality of polymorphisms. *Genetics* 74:175–95

71. Lewontin, R. C. 1974. *The Genetic Basis of Evolutionary Change*. New York: Columbia Univ. Press

72. Lewontin, R. C., Ginzburg, L. R., Tuljapurkar, S. D. 1978. Heterosis as an explanation for large amount of genic polymorphism. *Genetics* 88:149–70

73. Lewontin, R. C. 1985. Population Genetics. *Annu. Rev. Genet.* 19:81–102

74. Lipman, D. J., Wilbur, W. J. 1985. Interaction of silent and replacement changes in eukaryotic coding sequences. *J. Mol. Evol.* 21:161–67

75. MacIntyre, R. J., Collier, G. E. 1986. Protein evolution in the genus *Drosophila*. See Ref. 69, pp. 39–146

76. Martin, C. H., Meyerowitz, E. M. 1986. Characterization of the boundaries between adjacent rapidly and slowly evolving genomic regions in *Drosophila*. *Proc. Nat. Acad. Sci. USA* 83:8654–58

77. Mayr, E. 1963. *Animal Species and Evolution*. Cambridge, Mass.: Harvard Univ. Press

78. McDonald, J., Parson, P. A. 1973. Dispersal activities of the sibling species *Drosophila melanogaster* and *D. simulans*. *Behav. Genet.* 3:293–301

79. McLellan, T., Inouye, L. S. 1986. The sensitivity of isoelectric focusing and electrophoresis in the detection of sequence differences in proteins. *Biochem. Genet.* 24:571–78

80. Mettler, L. E., Voelker, R. A., Mukai, T. 1977. Inversion clines in natural populations of *D. melanogaster*. *Genetics* 87:169–76

81. Miyashita, N., Langley, C. H. 1988. Molecular and phenotypic variation of the white locus region in *Drosophila melanogaster*. *Genetics* 120:199–212

82. Montgomery, E. A., Langley, C. H. 1983. Transportable element in Mendelian populations. II. Distribution of three cupia-like elements in a natural population of *Drosophila melanogaster*. *Genetics* 104:473–83

83. Mueller, L. D. 1985. The evolutionary ecology of *Drosophila*. *Evol. Biol.* 19:37–98

84. Mukai, T. 1985. Experimental verification of the neutral theory. In *Population Genetics and Molecular Evolution*, ed.,

T. Ohta, K. Aoki, pp. 125–145. Tokyo: Jpn. Sci. Soc. Press

85. Nei, M. 1987. *Molecular evolutionary genetics.* New York: Columbia Univ. Press

86. Nigro, L. 1988. Natural populations of *Drosophila simulans* show great uniformity of the mitochondrial DNA restriction map. *Genetica* 77:133–6

87. Oakeshott, J. G., Chambers, G. K., Gibson, J. B., Willcocks, D. A. 1981. Latitudinal relationships of esterase-6 and phosphoglucomutase gene frequencies in *Drosophila melanogaster. Heredity* 47:385–96

88. Oakeshott, J. G., Gibson, J. B., Anderson, P. R., Knibb, W. R., Anderson, D. G., et al. 1982. Alcohol dehydrogenase and glycerol-3-phosphate dehydrogenase clines in *Drosophila melanogaster* on different continents. *Evolution* 36:86–96

89. O'Brien, S. J., MacIntyre, R. J. 1978. Genetic and biochemistry of enzymes and specific proteins of *Drosophila.* See Ref. 69, pp. 396–552

90. O'Farrell, P. H. 1975. High-resolution two-dimensional electrophoresis of proteins. *J. Biol. Chem.* 250:4007–21

91. Ohta, T. 1976. Role of very slightly deleterious mutations in molecular evolution and polymorphism. *Theor. Pop. Biol.* 10:254–75

92. Ohnishi, S., Kawanishi, M., Watanabe, T. K. 1983. Biochemical phylogenies of *Drosophila:* protein differences detected by two-dimensional electrophoresis. *Genetica* 61:55–63

93. Ohnishi, S., Leigh Brown, A. J., Voelker, R. A., Langley, C. H. 1982. Estimation of genetic variability in natural populations of *Drosophila simulans* by two-dimensional and starch gel electrophoresis. *Genetics* 100:127–36

94. Orr, H. A., Coyne, J. A. 1989. the genetics of post-zygotic isolation in the *Drosophila virilis* group. *Genetics* 121:527–537

95. Parsons, P. A. 1983. Ecobehavioral Genetics: Habitats and Colonists. *Annu. Rev. Ecol. Syst.* 14:35–55

96. Parsons, P. A. 1983. *The Evolutionary Biology of Colonizing Species.* New York: Cambridge Univ. Press

97. Parsons, P. A., Stanley, S. M. 1980. Spatial ecological studies—domesticated and widespread species. See Ref. 69, pp. 349–393

98. Raff, R. A., Kaufman, T. C. 1983. *Embryos, Genes and Evolution.* New York: Macmillan

99. Rhomberg, L. R., Singh, R. S. 1989. Evidence for a link between local and seasonal cycles in gene frequencies and latitudinal gene clines in a cyclic parthenogen. *Genetica* 78:73–79

100. Robertson, A. 1975. Gene frequency distributions as a test of selective neutrality. *Genetics* 81:775–85

101. Robertson, A., Briscoe, D. A., Louw, J. H. 1977. Variation in abdomen pigmentation in *Drosophila melanogaster* females. *Genetica* 47:73–76

102. Schaeffer, S. W., Aquadro, C. F., Langley, C. H. 1988. Restriction-map variation in the *Notch* region of *Drosophila melanogaster. Mol. Biol. Evol.* 5:30–40

103. Shah, D. M., Langley, C. H. 1979. Inter- and intra-specific variation in restriction maps of *Drosophila mitochondrial DNAs. Nature* 281:1–3

104. Schaffer, H. E., Johnson, F. M. 1974. Isozyme allelic frequencies related to selection and gene-flow hypothesis. *Genetics* 77:163–68

105. Shields, D. C., Sharp, P. M., Higgins, D. G., Wright, F. 1988. "Silent" sites in *Drosophila* genes are not neutral: Evidence of selection among synonymous codons. *Mol. Biol. Evol.* 5:704–16

106. Simmons, M. J., Crow, J. F. 1977. Mutations affecting fitness in *Drosophila* populations. *Annu. Rev. Genet.* 11:49–78

107. Singh, R. S. 1989. The genetic studies of species differences and their relevance to the problem of species formation. In *Electrophoretic Studies on Agricultural Pests,* ed. H. D. Loxdale, J. Den Hollander, pp. 17–49. Oxford: Clarendon

108. Singh, R. S. 1989. Patterns of species divergence and genetic theories of speciation. In *Topics in Population Biology and Evolution.* ed. K. Wohrmann, S. K. Jain. Berlin: Springer Verlag. In press

109. Singh, R. S., Choudhary, M., David, J. R. 1987. Contrasting patterns of geographic variation in cosmopolitan sibling species in *D. melanogaster* and *D. simulans. Biochem. Genet.* 25:27–40

110. Singh, R. S., Coulthart, M. B. 1982. Genic variation in abundant proteins of *Drosophila melanogaster* and *D. pseudoobscura. Genetics* 102:437–53

111. Singh, R. S., Hickey, D. A., David, J. R. 1982. Genetic differentiation between geographically distant populations of *Drosophila melanogaster. Genetics* 101:235–56

112. Singh, R. S., Rhomberg, L. R. 1987. A comprehensive study of genic variation in natural populations of *Drosophila melanogaster.* I. Estimates of geneflow

from rare alleles. *Genetics* 115:313–22

113. Singh, R. S., Rhomberg, L. R. 1987. A comprehensive study of genic variation in natural populations of *Drosophila melanogaster* II. Estimates of heterozygosity and patterns of geographic differentiation. *Genetics* 117:255–71

114. Skibinski, D. O. F., Ward, R. D. 1982. Correlations between heterozygosity and evolutionary rate of proteins. *Nature* 298:490–92

115. Solignac, M., Monnerot, M. 1986. Race formation, speciation, and introgression within the three homosequential species *Drosophila simulans, D. mauritiana,* and *D. sechellia* inferred from mitochondrial DNA analysis. *Evolution* 40:531–39

116. Stephens, J. C., Nei, M. 1985. Phylogenetic analysis of polymorphic DNA sequences at the *Adh* locus in *Drosophila melanogaster* and its sibling species. *J. Mol. Evol.* 22:289–300

117. Templeton, A. R. 1981. Mechanisms of speciation—A population genetics approach. *Annu. Rev. Ecol. Syst.* 12:23–48

118. Voelker, R. A., Cockerham, C. C., Johnson, F. M., Schaffer, H. E., Mukai, T., et al. 1978. Inversions fail to account for allozyme clines. *Genetics* 88:515–27

119. Wanner, L. A., Neel, J. V., Meisler, M.

H. 1982. Separation of allelic variants by two-dimensional electrophoresis. *Am. J. Hum. Genet.* 34:209–15

120. Ward, R. D., Skibinski, D. O. F. 1985. Observed relationships between protein heterozygosity and protein genetic distance and comparisons with neutral expectations. *Genet. Res.* 45:315–40

121. Watada, M., Ohba, S., Tobari, Y. N. 1986. Genetic differentiation in Japanese populations of *Drosophila simulans* and *D. melanogaster*. II. Morphological Variation. *Jpn. J. Genet.* 61:469–80

122. Watanabe, T. K., Watanabe, T. 1977. Enzyme and chromosomal polymorphism in Japanese natural populations of *Drosophila melanogaster*. *Genetics* 85:319–29

123. Wright, S. 1982. The shifting balance theory and macroevolution. *Annu. Rev. Genet.* 16:1–19

124. Yamazaki, T., Choo, J.-K., Watanabe, T. K., Takahata, N. 1986. Gene flow in natural populations of *Drosophila melanogaster* with special reference to lethal allelism rates and protein variation. *Genetics* 113:73–89

125. Zera, A. J., Koehn, R. K., Hall, J. G. 1984. Allozymes and biochemical adaptation. In *Comprehensive Insect Physiology, Biochemistry, and Pharmacology* 10:633–74. New York: Pergamon

Annu. Rev. Genet. 1989. 23:455–82

GENETIC REGULATION OF BACTERIAL VIRULENCE

Victor J. DiRita and John J. Mekalanos

Department of Microbiology and Molecular Genetics, Harvard Medical School, 200 Longwood Avenue, Boston, Massachusetts 02115

CONTENTS

> . . . what is actual is actual only for one time
> And only for one place
>
> -T. S. Eliot; *Ash-Wednesday*

INTRODUCTION

The question of how microbes cause disease can be approached from many viewpoints. For this reason, bacterial pathogens are often interesting to biochemists, immunologists, geneticists, toxicologists, and cell biologists. In

455

0066-4197/89/1215-0455$02.00

this review, we concentrate on genetic control of bacterial virulence. Common virulence determinants and strategies have been reviewed recently and are not extensively covered here (17, 48a, 84). We also limit this review to several systems that illustrate common molecular themes in control of virulence properties rather than presenting a comprehensive review of this extensive literature.

The genetic control of virulence can be divided into two broad categories: random and nonrandom. Random forms of control include cases in which a fraction of a given population exhibits a phenotype different from the rest of the population. Such phenomena are usually referred to as "phase" or "antigenic" variation and typically involve surface structures such as flagella, pili, outer membrane proteins, and capsules that enhance the colonization of host tissues or allow the microbe to evade phagocytosis (reviewed by Seifert & So 165). These structures are major targets of the host antibody response; therefore, the ability to rapidly change such surface features allows at least a fraction of the population to avoid the immune system. Alternatively, such variation may provide a benefit to the microbe before the onset of a specific immune response. For example, by expressing variant adhesion factors throughout the population, there is a better chance that a given cell will be successful in finding receptors on the mosaic of different tissues displayed by the host.

The second class of genetic control is nonrandom. Pathogenic bacteria, like their nonpathogenic counterparts, sense signals in the environment and respond accordingly by expressing gene products necessary for survival in the host (123). While the in vivo environment (i.e. that of host tissues) is clearly the most relevant, it has often been instructive to use appropriate laboratory (in vitro) models in attempting to understand mechanisms of signal transduction in pathogenesis. The signals likely to lead to virulence-gene expression can sometimes be deduced from the known lifestyle of the organism. Temperature, pH, and the presence or absence of specific ions or small molecules associated with certain environments are all important signals for various pathogens (123).

What follows is a discussion of various random and nonrandom means by which pathogenic microorganisms control the expression of virulence factors. We have concentrated on systems that have been well studied. This is by no means an exhaustive account. Wherever possible, we refer to relevant review articles to provide further depth or breadth to the discussion.

PHASE AND ANTIGENIC VARIATION

Type 1 Pilus Production in Escherichia coli

Most strains of *E. coli* produce type 1 pili (or fimbriae), which are made up of repeating, identical subunits and are approximately 1 μm in length and 7 nm in diameter (41, 97). *E. coli* expressing type 1 pili adhere to mannose-containing molecules on many types of eukaryotic cells (5, 11, 144). Adherent Fim$^+$ bacteria compete better for essential growth nutrients than do Fim$^-$ strains and a Fim$^+$ strain that elicits the *E. coli* heat labile toxin is more toxic to mouse Y1 adrenal cells than a Fim$^-$ strain (208). However, cells expressing these fimbriae are more susceptible to phagocytosis and liver clearance, perhaps accounting for the predominance of Fim$^-$ isolates from blood (5, 11, 107).

It has long been known that type 1 producing *E. coli* alternates between the Fim$^+$ and Fim$^-$ states (14, 179, 180) ("fim" and "pil" are both used to describe this phenotype and genes associated with it). Using *lacZ* fusions to a pilus gene, Eisenstein demonstrated that phase variation is under transcriptional control and that this RecA-independent event occurs at a rate of approximately one cell per 1000 per generation (40). The switch is a 314 bp segment, immediately preceding the *fimA* (pilin) gene [originally designated as *fimD* (57)], which is present in inverted orientation in the two phases (1, 56). The invertible DNA contains the promoter of the *fimA* gene so that the gene can be expressed when the promoter is in one orientation but not when it is inverted (1, 96). Upstream of the switch there is a consensus binding site for *E. coli* integration host factor (IHF), which is necessary for a number of diverse cellular processes and for integration of lambdoid phages (80, 105). It has recently been demonstrated that *E. coli* strains with lesions in either of the genes encoding the two subunits of IHF are incapable of phase variation (37, 42).

Mutational analysis of the cloned *pil* locus identified several Pil specific functions. In addition to genes involved in structure and function of the pilus, two loci (*fimB*, *fimE/hyp*) regulate piliation (96, 98, 145, 146, 148). The nucleotide sequences of *fimB* and *fimE* showed both the genes and the deduced amino acid sequences of the proteins to be quite homologous, indicating that they may have arisen as a result of a gene duplication event (96). FimB is necessary for switching in the OFF (Pil$^-$) to ON (Pil$^+$) direction, and FimE switches the DNA segment from the ON to the OFF orientation (96). Mutations in *fimE/hyp* affect switch frequency as well as the control of *fimA* expression, causing the expression of more pili per cell than seen in wild-type (147). This indicates that in addition to its role in switching, the *fimE/hyp* product represses *fimA* expression (147). Analysis of the *fimB* and *fimE/hyp* coding sequences has shown that these proteins share significant similarities to the integrase proteins of several bacteriophages (2, 37, 42, 93). Thus, FimB and FimE are probably involved in the site-specific recombination required for inversion of the *fimA* promoter.

The product of the *pilG* locus is a second "host" function, in addition to IHF, regulating phase variation (170). Mutations in this gene have switch frequencies up to 100 times the rate of normal (P. Orndorff, personal communication). The *pilG* gene product is involved in regulating the level of DNA supercoiling; mutations at this locus (designated as *osmZ* mutations) affect the expression of the osmotically regulated *proU* operon as well as other genes whose expression is influenced by the level of DNA supercoiling (74).

Phase and Antigenic Variation in Gonococcal Pilins

The ability of *Neisseria gonorrhoeae* to produce pili is essential to its virulence as these structures enable the organism to adhere to eukaryotic cell surfaces (20, 181). Pilus production in gonococci is subject to phase variation and antigenic variation, whereby successive isolates express either no pili or antigenically distinct pilin polypeptides, respectively (19, 169, 184, 187, 188). Rather than simply being a mechanism of avoiding the host immune response, antigenic variation in *Neisseria* may allow colonization of different host environments by assuring that the best adherence phenotype is expressed in at least a fraction of the population in all environments. That this is the case is suggested by the isolation from human volunteers of pilin variants of the inoculum strain before a detectable antibody response had been mounted (188). It is unclear, however, whether the gonococcal pilin is the actual adhesin or whether accessory proteins are involved (178).

There are several copies of *pil* sequences in the genome, although most of these (the *pilS* loci) represent only partial pilin genes because they lack sequences encoding the amino terminus of the pilin (66, 163, 186). In most cases, there is a single full-length expression locus, *pilE*, although at least one unusual strain has two functional *pilE* loci (121, 162, 163, 185, 186). Antigenic variants have different coding sequences at the *pilE* locus; segments of these coding sequences are also found at specific *pilS* loci. Antigenic variation is accomplished by RecA-mediated, nonreciprocal exchange of coding information (gene conversion) between the *pilS* and *pilE* loci (66, 70, 100, 121, 162, 188).

pilS loci harbor "minicassettes" of variable coding segments surrounded by DNA common to the *pilE* and *pilS* regions (66, 67). These conserved regions presumably mediate gene conversion between *pilS* and *pilE* loci and in functional pilins may encode portions of the molecule essential for overall structure, assembly, or adhesive properties (66, 188). These minicassettes represent the storage forms of variable coding information that allows different antigens to be expressed depending on which *pilS* copy converts the expression site (66, 188). In addition, the presence of repetitive sequences in only some of the *pilS* copies and in the *pilE* locus indicates that there may be a preferred sequence of transfer of information among silent loci and between silent and expressed loci (66).

Gene conversion between *pilS* and *pilE* is also responsible for phase variation. Revertible pilus⁻ strains contain copies of missense or nonsense coding sequences at the *pilE* locus that are replaced by pilin⁺ sequences in subsequent isolates (8, 186). Thus, the mechanism responsible for antigenic variation functions in phase variation as well. In some pilus⁻ gonococci, deletions within the expression locus have been observed; such strains are generally nonreverting, but can revert in the strain that harbors two expression sites (162, 185, 186). It is also possible that some Pil⁻ strains arise as a result of mutations in *pilA* or *pilB*, which encode *trans*-acting functions that regulate expression of the *pil* promoter (189).

It is not clear what role the pilus⁻ phase plays in the life of the gonococci. It may occasionally be an intermediate in antigenic variation (70), although phase variation is not a prerequisite for antigenic variation (8). It has also been suggested that the pilus⁻ phase may aid in dissemination of the organism (186).

Neisseria cultures undergo autolysis and are naturally competent for transformation (165). Seifert et al showed that a suitably marked *pilE* locus in one strain can convert the *pilE* locus in another during cocultivation of the two strains. This event is sensitive to DNase I and arises as a result of transformation (164). Antigenic variation may be the result of such transformation; the authors note, however, that because nonpiliated strains are not competent for transformation, the most likely way for a pilus⁻ strain to revert to pilus⁺ is through intracellular gene conversion as described above (164).

Variation in the Gonococcal P.II Protein

The P.II, or opacity, protein of *Neisseria gonorrhoeae* is a surface structure that may serve many functions for the cell. It has been implicated in resistance to cells of the immune system, adherence of gonococci to eukaryotic cells, survival in human serum, and the ability to colonize different microenvironments in host tissues (38, 86, 198). Like the *Neisseria* pilin, P.II undergoes both phase and antigenic variation with successive isolation both in vitro and in vivo (35, 86, 161, 182). Antigenic variation in vivo may allow differential outgrowth of variants selected by the host, as it has been demonstrated that variation occurs during a single infection and may be influenced by secretions in the host containing proteases or sex hormones (10, 38, 159, 161, 199). Changes in PII expression have also been observed to coincide with different phases of the menstrual cycle (86).

A single gonococcal isolate may produce from zero to several variants of the P.II protein, which share conserved domains but which also contain two regions of hypervariability in the coding sequence; these hypervariable regions are in areas of highest antigenicity (27, 165, 172, 183). Several genes encode P.II proteins in the gonococcal genome and, unlike the *pil* loci, each P.II gene (designated *opaE*) has a full-length copy (172).

As with the gonococcal pilin, antigenic variation of P.II is probably due to gene conversion between different copies. The conserved regions of the P.II coding sequences, which flank the hypervariable sequenes, could mediate these gene conversion events (27, 172).

There are no transcriptionally silent *opaE* loci (172). Thus, unlike pilin genes, the genes encoding the P.II proteins are all constitutively expressed; the number and type of P.II proteins is therefore determined posttranscriptionally. The hydrophobic core of the P.II signal sequence is encoded by a tandemly repeated sequence, CTCTT, which is present in variable copy number in different genes (136, 172). Although all genes are transcribed, the only transcripts successfully translated are those in which the open reading frame after the initiating ATG is maintained through the CTCTT-encoded signal sequence. The switch from P.II$^+$ to P.II$^-$ and back is thus associated with frameshifts due to variation in the number of CTCTT repeats within a P.II gene (136, 172). The gain or loss of one to several copies of this sequence is accomplished by slipped-strand mispairing of DNA in this region, perhaps during DNA replication (136).

Type b Capsule Expression in Haemophilus influenzae

Type b strains of *H. influenzae,* responsible for meningitis and other infections, produce a capsule made up of a ribose/ribitol-5-phosphate polymer that is a major determinant of virulence in this organism (135). Nonencapsulated strains of *H. influenzae* arise at a high frequency, suggesting that loss of capsule production may confer advantage to the organism under certain conditions (76).

The genes necessary for extracellular capsule production are in a locus termed *cap,* which contains a duplication of 17–18 kbp interrupted by a bridge region of 1.3 kbp (77, 102). Cap$^-$ variants of wild-type isolates occur at high frequency and are associated with resolution of the repeated DNA in a *rec*-dependent process (43, 102). Such deletions remove the bridge region, which encodes the *bexA* gene (*b* capsule *e*xpression A), whose product is required for transport of the capsule (101, 102). Thus, although these strains are phenotypically Cap$^-$, they accumulate capsule antigen intracellularly (101). The BexA protein shares similarity with several ATP-binding proteins involved in periplasmic transport functions, including the MalK protein of *E. coli,* and HisP and OppD of *Salmonella typhimurium.* Spontaneous deletion of *bexA* therefore removes a putative capsule transport function (101). Cap$^-$ *H. influenzae* adhere better to epithelial cells than Cap$^+$ strains, so there is an advantage to *H. influenzae* in becoming Cap$^-$ (104). Unlike other types of phase variation however, Cap$^-$ variants have apparently lost the ability to return to Cap$^+$. It is not known whether Cap$^-$ strains can reacquire *bexA* through transformation in vivo, as they can in vitro (135), although *H.*

influenzae is naturally competent, with a specialized system of recognizing and taking up its own DNA (173).

Variable Major Proteins in Borrelia

Relapsing fever caused by *Borrelia hermsii* features recurrent phases of bacteremia in which isolates from successive phases express surface antigens different from the previous phase (4). These serotype-specific surface proteins, termed variable major proteins (VMPs), are encoded by genes that reside on linear plasmids as either storage or expressed loci (151). The partial copies apparently do not contain sequences necessary for gene expression, while the expression sites harbor these sequences (151). Recombination between silent copies and expression sites leads to exchange of coding information resulting in expression of a new VMP, against which the host has yet to elicit an antibody response (120). As VMPs are evidently expressed only to allow the organism to evade the host antibody response, and not as colonization factors, the interaction between the host and *Borrelia* is very nearly a strict competition between two genetic rearrangement systems: That of the host, responsible for generating antibodies, and that of *Borrelia*, generating antigenic variants.

Streptococcal M Proteins

Size variation of the M protein, the antiphagocytic surface structure of Group A streptococci, is frequently observed among strains of the same serotype (51, 52). This variation results from intragenic recombination between highly homologous, tandem elements within the protein coding sequence (79). That these elements are not identical accounts for antigenic variation between the size variants (89). Streptococci thus engage a mechanism of antigenic variation of a surface protein using a single copy of the coding sequence, rather than several copies as seen in the systems described above. However, as with the gonococcal pilin, variation does not disrupt the essential structure of the M protein; its antiphagocytic function is therefore unaffected in variants (89).

ENVIRONMENTAL CONTROL OF GENE EXPRESSION

Agrobacterium tumefaciens and Crown Gall Tumorigenesis

Although much of the current review is devoted to human pathogens, perhaps the best understood bacterial pathogen is *Agrobacterium tumefaciens,* responsible for the formation of tumors on many plant species. The molecular biology of tumorigenesis has been recently reviewed (209). Briefly, virulent strains of *A. tumefaciens* harbor a large plasmid, the Ti-plasmid, a portion of which is integrated into the plant genome upon invasion of wound sites. This transferred portion (T-DNA) directs the synthesis of compounds that are catabolized by the invading organism using Ti-plasmid encoded functions (209 and references therein).

A region of the Ti-plasmid outside of the T-DNA directs the processing of the T-DNA prior to transfer to the plant cell (87, 196, 209). This *vir* region encodes six complementation loci *virA-E* and *virG* (95, 209). Two of these genes, *virA* and *virG,* are necessary for transcriptional activation of the other *vir* genes in response to phenolic compounds secreted by wounded plant cells (13, 171). The *virA* and *virG* products show similarity to a family of genes that make up the prokaryotic two component regulatory systems (106, 157, 203). In this system (reviewed by Albright et al, this volume), gene products similar to VirA, which are often membrane-associated, are responsible for sensing signals in the environment that, if present, lead to activation of genes by the VirG-homologous component (157). Sensory transduction in chemotaxis, nitrogen limitation, and outer-membrane protein expression in response to osmotic changes involves phosphorylation of the sensor (82) and phosphorylation of the regulator by the sensor (73, 92, 140, 141).

Two-component regulatory systems are involved in the virulence of many pathogens (123), including the regulation of secreted proteins in *Staphylococcus aureus* [by the *agr* locus; (156); S. Projan & R. Novick, personal communication); this locus has also been called *exp;* (134)], mucoidy in *Pseudomonas aeruginosa* [by the *algR* gene; (33)], and intracellular survival of *Salmonella* (by the *phoP* locus; see below). Aspects of this paradigm are also observed in *Bordetella pertussis* and *Vibrio cholerae* (see below). In addition, the promoters for pilin genes in *Pseudomonas aeruginosa, Neisseria spp.* and *Moraxella bovis* share consensus elements with those of genes regulated by the NtrB/NtrC two-component system (88). Although *P. aeruginosa* pilin gene expression is not controlled directly by these two gene products, it does require the alternative sigma subunit, RpoN, used in the nitrogen regulatory system, indicating that environmental signals may determine expression of this pilus as well (85).

Regulation of Virulence in Salmonella

Virulent species of *Salmonella,* responsible for many disease in man including typhoid fever, can invade epithelial cells and survive within macrophages. Attempts to construct a live vaccine against *Salmonella* have demonstrated that strains with defects in various properties are attenuated for virulence (108). These functions include purine biosynthesis (39, 143), aromatic amino acid biosynthesis (15, 39, 78), and lipopolysaccharide biosynthesis (53, 59). Virulence in *S. typhimurium* has been associated with functions encoded on a large plasmid; the role of this plasmid in virulence is not clear, although it does encode serum resistance (65, 68, 69). Catabolite control by adenylate cyclase and cAMP receptor protein (CRP) is important for virulence but so far no specific virulence determinants have been shown to be under this type of regulation in *S. typhimurium* (32). [In *E. coli*, production of heat-stable enterotoxin and of the PAP pilus, associated with pyelonephritis, are catabolite-controlled (3, 114)].

The *phoP* locus in *S. typhimurium* is essential for virulence, particularly for intracellular survival (46, 124). Mutations in *phoP* increase the LD_{50} of *S. typhimurium* for BALB/c mice by a factor of 10^5, yet *phoP* mutants are not noticeably growth-impaired on many different media in vitro (124). Tn*10* insertions into the *phoP* locus abrogate macrophage survival in *S. typhimurium*, presumably due to enhanced sensitivity to macrophage-derived cationic peptides called defensins (46).

The *phoP* locus from *S. typhimurium* has been cloned and sequenced (46, 124) and encodes two open reading frames, termed *phoP* and *phoQ*, that show similarity to the regulator and sensor prototypes, *virG* and *virA*, respectively, from *A. tumefaciens* (124; see above). Thus, a two-component regulatory system is required for virulence in *Salmonella*. The *phoQ* gene product has a periplasmic domain with which signals within the macrophage may interact (124).

A *phoP*-regulated gene required for *Salmonella* virulence has been identified (124). Strains with mutations in this gene, *pagC* (25 min. on the *S. typhimurium* chromosome), have LD_{50} levels approximately three orders of magnitude greater than wild-type, and both *phoP* and *pagC* mutants exhibit reduced survival in macrophages (S. Miller, J. Mekalanos, J. Swanson, unpublished observations). Since it is necessary for intracellular survival and has been identified as a secreted protein by Tn*phoA* analysis, PagC may be a surface-exposed protein that protects the organisms from toxic factors within the phagolysosome (124).

It has recently been shown that contact with epithelial cell monolayers in vitro stimulates synthesis in *S. cholerasuis* and *S. typhimurium* of at least three proteins during a 12 hour period (49). Induction of these proteins is correlated with an ability to invade polarized epithelial cells in vitro (49, 50), as avirulent Tn10 mutants defective for epithelial cell and macrophage invasion also did not synthesize these proteins (47, 49). Whether expression of these proteins is influenced by PhoP and PhoQ has not been reported.

Antigenic Modulation and Variation in Bordetella pertussis

The respiratory pathogen *B. pertussis* causes the disease whooping cough. This organism produces several virulence factors including pertussis toxin (Ptx), filamentous hemagglutinin (FHA), a hemolysin (Hly) and other toxins and potential colonization factors [recently reviewed by Weiss & Hewlett (201)]. Activation of virulence gene expression is controlled by the *vir* locus (200), which encodes three open reading frames (ORFs) of 209, 275, and 936

amino acids (*bvgA, bvgB,* and *bvgC,* respectively) (62). Expression of virulence factors in *B. pertussis* is subject to two forms of genetic control; phase
variation and antigenic modulation, which correspond to random and nonrandom mechanisms of control, respectively.

At a frequency of up to 10^{-3}, virulent organisms become avirulent with a
concomitant decrease in expression of virulence factors (201). Reversion back
to virulence occurs at an undetermined frequency. Stibitz and co-workers
have shown that this phase variation is the result of a frameshift mutation, in
the Vir$^-$ phase, within the largest of the three *vir*-encoded open reading
frames *(bvgC)* (62, 174). The protein encoded by this open reading frame
(BvgC) appears to be a one-component analogue of the two-component
regulatory model, exhibiting similarity to both the carboxy terminus of the
sensor class and the amino terminus of the regulatory class (62, 174). The
significance of these similarities is as yet only implied, but clearly BvgC is
indispensible to the Vir$^+$ phase.

Expression of *B. pertussis* virulence determinants is also under nonrandom
environmental control. This phenomenon, known as antigenic or phenotypic
modulation, is defined by the inability of *vir*-activated genes to be expressed
in the presence of $MgSO_4$, nicotinic acid, or at low temperature (22°C–27°C;
these are the commonly used modulators) [for an exhaustive analysis of
modulation, see (103)]. Using Tn*phoA* mutagenesis, Knapp & Mekalanos
identified both *vir*-activated genes *(vag),* which include *ptx* and *fha,* and
vir-repressed genes *(vrg)* (99). Unlike the *vag* genes, the *vrg* genes are
expressed only in the presence of the modulators or after inactivation of the
vir locus.

That phenotypic modulation is associated with the *vir* locus was demonstrated by Stibitz et al who showed that a cosmid clone containing *vir* and the
closely linked genes encoding the filamentous hemagglutinin *(fhaA, B, C)*
was sufficient for modulated expression of FHA antigen in *E. coli* (175).
Spontaneous mutations in a locus called *mod* lead to constitutive expression of
virulence determinants (99). In a Mod^c background, *vag* genes are maximally
expressed and *vrg* genes are minimally expressed, irrespective of the presence
of modulators (99). Thus, the *mod* gene product may be a sensor that
transmits the modulator status of the environment to a *vir*-encoded protein
that, in response, activates or represses expression from the appropriate
genes. Presumably, *mod* mutations map to one of the three *vir* ORFs described above.

Although the whole *vir* locus is required for modulated expression of FHA
in *E. coli,* the *vir* locus-encoded *bvgA* ORF, if present on a multicopy
plasmid, is sufficient to activate an *fhaB::lacZ* fusion in *E. coli* (123). The
primary sequence of BvgA is similar to the amino terminus of the regulator
class of two component activators (62, 123). Considering that BvgC also has

activator as well as sensor similarity, it is likely that modulated expression of virulence determinants in *B. pertussis* results from the interaction of several different *vir* locus products.

The previous unidentified *vrg* genes may encode products important in maintaining the human carrier state of the organism in in vivo environments rich in modulators (99). Avirulent phase *B. pertussis* can be isolated from long-term infected animals (91), suggesting that there is selection in vivo for such variants; whether these variants express *vrg* genes has not been determined. Peppler and coworkers have recently obtained evidence for the invasion and intracellular survival of *B. pertussis* and the closely related *B. parapertussis* in cultured mammalian cells and in infected animals (44, 204). Perhaps *vrg* genes are derepressed by modulators present in the cytoplasm of mammalian cells; if so, it is apparent that the *vir* regulon may recognize both intracellular and extracellular signals.

How *vir*-dependent activation and repression works is now being studied. Deletion analysis of the *ptx* promoter identified 170 bp upstream of the mRNA start site as necesary for *vir* dependent expression (63, 139). A direct repeat in this region is required for *trans* activation by Vir, as deletion into this repeat abolished expression of *ptx::CAT* fusions (63). Preliminary analysis of *vrg* genes indicates that repression of these involves elements downstream of transcription initiation (D. Beattie, S. Knapp, & J. Mekalanos, unpublished data); it will be interesting to see if the same *vir*-encoded protein(s) acts as an activator or a repressor depending on where its recognition site is (34).

Coordinate Regulation of Virulence Determinants in Vibrio cholerae

The intestinal pathogen responsible for Asiatic cholera, *V. cholerae*, has been the focus of much study for many years; this work has centered principally on the biochemistry and regulation of toxin production (reviewed in 36). Recent work has identified other virulence loci and, of continuing interest, the regulatory genes underlying their expression (125, 150, 191, 193).

Expression of several genes whose products are associated with virulence is under coordinate control of the *toxR* gene product (129, 130, 150, 193). In addition to the cholera toxin genes *(ctxAB)*, other genes in the *toxR* regulon include those required for elaboration of the TCP pilus, which is the major colonization factor (193), genes of the *acf* operon, which encode an as yet unidentified accessory colonization factor (150), two outermembrane proteins OmpT and OmpU, and several other genes yet to be characterized (150).

ToxR is a 32.5 kDa transmembrane protein that specifically binds to a repeated element within the *ctxAB* promoter (130). It is also the primary component in transduction of environmental signals that lead to virulence gene expression (130). Thus, ToxR apparently has both sensor and activator functions. We have recently described the *toxS* gene, whose product enhances ToxR activity (125). ToxS is a membrane-associated protein most of which resides within the periplasm. The results of experiments using ToxR-PhoA fusion proteins indicate that ToxS and ToxR interact in the periplasm; we postulate that this interaction leads to a conformational change of ToxR (perhaps oligomerization) that makes it competent for transcriptional activation (V. DiRita & J. Mekalanos, unpublished data). The role of this interaction in signal transduction is unclear at this time. Based on nucleotide sequence data, ToxR and ToxS do not appear to share the functional relationship of the two component regulators discussed above, although the amino terminus of ToxR is similar to putative DNA binding/transcriptional activation domains of the activators (130, 157). Missense mutations in conserved residues within this domain that lead to a ToxR⁻ phenotype have been identified (V. DiRita, K. Otteman, & J. Mekalanos, unpublished data). The specific effect of these mutations on ToxR function is currently under investigation.

A third gene involved in coordinate regulation has also been isolated. The product of this gene *(toxT)* can activate gene fusions that are dependent on *toxR* in *V. cholerae* but that cannot be activated in *E. coli* by cloned *toxR* alone (C. Parsot & J. Mekalanos, unpublished data). Cloned *toxT* expressed from a constitutive promoter can suppress some of the phenotypes of *ToxR* mutations in *V. cholerae* (V. DiRita, G. Jander, & J. Mekalanos, unpublished data), suggesting that *toxT* may be a transcriptional activator whose promoter is activated by ToxR. Two other genes encoding regulators of gene expression appear to be controlled by ToxR. These are *tcpH* and *tcpI,* whose products are positive and negative regulators of the genes required for synthesis and assembly of the TCP pilus (191; R. Taylor, personal communication). Whether ToxR controls these genes directly or through another gene product such as ToxT is unknown. However, it is apparent from these data that ToxR probably coordinately regulates the expression of some genes indirectly by regulating the expression of other regulatory genes. Such "cascading" of regulatory factors has been previously observed in bacterial developmental systems (109).

The Low Calcium Response of Yersiniae

Virulence of the three pathogenic *Yersinia* species, *Y. pestis, Y. pseudotuberculosis,* and *Y. enterocolitica* involves their ability to invade and grow within cells and to resist the bacteriocidal effects of serum (16, 128). Although virulence in these organisms is mainly plasmid encoded (45, 154), invasion and perhaps other virulence functions are chromosomally encoded phenotypes (83, 126, 127; Reviewed in 28, 128). Two in vitro signals mediating the virulence response, temperature and calcium, have received a great deal of attention.

Yersiniae require high levels of Ca^{2+} for growth at 37°C but not at 26°C (18). In the absence of Ca^{2+} at 37°C, there is a metabolic "stepdown" or restriction, accompanied by expression of genes whose products are associated with virulence. This complex virulence determinant, called the low calcium response (Lcr), is plasmid encoded and involves several gene products, including outer membrane proteins (encoded by the *yop* genes) (12, 31, 54, 122, 154), the V-antigen (encoded by the *lcrGVH* operon) (149) and regulatory genes (see 16, 18, 28 for a more complete discussion of the Lcr phenotype).

While it is clear that calcium affects expression of virulence related genes, the precise nature of this control has yet to be evinced (12, 54, 60, 176, 207). The calcium control region of the virulence plasmids in all three species is conserved and contains at least four transcriptional loci, *lcr(vir)A, B, C,* and *F* (30, 60, 153). The *lcrE* gene, a constituent of the *lcrA* locus in *Y. pestis,* may be involved in sensing the calcium status of the environment (207). Calcium concentration controls the action of a repressor of gene expression (12), and may also play a role in the processing of proteins destined for export during virulence (31). Although growth restriction coincident with virulence gene expression probably does not occur in vivo, genes associated with the Lcr are activated within macrophages and under conditions simulating the in vivo environment (R. Brubaker, personal communication; 152, 177).

Transcriptional activation of plasmid-encoded virulence determinants at 37°C requires the action of the *virF* gene product and, to a lesser extent, the products of *virA, B,* and *C* (29, 30, 60, 206). The deduced amino acid sequence of the VirF protein shows similarity at the C-terminus to the AraC protein, which regulates the arabinose operons of *E. coli* and *S. typhimurium* (29). Thermodependent expression of *virF* in *E. coli* indicates that the gene is autoregulated (29). Therefore, VirF is the primary factor in thermoregulation in *Yersiniae;* it has been suggested that the protein may act through temperature-sensitive conformational changes that affect its DNA-binding ability (29).

AraC similarity has also been seen in a virulence regulator of enterotoxinogenic *E. coli.* The *rns* gene, which controls the expression of the CS1 and CS2 adhesins synthesized by these strains, has been cloned and sequenced and the deduced amino acid sequence of the Rns protein revealed similarity to AraC in the same region as that observed within VirF (26). In the *E. coli* system, the regulatory gene resides on a plasmid and controls a chromosomally encoded phenotype (26); such an arrangement of a regulator and its target gene has not been observed before. Base composition analysis of *rns* indicates that its origin was in a species other than *E. coli* (26).

Temperature Control of Invasion by Shigella flexneri

The pathogenesis of *Shigella* is characterized by its ability to invade and replicate in epithelial cells of the colon (119). Strains that cannot invade are avirulent, as are strains that invade but do not reinvade contiguous cells (111, 115). The virulence mechanisms of this species have recently been reviewed, but recent work merits inclusion here (119).

Virulent strains of *Shigella* species harbor a 220 kbp plasmid that encodes several loci necessary for invasion and virulence. A 37 kbp region from this plasmid is sufficient to allow *Shigella flexneri* to invade HeLa cells (7, 115, 119, 160). This region contains the *ipa* genes, which encode surface proteins associated with virulence, as well as other unspecified invasion loci (6, 21, 71, 115, 197). Full virulence as measured by different invasion assays requires plasmid loci apparently not present within this 37 kbp (119). These include the *virG* locus, which is required for reinfection following in-tracellular growth, and a regulatory locus, *virF*, that encodes a 30 kDa activator of *virG, ipaB, C,* and *D* expression (111, 158).

Shigella virulence is strictly regulated by temperature so that strains grown at 30°C are noninvasive, while those grown at 37°C are fully invasive (116). This effect is at the level of gene expression (116, 117). The product of the *virR* gene represses expression of plasmid encoded loci at 30°C (118). Although virulence is substantially plasmid encoded, the *virR* gene resides on the chromosome (118). Recently, a gene capable of complementing a *VirR* defect in *Shigella flexneri* was cloned from *E. coli* K-12 (81). Since this strain of *E. coli* is not pathogenic, the *virR* product may regulate expression of other genes in the cell (81).

The relationship between *virR* and *virF* control is currently unclear. Regulation of virulence by both a chromosomally encoded repressor and a plasmid-encoded activator suggests that virulence gene expression is only part of a global response when the organisms are shifted to 37°C.

Gene Expression Controlled by Iron

The type of control seen in *Shigella* invasion, whereby extrachromosomally encoded determinants are expressed under conditions in which the cell presumably mounts other responses, has been seen in iron regulation of toxin gene expression in both *E. coli* and *Corynebacterium diphtheriae* (22–24, 137, 190). In both organisms, toxin genes (Shiga-like toxin and diphtheria toxin, respectively) are present on lysogenic phages and are maximally expressed in vitro in environments low in free iron, which is the likely environment of animal fluids rich in transferrin and lactoferrin (48). Many genes in *E. coli* whose products are required for low iron survival (both chromosomally and plasmid encoded) are repressed by the product of the *fur* locus during growth in high iron (72). The promoter for the Shiga-like toxin operon harbors a near consensus *fur* binding site and expression of these genes is repressed by the *fur* system during growth in high iron (23). Similarly, the hypothesis that iron acts as a corepressor for phage-encoded toxin production in *C. diphtheriae* (137) has recently been supported by the demonstration that repression of the cloned diphtheria toxin promoter in *E. coli* is dependent on *fur* (190).

Low iron expression of virulence determinants is not restricted to control by repressors. The exotoxin A gene of *Pseudomonas aeruginosa (toxA)* is positively regulated by the product of the *regA* gene (75) [this gene was originally identified as *toxR;* (205)]. Toxin production is nearly undetectable in media of high iron and this effect is due to the iron-regulated expression of the *regA* gene (55). This gene has two promoters; the T1 promoter is expressed early during low iron growth, followed by a decrease in *regA* mRNA. Late in low iron growth, the shorter T2 transcript is expressed. The biphasic nature of *regA* transcription is mirrored by *toxA* transcription, but extracellular toxin activity is not appreciably detectable until after accumulation of the T2 transcript (55). The nature of iron control of T2 and T1 expression has not been determined yet, but this system may involve repression as well as activation, which is also seen in other organisms described above; such fine-tuning of virulence gene expression is apparently a common motif in organisms that respond to environmental signals.

BUT WHAT DOES IT ALL MEAN?

In this article we have reviewed the genetic mechanisms used by pathogenic microorganisms to regulate the expression of virulence determinants. Because of space limitation, we have not covered all known examples of such phenomena. Recent reviews covering other important and well-studied virulence factors and their regulation have been written and are recommended (9, 110, 128, 142).

As biologists, we must consider pathogens that cause death of the host as being the most virulent. Death of the host, however, may properly be considered counterproductive to multiplication, dissemination, and persistence of the microbe in nature. The regulation of virulence factor expression may therefore play a role in balancing the pathogenic cycle to optimize the survival of the microbe within the individual host as well as within the population as a whole. Indeed, subclinical infections greatly outnumber those that cause overt disease for virtually all pathogens associated with their natural hosts. There are also classical examples (e.g. myxomytosis in Australia) where the reduction in virulence of the pathogen was the observed outcome of an artificially established host-parasite relationship in a natural setting (131). It is therefore important to consider all the possible benefits that a virulence factor (or its regulation) might contribute to the microbe, including those that pertain to survival within an infected individual, the entire host population (both sensitive and immune), transmission vectors, alternate host reservoirs, and even the environment. Thus, speculation on the reason(s) for the evolution of a particular virulence regulation strategy must be taken lightly since it is surely based on a narrow view of the host-parasite relationship.

The examples of phase and antigenic variation noted here demonstrate that pathogens of many different types have evolved the ability to genetically vary the expression of virulence properties. Such variation may actively change the properties of the microbe, allowing it to take advantage of new environments within or outside of the host. Alternatively, variation may act passively by allowing the host to set the rules of survival by selecting the variant that is most successful (and perhaps eliminating the others). We would like to coin the terms *active* and *passive random variation* to distinguish these two types of strategies in the control of microbial virulence properties.

These types of control are not restricted to bacteria; inversion of the G segment in bacteriophage Mu during lysogenic growth leads to expression of a different tail fiber that in turn broadens the host range of the viral progeny (64, 194). In the same way, phase variation of type I pili allows enteric bacteria to exist in two states, one of which is more appropriate for survival while adherent to the mucosal surface (Fim$^+$) and the other for survival in urine or in blood where motile, nonadherent Fim$^-$ forms of the organism might be more appropriate (107, 166). Thus, random variation can play an active role in determining host or tissue tropism and therefore the niche the organism occupies.

Random variation can also play a passive role in survival. For example, phase variation of the flagellins of *Salmonella* species (210) probably provides no new niche or survival advantage to variants until the host immune response is mounted against a given phase. Similarly, phase variation in the case of virulent *B. pertussis* yields Vir$^-$ organisms that are avirulent but may have passive survival advantage in an immune host late in the infection cycle when antibody responses to virulent phase antigens may effectively eliminate Vir$^+$ organisms (91). The Vir$^-$ phase might also actively enhance survival and persistence through the derepression of *vrg* genes (99). Clearly in gonococcus pilins, random variation of pilin expression may contribute both actively, in the fine-tuning of pilin-receptor binding properties, and passively, in avoidance of the immune response (188).

In cases of nonrandom or programmed responses, it is clear that environments within the host contain signals that induce pathogens to express different virulence determinants. In this review we have emphasized genetic control, but there are also examples of direct phenotypic changes induced by the environment (138).

That most bacterial pathogens do not constitutively express their virulence determinants underscores the importance that regulation must have in the overall success of pathogenic microorganisms. Thus, *when* a particular determinant is expressed may be more important to the organisms than that it *is* expressed at all. For example, since the toxin and TCP pili genes of *V. cholerae* are regulated (by the ToxR system), the environment in which these virulence factors have their effect (the epithelial cell surface) probably provides the relevant signals for the expression of this system. One might therefore predict that expression of toxin and TCP pili, which can clump the organism, may be deleterious to the organism if it occurs before the organisms are associated with the epithelium, because the resultant clumped organisms in the lumen could be washed out of the intestine by the toxin-induced secretory response. This might explain in part the reduced virulence of at least one strain of *V. cholerae* (569B) that is relatively unregulated (i.e. constitutive) in its production of toxin and TCP pili (125).

Regulation of virulence gene expression may be compared to the staged control of gene expression exhibited by organisms both less and more complex than many bacterial pathogens. Temperate bacteriophage exhibit a programmed pattern of gene expression where a series of decisions is based on the environment within the bacterial cell (155). Similarly, developmental control of gene expression by different bacterial (e.g. *Bacillus, Myxococcus,*) and protozoan species provide excellent examples of control in response to signals (starvation, heat shock) that these organisms encounter naturally (90, 94, 109, 195). Virulence gene expression is no doubt set up in stages so that organisms at one stage exhibit a phenotype different from that of organisms at another stage. Such staging can provide immediate benefit to the microbe by enhancing survival within the inducing environment or alternatively may prime the pathogen for exploiting a new host environment.

For example, gene products required for extracellular survival might not be expressed during intracellular growth. Thus, the genes for siderophore and Shiga toxin production are probably repressed by comparably high levels of iron during intracellular growth of shigellae but are probably immediately derepressed on exposure to reduced extracellular levels of iron, which are kept low by the action of transferrin and lactoferrin (48). However, it is also conceivable that passage through an intracellular environment prepares the organism, via the induction of appropriate genes, for entry into the extracellular milieu of the host. This may well be the case for pathogenic *Y. enterocolitica* since growth at low temperatures and high Ca^{++} enhances their adherence and invasive properties for epithelial cells, corresponding to exposure of the gut epithelium to environmentally grown organisms (113, 153a). On the other hand, growth at 37°C in low Ca^{++}, which are probable intracellular conditions, induces expression of antiphagocytic properties and serum resistance, corresponding to the emergence of the yersiniae into the lamina propria and blood stream (28, 113). *Salmonella* probably exhibits stage-specific virulence gene expression at the epithelial cell surface, during transcytosis, in the host extracellular fluids, and again intracellularly in professional phagocytes (49, 124).

However, such stage-specific virulence gene expression does not preclude some sort of environmental shift controlling gene expression at a gross level. Temperature control (as in the control of invasion by *S. flexneri*) would be a good example of this. Organisms that infect mammals are pathogens at 37°C, although they may originate in environments of lower temperature (e.g. insect vectors, contaminated food or water, etc). Thus, the shift from low to high temperature might set the stage for the next series of decisions concerning gene expression in the host, which could be more subtle and controlled by host microenvironments.

Environmental signals such as heat shock and starvation that elicit global control may be among the first to be utilized in the evolution of pathogenic microorganisms. Weakly pathogenic, saprophytic, free-living organisms (e.g. *Clostridium* species) may serve as examples of organisms at an early stage of evolution toward a more efficient host-parasite relationship. The development of potent neurotoxins and cytotoxins (complete with ganglioside receptor-binding B domains and highly vertebrate-specific, active, A domains) could not possibly have evolved by chance alone. Rather, it seems likely that expression of these toxins enhances the survival of these organisms in nature (even if they cannot survive for long in living animals) by simply increasing the chance that a hapless host will soon be converted to dead tissue (i.e. substrate) by even a mild infection with, for example, *C. tetani*. That some toxins of saprophytic organisms are frequently expressed during sporulation may reflect either a strategy that seeks to prepare the microbe for an eventual encounter with a toxin-sensitive host or, alternatively, linkage of the regulation of these toxins with hostile environmental conditions (spore inducing?) within the living host. In a sense, therefore, the ability of *B. thuringiensis* to elaborate spore toxins (202) is not unlike the ability of *Y. enterocolitica* to express invasive properties at ambient temperature (113), i.e. in each case, the organism has learned to prepare for a host environment that it encounters with reasonable frequency.

The conversion from an environmental organism (e.g. *Bacillus cereus*) to an *efficient* animal pathogen *(B. anthracis)* probably requires the acquisition of multiple virulence determinants, as well as a regulatory system controlling their expression (61). Thus, *Bacillus anthracis* may have made this transition simply by acquiring two plasmids, one encoding an antiphagocytic capsule and the other a pair of toxins; these plasmids also encode the ability to coordinately regulate the expression of their virulence factors in the presence of elevated levels of CO_2 (a condition that elegantly communicates the entry into the host tissues!) (5a, 112). We view the ability to regulate virulence gene expression as a critical step in the evolution of virulent microorganisms and ultimately an essential property for efficient maintenance of a pathogen within a host population.

Recent developments in the field of pathogenesis are encouraging. The widespread use of Tn*phoA* and other specialized mutagenesis agents to define virulence determinants, and construction of novel cloning and shuttle vectors have expanded analysis of virulence to previously recalcitrant organisms such as *Legionella, Mycobacterium, Listeria,* and *Brucella* (25, 58, 132, 133, 167, 168, 192). The important work on signal transduction being done on pathogenic as well as nonpathogenic organism should also provide practical benefits. It will be possible to use knowledge of the genetic control apparatus and its phenotypic consequences to design better vaccines and new antibiotics. In addition, as with the discoveries of immunoglobulins as the neutralizers of bacterial toxins, and DNA as the capsular transforming principle in *Streptococcus,* the study of bacterial pathogenesis will continue to yield exciting new insights into the function of not only bacterial cells but also higher organisms as well.

ACKNOWLEDGMENTS

We thank L. Csonka, A. Fouet, and the members of our laboratory for helpful comments on the manuscript. We also thank our colleagues who communicated results prior to publication, as noted in the text. This work was supported by Public Health Service grants AI-18045 and AI-26289 from the National Institutes of Allergy and Infectious Diseases. V. J. D. is supported by National Research Service Award AI-08050-01 from the National Institutes of Health.

Literature Cited

1. Abraham, J. M., Freitag, C. S., Clements, J. R., Eisenstein, B. I. 1985. An invertible. element of DNA controls phase variation of type 1 fimbriae of *Escherichia coli. Proc. Natl. Acad. Sci. USA* 82:5724–27

2. Argos, P., Landy, A., Abremski, K., Egan, J. B., Haggard-Ljunquist, E., et al. 1986. The integrase family of site specific recombinases: Regional similarities and global diversity. *EMBO J.* 5:433–40

3. Båga, M., Göransson, M., Normark, S., Uhlin, B. E. 1985. Transcriptional activation of a Pap pilus virulence operon from uropathogenic *Escherichia coli. EMBO J.* 4:3887–93

4. Barbour, A. G. 1988. Antigenic variation of surface proteins of *Borrelia* species. *Rev. Infect. Dis.* 10:S399–S402

5. Bar-Shavit, Z., Goldman, R., Ofek, I., Sharon, N., Mirelman, D. 1980. Mannose-binding activity of *Escherichia coli:* A determinant of attachment and ingestion of the bacteria by macrophages. *Infect. Immun.* 29:417–24

5a. Bartkus, J. M., Leppla, S. H. 1989. Transcriptional regulation of the protective antigen gene of *Bacillus anthracis. Infect. Immun.* In press

6. Baudry, B., Kaczorek, M., Sansonetti, P. J. 1988. Nucleotide sequence of the invasion plasmid antigen B and C genes

(*ipaB* and *ipaC*) of *Shigella flexneri.*
Microbial Pathog. 4:345–57

7. Baudry, B., Maurelli, A. T., Clerc, P., Sadoff, J. C., Sansonetti, P. J. 1987. Localization of plasmid loci necessary for the entry of *Shigella flexneri* into HeLa cells, and characterization of one locus encoding four immunogenic polypeptides. *J. Gen. Microbiol.* 133:3403–13

8. Bergstrom, S., Robbins, K., Koomey, J. M., Swanson, J. 1986. Piliation control mechanisms in *Neisseria gonorrhoeae. Proc. Natl. Acad. Sci. USA* 83:3890–94

9. Betley, M. J., Miller, V. L., Mekalanos, J. J. 1986. Genetics of bacterial enterotoxins. *Annu. Rev. Microbiol.* 40:577–605

10. Blake, M. S., Gotschlich, E. C., Swanson, J. 1981. Effects of proteolytic enzymes on the outer membrane proteins of *Neisseria gonorrhoeae. Infect. Immun.* 33:212–22

11. Blumenstock, E., Jann, K. 1982. Adhesion of piliated *Escherichia coli* strains to phagocytes: Differences between bacteria with mannose-sensitive pili and those with mannose-resistant pili. *Infect. Immun.* 35:264–69

12. Bolin, I., Wolf-Watz, H. 1988. The plasmid-encoded Yop2b protein of *Yersinia pseudotuberculosis* is a virulence determinant regulated by calcium and temperature at the level of transcription. *Mol. Microbiol.* 2:237–45

13. Bolton, G. W., Nester, E. W., Gordon, M. P. 1986. Plant phenolic compounds induce expression of the *Agrobacterium tumefaciens* loci needed for virulence. *Science* 232:983–85

14. Brinton, C. C. Jr. 1959. Non-flagellar appendages of bacteria. *Nature* 183: 782–86

15. Brown, A., Hormaeche, C. E., Demarco de Hormaeche, R., Winther, M., Dougan, G., et al. 1987. An attenuated *aroA Salmonella typhimurium* vaccine elicits humoral and cellular immunity to cloned β-galactosidase in mice. *J. Infect. Dis.* 155:86–92

16. Brubaker, R. R. 1981. Expression of virulence in *Yersiniae.* In *Microbiology-1979,* ed. D. Schlesinger, pp. 168–71. Washington, DC: Am. Soc. Microbiol.

17. Brubaker, R. R. 1985. Mechanisms of bacterial virulence. *Ann. Rev. Microbiol.* 39:21–50

18. Brubaker, R. R. 1986. Low calcium response of virulent *Yersiniae.* In *Microbiology-1986,* ed. L. Lieve, pp. 43–48. Washington, DC: Am. Soc. Microbiol.

19. Buchanan, T. M. 1975. Antigenic heterogeneity of gonococcal pili. *J. Exp. Med.* 141:1470–75

20. Buchanan, T. M., Pearce, W. A. 1976. Pili as mediators of the attachment of gonococci to human erythrocytes. *Infect. Immun.* 13:1483–89

21. Buysse, J. M., Stover, C. K., Oaks, E. V., Venkatesan, M., Kopecko, D. J. 1987. Molecular cloning of invasion plasmid antigen *(ipa)* genes from *Shigella flexneri:* Analysis of *ipa* gene products and genetic mapping. *J. Bacteriol.* 169:2561–69

22. Calderwood, S. B., Auclair, F., Donohue-Rolfe, A., Keusch, G. T., Mekalanos, J. J. 1987. Nucleotide sequence of the Shiga-like toxin genes of *Escherichia coli. Proc. Natl. Acad. Sci. USA* 84:4364–68

23. Calderwood, S. B., Mekalanos, J. J. 1987. Iron regulation of Shiga-like toxin expression in *Escherichia coli* is mediated by the *fur* locus. *J. Bacteriol.* 169:4759–64

24. Calderwood, S. B., Mekalanos, J. J. 1988. Confirmation of the Fur operator site by insertions of a synthetic oligonucleotide into an operon fusion plasmid. *J. Bacteriol.* 170:1015–17

25. Caparon, M. G., Scott, J. R. 1987. Identification of a gene that regulates expression of M protein, the major virulence determinant of group A streptococci. *Proc. Natl. Acad. Sci. USA* 84:8677–81

26. Caron, J., Coffield, L. M., Scott, J. R. 1989. A plasmid-encoded regulatory gene, *rns,* required for expression of the CS1 and CS2 adhesis of enterotoxinogenic *Escherichia coli. Proc. Natl. Acad. Sci. USA* 86:963–67

27. Connell, T. D., Black, W. J., Kawula, T. H., Barritt, D. S., Dempsey, J. A., et al. 1988. Recombination among protein II genes of *Neisseria gonorrhoeae* generates new coding sequences and increases structural variability in the protein II family. *Mol. Microbiol.* 2:227–36

28. Cornelis, G., Laroche, Y., Balligand, G., Sory, M. P., Wauters, G. 1987. *Y. enterocolitica,* a primary model for bacterial invasiveness. *Rev. Infect. Dis.* 9:64–87

29. Cornelis, G., Sluiters, C., Lambert de Rouvroit, C., Michielis, T. 1989. Homology between VirF, the transcriptional activator of the *Yersinia* virulence regulon, and AraC, the *Escherichia coli* arabinose operon regulator. *J. Bacteriol.* 171:254–62

30. Cornelis, G., Sory, M. P., Laroche, Y., Derclaye, I. 1986. Genetic analysis of the plasmid region controlling virulence

in *Yersinia enterocolitica* 0:9 by Mini-Mu insertions and *lac* gene fusions. *Microbial Pathog.* 1:349–59

31. Cornelis, G., Vanootegam, J.-C., Sluiters, C. 1987. Transcription of the *yop* regulon from *Y. enterocolitica* requires *trans* acting pYV and chromosomal genes. *Microbial Pathog.* 2:367–79

32. Curtiss, R. III, Kelly, S. M. 1987. *Salmonella typhimurium* deletion mutants lacking adenylate cyclase and cyclic AMP receptor protein are avirulent and immunogenic. *Infect. Immun.* 55:3035–43

33. Deretic, V., Dikshit, R., Konyecsni, W. M., Chakrabarty, A. M., Misra, T. K. 1989. The *algR* gene, which regulates mucoidy in *Pseudomonas aeruginosa*, belongs to a class of environmentally responsive genes. *J. Bacteriol.* 171:1278–83

34. Deuschle, U., Gentz, R., Bujard, H. 1986. *lac* repressor blocks transcribing RNA polymerase and terminates transcription. *Proc. Natl. Acad. Sci. USA* 83:4134–37

35. Diaz, J.-L., Heckels, J. E. 1982. Antigenic variation of outer membrane protein II in colonial variants of *Neisseria gonorrhoeae*. *J. Gen. Microbiol.* 128:585–91

36. DiRita, V. J., Peterson, K. P., Mekalanos, J. J. 1989. Regulation of cholera toxin synthesis. In *The Bacteria*, ed. B. H. Iglewski, V. L. Clark. Orlando: Academic. In press

37. Dorman, D. J., Higgins, C. F. 1987. Fimbrial phase variation in *Escherichia coli:* Dependence of integration host factor and homologies with other site-specific recombinases. *J. Bacteriol.* 169:3840–43

38. Draper, D. L., James, J. F., Brooks, G. F., Sweet, R. L. 1980. Comparison of virulence markers of peritoneal and fallopian tube isolates with endocervical *Neisseria gonorrhoeae* isolates from women with acute salpingitis. *Infect. Immun.* 27:882–88

39. Edwards, M. F., Stocker, B. A. D. 1988. Construction of (Δ)*aroA his* (Δ)*pur* strains of *Salmonella typhi*. *J. Bacteriol.* 170:3991–95

40. Eisenstein, B. I. 1981. Phase variation of type 1 fimbriae in *Escherichia coli* is under transcriptional control. *Science* 214:337–39

41. Eisenstein, B. I. 1988. Type 1 fimbriae of *Escherichia coli:* Genetic regulation, morphogenesis, and role in pathogenesis. *Rev. Infect. Dis.* 10:S341–44

42. Eisenstein, B. I., Sweet, D. S., Vaughn, V., Friedman, D. I. 1987. Integration

host factor is required for the DNA inversion that controls phase variation in *Escherichia coli. Proc. Natl. Acad. Sci. USA* 84:6506–10

43. Ely, S., Tippett, J., Kroll, J. S., Moxon, E. R. 1986. Mutations affecting expression and maintenance of genes encoding the serotype b capsule of *Haemophilus influenzae. J. Bacteriol.* 167:44–48

44. Ewanowich, C. A., Sherburne, R. D., Man, S. R. P., Peppler, M. S. 1989. *Bordetella parapertussis* invasion of HeLa 229 cells and human respiratory epithelial cells in primary culture. *Infect. Immun.* 57:1240–47

45. Ferber, D. M., Brubaker, R. R. 1981. Plasmids in *Yersinia pestis. Infect. Immun.* 31:839–41

46. Fields, P. I., Groisman, E. A., Heffron, F. 1989. A *Salmonella* locus that controls resistance to microbicidal proteins from phagocytic cells. *Science* 243:1059–62

47. Fields, P. I., Swanson, R. V., Haidaris, C. G., Heffron, F. 1986. Mutants of *Salmonella typhimurium* that cannot survive within the macrophage are avirulent. *Proc. Natl. Acad. Sci. USA* 83:5189–93

48. Finkelstein, R. A., Sciortino, C. V., McIntosh, M. A. 1983. Role of iron in microbe-host interactions. *Rev. Infect. Dis.* 5:S759–77

48a. Finlay, B. B., Falkow, S. 1989. Common themes in microbial pathogenicity. *Microbiol. Rev.* 53:210–30

49. Finlay, B. B., Heffron, F., Falkow, S. 1989. Epithelial cell surfaces induces *Salmonella* proteins required for bacterial adherence and invasion. *Science* 243:940–43

50. Finlay, B. B., Starnbach, M. N., Francis, C. L., Stocker, B. A. D., Chatfield, S., et al. 1988. Identification and characterization of Tn*phoA* mutants of *Salmonella* that are unable to pass through a polarized MDCK epithelial cell monolayer. *Mol. Microbiol.* 2:757–66

51. Fischetti, V. A., Jarymowycz, M., Jones, K. F., Scott, J. R. 1986. Streptococcal M protein size mutants occur at high frequency within a single strain. *J. Exp. Med.* 164:971–80

52. Fischetti, V. A., Jones, K. F., Scott, J. R. 1985. Size variation of the M protein in group A streptococci. *J. Exp. Med.* 161:1384–401

53. Formal, S. B., Baron, L. S., Kopecko, D. J., Washington, O., Powell, C., Life, C. A. 1981. Construction of a potential bivalent vaccine strain: In-

troduction of *Shigella sonnei* form I antigen genes into the *galE Salmonella typhi* Ty21a typhoid vaccine strain. *Infect. Immun.* 34:746–50

54. Forsberg, Å., Wolf-Watz, H. 1988. The virulence protein Yop5 of *Yersinia pseudotuberculosis* is regulated at the transcriptional level by plasmid pIB1-encoded *trans*-acting elements controlled by temperature and calcium. *Mol. Microbiol.* 2:121–33

55. Frank, D. W., Iglewski, B. H. 1988. Kinetics of *toxA* and *regA* mRNA accumulation in *Pseudomonas aeruginosa.* *J. Bacteriol.* 170:4477–83

56. Freitag, C. S., Abraham, J. M., Clements, J. R., Eisenstein, B. I. 1985. Genetic analysis of the phase variation control of expression of type 1 fimbriae in *Escherichia coli, J. Bacteriol.* 162:668–75

57. Freitag, C. S., Eisenstein, B. I. 1983. Genetic mapping and transcriptional orientation of the *fimD* gene. *J. Bacteriol.* 156:1052–58

58. Gaillard, J. L., Berche, P., Sansonetti, P. 1986. Transposon mutagenesis as a tool to study the role of hemolysin in the virulence of *Listeria monocytogenes. Infect. Immun.* 52:50–55

59. Germanier, R., Furer, E. 1975. Isolation and characterization of GalE mutant Ty21a of *Salmonella typhi:* A candidate strain for a live, oral typhoid vaccine. *J. Infect. Dis.* 131:553–58

60. Goguen, J. D., Yother, J., Straley, S. C. 1984. Genetic analysis of the low calcium response in *Yersinia pestis* Mu d1 (Ap *lac*) insertion mutants. *J. Bacteriol.* 160:842–48

61. Green, B. D., Battisti, L., Koehler, T. M., Thorne, C. B., Ivins, B. E. 1985. Demonstration of a capsule plasmid in *Bacillus anthracis. Infect. Immun.* 49:291–97

62. Gross, R., Aricò, B., Rappuoli, R. 1989. Genetics of pertussis toxin. *Mol. Microbiol.* 3:119–24

63. Gross, R., Rappuoli, R. 1988. Positive regulation of pertussis toxin expression. *Proc. Natl. Acad. Sci. USA* 85:3913–17

64. Grundy, F. J., Howe, M. M. 1984. Involvement of the invertible G segment in bacteriophage Mu tail fiber biosynthesis. *Virology* 134:296–317

65. Gulig, P. A., Curtiss, R. 1987. Plasmid associated virulence of *Salmonella typhimurium. Infect. Immun.* 55:2891–901

66. Haas, R., Meyer, T. F. 1986. The repertoire of silent pilus genes in *Neisseria gonorrhoeae:* Evidence for gene conversion. *Cell* 44:107–15

67. Haas, R., Schwarz, H., Meyer, T. F. 1987. Release of soluble pilin antigen coupled with gene conversion in *Neisseria gonorrhoeae. Proc. Natl. Acad. Sci. USA* 84:9079–83

68. Hackett, J., Kotlarski, I. K., Mathan, B., Francki, K., Rowley, D. 1988. The colonization of Peyer's patches by a strain of *S. typhimurium* cured of the cryptic plasmid. *J. Infect. Dis.* 153:1119–25

69. Hackett, J., Wyk, P., Reeves, P., Mathan, V. 1987. Mediation of serum resistance in *Salmonella typhimurium* by an 11 kilodalton polypeptide encoded by the cryptic plasmid. *J. Infect. Dis.* 155:540–49

70. Hagblom, P., Segal, E., Billyard, E., So, M. 1985. Intragenic recombination leads to pilus antigenic variation in *Neisseriae gonorrhoeae. Nature* 315:156–58

71. Hale, T. L., Oaks, E. V., Formal, S. B. 1985. Identification and antigenic characterization of virulence-associated plasmid-coded proteins of *Shigella* spp. and enteroinvasive *E. coli. Infect. Immun.* 50:620–29

72. Hantke, K. 1981. Regulation of ferric iron transport in *Escherichia coli* K12: isolation of a constitutive mutant. *Mol. Gen. Genet.* 182:288–92

73. Hess, J. F., Bourret, R. B., Simon, M. I. 1988. Histidine phosphorylation and phosphoryl group transfer in bacterial chemotaxis. *Nature* 336:139–43

74. Higgins, C. F., Dorman, C. J., Stirling, D. A., Waddell, L., Booth, I. R., et al. 1988. A physiological role for DNA supercoiling in the osmotic regulation of gene expression in S. typhimurium and E. coli. *Cell* 52:569–84

75. Hindahl, M. S., Frank, D. W., Hamood, A., Iglewski, B. H. 1988. Characterization of a gene that regulates toxin A synthesis in *Pseudomonas aeruginosa. Nucleic Acids Res.* 16:5699

76. Hoiseth, S. K., Connelly, C. J., Moxon, E. R. 1985. Genetics of spontaneous, high frequency loss of b capsule expression in *Haemophilus influenzae. Infect. Immun.* 49:389–95

77. Hoiseth, S. K., Moxon, E. R., Silver, R. P. 1986. Genes involved in *Haemophilus influenzae* type b capsule expression are part of an 18-kilobase tandem duplication. *Proc. Natl. Acad. Sci. USA* 83:1106–10

78. Hoiseth, S. K., Stocker, B. A. D. 1981. Aromatic-dependent *Salmonella typhimurium* are nonvirulent and effective as live vaccines. *Nature* 291:238–39

79. Hollingshead, S. K., Fischetti, V. A., Scott, J. R. 1987. Size variation in group A streptococcal M protein is generated by homologous recombination between intragenic repeats. *Mol. Gen. Genet.* 207:196–203

80. Hoyt, M. A., Knight, D. M., Das, A., Miller, H., I., Echols, H. 1982. Control of phage λ development by stability of cII protein: Role of the viral cIII and host *hflA, himA* and *himD* genes. *Cell* 31:565–73

81. Hromockyj, A. E., Maurelli, A. T. 1989. Identification of an *Escherichia coli* gene homologous to *virR*, a regulator of *Shigella* virulence. *J. Bacteriol.* 171:2879–81

82. Igo, M. M., Silhavy, T. J. 1988. EnvZ, a transmembrane environmental sensor of *Escherichia coli* K-12, is phosphorylated in vitro. *J. Bacteriol.* 170:5971–73

83. Isberg, R. R., Voorhis, D. L., Falkow, S. 1987. Identification of invasin: a protein that allows enteric bacteria to penetrate cultured mammalian cells. *Cell* 50:769–78

84. Isenberg, H. D. 1988. Pathogenicity and virulence: Another view. *Clin. Microbiol. Rev.* 1:40–53

85. Ishimoto, K. S., Lory, S. 1989. Formation of pilin in *Pseudomonas aeruginosa* requires the alternative sigma factor (RpoN) of RNA polmerase. *Proc. Natl. Acad. Sci. USA* 86:1954–57

86. James, J. F., Swanson, J. 1978. Studies of gonococcus infection. XIII. Occurrence of color/opacity colonial variants in clinical cultures. *Infect. Immun.* 19:332–40

87. Jayaswal, R. K., Veluthambi, K., Gelvin, S. B. 1987. Double stranded T-DNA cleavage and the generation of single stranded T-DNA molecules in *E. coli* by a *virD* encoded border specific endonuclease from *A. tumefaciens*. *J. Bacteriol.* 169:5035–45

88. Johnson, K., Parker, M. L., Lory, S. 1986. Nucleotide sequence and transcriptional initiation site of two *Pseudomonas aeruginosa* pilin genes. *J. Biol. Chem.* 261:15703–8

89. Jones, K. F., Hollingshead, S. K., Scott, J. R., Fischetti, B. A. 1988. Spontaneous M6 protein size mutants of group A streptococci display variation in antigenic and opsonogenic epitopes. *Proc. Natl. Acad. Sci. USA* 85:8271–75

90. Kaiser, D., Manoil, C., Dworkin, M. 1979. Myxobacteria: Cell interactions, genetics, and development. *Annu. Rev. Microbiol.* 33:595–639

91. Kasuga, T., Nakasi, Y., Ukishima, K., Takatsu, K. 1954. Studies on *Haemophilus pertussis*. V. Relation between the phase of bacilli and the progress of whooping cough. *Kitasato Arch. Exp. Med.* 27:57–62

92. Keener, J., Kustu, S. 1988. Protein kinase and phosphoprotein phophatase activities of nitrogen regulatory proteins NTRB and NTRC of enteric bacteria: Roles of the conserved amino-terminal domain of NTRC. *Proc. Natl. Acad. Sci. USA* 85:4976–80

93. Kikuchi, Y., Nash, H. A. 1978. The bacteriophage λ *int* gene product. *J. Biol. Chem.* 253:7149–57

94. Killeen, K. P., Nelson, D. R. 1988. Acceleration of starvation- and glycerol-induced myxospore formation by prior heat shock in *Myxococcus xanthus*. *J. Bacteriol.* 170:5200–7

95. Klee, H. J., White, F. F., Iyer, V. N., Gordon, M. P., Nester, E. W. 1983. Mutational analysis of the virulence region of an *Agrobacterium tumefaciens* Ti plasmid. *J. Bacteriol.* 153:878–83

96. Klemm, P. 1986. Two regulatory *fim* genes, *fimB* and *fimE*, control the phase variation of type 1 fimbriae in *Escherichia coli*. *EMBO J.* 5:1389–93

97. Klemm, P. 1985. Fimbrial adhesins of *Escherichia coli*. *Rev. Infect. Dis.* 7:321–40

98. Klemm, P., Jorgensen, B. J., van Die, I., de Ree, H., Bergmans, H. 1985. The *fim* genes responsible for synthesis of type 1 fimbriae in *Escherichia coli*, cloning and genetic organization. *Mol. Gen. Genet.* 199:410–14

99. Knapp, S., Mekalanos, J. J. 1988. Two *trans*-acting regulatory genes (*vir* and *mod*) control antigenic modulation in *Bordetella petussis*. *J. Bacteriol.* 170:5059–66

100. Koomey, M., Gotschlich, E. C., Robbins, K., Bergström, S., Swanson, J. 1987. Effects of *recA* mutations on pilus antigenic variation and phase transitions in *Neisseria gonorrhoeae*. *Genetics* 117:391–98

101. Kroll, J. S., Hopkins, I., Moxon, E. R. 1988. Capsule loss in *H. influenzae* type b occurs by recombination-mediated disruption of a gene essential for polysaccharide export. *Cell* 53:347–56

102. Kroll, J. S., Moxon, E. R. 1988. Capsulation and gene copy number at the *cap* locus of *Haemophilus influenzae* type b. *J. Bacteriol.* 170:859–64

103. Lacey, B. W. 1960. Antigenic modulation of *Bordetella pertussis*. *J. Hyg.* 58:57–91

104. Lampe, R. M., Mason, E. O. Jr., Kaplan, S. L., Umstead, C. L., Yow, M. D., et al. 1982. Adherence of

Haemophilus influenzae to buccal epithelial cells. *Infect. Immun.* 35:166–72

105. Leong, J. M., Nunes-Duby, S., Lesser, C. F., Youderian, P., Susskind, M. M., et al. 1985. The ϕ80 and P22 attachment sites: Primary structure and interaction with *Escherichia coli* integration host factor. *J. Biol. Chem.* 260:4468–77

106. Leroux, B., Yanofsky, M. J., Winans, S. C., Ward, J. E., Ziegler, S. F., et al. 1987. Characterization of the *virA* locus of *Agrobacterium tumefaciens:* a transcription regulator and host range determinant. *EMBO J.* 6:849–56

107. Leunk, R. D., Moon, R. J. 1982. Association of type 1 pili with the ability of livers to clear *Salmonella typhimurium. Infect. Immun.* 36:1168–74

108. Levine, M. M., Herrington, D., Morris, J. G., Losonsky, G., Murphy, J., et al. 1987. Safety, inactivity, immunogenicity, and in vivo stability of two attenuated auxotrophic mutant strains of *Salmonella typhi* 541Ty and 543Ty, used as oral vaccines in man. *J. Clin. Invest.* 79:888–902

109. Losick, R., Pero, J. 1981. Cascades of sigma factors. *Cell* 25:582–84

110. Mackman, N., Nicaud, J.-M., Gray, L., Holland, I. B. 1986. Secretion of haemolysin by *Escherichia coli. Curr. Top. Microbiol. Immunol.* 125:159–81

111. Makino, S., Sasakawa, C., Kamata, K., Kurata, T., Yoshikawa, M. 1986. A genetic determinant required for continuous reinfection of adjacent cells on a large plasmid in *Shigella flexneri* 2a. *Cell* 46:551–55

112. Makino, S., Sasakawa, C., Ucida, I., Terakado, N., Yoshikawa, M. 1988. Cloning and CO$_2$-dependent expression of the genetic region for encapsulation from *Bacillus anthracis. Mol. Microbiol.* 2:371–76

113. Martinez, R. J. 1983. Plasmid-mediated and temperature-regulated surface properties of *Yersinia enterocolitica. Infect. Immun.* 41:921–30

114. Martinez-Cadena, M. G., Guzman-Verduzco, L. M., Stieglitz, H., Kupersztock-Portnoy, Y. M. 1981. Catabolite repression of *Escherichia coli* heat-stable enterotoxin activity. *J. Bacteriol.* 145:722–28

115. Maurelli, A. T., Baudry, B., d'Hauteville, H., Hale, T. L., Sansonetti, P. J. 1985. Cloning of virulence plasmid DNA sequences involved in invasion of HeLa cells by *Shigella flexneri. Infect Immun.* 49:164–71

116. Maurelli, A. T., Blackmon, B., Curtiss, R. III. 1984. Temperature-dependent expression of virulence genes in *Shigella* species. *Infect. Immun* 43:195–201

117. Maurelli, A. T., Curtiss, R. III. 1984. Bacteriophage Mu d1 (Ap^r *lac*) generates *vir-lac* operon fusions in *Shigella flexneri* 2a. *Infect. Immun.* 45:642–48

118. Maurelli, A. T., Sansonetti, P. J. 1988. Identification of chromosomal gene controlling temperature regulated expression of *Shigella* virulence. *Proc. Natl. Acad. Sci. USA* 85:2820–24

119. Maurelli, A. T., Sansonetti, P. J. 1988. Genetic determinants of *Shigella* pathogenicity. *Annu. Rev. Micro.* 42:127–50

120. Meier, J. T., Simon, M. I., Barbour, A. G. 1985. Antigenic variation is associated with DNA rearrangements in a relapsing fever Borrelia. *Cell* 41:403–9

121. Meyer, T. F., Mlawer, N., So, M. 1982. Pilus expression in *Neisseria gonorrhoea* involves chromosomal rearrangement. *Cell* 30:45–52

122. Michiels, T., Cornelis, G. 1988. Nucleotide sequence and transcription analysis of *yop51* from *Yersinia enterocolitica* W22703. *Microbial Pathog.* 5:449–59

123. Miller, J. F., Mekalanos, J. J., Falkow, S. 1989. Coordinate regulation and sensory transduction in the control of bacterial virulence. *Science* 243:916–22

124. Miller, S. I., Kukral, A., Mekalanos, J. J. 1989. A two component regulatory system (*phoP* and *phoQ*) controls *Salmonella typhimurium* virulence. *Proc. Natl. Acad. Sci. USA* 86:5054–58

125. Miller, V. L., DiRita, V. J., Mekalanos, J. J. 1989. Identification of *toxS,* a regulatory gene whose product enhances ToxR-mediated activation of the cholera toxin promoter. *J. Bacteriol.* 171:1288–93

126. Miller, V. L., Falkow, S. 1988. Evidence for two genetic loci from *Y. enterocolitica* that can promote invasion of epithelial cells. *Infect. Immun.* 56:1242–48

127. Miller, V. L., Farmer, J. J. III, Hill, W. E., Falkow, S. 1989. The *ail* locus is found uniquely in *Yersinia enterocolitica* serotypes commonly associated with disease. *Infect. Immun.* 57:121–31

128. Miller, V. L., Finlay, B. B., Falkow, S. 1988. Factors essential for the penetration of mammalian cells by *Yersinia. Curr. Top. Microbiol. Immunol.* 138:15–39

129. Miller, V. L., Mekalanos, J. J. 1984. Synthesis of cholera toxin is positively regulated at the transcriptional level by *toxR. Proc. Natl. Acad. Sci. USA* 81:3471–75

130. Miller, V. L., Taylor, R. K., Mekalanos, J. J. 1987. Cholera toxin transcriptional activator ToxR is a transmembrane DNA binding protein. *Cell* 48:271–79

131. Mims, C. A. 1987. *The Pathogenis of Infectious Disease.* London: Harcourt Brace Jovanovich. 3rd ed.

132. Mintz, C. S., Chen, J., Shuman, H. A. 1988. Isolation and characterization of auxotrophic mutants of *Legionella pneumophila* that fail to multiply in human monocytes. *Infect. Immun.* 56:1449–55

133. Mintz, C. S., Shuman, H. A. 1987. Transposition of bacteriophage Mu in the Legionnaire's disease bacterium. *Proc. Natl. Acad. Sci. USA* 84:4645–49

134. Morfeldt, E., Janzon, L., Arvidson, S., Löfdahl, S. 1988. Cloning of a chromosomal locus *(exp)* which regulates the expression of several exoprotein genes in *Staphylococcus aureus. Mol. Gen. Genet.* 211:435–40

135. Moxon, E. R., Vaughn, K. A. 1981. The type b capsular polysaccharide of *Haemophilus influenzae:* Studies using clinical isolates and laboratory transformants. *J. Infect. Dis.* 143:517–24

136. Murphy, G. L., Connell, T. D., Barritt, D. S., Koomey, M., Cannon, J. G. 1989. Phase variation of gonococcal protein II: Regulation of gene expression by slipped-strand mispairing of a repetitive DNA sequence. *Cell* 56:539–47

137. Murphy, J. R., Bacha, P. 1976. Regulation of diphtheria toxin production. In *Microbiology-1976,* ed. D. Schlessinger. pp. 181–86 Washington, DC: Am. Soc. Microbiol.

138. Nairn, C. A., Cole, J. A., Patel, P. V., Parsons, N. J., Fox, J. E., Smith, H. 1988. Cytidine 5'-monophosho-*N*-acetylneuraminic acid or a related compound is the low M_r factor from human red blood cells which induces gonococcal resistance to killing by human serum. *J. Gen. Microbiol.* 134:3295–306

139. Nicosia, A., Rappuoli, R. 1987. Promoter of the pertussis toxin operon and production of pertussis toxin. *J. Bacteriol.* 169:2843–46

140. Ninfa, A. J., Magasanik, B. 1986. Covalent modification of the *glnG* product, NRI, by the *glnL* product, NRII, regulates the transcription of the *glnALG* operon in *Escherichia coli. Proc. Natl. Acad. Sci. USA* 83:5492–96

141. Ninfa, A. J., Ninfa, E. G., Lupas, A. N., Stock, A., Magasanik, B., et al. 1988. Crosstalk between bacterial chemotaxis signal transduction proteins and regulators of transcription of the Ntr regulon: Evidence that nitrogen assimilation and chemotaxis are controlled by a common phosphotransfer mechanism. *Proc. Natl. Acad. Sci. USA* 85:5492–96

142. Normark, S., Båga, M., Göransson, M., Lindberg, F. P., Lund, B., et al. 1986. Genetics and biogenesis of *Escherichia coli* adhesins. In *Microbial Lectins and Agglutinins,* ed. D. Mirelman, pp. 113–43. New York: Wiley.

143. O'Callaghan, D., Marshall, D., Liew, F. Y., Easman, C. S. F., Dougan, G. 1988. Characterization of aromatic and purine dependent *Salmonella typhimurium:* Attenuation, persistance, and ability to induce protective immunity in BALB/c mice. *Infect. Immun.* 56:419–23

144. Old, D. C. 1972. Inhibition of interaction between fimbrial haemagglutinins and erythrocytes by D-mannose and other carbohydrates. *J. Gen. Microbiol.* 71:149–57

145. Orndorff, P. E., Falkow, S. 1984. Organization and expression of genes responsible for type 1 piliation in *Escherichia coli. J. Bacteriol.* 159:736–44

146. Orndorff, P. E., Falkow, S. 1985. Nucleotide sequence of *pilA,* the gene encoding the structural component of type 1 pili in *Escherichia coli. J. Bacteriol.* 162:454–57

147. Orndorff, P. E., Falkow, S. 1984. Identification and characterization of a gene product that regulates type 1 piliation in *Escherichia coli. J. Bacteriol.* 160:61–66

148. Orndorff, P. E., Spears, P. A., Schauer, D., Falkow, S. 1985. Two modes of control of *pilA,* the gene encoding type 1 pilin in *Escherichia coli. J. Bacteriol.* 164:321–30

149. Perry, R. D., Harmon, P. A., Bowmer, W. S., Straley, S. C. 1986. A low-Ca^{2+} response operon encodes the V antigen of *Yersinia pestis. Infect. Immun.* 54:428–34

150. Peterson, K. M., Mekalanos, J. J. 1988. Characterization of the *Vibrio cholerae* ToxR regulon: Identification of novel genes involved in intestinal colonization. *Infect. Immun.* 56:2822–29

151. Plasterk, R. H. A., Simon, M. I., Barbour, A. G. 1985. Transposition of structural genes to an expression sequence on a linear plasmid causes antigenic variation in the bacterium *Borrelia hermsii. Nature* 318:257–63

152. Pollack, C., Straley, S. C., Klempner, M. S. 1986. Probing the phagolysosomal environment of human macrophages with a Ca^{++}-responsive operon fusion in *Yersina pestis. Nature* 322:834–36

153. Portnoy, D. A., Blank, H. F., Kingsbury, D. T., Falkow, S. 1983. Genetic analysis of essential plasmid determinants of pathogenicity in *Yersinia pestis*. *J. Infect. Dis.* 148:297–304

153a. Portnoy, D. A., Moseley, S. L., Falkow, S. 1981. Characterization of plasmids and plasmid-associated determinants of *Yersinia enterocolitica* pathogenesis. *Infect. Immun.* 31:775–82

154. Portnoy, D. A., Wolf-Watz, H., Bolin, I., Beeder, A. B., Falkow, S. 1984. Characterization of common virulence plasmids in *Yersinia* species and their role in the expression of outer membrane proteins. *Infect. Immun.* 43:108–14

155. Ptashne, M. 1986. *A Genetic Switch. Gene Control and Phage λ.* Cambridge/Palo Alto: Cell Press/Blackwell Sci.

156. Recsei, P., Kreiswirth, B., O'Reilly, M., Schlievert, P., Gruss, A., et al. 1986. Regulation of exoprotein gene expression in *Staphylococcus aureus* by *agr*. *Mol. Gen. Genet.* 202:58–61

157. Ronson, C. W., Nixon, B. T., Ausubel, F. M. 1987. Conserved domains in bacterial regulatory proteins that respond to environmental stimuli. *Cell* 49:579–81

158. Sakai, T., Saskawa, C., Yoshikaw, M. 1988. Expression of four virulence antigens of *Shigella flexneri* is positively regulated at the transcriptional level by the 30 kDa VirF protein. *Mol. Microbiol.* 2:589–97

159. Salit, E. 1982. The differential susceptibility of gonococcal opacity variants to sex hormones. *Can. J. Biochem.* 60:301–6

160. Sasakawa, C., Kamata, K., Sakai, T., Makino, S., Yamada, M., et al 1988. Virulence-associated genetic regions comprising 31 kilobases of the 230-kilobase plasmid in *Shigella flexneri* 2a. *J. Bacteriol.* 170:2480–84

161. Schwalbe, R. S., Sparling, P. F., Cannon, J. G. 1985. Variation of *Neisseria gonorrhoeae* protein II among isolates from an outbreak caused by a single gonococcal strain. *Infect. Immun.* 49:250–52

162. Segal, E., Billyard, E., So, M., Storzbach, S., Meyer, T. F. 1985. Role of chromosomal rearrangement in N. gonorrhoeae pilus phase variation. *Cell* 40:293–300

163. Segal, E., Hagblom, P., Seifert, H. S., So, M. 1986. Antigenic variation of gonococcal pilus involves assembly of separated silent gene segments. *Proc. Natl. Acad. Sci. USA* 83:2177–81

164. Seifert, H. S., Ajioka, R. S., Marchal, C., Sparling, P. F., So, M. 1988. DNA transformation leads to pilin antigenic variation in *Neisseria gonorrhoeae*. *Nature* 336:392–95

165. Seifert, H. S., So., M. 1988. Genetic mechanisms of bacterial antigenic variation. *Microbiol. Rev.* 52:327–36

166. Silverblatt, F. J., Ofek, I. 1978. Influence of pili on the virulence of *Proteus mirabilis* in experimental hematogenous pyelonephritis. *J. Infect. Dis.* 138:664–67

167. Smith, L. D., Heffron, F. 1987. Transposon Tn5 mutagenesis of *Brucella abortus*. *Infect. Immun.* 55:2774–76

168. Snapper, S. B., Lugosi, L., Jekkel, A., Melton, R. E., Kieser, T., et al. 1988. Lysogeny and transformation in mycobacteria: Stable expression of foreign genes. *Proc. Natl. Acad. Sci. USA* 85:6987–91

169. Sparling, P. F., Cannon, J. G., So, M. 1986. Phase and antigenic variation of pili and outer membrane protein II of *Neisseria gonorrhoeae*. *J. Infect. Dis.* 153:196–201

170. Spears, P. A., Schauer, D., Orndorff, P. E. 1986. Metastable regulation of type 1 piliation in *Escherichia coli* and isolation and characterization of a phenotypically stable mutant. *J. Bacteriol.* 168:179–85

171. Stachel, S. E., Zambryski, P. 1986. *virA* and *virG* control the plant-induced activation of the T-DNA and transfer process of *Agrobacterium tumefaciens*. *Cell* 46:325–33

172. Stern, A., Brown, M., Nickel, P., Meyer, T. F. 1986. Opacity genes in Neisseria gonorrhoeae: Control of phase and antigenic variation. *Cell* 47:61–71

173. Stewart, G. J., Carlson, C. A. 1986. The biology of natural transformation. *Annu. Rev. Microbiol.* 40:211–35

174. Stibitz, S., Aaronson, W., Monack, D., Falkow, S. 1989. Phase-variation in *Bordetella pertussis* by frameshift mutation in a gene for a novel two-component system. *Nature* 338:266–69

175. Stibitz, S., Weiss, A. A., Falkow, S. 1988. Genetic analysis of a region of the *Bordetella pertussis* chromosome encoding filamentous hemagglutinin and the pleiotropic regulatory locus, *vir*. *J. Bacteriol.* 170:2904–13

176. Straley, S. C., Bowmer, W. S. 1986. Virulence genes regulated at the transcriptional level by Ca^2 in *Yersinia pestis* include structural genes for outer membrane proteins. *Infect. Immun.* 51:445–54

177. Straley, S. C., Brubaker, R. R. 1981. Cytoplasmic and membrane proteins of

yersiniae cultivated under conditions simulating mammalian intracellular environment. *Proc. Natl. Acad. Sci. USA* 78:1224–28

178. Stromberg, N., Deal, C., Nyberg, G., Normark, S., So, M., et al. 1988. Identification of carbohydrate structures that are possible receptors for *Neisseria gonorrhoeae. Proc. Natl. Acad. Sci. USA* 85:4902–6

179. Swaney, L. M., Liu, Y.-P., Ippen-Ihler, K., Brinton, C. C. Jr. 1977. Genetic complementation analysis of *Escherichia coli* type 1 somatic pilus mutants. *J. Bacteriol.* 130:506–11

180. Swaney, L. M., Liu, Y.-P., To, C.-M., To, C.-C., Ippen-Ihler, K., et al. 1977. Isolation and characterization of *Escherichia coli* phase variants and mutants deficient in type 1 pilus production. *J. Bacteriol.* 130:495–505

181. Swanson, J. 1973. Studies on gonococcus infection. IV. Pili: Their role in attachment of gonococci to tissue culture cells. *J. Exp. Med.* 137:571–89

182. Swanson, J. 1978. Studies of gonococcus infection. XIV. Cell wall protein differences among color/opacity colony variants of *Neisseria gonorrhoeae. Infect. Immun.* 37:359–68

183. Swanson, J. 1980. ^{125}I-labeled peptide mapping of some heat-modifiable proteins of the gonococcal outer membrane. *Infect. Immun.* 28:54–64

184. Swanson, J., Barrera, O. 1983. Gonococcal pilus subunit size heterogeneity correlates with transition in colony piliation phenotype, not with changes in colony opacity. *J. Exp. Med.* 158:1459–72

185. Swanson, J., Bergstrom, S., Barrera, O., Robbins, K., Corwin, D. 1985. Pilus⁻ gonococcal variants: Evidence for multiple forms of piliation control. *J. Exp. Med.* 162:729–44

186. Swanson, J., Bergstrom, S., Robbins, K., Barrera, O., Crowin, D., et al. 1986. Gene conversion involving the pilin structural gene correlates with pilus⁺ ⇔ pilus⁻ changes in Neisseria gonorrhoeae. *Cell* 47:267–76

187. Swanson, J., Kraus, S. J., Gotschlich, E. C. 1971. Studies on gonococcus infection. I. Pili and zones of adhesion: Their relation to gonococcal growth patterns. *J. Exp. Med.* 134:886–906

188. Swanson, J., Robbins, K., Barrera, O., Corwin, D., Boslego, J., et al. 1987. Gonococcal pilin variants in experimental gonorrhea. *J. Exp. Med.* 165:1344–57

189. Taha, M. K., So, M., Seifert, H. S., Billyard, E., Marchal, C. 1988. Pilin expression in *Neisseria gonorrhoeae* is

under both positive and negative transcriptional control. *EMBO J.* 7:4367–78

190. Tai, S.-P. S., Holmes, R. K. 1988. Iron regulation of the cloned diphtheria toxin promoter in *Escherichia coli. Infect. Immun.* 56:2430–36

191. Taylor, R. K. 1989. Genetic studies of enterotoxin and other potential virulence factors of *Vibrio cholerae*. In *Genetics of Bacterial Diversity*, ed. D. A. Hopwood, K. F. Chater, pp. 309–29. London: Academic

192. Taylor, R. K., Manoil, C., Mekalanos, J. J. 1989. Broad host range vectors for delivery of Tn*phoA:* use in genetic analysis of secreted virulence determinants of *Vibrio cholerae. J. Bacteriol.* 171:1870–78

193. Taylor, R. K., Miller, V. L., Furlong, D. B., Mekalanos, J. J. 1987. Use of *phoA* gene fusions to identify a pilus colonization factor coordinately regulated with cholera toxin. *Proc. Natl. Acad. Sci. USA* 84:2833–37

194. Van de Putte, P., Cramer, S. Giphart-Gassler, M. 1980. Invertible DNA determines host specificity of bacteriophage Mu. *Nature* 286:218–22

195. Van der Ploeg, L. H. T., Giannini, S. H., Cantor, C. R. 1985. Heat shock genes: regulatory role for differentiation in parasitic protozoa. *Science* 228:1443–45

196. Veluthambi, K., Jayaswal, R. K., Gelvin, S. B. 1987. Virulence genes *A, G*, and *D* mediate the double-stranded border cleavage of T-DNA from the *Agrobacterium* Ti-plasmid. *Proc. Natl. Acad. Sci. USA* 84:1881–85

197. Venkatesan, M. M., Buysse, J. M., Kopecko, D. J. 1985. Characterization of invasion plasmid antigen genes (*ipaBCD*) from *Shigella flexneri. Proc. Natl. Acad. Sci. USA* 85:9317–21

198. Virji, M., Everson, J. S. 1981. Comparative virulence of opacity variants of *Neisseria gonorrhoeae. Infect. Immun.* 31:965–70

199. Ward, M. E., Watt, P. J., Robertson, J. N. 1974. The human fallopian tube: A laboratory model for gonococcal infection. *J. Infect. Dis.* 129:650–59

200. Weiss, A. A., Falkow, S. 1984. Genetic analysis of phase change in *Bordetella pertussis. Infect. Immun.* 43:263–69

201. Weiss, A. A., Hewlett, E. L. 1986. Virulence factors of *Bordetella pertussis. Annu. Rev. Microbiol.* 40:661–86

202. Whiteley, H. R., Schnepf, H. E. 1986. The molecular biology of parasporal crystal body formation in *Bacillus thuringiensis. Annu. Rev. Microbiol.* 40:549–76

203. Winans, S. C., Ebert, P. R., Stachel, S. E., Gordon, M. P., Nester, E. W. 1986. A gene essential for *Agrobacterium* virulence is homologous to a family of positive regulatory loci. *Proc. Natl. Acad. Sci. USA* 83:8278–82

204. Wood, D. E., Franklin, R., Cryz, S. J. Jr., Ganss, M., Pepper, M., et al. 1989. Development of a rat model for respiratory infection with *Bordetella pertussis*. *Infect. Immun.* 57:1018–24

205. Wozniak, D. J., Cram, D. C., Daniels, C. J., Galloway, D. R. 1987. Nucleotide sequence and characterization of *toxR*, a gene involved in exotoxin A regulation. *Nucleic Acids Res.* 15:2123–35

206. Yother, J., Chamness, T. W., Goguen, J. D. 1986. Temperature-controlled plasmid regulon associated with low calcium response in *Yersinia pestis*. *J. Bacteriol* 165:443–47

207. Yother, J., Goguen, J. D. 1985. Isolation and characterization of Ca^{2+}-blind mutants of *Yersinia pestis*. *J. Bacteriol.* 164:704–11

208. Zafriri, D., Oron, Y., Eisenstein, B. I., Ofek, I. 1987. Growth advantage and enhanced toxicity of *Escherichia coli* adherent to tissue culture cells due to restricted diffusion of products secreted by the cells. *J. Clin. Invest.* 79:1210–16

209. Zambryski, P. 1988. Basic processes underlying *Agrobacterium*-mediated DNA transfer to plant cells. *Annu. Rev. Genet.* 22:1–30

210. Zieg, J., Silverman, M., Hilman, M., Simon, M. 1977. Recombinational switch for gene expression. *Science* 196:170–72

Annu. Rev. Genet. 1989. 23:483–506

RHIZOBIUM GENETICS

Sharon R. Long

Stanford University, Department of Biological Sciences, Stanford,
California 94305-5020

CONTENTS

INTRODUCTION

Nitrogen fixation, the reduction of molecular dinitrogen to ammonia, is
exclusively a prokaryotic feature. Numerous plants nonetheless benefit
through symbiotic plant-prokaryote associations (141). One such association
is that between the *Rhizobiaceae* bacteria (*Rhizobium, Bradyrhizobium,* and
Azorhizobium), which infect plant hosts in a particular taxonomic family, the
Leguminosae. The rapid pace of study on this symbiosis is due largely to the
increasingly sophisticated genetics in the system. In turn, *Rhizobium* studies
have produced interesting surprises for geneticists, such as: the largest plasmids yet described which may carry required housekeeping functions (pSym a
and b, 1200 and 1500 kb, in *Rhizobium meliloti*); multiple genes for some

0066-4197/89/1215-0483$02.00

functions, such as two glutamine synthetases, one of which resembles the eukaryotic GS gene; and a founding member of a new family of positive gene activator proteins *(nodD)*. This review highlights recent progress in *Rhizobium* genetics, and aims to point out opportunities for genetic studies both on symbiosis and on fundamental questions of bacterial function and inheritance. Examples are chosen largely from papers since 1986, and to emphasize study of genes rather than physiological or developmental aspects of mutant analysis. Other reviews of *Rhizobium* present more extensive discussions of general genetics (10, 11, 98), plasmids (32, 142) and symbiosis (42, 112, 113, 146).

RHIZOBIUM BEHAVIOR AND CLASSIFICATION

The symbiosis of *Rhizobium* and its host requires recognition of the bacteria and the plant root. The *Rhizobium* bacteria associate with the host's epidermal root hairs, and usually penetrate by deformation of the hair and subsequent formation of a specialized invasion structure, the "infection thread." Mitoses and cell growth in the plant root cortex lead to the formation of a root nodule, in which bacteria infect host cells and differentiate into "bacteroids" that fix nitrogen. This is of considerable physiological benefit to the host plant in nitrogen-limited conditions.

The *Rhizobium* bacteria were formerly in one genus, but have been reclassified in three genera within the Rhizobiaceae, *Rhizobium*, *Bradyrhizobium*, and *Azorhizobium* (43, 88). These form nodules on roots (and sometimes on stems) of various host plants. Species or subspecies (biovars) within these genera are defined to an extent by host range. Taxonomic questions continue to be important in *Rhizobium* research. It is striking that the symbiotic interactions, which are highly conserved functions, are directed by diverse genes whose location and organization vary, and that are borne by numerous different bacteria. The major *Rhizobium* groups discussed in this review are presented in Table 1. When I am referring to all three genera of organisms as a group, I use "*Rhizobium*" with no specific epithet. The phenotype of a strain in the free-living state is designated in a standard way, e.g. Trp⁻. The phenotype for symbiotic gene changes are assigned by the properties of the plant and bacteria together: Nod⁻: no nodules; Fix⁻: nodules form, but no nitrogen fixation. Further refinements for naming symbiotic phenotypes have been proposed (133, 175).

RHIZOBIUM AS A GENETIC SYSTEM

Genetic analysis of all types depends upon finding or generating genetic variability, and upon having methods to move and map genes. Although there have been some innovations in the ten years since the use of broad host range

Table 1 Major groups of *Rhizobiaciae* bacteria used for genetic study

Bacterial species	Symbiotic host
Rhizobium leguminosarum	
biovar viciae	pea, vetch
biovar phaseoli	bean
biovar trifolii	clover
Rhizobium meliloti	alfalfa
Rhizobium fredii	soybean
Bradyrhizobium japonicum	soybean
Azorhizobium caulinodans	sesbania

R-factors and transposons gave *Rhizobium* genetics its initial push (10, 98), there remain several significant technical barriers to genetic analysis.

GENE TRANSFER AND VECTORS   Transfer of genes between *Rhizobium* cells by transduction or conjugation has been reviewed (11, 98, 112). Gene transfer into *Rhizobium* from *Escherichia coli* is carried out by conjugation, and useful stable vectors therefore must be mobilizable, and broad host range (e.g. inc-P, incQ, or native plasmid origins; see 98, 142, and 41, 63, 150). Modifications of *Rhizobium* so that it is infectible by λ phage has been described (115).

R-prime construction has special usefulness for symbiotic gene mapping in *Rhizobium,* since broad host range inc-P plasmids such as RP4 (R68, RK2) can stably bear extremely large inserts. Although more difficult to work with biochemically than the smaller cloning vectors derived from them, such as pRK290, the R-primes permit a wider survey of genetic material that is useful for dispersed sets of genes (24, 89, 127, 161).

MUTAGENESIS AND MARKERS   Chemical mutagens and UV mutagenesis have been widely used both for general genetic study and for symbiotic study. More commonly though, studies of *Rhizobium* genetics have used transposable element mutagenesis, particularly with Tn5 and derivatives such as Tn5-*mob* (162), which includes the transfer origin for RP4, and Tn5s with altered markers (34, 79) or with reporter fusions (38, 110). Other transposons have also been used (16a, 122, 123, 153), as have standard reporter fusions for study of gene expression including *lacZ* (92, 172); *lux* (103); and *uidA* (125). Transposon delivery methods typically involve unstable plasmids that cannot replicate, or that can be eliminated by incompatibility after transfer into *Rhizobium* (10, 89, 160, 163). Similar plasmids are used to introduce directed or random interrupted gene mutations through double crossover, cointegration into the genome, or transposon-mediated vector insertion (40, 72, 87, 151, 163).

Loss of vectors and cloned inserts during nodulation is a problem for functional studies in many strains of *Rhizobium* (41, 102). Marker exchange of insertions flanked by a neutral sequence resident in the chromosome is proposed as a method for introducing stable genomic copies of new or added genes. Among targets proposed for this job are the repeat sequences of *Bradyrhizobium japonicum* (1) and the *int-att* site for lysogenic phage 16-3 (137), which infects *R. meliloti* strain 41 and whose repressor bears an intriguing resemblance to that of phage λ, despite the difference in their target sites (26). In addition to families of repeat sequences (1, 73, 119), native transposable elements have been identified through insertions into symbiotic genes (44, 152, 178).

PLASMIDS   Most strains in the *Rhizobiaceae* have large plasmids and in *Rhizobium* and *Azorhizobium* these appear to be the primary sites of genes for symbiosis; this topic has been reviewed (11, 32, 98, 142). Current studies indicate diversity and fluidity of plasmid-genome arrangements (62, 75, 156, 165, 180). The arrangement of symbiotic genes differs in various species. Symbiotic plasmids sort into incompatibility groups, but having the same origins does not necessarily relate plasmids in terms of markers such as *sym* genes (reviewed in 32, 98; 17, 75). It is thus genetically possible to introduce two different p*Sym* into the same strain, and such strains are also recovered occasionally in natural populations (75, 176a); the symbiotic behavior of such strains is not stable, and often one of the plasmids is lost during plant infection (176a). *R. meliloti* carries genes necessary for symbiois on two extremely large plasmids, termed megaplasmids, of about 1200 and 1500 kilobases (18), designated p*Syma* and p*Symb*. These megaplasmids also carry genes for metabolic functions including thiamine biosynthesis *(thi)* (54) and dicarboxylic acid transport *(dct;* 49, 55, 177). By contrast, a single *Rhizobium leguminosarum* bv. *phaseoli* 410-Kb plasmid (pCFN299) bears all the genes needed to confer nodulation and nitrogen fixation on recipient *Agrobacterium tumefaciens* (117). Even tighter clustering is shown by *Rhizobium* bv. *trifolii*: a 32-Kd segment of the symbiotic plamid (pSym) has all the genes sufficient for host-specific nodulation and nitrogen fixation if placed in a pSym-cured *Rhizobium* background (86). This observation could provide a new impetus to defining a minimum set of symbiotic genes, as genetic studies of *nif* in free-living systems were accelerated by the initial discovery of a *nif*-sufficient *Klebsiella pneumoniae* clone (36).

BASIC MOLECULAR BIOLOGY   Few studies have been carried out on replication, recombination, transcription, and translation in *Rhizobium*. A *recA⁻ R.*

meliloti has been produced by cloning the *R. meliloti recA* gene through functional complementation of the *E. coli* locus, and by subsequent site-directed mutagensis (12). Other recombination-defective strains have been described (136). RNA polymerase activity has been obtained in vitro from strains of *B. japonicum* (144), *Rhizobium leguminosarum* bv. *viciae* (114), and *R. meliloti* (59, 129). In vitro coupled transcription-translation systems in *R. meliloti* have been used to define protein products from symbiotic genes (45, 61). Basic studies on transcription are needed to show what rules govern the general activity of promoters in *Rhizobium;* to date, transcripts have been characterized for relatively few constitutive genes (*nodD:* 59; *hemA:* 107, 121; *trpE:* 6; *glnA:* 21; *flaA* and *flaB:* 140), and little relationship can be seen among these promoters. It appears that many *Rhizobium* promoters do not function in *E. coli* (45, 59, 80), although enough genes have been cloned by complementation of *E. coli* mutants (*recA, glnA, cya*) that this approach should probably be tried routinely. It should be noted that these interspecific complementation tests sometimes yield very unexpected results, for example *glnT* (31).

Much room remains for genetic innovations that will facilitate analysis of *Rhizobium*. Restriction-deficient bacterial strains that allow direct transformation by DNA prepared from *E. coli* would be very useful, as they would permit use of non-conjugatable, smaller plasmid vectors. Transducing phage are still not available for most strains of *Rhizobium,* and the phage that exist would be more useful if temperature-sensitive or other conditional variants were obtained. Genetic and regulatory analyses would benefit from improved vectors, such as plasmids with regulatable promoters for conditional expression. Basic studies on and mutants in replication, transcription, translation (e.g. suppressors) and other processes are needed. For example, it is intriguing that some symbiotic genes are transcriptionally sensitive to gyrase inhibitors (30, 134). Such observations can be pursued in more detail when more analyses of basic molecular functions are carried out in *Rhizobium*.

METABOLIC GENES

The genetic study of *Rhizobium* metabolism is significant both because metabolism is directly involved in symbiotic processes, and because work with easily scored metabolic phenotypes permits rapid progress in analyzing underlying genetic processes. Our understanding of the genetics and regulation of *Rhizobium* symbiosis will be limited unless we develop a more extensive knowledge base about metabolism and general genetics. It would be a great boon to the field if more laboratories each took a so-far unelucidated pathway and thoroughly analyzed its genes and regulation.

Amino acid metabolism has been of interest on general principles, and because of the possible role of amino acids as biosynthetic precursors to plant hormones such as auxin and ethylene. Numerous studies have also found that certain amino acid pathway mutations affect ability to fix nitrogen sym- biotically, although these symbiotic effects vary (10, 95, 97, 154, 158). Such studies will be more easily interpreted when specific metabolic defects are defined.

The pathway for tryptophan biosynthesis has received most genetic scrutiny and is of particular interest because tryptophan is the metabolic precursor to the plant hormone indole-3-acetic acid (IAA). Trp$^-$ mutants of *R. l. viciae* were identified with pathway steps by Johnston and colleagues (reviewed in 98), by isolating *R. meliloti* R-primes that complemented each mutation, and sorting these according to complementation of known *Pseudomonas aerugi- nosa* mutants. The *R. meliloti trpE* gene identified in that study was cloned and sequenced by Crawford and colleagues (80); transcript analysis showed that transcription initiation occurs constitutively at a low rate, and that expression is regulated primarily by attenuation (6). Mutations in some but not all *trp* pathway steps cause symbiotic defects (G. Walker, personal communication).

Glutamine synthetase (GS) is the primary enzyme for ammonia assimila- tion in *Rhizobium,* as in other organisms. While enteric bacteria typically carry a single structural gene for GS, *glnA, Rhizobium* and *Bradyrhizobium* bacteria have two forms of GS, one of which resembles enteric GS *(glnA),* in that it functions as a 12-mer and is subject to adenylyl control, while the other GSII *(glnII)* resembles eukaryotic GS both in its gene structure and its function as an octomer, displaying no posttranslational regulation (27). The *glnA* and/or *glnII* genes have been studied in several organisms (21, 31, 51, 166). A third *R. meliloti*-DNA region able to complement glutamine syn- thetase function, *glnT,* has been identified; it appears to be a complex locus (31). In *R. meliloti,* both *glnA* and *glnII* are dispensable for normal symbiotic function, in that *glnA-glnII* auxotrophs still make Fix$^+$ nodules. What bacte- roids do to obtain amino acids during nitrogen fixation is obviously an interesting question, and relates to the models for metabolite exchange dis- cussed below. The various *gln* loci are differentially regulated in *B. japoni- cum* and *R. meliloti,* in that *glnA* is constitutively transcribed and subject to posttranslational regulation (21, 31), while *glnII* displays a typical "*ntr*"-type promoter and is transcriptionally regulated through the *ntrA-ntrC* circuitry (31) in response to nitrogen nutrition and oxygen level (2, 21, 31). *Azorhizo- bium caulinodans* is distinctive in having only one glutamine synthetase but two different glutamate synthases, one of which (the NADPH-dependent form, *glt*) has been cloned (38, 78) and is expressed independently of the *ntr*-regulatory circuit (78).

Some early mutant studies (10) proposed that purines and pyrimidines were essential for symbiotic function, which again might be suggested because of their relationship to plant hormones. Some purine-requiring *Rhizobium* have symbiotic defects (94, 97, 132). In *R. phaseoli,* seven purine auxotrophs made Ndv⁻ nodules. Pathway analysis showed that all seven could be rescued by IACAR, two steps from synthesis of inosine, and four were blocked early in the pathway. The bacteria with such lesions were able to provoke nodule initiation on plants, but were unable to invade (132).

Some strains of *Rhizobium,* along with other soil microbes, are able to synthesize melanin. The Mel⁺ phenotype sometimes is apparent only as colonies age; this dramatic result has provided the occasion for startling many a novice at *Rhizobium* genetics who revisits plates stored for a few weeks in the cold room. Melanin is derived from tyrosine, and the *mel* synthetic enzymes are typically encoded by *Rhizobium* plasmids (25). In a study of *R. l. phaseoli* (15), *mel* genes fell into two classes, one apparently synthetic (*mel* Class I); the other, *mel* class II, was regulatory on both *nif* and *mel* genes, and turned out to be *nifA* itself (A. W. B. Johnston, personal communiction). That *mel* is coregulated with some *nif* or *fix* functions implies that the pigment chemical or its derivative may have some symbiotic function.

The function of nitrogenase enzyme is dependent upon a low-oxygen environment, maintained in the symbiosis by a plant-encoded oxygen-buffering protein, leghemoglobin (Lb), which gives nodules a characteristic pink color. Genetic analysis has been used to track the source of the heme cofactor. The *R. meliloti hemA* gene, encoding amino levulinic acid synthase, the committed step leading to heme synthesis, is transcribed in both free-living cells and bacteroids from two different promoters with similar -10 sequences (106, 107). *R. meliloti hemA* mutants were Fix⁻, implying that *Rhizobium* heme synthesis was necessary for production of functional leghemoglobin.

By contrast, *B. japonicum hemA*::Tn5 mutants were Fix⁺ and leghemoglobin was formed at normal levels (69). A further complication was the observation that *B. japonicum* mutants defective in cytochrome and possibly in heme synthesis, in some cases also lacked leghemoglobin (135). These varied results leave unresolved the exact *Rhizobium* genetic requirement for support of leghemoglobin and at the very least imply that various host-*Rhizobium* pairs may differ.

Another peculiarity of the nitrogenase enzyme is its high ATP requirement, and its obligate evolution of molecular hydrogen (H_2) during nitrogen reduction, with consequent loss of energy and reducing equivalents (138, 141). Energetics is thus a particularly interesting topic in *Rhizobium*. Genes for adenylcyclase *(cya)* have been identified in *R. meliloti* and in *B. japonicum* (68, 96) by complementation of *E. coli* mutants or by mutagenesis. There may be redundant genes for *cya,* which has complicated efforts to test its symbiotic function. In some *Rhizobium,* an uptake hydrogenase enzyme is produced (Hup^+), which recycles the H_2 back to protons, regaining reducing power and saving some of the energy (141). The ability of *B. japonicum* to grow chemoautotrophically on CO_2 and H_2 through this uptake hydrogenase allows direct genetic study of Hup in this system (84, 101). Growth can also occur on CO_2 and formate, and genetic analysis of this trait has led to definition of a regulatory locus (120).

Genes for *hup* itself were cloned from *B. japonicum* by complementation of a Hup^- mutant. A 20-Kb DNA insert was sufficient to confer hydrogenase activity on *Rhizobium* and *Bradyrhizobium* strains with Hup^- background (101 and references therein). Deletion analysis of the cloned *hup* DNA indicated that a slightly different set of functions were required for Hup activity in free-living vs symbiotic *B. japonicum* cells, but functions for individual protein products (182) in this clone have not yet been biochemically demonstrated. Additional genes may be located elsewhere in the *B. japonicum* genome (84). Less is known about *hup* genes in other species of *Rhizobium*. Random mutagenesis and screening of a Hup^+ *R. l. viciae* strain yielded Hup^- mutants; all of these were phenotypically restored to Hup^+ by the cloned *B. japonicum* DNA, indicating functional and regulatory conservation of the trait (90). A general correlation is seen among numerous *R. l. viciae* strains between hydrogen uptake and DNA homology with the cloned *B. japonicum hup* DNA (109, 128).

NODULATION GENES

For phenotypic and genetic designation, nodulation is taken to mean emergence of nodules, functional or not. The bacterial genetics of nodulation has been most studied in *R. meliloti*, *R. leguminosarum* bv. *viciae* and bv. *trifolii*, and *B. japonicum* (reviewed in 42, 100, 112, 113, 146; additional references not covered in those reviews are indicated below). Mutations resulting in completely non-nodulating bacteria have been recovered in these and other *Bradyrhizobium*, *Rhizobium*, and *Azorhizobium* bacteria, and found to map in a cluster of four genes, *nodDABC* (see reviews; 118, 159, 164, 174, 179). Of these, *nodD* is read divergently and appears to be regulatory on *nodABC* and other *nod* operons. The *nodABC* genes function as allelic equivalents across some species with no effect on host range. Mutations in *nodA*, *nodB*, and *nodC* all have similar phenotype: almost no root hair curling or cell division in plants are caused by such bacteria.

Additional *nod* genes appear to influence nodulation efficiency on a given plant host, and to control host-plant selectivity (42, 100). Mutations in these genes are not complemented by symbiotic plasmids or clones from other *Rhizobium* species or biovars. Host-range determinants may act in combination with each other and with the common *nod* genes; they may function both positively (required for nodulation of a plant) and negatively to prevent nodulation of nonhosts (146). Mutant analyses imply the *nodABC* genes act epistatically to the host-range genes (50). Because mutations in some *nod* genes cause rather subtle changes in phenotype, many were not identified by random mutagenesis and screening, but by transposons and or deletion mutagenesis of DNA segments linked to *nodABC*, or otherwise shown to affect nodulation and host range (see reviews; 5, 65, 74, 108, 130, 170). Nucleotide sequencing of these cloned segments has revealed numerous genes. Downstream of *nodABC* are *nodI*, *nodJ*, and sometimes *nodX* or other genes. In *R. l. trifolii* and *viciae*, *nodFEL* and *nodMN* are downstream of *nodD*, and regulated by it (42). In *R. meliloti nodFE*, a downstream gene *nodG*, and a divergent gene, *nodH*, are present in a distinct cluster, instead of being adjacent to *nodD* (100, 113). Two newly defined genes, *nodP* and *nodQ*, are downstream from *nodFEG* (23, 157). Additional *nod* genes may be present in *R. meliloti* and *R. L. viciae/trifolii*, and are also being defined in *B. japonicum* and broad host range tropical *Rhizobium* strains (5, 65, 74, 108, 127, 130, 161, 174).

REGULATION   Although *nodD* is constitutive, most *Rhizobium nod* genes are not expressed in free-living cells. Exposure to plants or to plant exudates induces expression of these genes (reviewed in 42, 113); regulation occurs at least at the level of transcript abundance (61, 125). Regulation by plants may provide a route to discovery of new *nod* genes (7, 65, 149, 153). The factors in plant exudates that cause induction are small aromatic molecules, identified as various flavonoids (146 42). Induction of *nodABC* by plant exudates depends on *nodD;* the ability of *Rhizobium* to induce *nod* genes in response to specific flavonoid molecules, or to extracts of different plants, varies according to the species source of *nodD* gene placed in the strain (8, 83, 169). Some mutations in *nodD* create alleles that activate inducible *nod* gene expression in response to a broader range of plant-derived compounds, or at a higher basal level, or both (19, 167). In *R. l. viciae* and some other *Rhizobium*, the *nodD* gene product also regulates its own expression (42). Most *Rhizobium* strains have two or more loci showing strong physical homology to *nodD* probes (145). In a few cases, sequence analyses have been carried out that show these to be highly homologous to *nodD*, but only in one *Rhizobium* species, *R. meliloti*, have functional studies demonstrated that all the *nodD*-homologous loci display the NodD function of activating *nod* genes. In *R. meliloti*, there are three copies of *nodD* (71, 82, 125); the one linked to *nodABC* is *nodD1*, which activates transcription of inducible genes in response to luteolin; *nodD2* also responds to an unknown plant inducer distinct from luteolin (71, 82, 125). The *nodD3* product constitutively activates high rates of *nod* gene transcription, but is itself not expressed except under control of another *nodD*-like locus, *syrM* (125; M. Barnett, M. M. Yelton & S. Long, unpublished observations).

The inducible *nod* gene operons contain a highly conserved sequence, designated as the "*nod* box" (149, 168), which probably represents a transcriptional control region for the *nod* genes (reviewed in 42, 113). The *nod* box is located approximately 26–28 bases upstream of the mapped transcript start sites for *nodA, nodF* and *nodH* in *R. meliloti* (61, 125). NodD proteins bind to the promoters for inducible *nod* genes (60, 81) and establish a 50–60 base pair "footprint" from about -25 to -85 upstream of the $^{+}1$ mRNA start sites (58, 99). Neither overall binding nor the footprint has so far been observed to be affected by inducer.

GENETICS OF CELL SURFACE COMPONENTS

The *Rhizobium* cell surface has long been of interest because it was hypothesized to have a role in early recognition. Recent genetic studies complement biochemical and cellular analyses. The primary subjects of interest are the extracellular polysaccharides (of which there are several types), the chemotactic and motility apparatus, and other components of undetermined function, which may contribute to bacterial surface–plant surface interactions.

EXOPOLYSACCHARIDES   The categories and characteristics of cell surface carbohydrates in *Rhizobium* have been reviewed by Carlson (20). They are, briefly, the extracellular (or exo-) polysaccharide (EPS), capsular polysaccharide (CPS), cyclic β-glucans, and lipopolysaccharide (LPS). A protoype for genetic analysis of a *Rhizobium* surface is a series of studies by *R. meliloti* exopolysaccharides. Leigh et al (105) analyzed a series of *R. meliloti* Exo⁻ Tn5 mutants both by genetics and by NMR analysis of purified EPS-I, an acidic hetero-polysaccharide. Six complementation groups were initially defined, in which mutations abolished production of EPS-I (Exo⁻). These strains were able to cause nodule emergence, on the correct host, but were unable to invade (53, 105) (Ndv⁻, *n*odule *d*evelopment, phenotype; 133). Similar approaches have yielded mutants and mapped, cloned genes in other *Rhizobium* systems, although genetic analysis is more preliminary (14, 24, 124).

The number of *R. meliloti* genes known to affect the extracellular polysaccharides has been increased through further mutagenesis of DNA linked to the six original *exo* genes, and through further screening for altered structure or regulation of polysaccharides (52, 110, 111). Mutations in at least eight genes *(exoA, exoB, exoC, exoF, exoL, exoM, exoP, exoQ)* result in an exo⁻ phenotype, that is, no acidic extracellular polysaccharide EPS-I is produced (111). Two genes have been defined that affect acidic substitution, *exoH* for succinylation (104) and another gene needed for pyruvylation (124); defects in these genes also cause Ndv⁻ phenotype. Additional genes, *exoN*, *exoG*, *exoJ*, *exoD* and *exoK*, have quantitative effects on exopolysaccharide production, but do not in every case affect symbiotic properties (111; G. Walker, personal communication). *Exo* genes are expressed during nodulation (93). Control of polysaccharide genes appears to be critical, as regulatory mutants in *R. meliloti* and also in *R. l. phaseoli* are in some cases Fix⁻ (14, 37). Two negative regulatory elements in *R. meliloti* have been identified. The *exoR* locus is involved in the down-regulation of *exo* gene expression in response to nitrogen sufficiency; the *exoS* locus also affects *exo* gene expression, but to a lesser degree and its effect is not sensitive to nutritional status (37). Regulatory mutation *expR101* had the remarkable effect of derepressing synthesis of a second, structurally different exopolysaccharide termed EPS-II (64), through activation of at least seven genes, *expA*, *expC*, *expD*, *expE*, *expF*, *expG*, and the previously identified *exoB* (64; J. Leigh, personal communication). The second polysaccharide could functionally replace the EPS-I on one host plant, but not on several others.

LIPOPOLYSACCHARIDES   A series of noninfective Tn5 mutants of *R. l. phaseoli* were capable of deforming root hairs and stimulating cell divisions, but either did not invade or formed aborted infection threads (133). In one of these mutants, the lipopolysaccharide (LPS) structure was defective, in that the LPS lacked the O-antigen side chain. NTG mutagenesis yielded eight more strains with correlated defects in LPS and in symbiotic behavior. DNA segments containing two distinct sets of *R. l. phaseoli lps* genes have been cloned (22).

β-1, 2-GLUCANS   The genes encoding functions required for β-1,2-glucans synthesis were identified in *R. meliloti* DNA that was homologous to, and able to replace functionally, the *chvAB* virulence genes of *A. tumefaciens*. In *R. meliloti* two genes were identified as *ndvA* and *ndvB* (46). The *chvAB* and *ndvAB* loci are correlated with the ability of the bacteria to synthesize and export the neutral, cyclic Beta-1,2-glucan, which is characteristic of the Rhizobiaceae (20). *R. meliloti* mutated in *ndvA* or *ndvB* are noninfective, or form aborted infection threads (46). The *ndvB* gene product appears to be a 235-kDa inner membrane protein reported to be involved in the synthesis of the glucan (181); the *ndvA* gene product is homologous to the *E. coli* export protein HlyB, and is postulated to be involved in export of the glucan to the cell exterior (171).

CHEMOTAXIS   *Rhizobium* are chemotactic, and *R. meliloti* with mutations affecting flagellae, motility, and chemotaxis have been isolated and characterized (3). All such mutants are able to form normal nodules, perhaps with less competitive ability. Among chemotactic mutants, most were generally deficient in reponse to all stimuli; however, two had lost the ability to swim towards general attractants such as amino acids, but retained the ability to be attracted by plant exudates. The genetic analysis implies the existence of a second, plant-specific pathway for chemotaxis, which shares some components with the general chemotactic pathway (9). The *Rhizobium meliloti* flagellae are complex in morphology (140), and there are in fact two closely linked genes for flagellae, *flaA* and *flaB*. These two flagellins are 87% similar to each other, and show very little resemblance to other flagellins; it is proposed that the two distinct proteins assemble as heterodimers to produce the complex filament form of *R. meliloti* flagellae.

NIF, FIX, AND *DCT* GENES

Genes used in the final stages of symbiosis are generally identified by the observation that mutations in them make the *Rhizobium* unable to fix nitrogen, thus are Fix$^-$ (10). Where the lack of nitrogen fixation is due to a gross failure in the development of nodules, such as a defect in bacterial invasion into the nodule, the phenotype is described as Ndv$^-$, for nodule development (133). Although some genetic loci named *ndv* have been identified as affecting the bacterial surface (46; see above), there does not seem to be any reason why the phenotype designation could not also continue to be used for newly found genes, of initially unknown function, affecting gross developmental processes prerequisite to nitrogen fixation. The Fix$^-$ phenotype relates more particularly to the differentiation of bacteria into bacteroids, and to the functions of nitrogen fixation and interchange with the plant cytoplasm. Loci specifically affecting symbiotic nitrogen fixation (i.e. with no other metabolic effect) are designated by either of two names: those shown to be equivalent to a *Klebsiella pneumoniae nif* gene are assigned this same name; other genes are designated as *fix*.

The location and organization of *nif* and *fix* genes vary a great deal (16 and references therein). The structural *nif* genes of *R. meliloti,* and other *nif* and *fix* genes, are located in several clusters dispersed along more than 200 Kb of DNA on the megaplasmid pSym-a. The *R. leguminosarum* biovars appear to have more tightly linked *nif* and *fix* genes. In *Bradyrhizobium,* all the symbiotic genes appear to be located on the chromosome. The *nod* genes are typically linked to at least one cluster of *nif-fix* genes.

Nif genes have been identified through two main approaches: first, homology with the identified *nif* genes of *K. pneumoniae;* secondly, sequence analysis of DNA regions believed to carry *nif* genes because of linkage to known *nif* genes, or because of preliminary genetic evidence. In some systems it is possible to assay nitrogen fixation independently of the symbiotic state, and this can be used as a criterion to identify *nif* genes. For example, *A. caulinodans* can grow on the nitrogen that it fixes (33, 39, 139); *B. japonicum* derepresses *nif* enough to show enzyme activity (67). *R. meliloti* and other *Rhizobium* do not show colony growth or enzyme activity, but *nif* expression can be monitored by gene fusions (35, 172).

Of the more than 17 identified *nif* genes in *Klebsiella* (some counts now go as high as 21), at least 8 have been identified in various *Rhizobium* species (reviewed in 16, 30, 70, 76; see also 2, 33, 39, 66, 143, 155). The earliest to be studied were the genes for the nitrogenase enzyme peptides, *nifHDK*. These are found in all *Rhizobium*, *Azorhizobium* and *Bradyrhizobium* strains, although their organization varies in interesting ways. For example, the structural genes for nitrogenase, *nifHDK*, form an operon in *R. meliloti* and *R. leguminosarum*, but are split into two unlinked operons, *nifH* and *nifDK*, in *B. japonicum*. Other genes found in *R. meliloti*, *R. leguminosarum*, and *B. japonicum* are *nifN*, *nifE*, *nifB;* these are involved in the synthesis of the molybdenum-iron cofactor (FeMoco) required for activity of the nitrogenase component I, or molybdenum-iron protein. The *nifS* gene, identified in *B. japonicum*, has no currently proven function.

Fix genes have been identified in several ways: through random mutagenesis and screening; through mutagenesis and sequencing of DNA adjacent to the identified *nifHDK* or other *nif* genes; and by mapping DNA regions represented in differentiated bacteroid RNA (e.g. reference 29, 116, 151; and examples below). The functions of most *fix* genes have not yet been biochemically demonstrated, although sequence comparisons are suggestive. Some *fix* genes resemble bacterial ferredoxins or other metal-binding proteins; examples are *frxA* and *fixX* of *Bradyrhizobium* (48, 76), and *fixX* of *R. trifolii*, *R. meliloti* (47, 85). It has been proposed that these genes encode specialized components for electron transfer in the symbiosis, perhaps as assemblies with other *fix* gene products (47, 67). *FixGHIS* form an operon in *R. meliloti* (91); the sequences of all four indicate the proteins are likely to be membrane-bound. *FixG* is indicated to have an iron-sulfur center; the *fixI* sequence is similar to cation-transporting ATPases. A membrane-bound complex that couples an oxidation-reduction reaction to ion transport is proposed to be a likely function for the products of *fixGHIS* (91), and by DNA homology these genes appear to be widely conserved among different *Rhizobium*.

REGULATION   The study of regulation is often most revealing of the overall biology of a system, and symbiotic *nif* gene regulation has exemplified this nicely. The nitrogenase enzyme is expensive to operate, in that it takes up to 28 moles of ATP to reduce one mole of molecular nitrogen (141), and is very oxygen-labile. Nitrogen fixing bacteria thus typically regulate the expression of the *nif* genes in response to nitrogen availability, not using molecular nitrogen as a source except as a desperation measure, and even then only if ambient oxygen tension is appropriate.

In the general nitrogen regulatory circuit (70) transcription initiation at specific *ntr*-type promoters occurs only when nitrogen availability is low, and in the presence of the protein products of two genes, *ntrA* and *ntrC*. The *ntrA* gene product is a sigma factor interacting with *ntr*-type promoters. Regulatory circuits in *R. meliloti, B. japonicum* and *A. caulinodans* all involve the *ntrA* sigma factor. It has been directly shown to be required for *nif, nar* and *dct* expression in *R. meliloti* (148). The conservation of *ntrA*-type promoter structures for various *nif* and *fix* genes is striking; a consensus "−24/−12" promoter sequence (e.g. −26 CTGGYAYR-N_4-TTGCA-10; see 70) is found upstream of *nifHDK(E), FIXABCX, nifN, dctB,* and *ros* in *R. meliloti* (47, 126, 148, 172), upstream of *nifDKEN, nifS, nifBfrxA, nifH, fixBCX,* and *fixRnifAfixA* in *B. japonicum* (reviewed in 76), and upstream of *nifA* and *nifH* in *A. caulinodans* (30, 33). New work in progress on *ntr*-type genes in these and other systems is presented in (16).

Differences arise in various systems because the sigma factor encoded by *ntrA* works along with an upstream activator protein, and the identity and regulation of this activator may vary. The standard for comparison is the circuitry known for *K. pneumoniae* (70): the *ntrC* protein (NR-I) is activated, in nitrogen-limited conditions, by the kinase function of the *ntrB* gene product (NR-II). The *ntrC* and *ntrB* genes belong to a family of two-component regulatory systems (70, 131). The *ntrC*-type gene products are gene activators, which display domains for DNA binding and for interaction with the other, *ntrB*-type gene products, which function as sensors for environmental conditions (in the case of NR-II, response to nitrogen nutrition). In *K. pneumoniae,* the products of *ntrA* and *ntrC* together activate transcription of *nifA;* the *nifA* protein then replaces NR-I, as the *ntrA* and *nifA* proteins together activate expression of the other *nif* genes. Response to oxygen and possibly to other factors is mediated by the product of the nifL gene (70, 138).

Rhizobium systems have uncoupled to some extent the *ntrC-nifA* connection. The symbiotic *nif* and *fix* genes are, as expected, transcriptionally controlled by *nifA* and *ntrA*. However, the exact mechanisms for controlling *nifA* appear to vary among *Rhizobium, Bradyrhizobium* and *Azorhizobium.* In *R. meliloti,* the transcription of *nifA* itself is controlled by oxygen (35, 176), whose effect is mediated by two genes, *fixL* and *fixJ* located in a distantly linked cluster of symbiotic genes on pSym a (28). The sequences of *fixL* and *fixJ* suggest that they also belong to the family of two-component regulatory genes (*fixL* is *ntrB*-like; *fixJ* is *ntrC*-like). Furthermore, *R. meliloti* has a parallel regulatory system, in which *fixL* and *fixJ* activate a second regulatory gene, *fixK; fix* genes may then be turned on by the FixK-NtrA combination (77).

In *Azorhizobium*, the diazotrophic growth habit of the free-living cells allows parallel phenotypic analysis in culture and in nodules (33, 39, 139). Here, *nifA* is autoregulatory; *nifA* transcription is repressed by available nitrogen, but this is partly independent of *ntrC*, suggesting further regulatory genes (139). At least two more regulatory genes, *ntrYX*, have been identified at a site where a second site mutation rescues growth of an *hemA*::Tn5 mutant (30). Oxygen control may be mediated in part by one of these genes.

In *B. japonicum*, it is not known what exactly controls the expression of *nifA*, which is the second gene in the *fixRnifA* operon. The *nifA* locus is not autoregulatory (173); whether the *ntrC* locus plays a role in *nifA* control is not known. The ability of *nifA* to regulate other *nif* genes is decreased by oxygen, but the effect seems to occur as a result of direct oxygen inactivation of the NifA protein itself (57). The *B. japonicum* and *R. meliloti* NifA protein sequences include an inserted stretch of amino acids rich in cysteine, not found in other *NifA*s. Site-directed mutagenesis of the *B. japonicum nifA* gene showed that each of several cysteines was indispensible for NifA activity. This may represent a metal-binding and/or oxygen-sensitive domain (56). Action of the *R. meliloti nifA* protein is not as sensitive to oxygen (35, 56).

GENES FOR BACTEROID METABOLISM   The loci for transport of dicarboxylic acids such as fumarate, malate, and succinate appear to be strictly required for successful symbiotic function. Mutants in *dct* in diverse *Rhizobium* systems have been found to establish non-fixing nodules (for example, see 13, 49, 55, 147, 177); and previous work cited therein). Metabolic studies have shown that enzymes for most carbohydrate pathways are not present in bacteroids; Dilworth and colleagues have postulated that symbiotic bacteria are not exposed to the carbohydrate substrates that would induce the expression of these pathways; rather the plant-derived peribacteroid membrane allows only dicarboxylic acids to get through to the bacteria (34a). The *dct* loci of *R. l. viciae* include the structural gene *dctA*, and two tandem regulatory genes *dctB* and *dctD* that are read divergently from *dctA*. The *dctB-dctD* pair also belong to the two-component regulatory system family and *nifA* can substitute for *dctB* (148). The relevance of the *dct* loci to the symbiosis is supported by several recent discoveries: first, the genes for the *dct* proteins are encoded not on the chromosome, but on one of the pSym megaplasmids of *R. meliloti* (49, 55, 177). Second, the expression of these genes is controlled by the *ntrA* sigma factor, which also is required for the expression of *nif* genes (148). Further expression studies of the genes for both catabolic and synthetic carbohydrate and amino acid pathways may reveal more on the relationship of plant signals to the regulation of bacteroids. In turn, the observation that genes for *dct*, *mel*, and *ros* (a unique *Rhizobium* symbiotic metabolite) are subject to *ntrA* type expression control suggests that a global search for genes expressed by *ntrA* and *nifA* might be a useful approach to defining a more complete set of the genes needed for symbiosis.

ACKNOWLEDGMENTS

I wish to thank Dr. Graham Walker for encouragement and useful discussions; Dr. I. Robert Lehman for hospitality during a sabbatical leave that permitted much of the research and writing of this review; and Alexandra Bloom for skillful preparation of and assistance with the manuscript. I gratefully acknowledge research support from Department of Energy contract ASO3-82-ER12084, National Institutes of Health grant GM-30962, and a National Science Foundation Presidential Young Investigator award.

Literature Cited

1. Acuna, G., Alvarez-Morales, A., Hahn, M., Hennecke, H. 1987. A vector for the site-directed genomic integration of foreign DNA into soybean root-nodule bacteria. *Plant Mol. Biol.* 9:41–50
2. Adams, T. H., Chelm, B. K. 1988. Effects of oxygen levels on the transcription of *nif* and *gln* genes in *Bradyrhizobium japonicum*. *J. Gen. Microbiol.* 134:611–18
3. Ames, P., Bergman, K. 1981. Competitive advantage provided by bacterial motility in the formation of nodules by *Rhizobium meliloti*. *J. Bacteriol.* 148:728–29
4. Appelbaum, E. R., Thompson, D. V., Idler, K., Chartrain, N. 1988. *Rhizobium japonicum* USDA 191 has two *nodD* genes that differ in primary structure and function. *J. Bacteriol.* 170:12–20
5. Bachem, C. W. B., Kondorosi, E., Banfalvi, Z., Horvath, B., Kondorosi, A., et al. 1985. Identification and cloning of noduation genes from the wide host range *Rhizobium* strain MPIK-3030. *Mol. Gen. Genet.* 199:271–78
6. Bae, Y. M., Holmgren, E., Crawford, I. P. 1989. *Rhizobium meliloti* anthranilate synthase: Cloning sequence and expression in *Escherichia coli*. *J. Bacteriol.* 3471–78
7. Bassam, B. J., Djordjevic, M. A., Redmond, J. W., Batley, M., Rolfe, B. G. 1988. Identification of *nonD*-dependent locus in the *Rhizobium* strain NGR234 activated by phenolic factors secreted by soybeans and other legumes. *Mol. Plant-Microbe Int.* 1:161–68
8. Bender, G. L., Nayudu, M., Le Strange, K. K., Rolfe, B. G. 1988. The *nodD1* gene from *Rhizobium* strain NGR234 is a key determinant in the extension of host range to the nonlegume *parasponia*. *Mol. Plant-Microbe Int.* 1:259–66
9. Bergman, K., Gulash-Hofee, M., Hovestadt, R. E., Larosiliere, R. C., Ronco, P. G. II, et al. 1988. Physiology of behavioral mutants of *Rhizobium meliloti:* evidence for a dual chemotaxis pathway. *J. Bacteriol.* 170:3249–54
10. Beringer, J. E., Brewin, N. J., Johnston, A. W. B. 1980. The genetic analysis of *Rhizobium* in relation to symbiotic nitrogen fixation. *Heredity* 45:161–86
11. Beringer, J. E., Johnston, A. W. B., Kondorosi, A. 1987. Genetic maps of *Rhizobium leguminosarum, R. meliloti, R. phaseoli* and *R. trifolii*. In *Genetic Maps,* ed. S. J. O'Brien, 4:245–51. Cold Spring Harbor Press. 755 pp.
12. Better, M., Helinski, D. R. 1983. Isolation and characteriztion of the *recA* gene of *Rhizobium meliloti*. *J. Bacteriol.* 155:311–16
13. Bolton, E., Higgisson, B., Harrington, A., O'Gara, F. 1986. Dicarboxylic acid transport in *Rhizobium meliloti:* Isolation of mutants and cloning of dicarboxylic acid transport gene. *Arch. Microbiol.* 144:142–46

14. Borthakur, D., Barber, C. E., Lamb, J. W., Daniels, W. J., Downie, J. A., et al. 1986. A mutation that blocks exopolysaccharide synthesis prevents nodulation of peas by *Rhizobium leguminosarum* but not of beans by *Rhizobium phaseoli* and is corrected by cloned DNA from the phytophathogen *Xanthomonas*. *Mol. Gen. Genet.* 203:320–23

15. Borthakur, D., Lamb, J. W., Johnston, A. W. B. 1987. Identification of two classes of *Rhizobium phaseoli* genes required for melanin synthesis, one of which is required for nitrogen fixation and activates the transcription of the other. *Mol. Gen. Genet.* 207:155–60

16. Bothe, H., De Bruijn, F. J., Newton, W. E., eds. 1988. *Nitrogen Fixation: Hundred Years After*. Stuttgart/New York: Gustav Fischer. 878 pp.

17. Broughton, W. J., Heycke, N., Priefer, U., Schneider, G.-M., Stanley, J. 1987. Ecological genetics of *Rhizobium meliloti:* diversity and competitive dominance. *FEMS Microbiol. Lett.* 40:245–49

17a. Bullerjahn, G. S., Benzinger, R. H. 1984. Introduction of the mercury transposon Tn*501* into *Rhizobium japonicum* strains 31 and 110. *FEMS Microbiol. Lett.* 22:183–87

18. Burkhardt, B., Schillik, D., Püler, A. 1987. Physical characterization of *Rhizobium meliloti* megaplasmids. *Plasmid* 17:13–25

19. Burn, J., Rossen, L., Johnston, A. W. B. 1987. Four classes of mutations in the *nodD* gene of *Rhizobium leguminosarum* biovar *viciae* that affect its ability to autoregulate and/or activate other *nod* genes in the presence of flavonoid inducers. *Genes Dev.* 1:456–64

20. Carlson, R. W. 1982. Surface chemistry. In *Ecology of Nitrogen Fixation, Vol. 2: Rhizobium*, ed. W. J. Broughton, pp. 199–234. Oxford: Oxford Univ. Press. 353 pp.

21. Carlson, T., Martin, G. G., Chelm, B. K. 1987. Differential transcription of the two glutamine synthetase genes of *Bradyrhizobium japonicum*. *J. Bacteriol.* 169:5861–66

22. Cava, J. R., Elias, P. M., Turowski, D. A., Noel, K. D. 1989. *Rhizobium leguminosarum* CFN42 genetic regions encoding lipopolysaccharide structures essential for complete nodule development on bean plants. *J. Bacteriol.* 171:8–15

23. Cervantes, E., Sharma, S. B., Maillet, F., Vasse, J., Truchet, G., et al. 1989. The *Rhizobium meliloti* host-range *nodQ* gene encodes a protein which shares homology with translation elongation and initiation factors. *Mol. Microbiol.* 3:745–55

24. Chen, H., Gray, J. X., Nayudu, M., Djordjevic, M. A., Batley, M., et al. 1988. Five genetic loci involved in the synthesis of acidic exopolysaccharides are closely linked in the genome of *Rhizobium* sp strain NGR234. *Mol. Gen. Genet.* 212:310–16

25. Cubo, M. T., Buendia-Claveria, A. M., Beringer, J. E., Ruiz-Sainz, J. E. 1988. Melanin production by *Rhizobium* strains. *Appl. Environ. Microbiol.* 54:1812–17

26. Dallmann, G., Papp, P., Orosz, L. 1987. Related repressor specificity of unrelated phages. *Nature* 330:398–401

27. Darrow, R. A., Knotts, R. R. 1977. Two forms of glutamine synthetase in free-living root nodule bacteria. *Biochem. Biophys. Res. Commun.* 78:554

28. David, M., Daveran, M.-L., Batut, J., Dedieu, A., Domergue, O., et al. 1988. Cascade regulation of *nif* gene expression in *Rhizobium meliloti*. *Cell* 54:671–83

29. David, M., Domergue, O., Pognonec, P., Kahn, D. 1987. Transcription patterns of *Rhizobium meliloti* symbiotic plasmid pSym: identification of *nifA*-independent *fix* genes. *J. Bacteriol.* 169:2239–44

30. De Bruijn, F. J., Pawlowski, K., Ratet, P., Hilgert, U., Wong, C. H., et al. 1988. Molecular genetics of nitrogen fixation by *Azorhizobium caulinodans* ORS571, the diazotrophic stem nodulating symbiont of *Sesbania rostrata*. See Ref. 16, pp. 351–55

31. De Bruijn, F. J., Rossbach, S., Schneider, M., Ratet, P., Messner, S., et al. 1989. *Rhizobium meliloti* 1021 has three differentially regulated loci involved in glutamine biosynthesis, none of which is essential for symbiotic nitrogen fixation. *J. Bacteriol.* 171:1673–82

32. Denarie, J., Boistard, P., Casse-Delbart, F., Atherley, A. G., Berry, J. O., et al. 1981. Indigenous plasmids of *Rhizobium*. In *Biology of the Rhizobiaceae*, ed. K. Giles, A. Atherly, pp. 225–46. New York: Academic. 336 pp.

33. Denefle, P., Kush, A., Norel, F., Paquelin, A., Elmerich, C. 1987. Biochemical and genetic analysis of the *nifHDKE* region of *Rhizobium* ORS571. *Mol. Gen. Genet.* 207:280–87

34. De Vos, G. F., Walker, G. C., Signer, E. R. 1986. Genetic manipulations in *Rhizobium meliloti* utilizing two new transposon Tn*5* derivatives. *Mol. Gen. Genet.* 204:485–91

34a. Dilworth, M., Glenn, A. 1984. How does a legume nodule work? *Trends Biochem. Sci.* 9:519–23

35. Ditta, G., Virts, E., Palomares, A., Kim, C.-H. 1987. The *nifA* gene of *Rhizobium meliloti* is oxygen regulated. *J. Bacteriol.* 169:3217–23

36. Dixon, R., Postgate, J. 1972. Genetic transfer of nitrogen fixation form *Klebsiella pneumoniae* to *Escherichia coli*. *Nature* 237:102–4

37. Doherty, D., Leigh, J. A., Glazebrook, J., Walker, G. C. 1988. *Rhizobium meliloti* mutants that overproduce the *R. meliloti* acidic calcofluor-binding exopolysaccharide of *Rhizobium meliloti*. *J. Bacteriol.* 170:4249–56

38. Donald, R. G. K., Lapointe, J., Ludwig, R. A. 1988. Characterization of the *Azorhizobium sesbaniae* ORS571 genomic locus encoding NADPH-glutamate synthase. *J. Bacteriol.* 170:1197–204

39. Donald, R. G. K., Nees, D. W., Raymond, C. K., Loroch, A. I., Ludwig, R. A. 1985. Characterization of three genomic loci encoding *Rhizobium* spp. strain ORS571 N sub(2) fixation genes. *J. Bacteriol.* 165:72–81

40. Donald, R. G. K., Raymond, C. K., Ludwig, R. A. 1985. Vector insertion mutagenesis of *Rhizobium* sp. strain ORS571: Direct cloning of mutagenized DNA sequences. *J. Bacteriol.* 162:317–23

41. Donnelly, D. F., Birkenhead, K., O'Gara, F. 1987. Stability of IncQ and IncP-1 vector plasmids in *Rhizobium* spp. *FEMS Microbiol. Lett.* 42:141–45

42. Downie, J. A., Johnston, A. W. B. 1988. Nodulation of legumes by *Rhizobium*. *Plant Cell Environ.* 11:403–12

43. Dreyfus, B., Garcia, J. L., Gillis, M. 1988. Characterization of *Azorhizobium caulinodans* gene. nov., sp. nov., a stem-nodulating nitrogen-fixing bacterium isolated from *Sesbania rostrata*. *Int. J. Syst. Bacteriol.* 38:89–98

44. Dusha, I., Kovalenko, S., Banfalvi, Z., Kondorosi, A. 1987. *Rhizobium meliloti* insertion element ISRm2 and its use for identification of the fix-X gene. *J. Bacteriol.* 169:1403–9

45. Dusha, I., Schroeder, J., Putnoky, P., Banfalvi, Z., Kondorosi, A. 1986. A cell-free system from *Rhizobium meliloti* to study the specific expression of nodulation genes. *Eur. J. Biochem.* 160:69–76

46. Dylan, T., Ielpi, L., Stanfield, S., Kashyap, L., Douglas, C., et al. 1986. *Rhizobium meliloti* genes required for nodule development are related to chromosomal virulence genes in *Agrobacterium tumefaciens*. *Proc. Natl. Acad. Sci. USA* 83:4403–7

47. Earl, C. D., Ronson, C. W., Ausubel, F. M. 1987. Genetic and structural analysis of the *Rhizobium meliloti fixA, fixB, fixC* and *fixX* genes. *J. Bacteriol.* 169:1127–36

48. Ebeling, S., Noti, J. D., Hennecke, H. 1988. Identification of a new *Bradyrhizobium japonicum* gene (*frxA*) encoding a ferredoxin-like protein. *J. Bacteriol.* 170:1999–2001

49. Engelke, T., Jagadish, M. N., Puehler, A. 1987. Biochemical and genetical analysis of *Rhizobium meliloti* mutants defection in C_4-dicarboxylate transport. *J. Gen. Microbiol.* 133:3019–29

50. Faucher, C., Maillet, F., Vasse, J., Rosenberg, C., Van Brussel, A. A. N., et al. 1988. *Rhizobium meliloti* host range *nodH* gene determines production of an alfalfa-specific extracellular signal. *J. Bacteriol.* 170:5489–99

51. Filser, M. M. K., Moscatelli, C., Lamberti, A., Vincze, E., Guida, M., et al. 1986. Characterization and cloning of two *Rhizobium leguminosarum* genes coding for glutamine synthetase activities. *J. Gen. Microbiol.* 132:2561–69

52. Finan, T. M. 1988. Genetic and physical analyses of group E exo⁻ mutants of *Rhizobium meliloti*. *J. Bacteriol.* 170:474–77

53. Finan, T. M., Hirsch, A. M., Leigh, J. A., Johansen, E., Kuldau, G. A., et al. 1985. Symbiotic mutants of *Rhizobium meliloti* that uncouple plant from bacterial differentiation. *Cell* 40:869–77

54. Finan, T. M., Kunkel, B., De Vos, G. F., Signer, E. R. 1986. Second symbiotic megaplasmid in *Rhizobium meliloti* carrying exopolysaccharide and thiamin synthesis genes. *J. Bacteriol.* 167:66–72

55. Finan, T. M., Oresnik, I., Bottacin, A. 1988. Mutants of *Rhizobium meliloti* that are defective in succinate metabolism. *J. Bacteriol.* 170:3396–403

56. Fischer, H.-M, Bruderer, T., Hennecke, H. 1988. Essential and non-essential domains in the *Bradyrhizobium japonicum* NifA protein: identification of indispensable cysteine residues potentially involved in redox reactivity and/or metal binding. *Nucleic Acids Res.* 16:2207–24

57. Fischer, H.-M., Hennecke, H. 1987. Direct response of *Bradyrhizobium japonicum nifA*-mediated *nif* gene regulation to cellular oxygen status. *Mol. Gen. Genet.* 209:621–26

58. Fisher, R., Long, S. R. 1989. DNA footprint analysis of the transcriptional

activator proteins NodD1 and NodD3 on inducible *nod* gene promotors. *J. Bacteriol.* In press

59. Fisher, R. F., Brierley, H. L., Mulligan, J. T., Long, S. R. 1987. Transcription of *Rhizobium meliloti* nodulation genes: Identification of *nodD* transcription initiation site *in vitro* and *in vivo*. *J. Biol. Chem.* 262:6849–55

60. Fisher, R. F., Egelhoff, T. T., Mulligan, J. T., Long, S. R. 1988. Specific binding of proteins from *Rhizobium meliloti* cell-free extracts containing NodD to DNA sequences upstream of inducible nodulation genes. *Genes Dev.* 2:282–93

61. Fisher, R. F., Swanson, J., Mulligan, J. T., Long, S. R. 1987. Extended region of nodulation genes in *Rhizobium meliloti* 1021. II. Nucleotide sequence, transcription start sites, and protein products. *Genetics* 117:191–201

62. Flores, M., Gonzalez, V., Pardo, M. A., Leija, A., Martinez, E., et al. 1988. Genomic instability in *Rhizobium phaseoli*. *J. Bacteriol.* 170:1191–96

63. Gallie, D. R., Novak, S., Kado, C. I. 1985. Novel high and low-copy stable cosmids for use in *Agrobacterium* and *Rhizobium*. *Plasmid* 14:171–75

64. Glazebrook, J., Walker, G. C. 1989. A novel exopolysaccharide can function in place of the calcofluor-binding exopolysaccharide in nodulation of alfalfa by *Rhizobium meliloti*. *Cell* 56:661–72

65. Göttfert, M., Lamb, J. W., Gasser, R., Semenza, J., Hennecke, H. 1988. Mutational analysis of *Bradyrhizobium japonicum* common *nod* genes and further *nod* box-linked genomic DNA regions. *Mol. Gen. Genet.* In press

66. Groenger, P., Manian, S. S., Reilaender, H., O'Connell, M., Priefer, U. B., et al. 1987. Organization and partial sequence of a DNA region of *Rhizobium leguminosarum* symbiotic plasmid pRL6JI containing the genes *fixABC*, *hifA*, *nifB* and a novel open reading frame. *Nucleic Acids. Res.* 15:31–50

67. Gubler, M., Hennecke, H. 1986. *fixA, B C* genes are essential for symbiotic and free-living, microaerobic nitrogen fixation. *FEBS Lett.* 200:186–92

68. Guerinot, M. L., Chelm, B. K. 1984. Isolation and expression of the *Bradyrhizobium japonicum* adenylate cyclase gene *(cya)* in *Escherichia coli*. *J. Bacteriol.* 159:1068–71

69. Guerinot, M. L., Chelm, B. K. 1986. Bacterial Δ-aminolevulinic acid synthase activity is not essential for leghemoglobin formation in the soybean/ *Bradyrhizobium japonicum* symbiosis.

Proc. Natl. Acad. Sci. USA 83:1837–41

70. Gussin, G. N., Ronson, C. W., Ausubel, F. M. 1986. Regulation of nitrogen fixation genes. *Annu. Rev. Genet.* 20:567–91

71. Györgypal, Z., Iyer, N., Kondorosi, A. 1988. Three regulatory *nodD* alleles of diverged flavonoid-specificty are involved in host-dependent nodulation by *Rhizobium meliloti*. *Mol. Gen. Genet.* 212:85–92

72. Hahn, M., Hennecke, H. 1984. Localized mutagenesis in *Rhizobium japonicum*. *Mol. Gen. Genet.* 193:46–52

73. Hahn, M., Hennecke, H. 1987. Mapping of a *Bradyrhizobium japonicum* DNA region carrying genes for symbiosis and an asymmetric accumulation of reiterated sequences. *Appl. Environ. Microbiol.* 53:2247–52

74. Hahn, M., Hennecke, H. 1988. Cloning and mapping of a novel nodulation region from *Bradyrhizobium japonicum* by genetic complementation of a deletion mutant. *Appl. Environ. Microbiol.* 54:55–61

75. Harrison, S. P., Jones, D. G., Schünmann, P. H. D., Forster, J. W., Young, J. P. 1988. Variation in *Rhizobium leguminosarum* biovar *trifolii* sym plasmids and the association with effectiveness of nitrogen fixation. *J. Gen. Microbiol.* 134:2721–30

76. Hennecke, H., Fischer, H.-M., Gubler, M., Thöny, B., Anthamatten, D., et al. 1988. Regulation of *nif* and *fix* genes in *Bradyrhizobium japonicum* occurs by a cascade of two consecutive steps of which the second one is oxygen sensitive. See Ref. 16, pp. 339–44

77. Hertig, C., Li, R. Y., Louarn, A.-M., Garnerone, A.-M., David, M., et al. 1989. *Rhizobium meliloti* regulatory gene *fixJ* activates transcription of *R. meliloti nifA* and *fixK* genes in *Escherichia coli*. *J. Bacteriol.* 171:1736–38

78. Hilgert, U., Schell, J., De Bruijn, F. J. 1987. Isolation and characterization of Tn5-induced NADPH-glutamate synthase (GOGAT⁻) mutants of *Azorhizobium sesbaniae* ORS571 and cloning of the corresponding *glt* locus. *Mol. Gen. Genet.* 210:195–202

79. Hirsch, P. R., Wang, C. L., Woodward, M. J. 1986. Construction of a tn5 derivative determining resistance to gentamicin and spectinomycin using a fragment cloned from R1033. *Gene* 48:209–9

80. Holmgren, E., Crawford, I. P. 1982. Regulation of tyrptophan genes in *Rhizo-*

bium leguminosarum. J. Bacteriol. 149:1135–37

81. Hong, G.-F., Burn, J. E., Johnston, A. W. 1987. Evidence that DNA involved in the expression of nodulation *(nod)* genes in *Rhizobium* binds to the product of the regulatory gene *nodD. Nucleic Acids Res.* 15:9677–90

82. Honma, M. A., Ausubel, F. M. 1987. *Rhizobium meliloti* has three functional copies of the *nodD* symbiotic regulatory gene. *Proc. Natl. Acad. Sci. USA* 84:8558–62

83. Horvath, B., Bachem, C. W. B., Schell, J., Kondorosi, A. 1987. Host-specific regulation of nodulation genes in *Rhizobium* is mediated by a plant-signal interacting with the *nodD* gene product. *EMBO J.* 6:841–48

84. How, S. S. M., Novak, P. D., Maier, R. J. 1988. Transposon Tn5 generate *Bradyrhizobium japonicum* mutants unable to grow chemoautotrophically with H sub(2). *Appl. Environ. Microbiol.* 54:358–63

85. Iismaa, S. E., Watson, J. M. 1987. A gene upstream of the *Rhizobium trifolii nifA* gene encodes a ferredoxin-like protein. *Nucleic Acids Res.* 15:3180

86. Innes, R. W., Hirose, M. A., Kuempel, P. L. 1988. Induction of nitrogen fixing nodules on clover requires only 32 kilobase pairs of DNA from the *Rhizobium trifolii* symbiosis plasmid. *J. Bacteriol.* 170:3793–802

87. Jagadish, M. N., Bookner, S. D., Szalay, A. A. 1985. A method for site-directed transplacement of *in vitro* altered DNA sequences in *Rhizobium. Mol. Gen. Genet.* 199:249–55

88. Jordan, D. C. 1984. Rhizobiaceae. In *Bergey's Manual of Systematic Bacteriology*, ed. N. R. Krieg, J. G. Holt, 1:234–44. Baltimore/London: Williams & Wilkins, 965 pp.

89. Julliot, J. S., Dusha, I., Renalier, M. H. Terzaghi, B., Garnerone, A. M., et al. 1984. An RP4-prime containing a 285 kb fragment of *Rhizobium meliloti* pSym megaplasmid structural characterization and utilization for genetic studies on symbiotic functions controlled by pSym. *Mol. Gen. Genet.* 193:17–26

90. Kagan, B. A., Brewin, N. J. 1985. Mutagenesis of a *Rhizobium* plasmid carrying hydrogenase determinants. *J. Gen. Microbiol.* 131:1141–47

91. Kahn, D., David, M., Domergue, O., Daveran, M.-L., Ghai, J., et al. 1989. *Rhizobium meliloti fixGHI* sequence predicts involvement of a specific cation pump in symbiotic nitrogen fixation. *J. Bacteriol.* 171:929–39

92. Kahn, M. L., Timblin, C. R. 1984. Gene fusion vehicles for the analysis of gene expression in *Rhizobium meliloti. J. Bacteriol.* 158:1070–77

93. Keller, M., Müller, P., Simon, R., Pühler, A. 1988. *Rhizobium meliloti* genes for exopolysaccharide synthesis and nodule infection located on megaplasmid 2 are actively transcribed during symbiosis. *Mol. Plant-Microbe Int.* 1:267–74

94. Kerppola, T., Kahn, M. L. 1988. Symbiotic phenotypes of auxotrophic mutants of *Rhizobium meliloti* 104A14. *J. Gen. Microbiol.* 134:913–19

95. Kerppola, T. K., Kahn, M. L. 1988. Genetic analysis of carbamoylphosphate synthesis in *Rhizobium meliloti. J. Gen. Microbiol.* 134:921–29

96. Kiely, B., O'Gara, F. 1983. Cyclic 3'5'-adenosine monophosphate synthesis in *Rhizobium:* Identification of a cloned sequence from *Rhizobium meliloti* coding for adenyl cyclase. *Mol. Gen. Genet.* 192:230–34

97. Kim, C.-H., Kuykendall, D., Shah, K. S., Keister, D. L. 1988. Induction of symbiotically defective auxotrophic mutants of *Rhizobium fredii* HH303 by transposon mutagenesis. *Appl. Environ. Microbiol.* 54:423–27

98. Kondorosi, A., Johnston, A. W. B. 1981. The genetics of *Rhizobium.* See Ref. 32, pp. 191–224

99. Kondorosi, E., Gyuris, J., Schmidt, J., John, M., Duda, E., et al. 1989. Positive and negative control of *nod* gene expression in *Rhizobium meliloti* is required for optimal nodulation. *EMBO J.* 8:In press

100. Kondorosi, E., Kondorosi, A. 1986. Nodule induction on plant roots by *Rhizobium. Trends Biochem. Sci.* 11:296–99

101. Lambert, G. R., Cantrell, M. A., Hanus, F. J., Russell, S. A., Haddad, K. R., et al. 1985. Intra- and interspecies transfer and expression of *Rhizobium japonicum* hydrogen uptake genes and autotrophic growth capability. *Proc. Natl. Acad. Sci. USA* 82: 3232–36

102. Lambert, G. R., Harker, A. R., Cantrell, M. A., Hanus, F. J., Russell, S. A., et al 1987. Symbiotic expression of cosmid-borne *Bradyrhizobium japonicum* hydrogenase genes. *Appl. Environ. Microbiol.* 53:422–28

103. Legocki, R. P., Legocki, M., Baldwin, T. O., Szalay, A. A. 1986. Bioluminescence in soybean root nodules demonstration of a general approach to assay gene expression in-vivo by using

bacterial luciferase. *Proc. Natl. Acad. Sci. USA* 83:9080–84

104. Leigh, J. A., Reed, J. W., Hanks, J. F., Hirsch, A. M., Walker, G. C. 1987. *Rhizobium meliloti* mutants that fail to succinylate their calcofluor-binding exopolysaccharide are defective in nodule invasion. *Cell* 51:579–87

105. Leigh, J. A., Signer, E. R., Walker, G. C. 1985. Exopolysaccharide-deficient mutants of *Rhizobium meliloti* that form ineffective nodules. *Proc. Natl. Acad. Sci. USA* 82:6231–35

106. Leong, S. A., Ditta, G. S., Helinski, D. R. 1982. Heme biosynthesis in *Rhizobium* identification of a cloned gene coding for delta amino levulinic-acid synthetase from *Rhizobium-meliloti*. *J. Biol. Chem.* 257:8724–30

107. Leong, S. A., Williams, P. H., Ditta, G. S. 1985. Analysis of the 5′ regulatory region of the gene for δ-aminolevulinic acid synthetase of *Rhizobium meliloti*. *Nucleic Acids Res.* 13:5965–76

108. Lewin, A., Rosenberg, C., Meyer, H., Wong, C. H., Nelson, L., et al. 1987. Multiple host-specificity loci on the broad host-range *Rhizobium*-SP NGR234 selected using the widely compatible legume *Vigna unguiculata*. *Plant Mol. Biol.* 8:447–59

109. Leyva, A., Palacios, J. M., Mozo, G., Ruiz-Argüeso, T. 1987. Cloning and characterization of hydrogen uptake genes from *Rhizobium leguminosarum*. *J. Bacteriol.* 169:4929–34

110. Long, S., McCune, S., Walker, G. C. 1988. Symbiotic loci of *Rhizobium meliloti* identified by random Tn*phoA* metagenesis. *J. Bacteriol.* 170:4257–65

111. Long, S., Reed, J. W., Himawan, J., Walker, G. C. 1988. Genetic analysis of a cluster of genes required for synthesis of the calcofluor-binding exopolysaccharide of *Rhizobium meliloti*. *J. Bacteriol.* 170:4239–48

112. Long, S. R. 1984. Genetics of *Rhizobium* nodulation. In *Plant Microbe Interactions*, ed. T. Kosuge, E. Nester. pp. 265–306. New York: Macmillan. 444 pp.

113. Long, S. R. 1989. *Rhizobium*-legume nodulation: Life together in the underground. *Cell* 56:203–14

114. Lotz, W., Fees, H., Wohlleben, W., Burkardt, H. J. 1981. Isolation and characterization of the DNA-dependent RNA polymerase of *Rhizobium leguminosarum* 300. *J. Gen. Microbiol.* 125:301–9

115. Ludwig, R. A. 1987. Gene tandem-mediated selection of coliphage λ-receptive *Agrobacterium*, *Pseudomonas*, and *Rhizobium* strains. *Proc. Natl. Acad. Sci. USA* 84:3334–38

116. Maier, R. J., Graham, L., Keefe, R. G., Pihl, T., Smith, E. 1987. *Bradyrhizobium japonicum* mutants defective in nitrogen fixation and molybdenum metabolism. *J. Bacteriol.* 169:2548–54

117. Martinez, E., Palacios, R., Sánchez, F. 1987. Nitrogen-fixing nodules induced by *Agrobacterium tumefaciens* harboring *Rhizobium phaseoli* plasmids. *J. Bacteriol.* 169:2828–34

118. Marvel, D. J., Torrey, J. G., Ausubel, F. M. 1987. *Rhizobium* symbiotic genes required for nodulation of legume and nonlegume hosts. *Proc. Natl. Acad. Sci. USA* 84:1319–23

119. Masterson, R. V., Atherly, A. G. 1986. The presence of repeated DNA sequences and a partial restriction map of the pSym of *Rhizobium fredii* USDA193. *Plasmid* 16:37–44

120. McClung, C. R., Chelm, B. K. 1987. A genetic locus essential for formate-dependent growth of *Bradyrhizobium japonicum*. *J. Bacteriol.* 169:3260–67

121. McClung, C. R., Somerville, J. E., Guerinot, M. L., Chelm, B. K. 1987. Structure of *Bradyrhizobium japonicum* gene *hemA* encoding 5-aminolevulinic acid synthase. *Gene* 54:133–39

122. McGetrick, A., O'Regan, M., O'Gara, F. 1985. Expression and regulation of the lactose transposon Tn*951* in *Rhizobium* spp. *FEMS Microbiol. Lett.* 29:27–32

123. Miksch, G., Lentzsch, P. 1987. Marking of a Sym plasmid of *Rhizobium* with Tn7. *Mol. Gen. Genet.* 206:510–14

124. Müller, P., Hynes, M., Kapp, D., Niehaus, K., Pühler, A. 1988. Two classes of *Rhizobium meliloti* infection mutants differ in exopolysccharide production and in coinoculation properties with nodulation mutants. *Mol. Gen. Genet.* 211:17–26

125. Mulligan, J. T., Long, S. R. 1989. Extended nodulation gene region in *Rhizobium meliloti* 1021. III. Interaction of three copies of *nodD* and regulatory locus in the control of *Rhizobium meliloti* nodulation and exopolysaccharide genes. *Genetics* 85:6602–6

126. Murphy, P. J., Heycke, N., Trenz, S. P., Ratet, P., De Bruijn, F. J., et al. 1988. Synthesis of an opine-like compound, a rhizopine, in alfalfa nodules is symbiotically regulated. *Proc. Natl. Acad. Sci. USA* 85:9133–37

127. Nayudu, M., Rolfe, B. G. 1987. Analysis of R-primes demonstrates that genes for broad host range nodulation of *Rhi-*

zobium strain NGR-234 are dispersed on the Sym plasmid. *Mol. Gen. Genet.* 206:326–37

128. Nelson, L. M., Grosskopf, E., Tichy, H. V., Lotz, W. 1985. Characterization of hup-specific DNA in *Rhizobium leguminosarum* strains of different origin. *FEMS Microbiol. Lett.* 30:53–58

129. Nielsen, B. L., Brown, L. R. 1985. Purification and subunit characterization of *Rhizobium meliloti* RNA polymerase. *J. Bacteriol.* 162:645–50

130. Nieuwkoop, A. J., Banfalvi, Z., Deshmane, N., Gerhold, D., Schell, M. G., et al. 1987. A locus encoding host range is linked to the common nodulation genes of *Bradyrhizobium japonicum. J. Bacteriol.* 169:2631–38

131. Nixon, B. T., Ronson, C. W., Ausubel, F. M. 1986. Two-component regulatory systems responsive to environmental stimuli share strongly conserved domains with the nitrogen assimilation regulatory genes *ntrB* and *ntrC. Proc. Natl. Acad. Sci. USA* 83:7850–54

132. Noel, K. D., Diebold, R. J., Cava, J. R., Brink, B. A. 1988. Rhizobial purine and pyrimidine auxotrophs: Nutrient supplementation, genetic analysis, and the symbiotic requirement for de novo purine biosynthesis. *Arch. Microbiol.* 149:499–506

133. Noel, K. D., Vandenbosch, K. A., Kulpaca, B. 1986. Mutations in *Rhizobiumphaseoli* that lead to arrested development of infection threads. *J. Bacteriol.* 168:1392–401

134. Novak, P. D., Maier, R. J. 1987. Inhibition of hydrogenase synthesis by DNA gyrase inhibitors in *Bradyrhizobium japonicum. J. Bacteriol.* 169:2708–12

135. O'Brian, M. R., Kirshbon, P. M., Maier, R. J. 1987. Tn5-induced cytochrome mutants of *Bradyrhizobium japonicum* effects of the mutations on cells grown symbiotically and in culture. *J. Bacteriol.* 169:1089–94

136. Olasz, F., Dorgai, L., Páy, A., Orosz, L. 1983. Recombination deficient mutants of *Rhizobium meliloti 41. Mol. Gen. Genet.* 191:393–96

137. Olasz, F., Dorgai, L., Papp, P., Hermesz, E., Kósa, E., Orosz, L. 1985. On the site-specific recombination of phage *16-3* of *Rhizobium meliloti:* Identification of genetic elements and *int-att* recombinations. *Mol. Gen. Genet.* 201:289–95

138. Orme-Johnson, W. H. 1985. Molecular basis of biological nitrogen fixation. *Annu. Rev. Biophys. Biophys. Chem.* 14:419–59

139. Pawlowski, K., Ratet, P., Schell, J., De Bruijn, F. J. 1987. Cloning and characterization of *nifA* and *ntrC* genes of the stem nodulating bacterium ORS571, the nitrogen fixing symbiont of *Sesbania rostrata:* regulation of nitrogen fixation *(nif)* genes in the free living versus symbiotic state. *Mol. Gen. Genet.* 206:207–19

140. Pleier, E., Schmitt, R. 1989. Identification and sequence analysis of two related flagellin genes in *Rhizobium meliloti. J. Bacteriol.* 171:1467–75

141. Postgate, J. R. 1982. *The Fundamentals of Nitrogen Fixation.* New York: Cambridge Univ. Press. 252 pp.

142. Prakash, R. K., Atherly, A. G. 1986. Plasmids of *Rhizobium* and their role in symbiotic nitrogen fixation. *Int. Rev. Cytol.* 104:1–24

143. Quinto, C., De La Vega, H., Flores, M., Leemans, J., Cevallos, M. A., et al. 1985. Nitrogenase reductase: A functional multigene family in *Rhizobium phaseoli. Proc. Natl. Acad. Sci. USA* 82:1170–74

144. Regensburger, B., Hennecke, H. 1983. RNA polymerase from *Rhizobium japonicum. Arch. Microbiol.* 135:103–9

145. Rodriguez-Quinones, F., Banfalvi, Z., Murphy, P., Kondorosi, A. 1987. Interspecies homology of nodulation genes in *Rhizobium. Plant Mol. Biol.* 8:61–76

146. Rolfe, B. G., Gresshoff, P. M. 1988. Genetic analysis of legume nodule initiation. *Annu. Rev. Plant Physiol. Plant Mol. Biol.* 39:297–319

147. Ronson, C. W., Astwood, P. M., Downie, J. A. 1984. Molecular cloning and genetic organization of 4-carbon dicarboxylate transport genes from *Rhizobium leguminosarum. J. Bacteriol.* 160:903–9

148. Ronson, C. W., Nixon, B. T., Albright, L. M., Ausubel, F. M. 1987. *Rhizobium meliloti ntrA (rpoN)* gene is required for diverse metabolic functions. *J. Bacteriol.* 169:2424–31

149. Rostas, K., Kondorosi, E., Horvath, B., Simoncsits, A., Kondorosi, A. 1986. Conservation of extended promoter regions of nodulation genes in *Rhizobium. Proc. Natl. Acad. Sci. USA* 83:1757–61

150. Rubin, R. A. 1987. Genetic analysis of the gentamicin resistance region of pPH1JI and incorporation into a wide host range cloning vehicle. *Plasmid* 18:84–88

151. Ruvkun, G. B., Ausubel, F. M. 1981. A general method for site-directed mutagenesis in prokaryotes. *Nature* 289:75–78

152. Ruvkun, G. B., Long, S. R., Meade, H. M., Van Den Bos, R. C., Ausubel, F.

M. 1982. ISRm1 a *Rhizobium meliloti* insertion sequence that transposes preferentially into nitrogen fixation genes. *J. Mol. Appl. Genet.* 1:405–18

153. Sadowsky, M. J., Olson, E. R., Foster, V. E., Kosslak, R. M., Verma, D. P. S. 1988. Two host-inducible genes of *Rhizobium fredii* and characterization of the inducing compound. *J. Bacteriol.* 170:171–78

154. Sadowsky, M. J., Rosta, K., Sista, P. R., Bussey, H., Verma, D. P. S. 1986. Symbiotically defective histidine auxotrophs of *Bradyrhizobium japonicum*. *Arch. Microbiol.* 144:334–39

155. Schetgens, R. M. P., Hontelez, J. G. J., Van Den Bos, R. C., Van Kammen, A. 1985. Identification and phenotypical characterization of a cluster of *fix* genes, including a *nif* regulatory gene, from *Rhizobium leguminosarum* PRE. *Mol. Gen. Genet.* 200:368–74

156. Schofield, P. R., Gibson, A. H., Dudman, W. F., Watson, J. M. 1987. Evidence for genetic exchange and recombination of *Rhizobium* symbiotic plasmids in a soil population. *Appl. Environ. Microbiol.* 53:2942–47

157. Schwedock, J., Long, S. R. 1989. Nucleotide sequence and protein products of two new nodulation genes of *Rhizobium meliloti*, *nodP* and *nodQ*. *Mol. Plant-Microbe Int.* In press

158. Schwinghamer, E. A. 1977. Genetic aspects of nodulation and dinitrogen fixation by legumes: The microsymbiont. In *A Treatise on Dinitrogen Fixation*, ed. R. W. F. Hardy, W. S. Silver, 3:577–622. New York: Wiley. 675 pp.

159. Scott, D. B., Chua, K.-Y., Jarvis, B. D., Pankhurst, C. E. 1985. Molecular cloning of a nodulation gene from fast-growing and slow-growing strains of *Lotus* rhizobia. *Mol. Gen. Genet.* 201:43–50

160. Selvaraj, G., Hooper, I., Shantharam, S., Iyer, V. N., Barran, L., et al. 1987. Derivation and molecular characterization of symbiotically deficient mutants of *Rhizobium meliloti*. *Can. J. Microbiol.* 33:739–47

161. Shantharam, S., Engwall, K. S., Atherly, A. G. 1988. Symbiotic phenotypes of soybean root nodules associated with deletions and rearrangements in the symbiotic plasmid of *R. fredii* USDA191. *J. Plant Physiol.* 132:431–38

162. Simon, R. 1984. High frequency mobilization of gram-negative bacterial replicans by the *in vivo* constructed Tn5 mob transposon. *Mol. Gen. Genet.* 196:413–20

163. Simon, R., Priefer, U., Pühler, A.

1983. A broad host range mobilization system for *in vivo* genetic engineering transposon mutagenesis in gram-negative bacteria. *Biotechnology* 1:784–91

164. So, J.-S., Hodgson, A. L. M., Haugland, R., Leavitt, M., Banfalvi, A., et al. 1987. Transposon-induced symbiotic mutants of *Bradyrhizobium japonicum*: isolation of two gene regions essential for nodulation. *Mol. Gen. Genet.* 207:15–23

165. Soberón-Chávez, G., Nájera, R. 1989. Symbiotic plasmid rearrangement in a hyper-recombinant mutant of *Rhizobium leguminosarum* biovar *phaseoli*. *J. Gen. Microbiol.* 135:47–54

166. Somerville, J. E., Kahn, M. L. 1983. Cloning of the glutamine synthetase I gene from *Rhizobium meliloti*. *J. Bacteriol.* 156:168–76

167. Spaink, G. P., Wijffelman, C. A., Okker, R. J. H., Lugtenberg, B. E. J. 1989. Localization of functional regions of the *Rhizobium nodD* product using hybrid *nodD* genes. *Plant Mol. Biol.* In press

168. Spaink, H. P., Okker, R. J. H., Wijffelman, C. A., Pees, E., Lugtenberg, B. E. J. 1987. Promoters in the nodulation region of the *Rhizobium leguminosarum* Sym plasmid pRL1JI. *Plant Mol. Biol.* 9:29–37

169. Spaink, H. P., Wijffelman, C. A., Pees, E., Okker, R. J. H., Lugtenberg, B. J. J. 1987. *Rhizobium* nodulation gene *nodD* as a determinant of host specificity. *Nature* 328:337–40

170. Squartini, A., Van Veen, R. J. M., Regensburg-Tuink, T., Hooykaas, P. J. J., Nuti, M. P. 1988. Identification and characterization of the *nodD* gene in *Rhizobium leguminosarum* strain 1001. *Mol. Plant-Microbe Int.* 1:145–49

171. Stanfield, S., Ielpi, L., O'Brochta, D., Helinski, D. R., Ditta, G. S. 1988. The *ndvA* gene product of *Rhizobium meliloti* is required for Beta-(1-2)glucan production and has homology to the ATP binding export protein HlyB. *J. Bacteriol.* 170:3523–30

172. Szeto, W. W., Nixon, B. T., Ronson, C. W., Ausubel, F. M. 1987. Identification and characterization of the *Rhizobium meliloti ntrC* gene: *R. meliloti* has separate regulatory pathways for activation of nitrogen fixation genes in free-living and symbiotic cells. *J. Bacteriol.* 169:1423–32

173. Thöny, B., Fischer, H.-M., Anthamatten, D., Bruderer, T., Hennecke, H. 1987. The symbiotic nitrogen fixa-

tion regulatory operon *(fixR nifA)* of *Bradyrhizobium japonicum* is expressed aerobically and is subject to a novel, *nifA*-independent type of activation. *Nucleic Acids Res.* 15:8479–99

174. Van Den Eede, G., Dreyfus, B., Goethals, K., Van Montague, M., Holsters, M. 1987. Identification and cloning of nodulation genes from the stem-nodulating bacterium ORS571. *Mol. Gen. Genet.* 206:291–99

175. Vincent, J. M. 1980. Factors controlling the legume-*Rhizobium* symbiosis. In *Nitrogen Fixation,* ed. W. E. Newton, W. H. Orme-Johnson, 2:103–29. Baltimore: Univ. Park Press. 325 pp.

176. Virts, E. L., Stanfield, S. W., Helinski, D. R., Ditta, G. S. 1988. Common regulatory elements control symbiotic and microaerobic induction of *nifA* in *Rhizobium meliloti. Proc. Natl. Acad. Sci. USA* 85:3062–65

176a. Wang, C. L., Beringer, J. E., Hirsch, P. R. 1986. Host plant effects on hybrids of *Rhizobium leguminosarum* biovars *viceae* and *trifolii. J. Gen. Microbiol.* 132:2063–70

177. Watson, R. J., Chan, Y.-K., Wheatcroft, R., Yang, A.-F., Han, S. 1988. *Rhizobium meliloti* genes required for C_4-dicarboxylate transport and symbiotic nitrogen fixation are located on a megaplasmid. *J. Bacteriol.* 170:927–34

178. Wheatcroft, R., Watson, R. J. 1988. Distribution of insertion sequence IS*Rm1* in *Rhizobium meliloti* and other gram-negative bacteria. *J. Gen. Microbiol.* 134:113–21

179. Wilson, K. J., Anjaiah, V., Nambiar, P. T. C., Ausubel, F. M. 1987. Isolation and characterization of symbiotic mutants of *Bradyrhizobium-SP* arachis strain NC92 mutants with host-specific defects in nodulation and nitrogen fixation. *J. Bacteriol.* 169:2177–86

180. Young, J. P. W., Wexler, M. 1988. Sym plasmid and chromosomal genotypes are correlated in field populations of *Rhizobium leguminosarum. J. Gen. Microbiol.* 134:2731–39

181. Zorreguieta, A., Geremia, R. A., Cavaignac, S., Cangelosi, G. A., Nester, E. W., Ugalde, R. A. 1988. Identification of the product of an *Agrobacterium tumefaciens* chromosomal virulence gene. *Mol. Plant-Microbe Int.* 1:121–27

182. Zuber, M., Harker, A. R., Sultana, M. A., Evans, H. J. 1986. Cloning and expression of *Bradyrhizobium japonicum* uptake hydrogenase structural genes in *Escherichia coli. Proc. Natl. Acad. Sci. USA* 83:7668–72

Annu. Rev. Genet. 1989. 23:507–25

MOLECULAR GENETICS OF NEMATODE MUSCLE

Philip Anderson

Department of Genetics, University of Wisconsin, Madison, Wisconsin 53706

CONTENTS

INTRODUCTION

Striated muscle functions universally to generate biological force. It contains two major filament systems, the myosin-containing thick filaments and the actin-containing thin filaments. During contraction, hydrolysis of ATP by a myosin-linked ATPase causes thin filaments to slide past thick filaments, thus shortening the sarcomere and producing tension on the contractile unit (13, 40). During the past decade or two, the small nematode *Caenorhabditis elegans* has emerged as an important experimental organism for the study of muscle structure, assembly, and function (for a more comprehensive review see reference 78). Like muscle biologists in general, *C. elegans* investigators are concerned with the questions: What are the components of striated muscle? How are these components assembled within the cell? How do these components function during the contractile cycle?

Because *C. elegans* offers the opportunity to study these questions geneti-

507

0066-4197/89/1215-0507$02.00

cally as well as ultrastructurally and biochemically, traditional genetic analysis is a major theme of *C. elegans* muscle research. Genetics has been instrumental to our understanding of other problems of macromolecular function, such as bacteriophage morphogenesis (41, 85) and function of the ribosome (48). Genetics should be equally valuable for the study of muscle assembly and function. Genetic analyses of muscle development have been undertaken principally in *C. elegans* and *Drosophila* (17, 42, 64, 65). Within the past several years, traditional genetic analyses of *C. elegans* have been extended with a powerful array of new reagents and methodologies that make an integrated approach to muscle biology possible in this organism. Mutations can now be related to their genes and proteins; proteins and genes can be related to the corresponding mutants; and many of these components can be manipulated by design in vivo.

Much of this is made possible by a relatively trivial aspect of the nematode life cycle. *C. elegans* reproduces as a self-fertilizing hermaphrodite. Individual animals contain both eggs and sperm, and most reproduction occurs via self-fertilization. This has important consequences for genetics in general (10), but it is especially important for muscle genetics. Individual animals do not have to mate with each other in order to reproduce. Consequently, mutants that have even severe muscle defects are fertile. This facilitates their isolation, maintenance, and analysis. With mutants in hand, traditional genetic analysis follows in a straightforward manner, and the sophistication of *C. elegans* genetics makes the enterprise all the more powerful (37).

Muscle-defective mutants define genes and, hence, proteins needed for normal muscle function. The challenge to *C. elegans* investigators is to relate mutant strains to specific ultrastructural and biochemical defects. In doing so, an understanding of how those proteins function in the wild-type animal emerges. A variety of analytical tools are used to make these relationships. Polarized light microscopy provides a rapid, but relatively low resolution, examination of overall muscle structure. High resolution ultrastructure is determined by electron microscopy, a more laborious but more exact procedure. Coupled with these and other microscopic procedures, an increasing number of monoclonal antibodies and other reagents are available to investigate the presence and distribution within muscle cells of specific proteins. Biochemical methods are used to study those proteins and their genes in isolation. Many genes that have been identified by mutation can now be cloned. DNAs that have been manipulated in vitro can be stably transformed into the *C. elegans* germ line. Together with the traditional genetics, these methodologies have yielded new insights into the structure and function of nematode muscle. Although important structural and biochemical differences exist between vertebrate and invertebrate muscle, there is every reason to believe that principles emerging from investigations with *C. elegans* will be applicable to many different organisms.

STRUCTURE AND COMPOSITION OF WILD-TYPE MUSCLE

Muscle Cells and Their Ultrastructure

Of the approximately 1000 cells contained in the adult hermaphrodite, 135 are muscle cells. These include: 95 body-wall muscle cells, responsible for movement of the animal; 20 pharyngeal muscle cells, responsible for pumping food into the intestine; 16 vulval and uterine muscle cells, used to expel developing eggs; 2 intestinal muscles, responsible for shortening of the posterior intestine; and 2 anal muscles, which control defecation. Adult males do not contain vulval and uterine muscles, but instead have specialized muscle cells for mating. Body-wall and pharyngeal muscles have received the most attention, because they are the most significant in terms of cell number and mass.

Nematode body-wall muscle contains 95 rhomboid-shaped, mononucleate cells, which are located in four strips that run longitudinally along the body axis (see Figures 1B and 2). The controlled contraction and relaxation of dorsal/ventral groups of cells generate a sine wave that propagates along the length of the animal, thus propelling it forward or backward. Each muscle cell contains three distinct regions: (*a*) the cell body or "belly" region, which contains most cellular organelles, including the nucleus; (*b*) the muscle arm, a cellular projection that extends from the cell body to a nearby nerve cord and makes synapses with appropriate motor neurons; and (*c*) the spindle region, which contains the contractile filaments and their associated structures. The longitudinal axis of myofilaments is the same as that of the worm. In mature adults, the spindle region is about 1–2 microns deep and comprises a flat sheet near the surface of the worm, just underlying a thin hypodermis and the exterior cuticle. A basement membrane separates muscle cells from the hypodermis. Most investigations of *C. elegans* muscle are concerned with the assembly and function of this body-wall spindle region.

Like invertebrates, the basic contractile unit of *C. elegans* muscle is the sarcomere, in which myosin-containing thick filaments interdigitate with actin-containing thin filaments. The regularity of these interdigitations produces cross-striations that are especially prominent when viewed with polarized light optics. *C. elegans* muscle is *obliquely* striated (84), like that of many invertebrates (70). The structure of nematode muscle is diagrammed in Figure 1B and compared to that of the more familiar vertebrate striated muscle (Figure 1A). A polarized light micrograph is shown in Figure 2. Brightly birefringent A bands, which identify thick filament-containing regions of

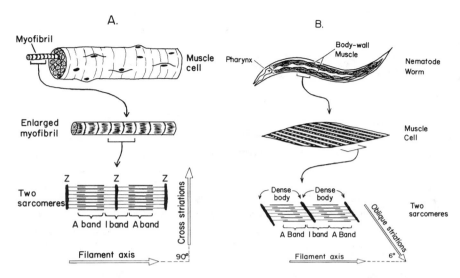

Figure 1   The structures of vertebrate cross-striated and *C. elegans* obliquely striated muscle. (A) Vertebrate striated muscle. A multinucleate vertebrate muscle cell is 10–50 microns in diameter and can be many millimeters long. Thin filaments are anchored in Z discs and interdigitate with thick filaments along a portion of their length. Thick filaments, contained within the A bands, are about 1.5 microns long. The cross striations are orthogonal to the filament axes. (B) *C. elegans* obliquely striated muscle. A *C. elegans* adult nematode is 50–80 microns in diameter and about one millimeter long. The animal contains 95 mononucleate body-wall muscle cells located in strips along the body axis (two strips are shown). Body-wall thin filaments are anchored in dense bodies, the analogs of vertebrate Z discs. Thick filaments are approximately 10 microns long. Birefringent striations are oriented at a 6° oblique angle relative to the filament axes, which are longitudinal with respect to the body axis. (In order to diagram more clearly the relationships between the striations and the filament axes, the angle of oblique striation shown is considerably greater than 6°.)

sarcomeres, alternate with non-birefringent I bands. The striations of *C. elegans* are oblique rather than orthogonal as in vertebrate striated muscle because filaments that are adjacent to each other around the circumference of the cell are offset from each other in the longitudinal direction by a constant amount. Filaments that are adjacent to each other along the radial axis of the worm (down through the lattice towards the interior of the cell) are aligned without offset. Thus, when viewed from the surface of the worm (as in the diagrams of Figure 1 and the polarized light micrographs of Figure 2), the striations are oriented at an oblique angle relative to that of the filaments, which are longitudinal. The stagger is such that striations subtend the longitudinal axis of the worm at an angle of approximately 6° (49). (In order to diagram more clearly the relationships between the striations and the filament

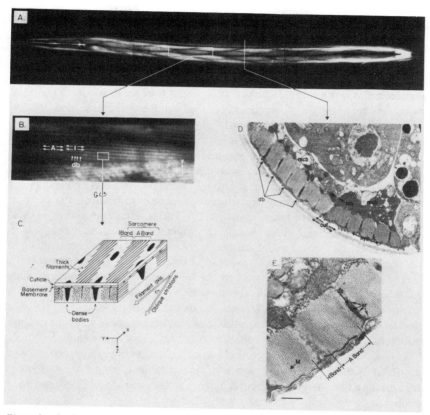

Figure 2   *C. elegans* body-wall muscle ultrastructure. (A) Polarized light micrograph of an adult male (dorsal view). Right and left dorsal muscle quadrants are visible; ventral quadrants are out of focus. (B) Higher magnification polarized light micrograph of a single muscle cell; four other muscle cells are partially in view. This cell is about 125 microns long and contains 9 sarcomeres. Brightly birefringent A bands alternate with nonbirefringent I bands (two of each are labeled with arrows). Birefringent dense bodies (db; four are indicated) are embedded within I bands. (C) Diagram showing the relationship between the myofilament axes and those of the oblique striations. Myofilaments are longitudinal with respect to the body axis (only thick filaments are shown). Birefringent striations intersect the filament axes at an oblique angle of 6°. (D) A low magnification electron micrograph of a transverse thin section of body wall muscle. One entire muscle quadrant, containing sections of two muscle cells and 11 sarcomeres, is visible. The plasma membranes (pm) that separate these two cells are arrowed. The muscle cell bodies (mcb) are located beneath the myofilament lattice, toward the interior of worm. Dense bodies (db; four are labeled) divide the spindle regions into 11 sarcomeres. The layers of cuticle (cut) and hypodermis (hyp) are indicated. Panel E shows one complete sarcomere and portions of two others at higher magnification. An A band, I Band, M line (M), dense body (db), and portions of the sarcoplasmic reticulum (sr) are marked. Measuring bar = 0.5 microns.

axis, the angle of oblique striation shown in Figure 1 and in the diagram of Figure 2 is considerably greater than 6°.)

Because filaments adjacent to each other in the circumferential direction are staggered, transverse sections reveal the ultrastructure of these filaments at many different points wthin the sarcomere. Thus, the organization of the sarcomere as a whole is displayed in a single transverse section. The relationships between the filament axes, the oblique striations, and the ultrastructure of a transverse section is shown in Figure 2.

C. elegans thin filaments are attached to dense bodies, structures analogous to the Z discs of vertebrate striated muscle. Dense bodies are proteinaceous, finger-like projections embedded within the I bands that extend from the cell surface deep into the myofilament lattice. Dense bodies align and anchor the thin filaments to which they are attached, and help to transmit the force of contraction to the overlying cuticle (31). Surrounding the dense bodies, extending towards and along the inside surface of the plasma membrane, is the sarcoplasmic reticulum, a network of membranous sacs assumed to be involved in the release of Ca^{2+} following excitation by motor neurons. Little is known about its components. In the interior of each sarcomere, thick filaments are arranged in an approximately hexagonal array, each thick filament surrounded by an average of about 12 thin filaments (84). The middle of each thick filament is associated with an amorphous, electron-dense structure that, by analogy to vertebrate muscle, likely represents the M line (31, 84). The M line appears to hold thick filaments in register and may be involved in establishing the position of thick filaments within the sarcomere by initiating their assembly. Both the dense bodies and the M lines are firmly anchored in the cell membrane, with electron-dense projections extending into the overlying basement membrane (31). The basement membrane is in turn attached through the hypodermis to the cuticle by a regular array of half-desmosomes and attached filaments (78). This series of attachments (beginning with the myofilaments and extending through the dense bodies, M lines, plasma membrane, basement membrane, and hypodermis to the cuticle) likely functions to transmit the force of contraction laterally, causing the cuticle to bend as needed. Force can probably also be transmitted between juxtaposed muscle cells via attachment plaques, membrane-associated structures found at specialized cell margins (31).

The second-most abundant muscle in C. elegans is the pharynx, which is located at the anterior end of the worm, connected to the outside. The pharynx is a two-lobed organ that contains 20 muscle cells and 42 other cells, including neurons. Coordinated contraction of specialized muscle groups is responsible for ingesting bacteria from the medium, concentrating them, breaking them open, and delivering the result to the intestine (3). Myofilaments of the procorpus, corpus, and isthmus are radially oriented, with half I bands anchored on one side of the cell to the cuticle forming the lumen of the pharynx, and on the other side of the cell to the muscle cell membrane and overlying basement membrane (3). Thick filaments occupy the central region of the sarcomere and interdigitate with thin filaments. Myofilaments of the terminal bulb are more complex, owing to specialized contractions of the grinder apparatus. Muscle cells of the terminal bulb contain a variety of radially, longitudinally, and obliquely oriented filaments. As elsewhere in the pharynx, these filaments are organized as single sarcomers.

Protein Components of Muscle

C. elegans muscle contains an expected variety of myofilament proteins, including actin, myosin heavy and light chains, paramyosin, tropomyosin, and probable troponins (23, 28, 35, 36, 71–73, 76, 79). These proteins were identified as such because of their similarities to well-described proteins from other muscle sources. *C. elegans* also contains some newly described muscle proteins, including components of thick filaments, dense bodies, and cell-to-cell or cell-to-cuticle attachment structures. These newly described proteins were discovered either by close inspection of thick filament substructure (22, 24), by identifying monoclonal antibodies that recognize specific muscle cell structures (31), or by cloning a mutant gene that in vivo causes muscle defects (62). With the development of reliable methods for cloning genes via transposon-tagging (for a review see reference 38), we can expect to see many more examples of this latter approach. In theory, all of the proteins that are defective in the muscle mutants can now be identified by cloning their genes. This is an especially exciting time, because genetic procedures have the ability to identify muscle components without regard to their abundance and without prior knowledge of their localization or biochemical properties. Textbook lists of the proteins of striated muscle will surely be expanded because of investigations with *C. elegans*.

THICK FILAMENTS   Nematode thick filaments differ from their vertebrate counterparts in a number of ways. They are longer, about 10 microns in body-wall muscle (49, 50, 84); they are thicker, tapering from about 33 nm at their centers to about 14 nm at their ends (24); they contain the filamentous protein paramyosin in addition to myosin (35, 79, 80); and they are organized about a "core" structure (24) whose protein components have been tentatively identified (22).

Nematodes contain four different myosin heavy chains (designated MHCs A, B, C, and D; 23, 28, 51, 82), which are encoded by four different genes (44, 59). MHCs C and D are restricted to the pharynx (5, 28, 51). MHCs A and B are found in all body-wall muscle cells and certain muscles other than the pharynx; MHCs A, B, C, and D constitute approximately 20%, 70%, 5%, and 5% of total myosin heavy chains, respectively (78, 82). The sequences of all four MHC genes have been determined (19, 43).

The *C. elegans unc-54* gene, which encodes MHC B, was the first myosin heavy chain gene (or protein) from any organism to be fully sequenced (43). This information led to a precise molecular model for the three-dimensional structure of the myosin rod segment (55, 56). The myosin rod is an alpha-helical, coiled coil and contains over 150 copies of a heptapeptide repeat. The 146 Å periodicity of positive and negative charge along the surface of the rod segment is responsible for aligning and stabilizing adjacent myosin dimers in the filament. All striated-muscle myosin rod segments that have been sequenced to date have this same basic structure (19, 77).

The locations of MHCs A and B within individual body-wall thick filaments are striking (27, 58; see Figure 3). MHC A occupies only the central 1.8 microns of each 10-micron long filament, while MHC B partially overlaps with MHC A and extends to the filament termini. This distribution suggests

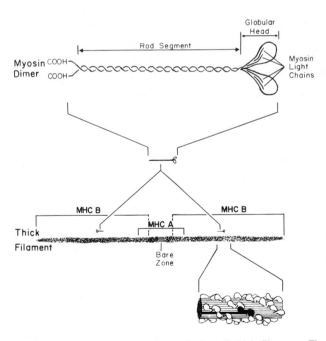

Figure 3 Diagram of myosin and of *C. elegans* body-wall thick filaments. The regions of individual thick filaments that contain either MHC A, MCH B, or both are indicated (58). The thickness of the filament and the size of the bare zone are exaggerated in proportion to the length of the filament, which is about 10 microns.

that MHC A and B have differing functions during filament assembly, a point that will be considered below.

Mutations affecting MHC A and MHC B have been identified. *unc-54* is the structural gene for MHC B. *unc-54* loss-of-function mutants are muscle-defective and almost paralyzed (28). *unc-54* mutants were among the first to be identified in *C. elegans* (10), and they have been extensively studied through the years. Indeed, the traditional and molecular genetics of *unc-54* are among the best to be found for any eukaryotic gene (4, 7, 18, 20, 28, 52–54, 63, 83). Until recently the only mutations affecting *myo-3,* the structural gene for MHC A, were duplications that elevate in vivo levels of MHC A (11, 57, 66, 69, 82). In fact, *myo-3* duplications were not recognized as such until genetic mapping of the cloned *myo-3* gene (2) showed it to be coincident with certain muscle-specific suppressor mutations (formerly called *sup-3;* 69, 82). *myo-3* loss-of-function mutations have since been isolated (R. H. Waterston, personal communication). They are inviable when homozygous due to a near total absence of body-wall thick filaments and an almost complete lack of movement. The differing roles that MHCs A and B might have for thick filament assembly are considered below.

C. elegans contains two electrophoretically separable myosin light chain (MLC) proteins, of 16,000 and 18,000 daltons (35, 72). The larger of these represents the regulatory MLCs (15). Regulatory MLCs are encoded by two closely linked genes (*mlc-1* and *mlc-2*), which specify virtually identical proteins (15). *mlc-2* is *trans*-spliced, but *mlc-1* is not. The in vivo functions of *mlc-1* are either unnecessary or redundant, because a deletion that removes *mlc-1* is phenotypically wild-type (C. Cummins, D. Albertson, J. Levin et al, unpublished observations). In the deletion mutant, *mlc-2* is the only regulatory MLC gene that can be detected in the genome. The 16,000 dalton MLCs presumably represent alkali MLCs (also called "essential" MLCs). An alkali MLC gene (*mlc-3*) has been cloned and sequenced, but it is unclear at present whether this represents the only alkali MLC gene in the genome (S. Sprunger & P. Anderson, unpublished observations).

C. elegans contains a single isoform of paramyosin (79), a filamentous protein found in many invertebrates at the core of thick filaments. *C. elegans* paramyosin is encoded by the *unc-15* gene (80) and is expressed in all muscle cells (5). Because they contain highly abnormal thick filaments, *unc-15* loss-of-function mutants are severely muscle defective and poorly viable. *unc-15* has been cloned and fully sequenced (H. Kagawa, personal communication). Like the rod segment of myosin, most of the paramyosin protein is an alpha-helical coiled coil. However, the periodicity of paramyosin's positive and negative surface charge (720 Å) is different from that of the myosin rod segment (146 Å), indicating that adjacent molecules of paramyosin are probably staggered more than those of myosin (H. Kagawa, personal communication).

Myosin and paramyosin can be removed from body-wall thick filaments either by mutation or by salt extraction of isolated wild-type filaments (24). The result is a smooth, filamentous, core structure approximately 15 nm in diameter. Several proteins that copurify with core structures have been identi-fied (22), but their relationships to each other and to the core are currently unknown. Mutations affecting the core structure have not been identified.

Cloning and molecular analysis of the muscle gene *unc-22* (62) has led to the identification of an intriguing protein that is likely to be associated with myosin (61). *unc-22* mutants have disorganized muscle, although the abnormalities can be subtle (84). The striking feature of *unc-22* mutants is that subcellular regions within individual muscle cells exhibit a spasmodic, con-tractile twitch (60, 84). Twitching within one region of the cell or animal occurs independently of that occurring elsewhere. This phenotype suggests that regulation of the contractile cycle is abnormal in *unc-22* mutants. Genetic analysis of *unc-22* suggests that the *unc-22* protein interacts with myosin B. Certain *unc-54* mutations located in subfragment 1 are suppressors of *unc-22*-induced twitching (18, 63). The *unc-22* gene has been cloned by transposon-tagging (62), and the clone has been used to identify and study the *unc-22* protein (61). *unc-22* protein is over 600 kilodaltons in size. It is localized in A bands of wild-type muscle, coincident with the distribution of MHC B. It is unclear whether *unc-22* protein is an integral component of thick filaments or whether it is localized to A bands in some other way. The phenotype of *unc-22* mutants, the location of *unc-22* protein within the cell, and the genetic interactions between *unc-22* and *unc-54* all suggest that *unc-22* interacts with myosin to regulate the contractile cycle. The sequence of the *unc-22* protein as translated from the genomic sequence is equally suggestive (G. Benian, J. E. Kiff, N. Neckelmann, et al, personal communication). *unc-22* contains a protein kinase domain of over 250 amino acids that is 52% identical to the kinase domain of chicken smooth muscle myosin light chain kinase (csmMLCK). The remainder of the *unc-22* protein is composed of repeated blocks of two different 100 amino acid motifs, copies of which are also present in csmMLCK. Thus, *unc-22* protein may regulate contraction by phosphorylating the regulatory MLCs. A *C. elegans* myofilament protein of the size expected for myosin regulatory light chains (15) is phosphorylated in vivo (L. Anderson & R. H. Waterston, personal communication), but it has not been established yet that *unc-22* is responsible for this.

THIN FILAMENTS   *C. elegans* thin filaments are about 6 microns long (84) and contain actin, tropomyosin, and presumably troponins (28). Actin iso-forms have not been extensively characterized, but the complete family of actin genes has been cloned, sequenced, and genetically mapped (2, 29, 39; M. Krause & D. Hirsh, personal communication). Four actin genes are present in the genome, each of which expresses a unique mRNA (45). Three of the genes (*act-1*, *act-2*, and *act-3*) are clustered within a 12-kb region of linkage group V; *act-4* is unlinked to this cluster. *act-1*, *act-2*, and *act-3* are *trans*-spliced, but *act-4* is not (46)

C. elegans actin genes exemplify the difficulties and some of the solutions to in vivo genetic analysis of multigene families. The original set of muscle genes was defined mostly by recessive, presumably loss-of-function, mutations (84). None of these proved to be actin mutants. Genetic analysis of actin followed, however, when a class of dominant, muscle-defective mutations proved to be genetically inseparable from the *act-1,2,3* gene cluster (81). Wild-type revertants of these mutants are frequent, and certain of the revertants contain insertions or deletions of *act-1* or *act-3* (47). The likely explanation for these results (47, 81) is that the dominant, muscle-defective mutants contain missense alleles of either *act-1* or *act-3*. These mutations yield a disruptive actin protein that interferes with sarcomere assembly. Revertants contain a second loss-of-function mutation within the same gene that eliminates the disruptive product without restoring wild-type protein. The remaining actins perform functions equivalent to those of the missing protein. This interpretation has important implications: (*a*) both *act-1* and *act-3* are muscle actins, and (*b*) the in vivo functions of *act-1*, *act-3*, and perhaps the remaining actin genes are redundant. Transformants that contain the *lacZ* gene fused to *act-4* promoter sequences strongly express the fusion protein in body-wall muscle, suggesting that *act-4* is a muscle actin as well (S. Stone & J. Shaw, personal communication).

Several additional muscle genes have dominant alleles that exhibit genetic behavior similar to the actin gene family (32–34, 67, 68). Without yet knowing the gene products, the redundancies and genetic relationships among the *unc-93/sup-9/sup-10/sup-11/sup-18* gene family have been clearly defined (32–34). Both *unc-93* and *sup-10* have now been cloned, so an understanding of the proteins will surely follow (J. Levin, personal communication; C. Cummins, D. Albertson, J. Levin, et al, unpublished observations).

DENSE BODIES AND MUSCLE ATTACHMENT STRUCTURES   The force of contraction occurs when thin filaments slide past thick filaments, but this force must be transmitted to the surface of the animal in order to bend the cuticle. Dense bodies, M lines, attachment plaques, and the specialized connections between muscle cells, hypodermal cells, and the overlying cuticle (see reference 31 and above) are responsible for this process. Although little is known about the proteins of these attachment structures, it is clear that many of the tools are now in hand for their identification and analysis. When nematodes are gently broken and extracted in high-salt buffer, many of the structures listed above remain insoluble, attached to the cuticle (31). Such cuticle fragments were used as immunogen to prepare monoclonal antibodies, and a rich variety of antibodies was obtained (31, 78). A molecular and genetic analysis of the antigens is just beginning, but the investigations promise new insights into the functions of muscle and cytoskeletal proteins. One class of monoclonal antibody recognizes p107a, a protein that is located throughout the dense body and appears to be the nematode homologue of alpha-actinin (31; L. Kleiman, R. Barstead & R. H. Waterston, personal communication). A second class of antibody recognizes a protein that is located at the base of the dense body (p107b/110) and whose cDNA sequence shows it to be nematode vinculin (6, 31). Other antibodies stain attachment plaques, dense body and M line extracellular matrix material, half desmosomes of the hypodermis, or trans-hypodermal filaments that connect these half desmosomes. Mutants have not yet been identified for any of the proteins defined by these antibodies, but there is every reason to expect that, like myosin and actin, precise genetic mapping of the gene clones will ultimately lead to the identification of mutants.

GENETIC ANALYSIS OF MUSCLE

Over the years, hundreds of muscle-defective mutants have been isolated, and these define about three dozen genes. Since most mutants have been identified as poorly motile animals, body-wall muscle has received the most attention. Certain selections or enrichments have aided the mutant hunts (9, 68, 86). Muscle mutants are distinguished from other types of uncoordinated mutants because they exhibit ultrastructural abnormalities of muscle cells. These criteria surely identify some mutants whose muscle defects are indirect, and they overlook many genes that function in muscle but which, for various reasons, do not yield a phenotype that is easily recognized as involving muscle cells. For example, certain mutants affecting muscle might not exhibit ultrastructural defects, or the defects might be too subtle to be recognized by polarized light microscopy. Other proteins might be essential for viability of the animal. Severe disruption of pharyngeal muscle would kill the animal because of an inability to feed. Other proteins might function in both muscle and nonmuscle cells, such that the mutational defects are not limited to muscle cells. Despite these difficulties and limitations, the available mutants already define about three dozen proteins needed for normal muscle structure, function, or position.

Muscle mutants have been classified into broad categories that in many cases indicate the sites or processes within the cell in which the wild-type proteins function. Genes have been defined that are involved in: thick filament number, morphology, or organization of A bands (*unc-15, unc-45, unc-54, unc-82, unc-89, myo-3*); thin filament morphology and I band organization (*unc-60, unc-78, unc-94, act-1, act-3*); regulation of contraction (*unc-22, unc-90, unc-93, unc-105, sup-10*); attachment of muscle cells to the cuticle (*unc-23, unc-52, unc-95, unc-97, mua-1, mua-2, mua-3*); and positioning of muscle cells within the body (*mup-1, mup-2, mup-3*) (78; E. Hedgecock, personal communication; P. Goh & T. Boggert, personal communication). The remaining genes have more general defects that do not strongly suggest modes of action. Waterston (78) provides a more thorough description and discussion of the mutants and their phenotypes.

For most of these genes the important questions to answer are: What is the protein product? Where is the protein located within the cell? What are the structural or biochemical functions of the protein? How do the available mutations lead to the observed muscle defects? It is often impossible to identify missing or altered proteins simply by biochemically analyzing the mutant strains. Indeed, the gene-protein relationships have probably already been established for all of the abundant, single-gene, nonessential muscle proteins that when eliminated cause a muscle-defective phenotype. Proteins for which mutants have not been identified could be completely dispensable (such that no mutant phenotype results from their absence); they could be redundantly encoded in the genome (such that the mutants are not completely missing a protein); or they might be essential for viability (such that the mutants have phenotypes difficult to recognize as being muscle-defective). Any one of these scenarios makes mutant isolation more difficult, but not impossible. For example, dominant mutations can produce phenotypes within multigene families (see above), and muscle mutants that are inviable when homozygous are only beginning to be analyzed (7, 81; B. Williams, L. Venolia & R. H. Waterston, personal communication).

Reversion analysis of muscle mutants is a valuable source of new mutations. Suppressors are identified by phenotypic criteria that are quite different from those of the mutant being reverted. Thus, unique and informative alleles can be isolated. For example, duplications of *myo-3* were first recovered as partial suppressors of *unc-54* mutations (69). Most of the genes in the *unc-93/sup-9/sup-10/sup-11/sup-18* gene family were identified by reversion of the muscle mutants *unc-93(e1500)* or *sup-10(n983)* (33, 34). Were it not for the revertants, we would probably not yet know about many of these genes. Reversion analysis can be an especially powerful way to identify proteins that physically interact. Mutations affecting one member of a pair of interacting proteins can compensate for those affecting the other. For example, the close associations between *unc-22* protein and MHC B (see above) were first indicated by the isolation of *unc-54* mutations that suppress the twitching of *unc-22* mutants (63). In addition to revealing these interactions, the *unc-54* mutations obtained in this way are unique among available alleles. They demonstrate that normal contractility per se is not required for sarcomere assembly (63).

THICK FILAMENT ASSEMBLY

The structure of individual thick filaments suggests differing roles for myosins A and B. Myosin A is restricted to the bare zone and nearby regions; myosin B overlaps with A and extends to the filament termini (58). This localization suggests that myosin A but not B is involved in initiating filament assembly, because filaments are believed to assemble from the bare zones outwards (16). The fact that *myo-3* loss-of-function mutations are lethal (R. H. Waterston, personal communication), but that those affecting *unc-54* are not, supports this idea.

The deduced three-dimensional structure of the myosin rod segment (see reference 55 and above) and the in vitro assembly properties of myosin and its proteolytic fragments (16) show unambiguously that the myosin rod segment is essential for thick filament assembly. Recent work with assembly-defective *unc-54* mutants demonstrates that filament assembly also requires functions of the myosin globular head. These mutants, designated *unc-54(d)*, exhibit a strongly dominant, muscle-defective phenotype (7, 54, 81). As heterozygotes [genotypes *unc-54(d)/+*], the mutant MHC B disrupts assembly of wild-type MHC B. Mutant MHC B also disrupts assembly of wild-type MHC A, the product of *myo-3*. Although certain *unc-54(d)* mutants accumulate near normal amounts of mutant MHC B (54, 81), typical *unc-54(d)* mutants accumulate steady-state amounts of mutant MHC B that are barely detectable, often less than 2% of wild-type levels (7). Thus, most of the mutant protein is degraded in vivo. Thirty one *unc-54(d)* mutations have been sequenced (8). All of them are located in the globular head of myosin. About half of the mutations are located in two tightly grouped clusters. One cluster affects amino acid residues of the highly conserved ATP binding site. It is clear from work with other enzymes that these mutations reduce or eliminate the myosin ATPase. The second cluster affects residues of the actin binding site, but it is unclear whether actin binding in the mutants is reduced, strengthened, or otherwise altered. These results demonstrate that functions of the myosin head, including the ATPase activity and actin binding, are required for thick filament assembly. The involvement of ATPase function in the assembly process suggests parallels between assembly of muscle thick filaments in *C. elegans* and assembly of smooth muscle and nonmuscle thick filaments in vertebrates (12).

Two other genes appear to influence directly the process of thick filament assembly. (*a*) *unc-82* mutants have reduced numbers of thick filaments (many of which are hollow, similar to those seen in *unc-15* mutants), and they accumulate large filamentous aggregates (84). The aggregates are at least partially composed of paramyosin, because they are eliminated in *unc-15;unc-82* double mutants (11, 84). The phenotypes of *unc-15;unc-82* double mutants depend on the alleles being used. These phenotypes and genetic interactions suggest that *unc-82* protein interacts with paramyosin during filament assembly. The recent demonstration that paramyosin is phosphorylated in vitro and in vivo (74) suggests a mechanism by which this control might operate. Some of the large aggregates present in *unc-82* mutants appear to be "multi-filament assemblages", in which multiple myosin-coated filaments extend from both ends of arrays of paracrystalline paramyosin (25). Such assemblages have been proposed to represent intermediates of normal assembly that accumulate in *unc-82* mutants because the assembly pathway is blocked (21, 25). It will be important to demonstrate that multi-filament assemblages are present at least transiently in the wild-type. Otherwise, they might represent unusual or aberrant structures that only occur in certain mutants. (*b*) *unc-45* mutants have drastically reduced numbers of thick filaments (26), and genetic evidence indicates that they are defective for myosin B assembly (78). Perhaps the *unc-45* protein modifies myosin B to promote thick filament assembly. Myosin heavy chains are phosphorylated in vitro (L. Schriefer & R. H. Waterston, personal communication), and the *unc-45* gene product would appear to be a good candidate for the kinase. *unc-45* protein must have other functions as well, however, because *unc-45* loss-of-function mutations and *unc-45; unc-54* loss-of-function double mutants are inviable when homozygous, but *unc-54* loss-of-function single mutants are not (L. Venolia & R. H. Waterston, personal communication).

FUTURE DIRECTIONS

The everyday activities of isolating, constructing, and analyzing mutant strains will continue to be an important part of *C. elegans* muscle research. Coupled with biochemical analyses of the affected proteins, these efforts should yield new insights into muscle structure and function. Within the past few years, important technical developments have given nematode muscle biologists several new tools. Reliable methods are now available for cloning genes via transposon-tagging (38). Thus, many of the approximately three dozen muscle genes can in theory be cloned. Such clones should identify new or unexpected components of striated muscle. These might, for example, be proteins that are present only in low abundance, or proteins that are needed for sarcomere assembly but which are absent from the finished structure. A physical map of the entire *C. elegans* genome is well along the way to completion (14). Ongoing efforts of the entire *C. elegans* community are making alignments between the physical and genetic maps increasingly precise. In many cases, accurate genetic mapping of a muscle mutation is sufficient to identify the corresponding wild-type clone. Such methods are especially valuable when transposon-induced alleles are unavailable for a particular gene. Muscle protein genes that have been cloned without prior knowledge of the mutant phenotype can be genetically mapped by comparing the clones to those of the ordered collection of genomic clones, by hybridization in situ to metaphase chromosomes (1, 2), or by genetic analysis of RFLPs (29). Knowing the precise map position of a cloned gene often allows realistic mutant hunts to be designed. Finally, reliable methods for transforming *C. elegans* by microinjection are now available (30, 75). Mutant rescue provides unequivocal proof that a cloned region corresponds to a particular mutant gene. More importantly, transformation allows muscle genes to be manipulated in vitro and reintroduced in vivo, where structure/function relationships can be systematically analyzed. For example, *cis*-acting sequences necessary for muscle-specific expression of the myosin heavy chain genes have been identified by this approach (A. Fire & S. Harrison, personal communication). When muscle mutations designed in vitro are subjected to reversion analysis in vivo, the interacting proteins will be identified in a manner not previously possible.

The basic premise for using *C. elegans* to study muscle biology is that an understanding of muscle assembly and function will emerge from a combination of genetic, ultrastructural, and biochemical investigations. Steady progress has been made towards these goals, but much remains to be done. The questions posed at the beginning of this article are by no means answered. Yet, the traditional genetic methods available in *C. elegans* coupled with new methodologies to manipulate genes both in vivo and in vitro should continue to provide important new insights into the biology of muscle.

Acknowledgments

I am grateful to the many *C. elegans* colleagues who shared with me their unpublished data. I especially thank Bob Waterston for helpful comments on the manuscript and for an electron micrograph of Figure 2.

Literature Cited

1. Albertson, D. G. 1984. Localization of the ribosomal genes in *Caenorhabditis elegans* chromosomes by *in situ* hybridization using biotin-labeled probes. *EMBO J.* 3:1227–34

2. Albertson, D. G. 1985. Mapping muscle protein genes by in situ hybridization using biotin-labeled probes. *EMBO J.* 4:2493–98

3. Albertson, D. G., Thomson, J. N. 1976. The pharynx of *C. elegans*. *Philos. Trans. R. Soc. London Ser. B* 275:299–325

4. Anderson, P., Brenner, S. 1984. A selection for myosin heavy-chain

mutants in the nematode *C. elegans*. *Proc. Natl. Acad. Sci. USA* 81:4470–74

5. Ardizzi, J. P., Epstein, H. F. 1987. Immunochemical localization of myosin heavy chain isoforms and paramyosin in developmentally and structurally diverse muscle cell types of the nematode *Caenorhabditis elegans*. *J. Cell Biol.* 105:2763–70

6. Barstead, R. J., Waterston, R. H. 1989. The basal component of the nematode dense-body is vinculin. *J. Biol. Chem.* 264:10177–85

7. Bejsovec, A., Anderson, P. 1988. Myosin heavy-chain mutations that dis-

rupt *Caenorhabditis elegans* thick filament assembly. *Genes Dev.* 2:1307–17

8. Bejsovec, A., Anderson, P. 1989. Functions of the myosin ATP and actin binding sites are required for *C. elegans* thick filament assembly. *Cell.* In press

9. Bejsovec, A., Eide, D., Anderson, P. 1984. Genetic techniques for analysis of nematode muscle. In *Molecular Biology of the Cytoskeleton,* ed. G. G. Borisy, D. W. Cleveland, D. B. Murphy, pp. 267–73. Cold Spring Harbor, NY: Cold Spring Harbor Lab.

10. Brenner, S. 1974. The genetics of *Caenorhabditis elegans. Genetics* 77:71–94

11. Brown, S. J., Riddle, D. L. 1985. Gene interactions affecting muscle organization in *C. elegans. Genetics* 110:421–40

12. Citi, S., Kendrick-Jones, J. 1987. Regulation of non-muscle myosin structure and function. *BioEssays* 7:155–59

13. Cooke, R. 1986. The mechanism of muscle contraction. *CRC Crit. Rev. Biochem.* 21:53–118

14. Coulson, A., Waterston, R., Kiff, J., Sulston, J., Kohara, Y. 1988. Genomic linking with yeast artificial chromosomes. *Nature* 335:184–86

15. Cummins, C., Anderson, P. 1988. Regulatory myosin light chain genes of *Caenorhabditis elegans. Mol. Cell. Biol.* 8:5339–49

16. Davis, J. S. 1988. Assembly processes in vertebrate skeletal thick filament formation. *Annu. Rev. Biophys. Biophys. Chem.* 17:217–39

17. Deak, I. I., Bellamy, P. R., Bienz, M., Dubuis, Y., Fenner, E., et al. 1982. Mutations affecting the indirect flight muscles of *Drosophila melanogaster. J. Embryol. Exp. Morphol.* 69:61–81

18. Dibb, N. J., Brown, D. M., Karn, J., Moerman, D. G., Bolten, S. L., Waterston, R. H. 1985. Sequence analysis of mutations that affect the synthesis, assembly and enzymatic activity of *unc-54* myosin heavy chain of *C. elegans. J. Mol. Biol.* 183:543–51

19. Dibb, N. J., Maruyama, I. N., Krause, M., Karn, J. 1989. Sequence analysis of the complete *Caenorhabditis elegans* myosin heavy chain gene family. *J. Mol. Biol.* 205:603–13

20. Eide, D., Anderson, P. 1985. The gene structures of spontaneous mutations affecting a *Caenorhabditis elegans* myosin heavy chain gene. *Genetics* 109:67–79

21. Epstein, H. F. 1988. Modulation of myosin assembly. *BioEssays* 9:197–200

22. Epstein, H. F., Berliner, G. C., Casey, D. L., Ortiz, I. 1988. Purified thick filaments from the nematode *Caenorhabditis elegans:* evidence for multiple proteins associated with core structures. *J. Cell Biol.* 106:1985–95

23. Epstein, H. F., Miller, D. M., Gossett, L. A., Hecht, R. M. 1982. Immunological studies of myosin isoforms in nematode embryos. In *Muscle Development, Molecular and Cellular Control,* ed. M. L. Pearson, H. F. Epstein, pp. 7–14. Cold Spring Harbor, NY: Cold Spring Harbor Lab.

24. Epstein, H. F., Miller, D. M., Ortiz, I., Berliner, G. C. 1985. Myosin and paramyosin are organized about a newly identified core structure. *J. Cell Biol.* 100:904–15

25. Epstein, H. F., Ortiz, I., Berliner, G. C. 1987. Assemblages of multiple thick filaments in nematode mutants. *J. Muscle Res. Cell Motil.* 8:527–36

26. Epstein, H. F., Thomson, J. N. 1974. Temperature-sensitive mutation affecting myofilament assembly in *Caenorhabditis elegans. Nature* 250:579–80

27. Epstein, H. F., Ortiz, I., Traeger-Mackinnon, L. A. 1986. The alteration of myosin isoform compartmentation in specific mutants of *Caenorhabditis elegans. J. Cell Biol.* 103:985–93

28. Epstein, H. F., Waterston, R. H., Brenner, S. 1974. A mutant affecting the heavy chain of myosin in *Caenorhabditis elegans. J. Mol. Biol.* 90:291–300

29. Files, J. G., Carr, S., Hirsh, D. 1983. Actin gene family of *Caenorhabditis elegans. J. Mol. Biol.* 164:355–75

30. Fire, A. 1986. Integrative transformation of *Caenorhabditis elegans. EMBO J.* 5:2673–80

31. Francis, G. R., Waterston, R. H. 1985. Muscle organization in *C. elegans:* Localization of proteins implicated in thin filament attachment and I-band organization. *J. Cell Biol.* 101:1532–49

32. Greenwald, I., Horvitz, H. R. 1986. A visible allele of the muscle gene *sup-10* X of *C. elegans. Genetics* 113:63–72

33. Greenwald, I. S., Horvitz, H. R. 1980. *unc-93(e1500):* A behavior mutant of *Caenorhabditis elegans* that defines a gene with a wild-type null phenotype. *Genetics* 96:147–64

34. Greenwald, I. S., Horvitz, H. R. 1982. Dominant suppressors of a muscle mutant define an essential gene of *Caenorhabditis elegans. Genetics* 101:211–25

35. Harris, H. E., Epstein, H. F. 1977. Myosin and paramyosin of *Caenorhabditis elegans:* Biochemical and structural

properties of wild-type and mutant proteins. *Cell* 10:709–19

36. Harris, H. E., Tso, M. W., Epstein, H. F. 1977. Actin and myosin-linked calcium regulation in the nematode *Caenorhabditis elegans*. Biochemical and structural properties of native filaments and purified proteins. *Biochemistry* 16:859–65

37. Herman, R. K. 1988. Genetics. In *The Nematode* Caenorhabditis elegans, ed. W. B. Wood, pp. 17–45. Cold Spring Harbor, NY: Cold Spring Harbor Lab.

38. Herman, R. K., Shaw, J. E. 1987. The transposable genetic element Tc1 in the nematode *C. elegans*. *Trends Genet.* 3:222–25

39. Hirsh, D., Files, J. G., Carr, S. H. 1982. Isolation and genetic mapping of the actin genes of *Caenorhabditis elegans*. See Ref 23, pp. 77–86

40. Huxley, H. E. 1969. The mechanism of muscle contraction. *Science* 164:1359–66

41. Jarvik, J., Botstein, D. 1973. A genetic method for determining the order of events in a biological pathway. *Proc. Natl. Acad. Sci. USA* 70:2046–50

42. Karlik, C. C., Coutu, M. D., Fyrberg, E. A. 1984. A nonsense mutation within the *act88F* actin gene disrupts myofibril formation in *Drosophila* indirect flight muscles. *Cell* 38:711–19

43. Karn, J., Brenner, S., Barnett, L. 1983. Protein structural domains in the *Caenorhabditis elegans* unc-54 myosin heavy chain gene are not separated by introns. *Proc. Natl. Acad. Sci. USA* 80:4253–57

44. Karn, J., Dibb, N. J., Miller, D. M. 1984. Cloning nematode myosin genes. In *Cell and Muscle Motility*, ed. J. Shay, 6:185–237. New York: Plenum

45. Krause, M., Hirsch, D. 1984. Actin gene expression in *Caenorhabditis elegans*. See Ref. 9, pp. 287–92

46. Krause, M., Hirsh, D. 1987. A trans-spliced leader sequence on actin mRNA in *C. elegans*. *Cell* 49:753–61

47. Landel, C. P., Krause, M., Waterston, R. H., Hirsh, D. 1984. DNA rearrangements of the actin gene cluster in *Caenorhabditis elegans* accompany reversion of three muscle mutants. *J. Mol. Biol.* 180:497–513

48. Lindahl, L., Zengel, J. M. 1986. Ribosomal genes in *Escherichia coli*. *Annu. Rev. Genet.* 20:297–326

49. Mackenzie, J. M., Epstein, H. F. 1980. Paramyosin is necessary for determination of nematode thick filament length in vivo. *Cell* 22:747–55

50. Mackenzie, J. M., Epstein, H. F. 1981.

Electron microscopy of nematode thick filaments. *J. Ultra. Res.* 76:277–85

51. Mackenzie, J. M., Jr., Schachat, F., Epstein, H. F. 1978. Immunocytochemical localization of two myosins within the same muscle cells in *Caenorhabditis elegans*. *Cell*. 15:413–19

52. MacLeod, A. R., Karn, J., Brenner, S. 1981. Molecular analysis of the unc-54 myosin heavy chain gene of *Caenorhabditis elegans*. *Nature* 291:386–90

53. MacLeod, A. R., Waterston, R. H., Brenner, S. 1977. An internal deletion mutant of a myosin heavy chain in *C. elegans*. *Proc. Natl. Acad. Sci. USA* 74:5336–40

54. MacLeod, A. R., Waterston, R. H., Fishpool, R. M., Brenner, S. 1977. Identification of the structural gene for a myosin heavy-chain in *C. elegans*. *J. Mol. Biol.* 114:133–40

55. McLachlan, A. D., Karn, J. 1982. Periodic charge distributions in the myosin rod amino acid sequence match cross-bridge spacings in muscle. *Nature* 299:226–31

56. McLachlan, A. D., Karn, J. 1983. Periodic features in the amino acid sequence of nematode myosin rod. *J. Mol. Biol.* 164:605–26

57. Miller, D. M., Maruyama, I. 1986. The sup-3 locus is closely linked to a myosin heavy chain gene in *C. elegans*. In *Molecular Biology of Development*, *UCLA Symp. Mol. Cell. Biol.*, ed. C. Emerson, D. Fischman, B. Nadal-Ginard, M. A. Q. Siddiqui, 29:629–38. New York: Liss

58. Miller, D. M., Ortiz, I., Berliner, G. C., Epstein, H. F. 1983. Differential localization of two myosins within nematode thick filaments. *Cell* 34:477–90

59. Miller, D. M., Stockdale, F. E., Karn, J. 1986. Immunological identification of the genes encoding the four myosin heavy chain isoforms of *Caenorhabditis elegans*. *Proc. Natl. Acad. Sci. USA* 83:2305–9

60. Moerman, D. G., Baillie, D. L. 1979. Genetic organization in *C. elegans*: Fine-structure analysis of the unc-22 gene. *Genetics* 91:95–104

61. Moerman, D. G., Benian, G. M., Barstead, R. J., Schriefer, L. A., Waterston, R. H. 1988. Identification and intracellular localization of the unc-22 gene product of *Caenorhabditis elegans*. *Genes Dev.* 2:93–105

62. Moerman, D. G., Benian, G. M., Waterston, R. H. 1986. Molecular cloning of the muscle gene unc-22 in *Caenorhabditis elegans* by Tc1 trans-

poson tagging. *Proc. Natl. Acad. Sci. USA* 83:2579–83

63. Moerman, D. G., Plurad, S., Waterston, R. H., Baillie, D. L. 1982. Mutations in the *unc-54* myosin heavy chain gene of *C. elegans* that alter contractility but not muscle structure. *Cell* 29:773–81

64. Mogami, K., Hotta, Y. 1981. Isolation of *Drosophila* flightless mutants which affect myofibrillar proteins of indirect flight muscle. *Mol. Gen. Genet.* 183:409–17

65. Mogami, K., O'Donnell, P. T., Bernstein, S. I., Wright, T. R. F., Emerson, C. P. Jr. 1986. Mutations of the *Drosophila* myosin heavy chain gene: Effects on transcription, myosin accumulation, and muscle function. *Proc. Natl. Acad. Sci. USA* 83:1393–97

66. Otsuka, A. J. 1986. *sup-3* suppression affects muscle structure and myosin heavy chain accumulation in *C. elegans.* See Ref. 30, pp. 619–28

67. Park, E.-C., Horvitz, H. R. 1986. Mutations with dominant effects on the behavior and morphology of the nematode *Caenorhabditis elegans. Genetics* 113:821–52

68. Park, E.-C., Horvitz, H. R. 1986. *C. elegans unc-105* mutations affect muscle and are suppressed by other mutations that affect muscle. *Genetics* 113:853–67

69. Riddle, D. L., Brenner, S. 1978. Indirect suppression in *C. elegans. Genetics* 89:299–314

70. Rosenbluth, J. 1965. Structural organization of obliquely striated muscle fibers in *Ascaris lumbricoides. J. Cell Biol.* 25:495–515

71. Schachat, F., Garcea, R. L., Epstein, H. F. 1978. Myosins exist as homodimers of heavy chains: Demonstration with specific antibody purified by nematode mutant myosin affinity chromatography. *Cell* 15:405–11

72. Schachat, F. H., Harris, H. E., Epstein, H. F. 1977. Two homogeneous myosins in body-wall muscle of *Caenorhabditis elegans. Cell* 10:721–28

73. Schachat, F. H., Harris, H. E., Epstein, H. F. 1977. Actin from the nematode *Caenorhabditis elegans* is a single electrofocusing species. *Biochim. Biophys. Acta* 493:304–9

74. Schriefer, L. A., Waterson, R. H. 1989. Phosphorylation of the N-terminal region of *Caenorhabditis elegans* paramyosin. *J. Mol. Biol.* 207:451–54

75. Stinchcomb, D. T., Shaw, J. E., Carr, S. H., Hirsh, D. 1985. Extrachromosomal DNA transformation of *C. elegans. Mol. Cell. Biol.* 5:3484–96

76. Tanii, I., Osafune, M., Arata, T., Inoue, A. 1985. ATPase characteristics of myosin from nematode *Caenorhabditis elegans* purified by an improved method. Formation of myosin-phosphate-ADP complex and ATP-induced fluorescence enhancement. *J. Biochem.* 98:1201–9

77. Warrick, H. M., Spudich, J. A. 1987. Myosin structure and function in cell motility. *Annu. Rev. Cell Biol.* 3:379–421

78. Waterston, R. H. 1988. Muscle. See Ref. 37, pp. 281–335

79. Waterston, R. H., Epstein, H. F., Brenner, S. 1974. Paramyosin of *Caenorhabditis elegans. J. Mol. Biol.* 90:285–90

80. Waterston, R. H., Fishpool, R. M., Brenner, S. 1977. Mutants affecting paramyosin in *Caenorhabditis elegans. J. Mol. Biol.* 117:679–97

81. Waterston, R. H., Hirsh, D., Lane, T. R. 1984. Dominant mutations affecting muscle structure in *Caenorhabditis elegans* that map near the actin gene cluster. *J. Mol. Biol.* 180:473–96

82. Waterston, R. H., Moerman, D. G., Baillie, D., Lane, T. R. 1982. Mutations affecting myosin heavy chain accumulation and function in the nematode *Caenorhabditis elegans.* In *Disorders of the Motor Unit,* ed. D. L. Shotland, pp. 747–59. New York: Wiley

83. Waterston, R. H., Smith, K. C., Moerman, D. G. 1982. Genetic fine structure analysis of the myosin heavy chain gene *unc-54* of *Caenorhabditis elegans. J. Mol. Biol.* 158:1–15

84. Waterston, R. H., Thomson, J. N., Brenner, S. 1980. Mutants with altered muscle structure in *Caenorhabditis elegans. Dev. Biol.* 77:271–302

85. Wood, W. B., King, J. 1979. Genetic control of complex bacteriophage assembly. In *Comprehensive Virology,* ed. H. Fraenkel-Conrat, R. Wagner, pp. 581–633. New York: Plenum

86. Zengel, J. M., Epstein, H. F. 1980. Identification of genetic elements associated with muscle structure in the nematode *Caenorhabditis elegans. Cell Motil.* 1:73–97

Annu. Rev. Genet. 1989. 23:527–77

ALTERNATIVE SPLICING IN THE CONTROL OF GENE EXPRESSION

Christopher W. J. Smith, James G. Patton, and Bernardo Nadal-Ginard

Laboratory of Molecular and Cellular Cardiology, Howard Hughes Medical Institute, Department of Cardiology, Children's Hospital and Department of Cellular and Molecular Physiology, Harvard Medical School, Boston, Massachusetts 02115

CONTENTS

0066-4197/89/1215-0527$02.00

INTRODUCTION

Since the discovery of pre-mRNA splicing a little over ten years ago, major advances have been made in understanding the biochemical basis of the splicing reaction (87, 154, 174, 225). At the same time, aspects of splicing such as the basis for the extremely high precision and fidelity of the splicing reaction, and the selective advantages and evolutionary origin of introns, have remained obscure. Alternative pre-mRNA splicing has emerged in recent years as a widespread device for regulating gene expression and generating isoform diversity. It has been a subject of major interest, both in its own right as an important biological regulatory mechanism, and also because it has provided insights into some fundamental aspects of splicing.

The number of known alternatively spliced genes has expanded so rapidly that it is no longer feasible to compile a comprehensive catalog, as it was in previous reviews (10, 29). Alternative splicing in most cases gives rise to protein isoforms sharing extensive regions of identity, and varying only in specific domains, thus allowing for the fine modulation of protein function. Through such changes, alternative splicing can affect almost all aspects of protein function, ranging from the determination of cellular and subcellular localization to the modulation of enzyme activity. The biological role of such changes can be dramatically amplified when the protein isoforms thus produced are themselves important regulatory molecules, such as transcription factors, hormone receptors, and ion channels. Alternative splicing is also used to quantitatively regulate gene expression, by giving rise to prematurely truncated open reading frames, or by regulating mRNA stability or translational efficiency via variability in the untranslated regions. Entire developmental pathways can be regulated by switching gene activity via alternative splicing early in the regulatory cascade. As a mechanism to regulate gene expression and generate protein diversity, alternative splicing complements, and has a number of advantages over, gene rearrangement and extensive multigene families. It allows for switches in protein isoforms in the absence of permanent changes in the cell's genetic content, and without changes in transcriptional activity. As such, it has been exploited from viruses to mammals as a genetically economical, posttranscriptional level of regulation. In evolutionary terms, alternative splicing appears to provide a mechanism for the exploitation of intragenic duplications, allowing the maintenance of the original open reading frame, while duplicated exon sequences can be allowed to drift under relaxed selection until their expression in a modified transcript becomes selectively advantageous.

The mechansims involved in the regulation of alternative splicing are the subject of intense experimental scrutiny. Though not presently well understood, several themes are emerging. Alternative splicing often appears to be achieved by subtle variations of the basic splicing mechanisms. The consensus splice-site elements involved in alternative splicing often do not appear to differ significantly from the elements involved in constitutive splicing (29), though in some cases the use of extreme variants in the organization and sequence of the *cis* elements is associated with differential splice-site utilization. In other cases, novel *cis* elements, unrelated to the consensus constitutive elements, are involved in regulating splice-site selection. Regulation also involves differences in cellular *trans* factors, as demonstrated by cell-specific and developmentally regulated splicing. Such cell-specific *trans* regulation is also required for the splicing of some apparently constitutive exons (31). For these reasons, an understanding of alternative splicing necessarily requires a thorough understanding of constitutive splicing. By the same token, the analysis of alternative splicing has already provided insights into the fundamental mechanisms of splicing, and is likely to yield many more.

In this review, we focus upon the mechanistic, functional, and evolutionary aspects of alternative splicing. The discussion of mechanisms attempts to be exhaustive, drawing not only upon direct experimental investigations of alternatively spliced genes, but also upon relevant themes derived from the study of constitutive splicing. However, only selected examples are used to illustrate the power and versatility of alternative splicing in regulating various biological functions. Finally, we discuss the advantages of alternative splicing as a level of qualitative and quantitative regulation of gene expression, and its potential implications for developmental and evolutionary biology.

MECHANISMS OF CONSTITUTIVE SPLICING

Eukaryotic pre-messenger RNAs undergo a series of nuclear processing events. In addition to 5' capping with 7-methyl guanosine (212) and the addition of a poly(A) tail to the 3' end of nascent transcripts (22), the noncoding intervening sequences must be precisely excised and the exons correctly ligated to avoid disrupting the open reading frame. The mechanisms whereby cells are able to accomplish this task are an area of intense research and have been the subject of recent reviews (87, 154, 174, 225). The understanding of this phenomenon is by no means complete, but a general picture is emerging from data derived from in vivo and in vitro splicing systems, coupled with the investigation of several self-splicing introns from fungal mitochondria.

The use of cell-free systems that accurately splice exogenous pre-mRNA substrates has allowed dissection of the splicing reaction (85, 130, 176, 208). Splicing is a two-step reaction involving successive transesterification reactions (225). Initially, cleavage occurs at the 5' donor splice site, and the

5'end of the intron is joined to an adenosine residue located within the intron, giving rise to a lariat form of the intron (Figure 1). The second transesterification reaction then occurs, resulting in ligation of the exons and release of the lariat intron. The conservation of phosphodiester bonds in these two steps points toward transesterification as the underlying mechanism. The splicing reaction takes place in a large complex, though whether catalysis is protein or RNA based is still not known. The existence of self-splicing introns (40, 41, 182) points to the potential for an RNA-catalyzed reaction (see below). In higher organisms the RNA catalysts could be the family of small nuclear RNAs (88). The experimental evidence and the high conservation of these RNAs suggests an important function, but their exact roles are not known.

How does the splicing machinery recognize 5', donor and 3' acceptor splice sites? An examination of eukaryotic genes has revealed conserved se-

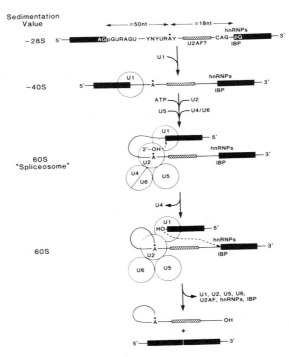

Figure 1 The mechanism of pre-mRNA splicing. The *cis* elements are shown as follows:-black boxes-exons, black line-intron, hatched box-polypyrimidine tract, splice sites and branch point consensus sequences are shown where Y is a pyrimidine and R is a purine. The branch point A is marked with a dot. The snRNPs are shown as circles. The precise order of binding and dissociation of the various factors is not fully established-(see text for details). The sedimentation values are those measured in nondisruptive conditions (68, 86).

quences at these sites (159). The 5' splice-site consensus sequence is (C/A)AG/GURAGU (where the intron-exon boundary is denoted by the vertical line) and the 3' splice-site consensus sequence is YAG/G. Of these bases, the two most highly conserved are the first and last two bases of the intron. Additional conserved sequence blocks include a very loosely defined branch-point sequence, YNYURAY (180, 195, 208), usually located about 30 nt upstream from the 3' splice site and stretch of pyrimidines typically found immediately adjacent to the 3' splice site. In yeast there are similar consensus sequences located at the 5' and 3' splice sites as well as the invariant UACUAAC branch-point sequence (141, 180, 188, 210, 228).

Although important, these consensus sequences are not sufficient to define efficient splice sites because similar sequences are found in positions other than actual splice sites. Some of these cryptic sites can be activated when the normal splice sites are mutated (4, 193, 207), and in some cases the bona fide splice site is not the sequence closest to the consensus present in the intron (172, 223). Somehow, the splicing machinery is able to distinguish the genuine splice sites from the cryptic. The idea that the family of small nuclear RNAs (snRNAs) might be involved in splice-site selection was based originally upon the complementarity between the 5' end of U1 RNA and pre-mRNA splice junctions (145, 200). Subsequently, U1, U2, U4, U5 and U6 have all been shown to be necessary for splicing (16, 23, 24, 42, 130, 131, 160, 177). The snRNAs are complexed with multiple proteins in ribonucleo-protein complexes (snRNPs). These snRNPs assemble with the pre-mRNA and additional proteins (45, 79, 209, 247, 249) to form a large 60S particle in metazoans (68, 86) and 40S in yeast (37). This so-called spliceosome contains the substrate, intermediates, and products of the splicing reaction. The interactions of the cis-acting elements with the trans-acting factors during the formation of spliceosomes are necessary for constitutive and alternative splicing. How these interactions and subsequent spliceosome formation are altered and/or regulated to allow alternative splicing is central to a complete understanding of both alternative and constitutive splicing.

SPLICE-SITE SELECTION

Modes of Alternative Splicing

A fundamental challenge in understanding the processes of both constitutive and alternative splicing is the basis of splice-site selection and pairing. This problem is all the more dramatic in alternative splicing where functional splice-sites that are selected in some circumstances are completely by-passed by the splicing machinery in others. A descriptive classification of the modes of alternative splicing is given below and summarized in Figure 2. This classification is based upon the various combinations of splice sites involved,

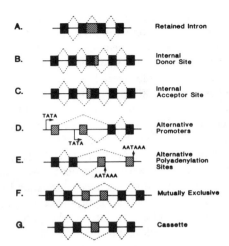

Figure 2   Modes of alternative splicing. Constitutively spliced exons are shown as black boxes and alternatively spliced exons as hatched boxes. Splicing pathways are shown by the diagonal lines. Promoters and poly(A) sites are denoted TATA and AATAAA, respectively.

in conjunction with other regulated processes such as differential promoter and poly(A) site selection, and does not correlate strictly either with the functional consequences of alternative splicing, or with underlying mechanisms.

RETAINED INTRONS   Perhaps the simplest form involves introns that can either be spliced or retained in the mature RNA. Failure to splice the intron can result in the insertion of a peptide segment if the open reading frame is maintained. Alternatively, premature termination of the reading frame or frameshift may result in functionally different products or may effectively shut off gene expression if the unspliced RNA produces no functional product. Examples of such retained introns are found in the genes for fibronectin (220), PDGF A chain (50, 252), and Drosophila P element transposase (142, 197).

ALTERNATIVE 5' DONOR SITES AND 3' ACCEPTOR SITES   The use of alternative donor or acceptor sites for a given exon results in the excision of introns of different lengths with complementary variations in exon size. Again, this can give rise to insertion of small alternative peptide segments, frameshifting, or premature termination of reading frames. Such use of competing splice sites is common in viral transcription units such as the adenovirus E1a (17) and SV40 T/t (268) units, but is also used in eukaryotic genes such as the Drosophila Ultrabithorax (14) and transformer (25) genes.

ALTERNATIVE PROMOTERS AND CLEAVAGE/POLYADENYLATION SITE   In some instances, alternative 5' and 3' end sequences arise through the differential utilization of promoters or cleavage/poly(A) sites, each associated with their own exons. This can lead to variability in the amino-acid sequence of the N or C terminal of the protein, respectively, if the variable segments extend into the protein-coding region. The variations in 5' and 3' untranslated sequences may lead to differential regulation of gene expression via effects on RNA stability, transport, or translational efficiency. Alternative promoter selection dictates the splicing pattern of the myosin light chain 1/3 (MLC1/3) gene (183, 241), while examples of alternative 3' end exons with associated poly(A) addition sites are found in the α and β tropomyosins (92, 204), the immunoglobulin μ heavy chain (6, 60, 198), and the calcitonin/CGRP (8) genes.

INTERNAL MUTUALLY EXCLUSIVE EXONS   Internal exons are sometimes used mutually exclusively. One member of the pair is always spliced into the mRNA, but never both. The two exons are thus never spliced together, nor are both skipped. These exons code for alternative peptide segments within an otherwise constant flanking amino-acid sequence. Such exons are found in the genes for the contractile proteins α and β tropomyosin (92, 204), troponin-T (31, 158), and myosin light chain 1/3 (183, 241), and in the pyruvate kinase gene (170).

CASSETTE EXONS   Some cassette exons can be included or excluded independently of other exons, and usually the same reading frame is maintained whether the exon is spliced in or out, though frameshift or premature termination can result from inclusion (15). When several cassette exons are present in a gene a very high degree of diversity can be generated, as in the NCAM (214) and troponin-T (31) genes. The observed hypervariability within the N-terminal region of troponin-T region is based on the presence of five cassette exons at the 5' end of the gene (31), these in conjuction with a pair of mutually exclusive exons (158) in theory allow the generation of up to 64 different Tn-T isoforms.

Regardless of the class in which an alternatively spliced gene resides, a thorough understanding of both constitutive and alternative splicing will necessarily involve knowing all the factors involved in determining efficient splice-site selection. Many *cis* and *trans* factors are already known to be involved in splice-site selection, some essential for splicing, others affecting the efficiency with which a splice site is utilized.

Cis *Requirements*

THE 5' SPLICE SITE   In yeast, the *cis*-splicing elements are highly conserved and mutations at most positions within the 5' splice site, /GUAUGU (the vertical line represents the exon-intron border), strongly inhibit splicing, with little or no activation of cryptic splice sites (167, 179, 188, 228). In contrast, with the exception of the GU dinucleotide at the intron boundary, the mammalian 5' splice-site consensus, (C/A)AG/GURAGU, is not so highly conserved. Only 5% of donor sites exactly match the strict consensus sequence, though the majority conform better than any other sequence in the flanking intron and exon (172, 223). Most point mutations within the 5' GU dinucleotide inhibit in vitro splicing but many do not abolish 5' cleavage (3, 4, 253). Only GU to GC mutations allow both steps in splicing to take place, consistent with some 5' splice sites containing a GC dinucleotide (116). Mutations of GU at the 5' splice site often cause activation of nearby cryptic splice sites in vivo (4). Variations of the 5' splice-site consensus outside of the GU do not absolutely impair splicing, but can affect the efficiency of splice-site utilization. Usually, mutation of 5' splice sites towards the consensus improves their competitive efficiency (3, 62, 149, 156, 184, 266, 267). This improvement likely reflects a higher degree of base pairing between U1 RNA and the splice site (266, 267). Nevertheless, correlation between the efficiency of U1 binding and 5' splice-site selection is not absolute (156, 166). No consistent differences within the 5' splice-site consensus sequences of constitutively and alternatively spliced exons have been detected (10).

THE 3' SPLICE SITE AND POLYPYRIMIDINE TRACT   The 3' splice-site consensus, usually written as Y_nNCAG/G (159, 172, 174), should be considered as two distinct elements, the polypyrimidine tract and the 3' splice site itself. The only absolutely conserved bases in either element are the AG at the intron-exon boundary. The effects of point mutations in, or deletions of, the AG dinucleotide are somewhat variable. Although the second step of splicing using the mutated 3' splice site is always inhibited, the effects upon 5' cleavage and lariat formation range from drastic inhibition to no effect at all (3, 19, 68, 138, 207, 210; C. W. J. Smith, J. G. Patton & B. Nadal-Ginard, unpublished observations). In some cases, AG to GG mutations, though inhibiting splicing to the mutated splice site, result in the utilization of the next downstream AG dinucleotide as the 3' splice site (3; C. W. J. S. J. G. P. & B. N.-G., unpublished observations). The importance of the polypyrimidine tract is less equivocal. Deletions into the polypyrimidine tract inhibit the 5' cleavage reaction (68, 194, 206, C. W. J. S., J. G. P. & B. N.-G., unpublished observations), binding of the splicing factors U2AF and U2snRNP (209) and splicing complex assembly (68; C. W. J. S., J. G. P. & B. N.-G., unpublished observations). The polypyrimidine tract is not always adjacent to the splice acceptor site and may be more correctly considered a branch-point association element. In at least two cases of branch points >100 nt upstream of the 3' splice site, the polypyrimidine tract is found adjacent to the branch point, not the 3' splice site (93, 232). These cases suggest that the 3' splice site itself has little specificity except as the first AG dinucleotide 3' of the branch point and polypyrimidine tract. This specification of the 3' splice site relative to the branch point probably underlies the relative proximity of the two elements in most introns, as intervening AG dinucleotides could be efficiently utilized as acceptor sites (237; C. W. J. S., J. G. P. & B. N-G., unpublished observations). The extent, nature, and location of the polypyrimidine tract may affect the efficiency of branch point/acceptor site utilization. Extensive polypyrimidine tracts are associated with several alternatively spliced acceptor sites (93, 232) and may be involved in making these sites more competitive. It should be noted, however, that not all introns contain a polypyrimidine tract, and that other sequence elements may be able to replace it. For instance, an intron in the apolipoprotein A1 gene has a repeating GU element that apparently fulfills the function of the polypyrimidine tract (227).

THE BRANCH-POINT SEQUENCE   In yeast, the invariant UACUAAC branch-point sequence is usually located within 50 nt of the 3' splice site (141, 188), with the final A being the point of branch formation. Most mutations in this consensus sequence drastically inhibit splicing (103, 167, 179), presumably because the base-pairing interaction with U2 RNA is disrupted (180). The mammalian consensus is less rigidly defined as YNYURAY (114, 119, 195, 208, 265), and is located to the 5' side of the polypyrimidine tract. The branch point is commonly found 20-40 nt upstream of the 3' splice site, though sometimes more than >150 nt upstream (93, 232). The most highly conserved position in the consensus is the A at the site of branch formation. Even this position is not absolutely conserved, however, and some branch points have been found to use U or C residues (91, 169). Point mutations of the branch-point sequence often have no effect upon splicing due to the activation of nearby cryptic branch points (175, 193, 207). The mammalian branch point sequence may also base pair with U5 RNA (91, 114, 180). Consistent with this hypothesis, mutation of mammalian branch-point sequences towards the mammalian or yeast consensus has recently been shown to improve their efficiency in a cis-competition assay (195, 267a). Even stronger evidence is provided by the preferential activation of non-consensus branch points by U2 mutants; these have compensatory mutations that restore branch-point sequence base pairing (A. M. Weiner, personal communication; J. L. Manley, personal communication). Some alternatively spliced introns contain multiple branch points associated with a single acceptor site (77, 93, 168, 169), although their functional significance in regulating alternative splicing is not known.

The branch point and 3' splice site are obligatorily linked. The early recognition by splicing factors appears to reside at the level of the polypyrimidine tract and branch point, with the 3' splice site being subsequently located relative to these elements. Thus, selection of 3' splice sites is probably determined by the branch point and polypyrimidine tract. For simplicity, we refer in subsequent discussion to 3' acceptor site selection, with the implicit understanding that this selection involves the associated branch point.

MINIMAL INTRON AND EXON LENGTHS   Apart from the specific sequence requirements discussed above, there are also nonsequence-specific requirements for minimal distances between the conserved elements. These minimal separation requirements likely reflect the steric constraints for the binding of splicing factors to the pre-mRNA substrate (44). In mammalian introns, the minimal distance between the branch point and 5' splice site is about 50 nt (70, 205, 232), and about 18 nt for the separation of branch point and 3' splice site (69, 87, 260), giving a minimal intron length of 66–70 nt (Figure 1). In nematodes and fireflies, introns as small as 31 nt are found (53, 58a, 110), probably reflecting differences in the spatial requirements of the splicing machinery in these species; such introns are not correctly excised in mammalian cells (110).

There appear to be no absolute restrictions upon exon length as exons of 3 (214), and 6 nt (52) have been described, both efficiently incorporated into mRNA. Minimum required exon lengths, determined by truncation of exons in single intron transcripts, probably reflect the requirement for minimal substrate sizes adjacent to the splice sites for factor binding (72, 178). Nevertheless, the proximity of the splice sites on either side of small exons introduces the potential for steric interference between bound splicing factors. Such interference has been implicated in the inefficient splicing of some small exons (2). Moreover, adjacent exon sequences do affect the efficiency of splice-site utilization (166, 194, 236; see below). A preferred order of intron removal could, therefore, result from steric effects, with the second splicing event being activated by the newly adjacent exon sequences. Such an effect is likely to play a role in the splicing of small cassette exons (32).

SPLICE-SITE PROXIMITY AND SPACING   Early models for mRNA splicing invoked simple scanning mechanisms to explain the high fidelity and accuracy of the splicing reaction in joining only the correct adjacent pairs of splice sites in pre-mRNA transcripts. It was envisaged that splicing factors would bind to a splice site and then process along the RNA molecule until the next splice site was encountered and recognized, either by primary sequence or by factors already bound at the site (224). However, even the proponents recognized that, in their simplest form, such scanning models could not accommodate all the available data (224). In the simple form, the major problem is the by-passing of otherwise functional splice sites in alternatively spliced transcripts. Moreover *cis*-competition tests with duplicated splice sites do not sustain the prediction that the nearest nighboring splice sites will always be joined (133, 140, 194). Finally, the requirement for a continuous RNA backbone is contradicted by the phenomena of *trans*-splicing. Two RNA molecules, one with a 5' splice site and one with a 3' splice site and branch point, can splice in vitro, provided the two molecules have a region of complementarity within the intron (120, 234). *Trans*-splicing has now also been demonstrated in chloroplasts (118), *Caenorhabditis Elegans* (132) and trypanosomes (163, 246), and probably proceeds via a mechanism very similar in most respects to that of conventional *cis*-splicing. Clearly then, simple RNA backbone scanning is not required for splice-site pairing.

Despite the inadequacies of simple scanning models, *cis*-competition assays show that when the duplicated splice sites have equivalent adjacent exon sequences, the proximal pair of sites is usually joined (133, 194). Thus, spatial proximity does promote splice-site pairing. This tendency may also have a temporal component. In vivo, splicing factors can bind to the nascent RNA transcript, and committed complexes between pairs of splice sites can be formed, thus limiting the available options for splice-site pairing to a 'first come, first served' principle (3, 18). However, in the immunoglobulin κ light chain J region where this principle has been tested, the exclusive use of the 5' most donor site does not require cotranscriptional commitment (112) (Figure 3).

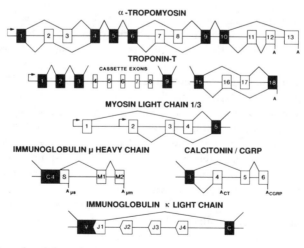

Figure 3 Examples of alternatively spliced genes. Constitutive exons are shown as black boxes, and alternative exons as white boxes. Poly(A) sites are marked by an A. Exons are numbered from the 5' end for all but the immunoglobulin exons, which are denoted C-constant region, S-secreted 3' end, M-membrane bound 3' end, V-variable region, J-joining region. The J segments have a splice donor site but no acceptor.

SPLICE-SITE SEQUENCE CONTEXT   Several studies have demonstrated that sequences outside the consensus splice-site elements can affect the efficiency of splice-site selection. The nature and extent of adjacent exon sequences affects the competitive efficiency of duplicated β globin 5' and 3' splice sites (194). A minimal exon sequence is essential for efficient splice-site utilisation, and non-exon sequences cannot be substituted. Utilization of 5' splice sites inserted into a β globin gene depends upon the position of the inserted splice site (166), its degree of match to the consensus (156, 166) and the ionic conditions of the in vitro reaction (156). Splice sites inserted into the exon are more efficiently used than those in the intron (166). Thus, the 'context' provided by adjacent sequences can affect splice-site utilization, although the basis for this effect is not clear. It may represent an as yet unidentified relaxed sequence specificity. RNAse-protection studies show that U1 and U2 snRNPs bind to regions far more extensive than their consensus binding elements (19, 43, 205), presumably via protein-RNA interactions. Therefore the adjacent sequences could feasibly affect the binding of splicing factors to the consensus elements.

SPLICE-SITE COMPATIBILITY   Factors such as the nature of the consensus splice-site elements and its sequence context determine the 'strength' of the splice site. Those with high match to the consensus and a favorable adjacent exon environment are strongly competitive in *cis*-competition assays. However, simple competition between splice sites, based upon a hierarchy of splice-site strengths, cannot explain all examples of alternative splicing. Rather, it appears that certain splice sites may have a high compatibility, irrespective of their intrinsic strength. For instance, selection between the mutually exclusive exons 3 and 4 of myosin light chain 1/3 (MLC1/3) in transfected constructs depends upon the upstream constitutive 5' splice site (M. E. Gallego & B. Nadal-Ginard, unpublished observations; 183, 241). Although exon 4 is usually spliced into mRNAs with the exclusion of exon 3, when exon 2 is upstream, exon 3 is spliced in. No other substituted upstream 5' splice site shows such a preferential selection of exon 3. A similar effect has been observed when the common 3' site/branch-point region of the adenovirus E1a transcription unit is substituted with a β globin branch point/3' splice site. This led to a switch from splicing predominantly to the 13S splice site to splicing mainly to the competing 9S splice site that lies further upstream (256). Similarly, changes in the composition of the polypyrimidine tract at the SV40 T and t common acceptor site affect discrimination between splicing to the alternative T and t donor sites (70). Increases in the pyrimidine content of this region increase the ratio of T:t splicing. In contrast, increasing the purine content reduces the overall efficiency of splicing, but at the same time decreases the T:t ratio (70).

Such results indicate that individual splice sites can discriminate specifically among an array of alternative competing sites, giving preferred combinations of sites. The biochemical basis for this effect is not clear. It could reflect a specific higher-affinity interaction between certain pairs of splice sites, or could arise from the differential binding of splice factor variants to donor and branch point-acceptor sites, from differences in the structure of factors upon binding to different sites, or directly from differing primary sequences of splice sites. Any of these effects could potentially give rise to preferred interactions between particular pairs of splice sites. It will be of interest to determine the generality of such preferred pairing of splice sites as it could potentially explain many alternative splicing phenomena.

Trans *Requirements*

Splicing of nuclear pre-mRNA transcripts cannot proceed in the absence of a large number of *trans*-acting factors that assemble with the transcript into a large splicing complex. Cases of alternative splicing, where identical pre-mRNA transcripts give rise to different mRNAs in a cell-type or developmental stage-specific manner, clearly demonstrate that the splicing environment of different cells varies for these transcripts (31, 259; and see below). Such variability in the splicing environment is likely determined by specific alternative splicing factors, whose nature in most cases, is by no means clear. However, the involvement of multiple RNA and protein factors in constitutive splicing, many of them present in variant forms, points to the possibility for regulation of alternative splice-site selection.

snRNAs   By a combination of antibody inhibition and RNase H mediated degradation experiments, U1 and U2, U4, U5 and U6 snRNAs have all been shown to be essential for splicing (16, 23, 24, 42, 130, 131, 160, 177). U1 and U2 snRNPs interact with the 5' splice site (131, 160) and branch point regions (23) respectively. Both interactions involve base pairing between the snRNA and the consensus *cis* sequence (267, 180, A. M. Weiner, personal communication, J. L. Manley, personal communication). In both cases, the base pairing, though limited in extent, is a determinant of efficient utilization, and is probably stabilized by protein-RNA and protein-protein interactions. U5 snRNP has been proposed to bind to the 3' splice site (42), though RNase-protection experiments suggest that, like the U4/U6 snRNP complex, it does not bind directly to the pre-mRNA (16, 24).

The genes for the human U snRNAs have been extensively investigated. The copy number for these genes ranges from 40 to 2000 per haploid genome (56). This discrepancy is possibly due to the variable number of pseudogenes detected for these RNAs. However, recent evidence suggests that not all of these so-called pseudogenes are inactive (151, 181). Heterogeneity of snRNAs may well be the rule rather than the exception. For example, while most human U1 RNAs consist of a single species, it has been recently shown that up to 15% of the total cellular U1 RNAs may be composed of variant sequences (151). In addition, in mouse and *Xenopus*, multiple U1 RNAs have been detected and appear to be developmentally regulated (69, 109, 152, 153). Likewise, U4 RNAs are developmentally and tissue-specifically regulated in chicken and *Xenopus* (123, 152). Sequence heterogeneity of U2 during development has not been detected but it is thought that posttranscriptional base modifications may be developmentally regulated and lead to heterogeneous U2 snRNPs (152). As for U5, little is known about possible developmental or tissue-specific regulation, although this RNA is represented by two major species (191). Heterogeneity of U6 RNA has not been described.

Heterogeneity of U RNAs could allow for differential splice-site recognition or compatibility. The very limited changes in U1 and U2 that can improve splicing efficiency in compensatory mutation experiments suggest that small changes in U1 or U2 populations can dramatically alter splicing. Even though none of the reported U RNA variants contains sequence heterogeneity in the regions that base pair with the pre-mRNA, this does not rule out their existence. Additionally, sequence variation at other sites in the RNA may also create functionally distinct snRNP complexes by altering either secondary structure of the RNA or the protein composition of the snRNP, or both. The U2 modificaton data (152) suggests that heterogeneity can also be generated without sequence variation. Functionally distinct, tissue-specific or developmentally regulated U snRNPs could allow cells to fine tune the existing splicing machinery and facilitate differential splice-site utilization.

snRNP PROTEINS   The family of small nuclear U RNAs do not exist as naked RNAs, but are found complexed with 8–10 different proteins (36, 96, 185). These include a core group of proteins reactive with autoantisera of the Sm class, as well as proteins specific to U1 and U2 snRNPs. cDNA clones have been generated for several of these proteins (89, 201, 229, 238, 239, 250, 263), and genomic clones have been isolated for the 70 kDa U1-specific protein and the Sm core E protein (237, 240). These genes contain introns and the 70 kDa protein apparently undergoes alternative splicing. The fact that proteins likely required for splicing contain introns themselves suggests possible feed-back loops in splicing regulation. Additional work has also suggested that the 70 kDa and E proteins are phosphoproteins (261). Thus, differentially spliced and modified snRNP protein components, as well as U RNA-sequence heterogeneity, may allow for functionally distinct snRNPs

OTHER PROTEINS   In addition to the snRNP proteins, other protein factors apparently required for splicing have been identified. Three hnRNP proteins (A1, C, and D) have been found to bind to the 3' end of introns and antibodies against the C protein inhibit splicing in vitro (45, 247). In addition, mutations in the 3' splice-site AG inhibit binding of these proteins, suggesting that they have a role in the 3' splice-site definition. Two additional unrelated proteins also bind to the 3' end of introns. The Intron Binding Protein (IBP or RBP) is a 70 kDa or 100 kDa protein thought to bind to the 3' splice site (79, 249). This protein is Sm-antibody reactive and has been proposed to be associated with U5, though the evidence is indirect. The second protein is U2AF, which is necessary for U2 snRNP binding to the branch-point region, and seems to bind to the polypyrimidine tract (209).

SPLICEOSOMES   All the *trans*-acting factors and pre-mRNA discussed above associate to form a large complex termed the spliceosome. This is a 60S or 40S particle in mammals and yeast respectively (73, 68, 86). The spliceosome minimally consists of the pre-mRNA with U1 snRNP bound to the 5' splice site, U2 snRNP bound to the branch-point region, perhaps U5 snRNP bound to the 3' bound splice site, and the U4/U6 snRNP associating via protein-protein interaction. The protein factors mentioned above are additional factors of this complex, as are other uncharacterized proteins (195). An assembly pathway for spliceosome formation has been proposed (20, 121, 122, 138, 187, 203) and is shown in Figure 1.

Glycerol and sucrose gradient data have suggested that spliceosome formation proceeds through the formation of at least three complexes (20, 68, 85). Initially, a 20–25S complex is formed consisting of the pre-mRNA bound by multiple RNA-binding proteins including the hnRNP proteins, cap-binding proteins, and perhaps IBP and U2AF (Figure 1). U1 snRNP then binds to the 5' splice site in an ATP-independent manner, most likely generating the 40S complex. Finally, a 60S complex is generated consisting of all the major known components of the mature spliceosome. Recent data suggest that the 60S spliceosome peak is composed of at least two distinct complexes (1, 195). Thus the scheme shown in Figure 1 should be considered a minimal pathway. Slightly different spliceosome assembly pathways have been proposed, based upon gel electrophoretic data (122, 138, 139).

Spliceosome assembly proceeds in multiple stages and results in the pairing of splice sites. At some stage, these pairs of sites become irreversibly committed to one another, such that they no longer have the option of interacting with any other splice sites. Commitment could occur as a consequence of the first interaction between factors stably bound at the 5' and 3' splice sites or branch points. It has been suggested that U1 snRNP may be the major early determinant of interaction between splice sites (143, 203, 222, 269). Alternatively, commitment could be the result of a conformational change within a 60S complex. The location of the commitment step in the splicing pathway, and the factors involved, are relevant to understanding both constitutive and alternative splicing, since it represents the final stage of splice-site selection. In alternatively spliced transcripts, the formation of different committed complexes could be determined by the more rapid binding of factors to some splice sites, possibly co-transcriptionally. Alternatively, higher-affinity interaction between factors bound at certain pairs of splice sites could give rise to specific committed complexes.

MECHANISMS OF ALTERNATIVE SPLICING

Default and Regulated Splicing Patterns; Cis- and Trans-Regulation

Even before any direct experimental investigations had been made, some conclusions could be drawn about the nature of the mechanisms of alternative splicing based upon the phenotype of alternatively spliced genes (10, 29). In cases of alternative splicing in which a gene is transcribed in different cell types, a nonspecific or default splicing pattern is found in most cells, while the regulated splicing pattern is found in a much more restricted set of cell types. For instance, the rat α tropomyosin (αTM) gene generates a large number of mRNAs via the use of two internal pairs of mutually exclusive exons, and a set of alternative 3' end exons with associated poly(A) addition sites (204, 259; Figure 3). Of these, exon 3 is used in all cells except smooth muscle where exon 2 is incorporated. Likewise, exon 8 is used in all but some transformed cells where exon 7 is used instead. Finally, exon 13 is used in all cells except for differentiated myogenic cells where exons 11 and 12 are incorporated. Exons 3, 8, and 13 are therefore the default choices, while exons 2, 7, and 11 and 12 each require different cell-specific environments for their incorporation (259). For genes that are not expressed in a wide variety of cell types, the default and regulated patterns can be distinguished by transfection of constructs into cell cultures or by the use of transgenic animals. Calcitonin was thus established as the default product of the calcitonin/CGRP gene in transgenic mice, and the production of CGRP occurs in neural tissue (54). Likewise, the regulated inclusion of some of the cassette exons of the rate troponin-T gene requires the environment of differentiated myogenic cells; expression of minigene constructs in nonmyogenic cells or undifferentiated myoblasts leads to the default exclusion of exons 4–8 (31). These observations imply that the default pattern will be produced by the general splicing machinery, while some other cell-specific *trans* contribution is needed for production of the regulated pattern. In cases of tissue or developmental stage-specific differential splicing of transcripts with the same 5' and 3' ends, such as the rat troponin-T and α-TM genes, *trans* regulation of the splicing patterns has to be invoked (31, 259). In the case of α-TM, the switching of splicing patterns within the three mutually exclusive groups of exons occurs in three different cell environments, suggesting that different *trans* factors are involved in regulating the splicing pattern in the different areas of the gene (259).

In contrast, alternative splicing of transcripts that have different 5' or 3' ends arising through the use of different promoters of poly(A) addition sites, or through DNA rearrangement, could be regulated entirely in *cis,* through the different structures of splicing substrates. In these cases, each specific transcript uses only one of various possible splicing pathways irrespective of the cell environment where it is expressed. This seems to be the case with the myosin light chain 1/3 gene (M. E. Gallego & B. Nadal-Ginard, unpublished observations; 76), in which the splicing pathway depends upon differential promoter activation. Similarly, splicing in the immunoglobulin κ light chain J region occurs only to the 5' splice site of the J segment that has been joined to the variable region by DNA rearrangement (111, 150).

Mutually Exclusive Splicing

Mutually exclusive selection occurs between pairs of exons at the 5' and 3' ends of transcripts, and between internal pairs of exons (Figures 2 and 3). In each case, an understanding of the alternative splicing event requires first an explanation for the mutually exclusive behavior and the lack of transcripts containing both exons, and second for the basis of exon selection in both default and regulated pathways.

There are obvious structural reasons for the mutually exclusive behavior of 5' and 3' end exons associated with their own promoters or poly(A) addition sites. The sequence separating a pair of exons, each with its own promoter, is not flanked by a pair of functional splice sites, but rather has a promoter to the 3' side. Likewise, 3' end exons are separated by a sequence with a 3' splice site at one end but a poly(A) addition site at the 5' end, and so cannot be spliced together. The basis for mutually exclusive splicing of internal pairs of exons is not so immediately apparent. Here, the intron between the exon pair is bounded by a pair of functional splice sites and, in all known cases, is above the minimal intron length required for splicing. Nevertheless, in some cases at least, simple structural constraints underlie this behavior. The mutually exclusive exons 2 and 3 of the rat α-TM gene have a 218 nt intron between them and yet they never splice to each other, even in transcripts containing no other splice sites, due to the unconventional location of the lariat branch point, positioned 177 nt upstream from the 3' splice site. The branch point is only 42 nt downstream of the exon 2 donor site, which is too

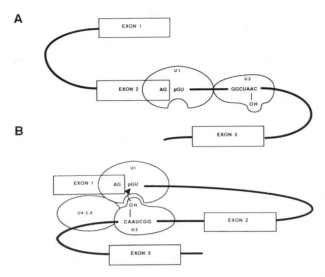

Figure 4 Enforcement of mutually exclusive splicing of α tropomyosin exons 2 and 3 by steric interference (231, 232). Exons are shown as boxes, introns as lines. In 'A' exons 2 and 3 are unable to splice as the factors bound to the exon 3 branch point and exon 2 donor site are too close to interact productively. In 'B' a productive spliceosome is formed between the branch point of exon 3 and the upstream exon 1 donor site.

close for splicing factors to be able to bind to each element and interact to form a productive spliceosome (232). As a result, exon 3 can splice to exon 1, but not to exon 2, and exon 2 can splice to exon 4, but not to exon 3 (Figure 4).

Such a strict mechanism for maintaining mutually exclusive splicing does not appear to be general. The mutually exclusive exons 3 and 4 of myosin light chain 1/3 (MLC1/3), and exons 7 and 8 of α-TM are mutually exclusive, and yet in single intron constructs both pairs are able to splice together (M. E. Gallego, D. Sengupta & B. Nadal-Ginard, unpublished observations). The branch points between exons 6 and 7 of rat β-TM are >140 nt from the 3' splice site, which is not close enough to the 5' splice site to absolutely prevent splicing (93). Nevertheless, the same general mechanism may underlie mutually exclusive behavior in these systems. An absolute block to splicing may not be necessary. A relatively low splicing efficiency between the adjacent splice sites, such that each one interacts more favorably with a distant splice site, could serve to enforce mutually exclusive splicing. α-TM exons 2 and 3 would then represent an extreme example of a more general mechanism.

The basis of exon selection has not yet been as well characterized. In transfected constructs containing α-TM exons 2 and 3, MLC1/3 exons 3 and 4 and troponin-T exons 16 and 17, mutually exclusive exon selection does not require extensive regions of the transcript, but rather is determined by sequences in the exons and their immediately flanking introns and exons (M. E. Gallego, B. Nadal-Ginard, unpublished observations; 231). Deletion of the exon that is usually selected allows incorporation of the normally excluded exon (exon 2 for α-TM and exon 3 for MLC1/3), suggesting that the splicing machinery can splice these exons, but that in wild-type transcripts they are out-competed by the default exons. Deletion of troponin-T exon 17 does not lead to inclusion of exon 16, however, suggesting that this exon is not available to the splicing machinery irrespective of competing exons (A. Andreadis & B. Nadal-Ginard, unpublished observations). In the case of MLC1/3 exons 3 and 4, substitution of the flanking exons can switch the selection from exon 4 to 3, or even in some cases to transcripts containing neither or both exons (M. E. Gallego & B. Nadal-Ginard, unpublished observations). While most 5' exons splice to exon 4, MLC1/3 exon 2 splices exclusively to exon 3, as it does in the native gene. Moreover, reversing the positions of the two exons in constructs that normally splice in exon 4 does not perturb exon selection. The results from the MLC1/3 gene suggest that interaction between some pairs of splice sites, such as MLC1/3 exons 2 and 3, may be specifically favored over other potential combinations of splice sites, even when the competing sites normally appear to be stronger. This may represent a very finely tuned form of *cis*-regulated alternative splicing, well-suited to a gene such as MLC1/3, where splicing is dictated in *cis* by differential promoter activation that gives rise to transcripts with different available 5' splice sites.

The mutually exclusive splicing of the rat β-TM exons 6 and 7 involves an ordered pathway in which neither the default exon 6 nor the skeletal-muscle specific exon 7 can be spliced to the upstream constitutive exon 5, until splicing to the downstream consitutive exon 8 has occurred (94). Thus, the splice to exon 8 determines exon selection. Exons 6 and 7 are both 76 nt long, so the requirement for prior splicing to the 3' side involves a relatively long-range interaction, and cannot be explained in terms of steric interference between splicing factors bound on either side of the exon (2). Such ordered pathways do not appear to be general for all mutually exclusive exons. Mutually exclusive exons 2, 3, 7 and 8 of α-TM can all splice to their respective upstream constitutive exon in vitro and with no requirement for splicing to the 3' constitutive exon (M. Mullen, D. Sengupta & B. Nadal-Ginard, unpublished observations)

Alternative 3' End Exons; Competition Between Splicing and 3' End Formation

Processing of RNAs with alternative 3' end exons with associated poly(A) sites has been extensively investigated for the immunoglobulin μ heavy chain (Igμ) and calcitonin/calcitonin gene-related peptide (CT/CGRP) genes. In both cases there is a choice between cleavage and poly(A) addition at an internal site, or splicing to a downstream exon (Figure 3). Regulated mutually exclusive 3' end exon selection involves a delicate balance between various processing events, including transcription termination, cleavage/poly(A) addition and splicing.

In $Ig\mu$ processing, the default pattern, as determined by nonlymphoid cells and *Xenopus* oocytes (254, 255), involves the use of the internal 3' end to produce the secreted $Ig\mu_s$ RNA. Only in pre-B lymphocytes is membrane-bound $Ig\mu_m$ produced, via splicing to the pair of downstream exons that code for the hydrophobic membrane-binding domain. The $Ig\mu_s$ RNAs arise in some cells as a result of transcription termination prior to the external poly(A) adenylation site (74, 115). In such prematurely terminated transcripts the only available processing pathway is cleavage and polyadenylation at the internal $Ig\mu_s$ site. However, in most $Ig\mu_s$-producing cells, transcription proceeds past the downstream $Ig\mu_m$ poly(A) site and alternative processing to $Ig\mu_s$ or $Ig\mu_m$ occurs from the same primary transcript (57, 115, 196, 199). In contrast to use of the internal poly(A) site, which enforces the default $Ig\mu_s$-processing pattern, cleavage and polyadenylation at the external site does not commit the transcript to either pathway (184). The fact that the internal poly(A) addition site is intrinsically less efficient than the external one (74, 184) may reflect its more central role in regulating the processing pathway, as a reaction competing with the splice to the downstream $Ig\mu_m$ exons. The regulated processing pattern, involving the splice to the $Ig\mu_m$ exons, only occurs to a significant extent in pre-B cells, apparently as a result of activating the splice to the $Ig\mu_m$ exon, rather than to down-regulating the competing polyadenylation at the $Ig\mu_s$ site, although this point remains a matter of controversy (75). Consistent with regulation of the $Ig\mu_m$ splice, mutational inactivation of the $Ig\mu_m$-splice donor has no effect upon the efficient use of the internal $Ig\mu_s$ poly(A) site in any cells (254), whereas its conversion to the donor consensus results in splicing to the $Ig\mu_m$ exons in cells that usually produce $Ig\mu_s$ (184). Mutational inactivation of the internal $Ig\mu_s$ poly(A) site, however, does not allow for splicing to the downstream exons in appropriate cell types, indicating that regulation of poly(A) site selection is not responsible for the switch in splicing patterns (254). Similar results were obtained by coinjection of $Ig\mu$ transcripts and mature B cell nuclei into *Xenopus* oocytes (255). These data strongly suggest that the nuclei of pre-B cells contain *trans*-acting factors that specifically stimulate splicing to the downstream $Ig\mu_m$ exon. In the absence of such factors, transcripts are cleaved and polyadenylated in the internal $Ig\mu_s$ site.

The regulation of processing in the calcitonin/CGRP gene appears to be quite similar. Use of the internal calcitonin poly(A) site is the default-processing pattern, except in neuronal cells where the downstream CGRP exons are used (54). Again, alternative processing occurs from a primary transcript containing both poly(A) addition sites (7, 26). Mutational inactivation of the internal calcitonin poly(A) site does not allow production of CGRP, except in neuronal cells that usually produce it (144). This suggests that, as with Igμ processing, the activation of the splice to the downstream exons causes the switch to the nondefault production of CGRP in neuronal cells.

Alternative 5' Splice Sites

In at least two genes there have been studies of alternative splicing of transcripts that contain multiple 5' splice sites competing for a common acceptor site. The alternative splicing of the SV40 T and t antigen cells 5' splice sites to their common acceptor has been intensively characterized and involves a balance between splice-site strength, proximity, and steric constraints. In most cell types the upstream T 5' splice site is selected. The major factor rendering the competing t splice inefficient is the minimal distance between the t 5' splice site and branch point that, at 48 nt, is at the lower limit to allow splicing. Deletion of only 2 nt between these elements eliminates t splicing, while insertion of spacer elements decreases the T:t ratio (69, 71). The proximity to the branch point of the t 5' splice site may be compensated somewhat by its sequence, which is closer to the consensus than the T 5' site. Mutation of the latter to the consensus further increases the T:t ratio. Splicing of SV40 large T involves the use of 6–9 branch sites at the common acceptor site (168, 169). In contrast, small t splicing uses only the two 3' most of these sites, which are the only ones distant enough from the small t donor site. It has been suggested that the multiple branch sites provide a region of high affinity for U2 snRNP that may facilitate splicing to small t even though this splice can only take place using the most downstream branch points (168, 169). The correlation between branch-point usage and 5' splice-site selection in this case is sterically enforced by the proximity of the small t donor site. Mutation of the major T branch points, leaving intact only the common T and t branch point, greatly decreases the T:t ratio in HeLa cells and suggests that the availability of multiple competing branch points favors the T splice. In the transformed 293 cell line, the ratio of T:t is greatly decreased compared to HeLa cells (69, 71). As yet the *trans*-acting factors responsible for this cell-specific switch have not been determined, though obvious possibilities consistent with the experimental data include negative regulators acting at the major T branch points or donor site, or activators of the t donor site.

Alternative selection of the immunoglobulin κ light chain J region 5' sites has also been studied in some detail, but the factors governing the strict splice-site selection have not been determined. In vivo, only the donor site of the 5' most J segment, to which the V region has been joined, is spliced to a common downstream acceptor (Figure 3). In vitro, all the J segment 5' splice sites are used with equal efficiency, using either HeLa or lymphocyte nuclear extracts (111, 112, 149, 150), though selection between sites can be altered by intron deletions and mutations of the donor sites. Nevertheless, in vivo the 5' most site is always selected, even in transcripts which in vitro favored selection of a 3' proximal donor. This selection of the 5' most site in vivo cannot be explained by a whole range of parameters that have been tested; it is not lymphoid-cell specific, does not depend upon the context provided by the adjacent variable region sequence, is not related to a hierarchy of splice-site strengths, is not apparently an artefactual result of selective degradation of transcripts utilizing downstream donor sites, nor an effect of co-transcriptional processing or commitment, since injection of transcripts into *Xenopus* oocytes still leads to selection of the 5' most site (112, 149). Selection of the 5' most site could be explained by proximity to the 5' cap structure. The components responsible for such a discriminating activity are presumably absent from the HeLa nuclear extract.

Nonsplice-Site Elements

The involvement of intronic elements other than the splice sites and branch point has been implicated in regulating splicing, most notably in viral RNAs. Many viruses produce primary transcripts used in both spliced and unspliced forms to produce different functional products. In these cases it is important that the initial transcript is not completely spliced, even though it is being processed by a standard eukaryotic splicing machinery and has correct functional splice sites. For instance, in influenza virus NS1 (136) and Rous Sarcoma virus gag transcription units (11) only 5–10% of the RNA is spliced. A negative regulatory sequence within the intron causes RSV splicing to be inefficient. Deletion or replacement of the element increases splicing to 70%, while its insertion into a heterologous intron severely inhibits splicing (11). The location of the inhibitory element in the influenza virus NS1 has not been determined though it is to the 3' side of the donor site, is not the 3' splice site or branch point, and apparently blocks splicing subsequent to the formation of a 60S spliceosome (5, 189). The means by which such elements inhibit splicing is not known.

Exon sequences have been identified that promote or inhibit splicing. Splicing of the cassette EDIIIA exon of human fibronectin is dependent upon exon sequences contained within an 81-bp fragment; deletion or substitution of the fragment results in constitutive skipping of the exon (155). The leukocyte common antigen gene has three cassette exons that are included in B cells but spliced out in thymocytes and other transfected cells (244). Linker scanning mutations within the cassette exon 4 show that at least 3 regions of the exon are required for its default exclusion. Mutation within these regions leads to the constitutive inclusion of the exon (242). Thus, cell-specific splicing of the transcript appears to require a B cell-specific factor to overcome the negative effects of the exon sequences. Sequences within the final exon of the bovine growth hormone gene are required for splicing of the preceding intron (90). This intron can be retained in pituitary cells and may generate an alternative form of growth hormone. Deletion of sequences within the exon results in an accumulation of RNA with the final intron retained. Splicing could be partially restored by substitution of a 10 nt palindromic sequence from the center of the deleted region.

Secondary Structure

Experiments with artificial constructs have suggested a potential role for RNA secondary structure in alternative splicing. Sequestration in vitro of an exon into the loop of a hairpin structure causes the production of some RNA in which the sequestered exon is skipped (233). However, in vivo this effect is minimal and requires perfectly base-paired stems of a length (>50nt) not found in natural mRNAs (235), leading to the suggestion that cellular factors would probably be required to stabilize potential structures in vivo. Skipping of splice donor sites sequestered within a potential stem structure occurs in vitro and in vivo, if the loop size is small (63). With larger loop sizes, the stems apparently fail to form in vivo and the splice site can be used efficiently (63). The relevance of such artifical constructs to real alternatively spliced transcripts is somewhat uncertain. Based upon genomic sequence analysis alone, sequestratioh of exons into stem and loop structures has been suggested to play a role in the mutually exclusive splicing of α-TM exons 2 and 3 (204), troponin-T exons 16 and 17 (31, 32), and the cassette exon 5 of chicken cardiac troponin-T (52). In all three cases, however, deletion of such elements has no effect upon splicing (51; B. Nadal-Ginard, C. W. J. Smith & A. Andreadis, unpublished observations). Sequence analysis of some mutually exclusive exon pairs, including exons 6 and 7 of β-TM, has revealed the potential for extensive secondary structures that are apparently conserved between the equivalent exon pairs in different species (64). As yet, however, we know of no direct evidence for the involvement of secondary structures in alternative splice site selection.

Trans *Regulation of Splicing Patterns*

As discussed above, the observation that identical primary transcripts are processed along alternative splicing pathways, according to cell type or stage of development, clearly demonstrates differences in the *trans*-splicing environments of these cells (30–32, 54, 69, 70, 144, 158, 254, 255, 259). With few exceptions, however, the nature of this difference in the splicing environment is not clear. It could reflect the presence of specific alternative splicing factors, such as the Drosophila Sex-lethal and suppressor of white apricot gene products (15, 46, 264), in cells expressing the regulated rather than the default splicing pathway. Alternatively, more subtle effects such as variations in the activities or levels of constitutive splicing factors could be the basis for the regulated alternative splicing patterns. The nature and mode of action of alternative splicing factors will depend upon, and to some extent can be predicted by, the nature of the *cis* elements that dictate default modes of splicing. Such factors could act directly at the splice sites or at other intron or exon *cis* elements, and could act in a positive or negative manner. Variants of the constitutive splicing factors (see above) would be expected to directly activate specific donor or branch point acceptor-sites. Thus, the putative pre-B cell-specific factor necessary for $Ig\mu_m$ production (184, 255) is likely to directly activate the $Ig\mu_m$ splice donor site, and so could be a variant component of the U1 snRNP. Important *cis* elements necessary for *trans* regulation, which are distant from the splice sites (90, 155, 242), are more likely to interact with novel factors rather than the constitutive splicing machinery. Such non-splice site factors could act by masking exons from the splicing machinery, or by stabilizing or destablizing secondary structures that limit the accessibility of the splicing machinery to certain splice sites.

The only specific alternative splicing factors thus far identified are the products of the Drosophila *Sex-lethal* (15), *transformer* (25), *transformer-2* (9) and *suppressor of white apricot* genes (46, 264). These genes were originally identified genetically by their regulatory effects upon downstream genes, and by the auto-regulation of *Sex-lethal* and *suppressor of white apricot* expression. They were only later characterized as alternative splicing factors (15, 46, 48, 264). Such fortuitous identification by genetic approaches is not likely in mammalian systems. The isolation and characterization of mammalian alternative splicing factors will require other strategies. The direct biochemical approach requires the development of in vitro splicing systems able to accurately reproduce cell-specific splicing patterns. This should be possible either by using nuclear extracts from different cell types, or by complementing HeLa nuclear extracts, but to date no such system has been developed. Moreover, even the in vivo default patterns of splice-site selection are not always faithfully reproduced in vitro (94, 111, 149). The only in vitro manipulations which have been shown to affect alternative splice-site selection are alterations in the ionic conditions, (94, 156, 217). However, such changes in K^+ and Mg^{2+} concentrations are unlikely to be the basis for switching of splicing patterns in vivo. Other strategies for the isolation of factors include coinjection of nuclei or nuclear extracts into *Xenopus* oocytes, along with the experimental transcript (112, 255). This would also allow purification of factors, the assay for activity being carried out in vivo rather than in vitro. An alternative to these biochemical approaches is to link an alternatively spliced locus to a drug-resistant selection marker, such that only expression of the regulated splicing pathway leads to drug resistance. Cotransfection with genomic or cDNA libraries would then allow the cloning of the regulatory factor. With one or more of these approaches, other alternative splicing factors will eventually be identified and their mode of action subsequently characterized. This remains one of the major goals in the study of alternative splicing mechanisms.

FUNCTIONAL CORRELATIONS OF ALTERNATIVE SPLICING

In the past few years the catalog of alternatively spliced genes has expanded enormously. The functional effects of alternative splicing often remain obscure, but increasingly correlations can be made between alternative splicing, molecular interactions and biological function. These cases have highlighted the importance of alternative splicing as a posttranscriptional level of gene regulation at which, like the transcriptional level, expression can be controlled both quantitatively and qualitatively. While transcriptional activity can determine levels of gene expression in a tissue-specific or developmentally regulated manner, alternative splicing can also introduce functional diversity into the products of a single gene. Moreover, in a few exceptional instances, alternative splicing has usurped the role of transcriptional regulation as the level at which gene expression is switched on and off, being used as the master switch for the expression of constitutively transcribed genes.

The functional effects of alternative splicing can be classified in to: (a) protein localization, (b) deletion of protein activity, (c) modification of protein activity (d) novel protein activities, and (e) RNA stability and translational efficiency. These categories are not mutually exclusive, but do provide a framework for discussion.

(a) Protein Localization

Regulated alternative RNA processing is often used to specify the localization of proteins. There are a number of well-characterized examples of proteins which are either membrane-bound or secreted. For instance, the immunoglobulin μ (Igμ) heavy chain gene is present in early B lymphocytes as a membrane-bound form, Igμ_m. Upon B-cell maturation, following antigen stimulation, Igμ_m levels are progressively decreased with concommitant elevation in the production of the secreted Igμ_s pentamer (6, 60, 198). This switch is achieved by the alternative use of 3' end exons, which code for the hydrophobic membrane-binding segment (discussed above; Figure 3).

Neural cell adhesion molecules (NCAMs) also ultilize alternative splicing to generate various membrane-bound and secreted forms. NCAMs belong to a family of glycoproteins involved in cell-cell interactions, and are expressed in the nervous system and skeletal muscle cells (55). The three major NCAM isoforms differ in their membrane binding or cytoplasmic domains but the extracellular domains are largely identical (55). The two transmembrane isoforms differ in the size of the cytoplasmic domain and a third isoform contains no transmembrane region but is anchored to the plasma membrane through a glycosyl-phosphatidylinositol linkage (55). A fourth secreted isoform is generated by the inclusion of a cassette exon containing a stop codon in the extracellular domain; a truncated NCAM is generated without the hydrophobic membrane-binding domain, and is therefore secreted (84). Additional tissue-specific and developmentally regulated variations within the extracellular, immunoglobulin-like domains have also been reported in brain and muscle NCAMS (59, 214, 230). Thus, the NCAM gene generates numerous functional products that vary both in their localization and potential for intra- and intercellular interactions.

Fibronectin is another cell-adhesion molecule with various binding activities; it exists in both insoluble and soluble forms present in the extracellular matrix and plasma, respectively (99). This diversity is generated via alternative splicing of cassette exons, alternative donors and acceptors, and retained introns (124, 220, 221). Incorporation of the ED type III cassette in fibroblasts produces cellular fibronectin, whereas its exclusion in liver generates plasma fibronectin (124, 125). Alternative splicing within the IIIcs region correlates with a melanoma-adhesion activity. Isoforms containing the REDV peptide encoded by the sequence between two alternative 3' splice sites posssess this activity (97, 125, 200, 220).

The gene for decay-accelerating factor (DAF) also gives rise to membrane-bound and secreted forms by the alternative excision or retention of the 3' most intron (39). DAF is a phosphatidylinositol-linked membrane glycoprotein that protects cells from the complement cascade system by binding activated complement fragments C3b and C4b. A soluble form of DAF is also found in body fluids, although its functon is presently unknown. The shorter, spliced isoform has a short hydrophobic C-terminal segment that makes the linkage to the phosphatidylinositol in the membrane-bound form. In the longer mRNA the open reading frame continues through the intron into the final exon and introduces a frameshift that results in a longer hydrophilic C-terminus giving rise to a secreted form of DAF (39). The gene for the platelet-derived growth factor (PDGF) A-chain also has a final intron that can either be excised or retained (50, 252). Retention of the intron in transformed cells produces a sequence with 15, predominantly acidic, additional amino acids before a termination codon is reached within the intron. This isoform dimerizes, is secreted and has mitogenic activity. In contrast, the product of the spliced RNA does not produce any secreted activity, presumably because it is unable to form homodimers or be secreted, requiring coexpression of the PDGF B-chain to become biologically active.

The actin filament-severing protein gelsolin exists as both a plasma and a cytoplasmic protein. The two mature proteins are identical except for an additional 25 N-terminal amino acids in the plasma isoform. The two isoforms are expressed from the same gene by the differential use of promoters and alternative splicing of 5' exons that code for the extra 25 aa of the plasma portein and a 27aa signal peptide that is cleaved from the mature plasma protein (134). No functional differences have been detected between the intra- and extracellular forms of gelsolin.

These examples show how proteins can be variously partitioned between cytoplasm, cell membrane, plasma, and extracellular matrix. Alternative splicing could also regulate the distributions of proteins between other subcellular compartments. The targetting of nuclear-encoded proteins to locations such as the mitochondrion involves the use of leader sequences that are removed from the mature protein during or after transportation. The distribution of proteins common to more than one cell compartment could thus be regulated by alternative splicing of such targetting sequences.

(b) Deletion of Function

Alternative splicing can be used to remove one or more functions from a protein product. Sometimes, only one of the splicing modes produces a functional mRNA. In such cases, the function of alternative splicing is to act as a simple on/off switch, in much the same way as transcription is used to control gene expression. The *Sex-lethal* (15), *transformer* (25), and *suppressor of white apricot* (264) genes of Drosophila all belong to this class. (These examples are discussed in the section on developmental regulation.)

The Drosophila P elements are one of the best-characterized families of eukaryotic mobile elements. Transposition of P elements is dependent upon a transposase function expressed only in germline cells and coded by a 2.9 kb autonomous P element. Germline-specific expression of the transposase activity arises from alternative splicing of the final intron of the transposase gene (142). Splicing out of the intron in germline cells gives rise to an active 87 kD transposase protein. In somatic cells the intron is retained and the resultant 66 kD protein has no sequences corresponding to the final exon and lacks transposase activity, although it still contains a DNA-binding bihelical motif, suggesting some functional role (142, 197).

The adenovirus E1a protein has many distinct activities, including activation of cellular DNA synthesis, transcriptional activation, and a distinct transcriptional repression activity correlating with cell-transforming activity. E1a isoforms encoded by the 13S RNA have all of these functions, but the isoform encoded by the 12S RNA lacks transcriptional activation yet retains the other functions. Transcription activation resides in a discrete 49 amino acid peptide, encoded by the sequence between the 12S and 13S 5' splice sites and incorporated only into mature 13S RNA (146, 147).

(c) Modulation of Function

The activities of protein isoforms are often subtly varied by alternative splicing. Correlations can be made between alternative splicing and the function of such diverse gene products as contractile proteins, glycolytic enzymes, ion channels and transcription factors. In these cases substitution of modular domains within an otherwise constant protein framework can be used to specifically modify sites of interaction with other proteins or ligands.

The alternative splicing of the five combinatorial cassette exons of fast muscle tyroponin-T (TnT) provides a good example where correlations can be made between isoforms arising from tissue-specific splicing, interprotein interactions and physiological function. These five cassette miniexons appear to be incorporated into mRNA independently of one another and could, in conjunction with the mutually exclusive exons 16 and 17 (158), potentially generate up to 64 different protein coding mRNAS (32). In fact, some TnT species are far more abundant than others and the isoforms that are produced in significant amounts in a wide range of adult muscles number between 5 and 10 (35). The N-terminal region of the protein for which these exons code interacts with the region in which tropomyosin dimers overlap to form a continuous regulatory strand along the muscle thin filaments (65). The activation of muscle contraction by the binding of Ca^{2+} to troponin-C and the subsequent movement of tropomyosin on the thin filament is a highly cooperative process, with the whole thin filament being switched on in a concerted manner (27). The degree of co-operativity and the Ca^{2+} concentration required for half-maximal tension development correlates with the combination of tropomyosin dimers, $\alpha\beta$ or $\alpha\alpha$, and the major TnT isoforms present (TnT1f, TnT2f, and TnT3f). In muscles that have the same tropomyosin content but differ only in the TnT isoforms, there is a direct correlation between the TnT isoforms and these parameters (33, 216). The variation between the TnT species is in the N-terminal region; TnT1f contains sequences corresponding to all 5 combinatorial exons while TnT2f lacks the sequences from exon 4 and TnT3f lacks those from exon 6 and 7 (33–35). Thus, the differences in the splicing pattern in this region of the TnT gene have specific effects upon the interaction of TnT with tropomyosin, that in turn gives rise to muscles with distinct mechanical parameters in their activation by Ca^{2+}.

Changes in enzymatic activity can be correlated with alternative splicing of the isoforms of the glycolytic enzyme pyruvate kinase (PK). PK has four isoforms, M1, M2, L, and R, each of which forms homotetramers. The M1 and M2 isozymes arise via substitution of an internal 45 aa segment that is coded for by a pair of mutually exclusive exons (170). This region of M2 is highly homologous to the equivalent regions of the L and R isozymes. The native M1 homotetramer shows Michaelis-Menten kinetics with respect to the substrate phosphoenolypyruvate, whereas the other isoforms are all allosterically regulated by phosphoenolypyruvate, fructose-1, 6-biphosphate, ATP, alanine, and pH, and have sigmoidal kinetics (100). In general, the M2, L and R isoforms are found in tissues where their regulation can be employed to prevent futile cycling of the common intermediates of glycolysis and gluconeogenesis. In contrast, the M1 isoform occurs in muscle where glycolysis predominates. The crystal structure of feline M1 PK has been determined by X-ray diffraction (244) and has been correlated with the amino acid sequence (162). The alternatively spliced exons of M1 and M2 encompass the regions involved in intersubunit contact, suggesting that the nature of the subunit interactions, determined by alternative splicing, dictates the appropriate tissue-specific kinetic and regulatory properties of the individual isozymes (170).

The Drosophila Shaker locus encodes the major structural component of the A type K^+ channel. The Shaker gene generates diverse transcripts via a complex process of alternative splicing, coding for at least nine different proteins (108, 219) that vary within the N and C terminal regions. Injection of two different Shaker mRNAs into Xenopus oocytes demonstrated that the different channels had distinct inactivation kinetics (251). It seems likely that the other K^+ channel isoforms encoded by Shaker will also prove to have distinct functional properties. The pattern of Shaker transcripts varies between head and body (219), so it appears that different tissue-specific electrophysiological properties can be established via alternative splicing.

A number of vertebrate transcription factors have recently been cloned and analysis of their expression by Northern blot has shown that many of them express a number of different size classes of RNAs. In a few instances it has been demonstrated that distinct isoforms are produced by differential RNA splicing. The CCAAT box DNA binding proteins CTF/NF-1, which are activators of both transcription and DNA replication, consist of a family of at least 3 different protein isoforms (215) arising through alternative splicing in at least two regions of the gene. However, initial in vitro analysis of the proteins indicated identical DNA binding, transcriptional activation, and DNA replication properties. In contrast, the rat α-thyroid hormone receptor (rTRα) gene generates a family of proteins with distinct biological properties. The rTRα gene belongs to a superfamily of steroid hormone receptors, the members of which are DNA-binding transcriptional regulators. At least two products of the rTRα gene, rTRα1 and rTRα2, which differ only in their carboxy terminal hormone binding domains, have been shown to have distinct properties (102). Although rTRα1 binds thyroid hormone (T3) and activates transcription of the thyroid hormone-sensitive α-myosin heavy chain gene (α-mhc), rTRα2 neither binds T3 nor activates α-mhc transcription. An extensive survey of potential ligands has revealed no hormone for which rTRα2 may be the receptor. However, rTRα2 has the same DNA-binding domain as rTRα1, binds in vitro to the αmhc promoter thyroid-responsive element (TRE), and in vivo has been shown to suppress activation of the growth hormone TRE by rTRα1 and rTRβ (117), although such an effect was not observed with the α-mhc TRE (S. Izumo, V. Madhavi, personal communication). Thus, rTRα2 may play a role in regulating gene expression either in response to an as yet unidentified hormone, as a negative regulator that can compete with activators such as rTRα1 for the TRE binding site, or as a component of an active or inactive heterodimer with rTRα1 or some other subunit. Alternatively, it may bind to DNA sequences quite distinct from the known TRE elements. Such promiscuous binding of transcriptional factors to apparently unrelated binding sites has been well documented.

The isolation of cDNA clones for factors that bind to the immunoglobulin κ light chain enhancer has demonstrated that two such distinct factors, which vary within the DNA-binding domain, are generated by alternative splicing (164). In vitro translated products of these two clones both bind to the κ enhancer DNA, but with different affinities (164). Thus, alternative splicing is used to vary both the ligand and DNA-binding domains of transcription factors in a modular fashion. A number of vertebrate homeobox-containing proteins have been identified in higher vertebrates, some of which are known transcription factors (95). The gene for the B-cell specific Oct-2 transcription factor generates at least two different mRNAs, which differ by an additional 12 aa C-terminal segment in one form (48). The functional significance of this variation is not clear. However, two closely related Oct-2 proteins have been detected that appear to interact differentially with the immunoglobulin heavy chain promoter and enhancer (218) and since Oct-2 shows heterogeneity of transcript size (48) it is likely that these functionally distinct isoforms could arise through differential RNA processing from the same gene. A homeobox-containing gene from *Xenopus laevis* generates transcripts containing and lacking the DNA-binding homeobox domain (262). The function of the homeoboxless isoform is unclear, though it may possibly be able to interact in a regulatory manner with other DNA-binding proteins.

(d) Products with Different Functions

In a limited number of cases alternative RNA processing gives rise to mature products with totally different functions. Here the alternatively spliced region codes for the major active site of the product. This type of alternative splicing is used in the generation of a number of peptide hormones. The calcitonin (CT) gene produces both CT and 'calcitonin gene-related peptide' (CGRP) by alternative splicing and poly(A) addition site usage (8, 105; Figure 2). Exons 1–3 are constitutive and contain the 5' untranslated and common coding regions. Exon 4 contains the CT coding sequence and a poly(A) addition site and is used in thyroid C cells. In neural tissue exon 3 is spliced instead to exons 5 and 6, which contain the CGRP coding and 3' untranslated sequences and poly(A) addition site. After proteolytic processing, the mature peptides are encoded entirely by the variable sequences. CT functions as a circulating Ca^{2+} homeostatic hormone while CGRP may have both neuromodulatory and trophic activities (66, 202). A similar mechanism is utilized by the R15 neuron of Aplysia to generate an overlapping set of peptides derived via alternative splicing and proteolytic processing (38).

(e) Effects upon RNA Turnover and Translational Efficiency

Many genes produce transcripts with heterogeneity within the 5' and 3' untranslated (UT) regions, often but not exclusively in conjunction with different coding regions. The UT sequences are known to be involved in regulating translational efficiency (127) and RNA stability (28, 190). Thus, alternative processing involving UT regions provides the potential for regulated expression of the same gene product in different cells, or appropriate levels of expression of different products arising from alternative splicing of coding regions. No clear demonstration has yet been made for such an effect on alternative splicing, but the increasing number of genes with regulated heterogeneity in only the UT regions is highly suggestive of such a role. The mRNA for HMG CoA synthetase, for intstance, has heterogeneity within the 5' UT sequences due to the presence of a cassette exon. It has been speculated that this difference could be used in the modulation of feedback-suppression of translation or in the selective cytoplasmic or mitochondrial targetting of the enzyme (80). The colony-stimulating factor-1 (CSF-1) gene produces two transcripts that are identical in the coding region but differ only in the 3' UT region. One 3' UT region contains a repetititve AUUUA element (135) found in a number of rapidly turning over mRNAs, and causes a dramatic destabilizing effect when inserted into the 3'UT region of β-globin mRNA (226). This suggests that CSF1 expression can be regulated at the level of mRNA turnover according to the selection of different 3'UT sequences (135). Another repetitive element has been found in the 3'UT region of a number of transcripts, increasing their stability in response to growth and viral transformation (83). This sequence element is present in an alternatively spliced 3' UT region of the L isoform of pyruvate kinase, raising the possibility of differential regulation of L-PK expression mediated by mRNA stability (148).

ALTERNATIVE SPLICING IN DEVELOPMENTAL CONTROL

The examples in the preceding section illustrate the role of alternative splicing in the final expression of the differentiated phenotypes of many cells. Such variants are the manifestations of tissue-specific phenotypes that allow the cell to carry out its particular function. Recently, however, it has emerged that alternative splicing also plays an important role further up in the hierarchy of developmental control, regulating the expression of genes responsible for initiating and maintaining different developmental programs. In a number of instances, alternative splicing has been shown to act as an on/off switch for gene function (reviewed in 21), whereas in others the activity of important regulatory genes such as *Ultrabithorax* and *Antennapedia* is modulated in a tissue-specific manner by alternative splicing.

Alternative Splicing as an On/Off Switch

The use of alternative splicing as an on/off switch has been well documented for a number of Drosopohila genes (21), including *Sex lethal* (15) and *transformer* (25), which are both important regulatory genes in sexual differentiation, and *suppressor of white apricot* (46, 264). The pathway involved in sexual differentiation of Drosophila illustrates particularly well the utility of alternative splicing as a developmental control mechanism (Figure 5).

The primary determinant of sex in Drosophila is the ratio of X chromosomes to autosomes (A) (for review, see 49). An X:A ratio of 1 leads to female development while a ratio of 0.5 results in male development. The response to this primary signal involves a hierarchy of regulatory genes that act in the initiation, expression, and subsequent maintenance of the determined sexual state (49). The final stage involves the development of sexually differentiated cells in both the soma and the germline, and X-chromosome dosage compensation that ensures that both male and female cells have similar transcriptional activity for the X-linked genes.

The epistatic hierarchy of genes in the pathway of somatic sexual differentiation has been well established (see Figure 5; 157, 165). It proceeds from the central control gene *Sex-lethal (Sxl)* which activates *transformer (tra)* (157). *tra*, in conjunction with *transformer-2 (tra-2)*, then acts to establish the male or female mode of the *double-sex (dsx)* gene. In females, the *dsx* female product acts in concert with *intersex (ix)* to repress the genes responsible for male differentiation, leading to female development. In contrast, the male *dsx* product acts alone to repress the genes for female development in males. At all levels of the pathway except for the bifunctional dsx, gene activity is required only for female development, and at all levels male development

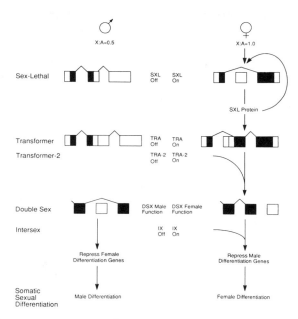

Figure 5   The regulatory cascade responsible for somatic sexual differentiation in Drosophila. The alternatively spliced *sex-lethal, transformer,* and *doublesex* genes are shown in the two splicing modes with the black boxes represented continuous open reading frames, and the white boxes noncoding regions or exons which are spliced out. Summarized from (13, 15, 25).

represents the default of gene expression. The molecular basis of the sex-specific expression of these genes has been characterized in some detail and, in the case of *Sxl, tra and dsx,* involves sex-specific alternative splicing.

The *Sxl* gene inhibits a complex sex-specific and developmentally regulated transcription pattern (15). Male and female specific transcripts differ by the inclusion in males of a specific exon that introduces an in-frame termination codon. This insertion results in the truncation of the major open reading frame of the transcript after only 42 or 48 codons. In females, the male specific exon is skipped and the mRNA encodes a 354 amino acid protein that has sequence similarities with a number of RNA-binding proteins (15). This provides a molecular rationale for the genetically observed autoregulation of the *Sxl* gene (49), which is responsible for continued maintenance of the committed state. In the absence of Sxl gene product, the gene transcript is spliced in the default male-specific pattern leading to the continued production of the male transcript, which does not encode a functional product. In female cells, in contrast, the Sxl product causes exclusion of the male-specific exon and production of functional mRNA, thus promoting its own further production. The *Sxl* protein could act by binding directly to the pre-mRNA and thus inhibit the incorporation of the male exon; alternatively it could specifically activate the female-specific splice. In either case, once a splicing pattern has been established, a positive feedback loop will stably maintain the same mode of expression. The processes involved in the initial establishment of the splicing mode are not clear, although the upstream genes daughterless *(da)* and sisterless *(sis)* are required for *Sxl* activity (49). In early embryogenesis the transcripts generated from the *Sxl* locus have structures that have not been determined but that are distinct from those present later in development (15, 213). It is likely that these early transcripts are involved in the initiation of the subsequent pattern of *Sxl* expression.

The downstream *tra* gene undergoes a similar process of sex-specific splicing to generate a functional mRNA only in female cells (25). In male cells, a 73 nt intron is spliced out between exons 1 and 2 and the longest open reading frame in the resultant RNA is only 48 amino acids long. In females, a downstream acceptor for exon 2 is used, resulting in the excision of a 248 nt first intron and the production of an mRNA with a single long open reading frame coding for the *tra* gene product. These splicing patterns are dependent upon *Sxl* expression (165). In the absence of *Sxl* activity only the default nonfunctional RNA is produced. Expression of the female-specific transcript in chromosomal males leads to female differentiation (157). It is likely that *Sxl* regulates *tra* splicing similarly to its own, probably by suppressing the male splice. The predicted amino acid sequence of the *tra* protein bears no resemblance to RNA-binding proteins. However, the product of the *tra-2* gene, which is also required for the regulation of activity of the downstream *dsx* gene, does have strong homology with other RNA-binding proteins (9) including the Sxl protein (15). *Dsx* gene expression is also controlled at the level of alternative splicing (13, 165), so its seems likely that *tra* and *tra-2* act in concert to regulate *dsx* splicing, with *tra-2* acting as an RNA-binding component.

dsx is the only gene in the pathway that is actively required in both males and females (12). This gene has active but opposite regulatory functions in males and females, acting in males to repress expression of the genes responsible for female sexual differentiation, and in females to repress the genes for male sexual differentiation (12). The molecular basis of the sex-specific activities of this gene has been shown to be due to the sex-specific alternative splicing of 3' end exons (13). As expected, loss of function mutations in *tra*, *tra-2* and *Sxl* that produce male development in chromosomal females all result in a switch from the female- to the male-specific *dsx* splicing pattern (165).

The *suppressor of white apricot (su(wᵃ))* locus provides another well-characterized example involving on/off switching and feedback auto-regulation by alternative splicing. The *su(wᵃ)* gene inhibits the expression of the hypomorphic mutant *white-apricot (w^a)* gene, increasing the use of a transcript termination site within a retrotransposon insertion into the second intron, at the expense of the normal wild-type splice (264). The *su(w^a)* product also down-regulates the production of its own mRNA by suppressing the excision of the first, and to a lesser extent the second, intron, resulting in RNAs that do not appear to contain functional long open reading frames (46, 264). The available data is consistent with a model in which the *su(w^a)* product interacts directly with its own and with *w^a* pre-mRNA to inhibit a specific splice, resulting in nonfunctional RNAs. This is in contrast to *Sxl*, which produced functional mRNAs by suppressing a splicing event. The predicted amino acid sequence of *su(w^a)* does not appear to contain the conserved ribonucleoprotein RNA-binding domain (46) found in *Sxl* and *tra-2*.

In these regulatory pathways alternative splicing is used as an on/off switch for the activity of *Sxl, tra, su(w^a)* and *w^a*, and as a switch between two alternative sets of functions in the case of *dsx*. In each case, the control of gene expression, which usually rests at the level of transcription of one or more (in the case of switching between activities) genes, resides at the level of differential splicing of a constitutively transcribed gene. The apparent waste inherent in the production of nonfunctional transcripts may represent the energetic cost of this mode of regulation, and in any case may not be too excessive if the RNAs are not translated, since this process represents the major energy-consuming phase in gene expression. In this regard, it will be of interest to determine whether the somatic cell product of the P-element transposase gene, which lacks transposase activity, does fulfill some other function (142, 197). Positive feedback regulation of alternative splicing appears to be a convenient way to establish and maintain a mode of expression, since the maximal required level of expression is set by the constant level of transcription and can not be exceeded. However, such feedback regulation of *Sxl* splicing may present a disadvantage to male cells if a small amount of the female transcript is inadvertently produced, since the effect will be rapidly amplified and *Sxl* expression switched to the inappropriate mode. The incidence of such events could be minimized if a significant threshold level of *Sxl* product is required to have an effect on the splicing process. In any case, cells that inappropriately express *Sxl* will rapidly be eliminated because of the consequent upset in X-linked gene expression.

Alternative Splicing of Morphogenetic Proteins

Unlike the preceding examples where whole regulatory cascades are controlled by differential splicing of constitutively expressed pre-mRNA at each level, there are cases where fine tuning of developmental regulation is achieved by the concerted involvement of both alternative splicing and transcriptional control.

The expression of the homeotic Antennapedia (*Antp*) and Ultrabithorax *(Ubx)* genes of Drosophila is now known to be regulated at the level of alternative splicing of the protein-coding regions of their transcripts. These genes are involved in both establishing segmental identity in the Drosophila larva and in subsequent tissue-specific development (78, 101, 186). Like many important genes involved in development, they are thought to act as transcriptional regulators in a cascade of interactions between genes (78, 101). The conserved homeo domain of these proteins contains a helix-turn-helix DNA-binding motif, and more recently the same conserved domain has been found to be part of the DNA-binding domain in known mammalian and yeast transcription factors (78, 95).

The *Ubx* gene has the capacity to produce 5–8 different RNAs, all of which vary within the protein-coding region (126, 171). This diversity is attained by the use of two internal 51 nt cassette exons, and by alternative splice donor sites for the common 5' exon (14, 126, 171). These different patterns of alternative exon selection also correlate with differential utilization of alternative poly(A) addition sites (126). The production of the different spliced mRNAs is time- and tissue-dependent (126, 171), and a mutant containing an in-frame termination codon within the second cassette exon is phenotypically similar to null mutations in its effects on epidermal development, but has no effects upon the development of the adult nervous system (257), presumably reflecting the fact that this exon is commonly skipped in neuronal, but not epidermal cells. On the basis of these and other data, it has been suggested that the overall levels of *Ubx* transcripts are responsible for the establishment of metameric identity via interactions with the promoters of homeotic genes, while the specific pattern of alternative splicing determines tissue-specific development by the modulation of interactions with other effector genes (126). The *Antp* gene has been shown to similarly utilize alternative splicing, with a cassette exon (exon 6) and alternative donor sites for exon 7 (243) allowing the production of 4 different proteins, all of which contain the conserved homeobox domain encoded by exon 8. The *Antp* RNAs are also produced in a developmental and tissue-specific fashion. The variable region within both *Ubx* and *Antp* is adjacent to both the homeobox and to other conserved elements such as the 'YPWM' box (14, 171, 243), and it seems likely that DNA binding, interactions with other proteins, and self-oligomerization could be modulated by such variable sequences. This could result in the recognition of different DNA-target sequences or it could alter the mode of regulation from activation to inhibition of transcription. Thus, in these examples, transcriptional activation of the gene is responsible for the determination of overall segmental identity, with alternative splicing allowing for more fine scale tissue-specific control of development.

ALTERNATIVE SPLICING AND EVOLUTION

Alternative splicing allows a high degree of protein diversity at low genetic cost   The evolution of complex biological systems has been accompied by increasing genomic complexity able to generate an array of protein variants, each adapted to function in a particular cell type, developmental pathway, or physiological state. Three main mechanisms have evolved to produce protein diversity at the level of primary structure: gene duplication, gene rearrangement, and alternative splicing. Of the three, gene duplication is the most costly in genetic terms because the coding of each isoform requires the duplication of the entire genetic sequence. Therefore, when genome size places strong evolutionary contraints, such as in very rapidly growing organisms and viruses, this method is of limited utility. Yet, the abundance of multigene families demonstrates the evolutionary success of this process. It is likely that the ease of generating gene duplications through polyploidy and DNA recombination can explain their widespread existence. In addition, the ability to place copies of the same gene under different developmental or metabolic regulatory pathways are likely to contribute further to the evolutionary importance of this process.

Gene rearrangement is a more efficient mechanism for the generation of protein diversity from single genes. The immunoglobulin and T-cell receptor genes, like alternatively spliced genes, contain intragenic duplications that correspond to the variable domains of the protein. Combinatorial rearrangement of a limited set of V, D, and J exons can produce an almost unlimited number of different antibodies and receptors. Yet surprisingly, given its tremendous power, this mechanism seems to be restricted to the immune system. A likely explanation is that DNA rearrangement appears to be a unidirectional process. It irreversibly changes the genetic content of the cell (173, 211) and, therefore, can be used only by cells in the process of terminal differentiation. Moreover, in order to be biologically significant, DNA rearrangement needs to occur in cells capable of clonal expansion. The DNA rearrangements leading to the production of unique immunoglobulin and T-cell receptor molecules would be of little value if these cells could not be selected and clonally expanded to produce a population of identical cells, sufficient to play a role in the immune response.

Alternative splicing offers many of the advantages and few of the limitations of DNA rearrangement and gene duplication. In contrast to DNA rearrangement, it does not change the genetic content of the cell nor is it irreversible in genetic terms. Splicing pathways need not be discarded in order to use different ones. This feature makes alternative splicing particularly suited to generate protein diversity during early developmental stages and in very long-lived and terminally differentiated cells. Isoforms generated by this mechanism can turn on differentiation pathways without restricting the genetic and phenotypic potential of the daughter cells (15, 25, 46, 142, 197, 264). The potential to reversibly regulate gene expression without transcriptional reprogramming plays a fundamental role in the ability of long-lived terminally differentiated cells to tailor their protein-isoform expression in a reversible manner.

Alternative splicing has the same quantitative potential to generate multiple protein isoforms from a single gene as DNA rearrangement. For a gene with n alternatively spliced exons, there are 2^n possible isoforms, assuming a combinatorial mechanism of exon usage. Thus, a gene with 20 or more alternatively spliced exons, such as recently found in the myosin heavy chain gene of Drosophila (C. Emerson, personal communcation), has the theoretical capacity to generate over a million different isoforms. This potential is further increased in some genes, such as the GTP-binding protein Gsα (128), by the ability to generate further diversity at exon borders by the use of duplicated donor and acceptor sites, in a manner very similar to the diversity created at the join region of the immunoglobulin genes.

The contribution of alternative splicing to phenotypic variability is increased many fold in the case of multiprotein complexes such as the sarcomere. In vertebrates, this organelle is produced by the assembly of seven major contractile proteins, each encoded by a multigene family with a minimum of four members (see 61). Assuming an average of 5 genes per family, in combination they have the capacity to generate $5^7 = 78125$ different sarcomeric types. This potential is significantly increased by alternative splicing. To date, more than 30 different exons or pairs of exons are known to be alternatively spliced in sarcomeric contractile protein genes (see 10, 29, 61, 113), raising to many billions the number of sarcomeres potentially produced by this limited set of genes. In fact, it is very likely that this maximum potential is never realized because not all genes are concurrently expressed in the same cell. However, even the limited subsets expressed in different muscle types and developmental stages have the potential to generate an impressive number of different sarcomere types.

Alternative splicing facilitates efficient exploitation of intragenic duplications and exon shuffling   DNA duplication followed by divergence has been a major source of genetic variability and has played an important role in protein evolution. Duplications span a continuum ranging from small intragenic regions, whole genes, chromosomes, and even the complete genome. Duplications of whole genetic loci or larger are likely to prove advantageous or neutral in the long run if they do not have immediate deleterious gene-dosage or position effects. On the other hand, intragenic duplications and/or trans-locations are expected to have neutral or deleterious effects. In constitutively spliced genes their effect will depend on the size and location of the dupli-cated region. Events limited to the introns might have little or no effect, whereas duplications or rearrangements affecting exons or splice sites could have considerable consequences. Even if translational frame shifts, stop codons, and splicing defects were avoided, the changes in primary protein sequence could easily disrupt protein folding and function. This scenario contrasts with the increasing evidence that exon duplication and shuffling has played a major role in protein evolution (58, 81, 82, 245). This apparent discrepancy between the genetic cost of most intragenic rearrangements and their evolutionary success suggests a role for alternative splicing. If the newly duplicated or translocated exons were alternatively spliced, the mutated gene could continue to produce the old gene product in addition to new ones. If these had a selective advantage they would eventually become fixed by evolving a constitutive or a regulated alternative pattern of splicing. If not advantageous, they would be deleted or converted into intronic sequences (104). The existence of alternatively spliced duplicated exons would acceler-ate protein evolution, since only one copy need be kept under rigid selection to maintain expression of the original gene product. The duplicated copy could then drift under relaxed selection until its expression in a different cellular environment imposes renewed selection.

Alternative splicing might be a predecesor of constitutive splicing Despite the fact that some introns are very recent and can be produced by retroinser-tion into the DNA (137, 248), there is compelling evidence that splicing is an old posttranscriptional process, since it is present in archeobacteria (107) and T-even phages (47), and intron positions of several genes have been con-served between unicellular organisms and mammals (106). It likely evolved from autocatalytic RNA ligation, as demonstrated by the existence of self-splicing introns (40, 41). It remains to be determined, however, whether alternative splicing is a predecessor or a refinement of constitutive splicing. The distinction between constitutive and regulated alternative spliced exons is relative and not determined by the intrinsic properties of the exon and its splice sites (31). This strongly argues that the two processes are part of a continuum and might have evolved simultaneously. If, as widely believed, most present-day genes have been assembled by a process of exon shuffling and duplication (58, 81, 82), it is likely that the primordial exons carried out some function that conferred a selective advantage. Therefore, it is possible that the new exon combinations were first alternatively spliced. If advantageous, these new combinations would have become fixed through mutations that rendered them constitutive. According to this scenario, stochastic alternative splicing would have been the primitive splicing form, while constitutive and regulated alternative splicing would represent its further refinement.

The evolutionary evidence for this or any alternative hypothesis is, unfortunately, not available. It is clear, however, that most alternatively spliced exons originated by exon duplication. Whether they were present in the ancestral gene or have arisen recently remains open to question. Structural comparisons among myosin light chain (183), troponin T (30) and α-TM, (204, 259) suggest that most of these duplications are ancient and were present in the ancestral genes, before the radiation of the insects more than 600 million years ago. Furthermore, the data indicate that, at least for these genes, a mechanism of alternative splicing was available at the time of the exon duplication event (259).

The preceding arguments point to the conclusion that alternative splicing arose prior to, or concomitantly with, constitutive splicing. Further, it appears to have played an important role in gene evolution, by facilitating the exploitation of exon duplications. Alternatively spliced duplicated exons would be retained when their differential expression optimizes the protein product to different cell environment. By contrast, genes whose expression is restricted to a single cell type, for which the gene product has already been optimized, would tend to be more streamlined, with constitutive exons only, and with reduced numbers of introns. The increased prevalence of alternative splicing in higher metazoans as compared with lower organisms is consistent with such a role of alternative splicing in gene evolution.

ACKNOWLEDGMENTS

We thank David Knaack, Maru Gallego, Dola Sengupta, Roger Breitbart, and Athena Andreadis for critical reading of the manuscript. This work was supported by grants from the National Institutes of Health and the Muscular Dystrophy Association.

Literature Cited

1. Abmayr, S. M., Reed, R., Maniatis, T. 1988. Identification of a functional mammalian spliceosome containing unspliced pre-mRNA. *Proc. Natl. Acad. Sci. USA* 85:7216–20
2. Adami, G. R., Carmichael, G. G. 1987. The length but not the sequence of the polyoma virus late leader exon is important for both late RNA splicing and stability. *Nucleic Acids Res.* 15:2593–611
3. Aebi, M., Hornig, H., Padgett, R. A., Reiser, J. Weissman, C. 1986. Sequence requirements for splicing of higher eukaryotic nuclear pre-mRNA. *Cell* 47:555–65
4. Aebi, M., Hornig, H., Weissman, C. 1987 5' Cleavage site in eukaryotic pre-mRNA splicing is determined by the overall 5' splice region, not by the conserved 5' GU. *Cell* 50:237–46
5. Agris, C. H., Nemeroff, M. E., Krug, R. M. 1989. A block in mammalian splicing occurring after formation of large complexes containing U1, U2, U4, U5, and U6 small nuclear ribonucleoproteins. *Mol. Cell. Biol.* 9:259–67
6. Alt, F. W., Bothwell, A. L. M., Knapp,

M., Siden, E., Mather, E., et al. 1980. Synthesis of secreted and membrane bound immunoglobulin m heavy chains is directed by mRNAs that differ at their 3' ends. *Cell* 20:293–302

7. Amara, S. G., Evans, R. M., Rosenfeld, M. G. 1984. Calcitonin/CGRP transcription unit: tissue specific expression involves selective use of alternative polyadenylation sites. *Mol. Cell Biol.* 4:2151–60

8. Amara, S. G., Jonas, V., Rosenfeld, M. G., Ong, E. S., Evans, R. 1982. Alternative RNA processing in calcitonin gene expression generates mRNAs encoding different polypeptide products. *Nature* 298:240–44

9. Amrein, M., Gorman, M., Nothiger, R. 1988. The sex-determining gene tra-2 of Drosophila encodes a putative RNA binding protein. *Cell* 55:1025–35

10. Andreadis, A., Gallego, M. E., Nadal-Ginard, B. 1987. Generation of protein isoform diversity by alternative splicing. *Annu. Rev. Cell Biol.* 3:207–42

11. Arrigo, S., Beemon, K. 1988. Regulation of rous sarcoma virus RNA splicing and stability. *Mol. Cell. Biol.* 8:4858–67

12. Baker, B. S., Ridge, K. 1980. Sex and the single cell: on the action of major loci affecting sex determination in Drosophila melanogaster. *Genetics* 94:383–423

13. Baker, B. S., Wolfner, M. F. 1988. A molecular analysis of doublesex, a bifunctional gene that controls both male and female sexual differentiation in Drosophila melanogaster. *Genes Dev.* 2:477–89

14. Beachy, P. A., Helfand, S. L., Hogness, D. S. 1985. Segmental distribution of bithorax complex proteins during Drosophila development. *Nature* 313:545–51

15. Bell, L. R., Maine, E. M., Schedl, P., Cline, T. W. 1988. Sex-lethal, a drosophila sex determination switch gene, inhibits sex-specific RNA splicing and sequence similarity to RNA binding proteins. *Cell* 55:1037–46

16. Berget, S. M., Robberson, B. L. 1986. U1, U2, and U4/U6 small nuclear ribonucleoproteins are required for in vitro splicing but not polyadenylation. *Cell* 46:691–96

17. Berk, A. J., Sharp, P. A. 1978. Structure of the adenovirus 2 early mRNAs. *Cell* 14:695–711

18. Beyer, A. L., Osheim, Y. N. 1988. Splice site selection, rate of splicing, and alternative splicing on nascent transcripts. *Genes Dev.* 2:754–65

19. Bindereif, A., Green, M. R. 1986. Ribonucleoprotein complex formation during pre-mRNA splicing in vitro. *Mol. Cell Biol.* 6:2582–92

20. Bindereif, A., Green, M. R. 1987. An ordered pathway of snRNP binding during mammalian pre-mRNA splicing complex assembly. *EMBO J.* 6:2415–24

21. Bingham, P. M., Chou, T-B., Mims, I., Zachar, Z. 1988. On/Off regulation of gene expression at the level of splicing. *Trends Genet.* 4:134–38

22. Birnstiel, M. L., Busslinger, M., Strub, K. 1985. Transcription termination and 3' processing: the end is in site. *Cell* 41:349–59

23. Black, D. L., Chabot, B., Steitz, J. A. 1985. U2 as well as U1 small nuclear ribonucleoproteins are involved in premessenger RNA splicing. *Cell* 42:737–50

24. Black, D. L., Steitz, J. A. 1986. Pre-mRNA splicing in vitro requires intact U4/U6 small nuclear ribonucleoprotein. *Cell* 46:697–704

25. Boggs, R. T., Gregor, P., Idriss, S., Belote, J. M., McKeown, M. 1987. Regulation of sexual differentiation in D. melanogaster via alternative splicing of RNA from the transformer gene. *Cell* 50:739–47

26. Bovenberg, R. A. L., van de Meerendonk, W. P. M., Baas, P. D., Steenbergh, P. H., Lips, C. J. M., Jansz, H. S. 1986. Model for alternative splicing in human calcitonin gene expression. *Nucleic Acids Res.* 14:8785–803

27. Brandt, P. W., Diamond, M. S., Schachat, F. H. 1984. The thin filament of vertebrate skeletal muscle cooperatively activates as a unit. *J. Mol. Biol.* 180:379–84

28. Brawerman, G. 1987. Determinants of messenger RNA stability. *Cell* 48:5–6

29. Breitbart, R. E., Andreadis, A., Nadal-Ginard, B. 1987. Alternative splicing; a ubiquitous mechanism for the generation of multiple protein isoforms from single genes. *Annu. Rev. Biochem.* 56:467–95

30. Breitbart, R. E., Nadal-Ginard, B. 1986. Complete nucleotide sequence of the fast skeletal troponin-T gene. *J. Mol. Biol.* 188:313–24

31. Breitbart, R. E., Nadal-Ginard, B. 1987. Developmentally induced, muscle specific trans factors control the differential splicing of alternative and constitutive troponin-T exons. *Cell* 49:793–803

32. Bretibart, R. E., Nguyen, H. T., Medford, R. M., Destree, A. T., Mahdavi, V., Nadal-Ginard, B. 1985. Intricate combinatorial patterns of exon splicing generate multiple regulated troponin T isoforms from a single gene. *Cell* 41:67–82

33. Briggs, M. M., Brandt, P. W., Schnurr, C. A., Schachat, F. H. 1989. Two alternative splicing events that generate functionally different fast troponin T isoforms. In *Cellular and Molecular Biology of Muscle Development*, ed. L. H. Kedes, F. E. Stockdale. pp. 597–607. New York: Liss

34. Briggs, M. M., Klevit, R. E., Schachat, F. H. 1984. Heterogeneity of contractile proteins. *J. Biol. Chem.* 259:10369–75

35. Briggs, M. M., Lin, J. J-C., Schachat, F. H. 1987. The extent of amino terminal heterogeneity in rabbit fast skeletal muscle troponin-T. *J. Muscle Res. Cell Motil.* 8:1–12

36. Bringmann, P., Lührmann, R. 1986. Purification of the individual snRNPs U1, U2, U5 and U4/U6 from HeLa cells and characterization of their protein constituents. *EMBO J.* 5:3509–16

37. Brody, E., Abelson, J. 1985. The "splicesome": yeast pre-messenger RNA associates with a 40S complex in a splicing-dependent reaction. *Science* 228:963–67

38. Buck, L. R., Bigelow, J. M., Axel, R. 1987. Alternative splicing in individual aplysia neurons generates neuropeptide diversity. *Cell* 51:127–33

39. Caras, I. W., Davitz, M. A., Rhee, L., Weddell, G., Martin, D. W., Nussenzweig, V. 1987. Cloning of decay-accelerating factor suggests novel use of splicing to generate two proteins. *Nature* 325:545–49

40. Cech, T. R. 1986. The generality of self-splicing RNA: relationship to nuclear mRNA splicing. *Cell* 44:207–10

41. Cech, T. R., Bass, B. L. 1986. Biological catalysis by RNA. *Annu. Rev. Biochem.* 55:599–629

42. Chabot, B., Black, D. L., LeMaster, D. M., Steitz, J. A. 1985. The 3' splice site of premessenger RNA is recognized by a small nuclear ribonucleoprotein. *Science* 230:1344–49

43. Chabot, B., Steitz, J. A. 1987. Multiple interactions between the splicing substrate and small nuclear ribonucleoproteins in spliceosomes. *Mol. Cell. Biol.* 7:281–93

44. Chabot, B., Steitz, J. A. 1987. Recognition of mutant and cryptic 5' splice sites by the U1 small nuclear ribonucleoprotein in vitro. *Mol. Cell. Biol.* 7:698–707

45. Choi, Y. D., Grabowski, P. J., Sharp, P. A., Dreyfuss, G. 1986. Heterogeneous nuclear ribonucleoproteins: role in RNA splicing. *Science* 231:1534–39

46. Chou, T-B., Zachar, Z., Bingham, P. M. 1987. Developmental expression of a regulatory gene is programmed at the level of splicing. *EMBO J.* 6:4095–104

47. Chu, F., Maley, G., Maley F., Belfort, M. 1984. Intervening sequence in the thymidylate synthase gene of bacteriophage T4. *Proc. Natl. Acad. Sci. USA* 81:3049–3053

48. Clerc, R. G., Corcoran, L. M., LeBowitz, J. H., Baltimore, D., Sharp, P. A. 1988. The B-cell-specific Oct-2 protein contains POU box- and homeo box-type domains. *Genes Dev.* 2:1570–81

49. Cline, T. W. 1985. Primary events in the determination of sex in *Drosophila melanogaster*. In *The Origin and Evolution of Sex*, ed. H. O. Halvorson, A. Monroy, pp. 301–27. New York: Liss

50. Collins, T., Bonthron, D. T., Orkin, S. H. 1987. Alternative RNA splicing affects function of encoded platelet derived growth factor A-chain. *Nature* 328:621–24

51. Cooper, T. A., Cardone, M. H., Ordahl, C. P. 1988. *Cis* requirements for alternative splicing of the cardiac troponin T pre-mRNA. *Nucleic Acids Res.* 16:8443–65

52. Cooper, T. A., Ordahl, C. P. 1985. A single cardiac toponin T gene generates embryonic and adult isoforms via developmentally regulated alternative splicing. *J. Biol. Chem.* 260:11140–48

53. Craig, S. P., Muralidhar, M. G., McKerrow, J. H., Wang, C. C. 1989. Evidence for a class of very small introns in the gene for hypoxanthine-guanine phosphoribosyl-transferase in *Schistosoma mansoni*. *Nucleic Acids Res.* 17:1635–48

54. Crenshaw, E. B., Russo, A. F., Swanson, L. W., Rosenfeld, M. G. 1987. Neuron specific alternative RNA processing in transgenic mice expressing a metallothionein-calcitonin fusion gene. *Cell* 49:389–98

55. Cunningham, B. A., Hemperly, J. J., Murray, B. A., Prediger, E. A., Brackenbury, R., Edelman, G. M. 1987. Neural cell adhesion molecule: structure, immunoglobulin like domains, cell surface modulation and alternative RNA splicing. *Science* 236:799–806

56. Dahlberg, J. E., Lund, E. 1988. In *Structure and Function of Major and Minor Small Ribonucleotprotein Particles*, ed. M. L. Birnstiel, pp. 38–70. Berlin: Springer-Verlag

57. Danner, D., Leder, P. 1985. Role of an RNA cleavage/poly(A) addition site in the production of membrane bound and secreted IgM mRNAs. *Proc. Natl. Acad. Sci. USA* 82:8658–62

58. Darnell, J. E. 1978. Implications of RNA-RNA splicing in the evolution of eukaryotic cells. *Science* 202:1257–60

58a. de Wet, J. R., Wood, K. V., deLuca,

M., Helsinki, D. R., Subramani, S. 1987. Firefly luciferase gene: Structure and expression in mammalian cells. *Mol. Cell Biol.* 7:725–37

59. Dickson, G., Gower, H. J., Barton, C. H., Prentice, H. M., Elson, V. L., et al. 1987. Human muscle neural cell adhesion molecule (N-CAM): identification of a muscle specific sequence in the extracellular domain. *Cell* 50:1119–30

60. Early, P., Rogers, J., Davis, M., Calame, K., Bond, M., et al. 1980. Two mRNAs can be produced from a single immunoglobulin m gene by alternative RNA processing pathways. *Cell* 20:313–10

61. Emerson, C., Fischman, D., Nadal-Ginard, B., Siddiqui, M. A. Q., eds. 1986. In *Molecular Biology of Muscle Development*. New York: Liss

62. Eperon, L. P., Estibeiro, J. P., Eperon, I. C. 1986. The role of nucleotide sequences in splice site selection in eukaryotic pre- messenger RNA. *Nature* 324:280–82

63. Eperon, L. P., Graham, I. R., Griffiths, A. D., Eperon, I. C. 1988. Effects of RNA secondary structure on alternative splicing of pre-mRNA: is folding limited to a region behind the transcribing RNA polymerase? *Cell* 54:393–401

64. Fiszman, M. Y., Libri, D., Savino, R., Lemonnier, M., Mouly, V., Meinnel, T. 1989. Expression of the tropomyosin genes during differentiation of avian muscle cells in vitro. See Ref. 33, pp. 607–20

65. Flicker, P. F., Philips, G. N., Cohen, C. 1982. Troponin and its interactions with tropomyosin. *J. Mol. Biol.* 162:495–501

66. Fontaine, B., Klarsfeld, A., Changeux, J. P. 1987. Calcitonin gene related peptide and muscle activity regulate acetylcholine receptor a-subunit mRNA levels by distinct intracellular pathways. *J. Cell Biol.* 105:1337–42

67. Forbes, D. J., Kirschner, M. W., Caput, D., Dahlberg, J. E., Lund, E. 1984. Differential expression of multiple U1 small nuclear RNAs in oocytes and embryos of xenopus laevis. *Cell* 38:681–89

68. Frendeway, D., Keller, W. 1985. The stepwise assembly of a pre-mRNA splicing complex requires U-snRNPs and specific intron sequences. *Cell* 42:355–67

69. Fu, X-Y., Colgan, J. D., Manley, J. L. 1988. Multiple *cis*-acting sequence elements are required for efficient splicing of simian virus 40 small-t antigen pre-mRNA. *Mol. Cell. Biol.* 8:3582–90

70. Fu, X-Y., Ge, H., Manley, J. L. 1988. The role of the polypyrimidine stretch at the SV40 early pre-mRNA 3' splice site in alternative splicing. *EMBO J.* 7:809–17

71. Fu, X-Y., Manley, J. L. 1987. Factors influencing alternative splice site utilization in vivo. *Mol. Cell. Biol.* 7:238–48

72. Furdon, P. J., Kole, R. 1986. Inhibition of splicing but not cleavage at the 5' splice site by truncating human β-globin pre-mRNA. *Proc. Natl. Acad. Sci. USA* 83:927–31

73. Deleted in proof

74. Galli, G., Guise, J. W., McDevitt, M. A., Tucker, P. W., Nevins, J. R. 1987. Relative position and strength of poly(A) sites as well as transcription termination are critical to membrane versus secreted m-chain expression during B-cell development. *Genes Dev.* 1:471–81

75. Galli, G., Guise, J., Tucker, P. W., Nevins, J. R. 1988. Poly(A) site choice rather than splice site choice governs the regulated production of IgM heavy chain RNAs. *Proc. Natl. Acad. Sci. USA* 85:2439–43

76. Garfinkel, L. I., Davidson, N. 1987. Developmentally regulated expression of a truncated myosin light-chain 1/3 gene. *Mol. Cell. Biol.* 7:3826–29

77. Gattoni, R., Schmitt, P., Stevenin, J. 1988. In vitro splicing of adenovirus E1A transcripts: characterization of novel reactions and of multiple branch points abnormally far from the 3' splice site. *Nucleic Acids Res.* 16:2389–408

78. Gehring, W. J. 1987. Homeo boxes in the study of development. *Science* 236:1245–52

79. Gerke, V., Steitz, J. A. 1986. A protein associated with small nuclear ribonucleoprotein particles recognizes the 3' splice site of premessenger RNA. *Cell* 47:973–84

80. Gil, G., Smith, J. R., Goldstein, J. L., Brown, M. S. 1987. Optional exon in the 5' untranslated region of 3-hydroxy-3-methylglutaryl coenzyme A synthase gene: conserved sequence and splicing pattern in humans and hamsters. *Proc. Natl. Acad. Sci. USA* 84:1863–66

81. Gilbert, W. 1978. Why genes in pieces? *Nature* 271:501–2

82. Gilbert, W. 1985. Genes-in-pieces revisited. *Science* 228:823–24

83. Glaichenhaus, N., Cuzin, F. 1987. A role for ID repetitive sequences in growth and transformation dependent regulation of gene expression in rat fibroblasts. *Cell* 50:1081–89

84. Gower, H. J., Barton, C. H., Elsom, V. L., Thompson, J., Moore, S. E., et al. 1988. Alternative splicing generates a secreted form of N-CAM in muscle and brain. *Cell* 55:955–64

85. Grabowski, P. J., Padgett, R. A., Sharp, P. A. 1984. Messenger RNA splicing in vitro: an excised intervening sequence and a possible intermediate. *Cell* 37:415–27

86. Grabowski, P. J., Seller, S. R., Sharp, P. A. 1985. A multicomponent complex is involved in the splicing of messenger RNA precursors. *Cell* 42:345–53

87. Green, M. R. 1986. Pre-mRNA splicing. *Annu. Rev. Genet.* 20:671–708

88. Guthrie, C., Patterson, B. 1988. Spliceosome snRNAs. *Annu. Rev. Genet.* 22:387–419

89. Habets, W. J., Sillekens, P. T. G., Hoet, M. H., Schalken, J. A., Roebroek, A. J. M., et al. 1987. Analysis of a cDNA clone expressing a human autoimmune antigen: full-length sequence of the U2 small nuclear RNA-associated B" antigen . *Proc. Natl. Acad. Sci. USA* 84:2421–25

90. Hampson, R. K., La Follette, L., Rottman, F. M. 1989. Alternative processing of bovine growth hormone mRNA is influenced by downstream exon sequences. *Mol. Cell. Biol.* 9:1604–10

91. Hartmuth, K., Barta, A. 1988. Unusual branch point selection in processing of human growth hormone pre-mRNA. *Mol. Cell. Biol.* 8:2011–20

92. Helfman, D. M., Cheley, S., Kuismanen, E., Finn, L. A., Yamawaki-Kataoka, Y. 1986. Nonmuscle amd muscle tropomyosin isoforms are expressed from a single gene by alternative RNA splicing and polyadenylation. *Mol. Cell Biol.* 6:3582–95

93. Helfman, D. M., Ricci, W. M. 1989. Branch point selection in alternative splicing of tropomyosin pre-mRNAs. *Nucleic Acids Res.* In press

94. Helfman, D. M., Ricci, W. M., Finn, L. A. 1988. Alternative splicing of tropomyosin pre-mRNAs in vitro and in vivo. *Genes Dev.* 2:1627–38

95. Herr, W., Sturm, R. A., Clerc, R. G., Corcoran, L. M., Baltimore, D., et al. 1988. The pou domain: a large conserved region in the mammalian pit-1, oct-1, oct-2, and *Caenorhabditis elegans* unc-86 gene products. *Genes Dev.* 2:1513–16

96. Hinterberger, M., Pettersson, I., Steitz, J. A. 1983. Isolation of small nuclear ribonucleoproteins containing U1, U2, U4, U5, and U6 RNAs. *J. Biol. Chem.* 258:2604–13

97. Humphries, M. J., Akiyama, S. K., Komoriya, A., Olden, K., Yamada, K. M. 1986. Identification of an alternatively spliced site in human plasma fibronectin that mediates cell type specific adhesion. *J. Cell Biol.* 103:2637–47

98. Deleted in proof

99. Hynes, R. 1985. Molecular biology of fibronectin. *Annu. Rev. Cell Biol.* 1:67–90

100. Imamura, K., Tanaka, T. 1982. Pyruvate kinase isozymes from rat. *Methods Enzymol.* 90:150–65

101. Ingham, P. W. 1988. The molecular genetics of embryonic pattern formation in Drosophila. *Nature* 335:25–34

102. Izumo, S., Mahdavi, V. 1988. Thyroid hormone receptor-α isoforms generated by alternative splicing differentially activate myosin HC gene-transcription. *Nature* 334:539–42

103. Jacquier, A., Rodriguez, J. R., Rosbash, M. 1985. A quantitative analysis of the effects of 5' junction and TACTAAC box mutants and mutant combinations on yeast mRNA splicing. *Cell* 43:423–30

104. Jaworski, C. J., Piatigorsky, J. 1989. A pseudo-exon in the functional human αA-crystallin gene. *Nature* 337:752–54

105. Jonas, V., Lin, C. R., Kawashima, E., Semon, D., Swanson, L. W., et al. 1985. Alternative RNA processing events in human calcitonin/calcitonin gene-related peptide gene expression *Proc. Natl. Acad. Sci. USA* 82:1994–98

106. Jung, G., Schmidt, C. J., Hammer, J. A. 1989. Cloning of a second Acanthamoeba myosin I heavy chain gene: Evidence for the existence of a third myosin I heavy chain isoform. *Gene.* In press

107. Kaine, B., Gupta, R., Woese, C. 1983. Putative introns in tRNA genes of prokaryotes. *Proc. Natl. Acad. Sci. USA* 80:3309–12

108. Kamb, A., Tseng-Crank, J., Tanouye, M. A. 1988. Multiple products of the Drosophila shaker gene may contribute to potassium channel diversity. *Neuron* 1:421–30

109. Kato, N., Harada, F. 1985. New U1 RNA species found in friend SFFV (spleen focus forming virus)-transformed mouse cells. *J. Biol. Chem.* 260:7775–82

110. Kay, R. J., Russnak, R. H., Jones, D., Mathias, C., Candido, E. P. M. 1987. Expression of intron containing C. Elegans heat-shock genes in mouse cells demonstrates divergence of 3' splice site recognition sequences between nematodes and vertebrates, and an inhibitory effect of heat shock on the mammalian splicing apparatus. *Nucleic Acids Res.* 15:3723–41

111. Kedes, D. H., Steitz, J. A. 1987. Accu-

rate 5' splice-site selection in mouse κ immunoglobulin light chain premessenger RNAs is not cell-type-specific. *Proc. Natl. Acad. Sci. USA* 84:7928–32

112. Kedes, D. H., Steitz, J. A. 1988. Correct in vivo splicing of the mouse immunoglobulin κ light-chain pre-mRNA is dependent on the 5' splice-site position even in the absence of transcription. *Genes Dev.* 2:1448–59

113. Kedes, L. H., Stockdale, F. E., eds. 1989. *Cellular and Molecular Biology of Muscle Development.* New York: Liss

114. Keller, E. B., Noon, W. A. 1984. Intron splicing: A conserved internal signal in introns of animal pre-mRNAs. *Proc. Natl. Acad. Sci. USA* 81:7417–20

115. Kelley, D. E., Perry, R. P. 1986. Transcriptional and posttranscriptional control of immunoglobulin mRNA production during B lymphocyte development. *Nucleic Acids Res.* 14:5431–47

116. King, C. R., Piatigorsky, J. 1983. Alternative RNA splicing of the murine αA-crystallin gene: protein-coding information with an intron. *Cell* 32:707–12

117. Koenig, R. J., Lazar, M. A., Hodin, R. A., Brent, G. A., Larsen, P. R., et al. 1989. Inhibition of thyroid hormone action by a non-hormone binding c-erbA protein generated by alternative mRNA splicing. *Nature* 337:659–61

118. Koller, B., Fromm, H., Galun, E., Edelman, M. 1987. Evidence for in vivo trans splicing of pre-mRNAs in tobacco chloroplasts. *Cell* 48:111–19

119. Konarska, M. M., Grabowski, P. J., Padgett, R. A., Sharp, P. A. 1985. Characterization of the branch site in lariat RNAs produced by splicing of mRNA precursors. *Nature* 313:552–57

120. Konarska, M. M., Padgett, R. A., Sharp, P. A. 1985. *Trans* splicing of mRNA precursors in vitro. *Cell* 42:165–71

121. Konarska, M. M., Sharp, P. A. 1986. Electrophoretic separation of complexes involved in the splicing of precursors to mRNAs. *Cell* 46:845–55

122. Konarska, M. M., Sharp, P. A. 1987. Interactions between small nuclear ribonucleoprotein particles in formation of spliceosomes. *Cell* 49:763–74

123. Korf, G. M., Botros, I. W., Stumph, W. E. 1988. Developmental and tissue-specific expression of U4 small nuclear RNA genes. *Mol. Cell. Biol.* 8:5566–69

124. Korhnblihtt, A. R., Umezawa, K., Vibe-Pedersen, K., Baralle, F. E. 1985. Primary structure of human fibronectin: differential splicing may generate at least 10 polypeptides from a single gene. *EMBO J.* 4:1755–59

125. Kornblihtt, A. R., Vibe-Pedersen, K., Baralle, F. E. 1984. Human fibronectin: cell specific alternative mRNA splicing polypeptide chains differing in the number of internal repeats. *Nucleic Acids Res.* 12:5853–68

126. Kornfeld, K., Saint, R. B., Beachy, P. A., Harte, P. J., Peattie, D. A., Hogness, D. S. 1989. Structure and expression of a family of Ultrabithorax mRNAs generated by alternative splicing and polyadenylation in Drosophila. *Genes Dev.* 3:243–58

127. Kozak, M. 1986. Regulation of protein synthesis in virus infected animal cells. *Adv. Virus. Res.* 31:229–92

128. Kozasa, T., Itoh, H., Tsukamoto, T., Kaziro, Y. 1988. Isolation and characterization of the human G$_s$α gene. *Proc. Natl. Acad. Sci. USA* 85:2081–85

129. Krainer, A. R., Maniatis, T. 1985. Multiple factors including the small nuclear ribonucleoproteins U1 and U2 are necessary from pre-mRNA splicing in vitro. *Cell* 42:725–36

130. Krainer, A. R., Maniatis, T., Ruskin, B., Green, M. R. 1984. Normal and mutant human b-globin pre-mRNAs are faithfully and efficiently spliced in vitro. *Cell* 36:993–1005

131. Kramer, A., Keller, W., Appel, B., Luhrmann, R. 1984. The 5' terminus of the RNA moiety of U1 small nuclear ribonucleoprotein particles is required for the splicing of messenger RNA precursors. *Cell* 38:299–307

132. Krause, M., Hirsh, D. 1987. A trans-spliced leader sequence on actin mRNA in *C. elegans. Cell* 49:753–61

133. Kuhne, T., Wieringa, B., Reiser, J., Weissmann, C. 1983. Evidence against a scanning model of RNA splicing. *EMBO J.* 2:727–33

134. Kwiatkowski, D. J., Stossel, T. P., Orkin, S. H., Mole, J. E., Colten, H. R., Yin, H. L. 1986. Plasma and cytoplasmic gelsolins are encoded by a single gene and contain a duplicated actin binding domain. *Nature* 323:455–58

135. Ladner, M. B., Martin, G. A., Noble, J. A., Nikoloff, D. M., Tal, R., et al. 1987. Human CSF-1: gene structure and alternative splicing of mRNA precursors. *EMBO J.* 6:2693–98

136. Lamb, R. A., Lai, C-J., Choppin, P. W. 1981. Sequences of mRNAs derived from genome RNA segment 7 of influenza virus: Colinear and interrupted

mRNAs code for overlapping proteins. *Proc. Natl. Acad. Sci. USA* 78:4170–74

137. Lambowitz, A. W. 1989. Infectious introns. *Cell* 56:323–26

138. Lamond, A. I., Konarska, M. M., Sharp, P. A. 1987. A mutational analysis of spliceosome assembly: Evidence for splice site collaboration during spliceosome formation. *Genes Dev.* 1:532–43

139. Lamond, A. I., Konarska, M. M., Grabowski, P. J., Sharp, P. A. 1988. Spliceosome assembly involves the binding and release of U4 small nuclear ribonucleoprotein. *Proc. Natl. Acad. Sci. USA* 85:411–15

140. Lang, K. M., Spritz, R. A. 1983. RNA splice site selection: evidence for a 5'-3' scanning model. *Science* 220:1351–55

141. Langford, C. J., Gallwitz, D. 1983. Evidence for an intron-contained sequence required for the splicing of yeast RNA polymerase II transcripts. *Cell* 33:519–27

142. Laski, F. A., Rio, D. C., Rubin, G. M. 1986. Tissue specificity of Drosophila P element transposition is regulated at the level of mRNA splicing. *Cell* 44:7–19

143. Legrain, P., Seraphin, B., Rosbash, M. 1988. Early commitment of yeast pre-mRNA to the spliceosome pathway. *Mol. Cell. Biol.* 8:3755–60

144. Leff, S. E., Evans, R. M., Rosenfeld, M. G. 1987. Splice commitment dictates neuron-specific alternative RNA processing in calcitonin/CGRP gene expression. *Cell* 45:517–24

145. Lerner, M. R., Boyle, J. A., Mount, S. M., Wolin, S. L., Steitz, J. A. 1980. Are snRNPs involved in splicing? *Nature* 283:220–24

146. Lillie, J. W., Green, M., Green, M. R. 1986. An adenovirus E1a protein region required for transformation and transcriptional repression. *Cell* 46:1043–51

147. Lillie, J. W., Loewenstein, P. M., Green, M. R., Green, M. 1987. Functional domains of adenovirus type 5 E1a proteins. *Cell* 50:1091–1100

148. Lone, Y. C., Simon, M-P., Kahn, A., Marie, J. 1986. Sequences complementary to the brain specific "identifier" sequence exist in L-type pyruvate kinase mRNA (a liver specific messenger) and in transcripts especially abundant in muscle. *J. Biol. Chem.* 261:1499–1502

149. Lowery, D. E., Van Ness, B. G. 1988. Comparison of in vitro and in vivo splice site selection in κ-immunoglobulin precursor mRNA. *Mol. Cell. Biol.* 8:2610–19

150. Lowery, D. E., Van Ness, B. G. 1987. In vitro splicing of Kappa-immunoglobulin precursor mRNA. *Mol. Cell. Biol.* 7:1346–51

151. Lund, E. 1988. Heterogeneity of human U1 snRNAs. *Nucleic Acids Res.* 16:5813–26

152. Lund, E., Dahlberg, J. E. 1987. Differential accumulation of U1 and U4 small nuclear RNAs during *Xenopus* development. *Genes Dev.* 1:39–46

153. Lund, E., Kahan, B., Dahlberg, J. E. 1985. Differential control of U1 small nuclear RNA expression during mouse development. *Science* 221:1271–74

154. Maniatis, T., Reed, R. 1987. The role of small nuclear ribonucleoprotein particles in pre-mRNA splicing. *Nature* 325:673–78

155. Mardon, H. J., Sebastio, G., Baralle, F. E. 1987. A role of exon sequences in alternative splicing of the human fibronectin gene. *Nucleic Acids Res.* 15:7725–33

156. Mayeda, A., Ohshima, Y. 1988. Short donor site sequences inserted within the intron of beta-globin pre-mRNA serve for splicing in vitro. *Mol. Cell. Biol.* 8:4484–91

157. McKeown, M., Belote, J. M., Boggs, R. T. 1988. Ectopic expression of the female transformer gene product leads to female differentiation of chromosomally male Drosophila. *Cell* 53:887–95

158. Medford, R. M., Nguyen, H. T., Destree, A. T., Summers, E., Nadal-Ginard, B. 1984. A novel mechanism of alternative RNA splicing for the developmentally regulated generation of troponin-T isoforms form a single gene. *Cell* 38:409–21

159. Mount, S. M. 1982. A catalogue of splice junction sequences. *Nucleic Acids Res.* 10:459–72

160. Mount, S. M., Pettersson, I., Hinterberger, M., Karmas, A., Steitz, J. A. 1983. The U1 small nuclear RNA-protein complex selectively binds a 5' splice site in vitro. *Cell* 33:509–18

161. Deleted in proof

162. Muirhead, H., Clayden, D. A., Barford, D., Lorimer, C. G., Fothergill-Gilmore, L. A., et al. 1986. The structure of cat muscle pyruvate kinase. *EMBO J.* 5:475–81

163. Murphy, W. J., Watkins, K. P., Agabian, N. 1986. Identification of a novel Y branch structure as an intermediate in trypanosome mRNA processing: evidence for trans splicing. *Cell* 47:517–25

164. Murre, C., Schonleber McCaw, P., Baltimore, D. 1989. A new DNA bind-

ing and dimerization motif in immunoglobin enhancer binding, daughterless, MyoD, and myc proteins. *Cell* 56:777–83

165. Nagoshi, R. N., McKeown, M., Burtis, K. C., Belote, J. M., Baker, B. S. 1988. The control of alternative splicing at genes regulating sexual differentiation in D. melanogaster. *Cell* 53:229–36

166. Nelson, K. K., Green, M. R. 1988. Splice site selection and ribonucleoprotein complex during in vitro pre-mRNA splicing. *Genes Dev.* 2:319–29

167. Newman, A. J., Lin, R-J., Cheng, S-C., Abelson, J. 1985. Molecular consequences of specific intron mutations on yeast mRNA splicing in vivo and in vitro. *Cell* 42:335–44

168. Noble, J. C. S., Pan, Z-Q., Prives, C., Manley, J. L. 1987. Splicing of SV40 early pre-mRNA to large-T and small-t mRNAs utilizes different patterns of lariat branch sites. *Cell* 50:227–36

169. Noble, J. C. S., Prives, C., Manley, J. L. 1988. Alternative splicing of SV40 early pre-mRNA is determined by branch site selection. *Genes Dev.* 2:1460–75

170. Noguchi, T., Inoue, H., Tunaka, T. 1986. The M1 and M2 type isozymes of rat ryruvate kinase are produced from the same gene by alternative RNA splicing. *J. Biol. Chem.* 261:13807–12

171. O'Connor, M. B., Binari, R., Perkins, L. A., Bender, W. 1988. Alternative RNA products from the Ultrabithorax domain of the bithorax complex. *EMBO J.* 7:435–45

172. Ohshima, Y., Gotoh, Y. 1987. Signals for the selection of a splice site in pre-mRNA. *J. Mol. Biol.* 195:247–59

173. Okazaki, K., David, D. D., Sakano, H. 1987. T-cell receptor β-gene sequences in the circular DNA of thymocyte nuclei: Direct evidence for intramolecular DNA deletion in V-D-J joining. *Cell* 49:477–85

174. Padgett, R. A., Grabowski, P. J., Konarska, M. M., Seiler, S., Sharp, P. A. 1986. Splicing of messenger RNA precursors. *Annu. Rev. Biochem.* 55:1119–1150

175. Padgett, R. A., Konarska, M. M., Aebi, M., Hornig, H., Weissmann, C., Sharp, P. A. 1985. Nonconsensus branch site sequences in the in vitro splicing of transcripts of mutant rabbit β-globin genes. *Proc. Natl. Acad. Sci. USA* 82:8349–53

176. Padgett, R. A., Konarska, M. M., Grabowski, P. J., Hardy, S. F., Sharp, P. S. 1984. Lariat RNAs as intermediates and

products in the splicing of messenger RNA precursors. *Science* 225:898–903

177. Padgett, R. A., Mount, S. M., Steitz, J. A., Sharp, P. A. 1983. Splicing of messenger RNA precursors is inhibited by antisera to small nuclear ribonucleoprotein. *Cell* 35:101–7

178. Parent, A., Zeitlin, S., Efstratiadis, A. 1987. Minimal exon sequence requirements for efficient in vitro splicing of mono-intronic nuclear pre-mRNA. *J. Biol. Chem.* 262:11284–11291

179. Parker, R., Guthrie, C. 1985. A point mutation in the conserved hexanucleotide at a yeast 5' splice site junction uncouples recognition, cleavage, and ligation. *Cell* 41:107–18

180. Parker, R., Siliciano, P. G., Guthrie, C. 1987. Recognition of the TACTAAC box during mRNA splicing in yeast involves base pairing to the U2-like snRNA. *Cell* 49:229–39

181. Patton, J. G., Wieben, E.D. 1987. U1 precursors: variant 3' flanking sequences are transcribed in human cells. *J. Cell. Biol.* 104:175–82

182. Peebles, C. L., Perlman, P. S., Mecklenburg, K. L., Petrillo, M. L., Tabor, J. H., et al 1986. A self splicing RNA excises an intron lariat. *Cell* 44:213–23

183. Periasamy, M., Strehler, E. E., Garfinkel, L. I., Gubits, R. M., Ruiz-Opazo, N., Nadal-Ginard, B. 1984. Fast skeletal muscle myosin light chains 1 and 3 are produced from a single gene by a combined process of differential RNA transcription and splicing. *J. Biol. Chem.* 259:13595–13604

184. Peterson, M. L., Perry, R. P. 1989. The regulated production of μm and μs mRNA is dependent on the relative efficiencies of μs poly (A) site usage and the Cμ4-to-M1 splice. *Mol. Cell. Biol.* 9:726–38

185. Pettersson, I., Hinterberger, M., Mimori, T., Gottlieb, E., Steitz, J. A. 1984. The structure of mammalian small nuclear ribonucleoproteins: identification of multiple protein components reactive with anti(U1)RNP and anti-Sm autoantibodies. *J. Biol. Chem.* 259:5907–14

186. Pfeifer, M., Karch, F., Bender, W. 1987. The bithorax complex: control of segmental identity. *Genes Dev.* 1:891–98

187. Pikielny, C. W., Rymond, B. C., Rosbash, M. 1986. Electrophoresis of ribonucleoproteins reveals an ordered assembly pathway of yeast splicing complexes. *Nature* 324:341–45

188. Pikielny, C. W., Teem, J. L., Rosbash,

M. 1983. Evidence for the biochemical role of an internal sequence in yeast nuclear mRNA introns: implications for U1 RNA and metazoan mRNA splicing. *Cell* 34:395–403

189. Plotch, S. J., Krug, R. M. 1986. In vitro splicing of influenza viral NS1 mRNA and NS1-β-globin chimeras: possible mechanisms for the control of viral mRNA splicing. *Proc. Natl. Acad. Sci. USA* 83:5444–48

190. Raghow, R., 1987. Regulation of messenger RNA turnover in eukaryotes. *Trends Biochem. Sci.* 12:122–66

191. Reddy, R. 1988. Compilation of small RNA sequences. *Nucleic Acids Res.* 16:r71–r85

192. Reed, R., Griffith, J., Maniatis, T. 1988. Purification and visualization of native spliceosomes. *Cell* 53:949–61

193. Reed, R., Maniatis, T. 1985. Intron sequences involved in lariat formation during pre-mRNA splicing. *Cell* 41:95–105

194. Reed, R., Maniatis, T. 1986. A role for exon sequences and splice-site proximity in splice-site selection. *Cell* 46:681–90

195. Reed, R., Maniatis, T. 1988. The role of the mammalian branchpoint sequences in pre-mRNA splicing. *Genes Dev.* 2:1268–76

196. Reuther, J. E., Maderious, A., Lavery, D., Logan, J., Man Fu, S., Chen-Kiang, S. 1986. Cell type specific synthesis of murine immunoglobulin m RNA from an adenovirus vector. *Mol. Cell. Biol.* 6:123–33

197. Rio, D. C., Laski, F. A., Rubin, G. M. 1986. Identification and immunochemical analysis of biologically active Drosophila P element transposase. *Cell* 44:21–32

198. Rogers, J., Early, P., Carter, C., Calame, K., Bond, M., et al. 1980. Two mRNAs with different 3' ends encode membrane bound and secreted forms of immunoglobulin μ chain. *Cell* 20:303–12

199. Rogers, J., Fasel, N., Wall, R. 1986. A novel RNA in which the 5' end is generated by cleavage at the poly(A) site of immunoglobulin heavy-chain secreted mRNA. *Mol. Cell. Biol.* 6:4749–52

199a. Rogers, J., Wall, R. 1980. A mechanism for RNA splicing. *Proc. Natl. Acad. Sci. USA* 77:1877–79

200. Rogers, S. L., Letourneau, P. C., Peterson, B. A., Furcht, L. T., McCarthy, J. B. 1987. Selective interaction of peripheral and central nervous system cells with two distinct cell-binding domains of fibronectin. *J. Cell. Biol.* 105:1435–42

201. Rokeach, L. A., Haselby, J. A., Hoch, S. O. 1988. Molecular cloning of a cDNA encoding the human Sm-D autoantigen. *Proc. Natl. Acad. Sci. USA* 85:4832–36

202. Rosenfeld, M. G., Amara, S. G., Evans, R. M. 1984. Alternative RMA processing: determining neuronal phenotype. *Science* 225:1315–20

203. Ruby, S. W., Abelson, J. 1988. An early hierarchic role of U1 small nuclear ribonucleoprotein in spliceosome assembly. *Science* 242:1028–1035

204. Ruiz-Opazo, N., Nadal-Ginard, B. 1987. A-Tropomyosin gene organization. *J. Biol. Chem.* 261:4755–65

205. Ruskin, B., Green, M. R. 1985. Specific and stable intron-factor interactions are established early during in vitro pre-mRNA splicing. *Cell* 43:131–42

206. Ruskin, B., Green, M. R. 1985. Role of the 3' splice site consensus sequence in mammalian pre-mRNA splicing. *Nature* 317:732–34

207. Ruskin, B., Greene, J. M., Green, M. R. 1985. Cryptic branch point activation allows accurate in vitro splicing of human beta-globin intron mutants. *Cell* 41:833–44

208. Ruskin, B., Krainer, A. R., Maniatis, T., Green, M. R. 1984. Excision of an intact intron as a novel lariat structure during pre-mRNA splicing in vitro. *Cell* 38:317–31

209. Ruskin, B., Zamore, P. D., Green, M. R. 1988. A factor, U2AF, is required for U2 snRNP binding and splicing complex assembly. *Cell* 52:207–19

210. Rymond, B. C., Rosbash, M. 1985. Cleavage of 5' splice site and lariat formation are independent of 3' splice site in yeast mRNA splicing. *Nature* 317:735–37

211. Sakano, H., Huppi, K., Heinrich, G., Tonegawa, S. 1979. Sequences at the somatic recombination sites of immunoglobulin light-chain genes. *Nature* 280:288–94

212. Salditt-Georgieff, M., Harpold, M., Chen-Kiang, S. 1980. The addition of 5' cap structures occurs early in hnRNA synthesis and prematurely terminated molecules are capped. *Cell* 19:69–78

213. Salz, H. K., Cline, T. W., Schedl, P. 1987. Functional changes associated with structural alternations induced by mobilization of a P element inserted in the Sex-Lethal gene of Drosophila. *Genetics* 117:221–31

214. Santoni, M. J., Barthels, D., Vopper, G., Boned, A., Goridis, C., Wille, W. 1989. Differential exon usage involving

an unusual splicing mechanism generates at least eight types of NCAM cDNA in mouse brain. *EMBO J.* 8:385–92

215. Santoro, C., Mermod, N., Andrews, P. C., Tijan, R. 1988. A family of human CCAAT-box-binding proteins active in transcription and expression of multiple cDNAs. *Nature* 334:218–24

216. Schachat, F. H., Diamond, M. S., Brandt, P. W. 1987. The effect of different troponin T-tropomyosin combinations on thin filament activation. *J. Mol. Biol.* 198:551–54

217. Schmitt, P., Gattoni, R., Keohavong, P., Stevenin, J. 1987. Alternative splicing of E1a transcripts of adenovirus requires appropriate ionic conditions in vitro. *Cell* 50:31–39

218. Schreiber, E., Matthias, P., Muller, M. M., Schaffner, W. 1988. Identification of a novel lymphoid specific octamer binding protein (OTF-2B) by proteolytic clipping bandshift assay (PCBA). *EMBO J.* 7:4221–29

219. Schwarz, T. L., Tempel, B. L., Papazian, D. M., Jan. Y. N., Jan, L. Y. 1988. Multiple potassium-channel components are produced by alternative splicing at the shaker locus in Drosophila. *Nature* 331:137–42

220. Schwarzbauer, J. E., Patel. R. S., Fonda, D., Hynes, R. O. 1987. Multiple sites of alternative splicing of the rat fibronectin gene transcript. *EMBO J.* 6:2573–80

221. Schwarzbauer, J. E., Tamkun, J. W., Lemischka, I. R., Hynes, R. O. 1983. Three different fibronectin mRNAs arise by alternative splicing within the coding region. *Cell* 35:421–31

222. Seraphin, B., Kretzner, L., Rosbash, M. 1988. A u1 snRNA:pre-mRNA base pairing interaction is required early in yeast spliceosome assembly but does not uniquely define the 5' cleavage site. *EMBO J.* 7:2533–38

223. Shapiro, M. B., Senapathy, P. 1987. RNA splice junctions of different classes of eukaryotes: statistics and functional implications in gene expression. *Nucleic Acids Res.* 15:7155–74

224. Sharp, P. A. 1981. Speculations on RNA splicing. *Cell* 23:643–46

225. Sharp, P. A. 1987. Splicing of messenger RNA precursors. *Science* 235:766–71

226. Shaw, G., Kamen, R. 1986. A conserved AU sequence from the 3' untranslated region of GM-CSF mRNA mediates selective mRNA degradation. *Cell* 46:659–67

227. Shelley, C. S., Baralle, F. E. 1987. Deletion analysis of a unique 3' splice site indicates that alternating guanine and thymine residues represent and efficient splicing signal. *Nucleic Acids Res.* 15:3787–99

228. Siliciano. P. G., Guthrie, C. 1988. 5' Splice site selection in yeast: genetic alterations in base-pairing with U1 reveal additional requirements. *Genes Dev.* 2:1258–67

229. Sillekens, P. T., Habets, W. J., Beijer, R. P., VanVenrooij, W. J. 1987. cDNA cloning of the human U1 snRNA-associated A protein: extensive homology between U1 and U2 snRNP-specific proteins. *EMBO J.* 6:3841–48

230. Small, S. J., Haines, S. L., Akeson, R. A. 1988. Polypeptide variation in an N-CAM extracellular immunoglobulin-like fold is developmentally regulated through alternative splicing. *Neuron* 1:1007–17

231. Smith, C. W. J., Gallego, M. E., Andreadis, A., Gooding, C., Nadal-Ginard, B. 1989. Investigation of the mechanism of mutually exclusive splicing in the alpha-tropomyosin, troponin T, and myosin light chain 1/3 genes. In *Cellular and Molecular Biology of Muscle Development,* ed. L. H. Kedes, F. E. Stockdale, pp. 573–84. New York: Liss

232. Smith, C. W. J., Nadal-Ginard, B. 1989. Mutually exclusive splicing of alpha-troponyosin exons enforced by an unudsual lariat branch point location; inplications for constitutive splicing. *Cell* 56:749–58

233. Solnick, D. 1985. Alternative splicing caused by RNA secondary structure. *Cell* 43:667–76

234. Solnick, D. 1985. *Trans* splicing of mRNA precursors. *Cell* 42:157–64

235. Solnick, D., Lee, S. I. 1987. Amount of RNA secondary structure required to induce an alternative splice. *Mol. Cell Biol.* 7:3194–98

236. Somasekhar, M. B., Mertz, J. E. 1985. Exon mutations that affect the choice of splice sites used in processing the SV 40 late transcripts. *Nucleic Acids Res.* 13:5591–5609

237. Spritz, R. A., Jagadeeswaran, P., Choudary, P. V., Biro, P. A., Elder, J. T. 1981. Base substitution in an intervening sequence of a β+-thalassemic human globin gene. *Proc. Natl. Acad. Sci.* 78:2455–59

238. Spritz, R. A., Strunk, K., Surowy, C. S., Hoch, S. O., Banon, D. E., Francke, U. 1987. The human U1-70K snRNP protein: cDNA cloning, chromosomal localization, expression,

alternative splicing and RNA-binding. *Nucleic Acids Res.* 15:10373–10392

239. Stanford, D. R., Perry, C. A., Holicky, E., Rohleder, A., Wieben, E. D. 1988. The snRNP E protein gene contains four introns and has upstream similarity to genes for ribosomal proteins. *J. Biol. Chem.* 263:17772–79

240. Stanford, D. R., Rohleder, A., Neiswanger, K., Wieben, E. D. 1987. DNA sequence of a human Sm autoimmune antigen. *J. Biol. Chem.* 262:9931–34

241. Strehler, E. E., Periasamy, M., Strehler-Page, M. E., Nadal-Ginard, B. 1985. Myosin light chain 1 and 3 gene has two structurally distinct and differentially regulated promoters evolving at different rates. *Mol. Cell. Biol.* 5:3168–82

242. Streuli, M., Saito, H. 1989. Regulation of tissue-specific alternative splicing: exon-specific cis-elements govern the splicing of leukocyte common antigen pre-mRNA. *EMBO J.* 8:787–96

243. Stroeher, V. L., Gaiser, C., Garber, R. L. 1988. Alternative RNA splicing that is spatially regulated: generation of transcripts from the antennapedia gene of Drosophila melanogaster with different protein-coding regions. *Mol. Cell. Biol.* 8:4143–54

244. Stuart, D. I., Levine, M., Muirhead, H., Stammers, D. K. 1979. Crystal structure of cat muscle pyruvate kinase at a resolution of 2.6 A. *J. Mol. Biol.* 134:109–42

245. Sudhof, T. C., Goldstein, J. L., Brown, M. S., Russell, D. W. 1985. The LDL receptor gene: a mosaic of exons shared with different proteins. *Science* 228:815–22

246. Sutton, R. E., Boothroyd, J. C. 1986. Evidence for trans splicing in trypanosomes. *Cell* 47:527–35

247. Swanson, M. S., Dreyfus, G. 1988. RNA binding specificity of hnRNP proteins: a subset bind to the 3' end of introns. *EMBO J.* 7:3519–29

248. Tani, T., Ohshima, Y. 1989. The gene for the U6 small nuclear RNA in fission yeast has an intron. *Nature* 337:87–90

249. Tazi, T., Alibert, C., Temsamani, J., Reveillaud, I., Cathala, G., et al. 1986. A protein that specifically recognizes the 3' splice site of mammalian pre-mRNA introns is associated with a small nuclear ribonucleoprotein. *Cell* 47:755–66

250. Theissen, H., Etzerodt, M., Reuter, R., Schneider, C., Lottspeich, F., et al. 1986. Cloning of the human cDNA for the U1 RNA-associated 70K protein. *EMBO J.* 5:3209–17

251. Timpe, L. C., Schwarz, T. L., Tempel, B. L., Papazian, D. M., Jan, Y. N., Jan, L. Y., 1988. Expression of functional potassium channels from shaker cDNA in Xenopus oocytes. *Nature* 331:143–45

252. Tong, B. T., Auer, D. E., Jaye, M., Kaplow, J. M., Ricca, G., 1987. cDNA clones reveal differences between human glial and endothelial cell platelet derived growth factor A-chains. *Nature* 328:619–21

253. Treisman, R., Orkin, S. H., Maniatis, T. 1983. Specific transcription and RNA splicing defects in five cloned β-thalassemia genes. *Nature* 302:591–96

254. Tsurushita, N., Avdalovic, N. M., Korn, L. J. 1987. Regulation of differential processing of mouse immunoglobulin m heavy chain mRNA. *Nucleic Acids Res.* 15:4603–15

255. Tsurushita, N., Ho, L., Korn, L. J. 1988. Nuclear factors in B-lymphoma enhance splicing of mouse membrane-bound μ-mRNA in Xenopus oocytes. *Science* 239:494–97

256. Ulfendahl, P. J., Kreivi, J-P., Akusjarvi, G. 1989. Role of the branch site/3'-splice site region in adenovirus-2 E1A pre-mRNA alternative splicing: evidence for 5'- and 3'-splice site cooperation. *Nucleic Acids Res.* 17:925–38

257. Weinzierl, R., Axton, J. M., Ghysen, A., Akam, M. 1987. Ultrabithorax mutations in constant and variable regions of the protein coding sequence. *Genes Dev.* 1:386–97

258. Deleted in proof

259. Wieczorek, D., Smith, C. W. J., Nadal-Ginard, B. 1988. The rat a-tropomyosin gene generates a minimum of six different mRNAs coding for striated, smooth and non-muscle isoforms by alternative splicing. *Mol. Cell Biol.* 8:679–94

260. Wieringa, B., Hofer, E., Weissmann, C. 1984. A minimal intron length but no specific internal sequence is required for splicing the large rabbit β-globin intron. *Cell* 37:915–25

261. Woppman, A., Rinke, J., Luhrmann, R. 1988. Direct cross-linking of snRNP proteins F and 70K to snRNAs by ultraviolet radiation in situ. *Nucleic Acids Res.* 16:10985–11004

262. Wright, C. V. E., Cho, K. W. Y., Fritz, A., Burglin, T. R., De Robertis, E. M. 1987. A Xenopus laevis gene encodes both homeobox-containing and homeobox-less transcripts. *EMBO J.* 6:4083–4094

263. Yamamoto, K., Miura, H., Moroi, Y., Yoshinoya, S., Goto, M., et al. 1988. Isolation and characterization of a complementary DNA expressing human U1 small nuclear ribonucleoprotein C polypeptide. *J. Immunol.* 140:311–17

264. Zachar, A., Chou, T-B., Bingham, P. M. 1987. Evidence that a regulatory gene autoregulates splicing of its transcript. *EMBO J.* 6:4105–4111

265. Zeitlin, S., Efstradiatis, A. 1984. In vivo splicing products of the rabbit β-globin gene. *Cell* 39:589–602

266. Zhuang, Y., Leung, H., Weiner, A. M. 1987. The natural 5' splice site of simian virus 40 large T antigen can be improved by increasing the base complementarity to U1 RNA. *Mol. Cell Biol.* 7:3018–20

267. Zhuang, Y., Weiner, A. M. 1986. A compensatory base charge in U1 snRNA suppresses a 5' splice site mutation. *Cell* 46:827–35

267a. Zhuang, Y., Goldstein, A. M., Weiner, A. M. 1989. UACUAAL is the preferred branch site for mammalian mRNA splicing. *Proc. Natl. Acad. Sci. USA* 86:2752–56

268. Ziff, E. B. 1982. Transcription and RNA processing by the DNA tumour viruses. *Nature* 287:491–99

269. Zillman, M., Rose, S. D., Berget, S. M. 1987. U1 small nuclear ribonucleoproteins are required early during spliceosome assembly. *Mol. Cell Biol.* 7:2877–83

Annu. Rev. Genet. 1989. 23:579–604

STRUCTURE AND FUNCTION OF TELOMERES

Virginia A. Zakian

Fred Hutchinson Cancer Research Center, 1124 Columbia Street, Seattle, Washington 98104

CONTENTS

TELOMERES DEFINED

Telomeres are the physical ends of eukaryotic chromosomes. The concept of the telomere derives from the pioneering studies in the 1930s and 40s of Hermann J. Müller (using *Drosophila melanogaster*) and Barbara McClintock (using *Zea mays*). By analyzing the frequency and types of chromosomal rearrangements arising after X-radiation, Müller deduced that terminal deletions and terminal inversions are exceptionally rare. From these data, he argued that chromosome stability requires a specialized terminal structure, which he called the telomere (90, 91). McClintock's studies advanced this concept by providing a description of the fate of an end produced by chromosome breakage. In her experiments, ends were produced by breakage of a

0066-4197/89/1215-0579$02.00

dicentric chromosome when the two centromeres attempted to move to opposite poles in meiotic anaphase. Whereas the broken end was reactive and
often fused with other broken ends (thereby re-forming a dicentric chromosome and continuing the breakage-fusion-bridge cycle), natural chromosome
ends were stable, fusing neither with broken ends nor with other telomeres
(83, 84). McClintock also demonstrated that in some tissues, broken ends
could be stabilized or "healed", presumably by the acquisition of a new
telomere.

Although telomeres were first defined in studies with multicellular organisms with visible chromosomes, much of our molecular understanding of
telomeres comes from studies carried out on single-celled eukaryotes. For the
purposes of this review, the termini on achromosomal (i.e. noncentromere-
bearing) DNA molecules of ciliates and other lower eukaryotes and on
artificially constructed linear yeast plasmids, as well as the ends of authentic
chromosomes, are called telomeres. The telomere, defined in molecular
terms, is the simple repeats found at these DNA termini and the proteins that
bind specifically to those sequences in vivo. DNA termini on viral,
mitochondrial, and nonnuclear plasmid DNA molecules are not considered in
this review.

TELOMERE FUNCTION

As first demonstrated by Müller and McClintock, the telomere provides a
protective "cap" for the end of the chromosome. In contrast to telomeres,
broken chromosomes and free DNA ends are susceptible to end-to-end fusions and exonucleolytic degradation (e.g. 53, 78, 83, 84, 112). By protecting against these events, telomeres deter formation of dicentric chromosomes
(which are usually unstable) and prevent loss of genetic information from
sub-telomeric regions of the chromosome.

Telomeres may also be involved in establishing the three-dimensional
architecture of the interphase nucleus. In many cells, individual telomeres are
positioned close to the nuclear envelope and 180° away from their centromeres (the "Rabl" orientation; reviewed in 44). Telomeres mediate transient
associations between both homologous and nonhomologous chromosomes.
For example, during prophase of the first meiotic division, chromosomes in
most species, assume the "bouquet" orientation in which telomeres are clustered together near the nuclear envelope (reviewed in 73). In polytene
chromosomes of Dipteran species, telomeres engage in ectopic pairing (associations between nonhomologous chromosomes) (58, 115). The role of telomere-telomere and telomere-nuclear envelope interactions is not known but it
is tempting to speculate that these interactions can influence gene expression,
chromosome behavior, or both. For example, in *Drosophila* (72) and *Saccharomyces* (D. Gottschling & V. Zakian, in preparation), transcription of a

gene can be repressed by placing it near a telomere. Conversely, in *Trypano-soma*, surface antigen genes are transcribed only when they are situated at a telomere (reviewed in 99). It remains to be determined if the effects of telomeres on transcription are related to the telomere's position within the nucleus rather than, for example, to a different chromatin structure arising from the telomere's association with specialized binding proteins.

Finally, telomeres in some way facilitate replication of the ends of chromosomes. Conventional DNA polymerases can synthesize DNA only in the 5' to 3' direction and cannot begin DNA synthesis de novo: they require a primer, which is usually RNA. After primer removal, gaps would remain at the 5' ends of newly replicated strands (137; Figure 1). Presumably telomeres in some way circumvent this dilemma and prevent the gradual loss of genetic information from the ends of chromosomes.

SEQUENCE AND STRUCTURE OF TELOMERES

The termini on the linear extrachromosomal ribosomal DNA (rDNA) molecules from the ciliated protozoan *Tetrahymena thermophila* were the first telomeres to be sequenced (15). These studies were carried out on native DNA termini by exploiting the fact that the *Tetrahymena* macronuclear rDNA is easily isolated because of its high copy number ($\sim 10^4$ copies/ macronucleus) and small size (21 kb) (*Tetrahymena* rDNA is reviewed in 142). These studies revealed that individual rDNA termini carry a variable

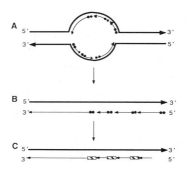

Figure 1   Replication of linear DNA molecules by conventional mechanisms leaves gaps at the 5' ends of newly replicated strands. Chromosomal DNA replication is primed by short stretches of RNA (● ●) and proceeds bi-directionally 5'→3' from an origin. Newly replicated DNA is indicated by thin solid lines, parental strands by thicker solid lines. After (or during) replication, RNA primers are removed leaving single-strand gaps. Because internal gaps have 3' hydroxyls (indicated by arrowheads), they can be repaired by DNA polymerase and ligation. Repair replication is indicated by hatched boxes (▱▱▱). A gap remains at 5' ends of newly replicated strands.

number ($\sim$120 to 420 bps) of C_4A_2/T_2G_4 repeats (15). The C_4A_2 strand runs 5' to 3' from the end of the rDNA molecule towards its center. Single base gaps are found in the terminal $\sim$100 bps of the C_4A_2 strand with gaps occurring every two to three repeats. These gaps make it possible to label *Tetrahymena* rDNA termini in vitro by nick translation with *Escherichia coli* DNA polymerase I without the prior addition of DNAse I.

The terminal sequence and structure of macronuclear DNA molecules from several species of *Oxytricha* and another hypotrichous ciliate, *Stylonchia*, were also determined by direct sequence analysis of native DNA termini (69, 95, 101). In these cases, sequence analysis was facilitated by the incredible abundance (for *Oxytricha*, $\sim$10^7 termini per macronucleus) and homogeneity of termini. All (or nearly all) of the macronuclear DNA molecules were found to have the same terminal sequence and structure:

$$5' \ C_4A_4C_4A_4C_4 \ . \ . \ .$$

$$3' \ OH \ G_4T_4G_4T_4G_4T_4G_4T_4G_4 \ . \ . \ .$$

The extension of the G-rich strand, first described in hypotrichs, may be a general feature of telomeres (see replication section).

In most organisms, telomeres are not sufficiently abundant to permit direct sequence analysis of native termini, and telomere sequencing must be carried out on cloned DNA. Cloning telomeres is complicated by the fact that fragments derived from the ends of chromosomes have only one restriction-enzyme generated terminus and are therefore rarely represented in a conventional DNA library. To create blunt ends from telomeres, genomic DNA is usually treated briefly either with the single-strand specific nuclease S1 or the (primarily) double-strand exonuclease Bal31. Both nucleases will remove 3' and 5' single-strand overhangs and both act at single-strand nicks. Since treatment with either enzyme eliminates the telomeric DNA distal to any nicks or gaps, cloned terminal restriction fragments obtained by S1 or Bal31 digestion are $\sim$50–500 bps shorter than the same restriction fragment in genomic DNA (e.g. 7, 119, 130). Alternatively, blunt-ended fragments can be produced from terminal restriction fragments by treatment with T4 DNA polymerase in the presence of dNTPs (102, 116, 136). The 3'→5' exonuclease associated with the T_4 DNA polymerase will remove 3' single-strand overhangs, but in the presence of dNTPs, the exonuclease does not act at nicks (79). Therefore, repair of termini with T4 DNA polymerase results in loss of very little telomeric DNA (S.-S. Wang & V. Zakian, unpublished results). Once a blunt end has been created near or at the telomere, a variety of strategies can be used to enrich for terminal restriction fragments (e.g. 41, 89, 110, 130). Finally, terminal restriction fragments can be isolated by their ability to act as telomeres in *Saccharomyces*. Terminal restriction fragments from *Tetrahymena* r DNA (34, 128), *Oxytricha* macronuclear DNA (100), and human chromosomal DNA (18, 32, 111) (as well as from *Saccharomyces* chromosomes; 128) can promote telomere formation when ligated to the ends of a yeast vector (see section on Formation of Telomeres). For at least some telomeric sequences, cloning in yeast also circumvents the problem of instability of telomeric DNA in *E. coli* (e.g. 110, 116, 119).

Once a putative telomeric fragment has been identified, a number of criteria can be used to demonstrate its terminal location on authentic chromosomes. The most definitive evidence for terminality is sensitivity to digestion by the exonuclease Bal31 (141, 144). Bal31 degrades the 5' and 3' strands of duplex DNA in a processive manner. Typically genomic DNA is subjected to Bal31 digestion for increasing amounts of time, then digested with a restriction enzyme and analyzed by gel electrophoresis and Southern hybridization. Whereas the sizes of terminal restriction fragments will decrease as a result of Bal31 digestion, the sizes of fragments derived from internal regions of the chromosome should not change. Mapping wth multiple restriction enzymes can also be used to demonstrate that a fragment is from the end of a chromosome by the fact that the telomere behaves like a "universal" recognition site for restriction enzyme cleavage (128, 141). For some organisms, in situ hybridization can be used to localize a sequence near telomeres (89).

Using these methods, telomeres have been cloned and sequenced from various lower eukaryotes, higher plants, and mammals (Table I). Telomeres carry multiple copies of simple repeats. Although the exact sequence of the repeat varies from organism to organism, most can be described by the consensus $5'-C_{(1-8)}^{A}T_{(1-4)}-3'$ (11). One version of this consensus, $5'-C_3TA_2-3'$ is found at telomeres in many distantly related organisms: the flagellated protozoan *Trypanosoma* (13, 130), an acellular slime mold (42), and all vertebrates tested to date (86, 89). Although many organisms, like *Tetrahymena* and *Oxytricha*, have a precise telomeric repeat (C_4A_2 and C_4A_4 respectively), in others the repeat is more heterogeneous (e.g. $C_{2-3}A(CA)_{1-3}$ in *Saccharomyces;* 119; $C_{1-8}T$ in *Dictyostelium;* 38). In *Dictyostelium*, the repeat unit may be considerably larger than the derived $C_{1-8}T$ consensus, since individual telomeres bear long tracts (>100 bps) of identical sequence (38). In many organisms, the telomeric sequence is also found at internal sites on the chromosome (e.g. *Saccharomyces;* 133: *Tetrahymena* macronuclear DNA; 141: acellular slime molds; 42). Telomeric sequences have a defined orientation in that the C-rich strand always runs 5' to 3' from the end towards the interior of the DNA molecule. In some cases, nicks or single base gaps in the C-rich strand were demonstrated directly (*Tetrahymena;* 15: *Physarum;* 65) and, in others, inferred from the ability of DNA polymerase to nick-translate telomeric sequences in vitro (*Trypanosoma;* 13: *Plasmodium;* 104: *Saccharomyces*; 128).

Table I   Repeat unit of telomere sequences[a]

Organism		Reference
PROTOZOANS		
Holotrichous ciliates		
Tetrahymena macronuclear DNA	5' C_4A_2 3'	15, 144
Paramecium macronuclear DNA	5' $C_3^CA_2$ 3'	3, 41
Hypotrichous ciliates		
Macronuclear DNA of Oxytricha,	5' C_4A_4 3'	69, 95, 101
Euplotes, and Stylonchia		
Oxytricha micronuclear DNA	5' C_4A_4 3'	35
Sporozoite		
Plasmodium	5' $C_3T_G^AAA$ 3'	104
Flagellates		
Trypanosoma	5' C_3TA_2 3'	13, 130
YEASTS		
Saccharomyces	5' $C_{2-3}A(CA)_{1-3}$ 3'	119
Schizosaccharomyces	5' $C_{1-6}G_{0-1}T_{0-1}GTA_{1-2}$ 3'	Sugawara & Szostak (un-published results) cited in 1, 81
SLIME MOLDS		
Dictyostelium	5' $C_{1-8}T$ 3'	38
Didymium	5' C_3TA_2 3'	42
VERTEBRATES	5' C_3TA_2 3'	86, 89
PLANTS		
Arabdopsis	5' C_3TA_3 3'	110

[a]The sequence of the strand running 5' $\rightarrow$ 3' from the end of the DNA molecule towards its interior is presented.

The amount of simple repeated DNA at individual telomeres can range from as little as 20 bps (for macronuclear DNA in hypotrichous ciliates like *Oxytricha*) up to several kilobases (e.g. *Oxytricha* micronuclear chromosomes; 35: *Arabadopsis;* 110: humans; 89). Moreover, for virtually all organisms, the amount of telomeric DNA varies from telomere to telomere (e.g. 118). As a result, terminal restriction fragments, even from a single chromosome, are heterogeneous in size and produce "fuzzy" bands after gel electrophoresis. In *Trypanosoma* (9) and *Tetrahymena* (70), terminal restriction fragments gradually increase in size (3–10 bps/generation) when cells are kept in continuous log-phase growth and can also suffer precipitous decreases in size. In *Saccharomyces*, the length of telomeric $C_{2-3}A(CA)_{1-3}$ tracts (hereafter abbreviated $C_{1-3}A$) do not increase during logarithmic growth (24, 116, 118) but can change in response to the addition of excess telomere DNA brought in on high copy number plasmids (116) or when cells are grown under limiting conditions for DNA polymerase I (23, 24). In *Saccharomyces*, the length of $C_{1-3}A$ tracts is under the control of multiple genes with naturally occurring variant alleles such that the average telomere length varies from strain to strain (61, 134). In some organisms, telomere length may vary during development: the length of terminal restriction fragments from human sex chromosomes are ~5 kb longer in sperm DNA than in DNA extracted from blood from the same person (30). Taken together, these results argue that telomeres are dynamic structures. The fact that telomeres can both grow and shrink argues that telomere length is determined by a balance of processes, some of which result in telomere elongation (e.g. replication or recombination) and some that result in telomere degradation (e.g. exonucleases, incomplete replication).

In summary, a general picture of telomere sequence and structure has emerged that applies to telomeres in organisms from single-celled eukaryotes to mammals and higher plants. That is, telomeres bear a variable number of simple repeats characterized by a defined polarity and the segregation of Gs and Cs largely into complementary strands, with nicks or gaps occurring in the C-rich strand. There are exceptions to this generalization. DNA termini on hypotrichous macronuclear DNA molecules are short, discrete, and lack nicks in the C-rich strand (69, 95, 101). The sequence deduced for telomeres of the fission yeast *Schizosaccharomyces pombe* is irregular and not easily related to the derived consensus (Table I). The termini of *Physarum* extrachromosomal rDNA may also bear sequences of a more complex nature than those found at most telomeres (7). *Physarum* rDNA termini were cloned after S1 treatment of isolated rDNA (a process that removed the terminal ~150 bps). These cloned termini bear 6–10 copies of a tandem, imperfect 140 ± 4 bps repeat, each of which contains internal inverted repetitions (palindromes). The sequence C_3TA (or its complement) is often found within the 140-bp repeat. These data could be reconciled with a more typical telomere structure if the very ends of chromosomes in *Physarum* and *Schizosaccharomyces* bear a short stretch of a conventional telomeric repeat that was lost during telomere cloning.

Many of the sequences that serve as telomeres in vivo have unusual physical properties in vitro, the most striking being their unusual base-pairing abilities. Nonidentical telomeric sequences can hybridize to each other or to telomere-like sequences in vitro (1, 86, 110). Some of the hybrids thus formed show remarkable stability: for example, poly $d(GT)\cdot d(CA)$ hybridizes at high stringency to *Saccharomyces* $C_{2-3}A$ $(CA)_{1-3}$ telomeric repeats (100, 135). Thus, cross-hybridization is not sufficient to argue that two organisms share the same telomeric repeat sequence. Pairing might also occur between nonidentical telomeric sequences in vivo: in *Saccharomyces*, recombination can occur between terminal stretches of C_4A_2 repeats and short ($\leqslant 28$ bps) terminal C_4A_4 tracts (102, 136). The G-rich strand from at least some telomeric sequences is able to engage in both intrastrand (55) and interstrand (96) associations in vitro. All tested G-rich telomeric strands $\geqslant 12$ bases in length displayed temperature-dependent changes in conformation due to intrastrand interactions as detected by altered mobility in nondenaturing polyacrylamide gels (55). NMR analysis of the structure of $(T_2G_4)_4$ is consistent with a structure held together primarily by non Watson-Crick guanine-guanine base pairs (55). The 16 base single-strand $(G_4T_4)_2$ tail on *Oxytricha* macronuclear DNA has been shown to mediate end-to-end associations of individual macronuclear DNA molecules in vitro. This association requires Na^+ for formation and can be stabilized by K^+, which is not typical behavior for B-form DNA (96). The authors hypothesize that the $(G_4T_4)_2$ tails associate to form triplex or quadraplex structures. Further evidence for an unusual structure for telomeric DNA is its hypersensitivity to nucleases (22, 56, 109). Although the exact structure of telomeric DNA is not known, its unusual behavior does not seem to be due to its assuming a left-handed (Z DNA) helical structure (22, 56). If all telomeric sequences formed a similar DNA structure, it might explain why different telomeric sequences can promote telomere formation in vivo in *Saccharomyces* (e.g. 102) and serve as substrates in vitro for *Tetrahymena* telomerase (50, 51).

SOLUTIONS FOR REPLICATION OF DNA TERMINI

A variety of solutions for replication of DNA termini have been proposed and can be grouped into three general categories (Figure 2): (*a*) models that invoke unusual DNA structures to permit replication via standard replication enzymes; (*b*) models that posit specialized recombination events to generate intermediates whose replication can be completed by standard replication enzymes; and (*c*) models that postulate novel enzymes for telomere replication.

The dilemma for replication of linear DNA molecules can be resolved if the structure of the telomere allows elimination of ends during DNA replication. The best example of end elimination is the circularization of bacteriophage lambda DNA by annealing the complementary 12 base single-strand tails at each end of its genome (reviewed in 146). Alternatively, ends can be effectively eliminated by extension of one of the two strands to form a terminal hairpin (Figure 2A). Terminal hairpin models predict palindromic sequences (self-complementary) at DNA termini and a site-specific endonuclease to cut at the boundary of (or within) the palindrome to allow dissociation of sister chromatids after replication (5, 25). Although some animal viruses (pox viruses, parvo viruses; see (4) and references therein), bacterial plasmids (2), and *Paramecium* mitochondrial DNA (107) have a terminal hairpin, there is little evidence for crosslinked termini or palindromic sequences at the ends of eukaryotic chromosomes (e.g. 127).

Because all telomeres in a given organism bear multiple copies of the same DNA sequence, telomeres should be able to recombine with each other. A

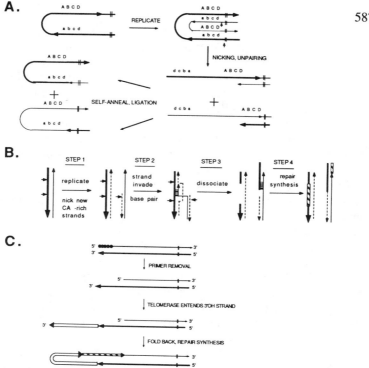

Figure 2 Models for replication of DNA termini. A. Terminal Hairpins: Based on the Bateman (5) modification of the Cavalier-Smith (25) model for telomere replication. During most of the cell cycle, the DNA helix is continuous (i.e. if denatured, the chromosome would be a very large single-strand circle). Parental strands are indicated by heavy solid lines, newly replicated DNA by thinner solid lines. Arrowheads indicate the 3' ends of the DNA backbone. Terminal sequences (indicated by letters) are palindromic. After replication, site-specific nicking occurs (position of nicks indicated by arrows), followed by unpairing to allow separation of sister chromatids. B. Recombination mediated replication: Picture based on model in (133) as described in (102). Each strand of a DNA duplex is represented by a line: solid lines are parental with the thinner line being the G-rich strand; dashed lines indicated newly replicated DNA. The 3' end of each strand is indicated by an arrowhead. Nicks in C-rich strand are indicated by horizontal arrows. Step 1: after DNA replication and RNA primer removal, a 5' terminal gap is created on each duplex (for simplicity, only one such gap is shown). Nicks are made on the newly replicated C-rich strand. Step 2: the free 3' tail (thin line) invades the sister duplex, base pairs in an out-of-register configuration with the complementary strand (thick line), displacing its complementary strand (dashed line). Step 3: dissociation of the newly base-paired region is facilitated by nicks present in the C-rich strand of the donor (thick line). Step 4: the internal gap and terminal 3' gap on each duplex can be filled by repair synthesis (hatched line, ▨) and remaining nicks sealed by ligase. C. Telomerase mediated replication: Picture based on model in (50, 51). The terminal region of a chromosome after replication is presented: thick lines indicate parental DNA; thinner lines indicate newly replicated DNA; circles indicate the RNA primer. Arrow heads indicate the 3' ends of the DNA strands. After replication and primer removal, the telomere has a single-strand 3'OH G-rich tail that can be extended by telomerase (open box). The extended tail folds back on itself, pairing by non Watson-Crick interactions (55), to form a primer that can be extended by repair replication (hatched line, ▨).

variety of recombination models for telomere replication have been proposed (7, 33, 40, 57, 87, 130, 133). Most models hypothesize formation and resolution of cross-strand exchanges in telomeric sequences (Figure 2B). These recombination events can occur between telomeric sequences located either on the same or at different telomeres or homologous sequences located elsewhere on the chromosome. Theoretically, recombination can move the gap at the 5' end of a newly replicated strand either to a more internal region of the chromosome or to a strand with a terminal 3'OH, both of which can be readily repaired by conventional enzymes (Figure 2B). There is some evidence for recombination at the ends of DNA molecules. Genetic and biochemical data indicate that replication of the ends of the bacteriophage T4 genome is initiated from recombinational intermediates formed when the 3' single strands remaining after primer removal invade and pair with homologous sequences on another T4 DNA molecule (reviewed in 40). In *Physarum*, electron microscopic studies reveal that ~2% of termini on native rDNA molecules interact with other DNA termini to form structures interpreted as being intermediates in recombinational replication (7). During formation of new telomeres in baker's yeast, the ends of linear DNA molecules recombine with telomeric sequences at other termini. This recombination involves the very ends of DNA molecules and is independent of the major pathway for mitotic recombination (102, 136). Models for recombination-dependent telomere replication are attractive because they can account for such unusual properties of telomeres as their ability to grow and shrink and the presence of single-strand nicks or gaps that are frequently found in telomeric sequences. However, even in *Saccharomyces* where telomere-telomere recombination has been demonstrated directly (102), there is no evidence that this recombination is exploited for replication.

Special proteins may allow replication of telomeres by novel mechanisms. Replication of some viruses such as the bacteriophage ϕ29 or the adeno group of animal viruses is accomplished by elongation of a protein-bound deoxymononucleotide (reviewed in 140). The result is that the 5' ends of newly replicated strands are covalently attached to the terminal protein and start with a base not encoded by the template strand. Although covalently bound proteins have also been detected at the 5' ends of linear plasmids in fungi (e.g. 67, 85) and mitochondrial plasmids in plants (39, 66), they are not found on the ends of achromosomal DNA molecules in ciliates (12, 48), or at the ends of *Saccharomyces* chromosomes or linear plasmids (V. Zakian, D. Gottschling, A. Pluta, K. Runge, unpublished results).

Telomerase, a telomere-specific terminal transferase first identified in ciliated protozoans, provides an excellent candidate for a telomere-specific polymerase (50, 148). (See also the Telomere Protein section). This activity can extend the 3' end of the G-rich strand of a variety of telomeric sequences in the absence of a DNA template (Figure 2C). The extended 3'OH strand can then serve as a template for primase and conventional polymerase-mediated replication of the complementary strand. Although RNA primer removal would still leave a gap at the end of the 5' strand, no sequence information would be lost. Alternatively, if the extended 3' OH strand can fold back on itself to form a terminal hairpin (as suggested by in vitro data; 55), it could serve as a primer to fill in the 5' gap left after RNA primer removal (Figure 2C). Replication via telomerase predicts that chromosomes should bear 3' G-rich single-strand tails at each end. It is well documented that termini on macronuclear DNA of hypotrichous ciliated protozoans like *Oxyticha* bear a 16-base $(G_4T_4)_2$ single-strand extension (69, 95, 101). Recently, sensitivity to the single-strand specific probe osmium tetraoxide was used to demonstrate that the G-rich strand is extended as a 12–16 base overhang on at least a subset of the extrachromosomal rDNA molecules in a ciliate *(Tetrahymena)* and a slime mold *(Didymium)* (54).

STRUCTURE OF SUBTELOMERIC REGIONS

In addition to the simple repeats found at the very ends of chromosomes, subtelomeric regions of chromosomes usually contain middle-repetitive elements called telomere-associated sequences (*Saccharomyces;* 27, 28: *Plasmodium;* 31, 98: *Drosophila;* 115:147: *Chironomus;* 117: *Secale;* 6: humans; 29, 124). In some species, telomere-associated sequences extend for several hundred kb (*Chironomus;* 117) and can be visualized as large blocks of heterochromatin (*Secale;* 6). For example, telomere-associated sequences account for > 10% of both the mass and length of *Secale cereale* chromosomes (6). In some species, copies of telomere-associated sequences are also present at interstitial sites (for example, in *Drosophila* in pericentric heterochromatin; 129, 147). In Diptera, telomere-associated sequences hybridize to the fibers that connect the telomeres of nonhomologous chromosomes engaged in ectopic pairing (115, 117, 147). Sequence analysis indicates that at least some telomere-associated sequences are internally repetitious (31, 38, 60, 98).

Telomere-associated sequences have been studied in greatest detail in *Saccharomyces* where there are two such classes of sequences, called Y' and X (27, 28) (see Figure 3 for the structure of a yeast telomere). Y' is a highly conserved element of 6.7 kb found in zero to four tandem copies per telomere (27, 28). X, a less conserved element that ranges in size from 0.3 to 3.7 kb, is found at most telomeres (27, 28, 149). When X and Y' are at the same telomere, Y' lies closer to the end of the chromosome (27). Tracts of $C_{1-3}A$ repeats, the same sequence found at the ends of yeast chromosomes, are found between X and Y' and between tandem copies of Y' (133). These internal stretches of $C_{1-3}A$ are shorter (50–130 bps) but otherwise similar to those at the ends of chromosomes. Likewise, in *Plasmodium*, telomere-associated sequences contain stretches of telomere or telomere-like sequences (98). Given the DNA requirements for forming new telomeres in *Saccharomyces* (92, 102; see next section), the internal tracts of $C_{1-3}A$ sequences could serve as "back up" telomeres: if chromosomes were degraded or broken so as to position these tracts near the end of the chromosome, these interstitial $C_{1-3}A$ repeats would be used efficiently as substrates for telomere formation.

X and Y' are both replicated late in the S phase (82). Moreover, X and Y' both contain an autonomously replicating sequence *ARS* (26). (*ARS*s are defined by their ability to promote self-replication of plasmid DNA molecules in yeast.) Since at least some *ARS*s serve as origins of replication on plasmids (17, 62) and chromosomes (63), telomere-associated *ARS*s could be specialized replication origins for chromosome ends. However, the X-associated *ARS* on the left arm of chromosome III does not appear to act as a replication origin (J. Huberman, personal communication). Moreover, 60-kb artificial chromosomes lacking Y' *ARS*s are replicated and segregated in mitosis and meiosis with the same efficiency as artificial chromosomes with Y' *ARS*s (138). Also, telomeric *ARS*s are not required to form a new telomere in *Saccharomyces* (92, 102, 138). Thus, in *Saccharomyces*, telomeric replication origins do not appear to be crucial for telomere replication or function.

Probably the most characteristic feature of telomere-associated sequences is their variability. For example, in *Saccharomyces* (149), *Drosophila* (147), and *Plasmodium* (98), the distribution of telomere-associated sequences on individual chromosomes varies from strain to strain. In humans, telomere-associated sequences on the sex chromosomes are highly polymorphic (29, 123, 124). It is in this region that obligate recombination between the X and Y chromosomes occurs (reviewed in 108). The often dramatic interstrain length variations characteristic of *Plasmodium* chromosomes are confined to

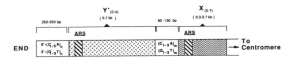

Figure 3   Structure of a yeast telomere. *Saccharomyces* chromosomes end in ~250–350 bps of $C_{1-3}A$ repeats. Individual telomeres have zero to four tandem copies of Y'; most telomeres have a copy of X. Between tandem Ys and at the Y'–X junction are 50–130 bps of $C_{1-3}A$ repeats. X and Y' both have *ARS*s sequences) (based on refs. 26–28, 133, 149).

regions near the ends of chromosomes (31). Variability in amount and kind of telomere-associated sequences is probably related to their high rate of recombination. This recombination is likely to be enhanced by the internally repetitious nature of telomere-associated sequences. Although the most detailed data are from *Saccharomyces* where rearrangement of Y' elements can be detected easily in both mitosis and meiosis (60, 61), data from other systems also indicate that subtelomeric regions of chromosomes, like telomeres themselves, are dynamic structures capable of rapid evolution.

Although widespread, telomere-associated sequences are not ubiquitous. In *Saccharomyces*, pulsed field gel analysis demonstrated that at least one chromosome in every strain examined lacked Y' completely, and in some strains, chromosome I, the smallest yeast chromosome, lacked detectable amounts of both X and Y' (149). Telomere-associated sequences are also apparently absent from some naturally occurring deletion derivatives of *Plasmodium* chromosomes (31, 103). Moreover, telomere-associated sequences can be deleted from *Saccharomyces* chromosome III without affecting mitotic chromosome stability (94). Taken together, these data argue against an essential role for telomere-associated sequences, although probably they will eventually be shown to serve an ancillary function in, for example, chromosome pairing during meiosis, repair of broken chromosomes after radiation damage, or protection of chromosome ends from degradation.

FORMATION OF NEW TELOMERES

Telomeres were first defined by the unstable behavior of chromosomes that lack telomeres. Thus, implicit in their definition was the notion that formation of new telomeres is a relatively rare event. However, for some organisms, there are developmental stages during which large-scale formation of new telomeres occurs.

The chromosomes of some nematodes, such as *Ascaris suum* and *Parascaris equorum*, undergo a process called chromatin diminution in those cells destined to form the somatic lineage of the animal (ref. 16 as described in 139). Chromatin diminution results in elimination of the heterochromatic ends of each chromosome and fragmentation of the rest of the chromosome to form multiple new chromosomes. Because the smaller chromosomes thus formed are stable, new telomeres are presumably formed at heretofore interstitial sites. In *A. suum*, $\sim$22% of the germline DNA is eliminated and the eliminated sequences consist predominately of many copies of two-related A+T-rich repeats of 125 and 131 bps (88, 113, 114, 126). After diminution, about 200–1000 copies of these repeats remain, although it is not known if they are localized at the telomeres of somatic chromosomes (114).

Chromosome fragmentation and telomere formation also occur during formation of the somatic nucleus in ciliated protozoa. In most ciliates, each cell contains one or more diploid micronuclei and one or more polyploid macronuclei. The micronucleus or germline nucleus contains typical sized eukaryotic chromosomes. The polyploid macronucleus is the site of transcription; it contains achromosomal DNA molecules that are derived from the micronuclear chromosomes by a series of events involving chromosome cleavage, DNA elimination, and telomere formation. In *Tetrahymena*, ~200 new telomeres are formed during development of the macronucleus. In hypotrichous ciliates like *Oxytricha*, macronuclear DNA molecules are very small (~0.4–20 kb) and abundant (~10^7 DNA molecules/macronucleus), and massive synthesis of new telomeres is required (reviewed in 143). In *Tetrahymena* and *Oxytricha*, regions of micronuclear DNA homologous to the ends of macronuclear DNA molecules have been sequenced (reviewed in 143). These studies reveal that telomeric repeats on macronuclear DNA molecules are not encoded in their micronuclear counterparts. Indeed, in only one case, the micronuclear sequence destined to be the 3' end of the *Tetrahymena* palindromic rDNA molecule, is there even a single copy of C_4A_2 (68). Therefore, during formation of the macronucleus, the telomeric repeats must be added de novo.

What are the rules for formation of new telomeres during macronuclear development? In *Tetrahymena*, the presence of the 15-bp chromosome breakage sequence (*cbs*) seems to be necessary and sufficient for fragmentation of micronuclear chromosomes (145; M.-C. Yao, personal communication). Since the *cbs* is eliminated during chromosome rearrangement, it probably does not serve as the signal for telomere-sequence addition. Indeed, sites of telomere formation, unlike sites of chromosome breakage, do not appear to be sequence-determined. For example, in *Paramecium*, the precise point of telomere addition for a given macronuclear chromosome can occur over a region of ~800 bps (3, 41). Only the 3' end of the *Tetrahymena* extrachromosomal rDNA molecule appears to be sequence-determined (143; M.-C. Yao, personal communication). In this case, the macronuclear addition of C_4A_2 repeats usually occurs at the single C_4A_2 repeat encoded in the micronuclear DNA. Other than the 3' end of the rDNA, the data suggest that any DNA terminus is a suitable substrate for telomere addition in the developing ciliate macronucleus. In *Paramecium*, even in mature macronuclei, there seem to be no sequence requirements for telomere formation (45, 46).

It seems likely that telomerase, an activity found in elevated amounts in developing macronuclei of *Tetrahymena* (50), is involved in forming new telomeres during macronuclear development. However, telomerase requires a telomeric primer for activity in vitro whereas most micronuclear fragments lack even a single telomere repeat. Moreover, a site shown by DNA sequence analysis to be a site for telomere addition in *Tetrahymena* in vitro does not serve as a substrate for *Tetrahymena* telomerase in vitro (125). Therefore, if telomerase is responsible for formation of telomeres during macronuclear development, its activity must be different in vivo and in vitro or a second activity must supply the primer.

In both nematodes and ciliates, massive genome rearrangement occurs in somatic but not germline nuclei. In nematodes, most of the eliminated DNA occurs as long heterochromatic blocks at the ends of germline chromosomes. Although for most ciliates the structure of micronuclear telomeres is not known, in *Oxytricha* micronuclear chromosomes end in stretches of C_4A_4 repeats far longer (up to 6 kb; 35) than the 36 bases of telomeric sequences characteristic of macronuclear DNA molecules. The ends of human chromosomes are also several kb longer in germline cells compared to somatic cells (30). Although these length differences could be due to trivial reasons, they may reflect a specialized role for telomeres in meiosis.

In *Saccharomyces,* new telomeres can be formed readily in vegetative cells. The assay for telomere formation is based on the observation that a plasmid can be maintained as a linear DNA molecule in yeast only if it has a functional telomere at each end. Because terminal restriction fragments from ciliates can be obtained easily, they were the first sequences tested for ability to act as telomeres. Terminal restriction fragments from either *Tetrahymena* macronuclear rDNA (34, 128) or total *Oxytricha* macronuclear DNA (100) were ligated to both ends of a plasmid and introduced into cells by transformation. Linearity was established by conventional gel electrophoresis and restriction enzyme mapping of plasmid DNA (34, 128) or by the behavior of plasmids during two-dimensional agarose gel electrophoresis under conditions that separate circular and linear DNA molecules (100). The functional cloning assay was also used to isolate an authentic yeast telomere (128). More recently, C_3TA_2 bearing terminal restriction fragments from human chromosomes have been shown to promote telomere formation in yeast (18, 32, 111).

Although the structure of linear plasmids with foreign termini was essentially that predicted from the structure of the input DNA, the terminal restriction fragments were about 200–1000 bps longer after propagation in yeast than when taken directly from *Tetrahymena* or *Oxytricha* (100, 128). Southern hybridization (100, 135) and DNA sequence analysis (119) indicate that the foreign termini were modified by the terminal addition of $C_{1-3}A$ repeats (the same sequence found at the ends of yeast chromosomes). This addition occurs in the absence of the *RAD52* gene product (36, 150). (*RAD52* is required for the majority of the recombination events in mitotic cells, presumably because it is necessary for repair of double-strand breaks; reviewed in 97). Thus, in yeast as in ciliates, formation of new telomeres involves the addition of telomeric repeats. However, in contrast to the situation in ciliates, in yeast telomere formation only occurs on substrates with a short stretch of telomeric DNA (see below).

Cloned telomeric DNA $>\sim100$ bps from *Tetrahymena* (bearing C_4A_2 repeats), *Oxytricha* (C_4A_4 repeats), or *Saccharomyces* ($C_{1-3}A$ repeats) can also efficiently promote telomere formation in yeast (92, 93, 102, 138). Thus, none of the in vivo modifications found at telomeres, such as nicks in the C-rich strand or a 3' single-strand tail, appears to be required for recognition in yeast. Telomere formation occurs on termini with as few as 28 bps of telomeric repeats although the efficiency of telomere formation decreases as the telomeric tract gets shorter (102). DNA termini bearing 16 bps (or fewer) of telomeric sequence (102) or telomeric DNA with the C-rich strand running $3' \rightarrow 5'$ (reverse of the in vivo orientation) (92, 102) do not act at detectable levels as substrates for telomere formation. Thus, telomeric repeats in their in vivo orientation are necessary and sufficient for telomere formation. Surprisingly, the telomeric repeats do not have to be at the very end of the molecule to support the addition of $C_{1-3}A$ repeats (92). During de novo telomere formation, DNA termini often recombine with other termini by a *RAD52*-independent pathway (102). Although the role of this recombination is not known, it is possible that the $C_{1-3}A$ repeats could also be added by recombination. Alternatively, if telomerase exists in yeast, it could explain yeast's ability to recognize and modify foreign telomeric sequences. Taken together, the data predict that de novo telomere formation will occur efficiently on natural yeast chromosomes if the chromosome is broken near a $5'C_{1-3}A3'$ or a $5'C_{1-3}A3'$-like tract of $>\sim20$ bps. Alternatively, a broken chromosome can be healed by *RAD52*-mediated gene conversion between the broken chromosome and an intact chromosome that has homology to the broken end (36, 53, 132).

PROTEINS THAT INTERACT WITH TELOMERES

The unusual replication requirements for DNA termini make it likely that specialized enzymes will interact with telomere DNA during DNA replication. Telomerase is an excellent candidate for a telomere-specific replication enzyme. Although telomerase was first isolated from and has been most extensively studied in extracts made from *Tetrahymena* developing macronuclei (50–52), it has also been detected in hypotrichous ciliates (*Oxytricha, Euplotes;* 120, 148) and HeLa cells (87). The *Tetrahymena* activity adds up to 30 T_2G_4 repeats onto the single-stranded DNA oligonucleotide $(T_2G_4)_4$ (50). The activity can also extend oligonucleotides comprised of the G-rich telomeric strand from *Saccharomyces* (50), *Trypanosoma, Oxytricha,* and *Dictyostelium* (51). Although the *Tetrahymena* activity can utilize different telomeric sequences as substrates for the addition reaction, it always adds G_4T_2 repeats to the substrate. In vivo, telomerase would be capable of extending a single-strand 3'-OH G-rich strand, the intermediate predicted to be left after replication and primer removal at the ends of linear DNA molecules (Figure 1). The activity of *Tetrahymena* (51) and *Oxytricha* (148) telomerase is sensitive to both protease and RNase. The single-copy gene encoding a 159 base RNA that copurifies with *Tetrahymena* telomerase contains the sequence 5'-$CA_2C_4A_2$-3' (52). Because complimentary DNA oligonucleotides made to this portion of the RNA decrease telomerase activity in vitro, it seems likely that the RNA provides the template that determines the sequence of the added DNA (52).

Telomeres are also likely to be associated with specialized structural proteins. In nuclei prepared from a variety of organisms, telomeric DNA is protected from degradation from nucleases or chemical cleavage agents. Although this protection is protease-sensitive, the size of the protected complex argues against its being a conventional nucleosome (*Tetrahymena,* 14; *Oxytricha,* 47; *Dictyostelium,* 37; *Physarum,* 75; *Saccharomyces,* D. Gottschling, J. Wright, V. A. Zakian, unpublished results). For example, in *Tetrahymena,* the protected telomeric DNA fragments are both larger and more heterogeneous (400–800 bps) than the DNA in nucleosomes (~200 bps) (14). In *Oxytricha,* the terminal ~100 bps on each molecule (including the 20 bps of C_4A_4/T_4G_4 duplex and the 16 base single-strand tail) are protected (47). In *Tetrahymena* (21), *Oxytricha* (47), and *Dictyostelium* (37), the nucleosomes in regions immediately adjacent to the telomeric repeats are precisely positioned (phased) relative to the underlying DNA sequence.

The telomeric DNA-protein complex at the ends of *Oxytricha* macronuclear DNA molecules is maintained even after chromatin is extracted with 2M salt (47). This behavior was exploited to facilitate the purification of two of the proteins in the telomeric DNA-protein complex (48). Two very basic proteins of 55 and 43 kDa copurify with 2M salt-extracted chromatin (48, 105, 106). The terminal location of these proteins in 2M salt-extracted chromatin was demonstrated by their ability to promote (*a*) retention of terminal restriction fragments onto nitrocellulose; (*b*) immunoprecipitation of terminal restriction fragments with antisera raised against telomere proteins; and (*c*) protection of terminal restriction fragments from digestion by Bal31 exonuclease (48, 49). Their ability to protect DNA termini from exonucleotytic degradation in vitro suggests that these proteins could also provide protection for telomeres in vivo.

Because the *Oxytricha* telomere proteins are associated so tenaciously with DNA, most procedures that remove them from DNA termini also destroy their DNA-binding ability (48, 105). Therefore, DNA-free telomere proteins were prepared by extensive micrococcal nuclease digestion of 2M salt-extracted chromatin. Telomere proteins prepared in this way efficiently and specifically associate with authentic DNA termini and with synthetic termini made by annealing deoxy-oligonucleotides of the appropriate sequence (48). By varying the sequence and structure of the synthetic termini, the 16 base $(G_4T_4)_2$ tail was shown to be essential for protein binding (48). Although the exact sequence of the duplex region adjacent to the 3' single-strand tail is not crucial, the most efficient substrate for the telomere proteins was DNA termini with the sequence and structure of authentic termini (48). Based on their in vitro binding properties, these proteins would be expected to bind efficiently in vivo only to natural termini. If similar proteins exist in organisms with more conventional chromosomes, they would provide an explanation for the cell's ability to distinguish between broken ends and true telomeres. The *Oxytricha* terminus binding proteins are probably the same proteins shown to mediate end-to-end association of *Oxytricha* macronuclear DNA molecules in vitro (74). Thus, these proteins are also candidates for mediating telomere-telomere associations in vivo.

In *Saccharomyces,* gel retardation assays were used to identify an activity that binds in vitro to $C_{1-3}A$ tracts either at telomeres or at internal loci; i.e. unlike the *Oxytricha* proteins, this activity does not require a DNA terminus for binding (8). The binding activity is the product of an essential gene *RAP1* (19, 20, 121). This protein also binds in vitro to many other sequences (20), including sites necessary for repression of transcription at the silent mating-type loci (19, 122) or for activation of transcription of some ribosomal protein genes (64, 131) and some glycolytic enzymes (19, 20). *RAP1,* an abundant protein ($>\sim4 \times 10^3$ molecules/cell; 20) fractionates efficiently with the nuclear scaffold and can promote formation of sequence-specific loops in vitro in some DNA molecules with *RAP1*-binding sites (59). Overexpression of *RAP1* affects telomere structure and chromosome stability (M. Conrad & V. Zakian, unpublished results). These data can be explained if interaction of *RAP1* with telomeres and with other *RAP1* binding sites is important for establishing the three-dimensional structure of the nucleus; if so, this structure is likely to be important for controlling gene expression and chromosome behavior.

In order to determine the binding properties of proteins that interact with *Saccharomyces* telomeres in vivo, a vast excess of potential binding sites for telomere proteins was introduced by transformation (116). Strains with 200–800 extra telomeres or with 100–200 circular plasmids that each bear a ~500 bp tract of $C_{1-3}A$ DNA were constructed (for comparison, haploid cells have 32 telomeres and ~17 kb of $C_{1-3}A$ DNA). This enormous excess of telomeric DNA had no detectable effect on viability, doubling time (116), or chromosome loss (K. Runge & V. Zakian, unpublished results). (In contrast, addition of ~5 extra centromeres is toxic to haploid yeast cells; 43). Increases in the number of termini or in the amount of $C_{1-3}A$ DNA did have the unexpected effect of increasing the lengths of terminal $C_{1-3}A$ tracts at all telomeres (116). These data suggest that in yeast telomere growth is inhibited by a limiting factor(s) that specifically recognizes $C_{1-3}A$ sequences and that this factor can be competed for by $C_{1-3}A$ sequences at telomeres or on circular plasmids.

Genetic analysis has also been used to identify genes whose products might interact with *Saccharomyces* telomeres. One approach was to screen for mutants with altered lengths of telomeric $C_{1-3}A$ tracts. Strains with mutations in *TEL1* and *TEL2* have telomeres 200–300 bps shorter than the starting strain but no other detectable phenotype (77). Although these mutants might be defective in processes that protect or replicate telomeres, mutations in a number of genes with general effects on DNA metabolism, such as the gene encoding the catalytic subunit of DNA polymerase I (23,24), also affect the length of telomeric $C_{1-3}A$ tracts. Thus, it is not clear if the products of *TEL1* or *TEL2* interact directly with telomeres. The phenotype of cells with mutations in the *EST1* gene (Ever Shorter Telomeres) argues for a more direct role for the *EST1* product in telomere biology (76). Null mutants in *EST1* display a progressive decrease in both telomere length and chromosome stability. Cell death, presumably due to chromosome loss, is concomitant with the loss of telomere DNA. *EST1* encodes a protein of 699 amino acids with no detectable similarity to other sequenced proteins. Based on the phenotype of *est1* mutants, the product of the *EST1* gene could be a structural protein, like the *Oxytricha* terminus binding proteins, required for protection of telomeres or an enzyme, like telomerase, required for telomere replication.

ARE TELOMERES REALLY ESSENTIAL?

The inability to recover terminal deletions after X-radiation of *Drosophila* chromosomes was crucial to the development of the concept of the telomere as an essential part of the chromosome (90, 91). However, two situations have now been described where terminal deletions can be obtained readily in *Drosophila*. In *mu-2* strains, terminal deletions can be recovered on all chromosomes after X-radiation of females (80). Once terminal deletions are recovered, they can be maintained in wild-type strains. However, in all strains, DNA is progressively lost from the end of the deleted chromosome at a rate of ~75 bp/fly generation (10). Likewise, when a P element transposon inserted in a subtelomeric region of a chromosome is induced to move, a process that, like X-radiation, induces chromosome breaks, terminal deletions of the chromosome bearing the P element are recovered. Again, these terminal deletions are unstable, with sequences being progressively lost from the end of the broken chromosome. Restriction enzyme mapping demonstrates that terminally deleted chromosomes produced by either method are capped by very little, if any, telomeric DNA. Therefore, the progressive loss of DNA is probably due to incomplete replication. This interpretation leads to the heretical conclusion that *Drosophila* chromosomes without telomeres are not particularly susceptible to degradation, end-to-end fusions, or loss.

FUTURE PROSPECTS

Molecular analysis reveals that many features of telomeric DNA are conserved among all eukaryotes. Among these conserved features are properties suggesting that telomeres assume an unusual DNA conformation. Continued analysis of telomeres provides an excellent opportunity to elucidate the biological function of a non-B form DNA. In vivo, telomeric DNA is complexed with nonhistone proteins that can protect the telomere from nucleases. In the ciliated protozoan *Oxytricha,* two of the proteins in the telomeric complex have been isolated. The in vitro behavior of these proteins can explain a number of the classic phenotypes of telomeres. It remains to be determined if terminus binding proteins are important for telomere structure or function in organisms with a more conventional chromosome structure and to elucidate the contributions of telomere proteins to the establishment of the three-dimensional organization of the nucleus. Determining the mechanism of telomere replication will continue to be a crucial area of research. Does recombination play a role in replication or maintenance of telomeres? Is telomerase required for formation of new telomeres and/or telomere replication in ciliates and other organisms? One of the most intriguing questions is the exact role of the telomerase RNA: is it a template, a ribozyme, or both? Finally, it needs to be demonstrated in multicellular organisms that telomeres are essential for chromosome integrity. Alternatively, somatic cells might have a phenotype similar to the *Saccharomyces est1* mutant: in this case, the progressive loss of DNA from the ends of chromosomes would be a normal and inevitable consequence of cell division that ultimately results in cell death.

ACKNOWLEDGMENTS

I thank my many colleagues who sent me reprints and preprints for this review. I am indebted to M. Conrad, R. Levis, K. Runge, S.-S. Wang, R. Wellinger, J. Wright, and M.-C. Yao for their comments on the manuscript. I thank R. Wellinger for Figure 3 and M. Conrad for help with the figures.

Research in my lab is supported by grant NP-574 from the American Cancer Society and GM-26938 from the National Institutes of Health.

Literature Cited

1. Allshire, R. C., Gosden, J. R., Cross, S. H., Cranston, G., Rout, D., et al. 1988. Telomeric repeat of *T. thermophila* cross hybridizes with human telomeres. *Nature* 332:656–59
2. Barbour, A. G., Garon, C. F. 1987. Linear plasmids of the bacterium *Borrelia burgdorferi* have covalently closed ends. *Science* 37:409–11
3. Baroin, A., Prat, A., Caron, F. 1987. Telomeric site position heterogeneity in macronuclear DNA of *Paramecium primaurelia*. *Nucleic Acids Res.* 15:1717–728
4. Baroudy, B. M., Venkatesan, S., Moss, B. 1982. Incompletely base-paired flip-flop terminal loops link the two DNA strands of the vaccinia virus genome into one uninterrupted polynucleotide chain. *Cell* 28:315–24
5. Bateman, A. J. 1975. Simplification of palindromic telomere theory. *Nature* 253:379–80
6. Bedbrook, J. R., Jones, J., O'Dell, M., Thompson, R. D., Flavell, R. B. 1980. A molecular description of telomeric heterochromatin in *Secale* species. *Cell* 19:545–60
7. Bergold, P. J., Campbell, G. R., Littau, V. C., Johnson, E. M. 1983. Sequence and hairpin structure of an inverted repeat series at termini of the Physarum extrachromosomal rDNA molecule. *Cell* 32:1287–99
8. Berman, J., Tachibana, C. Y., Tye, B-K. 1986. Identification of a telomere-binding activity from yeast. *Proc. Natl. Acad. Sci. USA* 83:3713–17
9. Bernards, A., Michels, P. A. M., Lincke, G. R., Borst, P. 1983. Growth of chromosomal ends in multiplying trypanosomes. *Nature* 303:592–97
10. Biessmann, H., Mason, J. M. 1988. Progressive loss of DNA sequences from terminal chromosome deficiencies in *Drosophila melanogaster*. *EMBO J.* 7:1081–86
11. Blackburn, E. H. 1984. Telomeres: do

the ends justify the means? *Cell* 37:7–8
12. Blackburn, E. H., Budarf, M. L., Challoner, P. B., Cherry, J. M., Howard, E. A., et al. 1983. DNA termini in ciliate macronuclei. *Cold Spring Harbor Symp. Quant. Biol.* 47:1195–207
13. Blackburn, E. H., Challoner, P. B. 1984. Identification of a telomeric DNA sequence in *Trypanosoma brucei*. *Cell* 36:447–57
14. Blackburn, E. H., Chiou, S. S. 1981. Non-nucleosomal packaging of a tandemly repeated DNA sequence at termini of extrachromosomal DNA coding for rRNA in *Tetrahymena*. *Proc. Natl. Acad. Sci. USA* 78:2263–67
15. Blackburn, E. H., Gall, J. G. 1978. A tandemly repeated sequence at the termini of the extrachromosomal rRNA genes in *Tetrahymena*. *J. Mol. Biol.* 120:33–53
16. Boveri, T. 1887. Über Differenzierung der Zellkerne während der Fürchung des Eies von *Ascaris megalocephala*. *Anat. Anz.* 2:688–93
17. Brewer, B. J., Fangman, W. L. 1987. The localization of replication origins on ARS plasmids in *S. cerevisiae*. *Cell* 51:463–71
18. Brown, W. R. A. 1989. Molecular cloning of human telomeres in yeast. *Nature* 338:774–76
19. Buchman, A. R., Kimmerly, W. J., Rine, J., Kornberg, R. D. 1988. Two DNA-binding factors recognize specific sequences at silencers, upstream activating sequences, autonomously replicating sequences & telomeres in *Saccharomyces cerevisiae*. *Mol. Cell Biol.* 8:210–25
20. Buchman, A. R., Lue, N. F., Kornberg, R. D. 1988. Connections between transcriptional activators, silencers, and telomeres as revealed by functional analysis of a yeast DNA-binding protein. *Mol. cell. Biol.* 8:5086–99
21. Budarf, M. L., Blackburn, E. H. 1986.

Chromatin structure of the telomeric region & 3'-nontranscribed spacer of *Tetrahymena* ribosomal RNA genes. *J. Biol. Chem.* 261:363–69

22. Budarf, M., Blackburn, E. H. 1987. S1 nuclease sensitivity of a double-stranded telomeric DNA sequence. *Nucleic Acids Res.* 15:6273–92

23. Carson, M. J. 1987. *CDC17, the structural gene for DNA polymerase I of yeast: mitotic hyper-recombination and effects on telomere metabolism.* PhD thesis. Univ. Wash., Seattle.

24. Carson, M. J., Hartwell, L. 1985. *CDC17:* an essential gene that prevents telomere elongation in yeast. *Cell* 42:249–57

25. Cavalier-Smith, T. 1974. Palindromic base sequences and replication of eukaryote chromosome ends. *Nature* 250:467–70

26. Chan, C. S. M., Tye, B-K. 1980. Autonomously replicating sequences in *Saccharomyces cerevisiae. Proc. Natl. Acad. Sci. USA* 77:6329–33

27. Chan, C. S. M., Tye, B-K. 1983. Organization of DNA sequences and replication origins at yeast telomeres. *Cell* 33:563–73

28. Chan, C. S. M., Tye, B-K. 1983. A family of *Saccharomyces cerevisiae* repetitive autonomously replicating sequences that have very similar genomic environments. *J. Mol. Biol.* 168:505–23

29. Cooke, H., J., Brown, W. R. A., Rappold, G. A. 1985. Hypervariable telomeric sequences from the human sex chromosomes are pseudoautosomal. *Nature* 317:687–92

30. Cooke, H. J., Smith, B. A. 1986. Variability at the telomeres of the human X/Y pseudoautosomal region. *Cold Spring Harbor Symp. Quant. Biol.* 51:213–19

31. Corcoran, L. M., Thompson, J. K., Walliker, D., Kemp, D. J. 1988. Homologous recombination within subtelomeric repeat sequences generates chromosome size polymorphisms in *P. falciparum. Cell* 53:807–13

32. Cross, S. H., Allshire, R. C., McKay, S. J., McGill, N. I., Cooke, H. J. 1989. Cloning of human telomeres by complementation in yeast. *Nature* 338:771–74

33. Dancis, B. M., Holmquist, G. P. 1979. Telomere replication & fusion in eukaryotes. *J. Ther. Biol.* 78:211

34. Dani, G. M., Zakian, V. A. 1983. Mitotic & meiotic stability of linear plasmids in yeast. *Proc. Natl. Acad. Sci. USA* 80:3406–10

35. Dawson, D., Herrick, G. 1984. Telomeric properties of C_4A_4-homologous sequences in micronuclear DNA of *Oxytricha fallax. Cell* 36:171–77

36. Dunn, B., Szauter, P., Pardue, M. L., Szostak, J. W. 1984. Transfer of yeast telomeres to linear plasmids by recombination. *Cell* 39:191–201

37. Edwards, C. A., Firtel, R. A. 1984. Site-specific phasing in the chromatin of the rDNA in *Dictyostelium discoideum. J. Mol. Biol.* 180:73–90

38. Emery, H. S., Weiner, A. M. 1981. An irregular satellite sequence is found at the termini of the linear extrachromosomal rDNA in *Dictyostelium discoideum. Cell* 26:411–19

39. Erickson, L., Beversdorf, W. D., Pauls, K. R. 1985. Linear mitochondrial plasmid in *Brassica* has terminal protein. *Curr. Genet.* 9:679–82

40. Formosa, T., Alberts, B. M. 1986. DNA synthesis dependent on genetic recombination: Characterization of a reaction catalyzed by purified bacteriophage T4 proteins. *Cell* 47:793–806

41. Forney, J. D., Blackburn, E. H. 1988. Developmentally controlled telomere addition in wild-type and mutant *Paramecia. Mol. Cell. Biol.* 8:251–58

42. Forney, J., Henderson, E. R., Blackburn, E. H. 1987. Identification of the telomeric sequence of the acellular slime molds *Didymium iridis* and *Physarum polycephalum. Nucleic Acids Res.* 15: 9143–52

43. Futcher, B., Carbon, J. 1986. Toxic effects of excess cloned centromeres. *Mol. Cell. Biol.* 6:2213–22

44. Gasser, S. M., Laemmli, U. K. 1987. A glimpse at chromosomal order. *Trends Genet.* 3:16–22

45. Gilley, D., Preer, J. R. Jr., Aufderheide, K. J., Polisky, B. 1988. Autonomous replication and addition of telomerelike sequences to DNA microinjected into *Paramecium tetraurela* macronuclei. *Mol. Cell. Biol.* 8:4765–72

46. Godiska, R., Aufderheide, K. J., Gilley, D., Hendrie, P., Fitzwater, T., et al. 1987. Transformation of *Paramecium* by microinjection of a cloned serotype gene. *Proc. Natl. Acad. Sci. USA* 84:7590–94

47. Gottschling, D. E., Cech, T. R. 1984. Chromatin structure of the molecular ends of *Oxytricha* macronuclear DNA: Phased nucleosomes and a telomeric complex. *Cell* 38:501–10

48. Gottschling, D. E., Zakian, V. A. 1986. Telomere proteins: Specific recognition and protection of the natural termini of

Oxytricha macronuclear DNA. *Cell* 47:195–205

49. Gottschling, D. E., Zakian, V. A. 1988. DNA-protein interactions at telomeres in ciliated protozoans. *Adv. Cell Biol.* 2:291–307

50. Greider, C. W., Blackburn, E. H. 1985. Identification of a specific telomere terminal transferase activity in *Tetrahymena* extracts. *Cell* 43:405–13

51. Greider, C. W., Blackburn, E. H. 1987. The telomere terminal transferase of *Tetrahymena* is a ribonucleoprotein enzyme with two kinds of primer specificity. *Cell* 51:887–98

52. Greider, C. W., Blackburn, E. H. 1989. A telomeric sequence in the RNA of *Tetrahymena thermophila* required for telomere repeat synthesis. *Nature* 337:331–37

53. Haber, J. E., Thorburn, P. C. 1984. Healing of broken linear dicentric chromosomes in yeast. *Genetics* 106:207–26

54. Henderson, E. R., Blackburn, E. H. 1989. An overhanging 3' terminus is a conserved feature of telomeres. *Mol. Cell. Biol.* 9:345–48

55. Henderson, E., Hardin, C. C., Walk, S. K., Tinoco, I. Jr., Blackburn, E. H. 1987. Telomeric DNA oligonucleotides form novel intramolecular structures containing guanine-guanine base pairs. *Cell* 51:899–908

56. Henderson, E., Larson, D., Melton, W., Shampay, J., Spangler, E., et al. 1988. Structure, synthesis, and regulation of telomeres. In *Eukaryotic DNA Replication*, pp. 453–61. Cold Spring Harbor, NY: Cold Spring Harbor Lab.

57. Heumann, J. M. 1976. A model for replication of the ends of linear chromosomes. *Nucleic Acids Res.* 3:3167–71

58. Hinton, T. 1945. A study of chromosome ends in salivary gland nuclei of *Drosophila*. *Biol. Bull.* 89:144–65

59. Hofmann, J. F-X., Laroche, T., Brand, A. H., Gasser, S. M. 1989. Purfied RAP-1 factor mediates DNA loop formation in vivo at the silent mating type locus, HML. *Cell* 57:725–37

60. Horowitz, H., Haber, J. E. 1984. Subtelomeric regions of yeast chromosomes contain a 36 base-pair tandemly repeated sequence. *Nucleic Acids Res.* 12:7105–21

61. Horowitz, H., Thornburn, P., Haber, J. E. 1984. Rearrangements of highly polymorphic regions near telomeres of *Saccharomyces cerevisiae*. *Mol. Cell Biol.* 4:2509–17

62. Huberman, J. A., Spotila, L. D., Nawotka, K. A., El-Assouli, S. M.,

Davis, L. R. 1987. The in vivo replication origin of the yeast $2\mu m$ plasmid. *Cell* 51:473–81

63. Huberman, J. A., Zhu, J., Davis, L. R., Newlon, C. S. 1988. Close association of a DNA replication origin and an *ARS* element on chromosome III of the yeast, *Saccharomyces cerevisiae*. *Nucleic Acids Res.* 16:6373–84

64. Huet, J., Sentenac, A. 1987. TUF, the yeast DNA-binding factor specific for UAS_{rpg} upstream activating sequences: Identification of the protein and its DNA-binding domain. *Proc. Natl. Acad. Sci. USA* 84:3648–52

65. Johnson, E. M. 1980. A family of inverted repeat sequences and specific single-strand gaps at the termini of the *Physarum* rDNA palindrome. *Cell* 22:875–86

66. Kemble, R. J., Thompson, R. D. 1982. S1 and S2, the linear mitochondrial DNAs present in a male sterile line of maize, possess terminally attached proteins. *Nucleic Acids Res.* 10:8181–90

67. Kikuchi, Y., Hirai, K., Hishinuma, F. 1984. The yeast linear DNA killer plasmids, pGKL1 & pGKL2, possess terminally attached proteins. *Nucleic Acids Res.* 12:5685–92

68. King, B. O., Yao, M.-C. 1982. Tandemly repeated hexanucleotide at *Tetrahymena* rDNA free end is generated from a single copy during development. *Cell* 31:177–82

69. Klobutcher, L. A., Swanton, M. T., Donini, P., Prescott, D. M. 1981. All gene-sized DNA molecules in four species of hypotrichs have the same terminal sequence and an unusual 3' terminus. *Proc. Natl. Acad. Sci. USA* 78:3015–19

70. Larsen, D. D., Spangler, E. A., Blackburn, E. H. 1987. Dynamics of telomere length variation in *Tetrahymena thermophila*. *Cell* 50:477–83

71. Levis, R. W. 1989. Viable deletions of a telomere from a Drosophila chromosome. *Cell.* 58:791–801

72. Levis, R., Hazelrigg, T., Rubin, G. M. 1985. Effects of genomic position on the expression of transduced copies of the *white* gene of *Drosophila*. *Science* 229:5 58–61

73. Lima-de-Faria, A. 1983. *Molecular Evolution and Organization of the Chromosome*, pp. 1186. Amsterdam, Neths.: Elsevier

74. Lipps, H. J., Gruissem, W., Prescott, D. M. 1982. Higher order DNA structure in macronuclear chromatin of the hypotrichous ciliate *Oxytricha nova*. *Proc. Natl. Acad. Sci. USA* 79:2495–99

75. Lucchini, R., Pauli, U., Braun, R., Koller, T., Sogo, J. M. 1987. Structure of the extrachromosomal ribosomal RNA chromatin of *Physarum polycephalum*. *J. Mol. Biol.* 196:829–43

76. Lundblad, V., Szostak, J. W. 1989. A mutant with a defect in telomere elongation leads to senescence in yeast. *Cell* 57:633–43

77. Lustig, A. J., Petes, T. D. 1986. Identification of yeast mutants with altered telomere structure. *Proc. Natl. Acad. Sci. USA* 83:1398–402

78. Mann, C., Davis, R. W. 1983. Instability of dicentric plasmids in yeast. *Proc. Natl. Acad. Sci. USA* 80:228–32

79. Masamune, Y., Richardson, C. C. 1971. Strand displacement during deoxyribonucleic acid synthesis at single strand breaks. *J. Biol. Chem.* 246:2692–701

80. Mason, J. M., Strobel, E., Green, M. M. 1984. *mu-2*: Mutator gene in *Drosophila* that potentiates the induction of terminal deficiencies. *Proc. Natl. Acad. Sci. USA* 81:6090–94

81. Matsumoto, T., Fukui, K., Niwa, O., Sugawara, N., Szostak, J. W., Yanagida, M. 1987. Identification of healed terminal DNA fragments in linear minichromosomes of *Schizosaccharomyces pombe*. *Mol. Cell. Biol.* 7:4424–30

82. McCarroll, R. M., Fangman, W. L. 1988. Time of replication of yeast centromeres and telomeres. *Cell* 54:505–13

83. McClintock, B. 1941. The stability of broken ends of chromosome in *Zea mays*. *Genetics* 26:234–82

84. McClintock, B. 1942. The fusion of broken ends of chromosomes following nuclear fusion. *Proc. Natl. Acad. Sci. USA* 28:458–63

85. Meinhardt, E., Kempken, F., Esser, K. 1986. Proteins are attached to the ends of a linear plasmid in the filamentous fungus *Ascobolus immersus*. *Curr. Genet.* 11:243–46

86. Meyne, J., Ratliff, R. L., Moyzis, R. K. 1989. Conservation of the human telomere sequence TTAGGG$_n$ among vertebrates. *Proc. Natl. Acad. Sci. USA.* In press

87. Morin, G. B., Cech, T. R. 1988. Mitochondrial telomeres: surprising diversity of repeated telomeric DNA sequences among six species of *Tetrahymena*. *Cell* 52:367–74

87a. Morin, G. B. 1989. The human telomere transferase enzyme is a ribonucleoprotein that synthesizes TTAGGG repeats. *Cell*. In press

88. Moritz, K. B., Roth, G. E. 1976. Complexity of germline and somatic DNA in *Ascaris*. *Nature* 259:55–57

89. Moyzis, R. K., Buckingham, J. M., Cram, L. S., Dani, M., Deaven, L. L., et al. 1988. A highly conserved repetitive DNA sequence, TTAGGG$_n$, present at the telomeres of human chromosomes. *Proc. Natl. Acad. Sci. USA* 85:6622–26

90. Müller, H. J. 1938. The remaking of chromosomes. *Collect. Net* 13:181–98

91. Müller, H. J., Herskowitz, I. H. 1954. Concerning the healing of chromosome ends produced by breakage in *Drosophila melanogaster*. *Am. Nat.* 88:177–208

92. Murray, A. W., Claus, T. E., Szostak, J. W. 1988. Characterization of two telomeric DNA processing reactions in *Saccharomyces cerevisiae*. *Mol. Cell. Biol.* 8:4642–50

93. Murray, A. W., Szostak, J. W. 1983. Construction of artificial chromosomes in yeast. *Nature* 305:189–93

94. Murray, A. W., Szostak, J. W. 1986. Construction and behavior of circularly permuted and telocentric chromosomes in *Saccharomyces cerevisiae*. *Mol. Cell. Biol.* 6:3166–72

95. Oka, Y., Shiota, S., Nakai, S., Nishida, Y., Okubo, S. 1980. Inverted terminal repeat sequence in the macronuclear DNA of *Stylonychia pustulata*. *Gene* 10:301–6

96. Oka, Y., Thomas, C. A., Jr. 1987. The cohering telomeres of *Oxytricha*. *Nucleic Acids Res.* 15:8877–98

97. Orr-Weaver, T. L., Szostak, J. W. 1985. Fungal recombination. *Microbiol. Rev* 49:33–58

98. Pace, T., Ponzi, M., Dore, E., Frontali, C. 1987. Telomeric motifs are present in a highly repetitive element in the *Plasmodium berghei* genome. *Mol. Biochem. Parsitol.* 24:193–202

99. Pays, E., Steinert, M. 1988. Control of antigen gene expression in African Trypanosomes. *Annu. Rev. Genet.* 22:107–26

100. Pluta, A. F., Dani, G. M., Spear, B. B., Zakian, V. A. 1984. Elaboration of telomeres in yeast: recognition and modification of termini from *Oxytricha* macronuclear DNA. *Proc. Natl. Acad. Sci. USA* 81:1475–79

101. Pluta, A. F., Kaine, B. P., Spear, B. B. 1982. The terminal organization of macronuclear DNA in *Oxytricha fallax*. *Nucleic Acids Res.* 10:8145–54

102. Pluta, A. F., Zakian, V. A. 1989. Recombination occurs during telomere formation in yeast. *Nature* 337:429–33

103. Pologe, L. G., Ravetch, J. V. 1988. Large deletions result from breakage and healing of *P. falciparum* chromosomes. *Cell* 55:869–74

104. Ponzi, M., Pace, T., Dore, E., Frontali, C. 1985. Identification of a telomeric DNA sequence in *Plasmodium berghei. EMBO J.* 4:2991–96

105. Price, C. M., Cech, T. R. 1987. Telomeric DNA-protein interactions of *Oxytricha* macronuclear DNA. *Genes Dev.* 1:783–93

106. Price, C. M., Cech, T. R. 1989. Properties of the telomeric DNA-binding protein from *Oxytricha nova. Biochemistry* 28:769–74

107. Pritchard, A. E., Cummings, D. J. 1981. Replication of linear mitochondrial DNA from *Paramecium:* Sequence and structure of the initiation-end crosslink. *Proc. Natl. Acad. Sci. USA* 78:7341–45

108. Pritchard, C., Goodfellow, P. N. 1985. The pseudoautosomal region and telomeres: the beginning of the end? *Trends Genet.* 1:289–90

109. Raibaud, A., Gaillard, C., Longacre, S., Hibner, U., Buck, G., et al. 1983. Genomic environment of variant surface antigen genes of *Trypanosoma equiperdum. Proc. Natl. Acad. Sci. USA* 80:4306–10

110. Richards, E. J., Ausubel, F. M. 1988. Isolation of a higher eukaryotic telomere from *Arabidopsis thaliana. Cell* 53:127–36

111. Riethman, H. C., Moyzis, R. K., Meyne, J., Burke, D. T., Olson, M. V. 1989. Cloning human telomeric DNA fragments into *Saccharomyces cerevisiae* using a yeast artificial-chromosome vector. *Proc. Natl. Acad. Sci. USA.* In press

112. Roth, D., Wilson, J. 1988. In *Genetic Recombination,* ed. R. Kucherlapati, G. R. Smith, pp. 621–53. New York: AMS Press

113. Roth, G. E. 1979. Satellite DNA properties of the germ line limited DNA and the organization of the somatic genomes in the nematodes *Ascaris suum* and *Parascaris equorum. Chromosoma* 74:355–71

114. Roth, G. E., Moritz, K. B. 1981. Restriction enzyme analysis of the germ line limited DNA of *Ascaris suum. Chromosoma* 83:169–90

115. Rubin, G. M. 1978. Isolation of a telomeric DNA sequence from *Drosophila melanogaster. Cold Spring Harbor Symp. Quant. Biol.* 42:1041–46

116. Runge, K. W., Zakian, V. A. 1989. Introduction of extra telomeric DNA sequences into *Saccharomyces cerevisiae* results in telomere elongation. *Mol. Cell. Biol.* 9:1488–97

117. Saiga, H., Edström, J. E. 1985. Long tandem arrays of complex repeat units in *Chironomus* telomeres. *EMBO J.* 4:799–804

118. Shampay, J., Blackburn, E. H. 1988. Generation of telomere-length heterogeneity in *Saccharomyces cerevisiae. Proc. Natl. Acad. Sci. USA* 85:534–38

119. Shampay, J., Szostak, J. W., Blackburn, E. H. 1984. DNA sequences of telomeres maintained in yeast. *Nature* 310:154–57

120. Shippen-Lentz, D., Blackburn, E. H. 1989. Telomere terminal transferase activity from *Euplotes crassus* adds large numbers of TTTTGGGG repeats onto telomeric primers. *Mol. Cell. Biol.* 9:2761–64

121. Shore, D., Nasmyth, K. 1987. Purification and cloning of a DNA binding protein from yeast that binds to both silencer and activator elements. *Cell* 51:721–32

122. Shore, D., Stillman, D. J., Brand, A. H., Nasmyth, K. A. 1987. Identification of silencer binding proteins from yeast: possible roles in SIR control & DNA replication. *EMBO J.* 6:461–67

123. Simmler, M. C., Johnsson, C., Petit, C., Rouyer, F., Vergnaud, G., Weissenbach, J. 1987. Two highly polymorphic minisatellites from the pseudoautosomal region of the human sex chromosomes. *EMBO J.* 6:963–69

124. Simmler, M. C., Royuer, F., Vergnaud, G., Nyström-Lahiti, N., Ngo, K. Y., et al. 1985. Pseudoautosomal DNA sequences in the pairing region of the human sex chromosomes. *Nature* 317:692–97

125. Spangler, E. A., Ryan, T., Blackburn, E. H. 1988. Developmentally regulated telomere addition in *Tetrahymena thermophila. Nucleic Acids Res.* 16:5569–85

126. Streeck, R. E., Mortiz, K. B., Beer, K. 1982. Chromatin diminution in *Ascaris suum:* Nucleotide sequence of the eliminated satellite DNA. *Nucleic Acids Res.* 10:3495–502

127. Szostak, J. W. 1982. Evolutionary conservation of the eucaryotic telomeres. *Recent Adv. Yeast Mol. Biol.* 1:76–92

128. Szostak, J. W., Blackburn, E. H. 1982. Cloning yeast telomeres on linear plasmid vectors. *Cell* 29:245–55

129. Traverse, K. L., Pardue, M. L. 1989. Studies of He-T DNA sequences in the pericentric regions of *Drosophila* chromosomes. *Chromosoma* 97:261–71

130. Van der Ploeg, L. H. T., Liu, A. Y. C.,

Borst, P. 1984. Structure of the growing telomeres of Trypanosomes. *Cell* 36: 459–68

131. Vignais, M-L., Woundt, L. P., Wassenaar, G. M., Mager, W. H., Sentenac, A., Planta, R. J. 1987. Specific binding of TUF factor to upstream activation sites of yeast ribosomal protein genes. *EMGO J.* 6:1451–57

132. Vollrath, D., Davis, R. W., Connelly, C., Hieter, P. 1988. Physical mapping of large DNA by chromosome fragmentation. *Proc. Natl. Acad. Sci. USA* 85:6027–31

133. Walmsley, R. M., Chan, C. S. M., Tye, B-K., Petes, T. D. 1984. Unusual DNA sequences associated with the ends of yeast chromosomes. *Nature* 310:157–60

134. Walmsley, R. M., Petes, T. D. 1985. Genetic control of chromosome length in yeast. *Proc. Natl. Acad. Sci. USA* 82:506–10

135. Walmsley, R. M., Szostak, J. W., Petes, T. D. 1983. Is there left-handed DNA at the ends of yeast chromosomes? *Nature* 302:84–86

136. Wang, S-S., Pluta, A. F., Zakian, V. A. 1989. DNA sequence analysis of newly formed telomeres in yeast. In *Aneuploidy: Mechanisms of Origin*, ed. M. Resnick. New York: Liss. In press

137. Watson, J. D. 1972. Origin of concatameric T7 DNA. *Nature* 239:197–201

138. Wellinger, R. J., Zakian, V. A. 1989. Lack of positional requirements for autonomously replicating sequence elements on artificial yeast chromosomes. *Proc. Natl. Acad. Sci. USA* 86:973–77

139. White, M. J. D. 1973. *Animal Cytology and Evolution*. Cambridge, UK: Cambridge Univ. Press

140. Wimmer, E. 1982. Genome-linked proteins of viruses. *Cell* 28:199–201

141. Yao, M.-C. 1981. Ribosomal RNA gene amplification in *Tetrahymena* may be associated with chromosome breakage and DNA elimination. *Cell* 24:765–74

142. Yao, M.-C. 1986. Amplification of ribosomal RNA genes. In *Molecular Biology of Ciliated Protozoa,* ed. J. G. Gall, pp. 179–201. New York: Academic

143. Yao, M.-C. 1988. Site-specific chromosome breakage and DNA deletion in cilites. In *Mobile DNA,* ed. D. Berg, M. Howe, pp. 715–34. Washington, DC: Am. Soc. Microbiol. Press

144. Yao, M.-C., Yao, C.-H. 1981. Repeated hexanucleotide C-C-C-C-A-A is present near free ends of macronuclear DNA of *Tetrahymena*. *Proc. Natl. Acad. Sci. USA* 78:7436–39

145. Yao, M.-C., Zheng, K., Yao, C.-H. 1987. A conserved nucleotide sequence at the sites of developmentally regulated chromosomal breakage in *Tetrahymena*. *Cell* 48:779–88

146. Yarmolinsky, M. B. 1971. Making & joining DNA ends. In *The Bacteriophage Lambda,* ed. A. D. Hershey, pp. 97–111. New York: Cold Spring Harbor Lab.

147. Young, B. S., Pession, A., Traverse, K. L., French, C., Pardue, M. L. 1983. Telomere regions in Drosophila share complex DNA sequences with pericentric heterochromatin. *Cell* 34:85–94

148. Zahler, A. M., Prescott, D. M. 1988. Telomere terminal transferase activity in the hypotrichous ciliate *Oxytricha nova* and a model for replication of the ends of linear DNA molecules. *Nucleic Acids Res.* 16:6953–85

149. Zakian, V. A., Blanton, H. M. 1988. Distribution of telomere-associated sequences on natural chromosomes in *Saccharomyces cerevisiae*. *Mol. Cell Biol.* 8:2257–60

150. Zakian, V. A., Blanton, H. M., Dani, G. M. 1985. Formation and stability of linear plasmids in recombination deficient strain of yeast. *Curr. Genet.* 9:441–45

Annu. Rev. Genet. 1989. 23:605–36

MECHANISM AND DEVELOPMENTAL PROGRAM OF IMMUNOGLOBULIN GENE REARRANGEMENT IN MAMMALS

T. Keith Blackwell and Frederick W. Alt

Howard Hughes Medical Institute and Departments of Biochemistry and Microbiology, Columbia University College of Physicians and Surgeons, 701 West 168 Street, New York, New York 10032

CONTENTS

0066-4197/89/1215-0605$02.00

INTRODUCTION

B-lineage cells mediate humoral immunity by producing specific antibodies. The basic structural subunit of an antibody (Ab) is an immunoglobulin (Ig) molecule; Ig molecules consist of a complex of two identical heavy (H) and two identical light (L) polypeptide chains (49). At the amino terminus of each H chain (HC) and L chain (LC) is a region that varies in amino acid sequence (variable region), with the remaining portion of HCs and LCs being relatively constant in amino acid sequence (constant region) (49). In an Ig molecule, HC and LC variable regions are juxtaposed to form the potential antigen-binding site (65). The genes that encode HC and LC variable regions are assembled somatically from segments of germline DNA during precursor B (pre-B) lymphocyte differentiation (143). Mammals have a nearly unlimited capacity for Ab diversity that derives in large part from this variable region gene assembly process (143). T lymphocytes mediate cellular immunity directly. The T cell antigen receptor (TCR) consists of one of two distinct heterodimers of polypeptides that each contain an Ig-related variable region (41). TCR variable region genes are similar in structure to Ig variable region genes and are assembled somatically in differentiating pre-T lymphocytes (41).

The variable region gene assembly process occurs independently from and prior to antigen recognition. In general, in differentiating pre-B lymphocytes HC variable region genes are assembled before those of LCs (Figure 1; discussed below). Each newly generated mature B lymphocyte expresses on its surface a homogeneous Ig species that functions as an antigen receptor (Figure 1; 38), and contains unique HC and LC variable region sequences. During mammalian fetal life B lymphocytes are produced in the liver, but at about the time of birth the site of differentiation shifts to the bone marrow (38). Contact of B lymphocytes with antigen usually occurs in the "peripheral" lymphoid organs, such as the spleen and lymph nodes. Peripheral B lymphocytes express a repertoire of antigen-binding specificities that is different from the "primary" repertoire expressed by newly generated B lymphocytes, suggesting that only a fraction of the B lymphocytes that are generated persist in the periphery (6). Recognition of antigen by the surface Ig species that is expressed by a B lymphocyte stimulates that cell to proliferate (Figure 1; 29). During these "antigen-dependent" stages of B lymphocyte differentiation, cells derived from the selected clone may then differentiate into plasma cells that produce the triggered Ig-specificity in a secreted form as Ab. The specificity of this clonal selection mechanism is dependent on the principle of allelic exclusion (31, 108), whereby each B lymphocyte expresses in its surface antigen receptor the products of only a single HC allele and a single LC allele. An expanding body of evidence, which is discussed below, indicates that allelic exclusion is imposed by regulation of the HC and LC variable region gene assembly process.

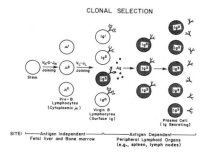

Figure 1   Clonal selection and B-lymphocyte differentiation. During pre-B lymphocyte differentiation, HC variable region gene assembly and μ HC expression generally precede LC gene assembly and expression. These antigen-independent differentiation stages generate independent B-lymphocyte clones, each of which expresses a surface receptor Ig with a unique antigen-recognition site. Binding of antigen by the surface receptor of a particular clone (Ig3) stimulates expansion of that clone and differentiation into plasma cells that secrete antibody molecules that have the same antigen-recognition site. Ag: antigen.

HC and LC constant (C) regions vary in amino acid sequence only among different isotypes that are each encoded by a different C region gene (132, 143). The class of an Ab molecule is defined by its HC isotype, which is designated by a corresponding Greek letter (i.e. IgM molecules contain mu (μ) H chains). HC isotypes mediate a variety of immunologic effector functions (100). Newly-generated B lymphocytes express surface IgM and can differentiate into cells that simultaneously express both IgM and IgD on their surface (38). IgM can be secreted in response to antigenic stimulation and is particularly effective at complement fixation (100). IgD is, in general, not secreted; its function is unknown (24, 94) but may be related to development of immunologic tolerance (53). The other HC isotypes can be secreted as Abs that perform different effector functions. For example, the IgG isotypes are crucial effectors of humoral responses, IgA mediates secretory immunity, and IgE is produced in response to parasitic infections and mediates allergic reactions (100). B lineage cells that express IgM can subsequently switch to production of one of these latter HC isotypes (HC class switching) by undergoing a DNA rearrangement event (class-switch recombination) that replaces the μ C region (Cμ) gene with the region that encodes the new isotype (132). Class-switch recombination is mechanistically distinct from variable region gene assembly (132). Class switching allows a B lymphocyte to produce an Ab that retains the original antigen-binding specificity provided by the variable region but performs a different immunologic effector function mediated by the new constant region. Class switching is influenced by factors that are produced by certain subsets of T lymphocytes; treatment of B cells with particular T-cell factors results in production of particular HC isotypes (33). Recent evidence suggests that these effects are mediated through modulation of the class-switching process itself (89). Control of HC class switching therefore appears to allow the immune system to tailor the Ab response so that Ab classes are produced that perform appropriate immune effector functions.

Both variable region gene assembly and HC class switching seem to be regulated processes. The mechanisms that control these events appear to be intimately related to the progression of B-lymphocyte differentiation (6, 89). In this review we describe current understanding of both the mechanism and regulation of these DNA rearrangement events.[1]

MODEL SYSTEMS FOR STUDYING IMMUNOGLOBULIN GENE REARRANGEMENT

Most studies have focused on murine model systems, but the general features of these DNA rearrangement processes seem to be analogous in humans (18). Tumor-cell analogs of differentiated Ig-producing cells were utilized for initial comparisons of rearranged Ig genes with their germline counterparts (143). These studies defined general aspects of the processes by which Ig genes are assembled and expressed (143). Pre-B cell lines can be generated by transformation with certain retroviruses (18). Of such lines, the most extensively studied are Abelson murine leukemia virus (A-MuLV) transformants, which represent most known phases of early pre-B cell maturation (5). Many A-MuLV transformants actively rearrange Ig variable region gene segments, and thus differentiate during culture (5). Analyses of A-MuLV transformants have allowed elucidation of many aspects of the developmental progression of Ig gene rearrangement events (5). Furthermore, these actively rearranging cell lines have been utilized as hosts for plasmid or viral vectors (recombination substrates) that contain cloned Ig or TCR variable region gene segments. In such lines recombination occurs by the normal mechanism within either chromosomally integrated (17, 81) or extrachromosomal (60) substrates. By manipulating recombination substrates—in particular, by altering the sequences that surround variable region gene segments, it has been possible to test various hypotheses concerning the mechanism and regulation of variable region gene assembly. Certain A-MuLV transformants also undergo HC class switching during culture, and therefore have been utilized both to study the modulation of class switching by various factors (89) and the substrates for class-switch recombination (106). In general, the conclusions derived from these studies of tumor-cell analogs agree with findings obtained using normal B-lineage cells (18, 38).

---

[1]Because of space limitations, we cite reviews wherever possible and individual papers only where the experimental findings are especially relevant to the discussion.

The use of transgenic mice has made it possible to study control of Ig variable region gene assembly and expression in a more physiological context. Assembled Ig genes introduced into the germline of mice are expressed in a generally tissue-specific manner (27, 55, 125), allowing analysis of the effect of expression of Ig polypeptides on B-lymphocyte differentiation. Studies in which recombination substrates are introduced as transgenes (28, 47, 52) hold additional promise for providing insight into both variable region gene assembly and HC class switching.

MOLECULAR BASIS OF ANTIBODY DIVERSITY

Immunoglobulin HC variable regions are encoded by variable (V), diversity (D), and joining (J) segments, LC variable regions by corresponding V_L and J_L segments, and TCR variable regions by analogous sets of gene segments (Figure 2; 41, 143). Within Ig-variable regions are three regions of greatest amino acid sequence variability (hypervariable regions) that interact to form the antigen-recognition site and are thus referred to as complementary-determining regions (CDRs) (65). The V segment encodes the bulk of the variable region domain, including CDRs 1 and 2 (143). CDR3 is encoded by sequences that are formed by the juxtaposition of V_H, D, and J_H (or V_L and J_L) segments (143). Diversity in CDRs 1 and 2 derives from sequence heterogeneity among multiple different germline-encoded V segments (65, 143). Further diversity in the antigen-recognition site is generated somatically, through combinatorial assortment among different V, (D), and J segments and by mechanisms (described below) that create nucleotide sequence heterogeneity where these segments are joined to each other (143). These latter mechanisms generate sequence heterogeneity within CDR3. Additional diversity may be derived from pairing of different HC and LC variable regions. Furthermore, a somatic hypermutation mechanism may operate on assembled variable region genes in a B lymphocyte that has been stimulated by an antigen (73). These mutations may alter the antigen-binding affinity of the expressed Ig molecule, and thus generate B-lymphocyte clones that produce Ig molecules with an enhanced affinity for the antigen (73). In mammals, gene conversion has been invoked as a generator of variable region diversity during evolution but has not been identified as a significant somatic generator of diversity (13, 18, 65). In contrast, in the chicken, gene conversion appears to be the primary somatic generator of diversity (150).

IMMUNOGLOBULIN GENE ORGANIZATION AND EXPRESSION

HC and LC Gene Organization

In the mouse, a cluster of four J_H segments is located about 7 kb 5' of the C_H genes (143), which lie in the 5' to 3' order: μ, δ, γ_3, γ_1, γ_{2b}, γ_{2a}, ϵ, and α

A: Variable Region Gene Assembly:

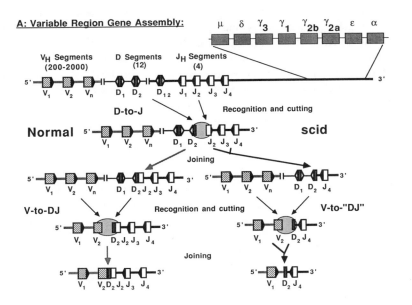

B: Heavy Chain Gene Expression:

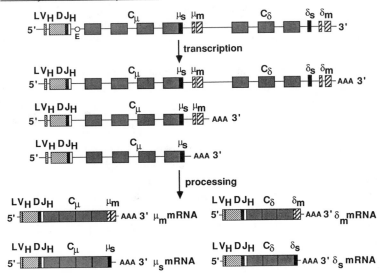

C: Light Chain Gene Organization:

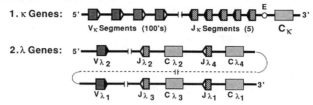

Figure 2 Organization and assembly of murine HC and LC variable region genes. A. Variable region gene assembly. Assembly of an HC gene is shown. Recombination recognition signals that are separated by a 23-bp spacer are indicated by black triangles, and those separated by a 12-bp spacer by open triangles. Shaded boxes denote C-region coding sequences. HC-variable region gene assembly is ordered, with D-to-J_H joining preceding V_H-to-DJ_H rearrangement. In *scid* mice the V(D)J recombinase activities that recognize and cut at signal sequences are normal, but joining of coding sequences to each other in the context of the V(D)J recombinase is impaired (see text). Aberrant *scid* chromosomal coding joins are generally formed by "rescue" of the resulting unrepaired double-strand break by an illegitimate recombination mechanism. However, the mechanisms that "target" variable region gene segments for rearrangement seem to function normally in *scid* mice—if the 5' set of recognition signals remains intact in an aberrant "DJ_H" rearrangement, this structure can serve as a substrate for attempted V_H-to-"DJ_H" joining (92). The V(D)J recombinase complex is represented by a filled oval, and the bold arrows shaded differently to indicate the differences in coding join formation between normal and *scid* mice. B. Heavy-chain gene expression. Expression of μ and δ HCs is controlled by modulation of RNA processing and termination. Coding sequences are represented as in A, except that the 5' leader and separate C_μ and C_δ exons are shown. The HC transcriptional enhancer is designated by E. The membrane-bound (m) and secreted (s) forms of HC termini are encoded by mini-exons that are shaded differently. The indicated primary transcripts can be processed to generate the mRNA species depicted below. C. Light-chain gene organization. Recombination recognition signals are shaded as in A, and the κ transcriptional enhancer indicated by E. Only four of the five J_κ segments are functional (143).

(133; Figure 2A). The murine V_H segments lie about 100 kb (M. Morrow & F. Alt, unpublished data) 5' of the J_H segments (Figure 2A). About 12 D segments lie in the intervening region between the D and J_H segments (Figure 2A). The number of V_H segments varies greatly among mouse strains (23) and is still a subject of debate (115), but has been estimated to be as high as 2000 in certain strains (87). The V_H segments identified so far have been grouped into families based on nucleotide sequence homology (115). In humans the organization of these loci is generally similar to that of the mouse (15, 74, 115).

Mammals produce two LC isotypes, kappa (κ) and lambda (λ), that are encoded in separate loci and for which no functional differences have been identified (143). The murine C_κ gene lies about 2 kb 3' of four functional J_κ segments, which in turn lie at an unknown distance 3' of about 200 V_κ segments (18; Figure 2C). In mice, two V_λ segments and four J_λ–C_λ units are organized as diagrammed in Figure 2C (130, 143); the location of a third V_λ segment, $V_\lambda X$ (42), has not been determined. These loci are organized in a generally analogous fashion in humans (18).

Expression of Assembled and Incompletely Assembled Ig Genes

A transcriptional promoter is located immediately upstream of each germline V_H and V_L segment. Transcription of an assembled HC or LC gene is initiated from the promoter that is associated with the rearranged V segment and is dependent upon the activity of a transcriptional enhancer element, which in HC and κ LC genes is located within the J-C intron (Figure 2; 14, 50, 98, 110). A corresponding enhancer element for λ LCs is presumed to exist (110) but has not yet been identified. Investigations into the mechanisms by which these regulatory sequences function have been reviewed (30). RNA processing mechanisms remove intervening sequences from HC and LC gene primary RNA transcripts and thus allow expression of V and C regions together in the same polypeptide (Figure 2B). Control of RNA polyadenylation and processing determines whether an HC is expressed in its membrane-bound (m) or secreted form (s) (7, 45, 123; Figure 2B), and also determines the relative levels of expression of μ and δ HCs (24, 94).

Germline and incompletely assembled Ig (and TCR) variable and C region loci are expressed in a tissue- and stage-specific pattern that seems to correlate with rearrangement of the corresponding variable region gene segments or C_H gene (reviewed in 18). For example, V_H gene segments in the largest family (V_HJ558) are expressed in early pre-B cells that undergo V_H-to-DJ_H rearrangement but not in mature B lymphocytes (153). D segments are flanked upstream by promoters from which DJ_H rearrangements are transcribed to generate a DJ_H-$C\mu$ transcript that can be expressed as a "$D\mu$" polypeptide (117). The germline J_K–C_K region is transcribed in mature B-lineage cell lines (144) and in certain pre-B cell lines that rearrange κ LC gene segments (21) but not in early stage pre-B cells that assemble HC variable region genes (97). These germline C_K transcripts are "sterile" (i.e. they do not encode obvious polypeptides) (144). Sterile transcripts of most germline C_H regions have also been identified and are discussed in detail below.

The pattern by which these various loci are expressed suggests that their transcription simply may be related directly to control of variable region gene assembly and HC class switching. However, some of these transcripts could encode identified or hypothetical protein products of an as yet undefined function. For example, proteins that are highly related to λ LC V and C regions are derived from λ-related genes that do not undergo rearrangement and are expressed early during murine and human B-lymphocyte differentiation (76, 77, 126). These proteins may be involved in intracellular compartmentalization of μ HCs (111). Similarly, D_μ or potential germline V_H protein products might also have a specific function during B-lymphocyte differentiation.

MECHANISM OF VARIABLE REGION GENE ASSEMBLY

The Joining Mechanism is Site-Specific and Nonreciprocal

Initial insights into the mechanism of variable region gene assembly came from comparisons of the DNA sequences of germline and assembled HC and LC variable region genes (143). All germline Ig variable region gene segments were found to be immediately flanked by a conserved, usually palindromic heptamer and an AT-rich nonamer separated by a relatively nonconserved spacer region of either $12+1$ or $23+1$ base pairs, or approximately one or two turns of the DNA helix (143: Figure 3A). The location and conservation of these "signal" sequences suggested that they are recognized by a site-specific recombination machinery (44, 127). Subsequent studies found highly related sequences adjacent to all germline TCR variable region gene segments (41), suggesting that they are assembled by the same or an analogous mechanism. It was noted that at each variable region locus segments of a given type (e.g. J_H segments) are flanked by signal sequences with spacers of the same length, and that segments that were joined to each other had been flanked in the germline by recognition elements separated by spacers of different length (Figure 2). These findings suggested that the possibilities for joining are determined by the length of the signal-sequence spacers (the 12/23 rule; 44, 127). Coding regions were often joined imprecisely, with potential coding bases lost in some joins but not in others (143), and many HC (and TCR) joins contained nucleotides at the junction that were not derived from known germline segments (78). Although joining between HC variable region gene segments generally deleted intervening sequences, many rearranged κ LC loci (and one HC gene rearrangement) retained the intervening DNA along with signal sequences fused back-to-back without base loss or addition (19).

Together, these observations suggested a general model for the recombination mechanism (referred to below as V(D)J recombination), in which coding and signal sequence ends are processed and joined together in multiple steps in a nonreciprocal fashion (Figure 3A; 4; reviewed in detail in 19). According to this model, a V(D)J recombination event is initiated by introduction of an endonucleolytic break precisely between the signal and coding sequences of both recombining segments. Bases may then be removed from coding but generally not from signal sequence ends, perhaps by an exonuclease activity. It was proposed that the extra nucleotides (referred to as "N" regions) are

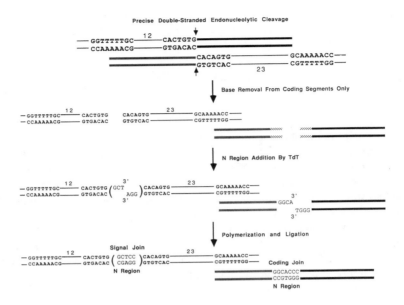

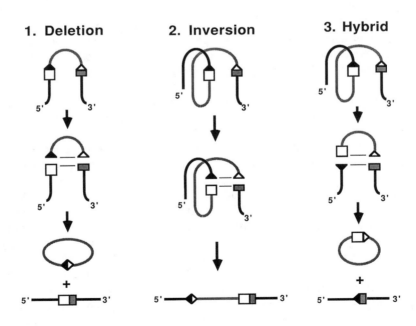

1. Deletion

2. Inversion

3. Hybrid

added de novo to coding-sequence ends, perhaps by the enzyme terminal deoxynucleotidyl transferase (TdT) (4). TdT is expressed in lymphoid cells and can add nucleotides to DNA ends in a 5'-to-3' direction without a template (4). Complementary bases would then be added to single-stranded regions and the signal and coding segment ends ligated to form "signal" and "coding" joins (Figure 3A). The imprecision inherent in coding-join formation appears to generate significant diversity in CDR3, and in the antigen-binding capability of Ig and TCR polypeptides (18).

This model predicts that V(D)J recombination can either delete or invert intervening DNA by the same mechanism, the outcome depending only on the relative orientation of the recombining segments (4; Figure 3B). If these segments are oriented with recognition signals pointed toward each other, intervening DNA is deleted as a circle. If the recognition signals are pointed in the same direction, the signal join is retained on the chromosome and the intervening DNA inverted. Joining by inversion was first demonstrated in an unusual D-to-J_H rearrangement that involved the 5' set of recognition signals flanking a D segment (4). A recombination substrate was utilized for the first unequivocal identification of both products of an inverted joining event (81). Subsequently, joining by inversion has been confirmed to occur frequently within endogenous Ig κ LC and TCR loci (19). The proposal that deletional joining events often generate circles containing a signal join has also been confirmed recently (19, 48, 104). Although signal-join formation is essential to maintain chromosomal integrity during joining by inversion, ligation of signal sequences within deleted DNA does not confer an obvious advantage and may simply reflect the common mechanism.

←——

Figure 3 Inversional and deletional V(D)J recombination events occur by a common mechanism. A. A model for the recombination mechanism. Shaded bars denote coding regions, with broken bars indicating removal of nucleotides. Consensus nonamer and heptamer sequences are shown, with thin lines representing spacer regions and flanking sequences. The efficiency at which N-regions are added by TdT to signal joins relative to coding joins is unknown (indicated by parentheses around the signal join N-region). Both N-regions are illustrated as having involved transient pairing of the 3' terminal A and T residues before polymerization of the opposite strand and ligation of the two ends, because evidence of such pairing is observed frequently (19). However, the exact mechanism of this stage of the recombination mechanism is unknown. B. Types of V(D)J recombination events. Recombination recognition signal sequences with different spacer lengths are indicated by black and open triangles, and coding segments denoted by boxes. V(D)J recombination that is mediated by signal sequences that are oriented as in (1) results in deletion of intervening DNA (including the signal join) as a circle. Joining by inversion (2) results in retention of the signal join and of intervening DNA in the chromosome. A hybrid join (3) is formed when these ends are paired inappropriately so that each coding segment is joined to the signal sequence that corresponds to the other partner. Hybrid joining therefore generates a deletion during recombination between two segments that would normally be joined together by inversion (or vice versa).

Recognition Sequence Requirements for V(D)J Recombination

Recombination substrate studies have shown that V(D)J recombination can be mediated by the consensus nonamer and heptamer elements in the absence of other sequences (2, 19) and have demonstrated that certain bases that are generally conserved within the heptamer (Figure 3A; 2, 115) are critical for the recombination process (2). These studies have also further confirmed the prediction that joining does not occur at high frequency between recognition elements that are separated by spacers of the same length (2). However, although the heptamer and nonamer signals are apparently sufficient to mediate V(D)J recombination, current evidence suggests that other elements and mechanisms (discussed below) may be important in vivo for directing the rearrangement process and enhancing the frequency of recombination.

The V(D)J recombinase apparently also can mediate joining between elements in which one partner consists of a heptamer not accompanied by a nonamer. For example, a set of recognition signals (RS element, κde in humans) that is located downstream of the C_K gene recombines at a significant frequency with a heptamer in the J_K-C_K intron (54, 130). In a V_H replacement event, the bulk of a rearranged V_H segment is replaced by an upstream germline V_H segment in a recombination event that is apparently mediated by a heptamer conserved within the coding region of most V_H segments (72, 119). However, V_H replacement does not appear to occur frequently (72, 119). Furthermore, in a recombination substrate in which the heptamer has been deleted from one partner, joining does not occur at a frequency significant relative to that of normal joining (S. Desiderio, personal communication). Thus, although it is not known whether additional sequences might influence the frequency of these events, the presence of an appropriately spaced nonamer seems to greatly increase the frequency at which V(D)J recombination can occur and may also act as restricting element in the context of the 12/23 rule (3).

Formation of Coding and Signal Joins

The proposal that N-regions are inserted by TdT is now supported by considerable but still indirect evidence. Consistent with the known activity of TdT, N-regions are usually GC-rich, and their presence in joined endogenous and introduced chromosomal elements closely correlates with TdT expression (19). During B-cell differentiation TdT is primarily expressed at earlier stages, when HC but not LC genes are assembled (4, 20, 79). Correspondingly, the major difference between analyzed HC and LC coding joins is the frequent appearance of N-regions in the former but not the latter—implying that these joins are generated by a common mechanism and that the difference between them simply reflects the presence or absence of TdT (4). Although the mechanisms that remove nucleotides from coding-segment ends do not act on signal-join partners (19), recent evidence indicates that signal-sequence ends can be substrates for N-region addition (Figure 3A). Presence of GC-rich N-regions in signal joins within extrachromosomal substrates correlates with TdT expression in host-cell lines (86). Chromosomal signal joins that were isolated in early studies did not contain inserts, but most were LC signal joins and therefore occurred in cells that probably did not express TdT (19). More recently, GC-rich inserts that are consistent with N-regions have been observed in an LC signal join and in a number of TCR signal joins (19). However, the relative frequency of N-region addition into chromosomal coding and signal joins is still unknown.

The V(D)J recombinase does not appear to be absolutely constrained with respect to juxtaposition of signal and coding sequence ends for joining. In "hybrid" joins a signal heptamer is joined to the coding end of the other partner (Figure 3B) (82, 95). Such joins may occur at a significant frequency (about 10%) relative to normal joins in both extrachromosomal and integrated substrates (82, 95), but it is not yet clear how often they occur in vivo and relatively few potential examples of endogenous hybrid joins have been identified (19). In "open-and-shut" joins, which have been observed at a low frequency in extrachromosomal substrates, signal and coding sequences have been cleaved then ligated back together after loss of bases from the coding sequence (82). Many of these joins appear to have derived from cleavage at only *one* set of signal sequences, suggesting that such events are possible even if cutting generally occurs on two sets of recognition signals simultaneously. It is not yet known whether the V(D)J recombinase can bind and cut at a single set of recognition sequences or whether signal sequences must normally be bound in pairs to target activity.

The scid *Mutation Results in Defective V(D)J Recombination*

Severe combined immune deficient *(scid)* mice (26) do not generate functional B or T lymphocytes because of a defect in V(D)J recombination (129). *scid* mice lack obvious signs of a general defect in DNA repair or ligation (26, 40), but pre-B and pre-T cells of *scid* mice assemble aberrant rearrangements at respective Ig and TCR loci (129). The coding joins produced by *scid* mice involve appropriate segments but generally result in deletion of either one or both recombining partners (21, 58, 70, 92, 106, 129: Figure 2A). Normal coding joins have been detected within integrated recombination substrates in *scid* pre-B cells, but at low frequency (20). In contrast, *scid* A-MuLV transformants can form extrachromosomal or endogenous signal joins that are normal in most respects (except for loss of a few bases in several) at an approximately normal efficiency (21, 84; Figure 2A) and can also form hybrid joins (84). These findings suggest that the *scid* mutation does not greatly impair the V(D)J recombinase activities that recognize, cut at, juxtapose, and ligate signal sequences. Instead, its impact is largely restricted to joining of coding sequences to each other. Therefore, these findings imply the existence of a component of the V(D)J recombinase that is required specifically for this process, perhaps an activity that binds to coding segment ends in a manner that is not sequence-specific and juxtaposes them for joining.

Because coding segment ends apparently cannot be joined to each other at a significant rate in the context of the *scid* V(D)J recombinase, it has been proposed that aberrant *scid* chromosomal coding joins derive from "rescue" of an otherwise lethal double-strand break by an illegitimate recombination mechanism (92). Similar aberrant coding joins have also been detected in B-lineage cells from normal mice (68, 101), and the structures of some chromosomal translocations that are characteristic of certain lymphoid malignancies are consistent with having been derived by this type of mechanism (12; Figure 4). Apparently, in normal lymphoid cells, coding segment ends may "escape" the V(D)J recombinase at a certain frequency and undergo recombination with other sequences by an alternative pathway.

V(D)J Recombinase Components

To date, an in vitro system for V(D)J recombination has not been developed, and efforts to isolate V(D)J recombinase components have focused on identification of endonuclease and sequence-specific DNA binding activities. Endonucleases that cut at signal sequences have been detected, but none appears to be lymphoid-specific and all lack cutting specificity (18). Factors that bind heptamer (1) and nonamer (56) sequences have been identified; isolation of the genes that encode them will make it possible to determine their relationship to V(D)J recombinase components. Recombination substrate assays have indicated that V(D)J recombinase activity is probably expressed only in pre-B and pre-T lymphocytes and, with a few exceptions, not in other cell types (22, 85, 128, 158). This notion is consistent with the observation that *scid* mice do not have obvious developmental defects in nonlymphoid lineages (40). Restriction of V(D)J recombinase activity to cell lineages in which it is needed might be important—it may minimize the chances that potentially harmful low-frequency rearrangement events, such as oncogene translocations, might occur. However, V(D)J recombinase activity has been generated in a fibroblast cell line by transfection of a limited segment of genomic DNA, suggesting transfer of a single gene (128). Whether the transfected region encodes a protein that along with factors expressed in the fibroblast can perform V(D)J recombination or, alternatively, encodes a regulatory factor that activates expression of V(D)J recombinase components, this experiment may represent a significant step towards isolation of genes that encode these components.

ABERRANT V(D)J RECOMBINATION EVENTS

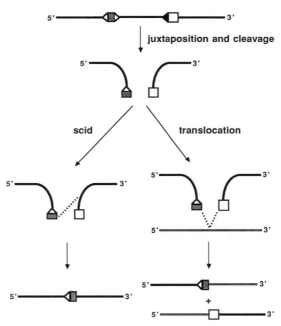

Figure 4   Aberrant V(D)J recombination events. *scid* coding joins and some chromosomal translocations that are characteristic of lymphoid tumors share certain common features. In both types of joins, appropriately cleaved coding segments are joined to DNA sequences that are not flanked by either "pseudo" recognition signals or by significant regions of homology. Furthermore, these joins may contain short sequence repeats that are characteristic of illegitimate recombination events (12, 19). In both cases a double-strand break that is otherwise lethal is repaired, but in the translocation this event involves invasion of another chromosome. Coding and signal sequences are represented as in Figure 3.

CONTROL OF IMMUNOGLOBULIN VARIABLE REGION GENE ASSEMBLY

"Regulated" Models of Allelic Exclusion and Ordered HC and LC Gene Assembly

The subject of how HC and LC allelic exclusion is imposed still remains controversial (18). However, the weight of current evidence most strongly supports "regulated" models. Such models propose that the polypeptide products that are expressed from assembled HC and LC genes effect allelic exclusion by influencing further rearrangement in the cell (5, 8, 10, 18). If the joining mechanism is randomly imprecise, it would be predicted that about one in three V_HDJ_H and V_LJ_L rearrangements would have the translational initiation codon of the V segment placed in the same triplet reading frame as the J and C regions, and therefore be "productive" and capable of encoding a complete Ig protein. Data from a number of experimental sources confirm that about one in three V_HDJ_H and V_LJ_L rearrangements are productive (18). These findings are not consistent with "stochastic" models that stipulate that the assembly process is extremely inefficient and not regulated (35). An efficient rearrangement process can theoretically generate B lymphocytes that violate allelic exclusion and express multiple HC and LC species in Ig molecules. Such cells arise in several lines of mice that express HC and LC transgenes (51, 52, 60, 99, 137), indicating that there is no inherent selection against their existence, as has been proposed as a model for allelic exclusion (147). The most direct conclusion to be drawn from these experiments is that allelic exclusion is mediated by regulation of the rearrangement process.

During B-lymphocyte differentiation, HCs are expressed before LCs (38). Evidence derived from pre-B cell lines (6, 18) and from normal B lymphocytes (34) indicates that HC variable region genes are assembled before those of LCs. HC variable region gene assembly is itself an ordered process that begins with assembly of DJ_H rearrangements, usually at both HC alleles (11). These DJ_H rearrangements may then either serve as a substrate for V_H-to-DJ_H joining, or at a similar or lower frequency might be replaced by joining of an upstream D and a downstream J_H segment (120). Direct V_H-to-D joining is rarely observed (11). The relative contributions of these rearrangements to antibody diversity have been reviewed in detail (6, 18). A striking feature of this process is that the most J_H-proximal V_H segments are used highly preferentially in V_H-to-DJ_H rearrangements, and that the relative frequency at which V_H segments are utilized is in roughly inverse proportion to their distance from the J_H region (120, 156, 157). The simplest model to account for this position dependence in V_H segment utilization is that the V(D)J recombinase operates by a one-dimensional tracking system and scans upstream from the DJH rearrangement for a V_H segment (152, 156). Tracking that initiates from the J_H region might also account for ordered rearrangement among V, D, and J segments. However, other explanations are possible and are discussed in the next section.

The μ HC polypeptide is thought to mediate HC allelic exclusion by signalling inhibition of further V_H-to-DJ_H rearrangement (5, 10). Consistent with this model, a significant percentage of normal surface Ig-positive B lymphocytes retain one HC allele in DJ_H configuration (11). Additional support has come from the finding that endogenous HC gene assembly is often inhibited in HC-transgenic mice (93, 102, 103, 125, 148). However, in these transgenic mice the "block" in rearrangement usually seems to be incomplete, and the extent to which the mature B-lymphocyte population expresses endogenous HC genes varies greatly among different lines (59, 93, 102, 103, 125, 137, 148). More recent analyses of LC-transgenic mice (discussed below) have demonstrated that B lymphocytes that express endogenous LC genes are selected to enter the population of peripheral B lymphocytes (99), presumably to increase diversity of the available repertoire. Apparently, such immune selective forces often greatly complicate the findings of Ig-transgenic studies.

The μ HC may also signal initiation of LC gene assembly (10, 118). In support of this "μ-signalling" model, pre-B cell lines that have assembled two nonproductive V_HDJ_H rearrangements usually will assemble LC genes only if they acquire μ HC expression, either from an introduced μ-expression vector or by having undergone a V_H-replacement event (121). It is not yet clear to what extent V_H-replacement, which is thought to occur infrequently, might "salvage" such cells in vivo. Data from cell lines (121) and from transgenic mice (93, 102, 103) suggest that the μm but not μs HC can perform these regulatory functions, presumably through an intracellular signal mediated by the transmembrane sequence. Accordingly, the δ HC has a similar carboxy terminus and also seems able to transmit this signal (63). However, certain B-lineage cell lines apparently have never produced an HC but have rearranged LC variable region gene segments (21, 75), and in A-MuLV transformants LC gene assembly is often not induced efficiently by introduction of a μm-expression vector (21, 121). Thus, if LC gene assembly is normally induced by μm-expression, this signal pathway might be influenced by aspects of transformation or by passage outside of a normal in vivo environment (21). Induction of LC gene assembly might also require other signals in addition to HC expression (121). Alternatively, it is possible that the timing of initiation of LC gene assembly does not directly involve HC gene expression, but instead is related to other events that occur during pre-B lymphocyte differentiation.

In both mice and humans, assembly of kappa LC variable region genes seems to be initiated before that of lambda (8, 61, 109). The mechanism that activates λ variable region gene assembly has not yet been identified, but may be related to rearrangements of the RS element into the κ locus and accompanying deletion of one or both C_κ genes (54, 109, 130). LC allelic and isotypic (κ versus λ) exclusion is thought to be effected by production of a "functional" LC that can complex with the cellular HC to form a complete Ig molecule (8). Separation of the HC and LC variable region gene assembly processes may be important to allow the pre-B lymphocyte to "test" for production of such an LC: nonfunctional $V_\kappa J_\kappa$ rearrangements can be replaced by joining events that utilize more 3' J_K segments (83, 131), or alternatively the cell may activate λ gene assembly.

Data from both κ- (122) and λ-transgenic mice (99) indicate that, at least in the majority of B lymphocytes, formation of Ig that contains either κ or λ LC results in a feedback inhibition of further LC gene assembly. However, coexpression of both κ and λ LC as Ig has been detected in a single Ly1$^+$ B-lineage cell line (57) and may occur frequently in some LC-transgenic mice (51, 99). It has been proposed that such cells represent a distinct B-lymphocyte lineage in which both κ and λ gene assembly occurs without feedback inhibition (51). However, in a study of λ-transgenic mice, κ- and λ-coexpression was observed in λ-transgenics but not in normal controls (99). Most B lymphocytes in newborn λ-transgenics expressed only λ LC, but the relative proportion of κ-producers increased dramatically with age (99). The latter findings imply the existence of mechanisms that select for B lymphocytes that express LCs (and in HC-transgenic mice, HCs) derived from endogenous genes. Such selection mechanisms may account for the existence of HC or LC double-producers in transgenic mice without requiring the proposal of hypothetical lineages.

It is not known whether LC allelic exclusion is effected by a specific inhibition of LC gene assembly, by termination of V(D)J recombinase activity, or by both mechanisms. In this regard, recombination substrate studies indicate that in general the V(D)J recombinase does not appear to be active in surface Ig-positive B lymphocytes (85, 158). Extrachromosomal substrate assays have also indicated that V(D)J recombinase activity is lower in pre-B cell lines that assemble LC genes than in less mature lines that assemble HC genes (85). However, it is not clear why recombinase activity should be lower during LC gene assembly. If these cells continued to divide, such a decrease might result in the accumulation of HC-expressing pre-B cells that had not yet assembled LC genes. More recent studies with chromosomally integrated substrates have not found a clear difference in recombinase activity between HC- and LC-gene rearranging pre-B cell lines (158). The apparent discrepancies among these assays might derive from a number of sources, including the nature of the substrates themselves or variation in V(D)J recombinase activity among cell lines.

Modulation of Recombinational "Accessibility" of Variable Region Loci

If Ig and TCR variable region genes are assembled by a common V(D)J recombinase, then the assembly process must be regulated by controlling the "accessibility" of the component segments to this recombinase. Initial support for the notion of "accessibility" control came from the finding that a number of Ig-variable region loci, most notably germline V_H segments (153), are transcribed when undergoing rearrangement. Direct experimental evidence for an "accessibility" model was subsequently provided by recombination substrate experiments. In these experiments, TCR D and J or Ig LC segments were introduced into an A-MuLV transformant that rearranges endogenous HC-variable region gene segments only (22, 154). Although the corresponding endogenous gene segments did not recombine in this cell line, the TCR and LC gene segments in the introduced substrates underwent recombination at high frequency. Presumably, the recombinational "accessibility" of the introduced segments derived from their proximity to an expressed selectable marker gene within the substrate, with the expressed gene substituting for normal regulatory factors (22, 154). However, the molecular mechanisms that determine "accessibility" to the V(D)J recombinase have not yet been elucidated.

In the context of recombinational "accessibility," the feedback regulation model described above would predict that μm-expression would signal "closing" of the V_H locus and "opening" of the κ LC locus. A number of experimental findings are consistent with this notion, but current evidence is incomplete. Early stage pre-B cell lines that assemble only HC-variable region gene segments can be induced to transcribe the C_κ locus by treatment with bacterial lipopolysaccharide (LPS), a polyclonal mitogen that activates the κ transcriptional enhancer (97). Accordingly, exposure to LPS may also induce κ gene assembly in such lines (P. Ammirati & F. Alt, unpublished data; M. Schlissel & D. Baltimore, personal communication). Certain pre-B cell lines that rearrange LC gene segments express germline C_κ transcripts (21) but not germline V_H segments (20, 153), although a larger number of lines should be examined before a firm correlation can be drawn. In general, the correlation between transcription and rearrangement at HC and LC loci should be interpreted with caution, and further investigations are necessary. Functional analyses of this correlation are not complete: transcripts of germline V_κ segments or of most V_H segment families have not been identified or even assayed for in appropriate lines and a role for potential protein products of these transcripts has not been eliminated.

Consistent with the notion that recombinational "accessibility" may be directly or indirectly related to germline variable region gene segment transcription, the frequency of V(D)J recombination within an introduced substrate is enhanced by transcription independently of enhancer activity or obvious chromatin changes (22). In other systems transcription increases the frequency of homologous recombination (142, 146). Any of a number of mechanisms might mediate a transcriptional enhancement of V(D)J recombination, including interactions between DNA-binding factors, or changes in chromatin structure or topology.

It may be possible to elucidate the molecular determinants of recombinational accessibility through utilization of transgenic recombination substrates (28, 47, 52). Transgenic experiments in which TCR β chain V, D, and J segments were linked to a $C\mu$ gene, either with or without sequences that contained the Ig HC enhancer, demonstrated that presence of the enhaner was both necessary and sufficient for D_β-to-J_β rearrangement to occur within the construct in both B and T lymphocytes (47; P. Ferrier, F. Alt, et al, submitted). However, V_β-to-DJ_β rearrangement within the substrate occurred at a significant frequency only in T lymphocytes (47; P. Ferrier, F. Alt, et al, submitted). Within an analogous substrate that was introduced into a pre-B cell line, D_β-J_β joining in the absence of V_β-DJ_β rearrangement was also observed (46). In this cell line the $D_\beta J_\beta$ but not V_β sequences appeared to be in an "active" chromatin structure (46). Thus, in these constructs, the enhancer element is apparently required to allow access of the locus for recombination, but an additional element associated with the V_β segment, potentially as a specific DNA-binding factor or transcription from its promoter, mediates the lineage-specificity of its rearrangement. In this context, sequential "activation" of J, D, and V segments for rearrangement might also be responsible for the ordered rearrangement observed at the Ig HC locus (11). Further manipulation of sequences within such substrates may ultimately allow elucidation of the elements that determine the order and tissue-specificity of rearrangement of these various types of gene segments.

HEAVY CHAIN CLASS SWITCHING

Mechanism of Heavy-Chain Class Switching

In general, B-lineage cell lines that produce HC isotypes other than μ or δ have undergone a class-switch recombination event that has placed the expressed C_H gene immediately downstream of the expressed variable region gene, with accompanying deletion of intervening C_H genes (Figure 5; 132). Class switch recombination is mediated by sets of repeated sequences (switch (S) regions), that are located upstream of every C_H gene except δ (Figure 5; 132). S-regions vary in length from 1 to 10 kb and are composed of groups of short homologous repeats that differ in structure among isotypes (89, 132). Class switch (S-region) recombination is usually intrachromosomal (89), and does not affect the translational reading frame of the HC gene because it occurs within introns (Figure 5).

S-region recombination differs from V(D)J recombination in that it is not strictly site-specific and occurs at various locations within the S-regions, most often near the conserved repeat elements (Reviewed in 43, 89). The mechanism of S-region recombination is not understood. S-regions resemble Chi sequences, which stimulate bacterial recombination (69), and may recombine in E. coli (66); this finding led to the proposal that the recombination mechanism is homology-mediated (132). However, S-region recombination occurs within an introduced substrate at a much greater frequency in a B-lineage than in a nonlymphoid cell line (106), suggesting that it probably is

Directed Heavy Chain Class Switching:

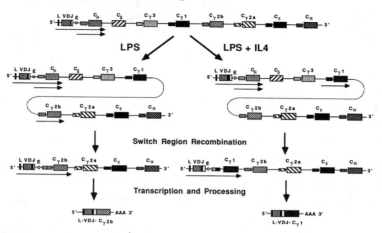

Figure 5 An "accessibility" model for directed HC class switching. The HC C-region genes are indicated by boxes with different shadings. Their corresponding upstream S-regions are represented by smaller boxes with the same shadings. Other sequences are indicated as in Figure 2. Arrows denote VDJ-C_μ, I_μ, and germline C-region transcripts. Only switching to γ2b and γ1 are shown. Activation by LPS treatment results in transcription of the germline γ2b region, which may be followed by class switching to γ2b. LPS treatment may also similarly induce switching to γ3. This pre-commitment to switch to γ2b (or γ3) is altered by exposure to IL4. Treatment with IL4 along with LPS results in production of a germline γ1 transcript, which may be followed by switching to γ1. Treatment with LPS plus IL4 may also induce switching to production of ε, although it has not yet been demonstrated that this latter event involves S-region recombination.

not mediated entirely by a generalized recombination pathway. S-region recombination events may generate point mutations and short duplications or deletions near the joining site, also suggesting that this recombination mechanism may involve DNA synthesis that might be error-prone (43). It also has been noted that certain S-region repeats can form specialized DNA conformations in vitro that have been implicated in recombination and gene regulation in other systems; it was proposed that such structures might be important for S-region recombination (37).

Directed Heavy-Chain Class Switching

The class of Ab that will be produced in response to a given antigen may be determined by helper T lymphocytes. These helper cells produce factors that direct the responding B lymphocytes to produce a particular HC isotype (33). The HC isotype-specificity of these responses can be mimicked in vitro by activation of splenocytes in the presence or absence of lymphokines that are produced by T lymphocytes. Stimulation by LPS polyclonally activates B lymphocytes in a manner that is presumably analogous to antigen stimulation. Culture of splenocytes with LPS results in production of IgM, IgG2b, and IgG3 (161), with class switching to the latter two isotypes involving S-region recombination (113). However, activation of B lymphocytes in the presence of lymphokines results in production of different isotypes. For example, treatment of splenocytes with interleukin 4 (IL4) along with LPS abrogates induction of IgG2b and IgG3 expression, and instead induces production of IgG1 and IgE (33, 64, 145). Treatment with different lymphokines similarly induces production of different Ab classes. For example, culture of splenocytes with γ interferon along with LPS results in production of IgG2a (135), and exposure to interleukin 5 (IL5) and LPS stimulates an enhancement of IgA production (33, 160).

These effects could derive from a random class-switching process followed by selection of cells that produce particular isotypes (36). However regardless of whether selective mechanisms operate, considerable evidence suggests that the class-switching mechanism is directed. For example, in murine B lymphocytes, class-switch recombination usually involves the same HC isotype on both productive and nonproductive alleles (62, 112, 151). Similarly, murine and rat plasmacytomas characteristically have translocations between the c myc gene and HC-switch regions; these translocations are generally present on the nonproductive allele and usually involve the S-region that corresponds to the expressed isotype (Reviewed in 39, 89). In contrast, in a series of human B-cell malignancies, switching had not necessarily occurred to the same isotype on the productive and nonproductive alleles (25, 149). This discrepancy could arise from a number of different sources, including the number of human cells examined and the possibility that multiple switches had occurred before transformation. However, the possibility that class switching may operate differently in humans and rodents has not been eliminated (25).

In further support of a directed model for HC class switching, B-lineage cell lines that undergo class switching during culture generally do so in an isotype-specific manner (89). For example, many A-MuLV transformants spontaneously switch to production of γ2b (reviewed in 89) or, in some instances, γ3 HCs (141). When these cell lines are treated with LPS, the frequency at which these isotype-specific switching events occur increases dramatically (90; P. Rothman et al, in preparation). The effect of LPS stimulation on these cultured pre-B cell lines is therefore generally analogous to its effect on splenocytes. The I29 B-cell lymphoma also undergoes isotype-specific switching in response to LPS treatment, but to production of α, γ2a, and ϵ HCs (139). Many of the effects of simultaneous IL4 plus LPS treatment are also observed in these cell lines; for example, when several A-MuLV transformants are exposed to IL4 plus LPS, they switch to expression of complete VDJ-C$_\epsilon$ mRNA (124; P. Rothman et al, manuscript in preparation).

An "Accessibility" Model for Control of Heavy-Chain Class Switching

A number of observations suggest that the mechanisms that direct HC class switching may be at least partially analogous to those that control V(D)J recombination, in that they involve modulation of the "accessibility" of the various HC isotypes to a common S-region recombinase machinery (90, 124, 138, 140, 155). In either A-MuLV transformants or cultured splenocytes, class switching to a particular isotype is generally preceded by expression of a germline C_H transcript corresponding to that isotype. For example, LPS stimulation of these cells results in expression of germline $\gamma 2b$ (90, 155,) and $\gamma 3$ transcripts (124), which are detected before appearance of VDJ-$\gamma 2b$ and VDJ-$\gamma 3$ transcripts. In an analogous fashion, LPS treatment of the I29 lymphoma induces appearance of transcripts of the germline α, $\gamma 2a$, and ϵ genes, which are demethylated in this cell line (140). As would be predicted for an "accessibility" model, exposure of pre-B cell lines to IL4 along with LPS treatment prevents expression of germline $\gamma 2b$ or $\gamma 3$ transcripts (90) and instead induces appearance of germline $\gamma 1$ (16, 124) and ϵ transcripts (Figure 5; 124; Rothman et al, in preparation).

The germline C_H transcripts characterized do date are generally analogous to each other in structure. The germline $\gamma 2b$ (88), $\gamma 3$ (P. Rothman et al, manuscript in preparation), $\gamma 1$ *(16), and* ϵ (124) transcripts all initiate upstream of the corresponding S-region. The germline $\gamma 2b$ (88), $\gamma 3$ (P. Rothman et al, manuscript in preparation), and ϵ (P. Rothman et al, manuscript in preparation) transcripts have been mapped in detail. All include a 5' exon that is derived from upstream of the S-region and is spliced to the corresponding C-region exons to generate a processed transcript (88; P. Rothman et al, manuscript in preparation). These processed transcripts are "sterile", because their 5' exons contain translation termination codons in the reading frame that corresponds to that of the downstream C-region (or in all three reading frames) (88; P. Rothman et al, in preparation). A germline transcript with a similar structure ($I\mu$ transcript; 80) is produced from the C_μ gene and is associated with transcriptional competence of that region (9, 80, 96).

The timing of appearance of these transcripts relative to isotype-specific switching suggests that germline C-region transcription may be indicative of "opening" of these loci to the S-region recombinase (Figure 5; 90, 124, 138, 140). Consistent with this " accessibility" model, in a cell line in which endogenous switching is detectable only following LPS treatment (and induction of germline C-region transcription), S-region recombination occurs at a significant frequency within an "accessible" substrate in the absence of LPS treatment (106). Alternatively, germline C-region transcription might mediate directed switching by a mechanism that is unrelated to potential changes in recombinational accessibility. Thus, either the transcripts themselves or hypothetical short polypeptides that they potentially could encode might play a role in this process.

A Model for Directed Heavy-Chain Class Switching

Normal pre-B lymphocytes that have undergone switching have not been detected (114). However, the capability of A-MuLV transformants to undergo directed class switch events suggests that pre-B lymphocytes may be capable of switching and are precommitted to switch to γ2b or γ3 production, but normally require activation to do so (89). Because of this precommitment, activation of these cells by LPS treatment results in production of germline γ2b and γ3 transcripts and in switching to these isotypes (90). The finding that LPS-stimulated splenocytes also switch to γ2b and γ3 (161) suggests that in general mature B lymphocytes inherit the precommitment to switch to these isotypes. However, the I29 lymphoma seems to be directed to switch to different isotypes (139, 140), indicating that this mature B lymphocyte might have been exposed to lymphokines earlier during its history. It is not known why the "baseline" state for HC class switching is a precommitment to switch to γ2b or γ3, or why exposure to lymphokines is generally required for B lymphocytes to switch to production of other isotypes. The predisposition to switch to γ2b or γ3 seems to be altered by lymphokine exposure in an analogous way in either pre-B or mature B lymphocytes. Thus, treatment of either pre-B or mature B lymphocytes with IL4 seems to "reprogram" these cells so that activation with LPS results in switching to γ1 and ϵ (see above). Consistent with the suggestion that lymphokine exposure alters a pre-existing commitment, pretreatment with IL4 directs splenocytes to switch to γ1 production when they are subsequently stimulated with LPS (136).

B Lymphocytes that Produce Multiple Heavy-Chain Isotypes Simultaneously

Numerous studies have identified B lymphocytes that appear to produce more than one HC isotype (excluding delta) simultaneously (32, 71, 91, 134, 159), leading to the proposal that class switching to a given isotype involves an intermediate stage during which the cell produces both μ and that isotype before DNA rearrangement (132). Some of this evidence is controversial (67), but an increasing number of examples of apparent double-producers have been identified, and it has been proposed that switching to IgE production normally does not involve S-region recombination (91). The model most commonly invoked to explain how double-isotype production occurs is that these cells generate long transcripts that are then spliced to generate the appropriate mRNA species. Such transcripts have not yet been identified, but there is some evidence for their existence (107). However, it is difficult to reconcile the long-transcript model with the notion that directed "accessibility" and germline C-region transcription are related—in the context of the latter model, the entire HC locus might be accessible in cells that produce long transcripts. An attractive alternative possibility that could reconcile a link between dual isotype production and directed switching would be the generation of the downstream isotype by trans-splicing between a VDJ-C_μ transcript and the germline transcript of the downstream isotype (89). However, there is as yet no experimental evidence for such a model.

PERSPECTIVES

The experiments described in this review have helped clarify many aspects of the V(D)J recombination mechanism and have expanded our understanding of how Ig-variable region gene assembly and H chain class switching are controlled during B lymphocyte differentiation. They have also left certain experimental questions unanswered and have defined new ones. For example, although a number of steps in the V(D)J recombination mechanism have been elucidated, almost nothing is known about the biochemical mechanisms and processes involved. To understand the V(D)J recombinase at such a level, it will be necessary to identify and purify its components and to develop an in vitro recombination system. Considerable evidence implicates Ig polypeptides in mediating control of the variable region gene assembly process, but it is not yet known how these regulatory signals might be effected or whether other factors might be involved. Expression of previously assembled transgenes that encode HC and LC polypeptides in *scid* mice may provide an invaluable system for addressing such questions. In *scid* Ig-transgenic mice it should be possible to better define the effects of expression of these polypeptides on B-lymphocyte differentiation —transgenic *scid* mice will be essentially unable to assemble endogenous Ig genes and may not be subject to the selection artifacts that may complicate studies involving normal Ig-transgenic mice. It is also of great interest to elucidate the mechanisms that seem to direct the action of the V(D)J recombinase. To understand these mechanisms it will be necessary to identify the molecular determinants of "accessibility," and of how the rearrangement process is ordered within a locus. Such questions should be approachable with existing experimental systems and perhaps with *scid* Ig-transgenic mice. With regard to H chain class switching, an apparent link has now been made between the regulatory action of T-lymphocytes and molecular events that seem to control S-region recombination. However, the mechanism by which S-region recombination occurs is still practically unknown. It has also not been determined whether germline C-region transcription may actually influence recombinational "accessibility" and/or may have another function. In this regard, it is of great interest to determine whether instances of apparent double HC isotype production derive from long transcripts or from splicing to germline C-region transcripts.

ACKNOWLEDGMENTS

We thank Paul Rothman and Gary Rathbun for critically reading this manuscript. This work was supported by the Howard Hughes Medical Institute and by NIH grants AI-200047 and CA-40427 to F. W. Alt. T. K. Blackwell is a fellow of the Howard Hughes Medical Institute.

Literature Cited

1. Aguilera, R. J., Akira, S., Okazaki, K., Sakano, H. 1987. A pre-B cell nuclear-protein that specifically interacts with the immunoglobulin V-J recombination sequences. *Cell* 51:909–17
2. Akira, S., Okazaki, K., Sakano, H. 1988. Two pairs of recognition signals are sufficient to cause immunoglobulin V-(D)-J joining. *Science* 238:1134–38
3. Alt, F. W. 1986. New mechanism revealed. *Nature* 322:772–73
4. Alt, F. W., Baltimore, D. 1982. Joining of immunoglobulin heavy chain gene segments: implications from a chromosome with evidence of three D-J_H fusions. *Proc. Natl. Acad. Sci. USA* 79: 4118–22
5. Alt, F. W., Blackwell, T. K., De Pinho, R. A., Reth, M. G., Yancopoulos, G. D. 1986. Regulation of genome rearrangement events during lymphocyte differentiation. *Immunol. Rev.* 89:5–30
6. Alt, F. W., Blackwell, T. K., Yancopoulos, G. D. 1987. Development of the primary antibody repertoire. *Science* 238:1079–87
7. Alt, F., Bothwell, A., Knapp, M., Siden, E., Mather, E., et al. 1980. Synthesis of secreted and membrane-bound immunoglobulin μ heavy chains is directed by mRNAs that differ at their 3' ends. *Cell* 20:293–301
8. Alt, F. W., Enea, V., Bothwell, A. L. M., Baltimore, D. 1980. Activity of multiple light chain genes in murine myeloma cells producing a single, functional light chain. *Cell* 21:1–12
9. Alt, F. W., Rosenberg, N., Enea, V., Siden, E., Baltimore, D. 1982. Multiple immunoglobulin heavy chain gene transcripts in Abelson murine leukemia virus transformed lymphoid cell lines. *Mol. Cell Biol.* 2:386–400
10. Alt, F. W., Rosenberg, N., Lewis, S., Thomas, E., Baltimore, D. 1981. Organization and reorganization of immunoglobulin genes in A-MuLV-

transformed cells: rearrangement of heavy but not light chain genes. *Cell* 27:381–90
11. Alt, F. W., Yancopoulos, G., Blackwell, T. K., Wood, C., Thomas, E., et al. 1984. Ordered rearrangement of immunoglobulin heavy chain variable region gene segments. *EMBO J.* 3:1209–19
12. Bakhshi, A., Wright, J. J., Graringer, W., Seto, M., Owens, J., et al. 1987. Mechanism of the t(14:18) chromosomal translocation: Structural analysis of both derivative 14 and 18 reciprocal partners. *Proc. Natl. Acad. Sci. USA*, 84:2396–400
13. Baltimore, D. 1981. Gene conversion: some implications for immunoglobulin genes. *Cell* 24:592–94
14. Banerji, J., Olson, L., Schaffner, W. 1983. A lymphocyte-specific cellular enhancer is located downstream of the joining region in immunoglobulin heavy chain genes. *Cell*:33:729–40
15. Berman, J. E., Mellis, S. J., Pollock, R. R., Smith, C. L., Suh, H., et al. 1988. Content and organization of the human Ig V_H locus: Definition of three new V_H families and linkage to the Ig C_H locus. *EMBO J.* 7:727–38
16. Berton, M. T., Uhr, J. W., Vitetta, E. S. 1989. Synthesis of germ-line $\gamma1$ immunoglobulin heavy-chain transcripts in resting B cells: Induction by interleukin 4 and inhibition by interferon γ. *Proc. Natl. Acad. Sci. USA* 86:2829–33
17. Blackwell, T. K., Alt, F. W. 1984. Site-specific recombination between immunoglobulin D and J_H segments that were introduced into the genome of a murine pre-B cell line. *Cell* 37:105–12
18. Blackwell, T. K., Alt, F. W. 1988. Immunoglobulin genes. In *Molecular Immunology,* ed. B. D. Hames, D. M. Glover, pp. 1–60. Oxford, UK: IRL Press. 248 pp.
19. Blackwell, T. K., Alt, F. W. 1989.

Molecular characterization of the Lymphoid V(D)J recombination activity. *J. Biol. Chem.* 264:10327–30

20. Blackwell, T. K., Ferrier, P., Malynn, B. A., Pollock, R. R., Covey, L. R., et al. 1989. The effect of the *scid* mutation on mechanism and control of immunoglobulin heavy and light chain gene rearrangement. *Curr. Top. Microbiol. Immunol.* In press

21. Blackwell, T. K., Malynn, B. A., Pollock, R. R., Ferrier, P., Covey, L., et al. 1989. Isolation of *scid* pre-B cells that rearrange kappa light chain genes: Formation of normal signal and abnormal coding joins. *EMBO J.* 8:735–42

22. Blackwell, T. K., Moore, M., Yancopoulos, G., Suh, H., Lutzker, S., et al. 1986. Recombination between immunoglobulin variable region segments is enhanced by transcription. *Nature* 324:585–89

23. Blankenstein, T., Krawinkel, U. 1987. Immunoglobulin V_H region genes of the mouse are organized in overlapping clusters. *Eur. J. Immunol.* 17:1351–57

24. Blattner, F., Tucker, P. 1984. The molecular biology of IgD. *Nature* 307:417–22

25. Borzillo, G., Cooper, M., Kubagawa, H., Landay, A., Burrows, P. 1987. Isotype switching in human B lymphocyte malignancies occurs by DNA deletion: Evidence for nonspecific switch recombination. *J. Immunol.* 139:1326–35

26. Bosma, G. C., Custer, R. P., Bosma, M. J. 1983. A severe combined immunodeficiency mutation in the mouse. *Nature* 301:527–30

27. Brinster, R. L., Ritchie, K. A., Hammer, R. E., O'Brien, R. L., Arp, B., Storb, U. 1983. Expression of a microinjected immunoglobulin gene in the spleen of transgenic mice. *Nature* 306:332–36

28. Bucchini, D., Reynaud, C.-A., Ripoche, M.-A., Grimal, H., Jami, J., Weill, J. C. 1987. Rearrangement of a chicken immunoglobulin gene occurs in the lymphoid lineage of transgenic mice. *Nature* 326:409–11

29. Burnet, F. M. 1957. A modification of Jerne's theory of antibody production using the concept of clonal selection. *Aust. J. Sci.* 20:67–69

30. Calame, K., Eaton, S. 1988. Transcriptional controlling elements in the immunoglobulin and T cell receptor loci. *Adv. Immunol.* 43:235–75

31. Cebra, J., Colberg, J. E., Dry, S. 1966. Rabbit lymphoid cells differentiated with respect to α, γ and μ-heavy polypeptide chains and to allotypic markers AA1 and AA2. *J. Exp. Med.* 123:547–58

32. Chen, Y.-W., Word, C., Dev, V., Uhr, J. W., Vitetta, E. S., Tucker, P. W. 1986. Double isotype production by a neoplastic B-cell line II. Allelically excluded production of μ and γ1 heavy chains without C_H gene rearrangement. *J. Exp. Med.* 164:562–79

33. Coffman, R., Seymour, B., Lebman, D., Hiraki, D., Christiansen, J., et al. 1988. The role of helper T cell products in mouse B cell differentiation and isotype regulation. *Immunol. Rev.* 102:5–28

34. Coffman, R., Weissman, I. 1983. Gene rearrangement in pre-B cells. *J. Mol. Cell. Immunol.* 1:33–41

35. Coleclough, C. 1983. Chance, necessity, and antibody gene dynamics. *Nature* 303:23–26

36. Coleclough, C. D., Cooper, D., Perry, R. 1980. Rearrangement of immunoglobulin heavy chain genes during B-lymphocyte development as revealed by studies of mouse plasmacytoma cells. *Proc. Natl. Acad. Sci. USA* 77:1422–26

37. Collier, D. A., Griffin, J. A., Wells, R. D., 1989. Non-B right-handed DNA confirmations of homopurine-homopyrimidine sequences in the murine immunoglobulin C_α switch region. *J. Biol. Chem.* 263:7397–405

38. Cooper, M. D., Burrows, P. 1989. B-cell differentiation. In *Immunoglobulin Genes* ed. T. Honjo, F. W. Alt, T. H. Rabbitts, ed. 1:1–21. New York: Academic.

39. Cory, S. 1986. Activation of cellular oncogenes in hemopoietic cells by chromosome translocation. *Adv. Cancer Res.* 47:189–234

40. Custer, R. P., Bosma, G. C., Bosma, M. J. 1985. Severe combined immunodeficiency (scid) in the mouse. *Am. J. Pathol.* 120:464–77

41. Davis, M. M., Bjorkman, P. J. 1988. T-cell antigen receptor genes and T-cell recognition. *Nature* 334:395–401

42. Dildrop, R., Gause, A., Muller, W., Rajewsky, K. 1987. A new V gene expressed in lambda-2 light chains of the mouse. *Eur. J. Immunol.* 17:731–34

43. Dunnick, W., Wilson, M., Stavnezer, J. 1989. Mutations, duplication, and deletion of recombined switch regions suggest a role for DNA replication in the immunoglobulin heavy chain switch. *Mol. Cell Biol.* In press

44. Early, P., Huang, H., Davis, M., Calame, K., Hood, L. 1980. An immunoglobulin heavy chain variable region gene is generated from three seg-

ments of DNA: V_H, D, and J_H. *Cell* 19:981–92

45. Early, P., Rogers, J., Davis, M., Calame, K., Bond, M., et al. 1980. Two mRNAs can be produced from a single immunoglobulin μ gene by alternate RNA processing pathways. *Cell* 20:313–19

46. Ferrier, P., Covey, L., Suh, H., Winoto, A., Hood, L., Alt, F. W. 1989. T cell receptor DJ but not VDJ rearrangement within a recombination substrate introduced into a pre-B cell line. *Int. Immunol.* 1:66–74

47. Ferrier, P., Krippl, B., Blackwell, K., Suh, H., Winoto, A., 1988. Control of VDJ recombinase activity: tissue specific assembly of transgenic T cell receptor variable region gene segments. In *Cellular Basis of Immune Modulation*, ed. J. G. Kaplan, D. R. Green, R. C. Bleackley, pp. 189–99. New York: Liss. 656 pp.

48. Fujimoto, S., Yamagishi, H. 1987. Isolation of an excision product of T-cell receptor α-chain gene rearrangement. *Nature* 327:242–44

49. Gally, J. A. 1973. Structure of immunoglobulins. In *The Antigens*, ed. M. Sela, 1:161–298. New York: Academic. 573 pp.

50. Gillies, S. D., Morrison, S. L., Oi, V. T., Tonegawa, S. 1983. A tissue-specific transcription enhancer element is located in the major intron of a rearranged immunoglobulin heavy chain gene. *Cell* 33:717–28

51. Gollahon, K. A., Hagman, J., Brinster, R. L., Storb, U. 1988. Ig γ-producing B cells do not show feedback inhibition of gene rearrangement. *J. Immunol.* 141:2771–80

52. Goodhart, M., Cavelier, P., Akimenko, M. A., Lutfalla, G., Babinet, C., Rougeon, F. 1987. Rearrangement and expression of rabbit immunoglobulin κ light chain in transgenic mice. *Proc. Natl. Acad. Sci. USA* 84:4229–33

53. Goodnow, C. C., Crosbie, J., Adelstein, S., Lavoie, T. B., Smith-Gill, S. J., et al. 1988. Altered immunoglobulin expression and functional silencing of self-reactive B lymphocytes in transgenic mice. *Nature* 334:676–82

54. Graininger, W. B., Goldman, P. L., Morton, C. C., O'Brien, S. J., Korsmeyer, S. J. 1988. The κ-deleting element: Germline and rearranged, duplicated and dispersed forms. *J. Exp. Med.* 167:488–501

55. Grosschedl, R., Weaver, D., Baltimore, D., Costantini, F. 1984. Introduction of a μ immunoglobulin gene into the mouse germ line: specific expression in lymphoid cells and synthesis of functional antibody. *Cell* 38:647–58

56. Halligan, B. D., Desiderio, S. V. 1987. Identification of a DNA binding protein that recognizes the nonamer recombinational signal sequence of immunoglobulin genes. *Proc. Natl. Acad. Sci. USA* 84:7019–23

57. Hardy, R. R., Dangl, J., Hayakawa, K., Jager, G., Herzenberg, L. A., Herzenberg, L. A. 1986. Frequent lambda light chain gene rearrangement and expression in a Ly-1 B lymphoma with a productive kappa chain allele. *Proc. Natl. Acad. Sci. USA* 83:1438–42

58. Hendrickson, E. A., Schatz, D. G., Weaver, D. T. 1988. The *scid* gene encodes a *trans*-acting factor that mediates the rejoining event of Ig gene rearrangement. *Genes Dev.* 2:817–29

59. Herzenberg, L. A., Stall, A. M., Braun, J., Weaver, D., Baltimore, D., et al. 1987. Depletion of the predominant B-cell population in immunoglobulin μ heavy-chain transgenic mice. *Nature* 329:71–73

60. Hesse, J. E., Lieber, M. R., Gellert, M., Mizuuchi, K. 1987. Extrachromosomal DNA substrates in pre-B cells undergo inversion or deletion at immunoglobulin V-(D)-J joining signals. *Cell* 49:775–83

61. Hieter, P., Korsmeyer, S., Waldmann, T., Leder, P. 1981. Human immunoglobulin κ light chain genes are deleted or rearranged in λ-producing cells. *Nature* 290:368–72

62. Hummel, M., Berry, J., Dunnick, W. 1987. Switch region content of hybridomas: Two spleen IgH loci tend to rearrange to the same isotype. *J. Immunol.* 138:3539–48

63. Iglesias, A., Lamers, M., Kohler, G. 1987. Expression of immunoglobulin delta chain causes allelic exclusion in transgenic mice. *Nature* 330:482–84

64. Isakson, P. C., Pure, E., Vitetta, E. S., Krammer, P. H. 1982. T cell-derived B cell differentiation factor(s). Effect on the isotype switch of murine B cells. *J. Exp. Med.* 155:734–48

65. Kabat, E. A. 1988. Antibody complementarity and antibody structure. *J. Immunol.* 141:525–36

66. Kataoka, T., Takeda, S.-I., Honjo, T. 1983. *Escherichia coli* extract-catalyzed recombination in switch regions of mouse immunoglobulin genes. *Proc. Natl. Acad. Sci. USA* 80:2666–70

67. Katona, I., Urban, J., Finkelman, F. 1985. B cells that simultaneously express surface IgM and IgE in *Nippo-*

strongylus brasilienis-infected SJA/9 mice do not provide a model for isotype switching without gene deletion. *Proc. Natl. Acad. Sci. USA* 82:511–15

68. Kelley, D. E., Wiedemann, L. E., Pittet, A.-C., Strauss, S., Nelson, K. J., et al. 1985. Nonproductive kappa immunoglobulin genes: Recombinational abnormalities and other lesions affecting transcription, RNA processing, and turnover. *Mol. Cell Biol.* 5:1660–75

69. Kenter, A., Birshtein, B. 1980. Chi, a promoter of generalized recombination in μ phage, is present in immunoglobulin genes. *Nature* 293:402–4

70. Kim, M.-G., Schuler, W., Bosma, M. J., Marcu, K. 1988. Abnormal recombination of Igh D and J gene segments in transformed pre-B cells of *scid* mice. *J. Immunol.* 141:1341–47

71. Kinashi, T., Godal, T., Noma, Y., Ling, N. R., Yaoita, Y., Honjo, T. 1987. Human neoplastic B cells express more than two isotypes of immunoglobulins without deletion of heavy chain constant region genes. *Genes Dev.* 1:465–70

72. Kleinfield, R., Hardy, R. R., Tarlinton, D., Dangl, J., Herzenberg, L. A., Weigert, M. 1986. Recombination between an expressed immunoglobulin heavy-chain gene and a germline variable gene segment in a Ly 1$^+$ B-cell lymphoma. *Nature* 322:843–46

73. Kocks, C., Rajewsky, K. 1989. Stable expression and somatic hypermutation of antibody V regions in B-cell differentiation. *Annu. Rev. Immunol.* 7:537–59

74. Kodaira, M., Kinashi, T., Umemura, I., Matsuda, F., Noma, T., et al. 1986. Organization and evolution of variable region genes of the human immunoglobulin heavy chain. *J. Mol. Biol.* 190:529–41

75. Kubagawa, H., Cooper, M. D., Carroll, A. J., Burrows, P. D. 1989. Light-chain gene expression before heavy-chain gene rearrangement in pre-B cells transformed by Epstein-Barr virus. *Proc. Natl. Acad. Sci. USA* 86:2356–60

76. Kudo, A., Melchers, F. 1987. A second gene, V_{preB} in the λ5 locus of the mouse, which appears to be selectively expressed in pre-B lymphocytes. *EMBO J.* 6:2267–72

77. Kudo, A., Sakaguchi, N., Melchers, F. 1987. Organization of the murine Ig-related λ5 gene transcribed selectively in pre-B lymphocytes. *EMBO J.* 6:103–7

78. Kurosawa, Y., Von Boehmer, H., Hass, W., Sakano, H., Traunecker, A., Tonegawa, S. 1981. Identification of D segments of immunoglobulin heavy chain genes and their rearrangement in T lymphocytes. *Nature* 290:565–70

79. Landau, N. R., Schatz, D. G., Rosa, M., Baltimore, D. 1987. Increased frequency of N-region insertion in a murine pre-B-cell line infected with a terminal deoxynucleotidyl transferase retroviral expression vector. *Mol. Cell Biol.* 7:3237–43

80. Lennon, G. G., Perry, R. P. 1985. C_μ-containing transcripts initiate heterogeneously within the IgH enhancer region and contain a novel 5'-nontranslatable exon. *Nature* 318:475–78

81. Lewis, S., Gifford, A., Baltimore, D. 1984. Joining of V_κ to J_κ segments in a retroviral vector introduced into lymphoid cells. *Nature* 308:425–28

82. Lewis, S. M., Hesse, J. E., Mizuuchi, K., Gellert, M. 1988. Novel strand exchanges in V(D)J recombination. *Cell* 55:1099–107

83. Lewis, S., Rosenberg, N., Alt, F., Baltimore, D. 1982. Continuing kappa-gene rearrangement in a cell line transformed by Abelson murine leukemia virus. *Cell* 30:807–16

84. Lieber, M. R., Hesse, J. E., Lewis, S., Bosma, G. C., Rosenberg, N., et al. 1988. The defect in murine severe combined immune deficiency: joining of signal sequences but not coding segments in V(D)J recombination. *Cell* 55:7–16

85. Lieber, M. R., Hesse, J. E., Mizuuchi, K., Gellert, M. 1987. Developmental stage specificity of the lymphoid V(D)J recombination activity. *Genes Dev.* 1:751–61

86. Lieber, M. R., Hesse, J. E., Mizuuchi, K., Gellert, M. 1988. Lymphoid V(D)J recombination: Nucleotide insertion at signal joints as well as coding joints. *Proc. Natl. Acad. Sci. USA* 85:8588–92

87. Livant, D., Blatt, C., Hood, L. 1986. One heavy chain variable region gene segment subfamily in the BALB/c mouse contains 500–1000 or more members. *Cell* 47:461–70

88. Lutzker, S. G., Alt, F. W. 1988. Structure and expression of germ line immunoglobulin γ2b transcripts. *Mol. Cell Biol.* 8:1849–52

89. Lutzker, S. G., Alt, F. W. 1988. Immunoglobulin heavy chain class switching. In *Mobile DNA*, ed. D. E. Berg, M. M. Howe, pp. 691–714. Washington, DC: Am. Soc. Microbiol. 972 pp.

90. Lutzker, S., Rothman, P., Pollock, R., Coffman, R., Alt, F. W. 1988. Mitogen- and IL-4-regulated expression of germ-line Ig γ_{2b} transcripts: evidence

for directed heavy chain class switching. *Cell* 53:177–84

91. MacKenzie, T., Dosch, H.-M. 1989. Clonal and molecular characteristics of the human IgE-committed B cell subset. *J. Exp. Med.* 169:407–30

92. Malynn, B. A., Blackwell, T. K., Fulop, G. M., Rathbun, G., Furley, A. J. W., et al. 1988. The *scid* defect affects the final step of the immunoglobulin VDJ recombinase mechanism. *Cell* 54:453–60

93. Manz, J., Denis, K., Witte, O., Brinster, R., Storb, U. 1988. Feedback inhibition of immunoglobulin gene rearrangement by membrane μ, but not by secreted μ heavy chains. *J. Exp. Med.* 168:1363–81

94. Mather, E. L., Nelson, K. J., Haimovich, J., Perry, R. P. 1984. Mode of regulation of immunoglobulin μ and δ-chain expression varies during B-lymphocyte maturation. *Cell* 36:329–38

95. Morzycka-Wroblewska, E., Lee, F. E.-H., Desiderio, S. V. 1988. Unusual immunoglobulin gene rearrangement leads to replacement of recombinational signal sequences. *Science* 242:261–63

96. Nelson, K. J., Haimovich, J., Perry, R. P. 1983. Characterization of productive and sterile transcripts from the immunoglobulin heavy-chain locus: Processing of μ_m and μ_s mRNA. *Mol. Cell Biol.* 3:1317–32

97. Nelson, K. J., Kelley, D. E., Perry, R. P. 1985. Inducible transcription of the unrearranged C_κ locus is a common feature of pre-B cells and does not require DNA or protein synthesis. *Proc. Natl. Acad. Sci. USA* 82:5305–9

98. Neuberger, M. S. 1983. Expression and regulation of immunoglobulin heavy chain gene transfected into lymphoid cells. *EMBO J.* 2:1373–78

99. Neuberger, M. S., Caskey, H. M., Pettersson, S., Williams, G. T., Surani, M. A. 1989. Isotype exclusion and transgene down-regulation in immunoglobulin-λ transgenic mice. *Nature* 338:350–52

100. Nisonoff, A., Hopper, J. E., Spring, S. B. 1975. Human immunoglobulins. In *The Antibody Molecule*, pp. 86–137. New York: Academic. 542 pp.

101. Nottenberg, C., St. John, T., Weissman, I. L. 1987. Unusual immunoglobulin DNA sequences from the nonexpressed chromosome of mouse normal B lymphocytes: implications for allelic exclusion and the DNA rearrangement process. *J. Immunol.* 139:1718–26

102. Nussenzweig, M. C., Shaw, A. C., Sinn, E., Campos-Torres, J., Leder, P. 1988. Allelic exclusion in transgenic mice carrying mutant human IgM genes. *J. Exp. Med.* 167:1969–74

103. Nussenzweig, M. C., Shaw, A. C., Sinn, E., Danner, D. B., Holmes, K. L., et al. 1987. Allelic exclusion in transgenic mice that express the membrane form of immunoglobulin μ. *Science* 236:816–19

104. Okazaki, K., Davis, D. D., Sakano, H. 1987. T cell receptor β gene sequences in the circular DNA of thymocyte nuclei: direct evidence for intramolecular DNA deletion in V-D-J joining. *Cell* 49:477–85

105. Okazaki, K., Nishikawa, S.-I., Sakano, H. 1988. Aberrant immunoglobulin gene rearrangement in *scid* mouse bone marrow cells. *J. Immunol.* 141:1348–52

106. Ott, D., Alt, F., Marcu, K. 1987. Immunoglobulin heavy chain switch recombination within a retroviral vector in murine pre-B cells. *EMBO J.* 6:577–84

107. Perlmutter, A., Gilbert, W. 1984. Antibodies of the secondary response can be expressed without switch recombination in normal B cells. *Proc. Natl. Acad. Sci. USA* 81:7189–93

108. Pernis, B. G., Chiappino, G., Kelus, A. S., Gell, P. G. H. 1965. Cellular localization of immunoglobulins with different allotype specificities in rabbit lymphoid tissue. *J. Exp. Med.* 122:853–76

109. Persiani, D. M., Durdik, J., Selsing, E. 1987. Active λ and κ antibody gene rearrangement in Abelson murine leukemia virus-transformed pre-B cell lines. *J. Exp. Med.* 165:1655–74

110. Picard, D., Schaffner, W. 1984. A lymphocyte-specific enhancer in the mouse immunoglobulin κ gene. *Nature* 307:80–82

111. Pillai, S., Baltimore, D. 1987. Formation of disulphide-linked $\mu_2\omega_2$ tetramers in pre-B cells by the 18K ω-immunoglobulin light chain. *Nature* 329:172–74

112. Radbruch, A., Muller, W., Rajewsky, K. 1986. Class switch recombination is IgG1 specific on the active and inactive IgH loci of IgG1-secreting B cell blasts. *Proc. Natl. Acad. Sci. USA* 84:3954–57

113. Radbruch, A., Sablitzky, F. 1983. Deletion of $C\mu$ gene in mouse B lymphocytes upon stimulation with LPS. *EMBO J.* 2:1929–35

114. Raff, M. C., Megson, M., Owen, J. J. T., Cooper, M. D. 1976. Early production of intracellular IgM by B-lymphocyte precursors in mouse. *Nature* 259:224–26

115. Rathbun, G., Berman, J., Yancopoulos, G., Alt, F. W. 1988. See Ref. 38, 2:63–90

116. Rathbun, G. A., Tucker, P. 1987. Conservation of sequences necessary for V gene recombination. In *Evolution and Vertebrate Immunity*, ed. G. Kelsoe, D. H. Schultze, pp. 85–115. Austin: Univ. Texas Press

117. Reth, M. G., Alt, F. W. 1984. Novel immunoglobulin heavy chains are produced from DJ_H gene segment rearrangements in lymphoid cells. *Nature* 312:418–23

118. Reth, M. G., Ammirati, P., Jackson, S., Alt, F. W. 1985. Regulated progression of a cultured pre-B-cell line to the B-cell stage. *Nature* 317:353–55

119. Reth, M., Gehrmann, P., Petrac, E., Wiese, P. 1986. A novel V_H to V_HDJ_H joining mechanism in heavy-chain-negative (null) pre-B cells results in heavy chain production. *Nature* 322:840–22

120. Reth, M. G., Jackson, S., Alt, F. W. 1986. V_HDJ_H formation and DJ_H replacement during pre-B differentiation: non-random usage of gene segments. *EMBO J.* 5:2131–38

121. Reth, M., Petrac, E., Wiese, P., Lobel, L., Alt, F. W. 1987. Activation of V_κ gene rearrangement in pre-B cells follows the expression of membrane-bound immunoglobulin heavy chains. *EMBO J.* 6:3299–305

122. Ritchie, K. A., Brinster, R. L., Storb, U. 1984. Allelic exclusion and control of endogenous immunoglobulin gene rearrangement in κ transgenic mice. *Nature* 312:517–20

123. Rogers, J., Early, P., Carter, C., Calame, K., Bond, M., et al. 1980. Two mRNAs with different 3' ends encode membrane-bound and secreted forms of immunoglobulin u chain. *Cell* 20:303–12

124. Rothman, P., Lutzker, S., Cook, W., Coffman, R., Alt, F. W. 1988. Mitogen plus interleukin 4 induction of C_E transcripts in B lymphoid cells. *J. Exp. Med.* 168:2385–89

125. Rusconi, S., Kohler, G. 1985. Transmission and expression of a specific pair of rearranged immunoglobulin μ and κ genes in a transgenic mouse line. *Nature* 314:330–34

126. Sakaguchi, N., Melchers, F. 1986. λ5, a new light-chain-related locus selectively expressed in pre-B lymphocytes. *Nature* 324:579–82

127. Sakano, H., Maki, R., Kurosawa, Y., Roeder, W., Tonegawa, S. 1980. Two types of somatic recombination are necessary for the generation of complete immunoglobulin heavy chain genes. *Nature* 286:676–83

128. Schatz, D. G., Baltimore, D. 1988. Stable expression of immunoglobulin gene V(D)J recombinase activity by gene transfer into 3T3 fibroblasts. *Cell* 53:107–15

129. Schuler, W., Weiler, I. J., Schuler, A., Phillips, R. A., Rosenberg, N., et al. 1986. Rearrangement of antigen-receptor genes is defective in mice with severe combined immune deficiency. *Cell* 46:963–72

130. Selsing, E., Durdik, J., Moore, M. W., Persiani, D. M. 1989. Immunoglobulin λ genes. See Ref. 38, 2:111–22

131. Shapiro, M. A., Weigert, M. 1987. How immunoglobulin V_κ genes rearrange. *J. Immunol.* 139:3834–39

132. Shimizu, A., Honjo, T. 1984. Immunoglobulin class switching. *Cell* 36:801–3

133. Shimizu, A., Takahashi, N., Yaoita, Y., Honjo, T. 1982. Organization of the constant region gene family of the mouse immunoglobulin heavy chain. *Cell* 29:499–506

134. Snapper, C. M., Finkelman, F. D., Stefany, D., Conrad, D. H., Paul, W. E. 1988. IL-4 induces co-expression of intrinsic membrane IgG1 and IgE by murine B cells stimulated with lipopolysaccharide. *J. Immunol.* 141:489–98

135. Snapper, C., Paul, W. 1987. Interferon-γ and B cell Stimulatory Factor-1 reciprocally regulate Ig isotype production. *Science* 236:944–47

136. Snapper, C., Paul, W. 1987. B cell Stimulatory Factor-1 (Interleukin-4) prepares resting murine B cells to secrete IgG1 upon subsequent stimulation with bacterial lipopolysaccharide. *J. Immunol.* 139:10–17

137. Stall, A. M., Kroese, F. G. M., Gadus, F. T., Sieckmann, D., Herzenberg, L. A., Herzenberg, L. A. 1988. Rearrangement and expression of endogenous immunoglobulin genes occur in many murine B cells expressing transgenic membrane IgM. *Proc. Natl. Acad. Sci. USA* 85:3546–50

138. Stavnezer, J., Radcliffe, G., Lin, Y.-C., Nietupski, J., Berggren, L., et al. 1988. Ig heavy chain switching may be directed by prior induction of transcripts from constant region genes. *Proc. Natl. Acad. Sci. USA* 85:7704–8

139. Stavnezer, J., Sirlin, S., Abbott, J. 1985. Induction of immunoglobulin isotype switching in cultured I.29 B lymphoma cells. *J. Exp. Med.* 161:577–601

140. Stavnezer-Nordgren, J., Sirlin, S. 1986. Specificity of immunoglobulin heavy chain switch correlates with activity of germline heavy chain genes prior to switching. *EMBO J.* 5:95–102
141. Sugiyama, H., Maeda, T., Akira, S., Kishimoto, S. 1986. Class switch from μ to $\gamma3$ or $\gamma2b$ production at pre-B cell stage. *J. Immunol.* 136:3092–97
142. Thomas, B. J., Rothstein, R. 1989. Elevated recombination rates in transcriptionally active DNA. *Cell* 56:619–30
143. Tonegawa, S. 1983. Somatic generation of antibody diversity. *Nature* 302:575–81
144. Van Ness, B. G., Weigert, M., Coleclough, C., Mather, E. L., Kelley, D. E., Perry, R. P. 1981. Transcription of the unrearranged mouse C_κ locus: Sequence of the initiation region and comparison of activity with a rearranged V_κ-C_κ gene. *Cell* 27:593–602
145. Vitetta, E. S., Ohara, J., Myers, C. D., Layton, J. E., Krammer, P. H., Paul, W. E. 1985. Serological, biochemical, and functional identity of B cell stimulatory factor-1 and B cell differentiation factor for IgG1. *J. Exp. Med.* 162:1726–31
146. Voelkel-Meiman, K., Keil, R., Roeder, G., 1987. Recombination stimulating sequences in yeast ribosomal DNA correspond to sequences regulating transcription by RNA polymerase I. *Cell* 48:1071–79
147. Wabl, M., Steinberg, C. 1982. A theory of allelic and isotypic exclusion. *Proc. Natl. Acad. Sci. USA* 79:6976–78
148. Weaver, D., Costantini, F., Imanishi-Kari, T., Baltimore, D. 1985. A transgenic immunoglobulin mu gene prevents rearrangement of endogenous genes. *Cell* 42:117–27
149. Webb, C., Cooper, M., Burrows, P., Griffin, J. 1985. Immunoglobulin gene rearrangements and deletions in human Epstein-Barr virus transformed cell lines producing IgG and IgA subclasses. *Proc. Natl. Acad. Sci. USA* 82:5495–99
150. Weill, J. C., Reynaud, C. A. 1987. The chicken B cell compartment. *Science* 238:1094–98
151. Winter, E., Krawinkel, U., Radbruch, A. 1987. Directed Ig class-switch recombination in activated B cells. *EMBO J.* 6:1663–71
152. Wood, C., Tonegawa, S. 1983. Diversity and joining segments of mouse Ig H genes are closely linked and in the same orientation. *Proc. Natl. Acad. Sci. USA* 80:3030–34
153. Yancopoulos, G. D., Alt, F. W. 1985. Developmentally controlled and tissue specific expression of unrearranged V_H gene segments. *Cell* 40:271–81
154. Yancopoulos, G. D., Blackwell, T. K., Suh, H., Hood, L., Alt, F. W. 1986. Introduced T cell receptor variable region gene segments recombine in pre-B cells: Evidence that B and T cells use a common recombinase. *Cell* 44:251–59
155. Yancopoulos, G., DePinho, R., Zimmerman, K., Lutzker, S., Rosenberg, N., Alt, F. 1986. Secondary rearrangement events in pre-B cells: V_HDJ_H replacement by a LINE-1 sequence and directed class switching. *EMBO J.* 5:3259–66
156. Yancopoulos, G. D., Desiderio, S. V., Paskind, M., Kearney, J. F., Baltimore, D., Alt, F. W. 1984. Preferential utilization of the most J_H-proximal V_H gene segments in pre-B-cell lines. *Nature* 311:727–33
157. Yancopoulos, G. D., Malynn, B. A., Alt, F. W. 1988. Developmentally regulated and strain-specific expression of murine V_H gene families. *J. Exp. Med.* 168:417–35
158. Yancopoulos, G. D., Nolan, G., Herzenberg, L., Alt, F. W. 1989. In *NATO Advanced Study Series: Vectors as Tools for the Study of Normal and Abnormal Growth and Differentiation*, ed. H. Lother, R. Dernick, W. Ostertag. Hamburg: Springer-Verlag. In press
159. Yaoita, Y., Kumagai, Y., Okumura, K., Honjo, T. 1982. Expression of lymphocyte surface IgE does not require switch recombination. *Nature* 297:697–99
160. Yokota, T., Coffman, R. L., Hagiwara, H., Rennick, D. M., Takebe, Y., et al. 1987. Isolation and characterization of lymphokine cDNA clones encoding mouse and human IgA-enhancing factor and eosinophil colony-stimulating factor activities: Relationship to interleukin 5. *Proc. Natl. Acad. Sci. USA* 84:7388–92
161. Yuan, D., Vitetta, E. S. 1983. Structural studies of cell surface and secreted IgG in LPS-stimulated murine B cells. *Mol. Immunol.* 20:367–75

Annu. Rev. Genet. 1989. 23:637–61

THE ISOCHORE ORGANIZATION OF THE HUMAN GENOME

Giorgio Bernardi

Laboratoire de Génétique Moléculaire, Institut Jacques Monod, 2 Place Jussieu, 75005 Paris, France

CONTENTS

INTRODUCTION

At a time when the human genome begins to be investigated on a massive scale, at least in the United States (19), it may be appropriate to review briefly what is already known about its organization. This information is relevant for

637

0066-4197/89/1215-0637$02.00

defining priorities and formulating strategies concerning future work, a subject that still is under debate (54, 107).

However fast our progress over the past few years, it is not generally appreciated that only about 600 human coding sequences are presently known in their primary structure. These represents 1% or 2% of all coding sequences if the total number of genes is 60,000 or 30,000, respectively, and if their average size is 1,000 base pairs and only about 0.02% of the human haploid genome, which comprises about 3 billion base pairs. Even less is known on the primary structure of noncoding, particularly intergenic, sequences. These account for 98–99% of the human genome (if the number and the size of coding sequences are those indicated above), and contain large families of repeated interspersed sequences, LINES and SINES (93), that are present in about 100,000 and 900,000 copies, respectively (44).

Consequently, it is not surprising that the recent advances in human molecular genetics have not shed much light on the general issue of genome organization, nor that many biologists still visualize the human genome, and the eukaryotic genome in general, as a "bean bag" (62), a collection of genes randomly scattered over vast expanses of "junk DNA" (74). In contrast, the investigations reviewed here indicate that the eukaryotic genome is an integrated structural, functional, and evolutionary system. This view arose from a comparative study of vertebrate genomes, centered on the analysis of their compositional patterns, namely of the compositional distributions of large DNA fragments, coding sequences, and introns (see Figures 1, 2, 5, for examples).

ISOCHORES AND GENOME ORGANIZATION

The Isochores

Equilibrium centrifugation in analytical CsCl density gradient shows that DNA preparations from warm-blooded vertebrates are characterized by a strong intermolecular compositional heterogeneity, whereas those from cold-blooded vertebrates exhibit a weak heterogeneity; moreover, the former reach GC levels that are not attained by the latter (101; see Figure 1).

To investigate the compositional distribution of DNA fragments from vertebrates in more detail, we developed an experimental approach (12, 22, 24, 29, 58, 59, 76, 82, 101, 115) derived from methods originally used for isolating satellite DNAs (15, 20, 21, 57, 65). This approach consists of fractionating DNA fragments by equilibrium centrifugation in preparative Cs_2SO_4 density gradients in the presence of sequence-specific DNA ligands, like Ag+ or BAMD, 3,6-bis (acetato-mercuri-methyl) dioxane; (netropsin was also used by others; 40). This approach, outlined in the legend of Figure 2, allowed the identification of a small number of families of DNA fragments

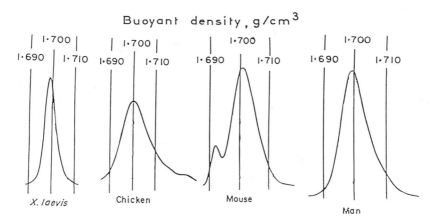

Buoyant density, g/cm^3

X. laevis   Chicken   Mouse   Man

Figure 1   CsC1 buoyant density profiles of DNA preparations from Xenopus, chicken, mouse and man. Band widths depend upon the intermolecular compositional heterogeneity (namely the spread of GC levels of DNA fragments) and molecular weight (lower molecular weights causing larger band widths because of the associated higher Brownian diffusion). After correction for differences in molecular weights (sedimentation coefficients of Xenopus, chicken, mouse and human DNAs were 35, 20, 24.5 and 25 S, respectively), band widths indicated a much lower intermolecular compositional heterogeneity (H; see Ref. 85) for Xenopus DNA (2.9% GC) than for the DNAs of chicken (5.1% GC) mouse (4.0% GC for main band) and man (4.8% GC). (Modified from ref. 101; H values are from G. Bernardi & G. Bernardi, paper in preparation) and refs. 24, 76.)

characterized by similar, though not identical, base compositions. These compositional families (or components) of DNA fragments comprise (*a*) major families, derived from "main band" DNA; (*b*) satellite families, derived from highly repeated, simple sequence satellite DNA(s); and (*c*) minor families, such as ribosomal DNA.

If this approach is applied to human DNA preparations in the 30–100 Kilobase (Kb) size range, one can see that the compositional distribution of DNA fragments (*a*) is characterized by a very broad GC range, and by the presence of two GC-poor major components, L1 and L2, representing about two thirds of the genome, and three GC-rich components, H1, H2, and H3, representing the remaining third (Figure 2; see refs. 12, 24, 101, 115); (*b*) is shared by the DNAs of all warm-blooded vertebrates studied, whereas the DNAs from cold-blooded vertebrates show much narrower distributions that do not extend as far in the high GC range (Figure 2); and (*c*) is largely independent of molecular size, at least between 3 Kb and over 300 Kb (this upper value was determined by lysing lymphocytes with sarkosyl directly in the analytical ultracentrifuge cell; the resulting very high molecular weight

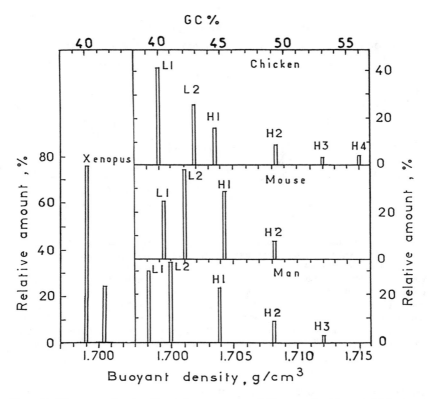

Figure 2 Histograms showing the relative amounts, modal buoyant densities and GC levels of the major DNA components from Xenopus, chicken, mouse and man. Xenopus data are from refs. 58, 101 and from G. Bernardi & G. Bernardi (paper in preparation). Chicken data (from ref. 22) include two components previously considered as minor (H3 and H4). Mouse and human data are from refs. 58, 82 and 115. Satellite and minor components are not shown. (Modified from ref. 12). The construction of the histograms involved the following steps: (*a*) the Cs₂SO₄ density gradient separation of DNA-ligand complexes on the basis of the sequence-specific, differential binding of the ligand (Ag+ or BAMD) to DNA fragments of different composition; (*b*) the analysis of the fractions in terms of relative DNA amounts and GC levels; (*c*) the resolution of the analytical CsCl band profiles of DNA fractions into Gaussian curves, and the identification of a small number of families of DNA fragments that are similar in composition (see ref. 58 for further details); (*d*) the pooling together of fractions containing the same families, after re-centrifugation in the case of very heterogeneous fractions; and (*e*) the assessment of the relative DNA amounts, modal buoyant densities and GC levels of the major DNA components. It should be noted that while the CsCl profiles of the isolated major components are centered on the modal buoyant densities shown in the Figure, they exhibit some degree of overlap with each other, particularly for the components that are most abundant and closer in density.

preparations exhibited multimodal profiles with maxima corresponding to the buoyant densities of major components; 58).

The independence of the compositional distribution of DNA fragments from molecular size indicates a remarkable compositional homogeneity over very long DNA stretches, called "isochores", (for "equal regions"; 12, 24). These gave rise to the DNA fragments that were actually fractionated, because of the occurrence of unavoidable mechanical and enzymatic breakage during DNA preparation. The existence of isochores is also indicated by the very low intermolecular heterogeneities of vertebrate DNAs and their components relative to their genome size and kinetic complexities, (24, 43, 79; G. Bernardi & G. Bernardi, in preparation; bacterial DNAs were used as references) as well as by other findings (see next section and Refs. 3, 45).

Satellite and minor components may also be considered as "isochores". They are not dealt with here, however, because they represent specialized genome compartments, corresponding to centromeres and telomeres, in the case of satellite DNAs, and to nucleoli, in the case of ribosomal genes.

Isochores and Genes

A number of genes from the genome of man and other warm-blooded vertebrates were localized in major components or compositional fractions by hybridization with appropriates probes (12, 23, 25, 82, 97, 115; see Figure 3). This research revealed three points:

1. In most cases, a given gene is embedded in DNA fragments that are within 1–2% of each other in GC content. This finding provides additional, independent evidence for the compositional homogeneity of isochores, as well as for the wide-spacing of isochore borders. Indeed, since the fragments making up DNA preparations are produced by random degradation, and since the genes probed can, therefore, be located anywhere on them, the narrow compositional range of the gene-carrying fragments indicates that they are very homogeneous in base composition over sizes roughly twice as large as the fragments themselves (i.e. up to 200 Kb). Incidentally, these observations are valid not only for isolated genes, but also for clustered genes, indicating again that isochores are large in comparison to the gene clusters explored (these varied in size from 4–40 Kb). Much less frequently, the gene is found in fractions covering a wider range of GC levels; obviously, if the DNA stretches around the probed gene are compositionally heterogeneous (as when the gene is close to isochore borders), random breakage produces gene-carrying fragments of different composition.

2. Gene distribution is strikingly nonuniform in the genome, most of the genes probed having been localized in component H3, which is the most heterogeneous component and only represents 3–5% of the genome (12).

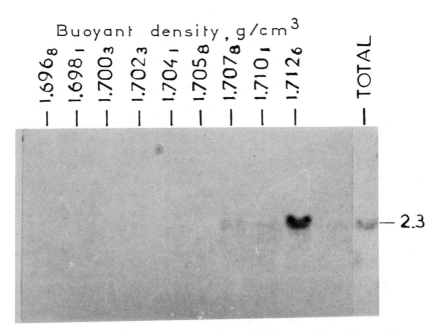

Figure 3 Localization of the *c-sis* oncogene in human DNA fractions obtained by centrifugation in a preparative Cs$_2$SO$_4$/BAMD density gradient. After dialysis to eliminate Cs$_2$SO$_4$ and BAMD, fractions were digested with EcoRI, electrophoresed on a 0.8% agarose gel, transferred to a nitrocellulose filter and hybridized with a *c-sis* probe. A hybridization band corresponding, as expected, to a 2.3 Kb EcoRI fragment was found in a fraction having a modal buoyant density of 1.7126 g/cm^3 and representing 2.3% of the human genome. (From ref. 115).

3. The GC levels of genes, introns, and exons (and their individual codon positions) are linearly correlated with those of the large DNA fragments embedding them; the GC levels of intergenic sequences are, however, systematically lower than those of exons and correspond to the unity slope passing through the origin (Figure 4). It should be noted that a correlation between GC levels of third codon positions and flanking sequences of some genes was independently reported (see 44a).

The linear relationships of Figure 4 are of interest in three respects. (*a*) Because genes represent only a minute amount of human DNA, isochores essentially consist of intergenic noncoding sequences. Thus the straight lines of Figure 4 correlate the GC levels of coding sequences with the GC levels of the noncoding sequences harboring them. (*b*) The GC levels of exons are higher by about 10% compared to those of the intergenic sequences flanking them; the slope of the exon plot is, however, equal to one. In contrast, introns, genes, and third codon position plots exhibit slopes higher than one

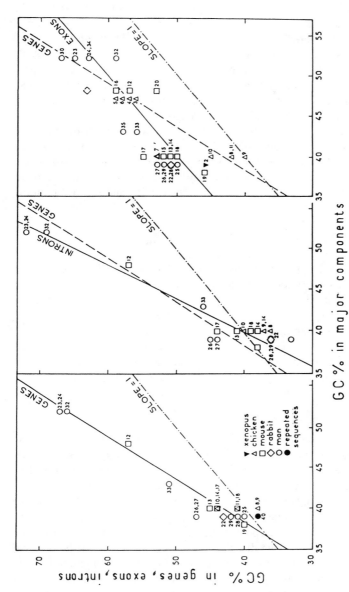

Figure 4   Plot of GC levels of genes (and mouse repeated sequence family L1), introns, and exons against the GC levels of DNA components in which they were localized. Gene localizations were made on either isolated major components or in compositional fractions (see ref. 12). In the latter case, gene localizations were assigned to the major component that was nearest in buoyant density, and this accounts for the vertical alignments of points. The numbers indicate genes and sequences (see ref. 12 for a list). Lines were drawn using the least-square method. The unit-slope lines correspond to the coincidence in GC levels of genes, introns and exons, respectively, and of the major components in which genes are located. (Modified from ref. 12).

(see Figure 4 and ref. 12). Interestingly, the different relationships just described can help in deciding whether a given anonymous probe, localized in compositional fractions, is a coding sequence or an intergenic sequence, since intergenic sequences fall on the unity slope line passing through the origin, whereas coding sequences are higher, by about 10% GC, than the DNA in which they are located. (c) The existence of such linear relationships suggests that the compositional pattern of a genome can also be studied by looking at the compositional distribution of coding sequences (and their different codon positions), and introns. The multimodal compositional patterns represented by these distributions (8–12, 70, 71, 78; see Figure 5) confirm the great predominance of GC-rich genes over GC-poor genes in the genome of man (and other warm-blooded vertebrates). In fact the concentration of genes found in the GC-richest and least abundant isochore family, H3, can be estimated as 5–10 times higher than that in the H1+H2 and L1+L2 families (70).

Isochores, CpG Doublets and "CpG Islands"

In vertebrate DNAs, CpGs are the only potential sites of methylation and the only doublets that are underrepresented relative to statistical expectations. Contrary to previous claims (80), the distribution of CpG doublets is strikingly nonuniform in the genome. Indeed, CpGs are very strongly avoided in genes and coding sequences located in GC-poor isochores, whereas they are hardly avoided at all in genes and coding sequences located in GC-rich isochores (7, 12). A similar situation is found in mammalian and avian viruses where GC-poor genomes avoid CpG, but GC-rich ones do not. Incidentally, the latter observation disposes of certain explanations for the different CpG shortage of small and large viral genomes (98, 99; see ref. 7 for further discussion).

"CpG islands" are sequences over 0.5 Kb in size, characterized by high GC levels, by clustered, unmethylated CpGs, by G/C boxes (GGGGCGGGGC and closely related sequences), and by clustered sites for rare-cutting restriction enzymes (these enzymes recognize GC-rich sequences that comprise one or two unmethylated CpG doublets; 13, 30, 33, 94, 102). CpG islands are associated with the 5' flanking sequences, exons and introns of all housekeeping genes and of many tissue-specific genes, and with the 3' exons of some tissue-specific genes (33). The distribution of CpG islands in the human and mouse genomes is also strongly nonuniform. This distribution parallels those of CpG doublets and of genes; CpG islands are rare in GC-poor isochores, but increasingly more frequent in GC-rich isochores; moreover, CpG islands are very rare in the genomes of cold-blooded vertebrates (B. Aissani, G. Bernardi & G. Bernardi, in preparation).

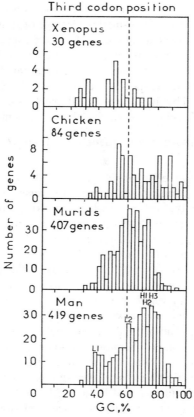

Figure 5 Compositional distribution of third codon positions from vertebrate genes. (This distribution is the most informative because of its wider spread in composition compared to coding sequences and first or second codon positions). The number of genes under consideration is indicated. The available gene sample of Xenopus was small, but the difference with the gene distribution from warm-blooded vertebrates was also found for homologous genes (see ref. 11), indicating that this difference is not due to the gene samples used. A 2.5% GC window was used. The broken line at 60% GC is shown to provide a reference. Approximate identifications of different compositional classes of coding sequences corresponding to the major components of the human genome (L1, L2, H1, H2 and H3) are indicated. The borders between L1-L2, H1-H2, and H3 can be tentatively estimated (70) as 67.5% and 77.5% GC, respectively. (Modified from ref. 11).

Isochores, "Single-Copy" and Middle-Repetitive Sequences

Reassociation kinetics of human major DNA components using hydroxyapatite chromatography (6, 14), revealed that "single copy" sequences decrease and middle-repetitive sequences increase from GC-poor to GC-rich components, indicating their different interspersion patterns (95); (these findings are not inconsistent with the increase of gene concentration in GC-rich isochores since genes represent a small percentage of single-copy sequences). Moreover, different reassociating classes from the same component showed very close GC levels (95), consistent with the compositional correlations of Figure 4. These results were confirmed and corroborated by the predominant location of LINES (93) in GC-pooor isochores, of SINES in the less abundant GC-rich isochores, and by the compositional match of these middle-repetitive sequences with the isochores in which they are present (66, 96).

Isochores and Chromosome Bands of Metaphase Chromosomes

G-bands (Giemsa positive or Giemsa dark bands; these are equivalent to Q-bands or Quinacrine bands), and R-bands (Reverse bands; these are equivalent to Giemsa negative, or Giemsa light bands) are produced by treating metaphase chromosomes with fluorescent dyes, proteolytic digestion, or differential denaturing conditions. GC-poor and GC-rich isochores largely correspond to the DNA of G- and R-bands, respectively. When this conclusion was proposed (24), it essentially rested on the parallelism between compositional heterogeneity of the genome, as seen at the DNA level, and chromosome banding in vertebrates. Indeed, we knew that genomes from cold-blooded vertebrates show a weak compositional heterogeneity (43, 79, 101; see Figures 1, 2, and 5), and we had realized (largely on the basis of unpublished data from other laboratories, quoted in ref. 24) that they show poor G- and R-bands, or no bands at all. In contrast, genomes from warm-blooded vertebrates exhibit a strong compositional heterogeneity (Figures 1, 2, and 5) and show distinct G- and R-bands. These findings were in agreement with the indirect cytogenetical evidence (18) suggesting that G- and R-bands correspond to GC-poor and GC-rich DNA sequences, respectively.

More recent data have confirmed this conclusion. (a) Additional evidence has been obtained for the link between compositional heterogeneity of DNA fragments and chromosome bands (63), as well as for the poor banding (1, 63, 86, 110), and the limited compositional heterogeneity of the genomes from cold-blooded vertebrates (G. Bernardi & G. Bernardi, in preparation). (b) G-bands replicate late, but R-bands replicate early (18), as do genes previously investigated in replication timing (31, 37) and localized in GC-poor and GC-rich isochores, respectively (12). (c) Genes are preferentially located in R-bands (37, 51), as well as in GC-rich isochores (11, 12, 70, 71). (d) Genes located in G- and R-bands are GC-poor and GC-rich, respectively (3, 45), as are genes located in GC-poor and GC-rich isochores (12). (e) G-bands can be produced by HaeIII degradation of metaphase chromosomes (55), because HaeIII splits GGCC sites that are much more frequent in GC-rich than in GC-poor isochores. In turn, R-bands can be produced by pancreatic DNase degradation of chromosomes protected by GC-specific binding of chromomycin A3 (90). (*f*) G-bands can be obtained in metaphase chromosomes by in situ hybridization of LINES (52, 60), that are mainly located in GC-poor isochores (66, 96), and R-bands by hybridizing Alu sequences (52, 60) that are predominantly distributed in GC-rich isochores (12, 96, 115). (*g*) Antibodies against Z-DNA produce R-bands (104), consistent with the expectation that Z-DNA structures are more frequent in GC-rich segments. (*h*) Antibodies against AT-rich triplex DNA have been localized in G-bands (16). (*i*) Unmethylated C, present in CpG islands (predominantly located in GC-rich isochores) is mainly found in R-bands (106). (*j*) Gene amplification leads to the formation of homogeneous staining regions in chromosomes (87); this is the result expected if the amplified genome segments are smaller in size than isochores, as is the case.

Points made in the following sections stress the fact that the general correlation between isochores and chromosome bands should only be considered a good approximation of the actual situation. Moreover, this correlation is not true for all organisms. Indeed, an isochore organization is also present in the genomes of plants (61, 81), in which case both low and high compositional heterogeneities have been found, and yet metaphase chromosomes do not exhibit G- or R-bands. This is probably because at metaphase plant chromosomes are 3.5–6.5 times more condensed than human chromosomes (39).

Isochores and the Fine Structure of Chromosome Bands

G- and R-bands not only differ in their overall isochore make-up, but also in their internal isochore structure, as indicated in the last section and by the following points:

(*a*) G-bands are remarkably homogeneous in DNA composition, because they are made up of GC-poor isochores that differ very little from each other in composition (see Figure 2). In contrast, R-bands are heterogeneous, since the corresponding GC-rich isochores encompass a wide GC range. This leads to both interband and intraband heterogeneity as shown by the compositional mapping of chromosome 21 (see the final section of this review). In fact, the interspersion of different GC-rich isochores within individual R-bands was already indicated by the finding that genes located in R-bands from many chromosomes (3, 45) are present in component H3. Since the latter only represents 3–5% of the genome, it cannot account for the totality of DNA of R-bands; other components must be present. A corollary of this conclusion is that, since gene concentration is highest in the isochores of the H3 family (10, 12, 70, 71), regions of high and low gene concentration should exist not only in R- and G-bands, respectively, but also within R-bands.

(*b*) GC-rich and GC-poor DNA components in the human genome are in a 1:2 ratio (24, 101), whereas R- and G-bands are in a 1:1 ratio (42). This discrepancy may mean that DNA concentration is lower in R-bands than in G-bands. In this connection, it is of interest (a) that such a lower DNA compaction, and the consequent greater accessibility of DNA, would affect chromosomal regions where the concentration of genes is higher, and (b) that DNase I sensitive chromosomal regions (corresponding to potentially active genes; 34, 109) usually correspond to Reverse bands, although not to all of them (48). An alternative explanation, not exclusive of the former, is that standard R-bands contain more GC-poor isochores (corresponding to the "thin" G-bands revealed by high resolution banding; 84, 105, 111, 112) than standard G-bands contain GC-rich isochores (corresponding to "thin" R-bands). This explanation appears to be supported by compositional mapping of chromosome 21 (see the final section) and other data (45).

ISOCHORES AND CHROMOMERES   Meiotic prophase chromosomes, especially at pachytene, show a characteristic pattern of chromomeres that represent centers of chromatin condensation along the chromosomes (28, 56, 75). This pattern can be seen because meiotic chromosomes are several times more extended than mitotic chromosomes and because meiotic pairing enhances the chromomeres (18). The chromomere-interchromomere patterns of pachytene bivalents strikingly resemble the high resolution G-banding patterns visualized in mid-prophase mitotic chromosomes (84, 105, 111, 112), namely in the high-resolution banding mentioned in the preceding section. Thus, it is generally accepted that G-bands of metaphase chromosomes correspond to individual or, more frequently, to closely spaced chromomeres, and that R-bands correspond to interchromomeric regions. Consequently, it may be speculated that chromomeres and interchromomeres correspond to GC-poor and GC-rich isochores.

ISOCHORES, CHROMATIN LOOPS AND NUCLEOSOMES   Levels of DNA organization lower than the chromomere-interchromomere patterns are known to exist in chromosomes. These range from the wrapping of the double helix arround histones to form nucleosomes, to the packaging of nucleosomes into chromatin fibers and to the folding of the fibers into chromatin loops. The latter consist of 30–100 Kb of supercoiled DNA and are fastened at their base by non-histone scaffold proteins, mainly topoisomerase II. The scaffold-associated regions of DNA (35) may serve as preferred sites at which DNA replication begins and tend to be close to promotor elements. The possible correlations of chromatin loops with isochores are very important questions that unfortunately cannot be answered at present. Yet another open question concerns the possibility of different nucleosome spacing in GC-poor and GC-rich isochores.

Isochores and Long-Range Physical Mapping

The correlations between isochores and chromosome band structure are now being confirmed in a number of genomic regions by long-range physical mapping, made possible by the advent of pulsed field gel electrophoresis (88, 89) and of rare-cutting restriction enzymes (72).

The partial physical map (32) of the long arm of human chromosome 21 has shown that (*a*) genes and unique sequences are clustered in R-band q22.3 whereas they are rare in G-band q21; and (*b*) rare-cutting restriction enzymes cut more frequently in R-band q22.3, producing small fragments, than in G-band q21, where they yield very large fragments. Similar observations are also being made about other mapped chromosomal regions. Needless to say, these data fit with expectations based on the distribution of genes and associated "CpG islands" discussed in previous sections.

Direct evidence has been found for the expected high gene concentration in a GC-rich region, the H-2K region of the mouse major histocompatibility complex (seven genes over 150 kb) (1).

ISOCHORES AND GENOME FUNCTIONS

Isochores and Integration of Mobile and Viral DNA Sequences

Stable integration of mobile and foreign DNA sequences is mostly found in isochores of matching composition. Mobile sequences that have been amplified by retrotranscription (108) and translocated to numerous loci of the human genome during mammalian evolution, such as LINES and SINES (93), are predominantly located in isochores of matching GC levels (12, 66, 77, 95, 96, 115). This indicates that reinsertion is targeted to matching genome environments, and/or that integration is more stable within such environments. Furthermore, such reinsertion may be a source of mutations if it occurs in genes (47, 69).

Similar behavior was observed for integrated viral sequences. In the four different viral systems studied so far, bovine leukemia (BLV), hepatitis B (HBV), mouse mammary tumor (MMTV) and Rous sarcoma (RSV), integration was found predominantly in isochores of matching GC contents (49, 83, 114; F. Kadi and associates, in preparation). These isochores are very GC-rich for BLV, HBV, and RSV, and very GC-poor for MMTV. In the case of integrated MMTV sequences, endogenous sequences, which have resided in the host genome for very many generations, are more concentrated in the GC-poorest isochores than the recently integrated exogenous sequences. This finding suggests that the broader compositional distribution of recent inserts is narrowed by the selection of more stable inserts. Needless to say, these observations are of interest in connection with the integration of foreign DNA into the genome of transgenic mammals. Here, too, an important, unresolved question concerns the effect of genomic compositional context on the expression of integrated sequences.

Isochores, Translocation Breakpoints and Fragile Sites

Translocation breakpoints are not randomly located on chromosomes (100). R-bands and G/R borders are the predominant sites of exchange processes, including spontaneous translocations, spontaneous and induced sister-chromatid exchanges, and the chromosomal abnormalities seen after X-ray and chemical damage. They also include the "hot spots" for the occurrence of mitotic chiasmata (41, 53, 68, 100,). Likewise, fragile sites tend to be more frequent in R-bands or near the border of R- and G-bands. For instance, 78 of the 89 fragile sites (88%) accepted at the Eighth International Workshop on Human Gene Mapping are situated in R-bands (41). A number of fragile sites have been located in R-bands near the border of G-bands (5, 113). Moreover, cancer-associated chromosomal aberrations are also nonrandom, with a limited number of genomic sites consistently involved and frequently associated with cellular oncogenes and fragile sites (67). These observations indicate that R-bands and G/R borders are particularly prone to recombination and raise the question of the role played in these phenomena by the compositional discontinuities at G/R borders and within R bands, as well as by the genomic distribution of Alu sequences, CpG islands and other recombinogenic sequences, such as minisatellites (46).

Chromosomal rearrangements have two important consequences; the activation of oncogenes by strong promotors that have been put upstream of them by the rearrangement (50), and the possibility that some chromosome rearrangements lead, in evolutionary time, to reproductive barriers and speciation (92). It may be speculated that the higher incidence of cancer and rate of speciation (17) shown by warm-blooded relative to cold-blooded vertebrates correlate with the higher propensity to chromosome rearrangements of the former due to the larger number of compositional discontinuities in their chromosomes.

Isochores and Replication/Condensation Timing in the Cell Cycle

As already noted, G-bands and genes located in GC-poor isochores replicate late in the cell cycle, whereas R-bands and genes located in GC-rich isochores replicate early (12, 18, 37). The correspondence of structural (G- and R-) banding and replication (BrdU) banding is, however, only found in warm-blooded vertebrates. Indeed, early and late replication occurs in cold-blooded vertebrates (2, 36, 110) in the absence or poorness of chromosomal banding.

Another feature distinguishing G- and R-bands is the condensation timing during mitosis. This occurs early in the cell cycle for G-bands and late for R-bands. In all probability, early and late condensation also occur in cold-blooded vertebrates in the absence of distinct G- and R-banding.

These findings suggest that replication and condensation timing are correlated with the basic chrommere-interchromomere organization of chromosomes, which is present in cold-blooded vertebrates, and only later in evolution became associated with the compositional differences that arose at the transition between cold-blooded and warm-blooded vertebrates.

That replication timing may be subject to more complex rules under certain circumstances, is indicated by the following observations: although "tissue-specific" genes, which are more frequent in G-bands (37), and in GC-poor isochores (70, 71), generally replicate late, they appear to replicate early in those cell types that express them (27); secondly, the two female X chromosomes have identical G- and R-band patterns, but different replication-band patterns (91).

Isochores, Genomic Distribution of Genes and Codon Usage

As previously indicated, gene concentration is much higher (by a factor of 5–10) in the GC-richest isochores of the human genome than in the other isochores. Moreover, housekeeping genes (including oncogenes) are preferentially distributed in GC-rich isochores (12, 70, 71) and R-bands (37; see also the next section), whereas tissue-specific genes are more abundant in GC-poor isochores and G-bands. These findings indicate that the higher gene concentration in the GC-richest isochores is likely to be underestimated at present, because housekeeping genes are still severely underrepresented in gene banks, and that the genomic distribution of genes is correlated, at least to some extent, with gene function.

The range of GC values (30–100%) in codon third positions of human genes (Figure 5) is almost as wide as that exhibited by the genes of all prokaryotes. This very extended range implies very large differences in codon usage for GC-poor and GC-rich genes of the same genome (8–12, 70, 71). In particular, at the high-GC end of the range, an increasing number (up to 50%) of codons are simply absent. In turn, the wide range of GC values in codon third positions is paralleled by the GC values in first and second positions, although to a more limited extent as would be anticipated. This wide range leads to very significant differences in the frequency of certain amino acids in GC-rich and GC-poor genes. For instance, the ratio of alanine+arginine to serine+lysine (namely, the two amino acids that contribute most to the thermodynamic stability of proteins over the two that do so the least; 4), increases by a factor of four between proteins encoded by the GC-poorest and the GC-richest coding sequences in the human genome (10).

These observations suggest that in warm-blooded vertebrates constitutively expressed, housekeeping genes tend to be the most biased in codon usage and to direct the synthesis of thermodynamically more stable proteins, compared to tissue-specific genes.

Isochores, CpG Doublets and "CpG Islands"

The distributions of CpG doublets and of CpG islands indicate that the distribution of methylation in the genome of man and other vertebrates is highly nonuniform, a point of interest in view of the functional role of DNA methylation and of the distribution of housekeeping and tissue-specific genes.

The results on CpG islands have an additional functional relevance. In GC-poor isochores, genes are usually endowed with a TATA or a CCAAT box and an upstream control region, whereas in the GC-rich isochores there is no TATA box, but promotors containing properly positioned "G/C boxes" that bind transcription factor Sp1, a protein that activates RNA polymerase II transcription (38, 64, 73, 103). These GC-rich promotors apparently are associated with all genes located in GC-rich isochores, and mainly arose with the appearance of warm-blooded vertebrates (B. Aissani, G. Bernardi & G. Bernardi, in preparation).

ISOCHORES AND GENOME EVOLUTION

Many differences and similarities in compositional patterns were found in vertebrate genomes (8–12, 22, 24, 29, 58, 59, 70, 71, 76, 78, 81, 101, 115); see Figures 1, 2, and 5. A comparison of these patterns sheds new light on genome evolution.

1. The genomes of the vast majority of cold-blooded vertebrates exhibit compositional distributions of both DNA fragments and coding sequences that are narrower and do not reach the GC levels of the GC-rich components of warm-blooded vertebrates and of the coding sequences contained in them. Moreover, a wide spread of GC compositions was found in DNAs from cold-blooded vertebrates (G. Bernardi & G. Bernardi in preparation).

2. In contrast, an overall similarity exists in the compositional patterns of all warm-blooded vertebrates, in spite of the fact that the genomes of birds and mammals not only differ in size by a factor of almost three (the avian genomes being smaller), but also arose separately in evolution. In birds, however, the compositional distributions of both DNA fragments and coding sequences attain higher GC values than in mammals.

3. Finally, two slightly different compositional patterns have been found in mammalian genomes (11, 70, 71, 82, 115; G. Sabeur, J. Filipski, F. Kadi & G. Bernardi, in preparation). The first one is very widespread in different orders of mammals, and includes the human genome, while the second one appears limited to three families of myomorpha (a sub-order of rodents),

namely murids (rat and mouse), cricetids (hamster) and spalacids (mole rat). The difference essentially consists in a narrower compositional distribution of DNA fragments and coding sequences in the myomorpha compared to other mammals.

These similarities and differences in the compositional patterns of vertebrate genomes define two modes of genome evolution (for a more detailed discussion, see 11). In the conservative mode prevailing in mammals (the relatively minor differences between myomorpha and other mammals are not covered here), the composition of DNA fragments and coding sequences is maintained in spite of a very high degree of nucleotide divergence (which may attain 50% in third codon positions, without correction for multiple hits). This compositional conservation, which can also be observed in banding patterns over a number of chromosomal regions (84), appears to require negative selection operating at the isochore level to eliminate any strong deviation from a presumably functionally optimal composition.

In contrast, in the transitional or shifting mode, parallel compositional changes are seen in both isochores and coding sequences. Investigations on the compositional changes occurring in cold-blooded vertebrates have shown that their extent is not correlated with evolutionary time and may be larger than that of synonymous changes indicating that the molecular clock (116) does not apply during compositional transitions (G. Bernardi & G. Bernardi, in preparation). Two typical compositional transitions are the GC increases that occurred between the genomes of reptiles on the one hand, and birds and mammals on the other (and that were accompanied by the replacement of TATA and CCAAT boxes by GC-rich promotors). These compositional changes are due to a directional fixation of point mutations caused by both negative and positive selection at isochore levels (10–12, 70, 71, 78). Selection appears to be for the higher thermal stabilities of proteins, RNA and DNA, that are required by the higher body temperature of warm-blooded vertebrates. Of course, selection implies functional differences and therefore supports the idea that isochores are functionally relevant structures. Moreover, the compositional relationships between coding and noncoding (particularly intergenic) sequences (Figure 4) indicate that the same compositional constraints apply to both kinds of sequences. The selection pressures underlying such constraints cannot be understood if noncoding sequences are "junk DNA" (74), with no biological function.

CONCLUSIONS: THE PALEOGENOME AND THE NEOGENOME

Isochores represent a new structural level in the organization of the genome of warm-blooded vertebrates that bridges the enormous size gap between the gene level, with its exon-intron systems and the corresponding regulatory sequences, and the chromosome level, with its banding patterns. These three levels are correlated with each other, since genes match compositionally the isochores in which they are harbored, while GC-poor and GC-rich isochores are DNA segments located in G- and R-bands, respectively.

The investigations that led to the discovery of isochores and of these two correlations (12, 24, 29, 57, 101) have firmly established the existence of differences in the base composition and gene concentration of DNA segments present in G- and R-bands. Moreover, they have revealed that these segments are characterized by strikingly different complexities, the isochores present in G-bands being very close in composition and characterized by a low gene concentration, whereas, the isochores present in R-bands belong to different compositional families, including those of the H3 family that have the highest concentration of genes and CpG islands. The R-band isochores might have a lower DNA compaction, and/or comprise a number of GC-poor isochores; they also appear to be accompanied by different higher order chromatin structures, as judged by DNase sensitivity. These results, as well as the compositional mapping data available so far (see next section), indicate that isochores correspond to a chromosome organization level lower than standard chromosomal bands, possibly to chromomeres and interchromomeres.

Although isochores from the genomes of warm-blooded vertebrates belong to a number of families characterized by large differences in base composition, this is not true for cold-blooded vertebrates. In this case, isochores exist, as shown by hybridization of gene probes and by the fact that the intermolecular compositional heterogeneities of cold-blooded vertebrates are close to those of bacteria in spite of a genome size two–three orders of magnitude higher and of a much larger kinetic complexity (24, 43, 79; G. Bernardi & G. Bernardi, paper in preparation). Isochores from cold-blooded vertebrates are characterized by much smaller differences in composition, which correspond to much weaker banding patterns in metaphase chromosomes. Needless to say, the existence of isochores raises the problem of how these compositionally homogeneous DNA stretches arose in evolution.

Isochores are, however, not only structural units, but also appear to play functional roles. Some of these, like the integration of mobile and viral sequences, recombination and chromosome rearrangements, are well established. In contrast, DNA replication timing, and chromosome condensation timing at mitosis seem rather to be correlated with the chromomere-interchromomere organization of chromosomes, independently of the composition of the corresponding DNA stretches. The observations on the gene distribution in the genome, the relationships of such distribution with gene functions (housekeeping, tissue-specific), with codon usage, and with different kinds of regulatory sequences are also indicative of functional roles for isochores. In contrast, the possible correlations between isochores and chromatin loops, replicons and transcription units still remain open questions.

Isochores are evolutionary units of vertebrate genomes. Their composition may be conserved in spite of enormous numbers of point mutations, or may undergo dramatic changes after relatively modest numbers of point mutations. In the case of the two independent compositional transitions from cold-blooded vertebrates to mammals and birds, compositional transitions seem to be largely associated with the optimization of genome functions following environmental body temperature changes. Interestingly, these transitions appear to be accompanied by very conspicuous changes in promotor sequences.

To sum up a number of points made in this review, two main compositional compartments can be distinguished in the human genome, and, more generally, in the genomes of warm-blooded vertebrates (see Table 1). The first compartment, the *paleogenome*, is characterized by its similarity to what it was, and still is, in cold-blooded vertebrates: the late-replicating, compositionally homogeneous, GC-poor isochores of early-condensing chromomeres contain relatively rare, GC-poor (largely tissue-specific) genes having

Table 1   The human genome[a]

The Paleogenome G-bands, GC-poor isochores	The Neogenome R-bands, GC-rich isochores	Ref.
Chromomeres	Interchromomeres	18
Late replication	Early replication	18
Early condensation	Late condensation	
AT-rich triplex DNA	Z-DNA	16,104
HaeIII digestion	DNase chromomycin A3	55, 90
Abundance of LINES	Abundance of SINES	52, 60, 66, 96, 115
Compositional homogeneity	Compositional heterogeneity	11(b)
Scarcity of genes	Abundance of genes (esp. in H3)	11, 12, 37, 51
GC-poor genes (esp. tissue-specific)	GC-rich genes (esp. housekeeping)	3, 11, 45
Scarcity of CpG islands	Abundance of CpG islands	(c), 106
TATA box promotors	G/C box promotors	26a
DNase I insensitivity	DNase I sensitivity	48
Less frequent recombination	More frequent recombination	

(a)It should be noted (i) that the early replication and late condensation of interchromomeres are neogenome features common to the corresponding compartment of the genome from cold-blooded vertebrates; (ii) that the abundance of genes in the neogenome may derive not only from a higher gene density in the early replicating compartment of cold-blooded vertebrates, but also from gene duplications and/or translocations; (iii) that only the most important properties are listed; secondary effects are not. For example, sites for rare-cutting restriction sites are more abundant where CpG islands are more abundant. Likewise, the abundance of SINES leads to a shorter interspersion pattern of repeats in GC-rich isochores.
(b) K. Gardiner, B. Aissani & G. Bernardi, paper in preparation.
(c) B. Aissani, G. Bernardi & G. Bernardi, paper in preparation.

TATA box promotors (CpG islands are scarce). The second compartment, the *neogenome,* is characterized by the fact that it changed its compositional features compared to what it was in cold-blooded vertebrates. In the *neogenome,* the ancestral, early-replicating, GC-poor isochores of late-condensing interchrommomeres were changed into compositionally heterogeneous, GC-rich isochores that contain abundant genes (perhaps including most housekeeping genes) having G/C box promotors (genes and CpG islands are particularly abundant in the GC-richest isochores). The *neogenome/ paleogenome* distinction supersedes the first proposal for a bipartite nature of the mammalian genome, that of an ontogenetic genome and a housekeeping genome. That proposal (37) was essentially based on the different distribution of tissue-specific and housekeeping genes, whereas the present proposal is based on the evolutionary history of the compartments.

PERSPECTIVES: COMPOSITIONAL MAPPING

The results discussed in this review suggest a novel experimental approach in human genome research: compositional mapping. Wherever long-range physical maps are available, compositional maps may be constructed by assessing GC levels around landmarks (localized on the physical maps) that can be probed. This simply requires the hybridization of the probes on DNA fractionated according to base composition. If DNA preparations of about 100 Kb in size are used, compositional mapping can define the base composition of DNA stretches of about 200 Kb around the sequence probed. When enough information is obtained, this approach can shed light on the base compositions and the approximate sizes of the isochores making up the genomic regions investigated. As far as chromosome structure is concerned, compositional mapping will provide the equivalent of a high-resolution banding, without the uncertainties of cytogenetics. The compositional map may well correspond to a chromomere-interchromomere map.

This approach has already been successfully tried for a set of 50 unique-sequence probes localized on the long arm of chromosome 21 and has provided a direct demonstration for the compositional homogeneity of G-bands and for the compositional heterogeneity of R-bands q22.1 and q22.3, the highest GC levels (corresponding to the H3 component) being in the telomere-proximal region of q22.3 (K. Gardiner, B. Aissani & G. Bernardi, in preparation). Incidentally, this latter observation agrees with cytogenetical evidence that telomeres almost always correspond to R-bands and that the terminal regions of 20 of them (including that of the long arm of chromosome 21) are the most denaturation-resistant regions of human chromosomes (26). This suggests that H3 may correspond to these denaturation-resistant telomeric regions and to some similar intercalary regions located on chromosomes 11, 19, and 22.

Needless to say, these regions, which contain a considerable fraction of all human genes, should be primary targets in genome sequencing projects. Interestingly, chromosome-specific regions of this kind can be isolated by taking advantage of the different compositional patterns of the genomes of man and myomorpha. Indeed, somatic cell hybrids carrying only one human chromosome (or a part of it) in a rodent background lend themselves to the preparation and cloning of the human GC-richest segments that have no equivalent in the genome of myomorpha and that are characterized by the highest gene concentration.

The modular nature of the isochore organization of the human genome and the properties of isochores from G- and R-bands, respectively, suggest that a detailed knowledge of isochores over a few genomic regions, obviously including the GC-richest ones, should greatly help to understand the organization of the human genome. Once attained, this understanding will not only concern the human genome, or the genomes of warm-blooded vertebrates but eukaryotic genomes in general.

Acknowledgments

Thanks are due to Toshimichi Ikemura and Alison Stewart for having prompted the writing of this paper; to Brahim Aissani, Giacomo Bernardi, Edwin Crouse, Bernard Dutrillaux, Dusko Ehrlich, Katheleen Gardiner, Christian Gautier, Richard Grantham, Farida Kadi, Julie Korenberg, Ladislav Pivec, Dominique Mouchiroud, Alla Rynditch, Georgette Sabeur, Vittorio Sgaramella, and, especially, to Maxine Singer for comments and criticism.

Literature Cited

1. Abe, K., Wei, J.-F., Wei, F.-S., Hsu, Y.-C, Uehara, H., et al. 1988. Searching for coding sequences in the mammalian genome: the H-2K region of the mouse MHC is replete with genes expressed in embryo. *EMBO J.* 7:3441–49
2. Almeida-Toledo, L. F., Viegas-Pequignot, E., Foresti, F., Toledo Filho, S. A., Dutrillaux, B. 1988. BrdU replication patterns demonstrating chromosome homologies in two species, Genus *Eigenmannia. Cytogenet. Cell Genet.* 48:117–20
3. Aota, S.-I., Ikemura, T. 1986. Diversity in G+C content at the third position of codons in vertebrate genes and its cause. *Nucleic Acids Res.* 14:6345–55
4. Argos, P., Rossmann, M. G., Grau, U. M., Zuber, H., Frank, G. 1979. Thermal stability and protein structure. *Biochemistry* 18:5698–703
5. Aurias, A., Prieur, M., Dutrillaux, B., Lejeune, J. 1978. Systematic analysis of 95 reciprocal translocations of autosomes. *Hum. Genet.* 45:259–82
6. Bernardi, G. 1965. Chromatography of nucleic acids on hydroxyapatite. *Nature* 206:779–83
7. Bernardi, G. 1985. The organization of the vertebrate genome and the problem of the CpG shortage. In *Chemistry, Biochemistry and Biology of DNA Methylation*, ed. G. L. Cantoni, A. Razin, pp. 3–10. New York: Liss
8. Bernardi, G., Bernardi, G. 1985. Codon usage and genome composition. *J. Mol. Evol.* 22:363–65
9. Bernardi, G., Bernardi, G. 1986. The human genome and its evolutionary context. *Cold Spring Harbor Symp. Quant. Biol.* 51:479–87
10. Bernardi, G., Bernardi, G. 1986. Compositional constraints and genome evolution. *J. Mol. Evol.* 24:1–11
11. Bernardi, G., Mouchiroud, D., Gautier, C., Bernardi, G. 1988. Compositional patterns in vertebrate genomes: conservation and change in evolution. *J. Mol. Evol.* 28:7–18
12. Bernardi, G., Olofsson, B., Filipski, J.,

658   BERNARDI

Zerial, M., Salinas, J., et al. 1985. The mosaic genome of warm-blooded vertebrates. *Science* 228:953–58

13. Bird, A. 1986. CpG-rich islands and the function of DNA methylation. *Nature* 321:209–13

14. Britten, R. J., Kohne, D. E. 1968. Repeated sequences in DNA. *Science* 161:529–40

15. Bünemann, M., Dattagupta, N. 1973. On the binding and specificity of 3,6-bis-(acetato-mercuri-methyl) dioxane to DNAs of different base composition. *Biochim. Biophys. Acta* 331:341–48

16. Burkholder, G. D., Latimer, L. J. P., Lee, J. S. 1988. Immunofluorescent staining of mammalian nuclei and chromosomes with a monoclonal antibody to triplex DNA. *Chromosoma* 97:185–92

17. Bush, G. L., Case, S. M., Wilson, A. C., Patton, J. L. 1977. Rapid speciation and chromosomal evolution in mammals. *Proc. Natl. Acad. Sci. USA* 74:3942–46

18. Comings, D. E. 1978. Mechanisms of chromosome banding and implications for chromosome structure. *Annu. Rev. Genet.* 12:25–46

19. Committee on Mapping and Sequencing the Human Genome. Board on Basic Biology. Commission on Life Sciences. Nat. Res. Counc. 1988. *Mapping and Sequencing the Human Genome*. Washington, DC: Natl. Acad. Press

20. Corneo, G., Ginelli, E., Soave, C., Bernardi, G. 1968. Isolation and characterization of mouse and guinea pig satellite DNA's. *Biochemistry* 7:4373–79

21. Cortadas, J., Macaya, G., Bernardi, G. 1977. An analysis of the bovine genome by density gradient centrifugation: fractionation in $Cs_2SO_4/3,6$ bis-(acetato-mercurimethyl) dioxane density gradient. *Eur. J. Biochem.* 76:13–19

22. Cortadas, J., Olofsson, B., Meunier-Rotival, M., Macaya, G., Bernardi, G. 1979. The DNA components of the chicken genome. *Eur. J. Biochem.* 99:179–86

23. Cuny, G., Macaya, G., Meunier-Rotival, M., Soriano, P., Bernardi, G. 1978. In *Genetic Engineering*, ed. H. W. Boyer, S. Nicosia, pp. 109–15. Amsterdam, Holland: Elsevier

24. Cuny, G., Soriano, P., Macaya, G., Bernardi, G. 1981. The major components of the mouse and human genomes. 1. Preparation, basic properties, and compositional heterogeneity. *Eur. J. Biochem.* 111:227–33

25. Dodemont, H. J., Soriano, P., Quax, W. J., Ramaeckers, F., Lenstra, J. A.,

et al. 1982. The genes coding for the cytoskeletal proteins actin and vimentin in warm-blooded vertebrates. *EMBO J.* 1:167–71

26. Dutrillaux, B. 1973. Nouveau système de marquage chromosomique: les bandes T. *Chromosoma* 41:395–402

26a. Dynan, W. S. 1986. Promoters for housekeeping genes. *Trends Genet.* 2:196–97

27. Epner, E., Forrestier, W. C., Groudine, M. 1988. Asynchronous DNA replication within the human α-globin gene locus. *Proc. Natl. Acad. Sci. USA* 85:8081–85

28. Ferguson-Smith, M. A., Page, B. M. 1973. Pachytene analysis in human reciprocal (10:11) translocation. *J. Med. Genet.* 10:283–86

29. Filipski, J., Thiery, J. P., Bernardi, G. 1973. An analysis of the bovine genome by Cs_2SO_4/AG^+ density gradient centrifugation. *J. Mol. Biol.* 80:177–97

30. Fischell-Ghodsian, N., Nicholls, R. D., Higgs, D. R. 1987. Long range genome structure around the human α-globin complex analysed by PFGE. *Nucleic Acids Res.* 15:6197–207

31. Furst, A., Brown, H., Braunstein, J. D., Schildkraut, C. T. 1981. α-globin sequences are located in a region of early-replicating DNA in murine erythroleukemia cells. *Proc. Natl. Acad. Sci. USA* 78:1023–27

32. Gardiner, K., Watkins, P., Münke, M., Drabkin, H., Jones, C., Patterson, D. 1988. Partial physical map of human chromosome 21. *Somatic Cell. Mol. Genet.* 14:623–38

33. Gardiner-Garden, M., Frommer, M. 1987. CpG islands in vertebrate genomes. *J. Mol. Biol.* 196:261–82

34. Garel, A., Axel, R. 1976. Selective digestion of transcriptionally active ovalbumin genes from oviduct nuclei. *Proc. Natl. Acad. Sci. USA* 73:3966–70

35. Gasser, S. M., Laemmli, U. K. 1987. A glimpse at chromosomal order. *Trends Genet.* 3:16–22

36. Giles, V., Thode, G., Alvarez, M. C. 1988. Early replication bands in two scorpion fishes, *Scorpaena porcus* and *S. notata* (order *Scorpaneiformes*). *Cytogenet. Cell Genet.* 47:80–83

37. Goldman, M. A., Holmquist, G. P., Gray, M. C., Caston, L. A., Nag, A. 1984. Replication timing of genes and middle repetitive sequences. *Science* 224:686–92

38. Goodfellow, P. J., Mondello, C., Darling, S. M., Pym, B., Little, P., Goodfellow, P. N. 1988. Absence of methylation of a CpG-rich region of the 5' end

of the MIC2 gene on the active X, the inactive X and the Y chromosome. *Proc. Natl. Acad. Sci. USA* 85:5605–9

39. Greilhuber, J. 1977. Why plant chromosomes do not show G-bands. *Theor. Appl. Genet.* 50:121–24

40. Guttman, T., Votavova, H., Pivec, L. 1976. Base composition heterogeneity of mammalian DNAs in CsC1-netropsin density gradient. *Nucleic Acids Res.* 3:835–45

41. Hecht, F. 1988. Enigmatic fragile sites on human chromosomes. *Trends Genet.* 4:121

42. Holmquist, G., Gray, M., Porter, T., Jordan, J. 1982. Characterization of Giemsa dark- and light-band DNA. *Cell* 31:121–29

43. Hudson, A. P., Cuny, G., Cortadas, J., Haschemeyer, A. E. V., Bernardi, G. 1980. An analysis of fish genomes by density gradient centrifugation. *Eur. J. Biochem.* 112:203–10

44. Hwu, H. R., Roberts, J. W., Davidson, E. H., Britten, R. J. 1986. Insertion and/or deletion of many repeated DNA sequences in human and higher ape evolution. *Proc. Natl. Acad. Sci. USA* 83:3875–79

44a. Ikemura, T. 1985. Codon usage and tRNA in unicellular and multicellular organisms. *Mol. Biol. Evol.* 2:13–34

45. Ikemura, T., Aota, S.-I. 1988. Global variation in G+C content along vertebrate genome DNA. Possible correlation with chromosome band structures. *J. Mol. Biol.* 203:1–13

46. Jeffreys, A. J., Wilson, V., Thein, S. L. 1985. Hypervariable "minisatellite" regions in human DNA. *Nature* 314:67–73

47. Kazazian, H. H. Jr., Wong, C., Youssoufian, H., Scott, A. F., Phillips, D. G., Antonarakis, S. E. 1988. Hemophilia-A resulting from *de novo* insertion of L1 sequences represents a novel mechanisms for mutation in men. *Nature* 332:164–66

48. Kerem, B.-S., Goitein, R., Diamond, G., Cedar, H., Marcus, M. 1984. Mapping of DNAase I sensitive regions of mitotic chromosomes. *Cell* 38:493–99

49. Kettmann, R., Meunier-Rotival, M., Cortadas, J., Cuny, G., Ghysdael, J., et al. 1979. Integration of bovine leukemia virus DNA in the bovine genome. *Proc. Natl. Acad. Sci. USA* 76:4822–26

50. Klein, G. 1983. Specific chromosomal translocations and the genesis of B-cell-derived tumors in mice and men. *Cell* 32:311–15

51. Korenberg, J., Engels, W. R. 1978. Base ratio, DNA content, and quinacrine-brightness of human chromosomes. *Proc. Natl. Acad. Sci. USA* 75:3382–86

52. Korenberg, J. R., Rykowski, M. C. 1988. Human genome organization: Alu, Lines, and the molecular structure of metaphase chromosome bands. *Cell* 53:391–400

53. Kuhn, E. M., Therman, E. 1986. Cytogenetics of Bloom's syndrome. *Can. Genet. Cytogenet.* 22:1–18

54. Lederberg, J. 1989. The genome project holds promise, but we must look before we leap. *The Scientist* 3:10

55. Lima-de-Faria, A., Isaksson, M., Olsson, E. 1980. Action of restriction endonucleases on the DNA and chromosomes of *Muntiacus muntjac*. *Hereditas* 92:267–73

56. Luciani, J. M., Morazzani, M. R., Stahl, A. 1975. Identification of pachytene bivalents in human male meiosis using G-banding technique. *Chromosoma* 52:275–82

57. Macaya, G., Cortadas, J., Bernardi, G. 1978. An analysis of the bovine genome by density-gradient centrifugation. *Eur. J. Biochem.* 84:179–88

58. Macaya, G., Thiery, J. P., Bernardi, G. 1976. An approach to the organization of eukaryotic genomes at a macromolecular level. *J. Mol. Biol.* 108:237–54

59. Macaya, G., Thiery, J. P., Bernardi, G. 1977. In *Molecular Structure of Human Chromosomes*, ed. J. J. Yunis, pp. 35–58. New York: Academic

60. Manuelidis, L., Ward, D. C. 1984. Chromosomal and nuclear distribution of the Hind III 1.9 Kb human DNA repeat segment. *Chromosoma* 91:28–38

61. Matassi, G., Montero, L. M., Salinas, J., Bernardi, G. 1989. The isochore organization and the compositional distribution of homologous coding sequences in the nuclear genome of plants. *Nucleic Acids Res.* 17:5273–290

62. Mayr, E. 1976. *Evolution and the Diversity of Life*, pp. 195, 309. Cambridge, Mass: Harvard Univ. Press

63. Medrano, L., Bernardi, G., Couturier, J., Dutrillaux, B., Bernardi, G. 1988. Chromosome banding and genome compartmentalization in fishes. *Chromosoma* 96:178–83

64. Melton, D. W., Konecki, D. S., Brennand, J., Caskey, C. T. 1984. Structure, expression and mutation of the hypoxanthine phosphoribosyltransferase gene. *Proc. Natl. Acad. Sci. USA* 81:2147–51

65. Meunier-Rotival, M., Cortadas, J., Macaya, G., Bernardi, G. 1979. Isola-

tion and organization of calf ribosomal DNA. *Nucleic Acids Res.* 6:2109–23

66. Meunier-Rotival, M., Soriano, P., Cuny, G., Strauss, F., Bernardi, G. 1982. Sequence organization and genomic distribution of the major family of interspersed repeats of mouse DNA. *Proc. Natl. Acad. Sci. USA* 79:355–59

67. Mitelman, F., Heim, S. 1988. Consistent involvement of only 71 of the 329 chromosomal bands of the human genome in primary neoplasia-associated rearrangements. *Cancer Res.* 48:7115–19

68. Morgan, W. F., Crossen, P. E. 1977. The frequency and distribution of sister chromatid exchanges in human chromosomes. *Hum. Genet.* 38:271–78

69. Morse, B., Rotherg, P. G., South, V. J., Spandorfer, J. M., Astrin, S. M. 1988. Insertional mutagenesis of the *myc* locus by a LINE-1 sequence in a human breast carcinoma. *Nature* 333:87–90

70. Mouchiroud, D., Fichant, G., Bernardi, G. 1987. Compositional compartmentalization and gene composition in the genome of vertebrates. *J. Mol. Evol.* 26:198–204

71. Mouchiroud, D., Gautier, C., Bernardi, G. 1988. The compositional distribution of coding sequences and DNA molecules in humans and murids. *J. Mol. Evol.* 27:311–20

72. Nelson, M., McClelland, M. 1987. The effect of site-specific methylation on restriction-modification enzymes. *Nucleic Acids Res.* 15:r219–r30

73. Nur, I., Pascale, E., Furano, A. V. 1988. The left end of rat L1 (L1Rn, long interspersed repeated) DNA which is a CpG island can function as a promoter. *Nucleic Acids Res.* 16:9233–51

74. Ohno, S. 1972. An argument for the genetic simplicity of man and other mammals. *J. Hum. Evol.* 1:651–62

75. Okada, T. A., Comings, D. E. 1974. Mechanisms of chromosome banding. III. Similarity between G-bands of mitotic chromosomes and chromomeres of mitotic chromosomes. *Chromosoma* 48:65–71

76. Olofsson, B., Bernardi, G. 1983. Organization of nucleotide sequences in the chicken genome. *Eur. J. Biochem.* 130:241–45

77. Olofsson, B., Bernardi, G. 1983. The distribution of CR1, an Alu-like family of interspersed repeats in the chicken genome. *Biochim. Biophys. Acta* 740:339–41

78. Perrin, P., Bernardi, G. 1987. Directional fixation of mutations in vertebrate evolution. *J. Mol. Evol.* 26:301–10

79. Pizon, V., Cuny, G., Bernardi, G. 1984. Nucleotide sequence organization in the very small genome of a tetraodontid fish *Arothron diadematus. Eur. J. Biochem.* 140:25–30

80. Russell, G. J., Walker, P. M. B., Elton, R. A., Subak-Sharpe, J. H. 1976. Doublet frequency analysis of fractionated vertebrate nuclear DNA. *J. Mol. Biol.* 108:1–23

81. Salinas, J., Matassi, G., Montero, L. M., Bernardi, G. 1988. Compositional compartmentalization and compositional patterns in the nuclear genomes of plants. *Nucleic Acids Res.* 16:4269–85

82. Salinas, J., Zerial, M., Filipski, J., Bernardi, G. 1986. Gene distribution and nucleotide sequence organization in the mouse genome. *Eur. J. Biochem.* 160:469–78

83. Salinas, J., Zerial, M., Filipski, J., Crépin, M., Bernardi, G. 1987. Non-random distribution of MMTV proviral sequences in the mouse genome. *Nucleic Acids. Res.* 15:3009–22

84. Sawyer, J. R., Hozier, J. C. 1986. High resolution of mouse chromosomes: banding conservation between man and mouse. *Science* 232:1632–35

85. Schmid, C. W., Hearst, J. E. 1972. Sedimentation equilibrium of DNA samples heterogeneous in density. *Biopolymers* 11:1913–18

86. Schmid, M., Guttenbach, M. 1988. Evolutionary diversity of reverse (R) fluorescent chromosome bands in vertebrates. *Chromosoma* 97:101–14

87. Schimke, R. T., ed. 1982. *Gene Amplification.* Cold Spring Harbor, NY: Cold Spring Harbor Lab.

88. Schwartz, D. C., Cantor, C. R. 1984. Separation of yeast chromosome-sized DNAs by pulsed field gradient gel electrophoresis. *Cell* 37:67–75

89. Schwartz, D. C., Koval, M. 1989. Conformational dynamics of individual DNA molecules during gel electrophoresis. *Nature* 338:520–22

90. Schweizer, D. 1977. R-banding produced by DNase I digestion of chromomycin-stained chromosomes. *Chromosoma* 64:117–24

91. Schweizer, D., Loidl, J., Hamilton, B. 1987. Heterochromatin and the phenomenon of chromosome banding. In *Structure and Function of Eukaryotic Chromosomes,* ed. W. Hennig, pp. 235–54. Heidelberg, Germany: Springer

92. Seuánez, H. N. 1979. *The Phylogeny of*

Human Chromosomes. Berlin, Germany: Springer-Verlag

93. Singer, M. F. 1982. SINES and LINES: highly repeated short and long interspersed sequences in mammalian genomes. *Cell* 28:433–34

94. Smith, D. I., Golembieski, W., Gilbert, J. D., Kizyma, L., Miller, J. O. 1987. Over abundance of rare-cutting restriction endonuclease sites in the human genome. *Nucleic Acids Res.* 15:1173–84

95. Soriano, P., Macaya, G., Bernardi, G. 1981. The major components of the mouse and human genomes. 2. Reassociation kinetics. *Eur. J. Biochem.* 115:235–39

96. Soriano, P., Meunier-Rotival, M., Bernardi, G. 1983. The distribution of interspersed repeats is non-uniform and conserved in the mouse and human genomes. *Proc. Natl. Acad. Sci. USA* 80:1816–20

97. Soriano, P., Szabo, P., Bernardi, G. 1982. The scattered distribution of actin genes in the mouse and human genomes. *EMBO J.* 1:579–83

98. Subak-Sharpe, H., Burk, R. R., Crawford, L. V., Morrison, J. M., Hay, J., Keir, A. M. 1966. An approach to evolutionary relationship of mammalian DNA viruses through analysis of the pattern of nearest neighbor base sequence. *Cold Spring Harbor Symp. Quant. Biol.* 31:737–48

99. Subak-Sharpe, J. H., Elton, R. A., Russell, G. J. 1974. Evolutionary implications of doublet analysis. *Symp. Soc. Gen. Microbiol.* 24:131–50

100. Sutherland, G. R., Hecht, F. 1985. *Fragile Sites on Human Chromosomes.* Oxford, England: Oxford Univ. Press

101. Thiery, J. P., Macaya, G., Bernardi, G. 1976. An analysis of eukaryotic genomes by density gradient centrifugation. *J. Mol. Biol.* 108:219–35

102. Tikocinski, M. L., Max, E. E. 1984. CG dinucleotide clusters in MHC gene and in 5' demethylated genes. *Nucleic Acids Res.* 12:4385–96

103. Valerio, D., van der Putten, H., Botteri, F. M., Hoegerbrugge, P. M. 1988. Activity of the adenosine deaminase promoter in transgenic mice. *Nucleic Acids Res.* 16:10083–97

104. Viegas-Pequignot, E., Derbin, C., Malfoy, B., Taillandier, E., Leng, M., Dutrillaux, B. 1983. Z-DNA immunoactivity in fixed metaphase chromosomes of primates. *Proc. Natl. Acad. Sci. USA* 80:5890–94

105. Viegas-Pequignot, E., Dutrillaux, B. 1978. Une méthode simple pour obtenir des prophase et des prométaphases. *Ann. Genet.* 21:122–25

106. Viegas-Pequignot, E., Dutrillaux, B., Thomas, G. 1988. Inactive X chromosome has the highest concentration of unmethylated Hha I sites. *Proc. Natl. Acad. Sci. USA* 85:7657–60

107. Weinberg, R. A. 1988. To sequence or not to sequence. *Sci. Am.* 259:112

108. Weiner, A. M., Deininger, P. L., Efstratiadis, A. 1986. Nonviral retroposons: genes, pseudogenes, and transposable elements generated by the reverse flow of genetic information. *Annu. Rev. Biochem.* 55:631–61

109. Weintraub, H., Groudine, M. 1976. Chromosomal subunits in active genes have an altered conformation. *Science* 93:848–56

110. Yonenaga-Yassuda, Y., Kasahara, S., Chu, T. H., Rodriguez, M. T. 1988. High-resolution RBG-banding pattern in the genus *Tropidurus (Sauria, Iguanidae). Cytogenet. Cell Genet.* 48:68–71

111. Yunis, J. J. 1976. High resolution of human chromosomes. *Science* 191: 1268–70

112. Yunis, J. J. 1981. Mid-prophase human chromosomes. The attainment of 2,000 bands. *Hum. Genet.* 56:291–98

113. Yunis, J. J., Soreng, A. L. 1984. Constitutive fragile sites and cancer. *Science* 226:1199–04

114. Zerial, M., Salinas, J., Filipski, J., Bernardi, G. 1986. Genomic localization of hepatitis B virus in a human hepatoma cell line. *Nucleic Acids Res.* 14:8373–86

115. Zerial, M., Salinas, J., Filipski, J., Bernardi, G. 1986. Gene distribution and nucleotide sequence organization in the human genome. *Eur. J. Biochem.* 160: 479–85

116. ZuckerKandl, E., Pauling, L. 1965. Evolutionary divergence and convergence in proteins. In *Evolving Genes and Proteins*, ed. V. Bryson, H. J. Vogel, pp. 97–165. New York: Academic

SUBJECT INDEX

A

Abelson murine leukemia virus
immunoglobulin gene
rearrangement and, 608
ACTH
See Adrenocorticotrophic hormone
Acute myelogenous leukemia
ETS2 oncogene and, 413-14
Additive gene action model, 342
Adenovirus E1A protein, 141-57
cellular transcription factors
and, 153-54
function at promoter, 154-57
structure and properties of,
143-45
transcription activation of,
553-54
Adenovirus E1B-21kd
polypeptide, 142
Adenovirus major late promoter,
145-47
Adenovirus major late transcription factor, 146-47
Adenovirus transcription
E1A protein-mediated activation of
mechanisms of, 153-57
targets of, 145-53
Adenylate cyclase
alcoholism and, 27-28
Salmonella virulence and, 462
Adoption studies
alcoholism and, 22-24
Adrenocorticotrophic hormone
adrenal steroidogenesis and,
371
alcoholism and, 28
Agrobacterium tumefaciens
crown gall tumorigenesis and,
461-62
Alcohol
metabolism of
oxidative pathway for, 28-
29
Alcohol abuse
antisocial behavior disorder
with, 23
Alcohol dehydrogenases
alcohol metabolism and, 28
Alcoholism, 19-32
adoption studies of, 22-24
biochemical markers of, 27-
29
electrophysiological markers
of, 26

family studies of, 24-25
male-limited, 23-24
mapping to human genome,
29-32
milieu-limited, 23-24
personality and, 25
twin studies of, 21-22
Aldehyde dehydrogenases
alcohol metabolism and, 28
Aldosterone
intravascular volume and
blood pressure and, 372
Algae
mating systems of
sexual recognition processes
in, 121-22
Alpha wave activity
alcoholics and, 26
Alternative splicing, 527-66
cis requirements in, 533-39
developmental control and,
558-63
evolution and, 563-66
functional correlations of,
551-57
mechanisms of, 542-51
modes of, 531-33
trans requirements in, 539-42
Alzheimer's disease
chromosome 21 and, 410-13
Amino acid metabolism
Rhizobium, 487-90
Androgens
effects of, 372-73
Androstenedione
effects of, 372-73
Anemia
Fanconi
homologous recombination
and, 213
Antibodies
diversity of
molecular basis of, 609
See also Immunoglobulins
Antigenic variation
bacterial virulence and, 456-
61
Antigens
HLA
alcoholic end-organ disease
and, 27
Antisocial behavior disorder
with alcohol abuse, 23
Arabidopsis thaliana
self-compatibility in, 136
Artificial selection
quantitative variation and, 340

Arylsulfatase A
alcoholism and, 28
Ascaris suum
chromatin diminution in, 591-
92
Ascomycetes
mating systems of
sexual recognition processes
in, 121-22
Atavistic polydactyly, 10
Ataxia telangiectasia
homologous recombination
and, 213
Auxin
biosynthetic precursors to
amino acids as, 487-88
habituation for, 395-404
Azorhizobium, 484, 486

B

Bacillus anthracis
virulence determinants in, 472
Bacillus subtilis
degradative exoenzymes in
regulation of, 322
Bacillus thuringiensis
spore toxins of, 472
Bacterial virulence, 455-73
antigenic variation and, 456-
61
gene expression and, 461-69
Bacteriocins
plasmid genes for, 60
Bacteriophage lambda
cII protein and, 188-90
growth in *Escherichia coli lon*
mutants, 173-74
lysogenic state of
cI gene product and, 402
protein turnover in, 165
tripartite operators in, 234-37
Barnase
destabilization in, 298
electrostatic interactions in,
301
Biogeography
historical
Drosophila melanogaster
and, 426-28
Biological markers
alcoholism and, 26-29
Biometric analysis
quantitative variation and, 340
Blood groups
MNS
alcoholism and, 30

CUMULATIVE INDEXES

CONTRIBUTING AUTHORS, VOLUMES 19–23

674

CHAPTER TITLES, VOLUMES 19–23

Annual Reviews Inc.

A NONPROFIT SCIENTIFIC PUBLISHER

4139 El Camino Way
P.O. Box 10139
Palo Alto, CA 94303-0897 • USA

Annual Reviews Inc. publications may be ordered directly from our office by mail, Telex, or use our Toll Free Telephone line (for orders paid by credit card or purchase order*, and customer service calls only); through booksellers and subscription agents, worldwide; and through participating professional societies. Prices subject to change without notice. ARI Federal I.D. #94-1156476

- **Individuals:** Prepayment required on new accounts by check or money order (in U.S. dollars, check drawn on U.S. bank) or charge to credit card—American Express, VISA, MasterCard.
- **Institutional buyers:** Please include purchase order number.
- **Students:** $10.00 discount from retail price, per volume. Prepayment required. Proof of student status must be provided (photocopy of student I.D. or signature of department secretary is acceptable). Students must send orders direct to Annual Reviews. Orders received through bookstores and institutions requesting student rates will be returned. You may order at the Student Rate for a maximum of 3 years.
- **Professional Society Members:** Members of professional societies that have a contractual arrangement with Annual Reviews may order books through their society at a reduced rate. Check with your society for information.
- **Toll Free Telephone orders:** Call 1-800-523-8635 (except from California) for orders paid by credit card or purchase order and customer service calls only. California customers and all other business calls use 415-493-4400 (not toll free). Hours: 8:00 AM to 4:00 PM, Monday-Friday, Pacific Time. ***Written confirmation** is required on purchase orders from universities before shipment.
- **Telex: 910-290-0275**

Regular orders: Please list the volumes you wish to order by volume number.
Standing orders: New volume in the series will be sent to you automatically each year upon publication. Cancellation may be made at any time. Please indicate volume number to begin standing order.
Prepublication orders: Volumes not yet published will be shipped in month and year indicated.
California orders: Add applicable sales tax.
Postage paid (4th class bookrate/surface mail) **by Annual Reviews Inc.** Airmail postage or UPS, extra.

ANNUAL REVIEWS SERIES	Prices Postpaid per volume USA & Canada/elsewhere	Regular Order Please send:	Standing Order Begin with:
Annual Review of ANTHROPOLOGY		Vol. number	Vol. number
Vols. 1-14	(1972-1985)................$27.00/$30.00		
Vols. 15-16	(1986-1987)................$31.00/$34.00		
Vol. 17	(1988) ....$35.00/$39.00		
Vol. 18	(avail. Oct. 1989) ....$35.00/$39.00	Vol(s). _____	Vol. _____
Annual Review of ASTRONOMY AND ASTROPHYSICS			
Vols. 1, 4-14, 16-20	(1963, 1966-1976, 1978-1982) . .$30.00/$30.00		
Vols. 21-25	(1983-1987)................$44.00/$47.00		
Vol. 26	(1988) ....$47.00/$51.00		
Vol. 27	(avail. Sept. 1989) ....$47.00/$51.00	Vol(s). _____	Vol. _____
Annual Review of BIOCHEMISTRY			
Vols. 30-34, 36-54	(1961-1965, 1967-1985).......$29.00/$32.00		
Vols. 55-56	(1986-1987)................$33.00/$36.00		
Vol. 57	(1988) ....$35.00/$39.00		
Vol. 58	(avail. July 1989)............$35.00/$39.00	Vol(s). _____	Vol. _____
Annual Review of BIOPHYSICS AND BIOPHYSICAL CHEMISTRY			
Vols. 1-11	(1972-1982)................$27.00/$30.00		
Vols. 12-16	(1983-1987)................$47.00/$50.00		
Vol. 17	(1988) ....$49.00/$53.00		
Vol. 18	(avail. June 1989) ....$49.00/$53.00	Vol(s). _____	Vol. _____
Annual Review of CELL BIOLOGY			
Vol. 1	(1985) ....$27.00/$30.00		
Vols. 2-3	(1986-1987)................$31.00/$34.00		
Vol. 4	(1988) ....$35.00/$39.00		
Vol. 5	(avail. Nov. 1989) ....$35.00/$39.00	Vol(s). _____	Vol. _____

ANNUAL REVIEWS SERIES	Prices Postpaid per volume USA & Canada/elsewhere	Regular Order Please send:	Standing Order Begin with:
		Vol. number	Vol. number

Annual Review of COMPUTER SCIENCE

Vols. 1-2	(1986-1987)............... $39.00/$42.00		
Vol. 3	(1988) .... $45.00/$49.00		
Vol. 4	(avail. Nov. 1989)........... $45.00/$49.00	Vol(s). _____	Vol. _____

Annual Review of EARTH AND PLANETARY SCIENCES

Vols. 1-10	(1973-1982)............... $27.00/$30.00		
Vols. 11-15	(1983-1987)............... $44.00/$47.00		
Vol. 16	(1988) .... $49.00/$53.00		
Vol. 17	(avail. May 1989)........... $49.00/$53.00	Vol(s). _____	Vol. _____

Annual Review of ECOLOGY AND SYSTEMATICS

Vols. 2-16	(1971-1985)............... $27.00/$30.00		
Vols. 17-18	(1986-1987)............... $31.00/$34.00		
Vol. 19	(1988) .... $34.00/$38.00		
Vol. 20	(avail. Nov. 1989)........... $34.00/$38.00	Vol(s). _____	Vol. _____

Annual Review of ENERGY

Vols. 1-7	(1976-1982)............... $27.00/$30.00		
Vols. 8-12	(1983-1987)............... $56.00/$59.00		
Vol. 13	(1988) .... $58.00/$62.00		
Vol. 14	(avail. Oct. 1989)........... $58.00/$62.00	Vol(s). _____	Vol. _____

Annual Review of ENTOMOLOGY

Vols. 10-16, 18 20-30	(1965-1971, 1973) (1975-1985)............... $27.00/$30.00		
Vols. 31-32	(1986-1987)............... $31.00/$34.00		
Vol. 33	(1988) .... $34.00/$38.00		
Vol. 34	(avail. Jan. 1989)........... $34.00/$38.00	Vol(s). _____	Vol. _____

Annual Review of FLUID MECHANICS

Vols. 1-4, 7-17	(1969-1972, 1975-1985)....... $28.00/$31.00		
Vols. 18-19	(1986-1987)............... $32.00/$35.00		
Vol. 20	(1988) .... $34.00/$38.00		
Vol. 21	(avail. Jan. 1989)........... $34.00/$38.00	Vol(s). _____	Vol. _____

Annual Review of GENETICS

Vols. 1-19	(1967-1985)............... $27.00/$30.00		
Vols. 20-21	(1986-1987)............... $31.00/$34.00		
Vol. 22	(1988) .... $34.00/$38.00		
Vol. 23	(avail. Dec. 1989)........... $34.00/$38.00	Vol(s). _____	Vol. _____

Annual Review of IMMUNOLOGY

Vols. 1-3	(1983-1985)............... $27.00/$30.00		
Vols. 4-5	(1986-1987)............... $31.00/$34.00		
Vol. 6	(1988) .... $34.00/$38.00		
Vol. 7	(avail. April 1989) .... $34.00/$38.00	Vol(s). _____	Vol. _____

Annual Review of MATERIALS SCIENCE

Vols. 1, 3-12	(1971, 1973-1982)........... $27.00/$30.00		
Vols. 13-17	(1983-1987)............... $64.00/$67.00		
Vol. 18	(1988) .... $66.00/$70.00		
Vol. 19	(avail. Aug. 1989)........... $66.00/$70.00	Vol(s). _____	Vol. _____

Annual Review of MEDICINE

Vols. 9, 11-15 17-36	(1958, 1960-1964) (1966-1985)............... $27.00/$30.00		
Vols. 37-38	(1986-1987)............... $31.00/$34.00		
Vol. 39	(1988) .... $34.00/$38.00		
Vol. 40	(avail. April 1989) .... $34.00/$38.00	Vol(s). _____	Vol. _____